# BASIC ELECTRONICS

Debashis De

*With contributions from*

Kamakhya Prasad Ghatak

Original edition, entitled *Basic Electronics,* , by *Ghatak & De* published by Pearson Education Limited, Copyright © 2010.

Indian Subcontinent Reprint



ISBN 978-81-317-1068-5

First Impression, 2010
Eighth Impression, 2024
**Ninth Impression, 2025**

Published by Pearson India Education Services Pvt. Ltd, CIN: U72200TN2005PTC057128

Head Office: 1st Floor, Berger Tower, Plot No. C-001A/2, Sector 16B, Noida – 201 301,Uttar Pradesh, India.

Registered Office: Featherlite, 'The Address' 5th Floor, Survey No 203/10B, 200 Ft MMRD Road, Zamin Pallavaram, Chennai - 600044
Website: in.pearson.com; Email:companysecretary.india@pearson.com

Digitally printed in India by Trinity Academy for Corporate Training Ltd, New Delhi in the year of 2025.

Dedicated to
Richard Phillips Feynman,
the father of nanotechnology

# BRIEF CONTENTS

# CONTENTS

# PREFACE

*Basic Electronics* has been written to serve as a textbook for core science and technology courses in engineering colleges and universities. This book is not only useful for students of science and technology, but also for readers from other fields aspiring to enhance their knowledge of electronics through self-study. The objective of the book is to develop the basic ability to understand electronics as a science that forms the core of engineering and technology in general. The basic prerequisite for those using this book is the alert mind. The required physics and mathematics has been developed throughout the book, and no prior knowledge of physical electronics has been assumed.

## Organization

Chapter 1 builds the foundation of electronics with a detailed analysis of crystalline materials, classification of materials, intrinsic and extrinsic semiconductors, electrical conduction phenomenon, continuity equation and the Hall Effect. The book then proceeds to focus on the major categories of electronic devices—the diode, bipolar junction transistor and field-effect transistor—with two chapters devoted to each device.

Chapter 8 deals with special semiconductor devices like SCR, TRIAC, DIAC, UJT and IGBT with emphasis on the constructional features, physical operations and characteristics. Chapter 9 discusses the conceptual development of feedback in amplifiers through block diagrams, properties of positive and negative feedback, calculations of open-loop gain, closed-loop gain, feedback factors, topologies of feedback amplifier, the effect of feedback on gain, input impedance and output impedance, practical implementation of the feedback topologies, sensitivity, bandwidth stability and the effect of positive feedback.

Chapter 10 deals with the fundamentals of integrated-circuit fabrication, which provides an insight into the world of nanotechnology where this basic idea of fabrication is important with respect to semiconductor device engineering. Chapter 11 studies the properties of the ideal operational amplifier, its terminal properties and real-life applications. Chapter 12 is devoted to a detailed examination of the working principle of an oscillator, the various classifications of oscillators along with a detailed examination of Barkhausen criteria.

Chapter 13 provides a comprehensive overview of number systems, binary codes, logic gates and the applications of digital circuits. The final chapter of the book, Chapter 14, deals with the key instruments of electronic measurement with special emphasis on the most versatile instrument of electronic measurement, the cathode-ray oscilloscope.

## Presentation

'Simplicity' is the key word when it is a question of building a foundation. The hallmark of this book is the presentation of the most complex theories in a manner that facilitates a lucid understanding of the fundamentals while establishing the platform for the next phase of subject development. Through simple language and clear-cut presentation, the book presents electronics intelligently.

There are numerous books on this subject, with a varying degree of depth and coverage that map the courses on electronics as offered by different institutions. This book emphasizes the importance of *observing* the applications by concentrating on the fundamentals. This approach will sustain an efficient assimilation of concepts in a coherent way. The primary aim of writing *Basic Electronics* is to establish a strong foundation in electronics and provide tools for self-assessment and further learning.

To complement the attempt at providing a student-friendly text, a comprehensive learning package has been combined with this book. Each chapter ends with: Important Formulae, Points to Remember, Objective Questions, Review Questions, Practice Problems and Suggested Readings. The companion Web site features a comprehensive lesson plan for instructors, which allows them to design and modify courses. An evaluation module for the students allows them to assess their progress. And a unique set of animated representations of key circuits and their functions, which makes the learning process simpler.

## Acknowledgements

The author is grateful to his colleagues, students and research scholars of the West Bengal University of Technology, University of Western Australia, University of Calcutta, and Bengal Engineering and Science University. I would like to thank my Professor K. P. Ghatak (University of Calcutta)—friend, philosopher and guide; Dr. Sitangshu Bhattacharya (IISc Bangalore), Anushka Roychowdhury (CTS). I would like to thank all the family members of Sadananda Bhavan for their constant support and encouragement. I would like to thank all the reviewers for their valuable feedback and inputs. I would also like to express my gratitude to Pearson Education who kindly agreed to publish this book. Special thanks to Rupesh Singh, Tapan Kumar Saha, Biswajit Banik, Thomas Rajesh, Yajnaseni Das and Preeta Priyamvada for their support and cooperation. I would be grateful to my readers for any constructive, positive criticisms and comments, which will definitely be addressed in the next edition. Please send your valuable feedback to us at **www.pearsoned.co.in/debashisde**.

**Debashis De**

# REVIEWERS

The final form in which *Basic Electronics* has been presented here would not have been possible without the feedback received from our reviewers. While we could not incorporate every suggestion from our reviewers, we do acknowledge that their feedback was invaluable in our attempt at creating the best possible basic electronics textbook.

### Consultant Board

Our Consultant Board provided us with a critical and unbiased analysis of each chapter. We would like to thank the following for their time and commitment:

**Kumar N Krishna Murthy**
*Faculty*
*Electronics and Communication Engineering*
*PES College of Engineering, Mysore*

**Asawari Dudwadkar**
*Lecturer*
*Vivekanand Education Society*
*Institute of Technology*

**Sitangshu Bhattacharya**
*Research Associate*
*Nanoscale Device Research Laboratory*
*Centre for Electronics Design and Technology*
*Indian Institute of Science, Banglore*

### Reviewers

The suggestions and thoughtful recommendations of many helped us further improve the quality of this book. We are grateful for the cooperation we received from the following reviewers:

**Abhijit Asati**
*Lecturer, EEE Group*
*BITS, Pilani*

**Dr. (Mrs) Rajeevan Chandel**
*Assistant Professor and Head*
*Department of Electronics and Communication Engineering*
*Coordinator SMDP-II*
*National Institute of Technology (NIT) Hamirpur*

**Aniruddha Chandra**
*Lecturer, ECE Department*
*NIT Durgapur*

**Professor Devaraju**
*Chairman, Electronics Department*
*Bangalore University*

**Sanjay Dhar Roy**
*Senior Lecturer, ECE Department*
*NIT Durgapur*

**Dr. M. K. Kowar**
*Dean (Faculty of Electronics, Electrical, Computer Science and Engineering)*
*Chhattisgarh Swami Vivekanad Technical University*

**Professor P. K. Sahu**
*NIT, Rourkela*

**Gagan Singh**
*Associate Professor and Head*
*EE Department, DIT Dehra Dun*

**Dr. Shree Prakash Singh**
*Division of ECE*
*NSIT, Dwarka*

**Sonika Singh**
*Assistant Professor*
*ECE Department, DIT Dehra Dun*

# THE AUTHOR AND THE CONTRIBUTOR

**Debashis De** studied radio physics & electronics and obtained his Ph.D. from Jadavpur University in 2005. He is currently Reader at the Department of Computer Science and Engineering, West Bengal University of Technology. With research interests in the field of semiconductor nanostructures, Dr De has more than 30 publications in international journals and conferences to his credit. In 2005, URSI awarded him with the International Young Scientists Award. In 2008, he was invited as a Post Doctoral Endeavour Research Fellow to the University of Western Australia, Perth, Australia, and the Australian Government awarded him with the Endeavour Award for research and professional development. He is currently an Adjunct Research Fellow at the University of Western Australia. Dr De has co-authored *Einstein Relation in Compound Semiconductors and Their Nanostructures* with Professor K. P. Ghatak. His current research topic includes theoretical and simulative investigation of different electronic and transport properties of armchair and zigzag nanotubes, nanoscale transistors, quantum wells, wires, dots, superlattices, *n-i-p-i* structures, and heavily doped systems under external controlled fields.

**Professor K. P. Ghatak** is the First Recipient of the Degree of Doctor of Engineering of Jadavpur University in 1991 since the inception of the University in 1955, and in the same year he received the prestigious Indian National Science Academy award. He joined as Lecturer in the Institute of Radio Physics and Electronics of the University of Calcutta in 1983, Reader in the Department of Electronics and Telecommunication of Jadavpur University in 1987 and Professor in the Department of Electronic Science of the University of Calcutta in 1994 respectively. His present research interest is nanostructured science and technology.

# 1

# Semiconductor Fundamentals

## Outline



## Objectives

This chapter equips the reader with the tools for studying the basic electronics namely, basics of crystal structures, the concept of energy bands, the density-of-states function of the carriers in semiconductors, the carrier statistics, and how the current in an almost filled band can be analysed using the vital concept of holes. Two types of transport mechanisms will be considered in this context. Firstly, the *drift* of carriers in the presence of an electric field, and secondly, the *diffusion* of carriers due to the presence of a concentration gradient will be discussed. Recombination mechanisms have been briefly discussed. The continuity equation has been formulated at the end—a useful tool in studying the current voltage characteristics of a junction diode. The chapter ends with a brief discussion of the Hall effect.

## 1-1 INTRODUCTION

The journey of electronics began in 1887 with the discovery of the *elementary particle electron* by the Nobel laureate British scientist Sir J. J. Thompson. Since the invention of the first amplifying device—the triode vacuum tube—in 1904 by John A. Fleming, electronics has evolved by leaps and bounds. The silicon diode was patented by Greenleaf Whittier Pickard in 1906, and this was followed by the invention of the first radio circuits using diodes and triodes between 1907 and 1927, the super heterodyne receiver by Major Edwin Howard Armstrong in 1920, the television in 1925, the field-effect devices in 1925, the concept of modulation by Armstrong in 1933 and the radar in 1940. The discovery of silicon transistor by John Bardeen, Walter Brattain and William Shockley in 1947 (the trio received Nobel Prize in 1956 for this wonderful discovery) marked the beginning of the era of solid-state electronics.

The next breakthrough came in 1956 with the development of the thyristor—the key device of power electronics. The first integrated circuit was developed in 1958 by Jack Kilby at Texas Instruments, and Robert Noyce and Gordon Moore at Fairchild Semiconductor, announcing the beginning of computer-based electronics.

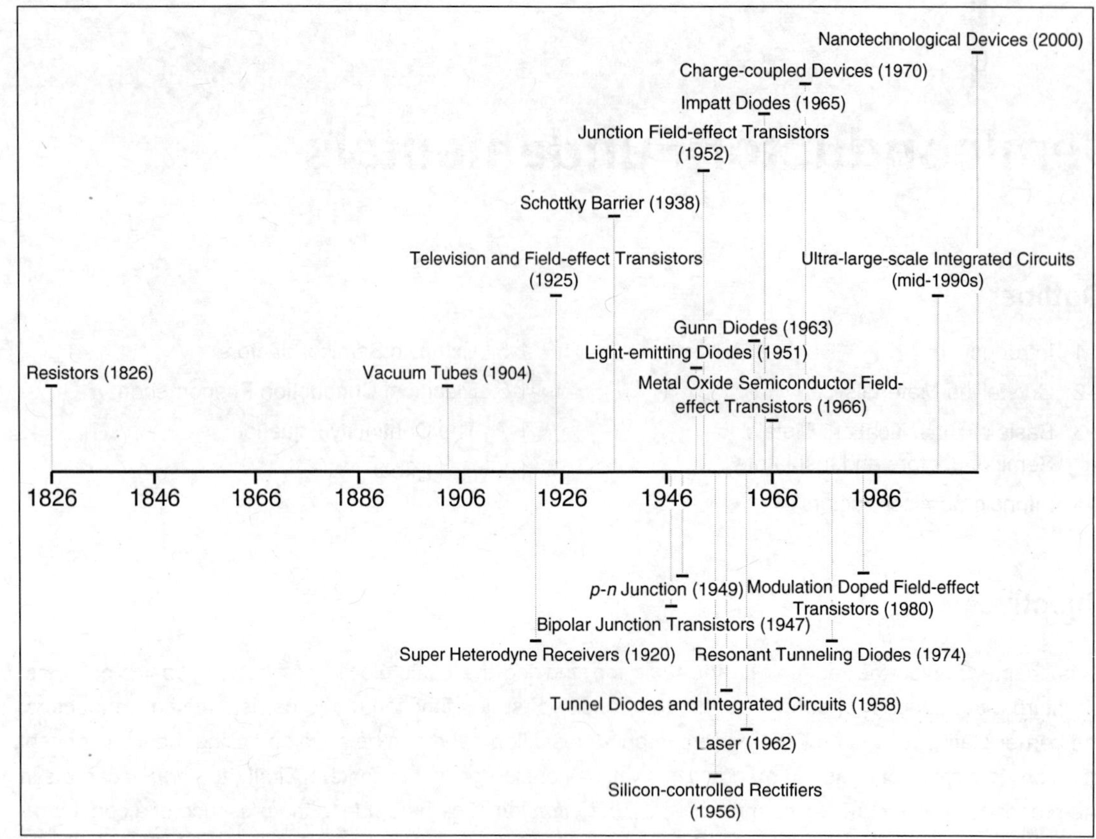


**Figure 1-1** Major milestones in the path of electronic revolution

Jack Kilby received the Nobel Prize in Physics in the year 2000. This discovery was followed by the 4004 microprocessor in 1971, the 8-bit microprocessor in 1972, and the gigabit memory chip in 1995, all by Intel. The ultra-large-scale integrated circuits, having more than $10^9$ components per chip, were developed in the mid-1990s.

All the aforementioned devices are made up of solid-state materials in general. The sound understanding of the functioning of these devices requires a detailed investigation into the electronic processes, which are different for different materials. Among the multidimensional aims of modern electronics, we need the choice of appropriate materials for different applications that can be monitored just by controlling the electrical behaviour. A majority of solid-state devices in the industry today are made of semiconductors. The recent developments in VLSI technology make electronic goods compact, cheaper and versatile so that it becomes popular in the global industrial market. Figure 1-1 provides a chronological view of the journey from resistors to nano-devices. It may be noted that 67 major devices and 110 related devices have been discovered within a period of less than 200 years.

## 1-2 CRYSTALLINE MATERIALS

Most of the materials used in electronic devices are crystalline in nature. The crystal is a systematic array of atoms. A three-dimensional lattice is defined by the three fundamental translation vectors such that the atomic arrangements look the same in every respect when viewed from two different directions. A pure crystal is constructed in such a way that it maintains directional invariance. Thus, the lattice is a regular periodic point in

**Table 1-1** Semiconductor materials

| *Classification* | *Example* |
|---|---|
| Gr – IV | Si, Ge |
| Gr III – V | GaAs, InP, InSb, GaN, GaSb |
| Gr II – VI | CdS, CdTe, CdSe, ZnO, ZnS |
| Alloys | $Al_xGa_{1-x}As$, $Ga_xIn_{1-x}As_{1-y}P_y$ |

space and is a mathematical abstraction. Such a structure yields a periodic potential throughout the material. An ideal crystal is constructed by the infinite repetition of identical structural unit in space. The crystals are made up of identical building blocks, each block being an atom or a group of atoms. It may be noted that *the periodicity of the crystal is the key property which controls the properties of the carriers inside the materials*.

The intrinsic property of a crystal is that the environment around the given atom or atoms is identical to that of other atom or atoms. Mathematically, the crystal structure is the sum of the lattice and the basis, where the basis is the site of the atom attached identically to every lattice point.

Two properties of crystals are of particular interest, since they are the ingredients required to formulate the total current in semiconductors. Firstly, we need to know how many fixed and how many mobile charges are present in the material. Secondly, we should understand the process of transport of the mobile carriers through the semiconductors. Table 1-1 illustrates a few elemental and compound semiconductors together with their alloys which find extensive applications in modern semiconductor industries.

It is important to note that the elemental semiconductor Silicon (Si) is preferred over another elemental semiconductor Germanium (Ge) for the following reasons:

(i) Higher temperature stability
(ii) Higher band gap
(iii) Lower leakage current
(iv) Higher breakdown voltage
(v) Technically easier fabrication process than germanium

In this context, it is worth remarking that the compound III-V semiconductor Gallium Arsenide (GaAs) is preferred over the elemental semiconductor Silicon (Si) for the following reasons:

(i) Higher temperature stability
(ii) Higher band gap
(iii) Higher cut-off voltage than silicon

Incidentally, there is one disadvantage; the fabrication of GaAs is technically more difficult than that of silicon. Some special microelectronic techniques are used to fabricate GaAs based devices.

## FOR ADVANCED READERS

### Semiconducting Materials

The choice of different semiconducting materials is discussed in Table 1-2 for various applications due to their different electronic behaviours such as tunneling efficiency, noise immunity, quantum efficiency etc.

**Table 1-2** Applications of different semiconducting materials

| *Device Types* | *Materials* | *Choice of Materials for Different Applications* |
|---|---|---|
| Tunnel diode | GaAs | Tunneling efficiency |
| Laser | GaAs, ZnO, CdSe, CdS, CdTe, $GaAs_xP_{1-x}$ | Direct band gap materials |

(Continued)

| Device Types | Materials | Choice of Materials for Different Applications |
|---|---|---|
| Light-emitting diode (LED)\ | GaP, $GsAs_{1-x}P_x$ | Quantum efficiency and the suitable band gap for proper emission of wavelength in visible region |
| Infrared detectors | InSb, PbS, PbSe, $Hg_{1-x}$, $Cd_xTe$ | Noise immunity and proper choice of band gap |
| Galvanometric devices | InSb, InAs, HgTe, HgSe | High electron mobility |
| Thermoelectric | PbS, $InAs_xP_{1-x}$ | Allowable operating temperature and high devices thermoelectric coefficients |
| Gunn Diode | GaAs, InP, CdTe | Transferred electron device has specific valley structure used in microwave generation |
| Acoustoelectric devices | CdS, $Cu_2S$, ZnO | Strong piezoelectric interaction |

## 1-2-1 Crystals and Crystal Structures

Crystal is a solid, composed of atoms arranged in an ordered array. Solid materials are classified on the basis of the arrangement of atoms. There are generally three types of solids.

**(i) Single crystal:** Materials in which the atoms are placed regularly in a long range order. Example: Silicon, Germanium, etc.

**(ii) Amorphous crystal:** Materials in which the atoms are placed randomly are called amorphous. Example: Common window glass, polystyrene, etc.

**(iii) Polycrystalline:** Materials in which atoms are placed in a high degree of short-range order and no long-range order with irregular orientation and size. Example: GaP, Ceramic, etc.

The crystal structure is called the atomic arrangement of atoms. In a crystal there is a regular arrangement of atoms in the form of points in space called *lattice points*. If all the atoms in the lattice points are identical, the lattice is called a Bravais lattice. A lattice is a mathematical abstraction. When the basis of atoms is attached identically to every lattice point, the resultant structure so formed is known as a crystal structure. Thus, we can write:

$$\text{Lattice} + \text{Basis} = \text{Crystal structure}$$

For a 2D system as shown in Fig. 1-2, the translation vector $R$ can be written as:

$$R = ar_1 + br_2$$

where $r_1$, $r_2$ are the fundamental translation vectors and $a$, $b$ are the arbitrary integers.

Similarly for 3D system:

$$R = ar_1 + br_2 + cr_3$$

where $r_1$, $r_2$ and $r_3$ are the fundamental translation vectors, and $a$, $b$ and $c$ are the arbitrary integers.

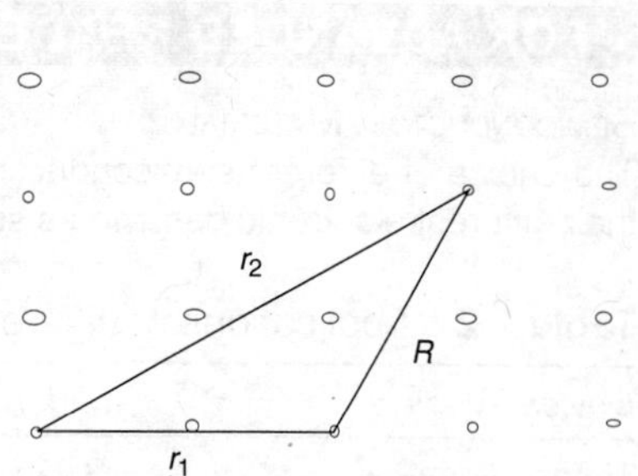


**Figure 1-2** A two-dimensional array of lattice points

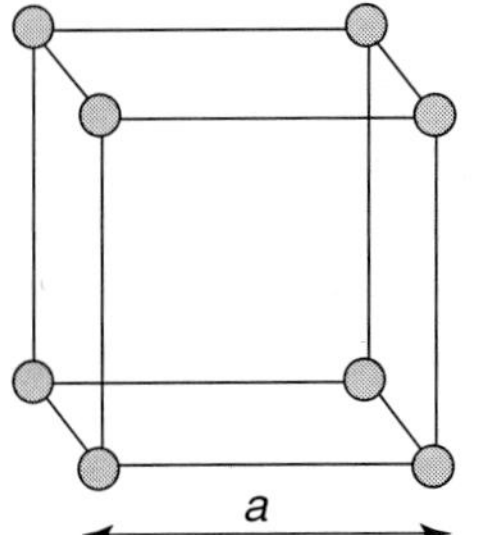


**Figure 1-3** Simple cubic lattices

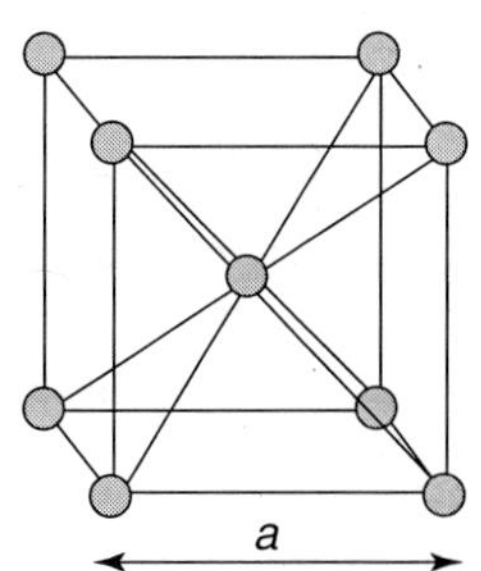


**Figure 1-4** Body-centred cubic lattice

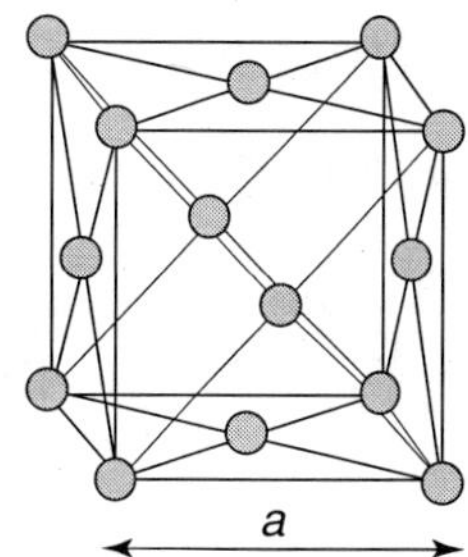


**Figure 1-5** Face-centred cubic lattice

A unit cell is a small portion of any given crystal that could be used to reproduce the crystal. While some crystals have a single atom placed at each lattice point, most crystals have a combination of atoms associated with each lattice point. The classification of lattices, the common semiconductor crystal structures and the growth of single-crystal semiconductors are discussed in the following sections.

The simple cubic lattice (SCL) consists of one lattice point at each corner of the cube, as shown in Fig. 1-3. It has one lattice point as $[(1/8) \times 8] = 1$

The body-centred cubic lattice (BCL) has one lattice point at the centre of the unit cell, as shown in Fig. 1-4. BCL has a total of two lattice points as $[(1/8) \times 8] + 1 = 2$

The face-centred cubic lattice (FCL) has lattice points on each of the faces of the cube, as shown in Fig. 1-5. FCL has a total of four lattice points as $[(1/8) \times 8] + [(1/2) \times 6] = 4$

### Growth of semiconductor crystals

The fabrication of semiconductor devices requires highly pure and mono-crystalline semiconductors. The Czochralski method, invented by the Polish scientist Jan Czochralski in 1916, is the preferred method for high-volume production of high quality Si single crystals. The silicon is pulled from a bath of molten silicon to form mono-crystalline silicon using a small seed crystal, as shown in Fig. 1-6. The semiconductor crystals can be obtained by cooling the molten semiconductor material. Incidentally, this procedure yields poly-crystalline material since crystals start growing in different locations with a different orientation. Single-crystal silicon can be grown by dipping one end of a seed crystal in the melt and by controlling the temperature difference between the seed crystal and the molten silicon. The output is large single-crystal silicon, cylindrical in shape. The diameter of the silicon will depend on the rotation of the seed crystal during the stages of its growth and partly on the cylindrical shape of the crucible containing the melt.

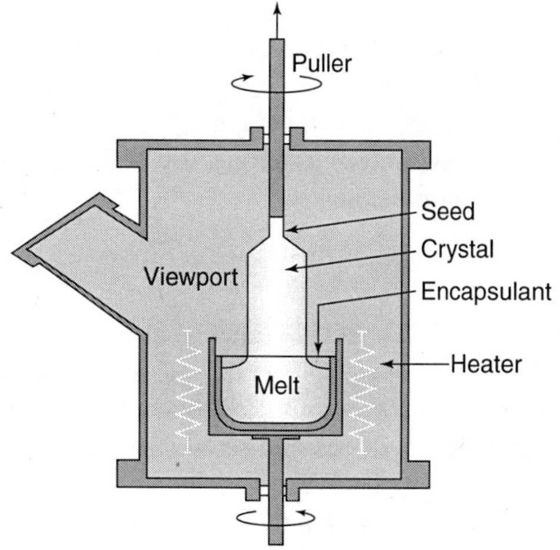


**Figure 1-6** Experimental setup of Czochralski growth of semiconductor crystals

The cylindrical silicon is then cut into smaller wafers with a diamond saw and again polished to yield the substrate material for the silicon device fabrication. The detailed steps of fabrication will be explained later in Chapter 10—'Fundamentals of Integrated Circuit Fabrication'.

## 1-2-2 Mechanical Properties

In the presence of stress, the tensorial nature of Hooke's law becomes prominent. Only the second- and third-order elastic constants are being affected in the presence of stress. In the *n*-type germanium, the shear elastic constant is being reduced by 8 per cent. In view of large changes of the elastic constants with the carrier density, the experiments on the velocity of sound involving shear mode as a function of carrier concentration exhibit the contribution of the electrons to the mechanical properties.

## 1-2-3 Energy Band Theory

In the energy band theory, we are only concerned with the conduction band and the valance band separated by a band gap. By using this theory, we can differentiate among insulators, metals and semiconductors. Another attribute of this theory is the formulation of the carrier energy spectrum, which in turn controls all the physical properties of semiconductors and semi metals in general.

### Energy levels in atoms

The sum of the kinetic and potential energies of an electron orbiting a nucleus can only assume a set of discrete values; those for the simplest atom (hydrogen) are shown in Fig. 1-7. This happens because the electrons exhibit wave-like properties.

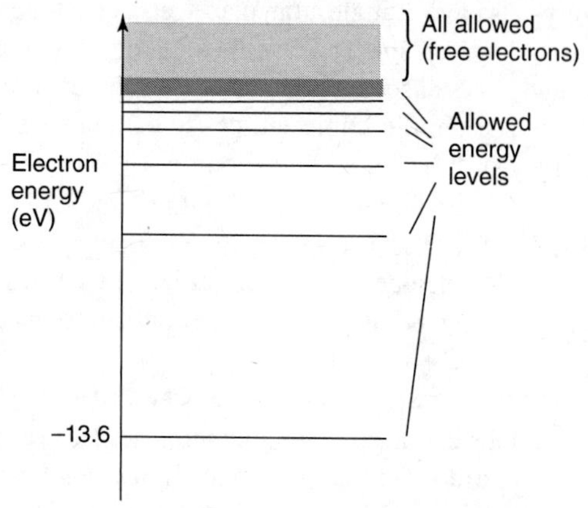


**Figure 1-7** Energy level diagram for a hydrogen atom

To be stable, the electron wave must repeat itself exactly at the same time when it is on a circular path around the nucleus. This is illustrated schematically in Fig. 1-8. It leads to the conclusion that in such a state an integral number of wavelengths must fit into its path. This state is termed *quantization condition.*

The electron's wavelength ($\lambda$) is inversely proportional to its momentum ($p$) as defined by de Broglie's relation of the wave-particle duality:

$$\lambda = \frac{h}{p} \tag{1-1}$$

The constant $h$ is called Planck constant, and has the value $6.6 \times 10^{-34}$ Js. The graph of Eq. (1-1) is rectangular hyperbolic in $p-\lambda$ plane. The quantization condition, thus, fixes the electron momentum to one of a set of discrete values, depending on the number of wavelengths fitting the circular path. The number, known as a *quantum number*, determines which set of discrete energies the electron has, since the energy is directly related to the electron's momentum. Four quantum numbers are needed to describe the electron in the hydrogen atom.

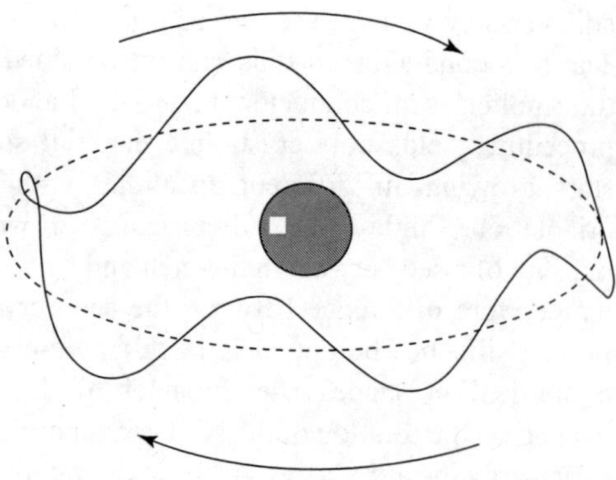

**Figure 1-8** Continuity of the electron wave around the nucleus resulting in an integral number of wavelengths fitting into any circular path

It should be noted that an electron freed from its parent nucleus can have any wavelength, and hence any kinetic energy. This explains the continuum of energy levels above the zero of energy, given in Fig. 1-7. When an atom contains many electrons, no more than two are found to exist simultaneously in the same state of nature. One of these electrons spins clockwise, the other anticlockwise. So the above rule can be expressed by stating, "In a closed system, no two electrons can occupy the same state." This is known as the Pauli exclusion principle.

In the simplest cases, when each state has a different energy we find electrons filling energy levels from the lowest up, one to each state. In the more general cases, it is possible for two or more orbits, or states, to have the same energy. Thus, a limited number of electrons are in the highest filled atomic energy levels, and these are the valence electrons, which take part in bonding an atom within a molecule or solid.

## Solved Examples

**Example 1-1** When a photon impinges upon a semiconductor it can take an electron from the valence band to the conduction band. The momentum is conserved in such transitions. Calculate the value of the wave vector that a 2.0 eV photon carries. The electron in the valence band can go into the conduction band with the momentum change of the photon.

**Solution:**

The relation gives the wavelength of a 2.0 eV photon:

$$\lambda = \frac{ch}{E_{\text{Photon}}} = \frac{(3 \times 10^{8}\,\text{m/sec})(6.64 \times 10^{-34}\,\text{Js})}{2.0 \times 1.6 \times 10^{-19}\,\text{J}} = 6.225 \times 10^{-7}\,\text{m}$$

$$= 6225\ \text{Å}$$

The magnitude of $k$-vector is:

$$|k| = \frac{2\pi}{\lambda} = 1.01 \times 10^{7}\,\text{m}^{-1}, \text{ which is an extremely small value.}$$

**Example 1-2** Identify the semiconductors with band gap large enough to emit photons with wavelength less than 0.5 μm.

**Solution:**

For light emission at wavelength of 0.5 μm or less, the material band gap $E_g$ has to be at least such that:

$$E_g = \frac{2\pi hc}{\lambda} = \frac{2\pi(1.05 \times 10^{-34}\,\text{Js})(3 \times 10^{8}\,\text{m/s})}{0.5 \times 10^{-6}\,\text{m}}$$

$$= 3.958 \times 10^{-19}\,\text{J} = 2.474\ \text{eV}$$

A number of semiconductors such as C, BN, GaN, etc. meet this criterion. The material SiC with a band gap of 2.4 eV is also quite close.

■

### Energy levels in solids

In comparison to isolated atoms, we expect the energies of electrons in solids to have only certain allowed values. If we were to push many atoms closer and closer together to form a regular crystalline solid, the electron orbits in each atom would be modified by the attractive forces exerted on them by the neighbouring nuclei. This causes the energy of each orbit to be shifted in such a way that the allowed energies form "bands" of closely spaced levels, separated in energy from other allowed energy bands, as shown in Fig. 1-9, by relatively large *forbidden energy gaps.*

Within each band, the allowed energies are so closely spaced that their separation is negligible ($<10^{-20}$ eV). This is because each level is derived from an energy level of one of the many atoms in the

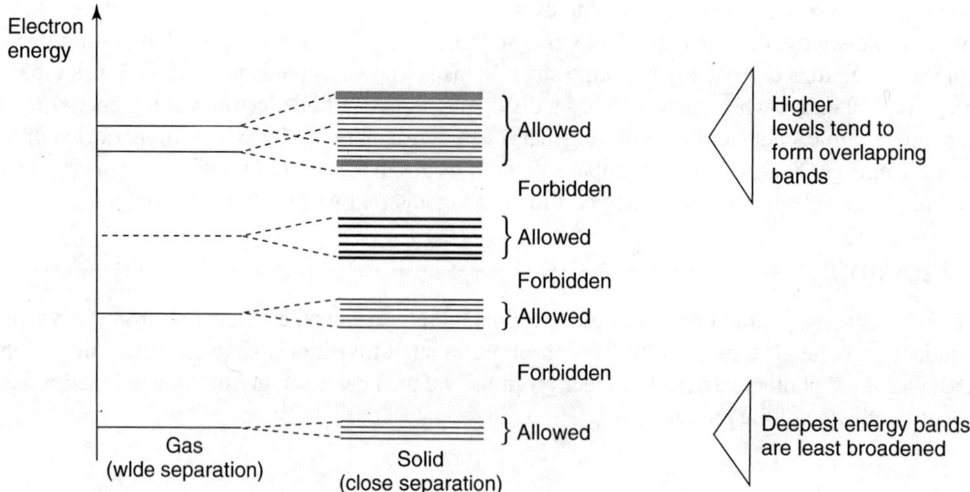


**Figure 1-9** Schematic energy level diagram of a solid, showing allowed bands and forbidden gaps

material. Thus, in a cubic centimetre, which contains about $10^{23}$ atoms, the spacing of levels in a single energy band with total width 1 eV, is about $10^{-23}$ eV.

The higher atomic energy levels are associated with the outermost orbits, and they are spread into wider bands than the lower levels, because in the lower levels the electrons are in tighter orbits around their parent nucleus and feel the influence of the neighbouring nuclei to a lesser extent. Hence, the upper bands of levels tend to overlap the neighbouring bands, causing the energy gaps to disappear there.

Pauli exclusion principle ensures that the levels are still filled from the bottom, and since there are finite number of energy levels in each band, the highest-filled level lies somewhere in one of the higher bands. Electrons in the higher bands move in "orbits", which are so far out from the parent atom that they cannot be identified exclusively with it. These electrons are in fact in motion throughout the solid and in these upper energy bands, the horizontal axis in the energy level diagram can be used to represent position. Thus, the valence electrons are shared, not just between the neighbouring atoms, but by the atoms throughout the crystal. These electrons are said to be *delocalized.*

## Bonding model

The isolated Si atom, or an Si atom not interacting with other atoms, was found to contain four valence electrons. Si atoms incorporated in the diamond lattice exhibit a bonding that involves an attraction between each atom and its four nearest neighbours. The implication is that in going from isolated atoms to the collective crystalline state the Si atoms come to share one of their valence electrons with each of the four nearest neighbours. This covalent bonding, or equal sharing of valence electrons with the nearest neighbours, and the fact that atoms in the diamond lattice have four nearest neighbours, which gives rise to the idealized semiconductor representation, called the bonding model, is shown in Fig. 1-10. Each circle in the bonding model represents the core of a semiconductor atom, while each line represents a shared valence electron. There are eight lines connected to each atom because any given atom not only contributes four shared electrons, but must also accept four shared electrons from adjacent atoms.

Two applications are presented in Fig. 1-11. We use the bonding model to picture a point defect, a missing atom in the lattice structure, as shown in Fig. 11-1(a). In Fig. 1-11(b) we visualize the breaking of an atom-to-atom bond and the associated release or freeing of an electron. Bond breaking (at $T > 0$ K) and defects occur naturally in all semiconductors, and hence the basic model of Fig. 1-10 is strictly valid for an entire semiconductor only at $T \rightarrow 0$ K when the semiconductor is devoid of defects and impurity atoms.

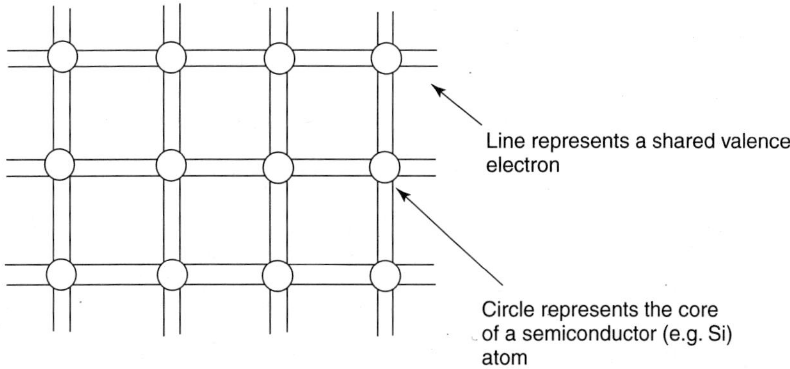


**Figure 1-10** The bonding model

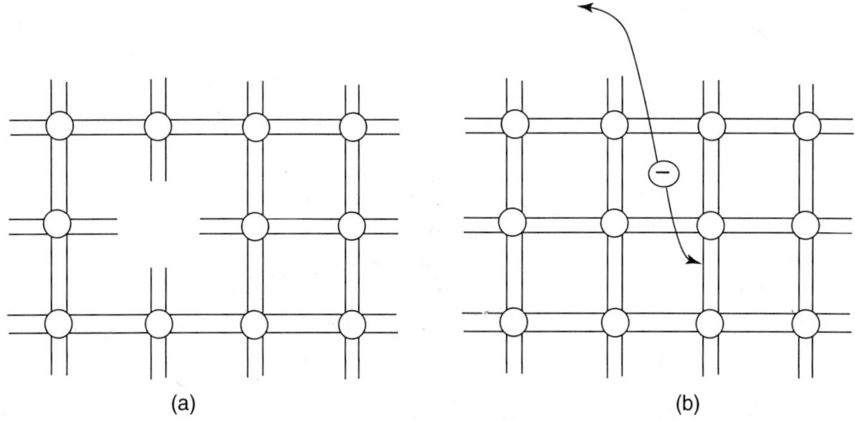


**Figure 1.11** Sample utilization of a bonding model: (a) visualization of a missing atom or point defect (b) breaking of an atom-to-atom bond and freeing of an electron

## Energy band model

Let us begin the conceptual path leading to the energy band model by recalling the situation inside an isolated Si atom. Ten of the 14 electrons inside an isolated Si atom are tightly bound to the nucleus and are unlikely to be significantly perturbed by normal atom–atom interactions. The remaining four electrons are rather weakly bound and, if unperturbed, occupy four of the eight allowed energy states immediately above the last core level. Moreover, it is implicitly understood that the electronic energy states within a group of $N$ number of Si atoms are far apart so that they are non-interacting.

The modification in the valence-electron energy states actually known to take place is summarized in Fig. 1-12.

While moving from N isolated Si atoms to an N-atom Si crystal, exactly half of the allowed states become depressed in energy and half increase in energy. The perturbation, moreover, causes a spread in allowed energies, forming two *ranges* or *bands* of allowed energy states separated by an intervening energy gap. The upper band of allowed states is called the *conduction band*; the lower band of allowed states, the *valence band*; and the intervening energy gap, the *forbidden gap*, or the *energy band gap*, or simply the *band gap*. In filling the allowed energy band states, electrons, of course, tend to gravitate to the lowest possible energies. Noting that electrons are restricted to single occupancy in allowed states (the Pauli exclusion

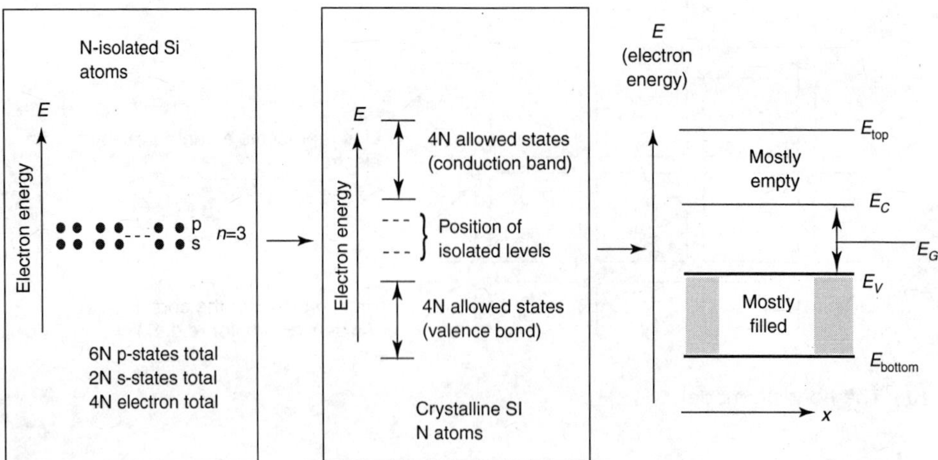


**Figure 1-12** Conceptual development of the energy band model starting with N isolated Si atoms on the left and concluding with a "dressed-up" version of the energy band model on the right

principle) and remembering that the 4N valence band states can just accommodate what were formerly 4N valence electrons, we typically find that the valence band is almost completely filled with electrons and the conduction band is all but devoid of electrons. Indeed, the valence band is completely filled and the conduction band completely empty at temperatures approaching $T \to 0$ K.

It should be mentioned that unlike the valence electrons in the isolated atom case, the band electrons in crystalline silicon are not tied to or associated with any one particular atom. The allowed electronic states are no longer atomic states, but are associated with the crystal as a whole. Independent of the point examined in a perfect crystal, one sees the same allowed-state configuration. We, therefore, conclude that for a perfect crystal under equilibrium conditions a plot of the allowed electron energies versus distance along any pre-selected crystalline direction (always called the *x*-direction) is as shown on the right-hand side of Fig. 1-12. The cited plot, a plot of allowed electron energy states as a function of position, is the basic energy band model. In Fig. 1-12, $E_C$ is the lowest possible conduction band energy, $E_V$ is the highest possible valence band energy, and $E_G = E_C - E_V$ is the band gap.

Finally, Fig 1-13 displays the form of the energy band model (for a perfect crystal under equilibrium conditions) actually employed in practice.

In this widely employed model, the line to indicate the top energy in the conduction band, the line to indicate the bottom energy in the valence band, the cross-hatching to indicate filled states, the *y*-axis or electron-energy axis, and the *x*-axis or position axis are all understood to exist implicitly, but are not known explicitly.

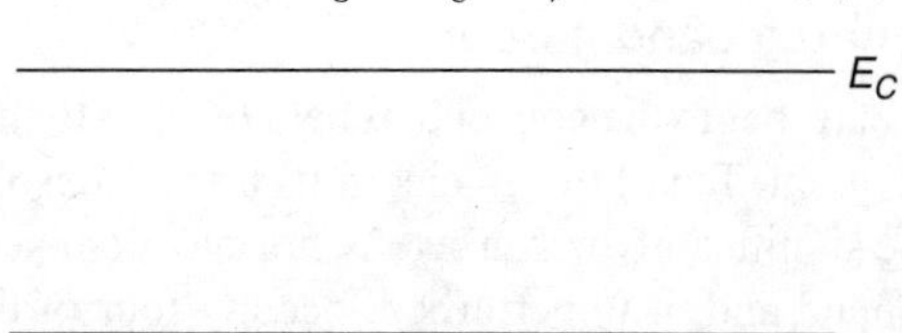


**Figure 1-13** The energy band diagram: a widely employed simplified version of the energy band model

## Carriers

Referring to Fig. 1-14(a), we note that if there are no broken bonds in the bonding model, or equivalently if in the energy band model the valence band is completely filled with electrons and the conduction band is devoid of electrons, then there are no carriers. Valence band electrons in the energy band model correspond to shared electrons in the bonding model and these electrons are not involved in charge transport.

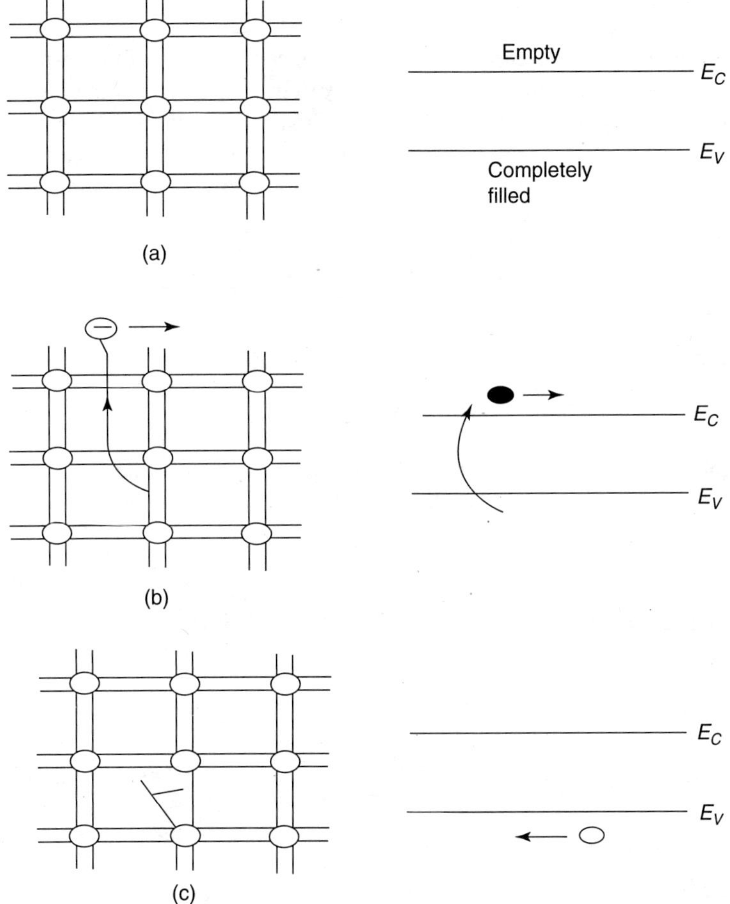


**Figure 1-14** Visualization of carriers using the bonding model (left) and the energy band model (right): (a) no carrier (b) visualization of an electron (c) visualization of a hole

The electrons involved in charge transport are visualized in Fig. 1-14(b). When the Si–Si bond is broken and the associated electron is free to move through the lattice, the released electron functions as a carrier. Equivalently, in terms of the energy band model, the excitation of valence band electrons into the conduction band creates carriers, i.e., the electrons in the conduction band are *carriers.*

In addition to releasing an electron, the breaking of the Si–Si bond also creates a missing bond or void in the bonding structure. The removal of an electron from the valence band creates an empty state in an otherwise vast sea of filled states. The empty state, like a bubble in a liquid, moves about rather freely in the lattice because of the cooperative motion of the valence band electrons. The missing bond in the bonding scheme, or the empty state in the valence band, is the second type of carrier found in semiconductors known as the *hole.* In brief, we can also say that the missing electron is known as the hole. The charge of the hole is opposite to that of the electron and therefore, the hole responds as if it has a positive charge. The mass of the hole is greater than that of the electron. Later on we shall learn that there are three types of holes: namely, heavy, light and split-off holes.

## Properties of carriers

**Effective mass.** Excluding the concept of charge, the concept of mass of the carriers occupies a central position in the field of solid-state electronics. This mass is different from that of the free carrier mass, and the free carrier mass needs to be replaced by the effective mass to account for the effects of crystalline force and the quantum mechanical properties so that the basic equations of the motion of the carriers in semiconductors remain unchanged The effective carrier mass, in general, can be expressed as follows. The effective carrier mass along a particular direction ($m^*$) is:

$$m^* = \frac{\text{momentum } (p) \text{ along this direction}}{\text{velocity } (v) \text{ along the same direction}} \tag{1-2}$$

From Eq. (1-1) we can write:

$$p = \frac{h}{\lambda} = \left(\frac{h}{2\pi}\right)\left(\frac{2\pi}{\lambda}\right) \tag{1-3}$$

The term $(h/2\pi)$ is called $\hbar$ and is called the normalized Planck constant or the *Dirac constant*, and the term $2\pi/\lambda$ is known as the carrier wave vector ($\vec{k}$). Therefore, Eq. (1-3) can be expressed as:

$$\vec{p} = \hbar\vec{k} \tag{1-4}$$

The velocity, as written in Eq. (1-2), must be the group velocity $\frac{\partial \omega}{\partial k}$ (frequency $\omega = E/\hbar$ in which $E$ is the total energy of the carrier) and not the phase velocity. Therefore, the velocity of the carrier is $\frac{\partial \omega}{\partial k}$. Thus, the mass of the carrier should, in general, be written as:

$$m^* = \frac{\hbar k}{\frac{\partial \omega}{\partial k}} = \frac{\hbar k}{\frac{1}{\hbar}\frac{\partial E}{\partial k}} \tag{1-5}$$

Then the effective mass of the carriers can be expressed as:

$$m^* = \hbar^2 k \frac{\partial k}{\partial E} \tag{1-6}$$

From Eq. (1-6), we observed that the effective mass changes with the slope of the *E–k* curve. This *E–k* relation is called the dispersion relation which changes from semiconductor to semiconductor, and consequently the $m^*$ also changes. Thus, mass can be a function of energy and will change with external physical conditions. It is very important to note that among the various definitions of effective carrier mass—acceleration effective mass, Faraday rotation effective mass—the momentum effective mass is the fundamental. From Newton's second law we can derive an expression for the acceleration effective mass. It can be shown that the acceleration effective mass and the momentum effective mass are the same for semiconductors having parabolic energy bands.

**Derivation.** From Newton's second law, we can write that the force $F$ on the carrier is given by:

$$F = \frac{dp}{dt} = \frac{d(hk)}{dt} = h\frac{dk}{dt} \quad (\text{since } p = \hbar k)$$

Also $F$ can be described as $F = m_f^* a$, where $a$ is acceleration of the carrier.

Thus,
$$a = \frac{\partial v_g}{\partial t} = \frac{1}{h}\frac{\partial}{\partial t}\left(\frac{\partial E}{\partial k}\right)$$

We know that $v_g = \frac{\partial E}{\partial k}$ is the group velocity. Combining the three equations, we get:

$$m_f^* = \frac{\hbar^2}{\frac{\partial^2 E}{\partial k^2}} \tag{1-6a}$$

The acceleration effective mass is also called the curvature effective mass. These two definitions yield the same result when $E \propto k^2$ i.e., the $E$–$k$ relation is parabolic. For any deviation from the parabolicity, these two definitions of the effective mass will not converge to the same expression. The effective momentum mass of the carriers, as given by Eq. (1-6a), affects all the properties of semiconductors, such as electronic heat capacity, diffusivity to mobility ratio, the Hall co-efficient, all types of transport co-efficient, and the changes due to electron concentration and other externally controllable parameters.

## The Fermi–Dirac statistics

The Fermi–Dirac statistics expresse the probability of the electron occupying the energy level $E$ and is extremely important in the field of solid-state electronics. In this section we present a very simplified treatment of the statistics with the following assumptions:

(i) A distribution function exists

(ii) The electron energy is the important specification of the states as far as this distribution function is concerned

(iii) In thermal equilibrium every identifiable physical process proceeds on an average at exactly the same rate as its own inverse. This is known as the principle of detailed balancing

(iv) The Pauli exclusion principle is valid

### FOR ADVANCED READERS

Simplified Derivation of the Fermi–Dirac Statistics

Let us consider a system of four states as shown in Fig. 1-15. All the notations of this figure are self-explanatory.

Let us assume that an electron goes from state 1 to state 3, and another electron goes from state 2 to state 4 simultaneously with the conservation of energy. The energy values are related by the equation:

$$E_1 + E_2 = E_3 + E_4 \tag{1-7}$$

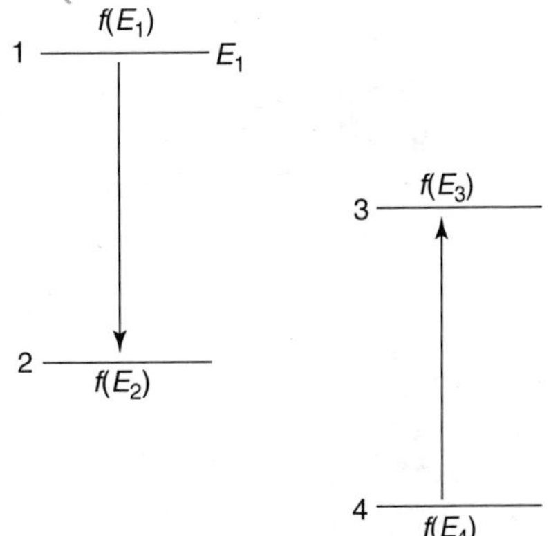


**Figure 1-15** Four states with the specified energy levels and the respective probability of occupation as used in the simplified formulation of the Fermi-level statistics.

The probability of occurrence of this reaction can take place if the states 1 and 2 are initially occupied. Thus, the probability of occurrence of this reaction must have $f(E_1) \times f(E_2)$ as a factor. Besides, due to the Pauli exclusion principle states 3 and 4 must be initially unoccupied for the reaction to proceed. The probability of both the states being unoccupied is $[1 - f(E_3)] \times [1 - f(E_4)]$ Therefore, the probability of interaction ($F$) taking place is:

$$F = A\{f(E_1) \times f(E_2) \times [1 - f(E_3)] \times [1 - f(E_4)]\} \tag{1-8}$$

where, $A$ is known as the *electronic interaction factor.*

The principle of detailed balancing requires the inverse process to take place with equal likelihood. In the inverse process, the electrons must start in the states 3 and 4 and finish in the states 1 and 2 respectively. Therefore, for the inverse process, we can write:

$$F = A\{f(E_3) \times f(E_4) \times [1 - f(E_1)] \times [1 - f(E_2)]\} \tag{1-9}$$

From Eqs. (1.8) and (1.9) we get:

$$A\{f(E_1) \times f(E_2) \times [1 - f(E_3)] \times [1 - f(E_4)]\} = A\{f(E_3) \times f(E_4) \times [1 - f(E_1)] \times [1 - f(E_2)]\} \tag{1-10}$$

Dividing both sides of Eq. (1-10) by $Af(E_1) \times f(E_4) \times f(E_2) \times f(E_3)$ (since the individual quantity is not equal to zero) yields:

$$\frac{[1 - f(E_3)] \times [1 - f(E_4)]}{f(E_3) \times f(E_4)} = \frac{[1 - f(E_1)] \times [1 - f(E_2)]}{f(E_1) \times f(E_2)}$$

$$\left[\frac{1}{f(E_3)} - 1\right]\left[\frac{1}{f(E_4)} - 1\right] = \left[\frac{1}{f(E_1)} - 1\right]\left[\frac{1}{f(E_2)} - 1\right] \tag{1-11}$$

From Fig. 1-15 we can write that the Eq. (1-7) will be satisfied if $E_1 = E_3 + x$ and $E_2 = E_4 - x$ are being satisfied simultaneously, where $x$ is the displacement in the energy scale. Substituting these values in Eq. (1-11) we get:

$$\left[\frac{1}{f(E_1 - x)} - 1\right]\left[\frac{1}{f(E_2 + x)} - 1\right] = \left[\frac{1}{f(E_1)} - 1\right]\left[\frac{1}{f(E_2)} - 1\right] \tag{1-12}$$

This equation is called the functional equation and can be satisfied if and only if we substitute $f(E)$ as:

$$f(E) = \frac{1}{1 + e^{(E - E_F)/k_BT}} \tag{1-13}$$

where, $E_F$ is the Fermi energy (a quantity defined later on), $k_B$ is the Boltzmann constant ($8.62 \times 10^{-5}$ eV/K) and $T$ is the absolute temperature in Kelvin (K). If Eq. (1-13) is the solution of Eq. (1-12) then, it will be satisfied by Eq. (1-13). The left hand side of the Eq. (1-12) can be written with the help of Eq. (1-13) as:

$$[1 + e^{(E1 - x - EF)/kBT} - 1]\,[1 + e^{(E_2 + x - E_F)/k_BT} - 1] = [1 + e^{(E_1 - E_F)/k_BT} - 1]\,[1 + e^{(E_2 - E_F)/k_BT} - 1] \tag{1-14}$$

$$\Rightarrow \quad e^{(E_1 + E_2 - 2E_F)/k_BT} \tag{1-15}$$

We find that the right hand sides of Eqs. (1-14) and (1-15) are same when $f(E)$, as given by Eq. (1-13), is being used. Hence Eq. (1-12) is verified. The most important thing to understand is the fact that excluding the form of $f(E)$, as given by Eq. (1-13), Eq. (1-12) will not be satisfied for any arbitrary form. The verification of this statement is left to the readers.

### Important Information

In Eq. (1-11), if $-1$ is being replaced by $+1$ then the resulting equation will be satisfied if, and only if we assume:

$$f(E) = \frac{1}{\{e \times b^{\,[(E - E_F)/k_BT]} - 1\}}$$

This equation is the well-known *Bose–Einstein distribution function*, and the particles with integral spin obeying this function are called *Bosons*, when $E \to E_F$, $f(E) \to \infty$. Thus, we infer that $E = E_F$ is a pole of the function mathematically, but physically this non-obedience of the Pauli exclusion principle leads to the concept of the *Bose–Einstein condensation*.

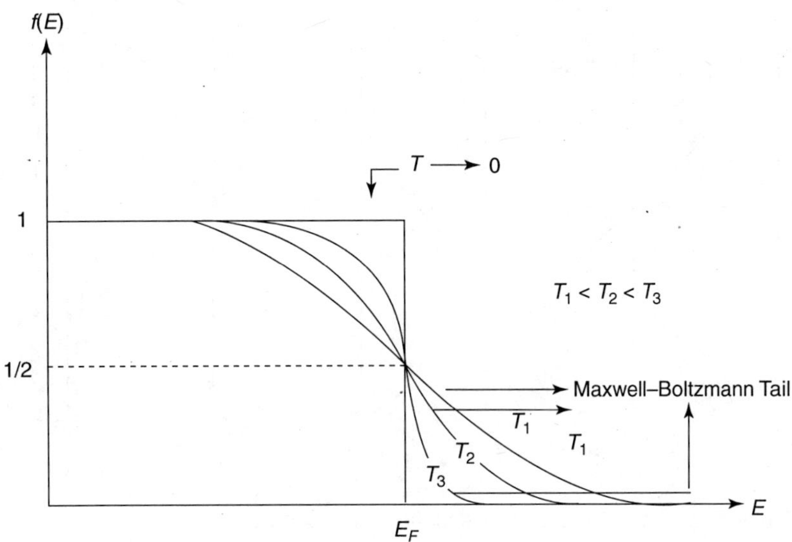


**Figure 1-16** The plot of $f(E)$ as a function of $E$ at different temperatures.

## Important Properties of the Fermi–Dirac Function

### The plot of $f(E)$ as a function of $E$

1. Figure 1-16 shows the plot of $f(E)$ as a function of $E$. Let us consider temperature $T > 0$ K, which has the following three consequences:
   **Case I**: If $E = E_F$, $f(E_F) = ½$, i.e., the probability of occupancy of the electron is ½ when its energy becomes equal to Fermi energy. Fermi level is the level where the occupancy of electron is ½.

   **Case II**: If $E > E_F$, more states above the Fermi level ($E_F$) will be empty.

   **Case III**: If $E < E_F$, most states below the Fermi level ($E_F$) will be filled.

2. As $T \to 0$,

   **Case I**: If $E = E_F$, then $f(E_F)$ indicates the transition point.

   **Case II**: If $E > E_F$, all states above Fermi energy are totally vacant, since in this case $f(E) \to 0$.

   **Case III**: If $E < E_F$, all states up to Fermi energy are totally filled up, since in this case $f(E) \to 1$.

Thus, we can infer that there are two definitions of the Fermi energy.

(i) Fermi energy is that energy which the electron can occupy with 50% energy.

(ii) Fermi energy is that energy up to which all states are filled up and above which all states are vacant, as $T \to 0$.

### Existence of a symmetry point

Figure 1-17 indicates that the point of symmetry exists. The probability of a state having energy $E$ being vacant (or equivalently being occupied by a hole) is:

$$f_h(E) = 1 - f(E) = 1 - \left[\frac{1}{1 - e^{(E - E_F)/k_BT}}\right] \quad \text{[substituting the value of } f(E) \text{ from Eq. (1-13)]}$$

$$= \frac{1 + e^{(E - E_F)/k_BT} - 1}{1 + e^{(E - E_F)/k_BT}} = \frac{e^{(E - E_F)/k_BT}}{1 + e^{(E - E_F)/k_BT}} = \frac{1}{1 + e^{(E_F - E)/k_BT}}$$

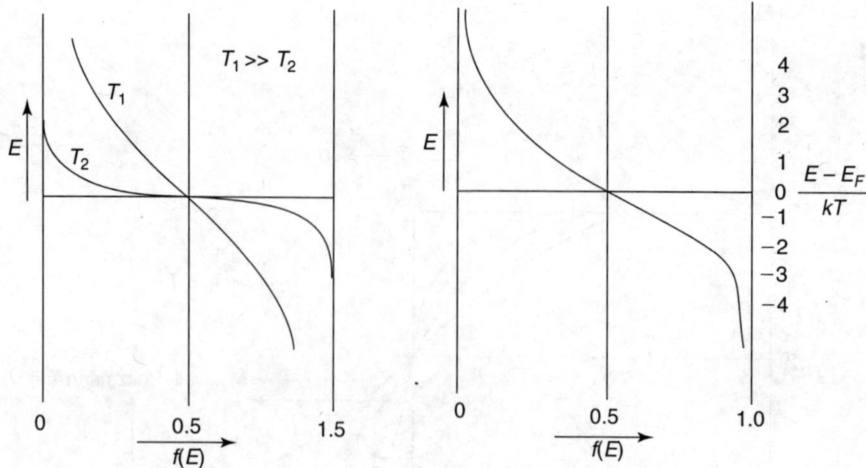


**Figure 1-17** *E* plotted along *y*-axis and *f(E)* is plotted along *x*-axis

Thus, $$f_h(E) = \frac{1}{1 + e^{(E_F - E)/k_BT}} \tag{1-16}$$

We can now conclude that the vacant states below the Fermi energy have the same characteristics as the filled states above Fermi energy with a reversal in the sense of energy. This symmetry proves to be very important in connection to the understanding of the dynamics of holes.

**Differentiating *f(E)* partially with respect to *E* and $E_F$**

Using Eq. (1-13) we get:

$$\frac{\partial F}{\partial E} = -\frac{\partial f(E)}{\partial E_F} \tag{1-17}$$

Thus, the slopes are equal in magnitude but opposite in signs.

**When $T \to 0$**

It can be proved that:

$$\frac{\partial f\,(E)}{\partial E} = -\partial'(E - E_F) \tag{1-18}$$

where, $\partial'$ is the Dirac's delta function.

This very important result leads to considerable simplifications of many tedious mathematical formulations.

Figure 1-18 shows the plot of $\left[-\frac{\partial f(E)}{\partial E}\right]$, which tends to the $E_F$ horizontally and the infinity vertically resembling the nature of the Dirac delta function when $T \to 0$.

**Under thermal equilibrium calculation of the term $\frac{\partial E}{\partial T}$**

We can write that:

$$\frac{\partial E}{\partial T} = \left[\frac{\partial f(E)}{\partial E}\right]^{-1}\left[\frac{\partial f\,(E)}{\partial T}\right] \tag{1-19}$$

From Eq. (1-19), differentiating partially with respect to *E* and *T*, we can write that:

$$\frac{\partial f(E)}{\partial E} = -\frac{e^{(E-E_F)/k_BT}}{\left[1+e^{(E-E_F)/k_BT}\right]^2}\frac{1}{k_BT} \tag{1-20}$$

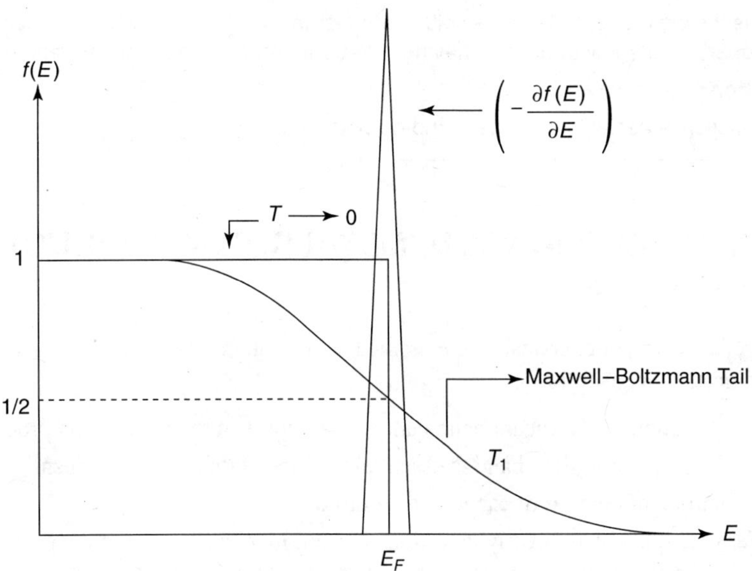


**Figure 1-18** Plot of $\left[-\frac{\partial f(E)}{\partial E}\right]$ versus $E$

$$\frac{\partial f(E)}{\partial E} = -\frac{e^{(E-E_F)/k_BT}}{\left[1+e^{(E-E_F)/k_BT}\right]^2}\left(-\frac{E-E_F}{k_BT^2}\right) \tag{1-21}$$

Substituting the values of Eqs. (1-19) and (1-20) in Eq. (1-21), we get:

$$\frac{\partial E}{\partial T} = \frac{E_F - E}{T} \tag{1-22}$$

Therefore, the change of carrier energy with respect to temperature is the difference of energy with respect to Fermi energy divided by temperature. Hence, we can conclude that there are three cases:

**Case I:** If $E_F > E$, then $\frac{\partial E}{\partial T} > 0$

**Case II:** If $E_F = E$, then $\frac{\partial E}{\partial T} = 0$

**Case III:** If $E_F < E$, then $\frac{\partial E}{\partial T} < 0$

**When** $\left(\frac{E - E_F}{k_BT}\right) >> 1$

In this case 1 is neglected with respect to $\exp\left(\frac{E-E_F}{k_BT}\right)$ in the Fermi–Dirac distribution function and Eq. (1-13) gets simplified as:

$$f(E) = \frac{1}{\exp\left(\frac{E-E_F}{k_BT}\right)} = A_o \exp\left(\frac{-E}{k_BT}\right)$$

where,
$$A_o = \exp\left(\frac{E_F}{k_BT}\right)$$

This equation is the well known Maxwell–Boltzmann distribution function, which is often used for studying the physical properties of semiconductors having non-degenerate carrier concentration.

**Fermi–Dirac function**

The Fermi–Dirac function is basically a no-zero and no-pole function.

---

## 1-3 BASIS OF CLASSIFICATION: METALS, SEMICONDUCTORS AND INSULATORS

Based on the energy band structure, crystals are classified as insulators, semiconductors and metals. The basic criteria for the classification are as follows:

(i) A fully filled-up or completely vacant band cannot participate in the conduction process
(ii) Presence of an incompletely filled band is essential for the conduction mechanism
(iii) Availability of sufficient number of carriers is required
(iv) Position of Fermi level, which directly determines the entire classification process, and is in turn determined by carrier density, temperature and other quantum parameters

A simplified description of the nature of energy band is as follows. The total energy $E$ of a conduction electron is given by:

$$E = \text{Kinetic energy} + \text{Potential energy}$$

$$= \frac{p^2}{2m_c^*} + \text{Potential energy}$$ (where, $m_c^*$ is the effective mass of the electron in the conduction band)

$$= \frac{p^2}{2m_c^*}$$ (assuming that the energy is measured from the edge of the conduction band $E_C$)

Since we have already proved that $p = \hbar k$, we can write that:

$$E = \frac{\hbar^2 k^2}{2m_c^*} \tag{1-23}$$

This indicates the parabolic dependence between the energy- and wave-vector, as shown in the Fig. 1-19, for constant effective mass $m_c^*$. This is known as the *electron parabola*. In Fig. 1-19 the horizontal line $E_C$ indicates the edge of the conduction band, the horizontal line $E_V$ is called the edge of valence band, the dotted horizontal line near $E_C$ is called the donor level $N_D$ (to be explained later on), the dotted horizontal line near $E_V$ is called the acceptor level $N_A$ (to be explained later on), $E_d$ is called the donor energy level, and $E_a$ is the acceptor energy level. The energy is measured from the edge of the conduction band $E_C$ in the vertical upward direction. The band gap $E_g$ is defined as $E_g = E_C - E_V$. For positive band gap, $E_C > E_V$; for negative band gap, $E_C < E_V$; and for zero band gap, $E_C = E_V$.

Below the line, $E_V$, the heavy hole, light hole and split-off hole parabolas have been drawn, since only three types of holes exist experimentally. The effective masses of the three types of holes are $m_{hh}$, $m_{lh}$ and $m_{soh}$ respectively.

### 1-3-1 Insulators ($E_g$ >> 4 eV)

For insulators, the forbidden energy gap is very large with both the bands being parabolic in nature under simplified assumptions. In such solids, at ordinary temperatures only a few electrons can acquire enough

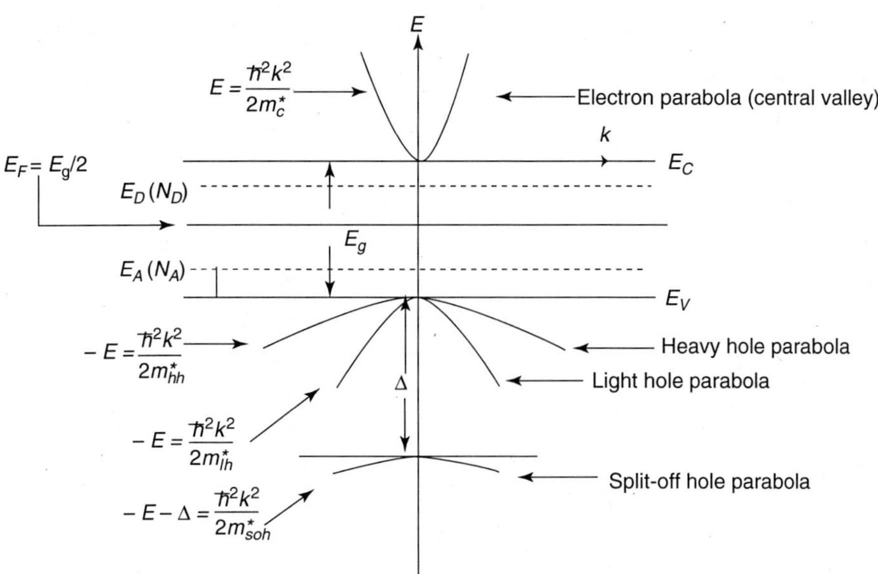


**Figure 1-19** Typical model of an energy band structure of the crystalline material (Δ is called spin orbit splitting constant of valence band)

thermal energy to move from the valence band into the conduction band. The valence band is completely filled and after a very large band gap the conduction band is completely empty. Thus, the insulator in principle possesses infinite resistivity. In this case the temperature plays the key role in the transfer process and not the movement of Fermi energy by doping. With only a few free electrons present in the conduction band, an insulator is a bad conductor of electricity. Diamond, with its forbidden gap of 6 eV, is a good example of an insulator. The energy band diagram of an insulator is shown in Fig. 1-20.

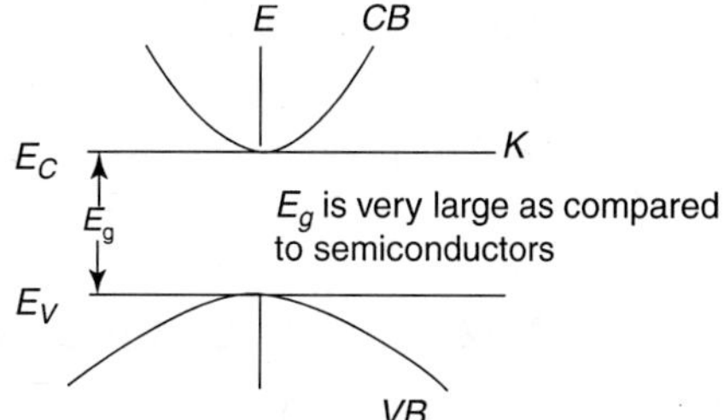


**Figure 1-20** Energy band diagram for insulators

## 1-3-2 Semiconductors (0 eV ≤ $E_g$ ≤ 4 eV)

The crystalline material with a ≤ 4 eV forbidden energy gap between the valence and the conduction band (approximately since the upper limit increases with the advance of modern fabrication technologies) is referred to as the semiconductor. Figure 1-19 shows a perfectly valid model for semiconductors in general. Germanium and silicon, which are the popular elemental semiconductors, have forbidden gaps of 0.78 and 1.2 eV respectively, at 0 K. The dispersion relation of the conduction electrons in elemental semiconductors is parabolic. As the forbidden gap is narrow, few of the valence electrons attain sufficient thermal energy to jump across into the empty conduction band. These electrons become free and can participate in the conduction process influenced by the applied electric field. The band gap is a function of the temperature also and hence the electrical properties may be tailored by fabrication of compound semiconductors (e.g. GaAs, InP, AlGaAs)—the criterion for composition being compatible lattice constants. The conduction electrons in III–V, II–VI and IV–VI semiconductors are defined by the Kane, Hopfield and Cohen models respectively, which take into account the various specialized band constants of the said semiconductors. Besides, semiconductors can be defined with respect to the position of Fermi energy, which can be controlled by doping. In semiconductors the effective mass is proportional to the band gap. The band gap $E_g$, in general,

**Table 1-3** Band gaps of commercially important semiconductors at 300 K

| *Name of Semiconductor* | *Energy Band Gap (eV)* |
|---|---|
| Mercury Telluride (HgTe) | 0 |
| Lead Sellinide (PbSe) | 0.15 |
| Indium Antimonide (InSb) | 0.24 |
| Indium Arsenide (InAs) | 0.35 |
| Germanium (Ge) | 0.66 |
| Silicon (Si) | 1.12 |
| Indium Phosphide (InP) | 1.35 |
| Gallium Arsenide (GaAs) | 1.43 |
| Gallium Phosphide (GaP) | 2.24 |
| Aluminium Arsenide (AlAs) | 2.94 |
| Silicon Carbide (SiC) | 2.99 |
| Gallium Nitride (GaN) | 3.36 |
| Zinc Sulphide ( ZnS ) | 3.84 |

is a function of the temperature, and in accordance with the Varshni law can be expressed as $E_g(T) = E_g(0) - [AT^2/(T + B)]$ where, $E_g(0)$ is the energy band gap at 0 K, and $A$ and $B$ are the constants of the semiconductor specimen. It may be noted that band gap increases with external pressure and magnetic field, whereas it decreases with an externally applied electric field. Table 1-3 exhibits the values of the band gap of a few commercially available semiconductors at 300 K.

## 1-3-3 Metals (Inter-penetrating Band Structure)

A crystalline solid is referred to as a metal when the conduction band is filled, as shown in Fig. 1-21. When the electric field is applied, these electrons acquire energy from the field and produce an electric current. This makes a metal a good conductor of electricity. The electrons in the conduction band are the *conduction electrons*. The holes at the top of the valence band predominate in the current flow as they reside at a higher energy level. The metals have a very large conductivity because of the very large number of free carriers participating in the conduction process. Thus, it is difficult to alter the conductivity of metals as a result of this.

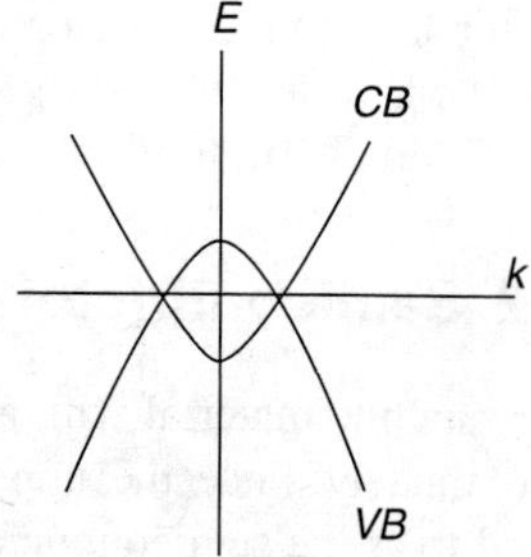


**Figure 1-21** Energy band diagram for metals

## Solved Examples

**Example 1-3** The effective mass of the conduction electron in a semiconductor at the band edge ($m_c^*$) is $0.1\ m_0$. Calculate the energy of the electron as measured from the edge of the conduction band in the vertically upward direction ($E$) corresponding to the wave-vector having the value 0.3/Å.

**Solution:**

The $E$–$k$ relation of the conduction electrons is given by $E = \hbar^2k^2/2m_c^*$

Substituting the data provided, we get:

$$E = \frac{(1.05 \times 10^{-34}\ \text{Js})^2\,(0.3 \times 10^{-10}\ \text{m}^{-1})^2}{2(0.1 \times 0.91 \times 10^{-30}\ \text{kg})} = 5.44 \times 10^{-19}\ \text{J}$$

$$= \frac{5.44 \times 10^{-19}}{1.6 \times 10^{-19}}\ \text{eV} = 3.4\ \text{eV}$$

**Example 1-4** Calculate the energies of electrons in GaAs and InAs conduction band with $k$-vectors (0.01, 0.01, 0.01) $\text{Å}^{-1}$. Refer the energies to the conduction band edge value and use the data you need.

**Solution:**

We know that $m^*_{c\text{GaAs}} = 0.067\ m_0$ and $m^*_{c\ \text{InAs}} = 0.01\ m_0$.

The energy of the electron in GaAs under parabolic approximation is:

$$E_{\text{GaAs}} = \frac{\hbar^2 k^2}{2m^*_{c\text{GaAs}}}$$

Substituting the data provided we get:

$$E_{\text{GaAs}} = \frac{(1.05 \times 10^{-34})^2\,[3(0.01 \times 10^{10})\ \text{m}^{-1}]^2}{2(0.067 \times 0.91 \times 10^{-30}\ \text{kg})} = 2.7 \times 10^{-21}\ \text{J} = 16.9\ \text{meV}$$

Similarly for InAs, we get:

$$E_{\text{InAs}} = \frac{(1.05 \times 10^{-34})^2\,[3(0.01 \times 10^{10})\ \text{m}^{-1}]^2}{2(0.01 \times 0.91 \times 10^{-30}\ \text{kg})} = 1.8 \times 10^{-21}\ \text{J} = 24.25\ \text{meV}$$

**Example 1-5** The effective mass of the conduction electron in $n$-GaAs at the band edge ($m^*_c$) is 0.067 $m_0$. Calculate the energy of the electron as measured from the edge of the conduction band in the vertically upward direction ($E$) corresponding to the wave-vector having the value $k$ = (0.1, 0.1, 0, 0)$\text{Å}^{-1}$. Comment on the answer.

**Solution:**

The $E$–$k$ relation of the conduction electrons is given by $E = \hbar^2k^2/2m^*_c$.

Substituting the data from the problem in the above equation we get,

$$E = \frac{(1.05 \times 10^{-34}\ \text{Js})^2\,[(0.1 \times 10^{10}\ \text{m}^{-1})^2 + (0.1 \times 10^{10}\ \text{m}^{-1})^2]}{2(0.067 \times 0.091 \times 10^{-30}\ \text{kg})}$$

$$= 1.8 \times 10^{-19}\ \text{J} = 1.125\ \text{eV}$$

**Comment:** At such a high-energy parabolic approximation of the dispersion relation used in this example is not at all good. The non parabolic $E$–$k$ dispersion relation is more appropriate.

**Example 1-6** An electron in the central Γ-valley of GaAs is to be transferred to the satellite $L$-valley. The energy separation between Γ and $L$ point is 0.3 eV. Using the parabolic expression for the band structure of GaAs along (111) direction, estimate the smallest $k$-vector along $x$-direction that is needed for this transition. The electron in the Γ-valley must have the energy equal to the position of the $L$-valley.

**Solution:**

To find the shortest $k$-vector needed to transfer the electron in the Γ-valley, we need to calculate the $k$-vector for an electron in the Γ-valley along the (111) direction and with energy of 0.3 eV. Using the parabolic expression for the energy we have along the (111) direction:

$$E = \frac{\hbar^2}{2m_c^*}\left(k_x^2 + k_y^2 + k_z^2\right) = \frac{3\hbar^2 k_x^2}{2m_c^*}$$

Therefore, $$k_x = \left(\frac{2m_c^* E}{3\hbar^2}\right)^{1/2}$$

Substituting $m_c^* = 0.067\ m_0$ and $E = 0.3$ eV, we get:

$$k_x = \left[\frac{2(0.067 \times 0.91 \times 10^{-30}\ \text{kg})(0.3 \times 1.6 \times 10^{-19}\ \text{J})}{3(1.05 \times 10^{-34}\ \text{Js})^2}\right]^{1/2}$$

Therefore, $$k_x = 4.2 \times 10^8\ \text{m}^{-1}$$

**Example 1-7** Calculate the energies of electrons in GaAs and InAs conduction band with $k$-vectors (0.01, 0.01, 0.01) Å$^{-1}$. Refer the energies to the conduction band edge value and use the data you need.

**Solution:**

We know that $m^*_{c\text{GaAs}} = 0.067m_0$ and $m^*_{c\text{InAs}} = 0.01m_0$.

The energy of the electron in GaAs under parabolic approximation is:

$$E_{\text{GaAs}} = \frac{\hbar^2 k^2}{2m^*_{c\text{GaAs}}}$$

Substituting the data, we get:

$$E_{\text{GaAs}} = \frac{(1.05 \times 10^{-34})^2[3(0.01 \times 10^{10})\ \text{m}^{-1}]^2}{2(0.067 \times 0.91 \times 10^{-30}\ \text{kg})} = 2.7 \times 10^{-21}\ \text{J} = 16.9\ \text{meV}$$

Similarly for InAs, we can get:

$$E_{\text{InAs}} = \frac{(1.05 \times 10^{-34})^2\,[3(0.01 \times 10^{10})\ \text{m}^{-1}]^2}{2(0.01 \times 0.91 \times 10^{-30}\ \text{kg})} = 1.8 \times 10^{-21}\ \text{J} = 24.25\ \text{meV}$$

■

## 1-4 INTRINSIC SEMICONDUCTORS

A semiconductor without impurities is termed as a pure or intrinsic semiconductor. We shall prove that $E_F = E_g/2$ in this case. The electron and hole concentration in an intrinsic semiconductor are equal because carriers within a very pure material are created in pairs (see Table 1-4).

**Table 1-4** Intrinsic concentration ($n_i$) at room temperature

| *Material* | *Concentration of Electrons (holes)* |
|---|---|
| GaAs | $2 \times 10^6$/cm$^3$ |
| Si | $1 \times 10^{10}$/cm$^3$ |
| Ge | $2 \times 10^{13}$/cm$^3$ |

## 1-5 EXTRINSIC SEMICONDUCTORS

Doped semiconductors whose properties are controlled by adding the impurity atoms are called extrinsic semiconductors. Doping increases the conductivity of a semiconductor.

### 1-5-1 Doping

Addition of impurity atoms in semiconductors in an appropriate ratio is called doping. Doping concentration varies according to the number of dopant atoms added, depending on the amount of doping required by a semiconductor. This is classified in Table 1-5.

**Table 1-5** Doping concentration

| *Type of doping* | *Ratio* |
|---|---|
| Heavily doped | $1:10^3$ |
| Moderately doped | $1:10^6$ |
| Low doped | $1:10^{10}$ |

### 1-5-2 Dopants

Specific impurity atoms, which are added to semiconductors in controlled amounts for the purpose of increasing the carrier concentration, are called *dopants* (see Table 1-6). Dopants have a remarkable effect on the electrical properties of the semiconductors.

**Table 1-6 Dopants**

| *Type of dopant* | *Elements used* |
|---|---|
| *n*-type | P, As, Sb |
| *p*-type | B, Ga, In, Al |

**Donor.** A donor is an impurity atom, which increases the electron concentration and is called an *n*-type dopant.

**Acceptor.** An acceptor is an impurity atom, which increases the hole concentration and is called a *p*-type dopant.

**Majority carriers.** These are the most abundant carriers in a given semiconductor, for example, electrons in an *n*-type material and holes in a *p*-type material.

**Minority carriers.** These are the least abundant carriers in a given semiconductor, for example, holes in an *n*-type material and electrons in a *p*-type material. The *n*-type semiconductors, as shown in Fig. 1-22, have the following properties:

(i) The donor energy level $E_D$ is situated near $E_c$ and lies within the band gap
(ii) The Fermi level is near $E_c$ and lies within the band gap under the condition of carrier non-degeneracy
(iii) The electron concentration is much greater than the hole concentration
(iv) The probability distribution function is given by:

$$f(E) = \frac{1}{1 + e^{(E - E_F)/k_B T}}$$

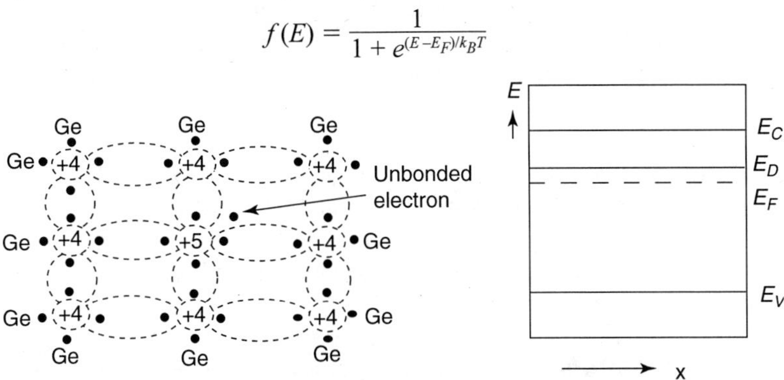


**Figure 1-22** The *n*-type semiconductor (Ge)

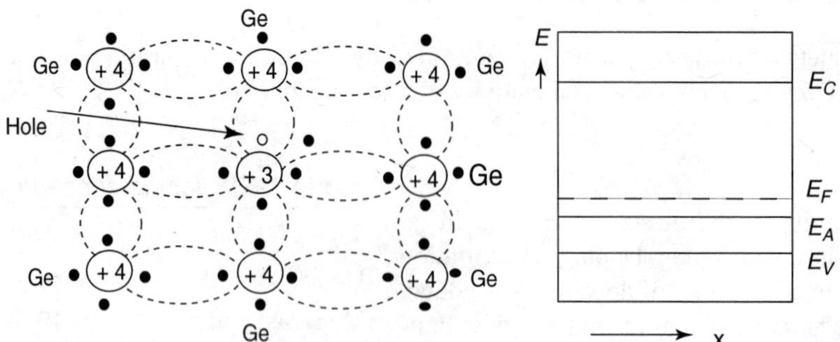


**Figure 1-23** The *p*-type semiconductor (Ge)

The *p*-type semiconductor, as shown in Fig. 1-23, has the following properties:

(i) The acceptor energy level $E_A$ is situated near $E_V$ and lies within the band gap

(ii) The Fermi level is near $E_v$ and lies within the band gap under the condition of carrier non-degeneracy

(iii) The electron concentration is much less than the hole concentration

(iv) The probability distribution function is given by:

$$f_h(E) = \frac{1}{1 + e^{(E_F - E)/k_B T}}$$

**Degeneracy in semiconductors.** The phenomenon of degeneracy in semiconductors is directly related to the movement of the Fermi energy. In intrinsic semiconductors, as noted already, $E_F = -E_g/2$. Let us discuss *n*-type semiconductors for which as a consequence of doping, $E_F$ will move toward $E_C$. When the Fermi energy comes within the energy interval $k_BT$, as measured in the edge of the conduction band in a vertically downward direction while the Fermi energy lies within the band gap, the degeneracy starts in. Thus, in the region $E_F < E_C - k_BT$, the semiconductor is known to be non-degenerate. When $E_F$ touches $E_C$ as a consequence of doping, the semiconductor is called critically degenerate. When Fermi energy is well above the conduction band $E_C$, the semiconductor is called highly degenerate. As a consequence of doping when $E_F > 5k_BT$, the semiconductor is called extremely degenerate and in this case $f(E) = 1$. When $E_F$ is well above $5\,k_BT$, we get the heavily doped semiconductors and as a consequence of heavy doping free carrier states exist in the band gap. When as a consequence of doping the $E_F$ even crosses the above limit, the semiconductor becomes amorphous.

**Compound semiconductors.** The III–V, ternary, quaternary, II–VI and IV–VI types of semiconductors are known as compound semiconductors, and their uses have been described in Table 1-1 and Table 1-2, respectively. They are basically narrow-gap semiconductors and through band gap engineering their band gap can be tailored.

**Zero band-gap semiconductors.** This is a special case of compound semiconductors where the band-gap is zero (e.g. *HgTe*), shown in Fig. 1-24. These materials are intrinsically degenerate.

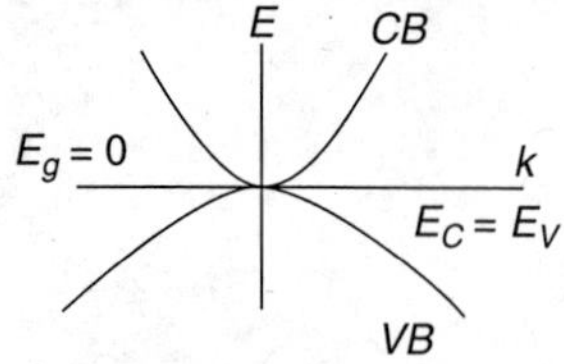


**Figure 1-24** Zero band-gap material (*HgTe*)

### 1-5-3 Carrier Statistics in *n*- and *p*-type Semiconductors

In this section we shall introduce a rather difficult, but very important concept called the density-of-state function.

#### Formulation of carrier statistics in *n*-type semiconductors

Carrier statistics refer to the relation between the electron concentration and the Fermi energy, which in turn depends on a very important concept known as the density-of-states function. The density-of-states function in three dimensions is defined as the number of carrier states per unit volume of wave-vector space per unit energy interval. The generalized formula of the density-of-state function is given by:

$$N(E) = \frac{2}{(2\pi)^3}\frac{dV(E)}{dE}\left(\frac{1}{eV\,m^3}\right) \tag{1-24}$$

where, $V(E)$ is the volume of $k$-space, and the valley degeneracy has been assumed as unity. The parabolic dispersion relation of the conduction electrons $V(E)$ should be determined from Eq.(1-23) in the following form:

$$k^2 = 2m_c^*\frac{E}{\hbar^2}$$

This is basically the equation of the electron parabola, since we have assumed that the simplified dispersion law of the conduction electron is parabolic for *n*-type semiconductors. Now, we can write:

$$k_x^2 + k_y^2 + k_z^2 = 2m_c^*\frac{E}{\hbar^2} \tag{1-25}$$

This indicates that the constant energy surface is a sphere in $k$-space whose volume $V(E)$ can be written as:

$$V(E) = \frac{4\pi}{3}\left(\frac{2m_c^*E}{\hbar^2}\right)^{3/2} \tag{1-26}$$

Using Eq.(1-24) and Eq.(1-26), we get:

$$N(E) = \frac{2}{8\pi^3}\frac{4\pi}{3}\left(\frac{2m_c^*}{\hbar^2}\right)^{3/2}\frac{3}{2}\sqrt{E}$$

Thus,

$$N(E) = 4\pi\left(\frac{2m_c^*}{h^2}\right)^{3/2}\sqrt{E}\,\frac{1}{eV\,m^3} \tag{1-27}$$

Equation (1-27) is known as the inverted parabolic dependence of the electronic density-of-states in parabolic *n*-type semiconductors, as shown in Fig. 1-25.

The density-of-state is used in calculating the carrier density, Hall coefficient, thermoelectric power, photo-emitted current density, mobility and almost all the transport parameters of semiconductor devices. The importance of the density-of-state function becomes apparent with the advent of nanotechnology. In two dimensions, the unit of the density-of-state is 1/eVm$^2$ and in one dimension the unit of the same function is 1/eVm.

#### Electron concentration in *n*-type semiconductors

Since $N_c(E)\,dE$ represents the number of conduction band states/cm$^3$ lying in the $E$ to $E + dE$ energy range, and $f(E)$ specifies the probability that an available state at an energy $E$ will be occupied by an electron, it then

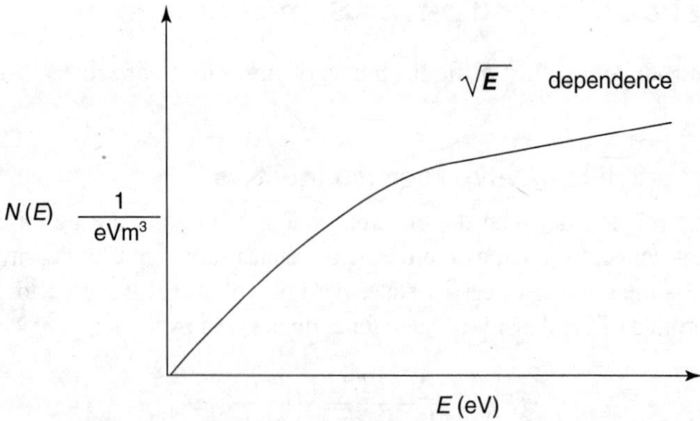


**Figure 1-25** Variation of density of states with energy E

follows that $N_c(E) f(E)\, dE$ gives the number of conduction band electrons/cm$^3$ lying in the $E$ to $E + dE$ energy range. Thus, the total electron concentration is given by:

$$n = \int_{E'}^{E_{top}} N_c(E) f(E)\, dE \tag{1-28}$$

where, $E'$ can be determined from the equation $N_c(E') = 0$. From the nature of variation of $f(E)$, $E_{top}$ can be replaced by infinity without introducing any appreciable error in the subsequent calculation.

Substituting the expressions of $N_c(E)$ and $f(E)$, from Eqs. (1-27) and (1-13) in Eq. (1-28) we get:

$$n_0 = \int_0^{\infty} 4\pi \left(\frac{2m_c^*}{h^2}\right)^{3/2} \frac{\sqrt{E}}{1 + \exp\left(\frac{E - E_F}{k_B T}\right)} dE \tag{1-29}$$

Since the lower limit of integral $E_c' = 0$, substituting $x = E/k_B T$ ($x$ is a new variable of normalized energy) and $\eta = E_F/k_B T$ (normalized Fermi energy) in Eq. (1.29) we get:

$$n_0 = 4\pi \left(\frac{2m_c^* k_B T}{h^2}\right)^{3/2} \int_0^{\infty} \frac{\sqrt{x}}{1 + \exp(x - \eta)} dx \tag{1-30}$$

This integral is a special case of the Fermi–Dirac integral of the order of $j$ and is given as:

$$F_j(\eta) = \frac{1}{\Gamma(j+1)} \int_0^{\infty} \frac{x^j}{1 + \exp(x - \eta)} dx \tag{1-31}$$

where, $\Gamma(j+1)$ is the Gamma function, $\Gamma(0) = 1$ and $\Gamma(1/2) = \sqrt{\pi}$.

The following table is for advanced readers but the results are important for any level of study of semiconductor electronics.

## FOR ADVANCED READERS

### Properties of the Fermi–Dirac Integral

A few important properties of the Fermi–Dirac integral are listed (without proof), as these help in obtaining direct results without cumbersome mathematics needed for the purpose of clear physical explanation.

$$\frac{d}{d\eta}\left[F_j(\eta)\right] = F_{j-1}(\eta) \qquad \text{(a)}$$

$$\int F_j(\eta)\,d\eta = F_{j+1}(\eta) \qquad \text{(b)}$$

$$F_{½}(\eta)\,(4/3\sqrt{\pi})\,\eta^{3/2}\left(1+\frac{\pi^2}{8\eta^2}\right), \quad \eta > 0 \qquad \text{(c)}$$

$$F_0(\eta) = \text{in}\,|1 + e^{\eta}| \qquad \text{(d)}$$

$$F_j(\eta) \approx e^{\eta},\ \eta < 0 \text{ for all } j \qquad \text{(e)} \qquad (1\text{-}32)$$

Among the five properties listed (a), (c) and (e) are often used. Hence, the electron concentration can be expressed as:

$$n_0 = N_c F_{½}(\eta) \qquad (1\text{-}33)$$

where, $N_c = 2(2\pi m_c^* k_B T/h^2)$ is termed as the effective number of states in the conduction band.

**Special cases**

**Case I:** When Fermi level lies within the band gap, the semiconductor becomes non-degenerate and $\eta < 0$. Under this condition, by using formula (e) of Eq. (1-32) and using Eq. (1-33) we can write:

$$n_0 = N_c \exp(\eta) \qquad (1\text{-}34)$$

**Case II:** When $E_F$ touches the edge of the conduction band then, $\eta \rightarrow 0$ and the electron concentration assumes the form:

$$n_0 = N_c F_{½}(0) \qquad (1\text{-}35)$$

**Case III:** Under degenerate electron concentration, using formula (c) of Eq. (1-32) and using Eq. (1-33) we can write:

$$n_0 = N_c\left[\frac{4}{3\sqrt{\pi}}\,\eta^{3/2}\left(1+\frac{\pi^2}{8\eta^2}\right)\right] \qquad (1\text{-}36)$$

From Eq. (1-36), using the binomial theorem and the theory of approximation of equations, it can be proved that:

$$E_F(T) = E_F(0)\left\{1 - \frac{\pi^2 k_B^2 T^2}{12[E_F(0)]^2}\right\} \qquad (1\text{-}37)$$

**Case IV:** Under the condition of extreme degeneracy the contribution of the second term of Eq. (1-36) is much less than that of the first term, and from Eq. (1-36) we can write:

$$n_0 = 2\left(\frac{2\pi m_c^* k_B T}{\hbar^2}\right)^{3/2} \times \frac{4}{3\sqrt{\pi}}\left(\frac{E_F}{k_B T}\right)^{3/2}$$

$$n_0 = \frac{8}{3\sqrt{\pi}}\left(\frac{2\pi m_c^*}{\hbar^2}\right)^{3/2}(E_F)^{3/2} \qquad (1\text{-}38)$$

From Eq. (1-38) we observe that under the condition of extreme degeneracy, the electron concentration is independent of temperature. In general, any electronic property of any semiconductor under the condition of extreme degeneracy will be temperature-independent. This is logical, because extreme degeneracy is obtained when $T \rightarrow 0$ and $f(E) \rightarrow 1$, together with the fact that all states are filled up to Fermi level and above the Fermi level all states are vacant. At $T \rightarrow 0$, let us designate the Fermi energy by $E_F(0)$. Therefore, from Eq. (1-38) we can write:

$$E_F(0) = b \times n_0^{2/3} \qquad (1\text{-}39)$$

where,

$$b = \left(\frac{h^2}{2\pi mh^*}\right)\left(\frac{3\sqrt{\pi}}{8}\right)^{2/3}$$

From Eq. (1-39) we will find that the Fermi energy in this case is proportional to the (electron concentration)$^{2/3}$.

## Hole concentration in *p*-type semiconductors

The hole concentration $p_0$ can be expressed as:

$$p_0 = \int_{-\infty}^{-E_g} N_h(E) f_h(E)\, dE \tag{1-40}$$

where, the energy $E$ is being measured from the edge of the conduction band in the vertically upward direction as noted already; $N_h(E)$ is the density-of-state function for the heavy holes and can be written following Eq. (1-27) as:

$$N_h(E) = 4\pi\left(\frac{2m^*_{hh}}{h^2}\right)^{3/2}\sqrt{-E - E_g} \tag{1-41}$$

and $f_h(E)$ is given by Eq. (1-16). Thus, substituting Eq. (1-16) and Eq. (1-41) in Eq. (1-40) we get:

$$p_0 = 4\pi\left(\frac{2m^*_{hh}}{h^2}\right)^{3/2}\int_{-\infty}^{-E_g}\frac{\sqrt{-E-E_g}}{1+\exp\left(\frac{E_F - E}{k_BT}\right)}dE \tag{1-42}$$

Let us substitute $y = -E - E_g/k_BT$ where, $y$ is a new variable. When $E \to -E_g$, $y \to 0$; and when $E \to -\alpha$, $y \to \alpha$. Besides $dE = -k_BTdy$.
Substituting in Eq. (1-42) we have:

$$p_0 = 4\pi\left(\frac{2m^*_{hh}}{h^2}\right)^{3/2}\int_{\infty}^{0}\frac{\sqrt{yk_BT}\,(-k_BT)}{1+\exp\left[(E_F + yk_BT + E_g)/k_BT\right]}dy$$

$$p_0 = 4\pi\left(\frac{2m^*_{hh}k_BT}{h^2}\right)^{3/2}\int_{0}^{\infty}\frac{\sqrt{y}}{1+\exp(y-\eta')}dy \tag{1-43}$$

where, $\eta' = \dfrac{-E_F - E_g}{k_BT}$

Thus, from Eq. (1-43), we get:

$$p_0 = N_v F_{1/2}\left(-\eta - \frac{E_g}{k_BT}\right) \tag{1-44}$$

where, $N_v = 2(2\pi\, m^*_{hh}\, k_BT/h^2)^{3/2}$ is termed as the effective number of states in the valence band. In this case for the purpose of simplicity we have neglected light hole and split-off hole concentrations respectively. Thus, the total hole concentration in this case is the concentration of heavy holes.

**Special case**

Under non-degenerate condition, we can write:

$$p_0 = N_v \exp\left(-\eta - \frac{E_g}{k_B T}\right) \tag{1-45}$$

**Electric neutrality equation**

The expressions of $n_0$ and $p_0$ can be used to find out the position of the Fermi level in intrinsic semiconductors. In general, the Fermi level may change when impurities responsible for localized states are introduced. The total charge of all charged particles both in the crystal as a whole and in any physically small volume should be zero, which is called the electric neutrality condition. Denoting the number of electrons and holes occupying donor and acceptor levels by $n_d$, $p_d$, $n_a$, $p_a$, we can write:

$$n_0 + n_d - p_0 - p_a = N_d - N_a \tag{1-46}$$

It can be proved that:

$$n_d = \frac{N_d}{\frac{1}{2}\exp\left(\frac{E_d - E_F}{k_B T}\right) + 1} \tag{1-47}$$

and,

$$p_a = \frac{N_a}{\frac{1}{2}\exp\left(\frac{E_F - E_a}{k_B T}\right) + 1} \tag{1-48}$$

where, $N_d$ is the donor concentration. Substituting the values of $n_0$, $p_0$, $n_d$ and $p_a$ from Eqs. (1-33), (1-44), (1-47) and (1-48), in Eq. (1-46) we get:

$$N_c F_{1/2}(\eta) + \frac{N_d}{\frac{1}{2}\exp\left(\frac{E_d - E_F}{k_B T}\right) + 1} - N_v F_{1/2}\left(-\eta - \frac{E_g}{k_B T}\right) - \frac{N_a}{\frac{1}{2}\exp\left(\frac{E_F - E_a}{k_B T}\right) + 1} = N_d - N_a \tag{1-49}$$

The general solution of Eq. (1-49) is very complicated although the said equation is written in the case of only one donor and one acceptor level.

**Special cases**

**Intrinsic semiconductors.** Since there are no impurities in this case, $n_0 = p_0$ is the form of charge neutrality equation. Therefore:

$$N_c F_{1/2}(\eta) = N_v F_{1/2}\left(-\eta - \frac{E_g}{k_B T}\right) \tag{1-50}$$

(i) Let us assume that $m_c^* = m_{hh}^*$. Therefore, $N_c = N_v$ and from Eq. (1-50) we can write:

$$\eta = -\eta - \frac{E_g}{k_B T} \quad \text{or,} \quad \eta = -\frac{E_g}{2k_B T}$$

Therefore, $E_F = \frac{-E_g}{2}$, i.e., the Fermi energy is independent of temperature and lies exactly at the middle of the band gap.

(ii) Let us assume that $m_c^* \neq m_{hh}^*$.

$$N_c \exp(\eta) = N_v \exp\left(-\eta - \frac{E_g}{k_B T}\right) \quad \text{or,} \quad 2\eta + \frac{E_g}{k_B T} = \ln\left|\frac{N_v}{N_c}\right|$$

Putting the value of $N_V$ and $N_c$ we have:

$$E_F = \frac{3k_BT}{4}\ln\left|\frac{m^*_{hh}}{m^*_c}\right| - \frac{E_g}{2} \tag{1-51}$$

We observe that the position of Fermi energy depends on temperature. As the temperature rises the Fermi energy also changes. Multiplying Eq. (1-34) and Eq. (1-45) we get:

$$n_0p_0 = \exp\left(-\frac{E_g}{k_BT}\right)(N_cN_v) = n_i^2 \tag{1-52}$$

where, $n_i$ is the intrinsic concentration. Equation (1-52) is known as the *law of mass action*. Multiplying Eq. (1-33) and Eq. (1-44) we get:

$$n_0\,p_0 = N_cN_vF_{½}\,(\eta)F_{½}\left(-\eta - \frac{E_g}{k_BT}\right) \neq n_i^2 \tag{1-53}$$

Thus, for degenerate parabolic energy bands, the law of mass action is not valid. Equation (1-52) is only valid for non-degenerate semiconductors having parabolic energy bands. Thus, in the non-degenerate semiconductors the product of electron and hole concentrations is independent of Fermi energy, and hence, of the doping of the semiconductors; it is equal to the square of the concentration of one type of carriers in the intrinsic semiconductors. Thus, if the electron concentration is known, then the hole concentration $p_0$ can be obtained from Eq. (1-52) as:

$$p_0 = \frac{n_i^2}{n_0} \tag{1-54}$$

Substituting the numerical values of the constants involved in the expressions for $N_c$ and $N_v$ in Eq. (1-52) we get:

$$n_i^2 = 2.31 \times 10^{31}\left(\frac{m^*_cm^*_{hh}}{m_0^2}\right)^{3/2} T^3 \exp\left(\frac{-E_g}{k_BT}\right) \tag{1-55}$$

Therefore,

$$\ln n_i = \text{const} - \left(\frac{3}{2}\ln\frac{1}{T}\right) - \left(\frac{E_g}{2k_B}\frac{1}{T}\right) \tag{1-56}$$

The second term in Eq. (1-56) is negligible with respect to the third term, therefore, the graphical dependence of $\ln(n_i)$ on inverse temperature will nearly be a straight line, which is shown in the plot of Fig. 1-26. The slope of the straight line is determined by the band gap through the equation

$$\tan\theta = \frac{E_g}{2k_B}$$

Therefore, $E_g = 2k_B\,|\tan\theta|$.

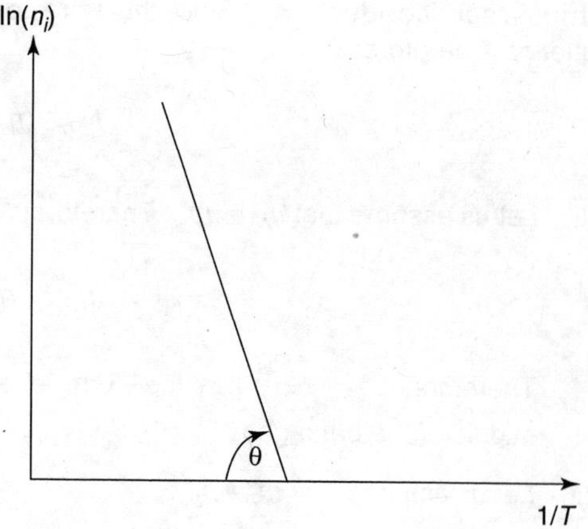

**Figure 1-26** The dependence of $\ln(n_i)$ on the inverse temperature in an intrinsic semiconductor

**Extrinsic semiconductors.** A semiconductor with impurities is termed as extrinsic semiconductor. Let us consider a semiconductor with impurity of one type, for which $N_d \neq 0$, $N_a = 0$. Therefore, the electric neutrality equation assumes the form:

$$n + n_d - p = N_d \tag{1-57}$$

At low temperatures, the impurity concentration plays the leading part and therefore, under the condition of non-degeneracy the above equation assumes the form:

$$N_c \exp\left(\frac{E_F}{k_B T}\right) = N_d \left[2 \exp\left(\frac{E_F}{k_B T}\right) \times \exp\left(\frac{E_d}{k_B T}\right) + 1\right]^{-1} \tag{1-58}$$

Substituting $x = \exp\left(\frac{E_F}{k_B T}\right)$ in Eq. (1-58), we have:

$$x^2 + \frac{1}{2} \exp\left(\frac{E_d}{k_B T}\right) \times x - \frac{N_d}{2N_c} \exp \frac{E_d}{k_B T} = 0 \tag{1-59}$$

Solving Eq. (1-59) we get:

$$x = \frac{1}{4} \exp\left(\frac{E_d}{k_B T}\right) \left[\pm \sqrt{1 + \frac{8N_d}{N_c} \exp\left(\frac{E_d}{k_B T}\right)} - 1\right] \tag{1-60}$$

Since $x > 0$, the minus sign before the radical should be omitted and Fermi energy can be expressed as:

$$E_F = k_B T \ln\left\{\frac{1}{4} \exp\left(\frac{E_d}{k_B T}\right)\left[\sqrt{1 + \frac{8N_d}{N_c} \exp\left(\frac{E_d}{k_B T}\right)} - 1\right]\right\} \tag{1-61}$$

Equation (1-62) is valid when the conduction electron is generated mainly as a result of donor impurity ionization.

Under the condition, $\frac{8N_d}{N_c} \exp\left(-\frac{E_d}{k_B T}\right) >> 1$, Eq. (1-60) is given by:

$$x = \frac{1}{4} \exp\left(\frac{E_d}{k_B T}\right)\left(\sqrt{\frac{8N_d}{N_c}}\right) \exp\left(-\frac{E_d}{2k_B T}\right) = \frac{1}{4} \exp\left(\frac{E_d}{2k_B T}\right)\left(\sqrt{\frac{8N_d}{N_c}}\right)$$

$$= \exp\left(\frac{E_F}{k_B T}\right) \tag{1-62}$$

Writing the expression for the position of Fermi level using Eq. (1-62) we get:

$$E_F = \frac{E_d}{2} + \frac{k_B T}{2} 1\text{n}\left(\frac{N_d}{2N_c}\right) \tag{1-63}$$

At $T \to 0$:

$$E_F = \frac{E_d}{2} \tag{1-64}$$

Thus, we can infer that the Fermi level lies midway between the bottom of the conduction band and the impurity level. The Fermi level rises as the temperature is increased, reaches a maximum at a certain temperature and then drops again.

Under the condition, $\frac{8N_d}{N_c} \exp\left(-\frac{E_d}{k_B T}\right) << 1$, Eq. (1-60) can be written as:

$$x = \frac{1}{4} \exp\left(\frac{E_d}{k_B T}\right)\left[1 + \frac{4N_d}{N_c} \exp\left(\frac{E_d}{k_B T}\right) + \ldots - 1\right] = \frac{N_d}{N_c} \tag{1-65}$$

The expression of Fermi level from the above equation is given as:

$$E_F = k_B T \ln\left(\frac{N_d}{N_c}\right) \tag{1-66}$$

Since it is valid for $N_c >> N_d$, the logarithm in Eq. (1-66) should be negative, and the Fermi level should sink with the rise in temperature. So, the electron concentration for this case is given as:

$$n_0 = N_c \exp\left(\frac{E_F}{k_B T}\right) = N_c \exp\left[\ln\left(\frac{N_d}{N_c}\right)\right] = N_d \tag{1-67}$$

Therefore, we can write that the electron concentration is independent of temperature and equal to the impurity concentration. This temperature range is termed *impurity depletion range*. The charge carriers are termed majority carriers when their concentration exceeds that of intrinsic carriers ($n_i$) at a given temperature. When their concentration is below $n_i$, they are termed as minority carriers. Thus, in *n*-type semiconductors electrons are the majority carriers. Besides, in the impurity depletion range, the majority carrier concentration remains constant; the concentration of the minority carrier on the other hand must change rapidly with temperature.

The hole concentration can be expressed from Eq. (1-54) as:

$$p_0 = \frac{n_i^2}{n_0} = \frac{n_i^2}{N_d} = \frac{N_c N_v}{N_d} \exp\left(-\frac{E_g}{k_B T}\right) \tag{1-68}$$

This expression is valid as long as the hole concentration remains much less than the electron concentration. Thus:

$$p_0 << n_0 = N_d$$

In the range of high temperature, the concentration of a hole increases and becomes comparable to the electron concentration. Hence, the electrical neutrality Eq. (1-57) can be expressed as,

$$n_0 = p_0 + N_d \tag{1-69}$$

For non-degenerate case, Eq. (1-69) can be written as:

$$n_0 = \frac{n_i^2}{n_0} + N_d \quad \text{or,} \quad n_0^2 - n_0 N_d - n_i^2 = 0 \tag{1-70}$$

On solving Eq. (1-70) we get:

$$n_0 = \frac{N_d}{2}\left(1 \pm \sqrt{1 + \frac{4n_i^2}{N_d^2}}\right) \tag{1-71}$$

Since the expression under the radical sign exceeds unity, and $n_0 > 0$, the minus sign should be omitted. Hence, the electron and hole concentrations are given by:

$$n_0 = \frac{N_d}{2}\left(1 + \sqrt{1 + \frac{4n_i^2}{N_d^2}}\right) \tag{1-72}$$

$$p_0 = \frac{2n_i^2}{N_d\left(1 \pm \sqrt{1 + \frac{4n_i^2}{N_d^2}}\right)} \tag{1-73}$$

Hence, the Fermi energy is given as:

$$E_F = k_B T \ln \frac{N_d}{2N_c}\left(1 + \sqrt{1 + \frac{4n_i^2}{N_d^2}}\right) \tag{1-74}$$

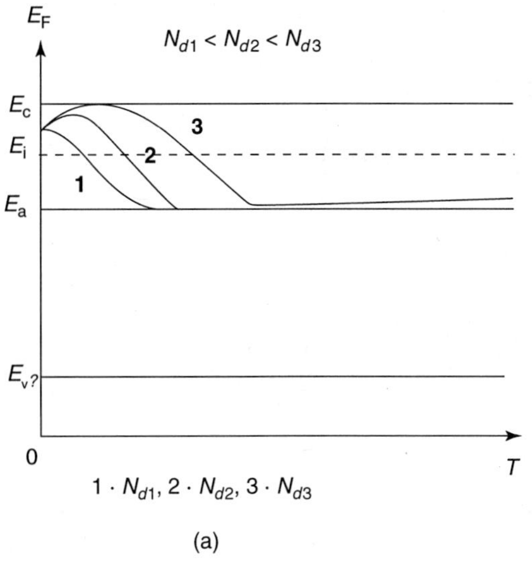


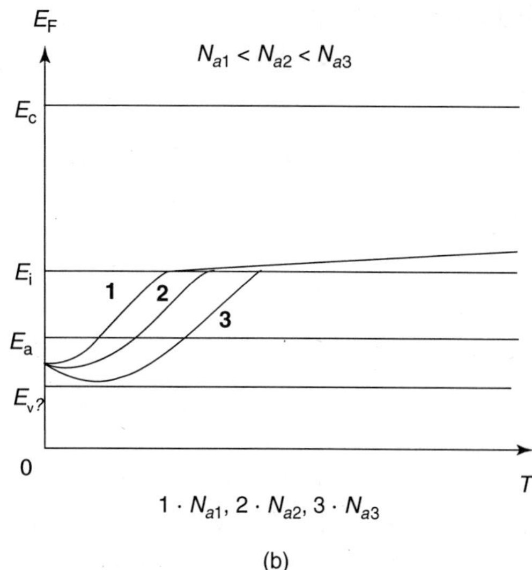


**Figure 1-27** (a) The temperature dependence of the Fermi level in a donor-doped *n*-type semiconductor, (b) the temperature dependence of the Fermi level in an acceptor-doped *p*-type semiconductor

If $\frac{4n_i^2}{N_d^2} << 1$, it follows from Eq. (1-72) and Eq. (1-74) that:

$$n_0 = N_d; \quad p_0 = \frac{n_i^2}{N_d}; \quad \text{and} \quad E_F = k_B T \ln\left(\frac{N_d}{N_c}\right) \tag{1-75}$$

If, on the other hand, $\frac{4n_i^2}{N_d^2} >> 1$, then:

$$n_0 = p_0 = n_i; \quad \text{and} \quad E_F = \frac{E_v}{2} + \frac{k_B T}{2} \ln\left(\frac{N_v}{N_c}\right) \tag{1-76}$$

as it should be for an intrinsic semiconductor.

Thus, in the temperature range there are two expressions to describe the position of the Fermi level in a non-degenerate semiconductor. The expression:

$$E_F = E_d = k_B T \ln\left[\frac{1}{4}\left(\sqrt{1 + \frac{8N_d}{N_c}\exp\left(-\frac{E_d}{k_B T}\right)} - 1\right)\right] \tag{1-77}$$

is valid from $T \rightarrow 0$ up to the depletion temperature $T_{dep}$. For temperatures in excess of $T_{dep}$ the expression is given by Eq. (1-74).

The temperature dependence of Fermi energy for *n*-type and *p*-type semiconductors is shown in Fig. 1-27.

## Solved Examples

**Example 1-8** A GaAs sample is doped *n*-type at $6 \times 10^{17}$ cm$^{-3}$. Find out the position of the Fermi level at 300 K assuming all donors are ionized. Given that $N_C = 4.45 \times 10^{17}$cm$^{-3}$

**Solution:**
Since all donors are ionized the electron concentration in the conduction band is:

$$n_0 = 6 \times 10^{17}\text{cm}^{-3} = N_D$$

Since the intrinsic carrier density is much less than the dopant density, the semiconductor is non-degenerate. Thus, using the Eq. (1-34) we get:

$$E_F = k_B T \ln\left(\frac{n_0}{N_c}\right)$$

$$= \left[0.026 \times \ln\left(\frac{6 \times 10^{17}}{4.45 \times 10^{17}}\right)\right] \text{ (At room temperature } k_B T = 0.026 \text{ eV)}$$

$$= 7.77 \times 10^{-3}$$

$$= 7.77 \text{ meV}$$

**Example 1-9** If the effective mass of an electron is equal to twice the effective mass of only the heavy hole (assuming only one type of hole) of a particular intrinsic semiconductor. Determine the position of the Fermi level at room temperature.

**Solution:**
From the question $m_e^* = 2m_h^*$ and $T = 300\,\text{K}$.

The Fermi level in an intrinsic semiconductor is given by:

$$E_F = \frac{3}{4} kT \ln\left(\frac{m_h^*}{m_e^*}\right) - \frac{E_G}{2}$$

$$= \frac{3}{4} \times 0.026 \ln\left(\frac{2m_h^*}{m_e^*}\right) - \frac{E_G}{2} \text{ (since at the room temperature } k_B T = 26 \text{ meV)}$$

$$= \left(-\frac{E_G}{2} - 0.014\right) \text{eV}$$

Thus the Fermi level is below the centre of the forbidden gap by 0.014 eV.

**Example 1-10** If the free electron concentration in $n$-GaAs is $10^{18}$ cm$^{-3}$ what is the position of the Fermi energy at 0 K as measured from the edge of the conduction band?

**Solution:**
At $T \to 0$ K, the semiconductor is extremely degenerate and the $n_0$ versus $E_F$ relation from Eq. (1-38) is given by:

Therefore,

$$E_F = \left(\frac{\hbar^2}{2m_c^*}\right)(3\pi^2 n_0)^{2/3} = \frac{(1.5 \times 10^{-34} \text{ Js})^2 \times [3 \times \pi^2 (10^{24} \text{ m}^{-3})]^{2/3}}{2(0.067 \times 0.91 \times 10^{-30} \text{ kg})}$$

$$= 8.3 \times 10^{-21} \text{ Joules} = 0.0518 \text{ eV}$$

Thus, the Fermi energy is placed 0.0518 eV above the edge of the conduction band and is within the conduction band.

■

# 1-6 ELECTRICAL CONDUCTION PHENOMENON

The concepts of mobility and scattering are basic ingredients of the electrical conduction phenomena in electronic materials and these are discussed in the following sections.

## 1-6-1 Mobility

Mobility is the essential property that characterizes the carrier transport. The carrier mobility ($\mu$) also plays a key role in characterizing the transport co-efficient in semiconductor devices. The mobility is defined as the drift velocity ($v_0$) per unit electric field, and can be written as:

$$\mu = \frac{v_0}{E_0} \tag{1-78}$$

where, $E_0$ is the applied electric field.

### Relationship to scattering

The force generated on a carrier of charge $e$ due to the application of the electric field $E_0$ is $eE_0$. The application of Newton's second law leads to the result that in a steady state $eE_0 = m^* \left(\frac{v_0}{\tau}\right)$ where, $\tau$ is known as the momentum relaxation time and has the significance that in the absence of an electric field, the momentum reduces to $e^{-1}$ of its initial value in the time $\tau$ sec. Thus, mobility is expressed as:

$$\mu = \frac{e\tau}{m^*} \tag{1-79}$$

In SI units, $e$ is in coulomb, $\tau$ is in second and $m^*$ is in kg. The dimensional substitution of these quantities in Eq. (1-79) leads to the fact that the dimension of mobility is $m^2/(V.s)$.

In general, $\tau$ and $m^*$ are the functions of carrier energy and the mobility for the carriers in isotropic energy bands can be expressed as:

$$\mu = e\left\langle \frac{\tau(E)}{m^*(E)} \right\rangle \tag{1-80}$$

where, $\langle \tau(E)/m^*(E) \rangle$ represents the average value of the bracketed quantity for all the carriers. Mobility is a measure of the ease of carrier motion within a semiconductor crystal. Carrier mobility varies directly with the momentum relaxation time when mass is constant, and inversely with the carrier effective mass when the momentum relaxation time is constant. The determination of mobility needs the expressions of $\tau(E)$ and $m^*(E)$ respectively. $\tau(E)$ can be determined through the solution of the Boltzmann transport equation (it is the integro-differential equation which generates the distribution function for a specific semiconductor under specific conditions) under different mechanisms of scattering and the specifications of the $E$–$k$ dispersion relation. The $m^*(E)$ can be derived from the specified dispersion relation of the carriers. Thus, ($\mu$) changes from semiconductor to semiconductor and also with respect to doping, temperature and other parameters.

### Effects of temperature and doping on mobility

There are various types of scattering mechanisms in semiconductors. The different types of scattering mechanisms are given in Figure 1-28.

Defect scattering includes scattering by ionized and neutral impurities, crystal defects and alloy scattering respectively. Phonon scattering occurs by the deformation potential (acoustic/optical) and polar scatterings (acoustic/optical). Among various scattering processes, two basic types of scattering mechanisms that influence electron and hole mobility are lattice scattering and impurity scattering respectively. In lattice scattering, a carrier moving through the crystal is scattered by a vibration of the lattice, caused by the temperature.

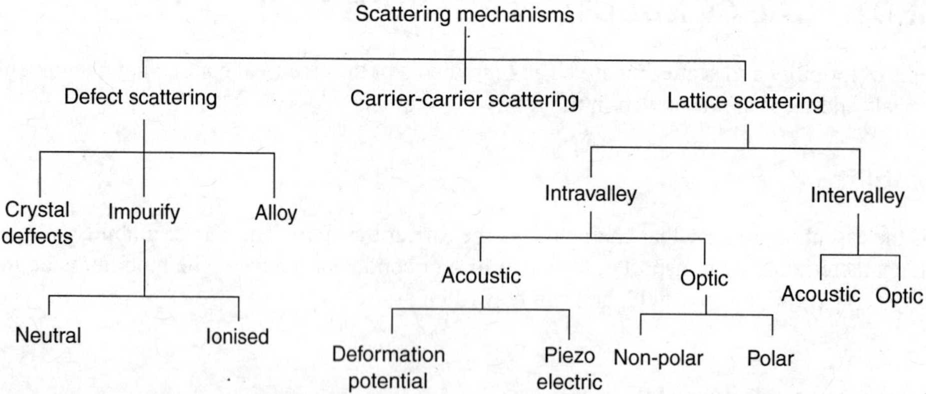


**Figure 1-28** Different scattering mechanisms in semiconductors

The frequency of such scattering events increases as temperature increases, since the thermal agitation of the lattice becomes greater. Therefore, we should expect the mobility to decrease as the sample is heated. On the other hand, scattering from crystal defects such as ionized impurities becomes the dominant mechanism at low temperatures. Since the atoms of the cooler lattice are less agitated, lattice scattering is less important; however, the thermal motion of the carriers is also slower. Since a slowly moving carrier is likely to be scattered more strongly by an interaction with a charged ion than a carrier with greater momentum, impurity scattering events cause a decrease in mobility with decreasing temperature. The approximate temperature dependencies are $T^{-3/2}$ for lattice scattering and $T^{3/2}$ for impurity scattering in the case of non-degenerate wide-gap isotropic parabolic semiconductors. Since the scattering probability is inversely proportional to the mean free time, and therefore, to mobility, the mobility due to two or more scattering mechanisms add inversely.

$$\frac{1}{\mu} = \sum_{i=1}^{n} \frac{1}{\mu_i} \tag{1-81}$$

Equation 1-81 is known as *Matthiessen's rule* and states that the resultant mobility may be deduced from the mobility due to each mechanism acting alone. It must be remarked that the Matthiessen's rule

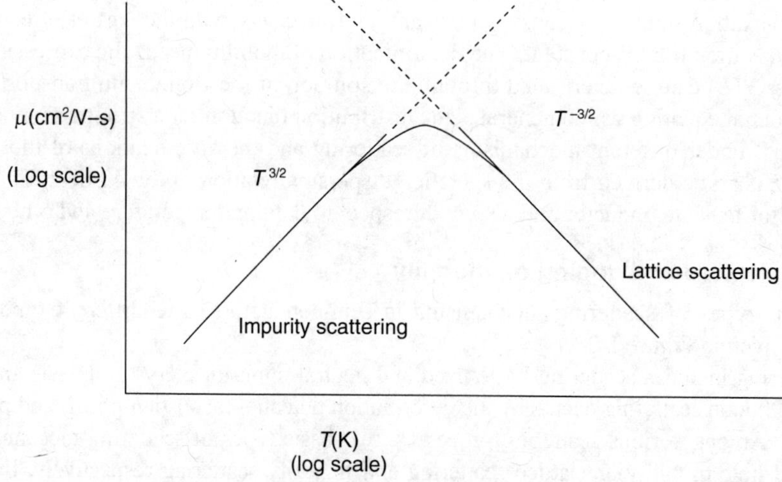


**Figure 1-29** Temperature dependence of mobility with both lattice and impurity scattering

**Table 1-7** Different scattering mechanisms and its dependence on temperature

| *Scattering Mechanisms* | *Temperature Dependence* |
|---|---|
| Acoustic phonon | $T^{-3/2}$ |
| Piezoelectric | $T^{-1/2}$ |
| Ionized impurity | $T^{3/2}$ |
| Neutral impurity | $T^{0}$ |
| Non- polar optical phonon | $T^{-3/2}$ |
| Intervalley phonon | $T^{-3/2}$ |
| Polar optical phonon | $T^{-1/2}$ |

applies only when the different scattering mechanisms have the same energy dependence. But actually it is not so, and thus, the use of Matthiessen's rule is rarely justified in practice. Nevertheless, it is commonly used because it is often easy to estimate the mobility for different scattering mechanisms independently but very difficult to do so when the processes occur simultaneously.

It may be noted that as the concentration of impurities increases, the effects of impurity scattering are felt at higher temperatures. In Fig. 1-29, the mobility of an isotropic wide-gap non-degenerate semiconductor has been shown as a function of temperature under impurity and lattice-dominated scattering mechanisms. The temperature dependence of momentum relaxation time for a few important scattering mechanisms are given in Table 1-7.

## 1-6-2 Conductivity

The expression between electrical conductivity ($\sigma$) and the current density ($J$) can be derived by considering a cylinder of area of cross section ($A$ m$^2$) as shown in Fig. 1-30. An electron having drift velocity $v$(m/s) travelling through this cylinder from one end traverses distance $v$ m in one second. The volume of the cylinder consequently becomes $Av$ m$^3$. If $n_0$ is the number of electrons per unit volume crossing this cylinder with the velocity $v$, then from the basic definition of current we can write:

$$I = n_0 \text{Ave} \tag{1-82}$$

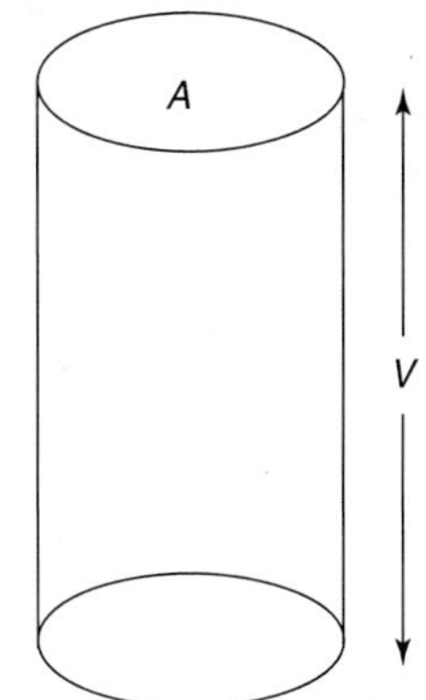


**Figure 1-30** Cylindrical model for the calculation of conductivity

from which the current density is the current per unit area and is given by:

$$J = \frac{\text{current } (I)}{\text{cross-sectional area perpendicular to the direction of flow } (A)}$$

$$J = \frac{n_0 \text{Ave}}{A} = n_0 \text{ev} \tag{1-83}$$

Substituting the expression of $v_0 = \frac{eE_0\tau}{m^*}$, as derived in the context of the derivation of the Eq. (1-79), in Eq. (1-83) we get:

$$J = \left(\frac{n_0 e^2 \tau}{m^*}\right) E_0 \tag{1-84}$$

The electrical conductivity ($\sigma$) is defined as the current density ($J$) per unit electric field ($E_0$) and can be written as:

$$\sigma = \frac{J}{E_0} \tag{1-85}$$

Comparing Eq. (1-84) and Eq. (1-85) we get:

$$J = \sigma E_0 \tag{1-86}$$

where,
$$\sigma = \frac{n_0 e^2 \tau}{m^*}$$

In general, $\tau$ and $m^*$ are functions of carrier energy and the conductivity for the carriers in isotropic energy bands, which can be expressed as:

$$\sigma = n_0 e^2 < \tau(E)/m^*(E) > \tag{1-87}$$

Using Eq. (1-80) we get:
$$\sigma = n_0 e \mu \tag{1-88}$$

Replacing $m^*$ by $m_c^*$ in Eq. (1-88), it can then be expressed as:

$$\sigma_n = n_0 e \mu_n \tag{1-89}$$

where, $\sigma_n$ and $\mu_n$ are the conductivity and mobility of the $n$-type semiconductors respectively.
Therefore, the current density for $n$-type semiconductors ($J_n$) is given by:

$$J_n = n_0 e \mu_n E_0 \tag{1-90}$$

Similarly, the conductivity for $p$-type semiconductors ($\sigma_p$) can be written as:

$$\sigma_p = p_0 e \mu_p \tag{1-91}$$

where, $\mu^p$ is the mobility of the $p$-type semiconductors.
Therefore, the current density for $p$-type semiconductors ($J_p$) is given by:

$$J_p = p_0 e \mu_p E_0 \tag{1-92}$$

The total current density $J$ in the semiconductor is the sum of $J_n$ and $J_p$. Thus:

$$J = \sigma E_0$$

where, $\sigma$ is the total conductivity. Therefore:

$$J = \sigma E_0 = J_n + J_p \tag{1-93}$$

Substituting the values of $J_n$ and $J_p$ from Eq. (1-90) and Eq. (1-92) in Eq. (1-93) we get:

$$\sigma = \sigma_n + \sigma_p = e(n_0 \mu_n + p_0 \mu_p) \tag{1-94}$$

Since this expression of conductivity is based on the concept of drift velocity, Eq. (1-94) may be called drift conductivity. For intrinsic semiconductors, $n_0 = p_0 = n_i$. Therefore, conductivity for intrinsic semiconductors ($\sigma_i$) is given by:

$$\sigma_i = e(\mu_n + \mu_p) n_i \tag{1-95}$$

The resistivity ($\rho$) is the reciprocal of conductivity. So, from Eq. (1-94), we can write:

$$\rho = \frac{1}{\sigma} = \frac{1}{e(n_0 \mu_n + p_0 \mu_p)} \tag{1-96}$$

Since $n_i$ increases with $T$, from Eq. (1-95) we can infer that $\sigma_i$ increases as $T$ increases. This property of semiconductors is utilized in thermistors, which are resistors with negative temperature co-efficient.

## 1-6-3 Diffusion of Carriers

In addition to conduction of the current by drift velocity, the transport of charge carriers in semiconductors may be accounted for by a mechanism called diffusion. Diffusion current is the net flow of the randomly moving electrons and holes from a region of high carrier concentration to regions of lower carrier density. It is the same process by which molecules in a gas arrange themselves to establish a uniform pressure. Thus, it is a process of redistribution of carriers from non-uniformity to uniformity and is analogous to *Fick's law of classical thermodynamics*. The motion of the carrier by diffusion plays a vital role in the operation of junction diodes and transistors.

An illustration of the motion of the carrier by diffusion is shown in Fig. 1-31. If we choose a reference plane at $x$ we observe that the electrons diffuse to the right because the electron density in the right-hand side is lower than that in the left-hand side. In the process of re-distribution some carriers are lost by recombination (an important process described later on) and thus, the average electron concentration decreases. At equilibrium, the electron concentration becomes uniform and the motion of the electrons by diffusion does not exist. It must be remembered that the cause of the diffusion current is the presence of non-zero rate-of-change of carrier concentration with respect to the distance (known as concentration gradient). The higher the magnitudes of the concentration gradient, the higher are the numerical values of the diffusion current. Therefore, for electrons, the diffusion-current density is given by:

$$J_n = eD_n \nabla n \tag{1-97}$$

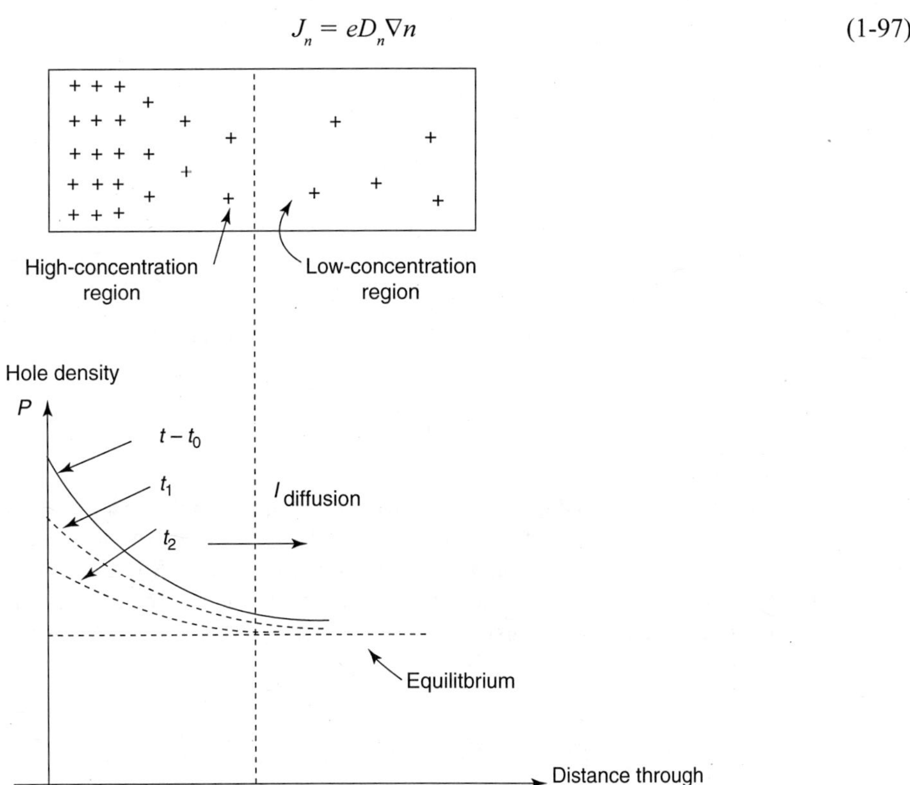


**Figure 1-31** A plot of carrier concentration through a crystal, illustrating the diffusion process

Where $\nabla n$ is the gradient of electron concentration and $D_n$ is the electron diffusion constant. The diffusion constant is a function of the type of the carrier, the specific band structure, the hot electron effects in case of high-field transport, and the other external parameters. Thus, the motion of the carriers is guided by the conduction mechanism through drift and diffusion respectively. The total current due to the motion of holes by drift and diffusion is:

$$J_p = e(\mu_p p_0 E_0 - D_p \nabla p) \tag{1-98}$$

and for electrons:

$$J_n = e(\mu_n n_0 E_0 + D_n \nabla n) \tag{1-99}$$

We observe that in Eq. (1-98) and Eq. (1-99) the sign of the charge for the particles is included, so that $e$ is a positive quantity. A comparison of Eq. (1-98) and Eq. (1-99) with the conduction current for vacuum tubes, where the carriers are usually only electrons moving in vacuum, shows that the situation is somewhat more complicated in solids. In semiconductors, there are usually four components of currents, two due to the motion of electrons and holes by drift, and two due to the diffusive flow of electrons and holes.

## 1-6-4 Einstein Relation

The diffusivity to the mobility ratio of the carriers is called the *Einstein relation*. This was investigated by Albert Einstein in 1905. This relation was first determined in connection to the diffusion of gas particles and is usually termed as the Nernst–Townsend–Einstein relation or simply the Einstein relation.

Under the condition of equilibrium Eq.(1-99) can be expressed as:

$$J_n = 0 = n_0 e \mu_n E_0 + D_n e \frac{\partial n_0}{\partial x}$$

or,

$$-n_0 \mu_n \left(-\frac{\partial V}{\partial x}\right) = -D_n e \frac{\partial n_0}{\partial E_F} \frac{\partial E_F}{\partial x}$$

or,

$$n_0 \mu_n \frac{\partial V}{\partial x} = D_n e \frac{\partial V}{\partial x} \frac{\partial n_0}{\partial E_F}$$

Therefore, the Einstein relation for the electrons can, in general, be expressed as:

$$\frac{D_n}{\mu_n} = \frac{1}{e} \frac{n_0}{\dfrac{\partial n_0}{\partial E_F}} \tag{1-100}$$

But due to the presence of three types of holes, no such simple relation exists for holes in general. From Eq. (1-100) it appears that the Einstein relation depends only on the electron statistics, which in turn is determined by the specific dispersion relation of a particular semiconductor. Since the electron energy spectrum changes for various types of semiconductors and also changes under different external physical conditions, the Einstein relation consequently assumes different expressions in each specific case.

Differentiating Eq.(1-33) with respect to $E_F$, we have:

$$\frac{\partial n_0}{\partial E_F} = N_c \frac{\partial}{\partial E_F}[F_{1/2}(\eta)] = N_c \frac{\partial}{\partial E_F}[F_{1/2}(\eta)] \frac{\partial \eta}{\partial E_F}$$

Therefore,

$$\frac{\partial n_0}{\partial E_F} = N_c \frac{1}{k_B T} F_{-1/2}(\eta) \tag{1-101a}$$

Since $\frac{\partial}{\partial \eta}[F_{1/2}(\eta)] = F_{-1/2}(\eta)$ (from Eq. (1-32a) under the condition $j = 1/2$)

and,
$$\frac{\partial \eta}{\partial E_F} = \frac{\partial}{\partial E_F}\left(\frac{E_F}{k_B T}\right) = \frac{1}{k_B T}$$

Using Eq. (1-33), Eq. (1-101a) and Eq. (1-100), we get:

$$\frac{D_n}{\mu_n} = \frac{k_B T}{e}\left[\frac{F_{1/2}(\eta)}{F_{-1/2}(\eta)}\right] \tag{1-101b}$$

Under the condition of degeneracy using Eq. (1-32) and Eq. (1-101), we can write:

$$F_{1/2}(\eta) = \frac{4}{3\sqrt{\pi}}\left(\eta^{3/2} + \frac{\pi^2}{8\sqrt{\eta}}\right) \tag{1-101c}$$

and,
$$F_{-1/2}(\eta) = \frac{4}{3\sqrt{\pi}}\left(\frac{3}{2}\eta^{1/2} - \frac{\pi^2}{16\eta^{3/2}}\right) \tag{1-101d}$$

Substituting these values in Eq. (1-101b), we have:

$$\frac{D_n}{\mu_n} = \frac{2}{3}E_F \frac{\left(1 + \frac{\pi^2}{8\eta^2}\right)}{\left(1 - \frac{\pi^2}{24\eta^2}\right)} \tag{1-102}$$

Now, applying binomial theorem in the denominator we get:

$$\frac{D_n}{\mu_n} = \frac{2}{3}E_F\left(1 + \frac{\pi^2}{8\eta^2}\right)\left(1 - \frac{\pi^2}{24\eta^2}\right)^{-1}$$

Neglecting high powers of $\eta$, since $\eta >> 1$, we get:

$$\frac{D_n}{\mu_n} = \frac{2}{3}E_F\left(1 + \frac{\pi^2}{8\eta^2}\right)\left(1 + \frac{\pi^2}{24\eta^2}\right)$$

$$\frac{D_n}{\mu_n} = \frac{2}{3}E_F\left(1 + \frac{\pi^2}{6\eta^2}\right) \tag{1-103}$$

Equation (1-103) infers that the Einstein relation for the electrons increases with increasing electron concentration for constant temperature and increases with temperature for constant electron concentration.

Under the condition, $T \to 0$ Eq. (1-103) reduces to the form:

$$\frac{D_n}{\mu_n} = \frac{2}{3}\frac{E_F}{e} \tag{1-104}$$

From Eq. (1-104) we can conclude that the Einstein relation under the case of extreme degeneracy is a function of only electron concentration and is independent of temperature. Besides, electron concentration versus $D_n/\mu_n$ graph is a replica of $n_0$ versus $E_F$ graph.

Under the condition of non-degeneracy using Eq. [1-32(e)] and Eq. [1-101(b)] we get:

$$\frac{D_n}{\mu_n} = \frac{k_B T}{e}\left(\frac{e^{\eta}}{e^{\eta}}\right) = \frac{k_B T}{e} \tag{1-105}$$

Equation (1-105) states that the Einstein relation in $n$-type semiconductors is directly proportional to the temperature only and this result is valid for all semiconductors having arbitrary dispersion laws. From the slope of the $D_n/\mu_n$ versus temperature graph (which is of course a straight line but not valid for

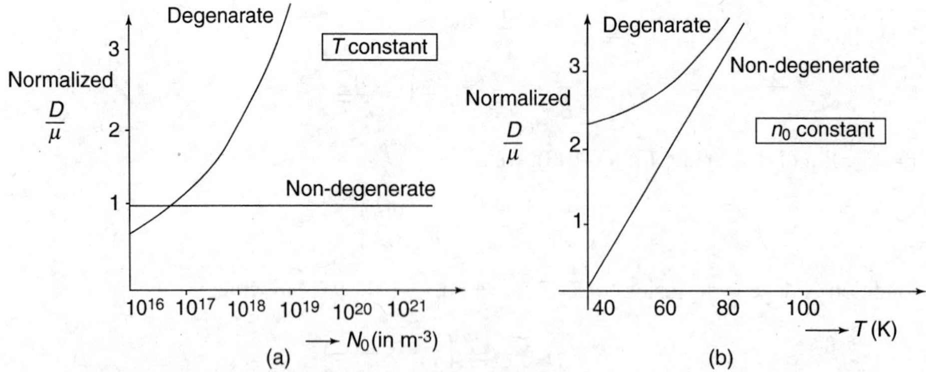


**Figure 1-32** Normalized plot of the diffusivity to mobility ratio in parabolic semiconductors as functions of (a) electron concentration ($T$ constant) and (b) temperature ($n_0$ constant) for both degenerate and non-degenerate cases respectively

low temperatures) we can determine either $e$ for the known value of $k_B$; or $k_B$, assuming $e$ as a known quantity. Thus, the Einstein relation under the condition of non-degeneracy not only provides a general result valid for all semiconductors but also determines the two fundamental constants namely, $k_B$ and $e$. The plots of the Einstein relation are shown in Fig. 1-32.

## 1-6-5 Recombination and Generation Processes

Generation is a process whereby electrons and holes are created. Recombination is a process where electrons and holes are annihilated or destroyed.

### Recombination

There are three types of common recombination processes. These are as follows:

**(i) Band-to-Band Recombination:** In the case of high temperatures, electron-hole pairs (EHPs), generated as valence-band electrons, are excited thermally across the band gap to the conduction band. These EHPs are the only charge carriers in intrinsic material. The generation of EHPs can be explained qualitatively by considering the breaking of covalent bonds in the crystal lattice. If one of the Si valence electrons is broken away from its position in the bonding structure such that it becomes free to move about in the lattice, a conduction electron is being generated and a broken bond (hole) is left behind. The energy required to break the bond is the band gap energy $E_g$. The band-to-band recombination process is shown in Fig. 1-33.This model helps in visualizing the physical mechanism of EHP creation, but the energy band model is more productive for purposes of quantitative calculation. One important difficulty in the "broken bond" model is that the free electron and the hole seem to be deceptively localized in the lattice. Actually, the positions of the free electron and the holes are spread out over several lattice spacings and should be considered quantum mechanically by probability distributions.

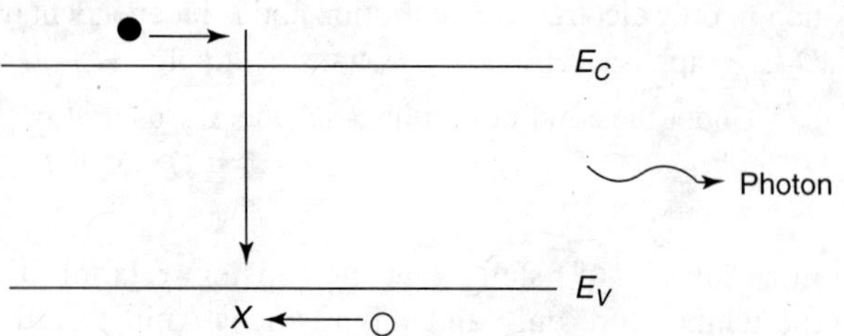


**Figure 1-33** Band-to-band recombination process

Since the electrons and holes are created in pairs, the conduction-band electron concentration $n$ (electrons per cm$^3$) is equal to the concentration of holes $p$ (holes per cm$^3$) in the valence band. Each of these intrinsic carrier concentrations is commonly referred to as $n_i$. Thus, for intrinsic material:

$$n = p = n_i$$

At a given temperature, there is a certain concentration of electron-hole pairs $n_i$. If a steady state carrier concentration is maintained, there must be recombination of EHPs at the same rate at which they are generated. Recombination occurs when an electron in the conduction band makes a transition (direct or indirect) to an empty state (hole) in the valence band, thus annihilating the pair. If we denote the generation rate of EHPs as $g_i$ (EHP/cm$^3$-s) and the recombination rate as $r_i$, equilibrium requires that:

$$r_i = g_i \tag{1-106}$$

Each of these rates is temperature-dependent. For example, $g_i(T)$ increases when temperature is raised, and a new carrier concentration $n_i$ is established such that the higher recombination rate $r_i(T)$ just balances generation. At any temperature, we can predict that the rate of recombination of electrons and holes $r_i$ is proportional to the equilibrium concentration of electrons $n_0$ and the concentration of holes $p_0$:

$$r_i = k_r n_0 p_0 = k_r n_i^2 = g_i \tag{1-107}$$

The factor $k_r$ is a constant of proportionality which depends on the particular mechanism by which recombination takes place.

**Direct recombination of electrons and holes.** Electrons in the conduction band of a semiconductor may make transitions to the valence band, i.e., recombine with holes in the valence band either directly or indirectly. In direct combination, an excess population of electrons and holes decays with electrons falling from the conduction band to empty states (holes) in the valence band. Energy lost by an electron in making the transition is given up as a photon. Direct recombination occurs spontaneously; that is, the probability that an electron and a hole will recombine is constant in time. As in the case of carrier scattering, this constant probability leads us to expect an exponential solution for the decay of the excess carriers. In this case the rate of decay of electrons at any time $t$ is proportional to the number of electrons remaining at $t$ and the number of holes, with some constant of proportionality for recombination, $k_r$. The net rate of change in the conduction-band electron concentration is the thermal generation rate $k_r n_i^2$ minus the recombination rate:

$$\frac{dn(t)}{dt} = k_r n_i^2 - k_r n(t)p(t) \tag{1-108}$$

Let us assume that the excess electron-hole population is created at $t = 0$, for example, by a short flash of light, and the initial excess electron and hole concentrations $\Delta n$ and $\Delta p$ are equal. Then as the electrons and holes recombine in pairs, the instantaneous concentrations of excess carriers $\delta n(t)$ and $\delta p(t)$ are also equal. Thus, we can write the total concentrations of Eq. (1-108) in terms of the equilibrium values $n_0$ and $p_0$ and the excess carrier concentrations $\delta n(t) = \delta p(t)$.Using, $n_0 p_0 = n_i^2$, we have:

$$\begin{aligned}\frac{d\delta n(t)}{dt} &= k_r n_i^2 - k_r\left[n_0 + \delta n(t)\right]\left[p_0 + \delta p(t)\right] \\ &= -k_r\left[(n_0 + p_0)\delta n(t) + \delta n^2(t)\right]\end{aligned} \tag{1-109}$$

If the excess-carrier concentrations are small, we can neglect the $\delta n^2$ term. Furthermore, if the material is extrinsic, we can usually neglect the term representing the equilibrium minority carriers. For example, if the material is $p$-type ($p_0 \approx n_0$), Eq. (1-109) becomes:

$$\frac{d\delta n(t)}{dt} = -k_r p_0 \delta n(t) \tag{1-110}$$

The solution to this equation is an exponential decay from the original excess carrier concentration $\Delta n$:

$$\delta n(t) = \Delta n \exp(-k_r p_0 t) = \Delta n \exp\left(-\frac{t}{\tau_n}\right) \tag{1-111a}$$

Excess electrons in a $p$-type semiconductor recombine with a decay constant $\tau_n = (k_r p_0)^{-1}$, called the recombination lifetime. Since the calculation is made in terms of the minority carrier, $\tau_n$ is often called the minority-carrier lifetime. The decay of excess holes in $n$-type material occurs with $\tau_p = (k_r n_0)^{-1}$. In the case of direct recombination, the excess majority carriers decay at exactly the same rate as the minority carriers.

**(ii) Auger Recombination:** Auger recombination involving two electrons and one hole is shown in Fig. 1-34. When two electrons collide, one electron is sent to a higher energy state and the other electron falls to the valence band and recombines with a hole there. The surviving electron will fall back to the edge of the conduction band because it loses energy through collisions with the semiconductor lattice. The released energy is transferred to heating the lattice thus causing the lattice to vibrate with the creation of phonons in turn. Since the chance of collision increases with increasing free-carrier concentration, Auger recombination becomes more important in $n^+$ or $p^+$ semiconductors.

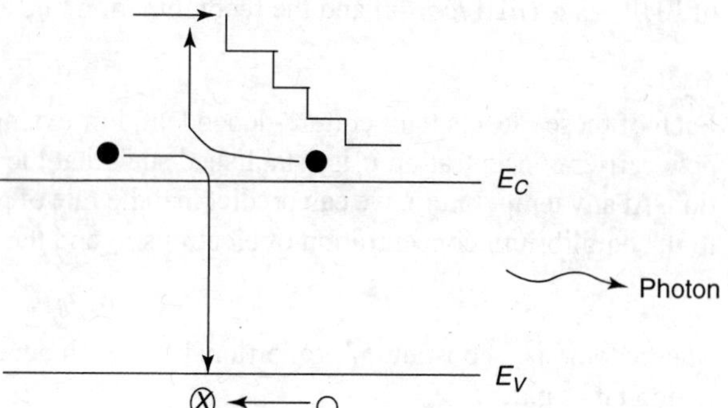


**Figure 1-34** Auger recombination process

**(iii) Shockley–Reed–Hall (SRH) Recombination:** SRH recombination describes the recombination processes assisted by the localized states in the energy band gap. The localized states, which are recombination–generation (R-G) centres, can result from impure atoms or crystal defects in general. This particular mechanism as shown in Fig. 1-35 is the most important recombination in a lightly-doped indirect band gap semiconductor like silicon. The possible electron and hole transitions through a single localized energy state ($E_T$) are:

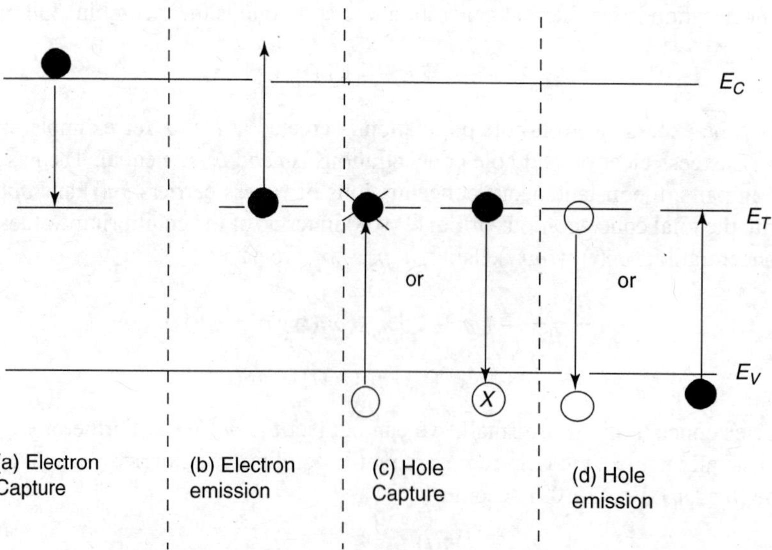


**Figure 1-35** Shockley-Reed-Hall (SRH) recombination

1. Electron capture at an R-G centre ($P$)
2. Electron emission from an R-G centre ($Q$)
3. Hole capture at an R-G centre ($R$)
4. Hole emission from an R-G centre

The net maximum SRH recombination rate ($U$) can be expressed as:

$$U = \frac{(pn - n_i^2)}{[\tau_0 (p + n + 2n_i)]} \tag{1-111b}$$

where, $p$ is the hole concentration, $n$ is the electron concentration, $n_i$ is the intrinsic carrier concentration. And, $\tau_0 = 1/N_T\sigma V_{th}$, where, $N_T$ is the trapping density at $E_T$, $\sigma$ is the carrier capture cross-section and $V_{th}$ is the carrier thermal velocity.

**Generation.** The various recombination mechanisms are important in a sample at thermal equilibrium or with a steady rate EHP generation-recombination. For example, a semiconductor at equilibrium experiences thermal generation of EHPs at a rate $g(t) = g_i$. This generation is balanced by the recombination rate so that the equilibrium concentrations of carriers $n_0$ and $p_0$ are maintained:

$$g(t) = k_r n_i^2 = k_r n_0 p_0 \tag{1-112}$$

This equilibrium rate balance can include generation from defect centres as well as band-to-band generation. When a semiconductor sample is exposed to photons, an optical generation rate $g_{op}$ will be added to the thermal generation, and the carrier concentrations $n$ and $p$ will increase to new steady state values. We can write the balance between generation and recombination in terms of the equilibrium carrier concentrations and the departures from equilibrium $\delta n$ and $\delta p$ as:

$$g(T) + g_{op} = k_r (n_0 + \delta n)(p_0 + \delta p) \tag{1-113}$$

For steady state recombination and no trapping, $\delta n = \delta p$, Eq. (1-113) becomes:

$$g(T) + g_{op} = k_r n_0 p_0 + k_r [(n_0 + p_0)\,\delta n + \delta n^2] \tag{1-114}$$

The term $k_r n_0 p_0$ is just equal to the thermal generation rate $g(T)$. Thus, neglecting the $\delta n^2$ term for low-level excitation, we can rewrite Eq. (1-114) as:

$$g_{op} = k_r (n_0 + p_0)\,\delta n = \frac{\delta n}{\tau_n} \tag{1-115}$$

The excess carrier concentration can be written as:

$$\delta n = \delta p = g_{op}\tau_n \tag{1-116a}$$

In the context of excess carriers in semiconductors, the concept of recombination and generation of carriers has been created. With respect to various types of profiles, various types of solutions can be generated depending on the nature of the initial conditions.

**Boltzmann transport equation.** After discussing the carrier statistics and recombination–generation mechanisms in semiconductors, it is appropriate to introduce the most fundamental equation for semiconductor devices: the Boltzmann transport equation (BTE). It is a semi-classical equation and describes the time-dependent position and momentum of electrons and holes in semiconductors in terms of a distribution function $f(\vec{r}, \vec{p}, t)$, where, $\vec{r}$ is the position, $\vec{p}$ is the momentum, and $t$ is the time. The BTE can be obtained from the conservation principle that the change in $f(\vec{r}, \vec{p}, t)$ with respect to time is the sum of the change in $f$ with respect to the space, the change in $f$ with respect to the momentum, and the change in $f$ with respect to time due to generation–recombination ($G$–$R$) processes. In one-dimensional analysis, the time-dependent BTE can be expressed as:

$$\frac{\partial f}{\partial t} = \left(-\frac{v\partial f}{\partial r}\right) - \left(\frac{F\partial f}{\partial p}\right) + \left(\left(\frac{\partial f}{\partial t}\right)_{\text{collision}}\right) + G - R \qquad (1\text{-}116b)$$

where, $v$ is the carrier velocity along $x$-axis, $F$ is the force along $x$-axis, $G$ is the generation rate and $R$ is the recombination rate.

It may be noted that the most important single equation in the field of semiconductor science excluding band-structure is the BTE as given in the simplified form of Eq. (1-116b). This equation is the key to solving all types of transport coefficients in general and the carrier motion in semiconductor devices.

## Solved Examples

**Example 1-11** Consider an electron in the bottom of the conduction band in GaAs. An electric field of $10^4$ V/cm is applied to the materials of the $x$-direction. Calculate the time it takes the electron to reach the Brillouin zone. Use the data you need.

**Solution:**
The equation for motion of an electron in a periodic structure with no scattering is $\hbar \frac{dk}{dt} = eE_0$ where, $E_0$ is the external electric field.
From the previous equation we get:

$$\hbar \int_0^{k_B} dk = eE_0 \int_0^{\tau} dt$$

where, $k_B$ is the wave-vector at Brillouin zone along $x$-axis and $\tau$ is the time taken for the electron to reach the Brillouin zone along $x$-axis starting from the $\Gamma$ point. Therefore, $\tau = \frac{\hbar k_B}{eE_0}$

The Brillouin zone edge along the $x$ direction is:

$k_B = \frac{2\pi}{a} = \frac{2\pi}{5.65 \times 10^{-8}} = 1.112 \times 10^8\ \text{cm}^{-1}$ (since the lattice constant a of $n$-GaAs is $5.62 \times 10^{-8}$ cm)

Substituting $k_B = 1.112 \times 10^8\ \text{cm}^{-1}$ and $E_0 = 10^4$ v/cm we get:

$$\tau = \left(\frac{(1.05 \times 10^{-34}\ \text{Js})\ (1.12 \times 10^8\ \text{cm}^{-1})}{(1.6 \times 10^{-19}\ \text{C})\ (10^4\ \text{V/cm})}\right) = 7.297\ \text{ps}$$

**Example 1-12** Consider the sample of GaAs with electron effective mass of $0.067m_0$. If an electric field of 1 KV/cm is applied, calculate the drift velocity produced if:

(a) $\tau_{sc} = 10^{-13}$ s (b) $\tau_{sc} = 10^{-12}$ s (c) $\tau_{sc} = 10^{-11}$ s

**Solution:**
The drift velocity is related to the applied field and the scattering time by the equation:

$$v_0 = \frac{e\tau_{sc}E_0}{m_c^*}$$

(a) Substituting the given data in this equation we get:

$$v_0 = \frac{(1.6 \times 10^{-19}\ \text{C})(10^{-13}\ \text{s})(10^5\ \text{kV/m})}{0.067 \times 0.91 \times 10^{-30}\ \text{kg}} = 2.62 \times 10^4\ \text{m/s} = 2.62 \times 10^6\ \text{cm/s}$$

(b) Substituting $\tau_{sc} = 10^{-12}$ s we get:

$$v_0 = \frac{(1.6 \times 10^{-19}\ \text{C})(10^{-12}\ \text{s})(10^5\ \text{kV/m})}{0.067 \times 0.91 \times 10^{-30}\ \text{kg}} = 2.62 \times 10^7\ \text{cm/s}$$

(c) Substituting $\tau_{sc} = 10^{-11}$ s we get:

$$v_0 = \frac{(1.6 \times 10^{-19}\,\text{C})\,(10^{-11}\,\text{s})\,(10^5\,\text{kV/m})}{0.067 \times 0.91 \times 10^{-30}\,\text{kg}} = 2.62 \times 10^8\ \text{cm/s}$$

**Example 1-13** A specimen of metal has $7.87 \times 10^{28}$ free electrons per cubic meter. The mobility of electrons in the metal is 35.2 cm²/vs (a) Compute the conductivity of the metal (b) If an electric field of 30 V/cm is applied across the specimen, find the drift velocity of free electrons and the current density.

**Solution:**

(a) Mobility $= \mu = 35.2 \times 10^{-4}$ m²/vs

Conductivity $\sigma = n_0 e\mu = 7.87 \times 10^{28} \times 1.6 \times 10^{-19} \times 35.2 \times 10^{-4}$

$= 44.32 \times 10^6$ s/m

(b) Electric field $E_0 = 30 \times 10^2$ V/m

Hence, drift velocity of free electrons is $V_0 = \mu E_0 = 10.44$ m/s

Current density $J = \sigma E_0 = 13.3 \times 10^{10}$ A/m³

**Example 1-14** Calculate the drift velocity of free electrons in a copper conductor of cross-sectional area ($A$) $10^{-5}$ m² and in which there is a current ($I$) of 100 A, assuming the free electron concentration ($n_0$) of copper to be $8.5 \times 10^{28}$ m⁻³.

**Solution:**

The current $I$ through a conductor of cross-section area $A$ and free electron concentration $n_0$ is given by:

$$I = n_0 A v_d e$$

where, $v_d$ is the drift velocity.

From the question, $A = 10^{-5}$ m², $I = 100$ A and $n_0 = 8.5 \times 10^{28}$ m⁻³.

Therefore, the drift velocity $v_d = \dfrac{I}{dAe} = \dfrac{100}{8.5 \times 10^{28} \times 10^{-5} \times 1.6 \times 10^{-19}}$

$= 0.735 \times 10^{-3}$ ms⁻¹

**Example 1-15** Calculate the drift velocity of free electrons in a copper conductor of cross-sectional area ($A$) $10^{-5}$ m² and in which there is a current ($I$) of 100 A, assuming the free electron concentration ($n_0$) of copper to be $8.5 \times 10^{28}$ m⁻³.

**Solution:**

The current $I$ through a conductor of cross-section area $A$ and free electron concentration $n_0$ is given by:

$$I = n_0 A v_d e \text{ where } v_d \text{ is the drift velocity.}$$

By the question $A = 10^{-5}$ m², $I = 100$A and $n_0 = 8.5 \times 10^{28}$ m⁻³

Therefore, the drift velocity

$$v_d = \frac{I}{dAe} = \frac{100}{8.5 \times 10^{28} \times 10^{-5} \times 1.6 \times 10^{-19}} = 0.735 \times 10^{-3}\ \text{ms}^{-1}$$

**Example 1-16** A potential difference of 1V is applied across a uniform wire of length ($L$) 10m. Calculate the drift velocity ($v_0$) of electrons through the copper. Given that the relaxation time ($\tau$) is $10^{-14}$ s and the effective mass of the electron is $0.02 \times 9.1 \times 10^{-31}$ kg.

**Solution:**

By the question the applied potential difference is 1V, L = 10 m, $\tau = 10^{-14}$ s

Therefore the electric field, $E_0 = \frac{1}{10} = 0.1\text{Vm}^{-1}$

The drift velocity, $v_0 = \frac{eE_0\tau}{m^*} = \frac{1.6 \times 10^{-19} \times 0.1 \times 10^{-14}}{0.02 \times 9.1 \times 10^{-31}}$

$= 0.176 \times 10^{-3}\ \text{ms}^{-1}$

$= 0.0088\ \text{ms}^{-1}$

**Example 1-17** A Si sample is doped with $10^{17}$ As atoms/cm³. What is the equilibrium hole concentration $p_0$ at 300 K?

**Solution:**

Since $N_d >> n_i$, we can approximate $n_0 = N_d$, and:

$$p_0 = \frac{n_i^2}{n_0} = \frac{2.25 \times 10^{20}}{10^{17}} = 2.25 \times 10^3\ \text{cm}^{-3}$$

**Example 1-18** Consider an electron in the bottom of the conduction band in GaAs. An electric field of $10^4$ V/cm is applied to the materials of the $x$-direction. Calculate the time it takes the electron to reach the Brillouin zone. Use the data you need.

**Solution:**

The equation of motion for an electron in a periodic structure with no scattering is $\hbar \frac{dk}{dt} = eE_0$ where $E_0$ is the external electric field.

From the above equation, we get:

$$\hbar \int_0^{k_B} dk = eE_0 \int_0^{\tau} dt$$

where, $k_B$ is the wave-vector at Brillouin zone along $x$-axis and $\tau$ is the time taken for the electron to reach the Brillouin zone along $x$-axis starting from the Γ point. Therefore,

$$\tau = \frac{\hbar k_B}{eE_0}$$

The Brillouin zone edge along the $x$-direction is:

$$k_B = \frac{2\pi}{a} = \frac{2\pi}{5.65 \times 10^{-8}} = 1.112 \times 10^8\ \text{cm}^{-1}$$

(since the lattice constant $a$ of $n$-GaAs is $5.62 \times 10^{-8}$ cm)

Substituting $k_B = 1.112 \times 10^8\ \text{cm}^{-1}$ and $E_0 = 10^4$ v/cm we get:

$$\tau = \left(\frac{(1.05 \times 10^{-34}\ \text{Js})(1.12 \times 10^8\ \text{cm}^{-1})}{(1.6 \times 10^{-19}\ \text{C})(10^4\text{V/cm})}\right) = 7.297\ \text{ps}$$

**Example 1-19** An electron in the central Γ-valley of GaAs is to be transferred to the satellite L – valley. The energy separation between Γ and L point is 0.3eV. Using the parabolic expression for the band structure of GaAs along (111) direction, estimate the smallest $k$-vector along $x$-direction that is needed for this transition. The electron in the Γ valley must have an energy equal to the position of the L-valley.

**Solution:**

To find the shortest $k$-vector needed to transfer the electron in the Γ-valley, we need to calculate the $k$-vector for an electron in the Γ valley along the (111) direction and with an energy of 0.3 ev. Using the parabolic expression for the energy we have along the (111) direction

$$E = \frac{h^2}{2m_c^*}(k_x^2 + k_y^2 + k_z^2) = \frac{3h^2k_x^2}{2m_c^*} \quad \text{or,} \quad k_x = \left(\frac{2m_c^*E}{3h^2}\right)^{½}$$

Substituting $m_c^* = 0.067\ \mathrm{m_0}$ and $E = 0.3\mathrm{eV}$, we get:

$$k_x = \left[\frac{2(0.067 \times 0.91 \times 10^{-30}\ \mathrm{kg})(0.3 \times 1.6 \times 10^{-19}\ \mathrm{J})}{3(1.0510^{-34}\ \mathrm{Js})^2}\right]^{½}$$

Therefore, $k_x = 4.2 \times 10^8\ \mathrm{m^{-1}}$

**Example 1-20** Show that the minimum conductivity of a Si sample occurs when it is slightly $p$-type. Calculate the electron and hole concentration when the conductivity is minimum, given that $\mu_n = 1350\ \mathrm{cm^2/vs}$, $\mu_p = 450\mathrm{cm^2/v\text{-}s}$ and the intrinsic carrier concentration $n_i = 1.5 \times 10^{10}\ \mathrm{cm^{-3}}$.

**Solution:**

The total conductivity is given by $\sigma = e(n_0\,\mu_n + p_0\,\mu_p)$ (1)

The law of mass action is given by $n_i^2 = n_0 p_0$ (2)

Eliminating $p_0$ between (1) and (2) we get

$$\sigma = e(n_0\mu_n + \mu_p n_i^2/n_0) \quad (3)$$

Differentiating both sides of (3) w.r.t. $n_0$, we get:

$$\frac{\partial\sigma}{\partial n_0} = e\left(\mu_n - \frac{\mu_p n_i^2}{n_0^2}\right) \quad (4)$$

$$\frac{\partial^2\sigma}{\partial n_0^2} = e\left(\frac{2\mu_p n_i^2}{n_0^3}\right) \quad (5)$$

For extremum $\frac{\partial\sigma}{\partial n_0} = 0$ which leads to:

$$e(\mu_n - (n_i^2/n_0^2)\mu_p) = 0$$

therefore, $n_0 = n_i\sqrt{\frac{\mu_p}{\mu_n}}$

Substituting this value of $n_0$ in Eq. (5) we get: $\frac{\partial^2\sigma}{\partial n_0^2} > 0$

Therefore, $\sigma$ is minimum when $n_0 = n_i\sqrt{\frac{\mu_p}{\mu_n}}$

Substituting this value of $n_0$ in Eq. (2) we get: $P_0 = n_i\sqrt{\frac{\mu_n}{\mu_p}}$

$\mu_n > \mu_p \therefore p_0 > n_0$

Therefore, the $Si$ sample is $p$-type.

Thus, the corresponding electron and hole concentrations are given by:

$$n_0 = 1.5 \times 10^{10}\sqrt{\frac{450}{1350}} = 8.66 \times 10^9\ \mathrm{cm^{-3}}$$

$$p_0 = 1.5 \times 10^{10}\sqrt{\frac{1350}{450}} = 2.6 \times 10^{10}\ \mathrm{cm^{-3}}$$

**Example 1-21** In $n$-type Ge, the donor concentration corresponds to 1 atom per $10^7$ Ge atoms. Assume that the effective mass of the electron equals one half the free electron mass. At room temperature, how far from the edge of the conduction band is the Fermi level?

**Solution:**
The concentration of Ge atom is = [$6.02 \times 10^{23}$ (atom/mole) × (1 mole/72.6 gm) × (5.32 g/cm$^2$)]
= $4.41 \times 10^{22}$ cm$^{-3}$

Therefore, $N_D = 4.41 \times 10^{22}\ \text{cm}^{-3} \times 1/10^7 = 4.41 \times 10^{15}\ \text{cm}^{-3}$
Assuming full ionization $n_0 = N_D$

$$N_c = 2\left(\frac{2\pi m_c^* k_B T}{h^2}\right)^{3/2} = 4.82 \times 10^{15}\left[\left(\frac{m_c^*}{m_0}\right)T\right]^{3/2}\text{cm}^{-3}$$

By the question, T = 300 K and $m_c^* = 1/2m_0$
Therefore, $N_c = 4.82 \times 10^{15} \times 1/\sqrt{8} \times (300)^{3/2}$
$= 8.87 \times 10^{18}\ \text{cm}^{-3}$

We know $E_F = K_B T \ln\left(\frac{n_0}{N_c}\right)$

Substituting the values of $n_0$ and $N_C$, we get:

$$E_F = 0.026 \ln\left(\frac{4.41 \times 10^{15}}{8.87 \times 10^{18}}\right)$$
$$= -0.1977$$

Therefore, $E_F$ is below $E_C$.

**Example 1-22** Compute the mobility of the free electrons in aluminium for which the density ($d$) is 2.70 g/cm$^3$ and the resistivity ($\rho$) is $3.44 \times 10^{-6}$ Ω-cm. Assume that aluminium has three valence electrons per atom.

**Solution:**
We have proved that:

$$n_0 = \frac{dv}{AM}$$

Therefore, substituting the data for this problem we get:

$$n_0 = \left(2.7\frac{\text{g}}{\text{cm}^3}\right)\left(3\frac{\text{electrons}}{\text{atom}}\right)\left(10^{-3}\frac{\text{kg}}{\text{g}}\right)\left(\frac{1\ \text{atom}}{26.97 \times 1.66 \times 10^{-27}\,\text{kg}}\right)\left(10^6\frac{\text{cm}^3}{\text{m}^3}\right)$$
$$= 18 \times 10^{28}\ \text{m}^{-3}$$

We know that: $\mu = \frac{\sigma}{n_0 e} = \frac{1}{n_0 e p}$ (1)

Substituting the data in we get:

$$\mu = \frac{1}{\left(18 \times 10^{28}\frac{\text{electrons}}{\text{m}^3}\right)\left(1.6 \times 10^{-19}\frac{\text{C}}{\text{electron}}\right)} \cdot \frac{10^2\ \text{cm/m}}{3.44 \times 10^{-6}\ \text{ohm} - \text{cm}}$$
$$= 10^{-3}\ \text{m}^2/\text{V-sec}$$

**Example 1-23** Consider intrinsic germanium at room temperature (300 K). By what percent does the intrinsic carrier concentration increase per degree rise in temperature? (Use the band gap Eg in this case as 0.785 eV)

**Solution:**
From Eq. (1-56) we get:

$$\ln n_i = \text{const} - \left(\frac{3}{2}\ln\frac{1}{T}\right) - \left(\frac{E_g}{2k_B}\frac{1}{T}\right)$$
$$= \text{const} - \left(\frac{3}{2}\right)(\ln(1) - \ln(T)) - \left(\frac{E_g}{2k_B}\frac{1}{T}\right)$$

Taking differentials we get:

$$\frac{dn_i}{n_i} = \frac{3}{2}\frac{dT}{T} + \frac{E_g}{2kT^2}dT = \left(\frac{3}{2} + \frac{E_g}{2kT}\right)\frac{dT}{T}, \quad \text{at} \quad T = 300\text{ K}, KT = 0.026$$

$$T = 300\text{ K}, \frac{dT}{T} = \frac{1}{300}$$

Therefore, $100\left(\frac{dn_i}{n_i}\right) = \left(1.5 + \frac{0.785}{0.052}\right)\left(\frac{1}{300}\right)(100\%) = 5.5\%$ per degree

**Example 1-24** Redo Example 1-23 for intrinsic Silicon. (Use the band gap $E_g$ in this case as 1.21 eV)

**Solution:**
Using the result of Example 1-23 and $E_g$ of Si =1.21eV at 0 K:

$$100\frac{dn_i}{n_i} = \left(1.5 + \frac{1.21}{0.052}\right)\left(\frac{1}{300}\right)(100\%) = 8.3\% \text{ per degree}$$

**Example 1-25** The resistance of No. 18 copper wire (diameter =1.03 mm) is 6.51 Ω per 1,000 ft. The concentration of electrons in copper is $8.4 \times 10^{27}$ electrons/m$^3$. If the current is 2A, find the (a) drift velocity, (b) mobility, (c) conductivity.

**Solution:**
The cross section of the wire is $A = \frac{\pi}{4}d^2 = 0.835 \times 10^{-6}$ m$^2$
The current density $J = I/A = (2/0.835) \times 10^6 = 2.4 \times 10^6$ A/m$^2$
Therefore, the drift velocity is given by:

$$v_0 = J/n_0e = \frac{2.4 \times 10^6}{8.4 \times 10^{27} \times 1.6 \times 10^{-19}} = 1.78 \times 10^{-3}\text{ m/s (since } n_0 = 8.4 \times 10^{27}\text{ m}^{-3})$$

The resistance per meter is $\frac{6.51\ \Omega}{10^3\text{ ft} \times 0.304\text{ m/ft}} = 0.0214\ \Omega/\text{m}$
and the electric field $E_0 = 0.0214 \times 2 = 0.0428$ V/m.
Since mobility $\mu = v_0/E_0$ (from Eq. (1-78)). Therefore, mobility in this case is:

$$\mu = \frac{1.78 \times 10^{-3}\text{ m/s}}{0.0428\text{ V/m}} = 4.16 \times 10^{-2}\text{ m}^2/\text{V-s}$$

(c) From Eq. (1-88), $\sigma = n_0e\mu$. Therefore the conductivity $\sigma$ is equal to:

$$(8.4 \times 10^{28})(1.64 \times 10^{-19})(4.16 \times 10^{-2}) = 5.61 \times 10^8\ (\Omega\text{-m})^{-1}$$

**Example 1-26** (a) Determine the concentration of free electrons and holes in a sample of germanium at 300 K which has concentration of donor atoms equal to $2 \times 10^{14}$ atoms/cm$^3$ and concentration of acceptor atoms equal to $3 \times 10^{14}$ atoms/cm$^3$. Is this *p*- or *n*-type Germanium? In other words, is the conductivity due primarily to holes or to electrons?

(b) Repeat part (a) for equal donor and acceptor concentrations of $10^{15}$ atoms/cm$^3$. Is this $p$- or $n$-type Germanium?

(c) Repeat part (a) for donor concentration of $10^{16}$atoms/cm$^3$ and acceptor concentration $10^{14}$ atoms/cm$^3$. It is given that $n_i$ (at 300 K) = $2.5 \times 10^{19}$ m$^{-3}$

**Solution:**

We know that $n_0 \cdot p_0 = n_i^2$ and $N_A + n_0 = p_0 + N_D$

Therefore, $N_A + n_0 = n_i^2/n_0 + N_D$

Solving the above equation for $n_0$, we get:

$$n_0 = -\frac{N_A - N_D}{2} + \left[\left(\frac{N_A - N_D}{2}\right)^2 + \frac{4}{4} n_i^2\right]^{1/2}$$

Similarly, eliminating $n_0$, we can write: $\quad N_A + \dfrac{n_i^2}{p_0} = p_0 + N_D$

Solving the above equation for $p_0$, we obtain:

$$p_0 = -\frac{N_D - N_A}{2} + \left[\left(\frac{N_D - N_A}{2}\right)^2 + \frac{4}{4} n_i^2\right]^{1/2}$$

We have used only the + sign in the solution of quadratic equation because we know that $n_0 > 0$ and $p_0 > 0$. Substituting the data $N_D = 2 \times 10^{14}$ cm$^{-3}$, $N_A = 3 \times 10^{14}$ cm$^{-3}$ and $n_i = 2.5 \times 10^{19}$ m$^{-3}$ in the above equations, we get:

$$n_0 = -\frac{1}{2} \times 10^{14} + (0.25 \times 10^{28} + 6.25 \times 10^{26})^{1/2} = 0.058 \times 10^{14} \text{ cm}^{-3}$$

$$p_0 = \frac{1}{2} \times 10^{14} + (.25 \times 10^{28} + 6.25 \times 10^{26})^{1/2} = 1.058 \times 10^{14} \text{ cm}^{-3}$$

Since $p_0 > n_0$, therefore the sample is $p$-type.

By the question $N_A = N_D = 10^{15}$cm$^{-3}$.

Substituting this condition in the charge neutrality equation:

$$N_A + n_0 = p_0 + N_{D,}$$

we get, $n_0 = p_0$ (which is the condition for intrinsic semiconductor)

Thus we realize that a semiconductor can be made intrinsic either by without doping or by equal amount of donor and acceptor doping. Thus, the germanium sample in this question is intrinsic by compensation.

$N_A << N_D$, therefore, $n_0 \approx N_D = 10^{16}$ cm$^{-3}$.

$$p_0 = \frac{6.25 \times 10^{26}}{10^{16}} = 6.25 \times 10^{10} \text{ /cm}^{-3}$$

Since $n_0 > p_0$, the germanium sample in this question is $n$-type.

**Example 1-27** (a) Find the concentration of holes and electrons in $p$-type germanium at 300 K if the conductivity is 100($\Omega$-cm)$^{-1}$.

(b) Repeat part a for $n$-type silicon if the conductivity is 0.1($\Omega$-cm)$^{-1}$.

Given that $\mu_p$ = 1800 cm$^2$/V-sec for $p$-type germanium at 300 K, $\mu_n$ = 1300 cm$^2$/V-sec for $n$-type silicon at 300 K. The intrinsic concentration $n_i = 2.5 \times 10^{13}$ cm$^{-3}$ for germanium and $n_i = 1.5 \times 10^{10}$ cm$^{-3}$ for silicon at 300 K.

**Solution:**

(a) We know that $\sigma = p_0 e\mu_p$, since $p_0 >> n_0$.

Thus, $$p_0 = \frac{\sigma}{e\mu_p} = \frac{100}{1.6 \times 10^{-19} \times 1800} = 3.47 \times 10^{17}\,\text{cm}^{-3}$$

We have, $$n_0 = \frac{n_i^2}{P_p} = \frac{(2.5 \times 10^{13})^2}{3.47 \times 10^{17}} = 1.8 \times 10^{15}\text{m}^{-3}$$

We have $\sigma = n_0 e\mu_0$, since $n_0 >> p_0$

$$n_0 = \frac{\sigma}{\mu_n e} = \frac{0.1}{1300 \times 1.6 \times 10^{-19}} = 4.81 \times 10^{14}\,\text{cm}^{-3}$$

We have,

$$p_0 = \frac{n_i^2}{n_0} = \frac{(1.5 \times 10^{10})^2}{4.81 \times 10^{14}} = 4.68 \times 10^{5}\,\text{cm}^{-3} = 4.68 \times 10^{11}\,\text{m}^{-3}$$

**Example 1-28** Show that the resistivity of intrinsic germanium at 300 K is 45 Ω-cm.

**Solution:**

We know that $\sigma = e(n_0\mu_n + p_0\mu_p)$. Since the germanium is intrinsic $n_0 = p_0 = n_i$ or $\sigma = en_i(\mu_n + \mu_p)$ as $n_i = 2.5 \times 10^{13}\,\text{cm}^{-3}$, $\mu_n + \mu_p = 5600\,\text{cm}^2/\text{V-sec}$, hence $\sigma = 2.5 \times 10^{13}\,\text{cm}^{-3} \times 5600\text{cm}^2/\text{V-sec} \times 1.6 \times 10^{-19}\,\text{C/electrons} = 0.0232\,(\text{s/cm})^{-1}$ or $\rho = 44.6$ Ω-cm

**Example 1-29** (a) Find the conductivity of intrinsic Germanium at 300 K.

(b) If donor-type impurity is added to the extent of 1 impurity atom in $10^7$ Ge atoms, find conductivity (c) If acceptor is added to the extent of 1 impurity atom in $10^7$ Ge atoms, find the conductivity. Given that $n_i$ at 300 K is $2.5 \times 10^{13}/\text{cm}^3$, $\mu_n$ and $\mu_p$ in Ge are 3800 and 1800 cm²/v-s respectively and the concentration of Ge atoms $= 4.41 \times 10^{22}\,\text{cm}^{-3}$.

**Solution:**

(a) For intrinsic semiconductors $n_0 = p_0 = n_i$

Therefore the intrinsic conductivity $\sigma = n_0 e(\mu_n + \mu_p)$

Substituting the given data in the above equation:

$$\sigma = (2.25 \times 10^{13})(1.6 \times 10^{-19})(3800 + 1800)\,\text{s/cm}$$
$$= 0.0224\,\text{s/cm}$$

(b) By the question, the concentration of Ge atoms $= 4.41 \times 10^{22}\,\text{cm}^{-3}$ and it is given that the donor-type impurity is added to the extent of 1 impurity atom in $10^7$ Ge atoms.

$$\therefore \text{ the concentration of donor atoms } N_D = 4.41 \times \frac{10^{22}}{10}\,\text{cm}^{-3}$$
$$= 4.41 \times 10^{15}\,\text{cm}^{-3}$$

Besides $n_0 = N_D$

∴ the hole concentration $p_0 = n_i^2/N_D$

$$= \frac{(2.5 \times 10^{13})^2}{4.41 \times 10^{15}} = 1.41 \times 10^{11}\,\text{cm}^{-3}$$

Since $n_0 >> p_0$, we may neglect $p_0$ in calculating conductivity

$$\sigma = n_0 e\mu_n = (4.41 \times 10^{15}) \times (1.6 \times 10^{-19}) \times (3800)$$
$$= 2.68 \text{ s/cm}$$

(c) With the given acceptor impurity of one acceptor atom per $10^7$ Ge atoms:

$$N_A = 4.41 \times 10^{15} \text{ atoms/cm}^3$$

Further $p_0 = N_A$

Therefore, $n_0 = \dfrac{n_i^2}{N_A} = \dfrac{(2.5 \times 10^{13})^2}{4.41 \times 10^{15}} = 1.42 \times 10^{11}$ electrons/cm$^3$.

Since $p_0 >> n_0$, we may neglect $n_0$ in calculating conductivity. Therefore, from the given data:

$$\sigma = p_0 e\mu_p = 4.41 \times 10^{15} \times 1.6 \times 10^{-19} \times 1800 = 1.275 \text{ s/cm}$$

■

## 1-7 THE CONTINUITY EQUATION

The basic equation for the analysis of current flow in a *p–n* junction diode is provided by the continuity equation. The continuity equation states that the rate of increase or decrease in the density of carriers in a volume is equal to the difference between the current flowing into the volume and the current leaving the volume.

The continuity equation applies for holes and electrons; however, in order to avoid duplication of similar equations, we shall consider holes only. Consider a one-dimensional flow of carriers, as indicated in the Fig. 1-36. The continuity equation is based on the following assumptions:

(i) Only one-dimensional flow of current will be assumed
(ii) Surface effects are neglected
(iii) Transition region is very small with respect to diffusion length
(iv) All the applied voltage appears across the transition region only
(v) The *x*-dimensions of *n* and *p* regions are large compared to the diffusion length of the respective regions
(vi) The external circuit is connected to the *p–n* diode through ohmic contacts

The rate of increase in holes in volume = (holes generated minus holes recombined per unit volume per unit time) + (difference between hole current entering and leaving volume). Therefore, mathematically we can write:

$$\underbrace{eA\Delta\frac{\partial p(x,t)}{\partial t}}_{\text{The rate of increase in holes in volume}} = \underbrace{eA\Delta x\left[g_p(x,t) - r_p(x,t)\right]}_{\text{Holes generated minus holes recombined per unit volume per unit time}} + \underbrace{A\left[J_p(x) - J_p(x+\Delta x)\right]}_{\text{Difference between hole currents into and leaving volume}} \tag{1-117}$$

Equation (1-117) can be written as:

$$\frac{\partial p(x,t)}{\partial t} = \left[g_p(x,t) - r_p(x,t)\right] + \frac{1}{e}\nabla J_p(x,t) \tag{1-118}$$

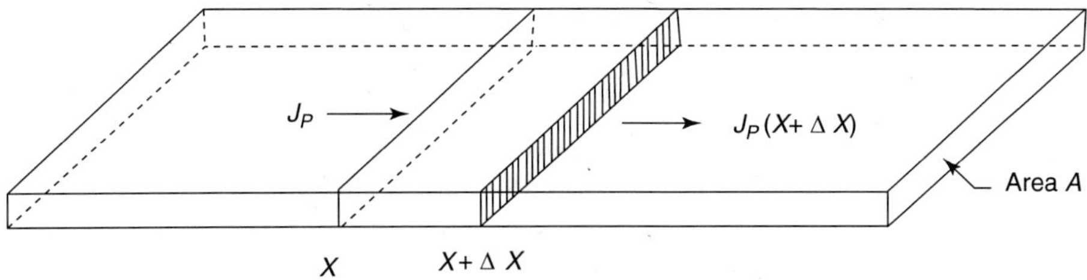


**Figure 1-36** One-dimensional flow of holes in an n-type crystal used to illustrate the development of the continuity equation

where, $g_p(x, t)$ stands for the holes generated per unit time and is given by:

$$g_p(x, t) = \frac{p_n}{\tau_p}$$

($p_n$ is the equilibrium value of the hole concentration for *n*-type semiconductors and $\tau_p$ is the hole life time), $r_p(x, t)$ is the time rate of recombination of holes and can be written as $p/\tau_p$ ($p$ is the instantaneous hole concentration in *n*-type semiconductors), $\nabla$ denotes the gradient. Thus, using these concepts from Eq. (1-2) we get:

$$\frac{\partial p}{\partial t} = \frac{p - p_n}{\tau_p} - \frac{1}{e}\nabla J_p \tag{1-119}$$

Similarly for electrons:

$$\frac{\partial n}{\partial t} = \frac{n - n_p}{\tau_n} + \frac{1}{e}\nabla J_n \tag{1-120}$$

The total hole current is given by:

$$J_p = e\mu_p p E_0 - eD_p \nabla p \tag{1-121}$$

From Eq. (1-119) and Eq. (1-121) we obtain:

$$\frac{\partial p(x, t)}{\partial t} = -\frac{p - p_n}{\tau_p} + D_p \frac{\partial^2 p(x, t)}{\partial x^2} - \mu_p E_0 \frac{\partial p(x, t)}{\partial x} \tag{1-122}$$

Equation (1-122) is called the continuity equation, which is also known as charge conservation equation, or diffusion equation. This equation is extremely important in the study of diodes and transistors. Equation (1-122) is a partial differential equation containing two variables namely, $x$ and $t$. It is first order in time and both second and first order in space. Thus, many special cases of physical interests are possible. Equation (1-122) is written in one dimension, although its extension in two and three dimensions is possible.

## 1-8 HALL EFFECT

The motion of carriers in the presence of electric and magnetic fields gives rise to a number of galvanometric effects. The most important of these effects is the Hall effect.

When a semiconductor sample carrying a current $I$ is placed in a transverse magnetic field $B$, then an electric field $E_0$ is induced in the specimen, in the direction perpendicular to both $B$ and $I$.

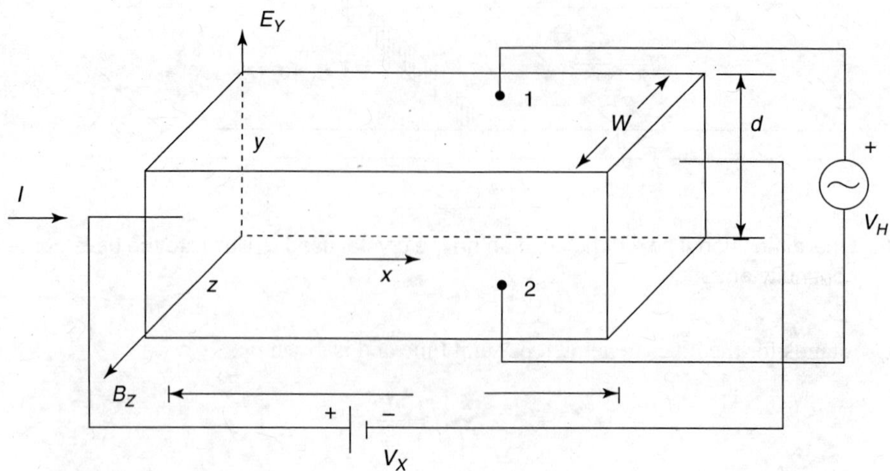


**Figure 1-37** Schematic diagram of Hall effect. The carriers (electrons or holes) are subjected to a magnetic force in the negative *y*-direction.

This phenomenon is called the Hall effect. The Hall effect may be used for determining whether a semiconductor is *n*-type or *p*-type by finding the carrier concentration and calculating the mobility $\mu$, by measuring the conductivity $\sigma$.

Let us consider a rectangular bar of *n*-type semiconductor of length *L*, width *W* and thickness *d*, as shown in Fig. 1-37.

If a current *I* is applied in the positive *x*-direction and a magnetic field *B* is applied in the positive *z*-direction, a force will be exerted in the negative *y*-direction of the current carriers. The current is carried by electron from side 1 to side 2, if the semiconductor is *n*-type. Therefore Hall voltage $V_H$ appears between surfaces 1 and 2. The electric field developed in *y*-direction $E_y$ is given by:

$$\frac{V_H}{d} = E_y \tag{1-123}$$

where, *d* is the distance between surfaces 1 and 2. In the equilibrium state the electric field $E_y$ due to the Hall effect must exert a force on the carrier, which just balances the magnetic force, and we can write:

$$eE_y = Bev_0 \tag{1-124}$$

where, *e* is the magnitude of the charge of the carriers, $v_0$ is the drift speed.
Therefore current density *J* is given by:

$$J = \rho v = \frac{I}{Wd} \tag{1-125}$$

where, $\rho$ is the charge density and *W* is the width of the specimen in the direction of the magnetic field. Combining Eq. (1-123), Eq. (1-124) and Eq. (1-125) we find:

$$V_H = E_y d = Bv_0 d = \frac{BJd}{\rho} = \frac{BI}{\rho W} \tag{1-126}$$

If $V_H$, $B$, $W$ and $I$ are measured, the charge density $\rho$ can be measured from Eq. (1-126). If the polarity of $V_H$ is positive at terminal 1 then the carrier must be an electron and $\rho = n_0 e$ where $n_0$ is the electron concentration. If the terminal 2 becomes positively charged with respect to terminal 1 the semiconductor must be of $p$-type and $\rho = p_0 e$ where $p_0$ is the hole concentration:

$$R_H = \frac{l}{\rho} \tag{1-127}$$

where, $R_H$ is the Hall coefficient.

$$R_H = \frac{V_H w}{BI} \tag{1-128}$$

Conductivity $\sigma$ is related to mobility $\mu$ by: $\sigma = n_0 e\mu$

$$\sigma = p_0 \mu$$

where, $\rho = n_0 . e$ (1-129)

If the conductivity is measured with Hall coefficient, mobility $\mu$ can be determined by:

$$J = \rho\mu$$

$$\frac{1}{\rho} = \frac{\mu}{\sigma} = R_H \tag{1-130}$$

From Eq. (1-127) and Eq. (1-130) we get:

$$R_H = \frac{\mu}{\sigma} \tag{1-131}$$

$$\mu = \sigma R_H$$

In the presence of scattering the mobility can approximately be written as:

$$\mu = \frac{8\sigma}{3\pi} R_H \tag{1-132}$$

## Applications of Hall Effect

1. Experimental determination of carrier concentration: From the basic formula of Hall effect, the electron and hole concentrations can be determined by using the experimental values of Hall coefficient.
2. By using Hall effect, the type of the semiconductor can be determined as follows: $R_H > 0$ for $p$-type semiconductors and $R_H < 0$ for $n$-type semiconductors.
3. Determination of the mobility: The equation $\mu = [\sigma R_H]$ determines the mobility by using the experimental values of $R_H$.
4. Hall effect multiplier: If the magnetic field $B$ produces current $I_1$ then Hall voltage $V_H \infty II_1$. The Hall effect multiplier generates an output proportional to the product of two signals. Thus, Hall effect can be used for analogue multiplication.
5. The power flow in an electromagnetic wave can be measured by the help of Hall effect.
6. Experimental determination of the magnetic field: By knowing the values of $V_H$, $I$, $\rho$ and $W$, we can determine the value of $B$ experimentally.

## POINTS TO REMEMBER

1. Lattice plus basis is equal to crystal structure.
2. The bravais lattices are distinct lattice types, which when repeated can fill the whole space.
3. In two dimensions there are five distinct bravais lattices while in three dimensions there are fourteen.
4. Electrons, which revolve in the outermost orbit of an atom, are called valence electrons.
5. Electrons, which are detached from the parent atoms and move randomly in the lattice of a metal, are called free electrons.
6. The range of energies possessed by electrons in a solid is known as energy band.
7. The range of energies possessed by valance electrons is called the valance band.
8. The range of energies possessed by free electrons is called the conduction band.
9. The separation between conduction and valence band on the energy level diagram is called the forbidden energy gap.
10. The semiconductors are a class of materials whose electrical conductivity lies between those of conductors and insulators. Germanium and silicon are called elemental semiconductors. The III-V, II-VI, IV-VI, ternaries and quaternaries are called compound semiconductors.
11. There are three types of semiconductors namely *n*-type semiconductors (when electron concentration is much greater than hole concentration), *p*-type semiconductors (when electron concentration is much less than hole concentration) and intrinsic semiconductors (when electron concentration is equal to hole concentration).
12. A semiconductor without any impurities is called an intrinsic semiconductor and a semiconductor with impurities is called extrinsic semiconductor.
13. When a covalent bond is broken, an electron vacates an energy level in valence band. This vacancy may be traced as a particle called hole.
14. The resistivity of pure semiconductor is of the order of $10^3$ Ωm. The energy gap of pure silicon is 1.12 eV and of germanium is 0.72 eV.
15. Thermal process is the only process that generates carriers in an intrinsic semiconductor.
16. The additional velocity acquired by the charge carriers in the electric field is called the drift velocity.
17. Mobility is defined as the drift velocity acquired per unit electric field strength.
18. Conductivity of a material is a measure of the material's ability to allow charge carriers to flow through it.
19. The process of adding impurities to a pure semiconductor is called doping. A doped semiconductor is called an extrinsic semiconductor. The dopants are usually trivalent or pentavalent impurities for a tetravalent (Si or Ge) semiconductor material.
20. An *n*-type semiconductor is obtained by adding pentavalent impurity to a pure Si or Ge semiconductor. In an *n*-type semiconductor electrons are majority carriers and holes are minority carriers.
21. A *p*-type semiconductor is obtained by adding trivalent impurity to a pure Si or Ge semiconductor. In a *p*-type semiconductor holes are majority carriers and electrons are minority carriers. The energy gap of extrinsic silicon semiconductor is 0.7 eV and of germanium is 0.3 eV.
22. An extrinsic semiconductor is electrically neutral.
23. In an *n*-type material, the free electron concentration is approximately equal to the density of donor atoms. Thus, $n_n \cong N_A$
24. In an extrinsic parabolic semiconductors, Fermi level lies exactly midway between valance and conduction bands at $T \to 0$.
25. The product of $n$ and $p$ is a constant and is known as the law of mass action.
26. The flow of current through a semiconductor material is normally referred to as one of two types: drift or diffusion.

27. The combined effect of the movement of holes and electrons constitute an electric current under the action of an electric field and is called a drift current.

28. The directional movement of charge carriers due to their concentration gradient produces a component of current known as diffusion current.

29. The total current is equal to the drift part of the current plus diffusion part of the current.

30. The distance that free carrier travels before recombining is called the diffusion length.

31. The average time an electron or hole can exist in the free state or the average time between the generation and recombination of a free electron is called lifetime.

32. If a metal or semiconductor carrying a current is placed in a transverse magnetic field, an electric field is induced in the direction perpendicular to both the current and magnetic field. This phenomenon is called Hall effect.

33. Hall effect measurements help us in identifying the type of majority carriers in determining carrier concentration and carrier mobility.

## IMPORTANT FORMULAE

1. De-Broglie's relation of the wave particle is given by: $\lambda = h/p$

2. The effective momentum mass of the carriers is given by: $m^* = h^2 k \dfrac{\partial k}{\partial E}$

3. Energy gap between conduction and valence band is: $E_g = E_c - E_v$

4. The density-of-state function of the conduction electron in parabolic $n$-type semiconductors is given by:

$$N(E) = 4\pi\left(\frac{2m_c^*}{h^2}\right)^{3/2}\sqrt{E}\,\frac{1}{eV.m^3}$$

5. The Fermi–Dirac integral of order $j$ is given by:

$$F_j(\eta) = \frac{1}{\lfloor j+1}\int_0^\infty \frac{x^j}{1+\exp(x-\eta)}\,dx$$

which obeys the following properties:

1. $\dfrac{d}{d\eta}\left[F_j(\eta)\right] = F_{j-1}(\eta)$
2. $\int F_j(\eta)\,d\eta = F_{j+1}(\eta)$
3. $F_{1/2}(\eta) \approx \dfrac{4}{3\sqrt{\pi}}\eta^{3/2}\left(1+\dfrac{\pi^2}{8\eta^2}\right),\ \eta > 0$
4. $F_0(\eta) = \ln|1+e^\eta|$
5. $F_j(\eta) \approx e^\eta,\ \eta < 0$ for all $j$.

6. The electron concentration in $n$-type parabolic semiconductors is given by: $n_0 = N_c F_{1/2}(\eta)$

(a) Under non-degenerate electron concentration $n_0 = N_c \exp(\eta)$

(b) The expression of the critical electron concentration when $E_F$ touches $E_C$ is given by: $n_0 = N_c F_{1/2}(0)$

(c) The variation of Fermi energy with temperature for relatively low values of temperatures in $n$-type parabolic semiconductors is given by:

$$E_F(T) = E_F(0)\left[1 - \frac{\pi^2 k_B^2 T^2}{12(E_F(0))^2}\right]$$

(d) Under the condition of extreme carrier degeneracy

$$n_0 = \frac{8}{3\sqrt{\pi}}\left(\frac{2\pi m_c^*}{h^2}\right)^{3/2}(E_F)^{3/2}$$

7. The concentration of heavy holes in parabolic $p$-type semiconductors is given by:

$$p_0 = N_v F_{1/2}\left(-\eta - \frac{E_g}{K_B T}\right)$$

8. Under the condition of non-degenerate hole concentration we can write the above equation reduces to:

$$p_0 = N_v \exp\left(-\eta - \frac{E_g}{K_B T}\right)$$

9. For intrinsic semiconductors $E_F$ is given by:t

$$E_F = \frac{3K_B T}{4}\ln\left|\frac{m_{hh}^*}{m_c^*}\right| - \frac{E_g}{2}$$

10. The law of mass action for non-degenerate parabolic semiconductors is given by:

$$n_0 p_0 = \exp\left(\frac{E_g}{K_B T}\right)(N_c N_v) = n_i^2$$

11. The intrinsic carrier concentration is given by:

$$n_i^2 = 2.31 \times 10^{31}\left(\frac{m_c^* m_{hh}^*}{m_0^2}\right)^{3/2} T^3 \exp\left(-\frac{E_g}{K_B T}\right)$$

12. For extrinsic non-degenerate parabolic semiconductors with $N_d \neq 0, N_a = 0$ and $\frac{8N_d}{N_c}\exp\left(-\frac{E_d}{K_B T}\right)$ >> 1, the Fermi energy can be written as:

$$E_F = \frac{E_d}{2} + \frac{K_B T}{2}\ln\frac{N_d}{2N_c}$$

13. For extrinsic non-degenerate parabolic semiconductors with $N_d \neq 0, N_a = 0$

and $\frac{8N_d}{N_c}\exp\left(-\frac{E_d}{K_B T}\right) << 1$, the Fermi energy can be written as:

$$E_F = K_B T \ln\left(\frac{N_d}{N_c}\right)$$

14. In the zone of very high temperatures the Fermi energy for non-degenerate parabolic semiconductors when $N_A = 0$ is given by:

$$E_F = K_B T \ln\left[\frac{N_d}{2N_c}\left(1 + \sqrt{1 + \frac{4n_i^2}{N_d^2}}\right)\right]$$

15. The electron mobility ($\mu$) is given by:

$$\mu = \frac{v_0}{E_0} = \frac{e\tau}{m^*}$$

In general, the mobility for isotropic bands is expressed as:

$$\mu = e < \tau(E)/m^*(E) >$$

16. Mathiessen's rule is given by:

$$\frac{1}{\mu} = \frac{1}{\mu_1} + \frac{1}{\mu_2} + \frac{1}{\mu_3} + \cdots$$

17. The conductivity for mixed conduction is given by: $\sigma = \sigma_n + \sigma_p = e\,(n_0\mu_n + p_0\mu_p)$

18. The electron and hole current densities in the presence of drift and diffusion are given by:

$$J_p = e(\mu_p p_0 E_0 - D_p \nabla p)$$
$$= e\left(p\mu_p E - D_p \frac{dp}{dx}\right)$$
$$J_n = e(\mu_n n_0 E_0 + D_n \nabla n)$$
$$= e\left(n\mu_n E + D_n \frac{dn}{dx}\right)$$

19. The generalized expression of the Einstein relation for electrons in $n$-type semiconductors is given by:

$$\frac{D_n}{\mu_n} = \frac{1}{e}\frac{n_0}{\frac{\partial n_0}{\partial E_F}}$$

(a) Under the condition of extreme degeneracy, the Einstein relation in $n$-type parabolic semiconductors is given by:

$$\frac{D_n}{\mu_n} = \frac{2}{3}\frac{E_F}{e}$$

(b) Under non-degenerate electron concentration, the Einstein relation for $n$-type semiconductors is given by:

$$\frac{D_n}{\mu_n} = \frac{k_B T}{e}$$

$$\frac{D_p}{\mu_p} = \frac{D_n}{\mu_n} = V_T = \frac{kT}{e}$$
$$= \frac{T}{11{,}600} = 26\text{ mV}$$

20. The relation between the rate of recombination of electrons and holes and equilibrium concentration of electrons and holes is given by:

$$r_i = k_r n_0 p_0 = k_r n_i^2 = g_i$$

21. The time dependence of excess carrier concentration for a $p$-type material is given by:

$$\delta n(t) = \Delta n \exp(-k_r p_0 t) = \Delta n \exp\left(-\frac{t}{\tau_n}\right)$$

22. The continuity equation for holes for $n$-type semiconductors is given by:

$$\frac{\partial p\,(x,t)}{\partial t} = \frac{p - p_n}{\tau_p} + D_p \frac{\partial^2 p(x,t)}{\partial x^2}$$
$$-\mu_p E_0 \frac{\partial_p(x,t)}{\partial x}$$

23. The transverse voltage produced in Hall effect is given by:

$$V_H = \frac{BI}{\rho t} = R_H \frac{BI}{t}$$

24. The Hall coefficient: $R_H = \frac{1}{\rho} = \frac{1}{ne} = \frac{1}{pe}$

25. Charge density: $\rho = \frac{BI}{V_H t}$

26. Mobility: $\mu = \sigma R_H$

27. The transverse voltage produced in Hall effect is given by:

$$V_H = \frac{BI}{\rho t} = R_H \frac{BI}{t}$$

28. The Hall coefficient: $R_H = \frac{1}{\rho} = \frac{1}{ne} = \frac{1}{pe}$

29. Charge density: $\rho = \frac{BI}{V_H t}$

## OBJECTIVE QUESTIONS

1. A silicon sample is uniformly doped with $10^{16}$ phosphorus atoms/cm$^3$ and $2 \times 10^{16}$ boron atoms/cm$^3$. If all the dopants are fully ionized, the material is:
   (a) $n$-type with carrier concentration of $3 \times 10^{16}$/cm$^3$
   (b) $p$-type with carrier concentration of $10^{16}$/cm$^3$
   (c) $p$-type with carrier concentration of $4 \times 10^{16}$/cm$^3$
   (d) Intrinsic
2. $n$-type semiconductors are:
   (a) Negatively charged
   (b) Produced when Indium is added as an impurity to Germanium
   (c) Produced when phosphorous is added as an impurity to silicon
   (d) None of the above
3. The probability that an electron in a metal occupies the Fermi-level, at any temperature ($>$ 0 K) is:
   (a) 0 (b) 1
   (c) 0.5 (d) None of the above
4. Measurement of Hall coefficient enables the determination of:
   (a) Mobility of charge carriers
   (b) Type of conductivity and concentration of charge carriers
   (c) Temperature coefficient and thermal conductivity
   (d) None of the above
5. If the energy gap of a semiconductor is 1.1 eV it would be:
   (a) Opaque to the visible light
   (b) Transparent to the visible light
   (c) Transparent to the ultraviolet radiation
   (d) None of the above
6. The conductivity of an intrinsic semiconductor is given by (symbols have the usual meanings):
   (a) $\sigma_i = en_i^2(\mu_n - \mu_p)$ (b) $\sigma_i = en_i(\mu_n - \mu_p)$
   (c) $\sigma_i = en_i(\mu_n + \mu_p)$ (d) None of the above
7. Consider the following statements: Compared to Silicon, Gallium Arsenide (GaAs) has:
   1. Higher signal speed since electron mobility is higher
   2. Poorer crystal quality since stoichiometric growth difficult
   3. Easier to grow crystals since the vapour pressure Arsenic is high
   4. Higher optoelectronic conversion efficiency

   Of these statements:
   (a) 1, 2, 3 and 4 are correct
   (b) 1, 2 and 3 are correct
   (c) 3 and 4 are correct
   (d) None of the above
8. In an intrinsic semiconductor, the mobility of electrons in the conduction band is:
   (a) Less than the mobility of holes in the valence band
   (b) Zero
   (c) Greater than the mobility of holes in the valence band
   (d) None of the above
9. The Hall coefficient of sample (A) of a semiconductor is measured at room temperature. The Hall coefficient of (A) at room temperature is $4 \times 10^{-4}$ m$^3$ coulomb$^{-1}$. The carrier concentration in sample (A) at room temperature is:
   (a) ~$10^{21}$ m$^{-3}$ (b) ~$10^{20}$ m$^{-3}$
   (c) ~$10^{22}$ m$^{-3}$ (d) None of the above

10. In a semiconductor, $J$, $J_p$ and $J_n$ indicate total diffusion current density hole current density and electron current density respectively, $\frac{\partial n}{\partial x}$ and $\frac{\partial p}{\partial x}$ are the electron and hole concentration gradient respectively in $x$-direction and $D_p$ and $D_n$ are the hole and electron diffusion constants respectively. Which one of the following equations is correct? ($e$ denotes charge of electron).

    (a) $J_n = -e\, D_n \frac{\partial n}{\partial x}$ for electrons

    (b) $J = -e\, D_p \frac{\partial p}{\partial x}$ for holes

    (c) $J_p = -e\, D_p \frac{\partial p}{\partial x} - e\, D_n \frac{\partial n}{\partial x}$

    (d) None of the above

11. If the drift velocity of holes under a field gradient of 100 v/m is 5 m/s, the mobility (in the same SI units) is:

    (a) 0.05 (b) 0.55

    (c) 500 (d) None of the above

12. The Hall effect voltage in intrinsic silicon is:

    (a) Positive (b) Zero

    (c) Negative (d) None of the above

13. The Hall coefficient of an intrinsic semiconductor is:

    (a) Positive under all conditions

    (b) Negative under all conditions

    (c) Zero under all conditions

    (d) None of the above

14. Consider the following statements: Pure germanium and pure silicon are examples of:

    1. Direct band-gap semiconductors

    2. Indirect band-gap semiconductors

    3. Degenerate semiconductors

    Of these statements:

    (a) 1 alone is correct

    (b) 2 alone is correct

    (c) 3 alone is correct

    (d) None of the above

15. Assume $n_e$ and $n_h$ are electron and hole densities, and $\mu_e$ and $\mu_n$ are the carrier mobilities; the Hall coefficient is positive when:

    (a) $n_h \mu_h > n_e \mu_e$ (b) $n_h \mu_h^2 > n_e \mu_e^2$

    (c) $n_h \mu_h < n_e \mu_h$ (d) None of the above

16. A long specimen of $p$-type semiconductor material:

    (a) Is positively charged

    (b) Is electrically neutral

    (c) Has an electric field directed along its length

    (d) None of the above

17. The electron and hole concentrations in an intrinsic semiconductor are $n_i$ and $p_i$ respectively. When doped with a $p$-type material, these change to $n$ and $P$ respectively. Then:

    (a) $n + p = n_i + p_i$

    (b) $n + n_i = p + p_i$

    (c) $np = n_i p_i$

    (d) None of the above are applicable

18. If the temperature of an extrinsic semiconductor is increased so that the intrinsic carrier concentration is doubled, then:

    (a) The majority carrier density doubles

    (b) The minority carrier density doubles

    (c) Both majority and minority carrier densities double

    (d) None of the above

19. At room temperature, the current in an intrinsic semiconductor is due to:

    (a) Holes

    (b) Electrons

    (c) Holes and electrons

    (d) None of the above

20. A semiconductor is irradiated with light such that carriers are uniformly generated throughout its volume. The semiconductor is $n$-type with $N_D = 10^{19}$ per cm$^3$. If the excess electron concentration in the steady state is $\Delta n = 10^{15}$ per cm$^3$ and if $\tau_p = 10\ \mu$sec. (minority carrier life time) the generation rate due to irradiation is:

(a) $10^{22}$ e-h pairs/cm$^3$/s

(b) $10^{10}$ e-h pairs/cm$^3$/s

(c) $10^{24}$ e-h pairs/cm$^3$/s

(d) None of the above

21. A small concentration of minority carriers is injected into a homogeneous semiconductor crystal at one point. An electric field of 10 V/cm is applied across the crystal and this moves the minority carriers a distance of 1 cm is 20 $\mu$sec. The mobility (in cm$^2$/volt.sec) is:

(a) 1,000 (b) 2,000

(c) 50 (d) None of the above

22. The mobility is given by:

(a) $\mu = \frac{V_0}{E_0}$ (b) $\mu = \frac{V_0^2}{E_0}$

(c) $\mu = \frac{V_0}{E_0^2}$ (d) None of the above

23. Hall effect is observed in a specimen when it (metal or a semiconductor) is carrying current and is placed in a magnetic field. The resultant electric field inside the specimen will be in:

(a) A direction normal to both current and magnetic field

(b) The direction of current

(c) A direction anti parallel to magnetic field

(d) None of the above

24. In a *p*-type semiconductor, the conductivity due to holes ($\sigma_p$) is equal to: (*e* is the charge of hole, $\mu_p$ is the hole mobility, $p_0$ is the hole concentration):

(a) $p_0 e/\mu_p$ (b) $\mu_p/p_0 e$

(c) $p_0 e \mu_p$ (d) None of the above

25. The difference between the electron and hole Fermi energies of a semiconductor laser is 1.5 eV and the band gap of the semiconductor is 1.43 eV. The upper and lower frequency limits of the laser will be respectively:

(a) $3.3 \times 10^{15}$ and $9.9 \times 10^{13}$ Hz

(b) $3.7 \times 10^{16}$ and $3.5 \times 10^{14}$ Hz

(c) $6.28 \times 10^{17}$ and $3.1 \times 10^{13}$ Hz

(d) None of the above

26. A sample of *n*-type semiconductor has electron density of $6.25 \times 10^{18}$/cm$^3$ at 300 K. If the intrinsic concentration of carriers in this sample is $2.5 \times 10^{13}$/cm$^3$ at this temperature, the hole density becomes:

(a) $10^{16}$/cm$^3$ (b) $10^7$/cm$^3$

(c) $10^{17}$/cm$^3$ (d) None of the above

27. The intrinsic carrier density at 300K is $1.5 \times 10^{10}$/cm$^3$ in silicon. For *n*-type silicon doped to $2.25 \times 10^{15}$ atoms/cm$^3$ the equilibrium electron and hole densities are:

(a) $n_0 = 1.5 \times 10^{16}$/cm$^3$, $p_0 = 1.5 \times 10^{12}$/cm$^3$

(b) $n_0 = 1.5 \times 10^{10}$/cm$^3$, $p_0 = 2.25 \times 10^{15}$/cm$^3$

(c) $n_0 = 2.25 \times 10^{17}$/cm$^3$, $p_0 = 1.0 \times 10^{14}$/cm$^3$

(d) None of the above

28. In a *p*-type silicon sample, the hole concentration is $2.25 \times 10^{15}$/cm$^3$. If the intrinsic carrier concentration $1.5 \times 10^{10}$/cm$^3$ the electron concentration is:

(a) $10^{21}$/cm$^3$ (b) $10^{10}$/cm$^3$

(c) $10^{16}$/cm$^3$ (d) None of the above

29. A good ohmic contact on a *p*-type semiconductor chip is formed by introducing:

(a) Gold as an impurity below the contact

(b) A high concentration of acceptors below the contact

(c) A high concentration of donors below the contact

(d) None of the above

30. Measurement of Hall coefficient in a semiconductor provides information on the:

(a) Sign and mass of charge carriers

(b) Mass and concentration of charge carriers

(c) Sign of charge carriers alone

(d) Sign and concentration of charge carriers

## REVIEW QUESTIONS

1. Briefly discuss the basic developments in the study of electronics.
2. What are crystalline materials?
3. Give three examples of Group III-V semiconductors.
4. What do you mean by pure crystals?
5. Why is Si preferred over Ge?
6. Why is GaAs preferred over Si?
7. Give an example of the constituent material of Gunn Diode.
8. What is Bravais lattice? Discuss briefly.
9. What are unit cells and lattice constants?
10. Explain the differences among simple cubic, body-centred cubic and face-centred cubic lattices respectively.
11. Explain Czochralski growth of the semiconductor crystal in detail.
12. Explain the wave particle duality principle.
13. State Pauli exclusion principle.
14. What is degenerate energy level?
15. What is energy band gap?
16. What do you mean by covalent and electrovalent bonds? How these are affected by the temperature?
17. Explain thermal equilibrium.
18. What is valence band and conduction band?
19. What are conduction band carriers?
20. Explain the existence of hole.
21. What is momentum effective mass of the carriers? What is its difference with acceleration effective mass?
22. Explain indistinguishability between the particles. What should be the value of the spin of the particles obeying Fermi–Dirac statistics?
23. Explain the term "occupation probability".
24. What is bonding model?
25. Explain the energy band models of semiconductors.
26. Write the assumptions behind Fermi–Dirac statistics. Give a very simple proof of the same statistics.
27. Write the five properties of the Fermi–Dirac function.
28. Define Fermi energy. How does Fermi energy vary with temperature?
29. Write the basic criterion for the classification of metals, semiconductors and insulators.
30. Draw the model energy band structure diagram of a semiconductor in general.
31. The split-off hole parabola is more flattened than the light hole parabola. Justify your answer very briefly.
32. What are intrinsic semiconductors?
33. What are extrinsic semiconductors?
34. Explain donor ion, acceptor ion, majority carriers, minority carriers, doping and dopants in a semiconductor.
35. What do you mean by the term "carrier degeneracy" of a semiconductor? Explain in detail.
36. What are compound semiconductors? Explain the uses of compound semiconductors.
37. Explain the difference between metals, insulators and semiconductors with the help of band structure model.
38. Define the term "density-of-state function". Derive an expression for the density-of-state function of the conduction electrons in a semiconductor having parabolic energy bands. Draw the graph of the density of state function versus energy and explain the graph.
39. Derive an expression of electron concentration in $n$-type semiconductors. Discuss all the special cases.
40. Derive an expression of electron concentration in $p$-type semiconductors. Discuss all the special cases.
41. What is charge density equation?
42. Show that:

$$E_F = K_B T \ln\left[\frac{N_d}{2N_c}\left(1+\sqrt{1+\frac{4n_i^2}{N_d^2}}\right)\right]$$

where, the notations mean as usual.

43. Derive the law of mass action. Is it valid for degenerate semiconductors? Give reasons.
44. Does the band gap vary with temperature? Give reasons.
45. How does concentration vary with temperature?
46. What is mobility?
47. What is relaxation time?
48. What is conductivity?
49. What is Mathiessen's rule?
50. What is diffusion?
51. What is recombination?
52. What is generation?
53. Prove that: $$\frac{D_n}{\mu_n} = \frac{K_B T}{e}\left[\frac{F_{\frac{1}{2}}(\eta)}{F_{-\frac{1}{2}}(\eta)}\right]$$
54. Write the assumptions of continuity equation and derive continuity equation.
55. Explain physically the continuity equation.
56. What is Hall effect?
57. Derive the relation between mobility and Hall coefficient.
58. What are the applications of Hall effect?
59. Find out an expression for Hall coefficient in a semiconductor when both carriers contribute to the current.

## PRACTICE PROBLEMS

1. The energy spectrum of the conduction electrons of III-V compound semiconductors can be expressed as $E\left[1 + (E/E_g)\right] = \frac{\hbar^2 k^2}{2m_c^*}$ where the notations have their usual meaning. Find out the momentum effective mass and the acceleration effective mass of the conduction electrons respectively. Draw the plots in two cases on the same graph paper with the independent variable as energy by taking the example of $n$-GaAs. Interpret the results. Explain the results for $E_g \to \infty$.
2. The dispersion relation of the carriers in a semiconductor is approximately given by
   $E = E_0 - A\cos(\alpha k_x) - B[\cos(\beta k_y) + \cos(\beta k_z)]$
   where $E_0$, $A$, $B$, $\alpha$ and $\beta$ are constants. Develop an expression of the density-of-states function for small $k$.
3. (a) Calculate the coordinates of three points on the Fermi–Dirac function and enter the coordinates in this table.

| $\frac{\xi - \xi_F}{kT}$ | $P(\xi)$ |
|---|---|
| | |
| | |
| | |

   (b) Plot the same three points in the following axis and sketch the Fermi–Dirac function with reasonable accuracy.

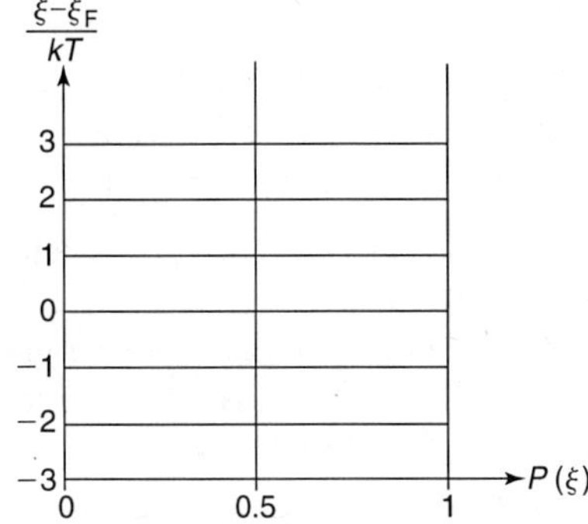


4. In a certain silicon sample at equilibrium, the Fermi level resides at 0.500 eV above the centre of the band gap.
   (a) Calculate the occupancy probability for a lone isolated state located right at the centre of the band gap.
   (b) This sample contains donor impurities and no acceptor impurities. The donor states are situated 0.045 eV below the conduction-band edge. Find the occupancy probability of the donor states.
   (c) Comment on the validity of the assumption of 100 per cent ionization of the donor states in the present situation.

(d) Derive an approximate form of the Fermi–Dirac probability function that could be applied in (b) with reasonable validity.

(e) Use your approximate expression to recalculate the probability in (b) and find the percentage difference in the exact and approximate results.

(f) Sketch the exact probability function accurately by plotting several points. Superimpose on the same diagram an accurate sketch of the approximate expression of (d).

(g) Comment on conditions where the use of the approximate expression is justified.

(h) How much error results when the approximate expression is used to calculate occupancy probability for a state located right at the Fermi level?

5. The conduction band can be characterized by a state density (number of states per cm$^3$) of $N_c = 3.75 \times 10^{19}$/cm$^3$, with these states assumed to be situated right at the conduction-band edge.

(a) Using this assumption, calculate the conduction-electron density $n_0$ (number of electrons per cm$^3$) for the conditions of Problem 4.

(b) The valence band can be characterized by a state density of $N_v \cong N_c$, with these states assumed to be situated right at the valence-band edge. Using this assumption, calculate the hole density $p_0$.

(c) Calculate the *p*–*n* product using the results from (a) and (b).

6. Determine the approximate density of donor states $N_D$ (number of donor states per cm$^3$) for the silicon sample of Problems 4 and 5. Provide reasons.

7. Derive an expression relating the intrinsic level $E_i$ to the centre of the band gap $E_g/2$. Calculate the displacement of $E_i$ from $E_g/2$ for Si at 300 K, assuming the effective mass values for electrons and holes are 1.1 $m_0$ and 0.56 $m_0$, respectively.

8. (a) Explain why holes are found at the top of the valence band, whereas electrons are found at the bottom of the conduction band.

(b) Explain why Si doped with $10^{14}$ cm$^{-3}$ Sb is *n*-type at 400 K but similarly doped Ge is not.

9. Calculate the $N_c$ and $N_v$ for the conduction and valence bands of Si and GaAs at 100 K, 200 K and 400 K respectively by assuming the data you require.

10. A certain uniformly doped silicon sample at room temperature has $n_0 = 10^6$/cm$^3$ and $N_A = 10^{15}$/cm$^3$.

(a) Find $p_0$.

(b) Find $N_D$.

11. Using the Boltzmann approximation to the Fermi–Dirac probability function (obtained by dropping the unity term in its denominator), find the Fermi-level position relative to the conduction band edge for a sample having $n_0 = 3 \times 10^{15}$/cm$^3$ at room temperature and at equilibrium.

12. Given that the majority impurity in the foregoing problem is phosphorus, find the occupancy probability at the donor level. Comment on the assumption of 100 per cent ionization in this case.

13. A silicon sample has $N_A = 10^{15}$/cm$^3$ and $N_D = 0$. Find the:

(a) Majority-carrier density

(b) Minority-carrier density

(c) Conductivity

14. A certain silicon sample has $p_0 = 2.5 \times 10^{10}$/cm$^3$. Find the:

(a) Electron density $n_0$

(b) Resistivity $\rho$

15. Calculate the position of the intrinsic Fermi level measured from the midgap for InAs.

16. Calculate and plot the position of the intrinsic Fermi level in Si between 80 K and 400 K.

17. Calculate the density of electrons in silicon if the Fermi level is 0.45 eV below the conduction bands at 290 K. Compare the results by using the Boltzmann approximation and the Fermi–Dirac integral.

18. In a GaAs sample at 310 K, the Fermi level coincides with the valence band-edge. Calculate the hole density by using the Boltzmann approximation. Also calculate the electron density using the law of mass action.
19. The electron density in a silicon sample at 310 K is $10^{15}$ cm$^{-3}$. Calculate $E_c - E_F$ and the hole density using the Boltzmann approximation.
20. A GaAs sample is doped $n$-type at $4 \times 10^{18}$ cm$^{-3}$. Assume that all the donors are ionized. What is the position of the Fermi level at 300 K?
21. Consider a $n$-type silicon with a donor energy 60 meV below the conduction band. The sample is doped at $10^{15}$ cm$^{-3}$. Calculate the temperature at which 30 per cent of the donors are not ionized.
22. Consider a GaAs sample doped at $N_d = 10^{15}$ cm$^{-3}$ where the donor energy is 5 meV. Calculate the temperature at which 80 per cent of the donors are ionized.
23. Estimate the intrinsic carrier concentration of diamond at 700 K (you can assume that the carrier masses are similar to those in Si). Compare the results with those for GaAs and Si.
24. A Si device is doped at $2.5 \times 10^{16}$ cm$^{-3}$. Assume that the device can operate up to a temperature where the intrinsic carrier density is less than 10 per cent of the total carrier density. What is the upper limit for the device operation?
25. Estimate the change in intrinsic carrier concentration per K change in temperature for Ge, Si, and GaAs and InSb at 300 K.
26. A certain silicon sample has $N_D = 5.30 \times 10^{15}$/cm$^3$ and $N_A = 4.50 \times 10^{15}$/cm$^3$.

    Find the:
    (a) Majority-carrier density
    (b) Minority-carrier density
    (c) Conductivity $\sigma$, using $\mu_n = 400$ cm$^2$/Vs, and $\mu_p = 300$ cm$^2$/Vs
    (d) $n_0$
    (e) $\sigma$ using $\mu_n = 500$ cm$^2$/V.s and $\mu_p = 300$ cm$^2$/V.s
    (f) $\rho$
27. A silicon samples is doped with $2.5 \times 10^{15}$ phosphorus atoms per cubic centimetre and $0.5 \times 10^{15}$ boron atoms per cubic centimetre, and is at equilibrium. Find the:
    (a) Majority-carrier density.
    (b) Minority-carrier density.
    (c) Calculate the sample's conductivity, using $\mu_n = 1020$ cm$^2$/Vs.
    (d) Calculate its resistivity.
28. A silicon sample has $N_D = 10^{16}$/cm$^3$ and $N_A = 0$.
    (a) Find $\rho$.
    (b) What is the conductivity type of the sample in (a)?
    (c) Another sample has $N_D = 0.5 \times 10^{14}$/cm$^3$ and $N_A = 10^{16}$/cm$^3$. Find $p_0$.
    (d) Find $n_0$ in the sample of c.
    (e) Another sample has $N_D = 1.5 \times 10^{15}$/cm$^3$ and $N_A = 10^{15}$/cm$^3$. Find $p_0$.
    (f) Find $n_0$ in the sample of (e).
    (g) Another sample has $N_D = 0.79 \times 10^{14}$/cm$^3$ and $N_A = 10^{15}$/cm$^3$. Find $p_0$.
    (h) Find $n_0$ in the sample of (g).
29. A Silicon specimen in the form of circular cylinder (Length $L = 20$ mm, area of cross-section $A = 2$ mm$^2$ and resistivity $\rho = 15$ ohm cm is placed in series with an ideal battery of 2 V in a complete circuit. Answer the following questions.
    (a) Calculate hole current density $J_p$.
    (b) Calculate electron current density $J_n$.
    (c) How $J_n$ is affected by doubling sample length $L$ and keeping $A$, $\rho$ and applied voltage $V$ the same as in part (a)?
    (d) How is $J_n$ affected by doubling cross-sectional area $A$ and keeping $\rho$, applied voltage $V$, and sample length $L$ the same as in part (a)?
30. Ohmic contacts are made to the ends of a silicon resistor having a length $L = 0.5$ cm. The resistor has a cross-sectional area $A$ that is given by the product of its width $W$ and thickness $X$, where $W = 1$ mm and $X = 2\ \mu_m$. The silicon is uniformly doped with $N_A = 2.5 \times 10^{15}$ cm$^3$ and $N_D = 2.5 \times 10^{15}$/cm$^3$. For this doping density, $\mu_p = 250$ cm$^2$/Vs and $\mu_n = 400$ cm$^2$/Vs. Find the resistance $R$.

31. A silicon sample contains $3.5 \times 10^{16}/\text{cm}^3$ of one impurity type and a negligible amount of the opposite type. It exhibits a resistivity of 0.22 Ω–cm at room temperature.
    (a) Determine majority-carrier mobility.
    (b) Is the sample *n*-type or *p*-type? Explain your reasoning.
32. (a) Given an extrinsic but lightly doped *n*-type silicon resistor *R* of length *L* and cross-sectional area *A*, derive an expression for its net impurity density.
    (b) For the resistor with, $R$ = 1 kilo-ohm, $L$ = 5 mm, and $A$ = 1.5 mm². Evaluate the expression derived in a. What is the probable majority impurity in the resistor?
33. Minority carriers in a particular silicon sample drift 1 cm in 100 μs when $E_0$ = 15 V/cm.
    (a) Determine drift velocity $V_0$.
    (b) Determine minority-carrier diffusivity *D*.
    (c) Determine the conductivity type of the sample and explain your reasoning. A sample of heavily doped *p*-type silicon has a drift-current density of 100 A/cm². Hole drift velocity is 50 cm/s. Find hole density $p_0$.
35. Calculate the intrinsic carrier concentration of Si, Ge and GaAs as a function of temperature from 4 K to 600 K. Assume that the band gap is given by:

$$E_g(T) = E_g(0) - \frac{\alpha T^2}{T+\beta}$$

where $E_g(0)$, $\alpha$, $\beta$ are given by

Si: $E_g(0) = 1.17$ eV, $\alpha = 4.37 \times 10^{-4}$ K$^{-1}$,
$\beta = 636$ K

Ge: $E_g(0) = 0.74$ eV, $\alpha = 4.77 \times 10^{-4}$ K$^{-1}$,
$\beta = 235$ K

GaAs: $E_g(0) = 1.519$ eV, $\alpha = 5.4 \times 10^{-4}$ K$^{-1}$,
$\beta = 204$ K

36. The resistance of No. 18 copper wire (having a diameter $d$ = 1.15 mm) is 6.5 ohm/1000 ft. The density of conduction electrons in copper is $n_0 = 8.3 \times 10^{22}/\text{cm}^3$.
    (a) Given that the current in the wire is 2 A, calculate the current density *J*.
    (b) Find the magnitude of the drift velocity $V_0$ of the electrons in cm/s and cm/hr.
    (c) Calculate the resistivity $\rho$ of the wire.
    (d) Calculate the electric field $E_0$ in the wire.
    (e) Calculate the mobility magnitude $|\mu_n|$ of the electrons in copper.
37. Assume complete ionization.
    (a) Combine the neutrality equation and the law of mass action to obtain an accurate expression for $n_0$ in near-intrinsic *n*-type silicon.
    (b) Use the expression obtained in a. to calculate $n_0$ in a sample having $N_D = 0.9 \times 10^{14}/\text{cm}^3$ and $N_A = 1.0 \times 10^{14}/\text{cm}^3$.
    (c) Comment on the accuracy of the approximate equation $n_0 \approx N_D - N_A$ for ordinary doping values.
38. Given a hole concentration gradient of $-10^{20}/\text{cm}^4$ in a lightly doped sample, calculate the corresponding hole diffusion-current density.
39. A lightly doped field-free sample of silicon that is 1 μm thick in the *x* direction exhibits a majority-electron gradient of $(dn/dx) = -10^{15}/\text{cm}^4$.
    (a) Calculate the diffusion current density for electrons in this case.
    (b) If the electron density at the higher-density (left) face is $10^{14}/\text{cm}^3$, what is it at the right face?
    (c) In a thought experiment, we supply an additional population of electrons throughout the sample in the amount of $10^{14}/\text{cm}^3$. Repeat the calculation of (a).
    (d) Return to (a) and (b). Suppose that without any other changes an electric field of $E_0$ = 0.045 V/cm is superimposed on the sample in the positive *x* direction. Calculate the diffusion current density due to electrons approximately.
    (e) Calculate the total current density for the electrons.
40. In a certain lightly doped silicon sample there exists a hole current-density value due to drift equal to −1.075 A/cm². Calculate the corresponding density gradient.

41. A certain silicon sample has negligible acceptor doping and a donor doping that is linearly graded from $1.5 \times 10^{15}/\text{cm}^3$ at the left surface to $2.5 \times 10^{14}/\text{cm}^3$ at the right surface. The two surfaces are 100 $\mu$m apart.

 (a) Assuming that $n(x) \approx N_D(x)$ throughout, calculate diffusion current density due to electrons at the middle of the sample.

 (b) In a thought experiment, we add $1.5 \times 10^{16}/\text{cm}^3$ of donors uniformly to the sample of (a); recalculate diffusion current density due to electrons.

 (c) In a second thought experiment we add $1.5 \times 10^{15}/\text{cm}^3$ of acceptors to the sample of (a); recalculate diffusion current density due to electrons.

42. A thin silicon sample receives steady-state radiation that produces excess carriers uniformly throughout the sample in the amount $\Delta p_0 = \Delta n_0 = 10^{10}/\text{cm}^3$. Excess-carrier lifetime in the sample is 1 $\mu$s. At $t = 0$, the radiation source is turned off. Calculate the excess-carrier density and recombination rate at (a) $t = 1$ $\mu$s; (b) $t = 1.5$ μs; (c) $t = 4$ $\mu$s.

43. A thin $n$-type sample of silicon having an equilibrium minority-carrier density $p_0$ is subjected to penetrating radiation with the radiation source turned on at $t = 0$. At $t = \infty$, $p = p(\infty)$.

 (a) Write the differential equation appropriate to this situation.

 (b) Find the solution for the differential equation of (a) under the given boundary conditions.

44. Uniform, steady-state ultraviolet radiation impinges on the surface of a semi-infinite silicon sample in which $n_0 = 10^{15}/\text{cm}^3$, producing an excess-carrier density at the surface of $\Delta p_0(0) = \Delta n_0(0) = 10^{11}/\text{cm}^3$. Given further that $\tau = 1$ $\mu$s, and that the spatial origin is at the irradiated surface.

 (a) Calculate diffusion current density due to holes at $x = 0$.

 (b) Calculate diffusion current density due to electrons at $x = 0$.

 (c) Calculate diffusion current density due to holes and diffusion current density due to electrons at $x = L_p$. Sketch vectors to scale representing these current-density components.

 (d) Since the sample is open-circuited, the total current density at $x = L_p$ must be zero, $J = 0$. In fact, a tiny electric field accounts for the "missing" current density component. Speculate on the cause of the electric field.

 (e) Calculate the magnitude and direction of the field cited in (d) at the position $x = L_p$.

 (f) Does the presence of the field destroy the validity of the current density due to holes calculations that were based upon a pure-diffusion picture? Explain.

 (g) Derive an approximate expression for $E_0$. A thin $n$-type silicon sample with $\tau = 5$ $\mu$ sec is subjected to infrared radiation for a long period. At $t = 0$, the radiation source is turned off. At $t = 020$ $\mu$ sec, excess – carrier recombination rate is observed to be $4.5 \times 10^{13}/\text{cm}^3/\text{s}$. Calculate the steady – state generation rate caused by the radiation.

46. (a) A Si bar 0.5 cm long and 120 $\mu\text{m}^2$ in cross-sectional area is doped with $10^{17}\text{cm}^{-3}$ phosphorus. Find the current at 300 K with 10 V applied. Repeat for a Si bar 1 $\mu$m long.

 (b) How long does it take an average electron to drift 1 $\mu$m in pure Si at an electric field of 500 V/cm? Repeat for 10 V/cm.

47. A Si sample is doped with $6.5 \times 10^{15}$ $\text{cm}^{-3}$ donors and $2.5 \times 10^{15}$ $\text{cm}^{-3}$ acceptors.

 Find the position of the Fermi level with respect to $E_i$ at 300 K. What is the value and sign of the Hall coefficient?

48. Find out the expressions of the density-of-state functions for the dispersion relation of the conduction electrons as given in Problem 1 under the conditions (a) $E/E_g << 1$ and (b) $E/E_g >> 1$ respectively. Draw the graphs and explain the result physically.

49. Find the expressions for $n_0$ under the conditions as stated in Problem 47. Draw the graphs and explain the result physically.

50. Find the expression of the average energy of electrons in semiconductors having parabolic energy bands for the conditions of both non degenerate and degenerate electron concentrations.

51. Find out a simple expression of photo emitted current density from semiconductors having parabolic energy bands.

## SUGGESTED READINGS

1. Pierret, R. F. and G.W. Neudeck. 1989. *Modular Series on Solid State Devices.* Boston M.A.: Addison Wesley.
2. Singh, J. 1994. *Semiconductor Devices: An Introduction.* New York, NY: McGraw-Hill.
3. Millman, Jacob and Christos C. Halkias. 1986. *Integrated Electronics: Analog and Digital Circuits and Systems*. New Delhi: McGraw-Hill.
4. Streetman, B.G. and S. Banerjee. 2000. *Solid State Electronic Devices*. New Delhi: Pearson Education.

# 2

# Diode Fundamentals

## Outline



## Objectives

This chapter introduces the reader to the first non-linear element in the whole field of electronics, namely the diode. William Shockley first demonstrated the diode in his famous paper on this topic, and discussed the formation of the *p–n* junction and the energy band diagram together with the basic concept of built-in potential. In analog electronics the modes of operation of the *p–n* junction is very important. The fundamental concepts of the junction capacitance, the I–V characteristics and linear piecewise models have been introduced for better understanding. Breakdown diodes that are very important components in power electronics have been discussed along with Zener and Avalanche breakdown. The degenerate semiconductors that were the first to enjoy device applications through tunnel diodes, the first element to produce negative resistance in microwaves, have also been discussed. The chapter ends with an examination of optoelectronic devices like light-emitting diode, photo detector diode, etc.

## 2-1 INTRODUCTION

The origin of a wide range of electronic devices being used can be traced back to a simple device, the *p–n* junction diode. *The p–n junction diode is formed when a p-type semiconductor impurity is doped on one side and an n-type impurity is doped on the other side of a single crystal.* All the macro effects of electronic devices, i.e., wave shaping, amplifying or regenerative effects, are based on the events occurring at the junction of the *p–n* device. Most modern devices are a modification or amalgamation of *p–n* devices in various forms. Prior to the era of semiconductor diodes, vacuum tubes were being extensively used. These were bulky, costly and took more time to start conducting because of the thermo-ionic emission. The semiconductor diodes and the allied

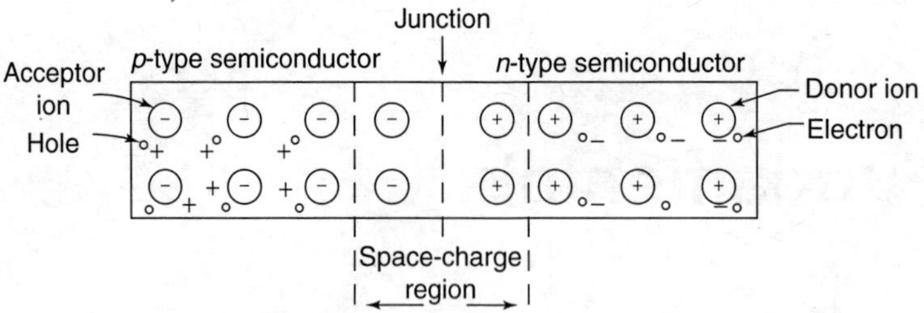


**Figure 2-1** A semiconductor *p–n* junction

junction devices solved all these problems. The *p–n* junction forms the basis of the semiconductor devices in general. This chapter starts with a discussion on the formation of the *p–n* junction diode and the corresponding energy band diagrams explicating their fundamental importance in the field of solid state electronics.

## 2-2 FORMATION OF THE *p–n* JUNCTION

When donor impurities are introduced into one side and acceptors into the other side of a single crystal semiconductor through various sophisticated microelectronic device-fabricating techniques, a *p–n* junction is formed. The *p–n* junction is thus formed by an intimate contact, as shown in Fig. 2-1. The presence of a concentration gradient between two materials in such intimate contact results in a diffusion of carriers that tends to neutralize this gradient. This process is known as the *diffusion process*.

The nature of the *p–n* junction so formed may, in general, be of two types:

(i) A step-graded junction (ii) A linearly-graded junction

In a step-graded semiconductor junction, the impurity density in the semiconductor is constant. In a linearly-graded junction, the impurity density varies linearly with distance away from the junction.

## 2-3 ENERGY BAND DIAGRAMS

The discussion in this section is based on the realistic assumption that a junction is made up of uniformly doped *p*-type and *n*-type crystals forming a step-graded junction. The variations in the expressions due to the other types of junctions will be mentioned later on.

### 2-3-1 The *p–n* Junction at Thermal Equilibrium

It is assumed that the discontinuity at the junction surfaces is confined within a narrow region of length $L$, as shown in Figs. 2-2(b) and 2-2(c), and that on either side of the junction the non-degenerate carrier concentrations are uniform, on planes parallel to the junction surface. Figure 2-2(a) exhibits the location of Fermi energy in *p*- and *n*-type semiconductors just before contact. Figure 2-2(b) shows the formation of the *p–n* junction after contact, which is based on the fact that the Fermi energy becomes invariant with respect to distance.

Figure 2-2(c) shows the energy band diagram of the *p–n* junction under the condition of thermal equilibrium. A *p–n* junction is said to be in thermal equilibrium when it is at a uniform temperature and no external disturbances, such as light or a bias voltage, are acting on it leading to a steady state charge and potential

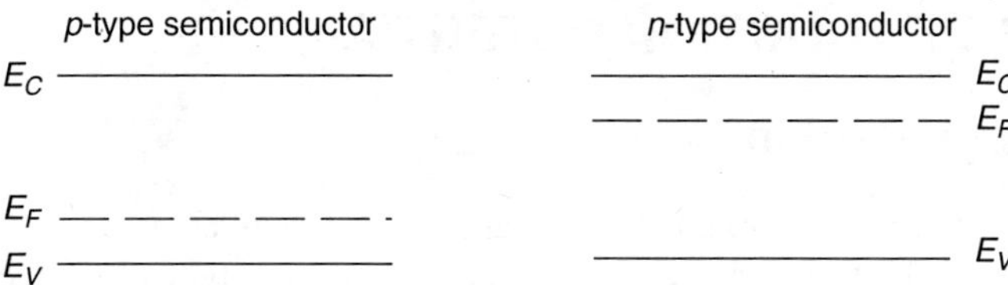


**Figure 2-2(a)** *p*-type and *n*-type semiconductors just before contact

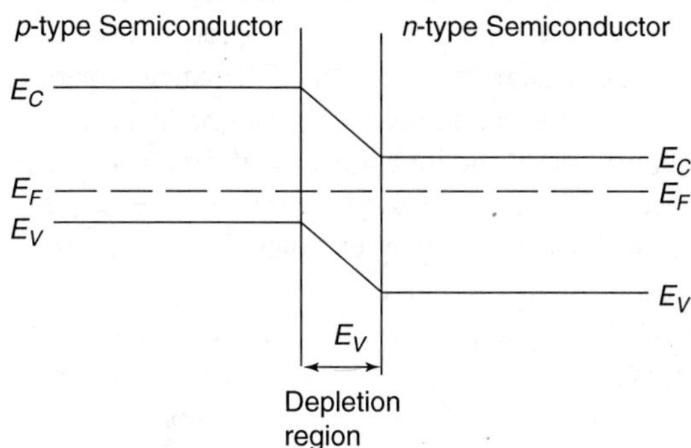


**Figure 2-2(b)** Band structure of *p–n* junction

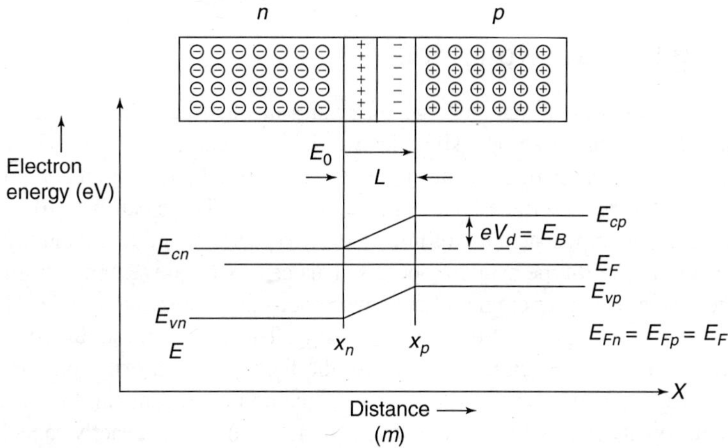


**Figure 2-2(c)** The energy band diagram of a *p–n* junction under the condition of thermal equilibrium

distribution. From the discussion of the law of mass action (see Chapter 1), the carrier concentrations on either side away from the junction are given by:

$$n_n \approx N_d \qquad p_n \approx \frac{n_i^2}{N_d} \qquad \text{(For } n\text{-type)} \tag{2-1a}$$

$$p_p \approx N_a \qquad n_p \approx \frac{n_i^2}{N_a} \qquad \text{(For } p\text{-type)} \tag{2-1b}$$

where $p_n$ is the hole concentration in *n*-type semiconductors, $n_p$ is the electron concentration in *p*-type semiconductors; $n_n$ and $p_p$ are the electron and hole concentrations in *n*- and *p*-type semiconductors respectively. The behaviour of these carriers in a junction is intimately related to the potential formulation of the junction.

# 2-4 CONCEPTS OF JUNCTION POTENTIAL

## 2-4-1 Space-Charge Region

The non-uniform concentration of holes and electrons at the junction gives rise to a diffusive flow of carriers. Since the electron density is higher in the *n*-type crystal than in the *p*-type crystal, electrons flow from the *n*-type to the *p*-type and simultaneously, due to reversibility, the holes flow from the *p*-type to the *n*-type. The result of this migration of carriers is that the region near the junction of the *n*-type is left with a net positive charge (only ionized donor atoms) while that of the *p*-type is left with a net negative charge (only ionized acceptor atoms). This diffusive mechanism of migration of the carriers across the junction creates a region devoid of free carriers, and this region is called the *space-charge region*, the depletion region or the transition region. Thus, in the neighbourhood of the junction we will find a space charge that is a net charge. It is to be noted that the space charge, both negative and positive, must be equal, which effectively means that the opposite they must be numerically equal. It may be noted that the number of free carriers is particularly significant near the edges of the space-charge region.

The junction, as noted above, has three major properties: (i) There is a space charge and an electric field across the junction, which in turn indicates that the junction is pre-biased (i.e., there exists a built-in potential, a very important concept, which will be discussed shortly); (ii) The impure atoms maintaining the space charge are immobile in the temperature range of interest (at very high temperatures, the impurities become mobile). The pre-biased condition can be maintained indefinitely; (iii) The presence of any free electron or hole is strictly forbidden.

## 2-4-2 Built-in and Contact Potentials

This diffusive flow process results in a space-charge region and an electric field. The resulting diffusion current cannot build up indefinitely because an opposing electric field is created at the junction. The homogeneous mixing of the two types of carriers cannot occur in the case of charged particles in a *p*–*n* junction because of the development of space charge and the associated electric field $E_0$. The electrons diffusing from the *n*-type to the *p*-type leave behind uncompensated donor ions in the *n*-type semiconductor, and the holes leave behind uncompensated acceptors in the *p*-type semiconductors. This causes the development of a region of positive space charge near the *n*-side of the junction and negative space charge near the *p*-side. The resulting electric field is directed from positive charge towards negative charge. Thus, $E_0$ is in the direction opposite to that of the diffusion current for each type of carrier. Therefore, the field creates a drift component of current from *n* to *p*, opposing the diffusion component of the current. Since no net current can flow across the junction at equilibrium, the current density due to the drift of carriers in the $E_0$ field must exactly cancel the current density due to diffusion of carriers. Moreover, since there can be no net build-up of electrons or holes on either side as a function of time, the drift and diffusion current densities must cancel for each type of carrier.

Therefore, the electric field $E_0$ builds up to the point where the net current density is zero at equilibrium. The electric field appears in the transition region of length $L$ about the junction, and there is an equilibrium potential difference $V_0$ across $L$ (known as contact potential). In the electrostatic potential diagram, there is a gradient in potential in the direction opposite to $E_0$. In accordance with the following fundamental relation:

$$E_0(x) = -\frac{dV(x)}{dx}$$

the electric field is zero in the neutral regions outside $L$. The contact potential appearing across $L$ under condition of zero external bias is a built-in potential barrier, in that it is necessary for the maintenance of equilibrium at the junction. It does not imply any external potential. $V_0$ is an equilibrium quantity, and no net current can result from it. In general, the contact potential is the algebraic sum of the built-in potential and the applied voltage. The variations in the contact potential under the condition of applied bias are given in the subsequent sections.

The field is in a direction that opposes further migration of carriers. Since there can be no current flow in the steady state equilibrium condition, the tendency for the diffusion current density is exactly balanced by the tendency for drift current density due to the transition region field. The presence of the field $E_0$ in the transition region (space-charge region) implies the formation of a potential barrier $V_d$ for the charge carriers. Assuming that the field is confined within the space-charge region $L$, the potential barrier $V_d$ and the field $E_0$ are related by:

$$V_d = \left| \int_{x_n}^{x_p} E_0 dx \right| \quad (2\text{-}2)$$

It should be noted that a voltmeter cannot measure this electrostatic potential since the internal field is set up to oppose the diffusion current and also since the built-in potential is cancelled exactly by the potential drop across the contact. The barrier energy corresponding to barrier potential $V_d$ is expressed as $E_B = eV_d$, as shown in Fig. (2-2c). The value of $E_B$ can be changed by doping change. The value of $E_B$ is different for different semiconductors.

## 2-4-3 Effect of Doping on Barrier Field

The width of the depletion region is inversely proportional to the doping strength, as a larger carrier concentration enables the same charge to be achieved over a smaller dimension. It should be noted that the depletion charge for different doping is not constant. The barrier field is normally independent of the doping concentration except under conditions of heavy doping, which may alter the band-gap itself, thereby modifying the barrier field. The value of $V_d$ in terms of the hole and electron concentrations can be derived in the following manner.

At thermal equilibrium, the non-degenerate electron concentrations for the $n$-type and $p$-type can be written as:

$$n_n = N_c e^{-(E_{cn} - E_{fn})/k_B T} \quad (2\text{-}3a)$$

$$n_p = N_c e^{-\left(E_{cp} - E_{fp}\right)/k_B T} \quad (2\text{-}3b)$$

where $E_{cn}$, $E_{cp}$, $E_{fn}$, and $E_{fp}$ are the conduction and Fermi level energies of the $n$-type and $p$-type semiconductors, respectively, and $N_c$ is the effective density-of-states. From Eqs. (2-3) and (2-1), the Fermi levels are given by:

$$E_{fn} = E_{cn} - k_B T \ln \frac{N_c}{N_d} \quad (2\text{-}4a)$$

$$E_{fp} = E_{cp} - kT \ln \frac{N_c N_a}{n_i^2} \quad (2\text{-}4b)$$

At equilibrium condition, the Fermi level must be constant throughout the entire crystal. Otherwise, because of the availability of lower energy levels, a flow of carriers would result. The Fermi levels, therefore, must line up at the equilibrium. This extremely important theorem, which is the key in drawing the energy-band diagrams of various junction semiconductor devices, will now be proved.

**FOR ADVANCED READERS**

INVARIANCE OF FERMI LEVEL AT THERMAL EQUILIBRIUM

Let us consider the existence of the energy level $E$ that is identical in the two materials, 1 and 2, which are in physical contact with each other as shown in Fig. 2-3. Therefore, the electron in material 1 with energy $E$ are able to move in material 2 with the same energy $E$ without any expenditure of energy. Besides, the electrons must be available at energy $E$ in material 1, and there must be allowed empty energy states at energy $E$ in material 2.

Under the condition of a thermal equilibrium there is no current. Therefore, there is no net charge transfer and consequently there is also no net transfer of energy. Thus, for each energy $E$ the opposite transfer of electrons from material 2 to material 1 must exactly balance any transfer of electrons from material 1 to material 2.

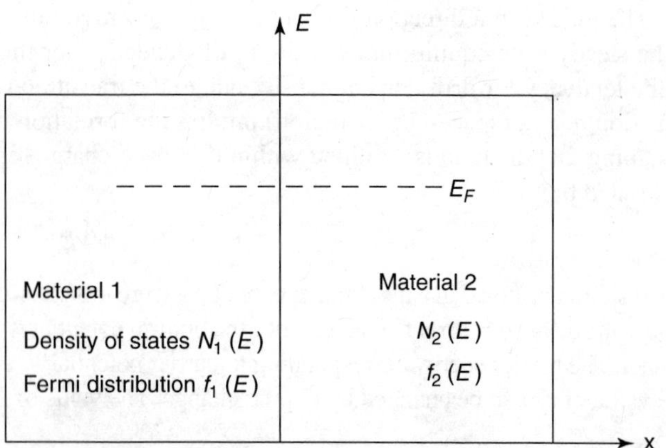


**Figure 2-3** The contact equilibrium diagram of two materials

At energy $E$, the rate of transfer of electrons from material 1 to material 2 is proportional to the number of filled states at $E$ in material 1 multiplied by the number of empty states at in material 2. By using the product law of probability, the rate of transfer of electrons ($R_{12}$), from material 1 to material 2 is given by:

$$R_{12} = [C_1N_1(E)\,f_1(E)] \times \{N_2(E)[1 - f_2\,(E)]\} \tag{2-5}$$

where, $C_1$ is a constant and $f(E)$ is the well-known Fermi–Dirac distribution function as given by Eq. (1-13). Similarly, the rate of transfer of electrons from material 2 to material 1 ($R_{21}$) can be expressed as:

$$R_{21} = [C_1N_2(E)f_2(E)] \times \{N_1(E)[1 - f_1(E)]\} \tag{2-6}$$

At the thermal equilibrium, these rates must be equal and thus we can write:

$$[N_1(E)f_1(E)] \times \{N_2(E)[1 - f_2(E)]\} = [N_2(E)f_2(E)] \times \{N_1(E)[1 - f_1(E)]\} \tag{2-7a}$$

Dividing both sides by $N_1(E)N_2(E)f_1(E)f_2(E)$ we get:

$$\frac{1}{f_2(E)} - 1 = \frac{1}{f_1(E)} - 1 \tag{2-7b}$$

Since $N_1(E)N_2(E)f_1(E)f_2(E)$ is not equal to zero and results in $f_1(E) = f_2(E)$:

$$[1 + e^{(E-E_{F1})/k_BT}]^{-1} = [1 + e^{(E-E_{F2})/\,k_BT}]^{-1} \tag{2-8}$$

We can now conclude that $E_{F1} = E_{F2}$, which indicates that there is no discontinuity in the equilibrium Fermi level across materials in intimate contact. Thus, mathematically $E_F$ is a constant with respect to distance in this case. From differential calculus, we know that if C is a constant with respect to a particular variable (say $\beta$) then:

$$\frac{dC}{d\beta} = 0$$

Then, in this case we can write that:

$$\frac{dE_F}{dx} = 0 \tag{2-9}$$

## 2-4-4 Formulation of Built-in Potential

For the *n*-type material, $E_F$ is near $E_c$, while for the *p*-type material, $E_F$ is near $E_v$. Therefore, in order for $E_F$ to satisfy these two conditions simultaneously while keeping itself invariant, the energy bands must get themselves modified as shown in Figs. 2-2(b) and 2-2(c) respectively. Thus:

$$E_{Fn} = E_{Fp} = E_F \tag{2-10}$$

and,
$$eV_d = E_{cp} - E_{cn} \tag{2-11}$$

At equilibrium, the ratio of Eqs. (2-3a) to (2-3b) yields:

$$\frac{n_n}{n_p} = e^{(E_{cp}-E_{cn})/k_BT} = e^{eV_d/k_BT} \tag{2-12}$$

Similarly, the ratio of the holes in the two regions is:

$$\frac{p_p}{p_n} = e^{eV_d/k_BT} \tag{2-13}$$

Equations (2-12) and (2-13) are referred to as Boltzmann equations and from them $V_d$ can be determined in terms of the electron and hole concentrations:

$$V_d = \frac{k_BT}{e}\ \ln\frac{n_n}{n_p} = \frac{k_BT}{e}\ \ln\frac{p_p}{p_n} \tag{2-14}$$

From Eqs. (2-14), and (2-1):
$$V_d \approx \frac{k_BT}{e}\ \ln\frac{N_aN_d}{n_i^2} \tag{2-15a}$$

The barrier potential can be further expressed in terms of the conductivities of the *p*-type and *n*-type semiconductors:

$$V_d \approx \frac{k_BT}{e}\ \ln\frac{\sigma_n\sigma_p}{\mu_n\mu_p e^2 n_i^2} \tag{2-15b}$$

where, $N_d e\mu_n = \sigma_n$ and $N_a e\mu_p = \sigma_p$ have been used.

From Eq. (1-105), assuming only the presence of heavy holes, under the condition of non-degenerate hole concentration, the Einstein relation for holes can be written as a logical extension of Eq. (1-105) as:

$$\frac{\mu_p}{D_p} = \frac{e}{k_BT} \tag{2-16}$$

In order to appreciate the order of magnitude of the diffusion and drift currents in *p–n* junctions at thermal equilibrium, consider a typical junction where the electron density may drop from $n_n = 10^{17}$ cm$^{-3}$ to $n_p = 10^{10}$ cm$^{-3}$ within a transition region $L = 10^{-4}$ cm. The electron gradient $dn/dx = (10^{17} - 10^{10})/10^{-4}$ cm$^{-4} \cong 10^{17}/10^{-4}$ cm$^{-4} \approx 10^{21}$ cm$^{-4}$, and hence, the electron diffusion current density for germanium is:

$$J_n = -eD_n\frac{dn}{dx} = 1.6 \times 10^4 \text{ amp/cm}^2 \tag{2-17}$$

This very large magnitude of electron current is almost completely cancelled by an equal electron drift current in the opposite direction. A similar state of affairs exists for the holes. Since such an enormous current cannot exist inside the junction, the balance of the built-in field and the concentration gradient prevents the net motion of carriers at equilibrium. For small currents in the non-equilibrium case we should expect that this particular balance would remain essentially the same at the junction.

## Solved Examples

**Example 2-1** (a) The resistivities of the two sides of a step-graded germanium diode are 1.5 Ω-cm (*p*-side) and 1 Ω-cm (*n*-side). Calculate the height of the potential-energy barrier. (b) Repeat the part (a) for silicon *p–n* junction.

**Solution:**

(a)
$$\rho = \frac{1}{\sigma} = \frac{1}{N_A e \mu_p} = 2\ \Omega\text{-cm}$$

or,
$$N_A = \frac{1}{1.5 \times 1.6 \times 10^{-19} \times 1800} = 2.31 \times 10^{15}/\text{cm}^3$$

Similarly,
$$N_D = \frac{1}{1 \times 1.6 \times 10^{-19} \times 3800} = 1.65 \times 10^{15}/\text{cm}^3$$

The height of the potential energy barrier is:

$$V_0 = 0.026 \times \ln \frac{1.65 \times 10^{15} \times 2.31 \times 10^{15}}{(2.5 \times 10^{13})^2} = 0.226\ \text{eV}$$

(b)
$$N_A = \frac{1}{1.5 \times 1.6 \times 10^{-19} \times 500} = 8.33 \times 10^{15}/\text{cm}^3$$

$$= \frac{1}{2 \times 1.6 \times 10^{-19} \times 1300} = 2.4 \times 10^{15}/\text{cm}^3$$

Then,
$$V_0 = 0.026 \times \ln \frac{2.4 \times 10^{15} \times 8.33 \times 10^{15}}{(1.5 \times 10^{10})^2} = 0.655\ \text{eV}$$

## 2-5 MODES OF THE *p–n* JUNCTION

There are two modes of switching of a *p–n* junction diode.

**(i) Forward-biased *p–n* junction:** When the positive terminal of a battery is connected to the *p*-type side and the negative terminals to the *n*-type side of a *p–n* junction, the junction allows a large current to flow through it due to the low resistance level offered by the junction. In this case the junction is said to be forward-biased. A forward-biased *p–n* junction is shown in the Fig. 2-4(a). This shows the decrease of barrier energy by $eV_a$ amount.

**(ii) Reverse-biased *p–n* junction:** When the terminals of the battery are reversed i.e., when the positive terminal is connected to the *n*-type side and the negative terminal is connected to the *p*-type side, the junction allows a very little current to flow through it due to the high resistance level offered by the junction. Under this condition, the *p–n* junction is said to be reverse-biased. A reverse-biased *p–n* junction is shown in Fig. 2-4(b).

In this case, the voltage applied to the junction causes the holes in the *p*-type side and the electrons in the *n*-type side to move away from the junction. This increases the width of the depletion region and the barrier energy (eV) gives the amount of increase in the barrier energy where $V_a$ is the magnitude of the applied voltage. Due to the increase in the barrier energy, a negligible number of majority carriers will be able to cross the junction and the current will be practically zero. But the minority carriers, which travel down the potential barrier, remain unaffected and give a small current. This current is called the reverse saturation current ($I_S$). The reverse saturation current increases with the temperature of the diode, but is independent of the applied reverse voltage, the increasing temperature breaks the covalent bonds in the semiconductor.

### 2-5-1 The *p–n* Junction with External Applied Voltage

If an external voltage $V_a$ is applied across the *p–n* junction, the height of the potential barrier is either increased or diminished as compared to $V_a$, depending upon the polarity of the applied voltage. The energy

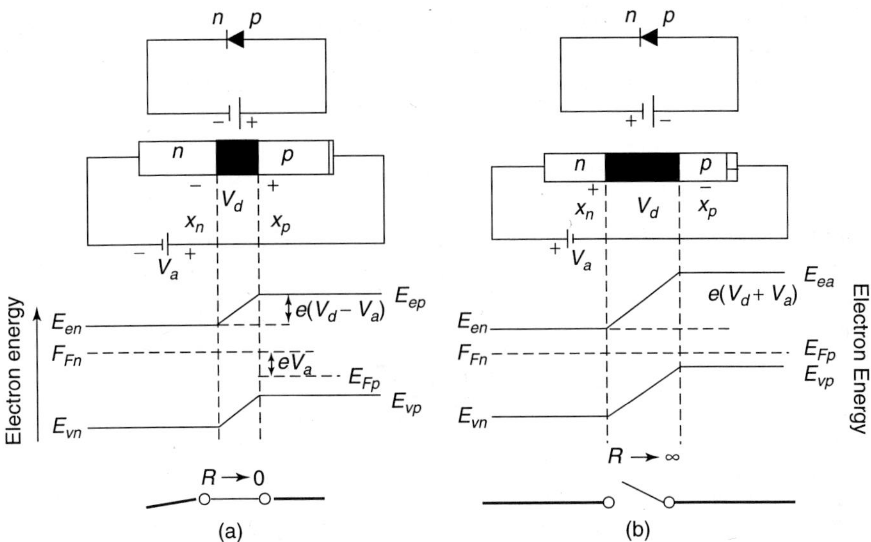


**Figure 2-4** Energy band diagram of a *p–n* junction under an externally applied voltage: (a) forward-biased condition (b) reverse-biased condition

band distribution, with applied external voltage, is shown in Fig. 2-4. For these non-equilibrium conditions, the Fermi level can no longer be identified. In order to describe the behaviour of the *p–n* junction, quasi-Fermi levels are introduced as shown by the dashed lines in Fig. 2-4.

If the polarity of the applied voltage is such that the *p*-type region is made positive with respect to the *n*-type, the height of the potential barrier is reduced, thereby making it relatively easier for the majority carriers, holes or electrons to surmount the barrier. The reduction in the height of the potential barrier is equal to the applied voltage as shown in Fig. 2-4(a), under the condition that the voltage drop across the body of the semiconductor and the ohmic drop at the contacts are negligible. Usually the majority carrier densities are much higher than the intrinsic density in both the *n*- and *p*-types making the injected minority carrier concentration much less than the majority carrier. These injected carriers, which become minority carriers, have a much higher concentration in the neighbourhood of the junction than the minority carriers present far away from the junction. The gradient in the minority-carrier concentration causes a diffusive flow of carriers and hence, an electric current under the forward-biased condition. The magnitude of the current clearly increases with increase in the forward applied voltage.

## 2-5-2 Rectifying Voltage–Current Characteristics of a *p–n* Junction

If the polarity of the applied voltage is such that the *p*-type region is made negative with respect to the *n*-type, the height of the potential-barrier is increased. Under this reverse-biased condition, it is relatively harder for the majority of the carriers to surmount the potential-barrier. The increase in the potential-barrier height is essentially equal to the applied voltage as shown in Fig. 2-4(b).

Because of the increase in barrier height, the increased field helps in the motion of the minority carriers from the *p*-type to the *n*-type, and vice versa. Hence, the minority-carrier concentration near the junction is reduced from its equilibrium value. Since the minority-carrier concentration in this case is much less than the equilibrium value, the increase in applied reverse voltage has a negligible effect on the magnitude of the reverse current.

Under an external applied voltage, the carrier concentrations near the junction corresponding to Eqs. (2-12) and (2-13) are:

$$\frac{n_n}{n_p} = e^{(e/kBT)\,(Vd \pm Va)} \tag{2-18a}$$

$$\frac{p_p}{p_n} = e^{(e/kBT)\,(Vd \pm Va)} \tag{2-18b}$$

where, the sign of the applied voltage is included in Eq. (2-18). In other words, the plus and minus signs are for the reverse-biased and the forward-biased conditions, respectively (see Fig. 2-4).

From Eqs. (2-12), (2-13) and (2-18), the injected or extracted minority-carrier concentrations near the junction can be written as:

$$p_n(x_n) = p_p e^{-(e/k_BT)(V_d \mp V_a)} = n_p e^{\mp (e/kBT)\, V_a} \tag{2-19}$$

$$n_p(x_p) = n_n e^{-(e/k_BT)(V_d \mp V_a)} = n_p e^{\mp (e/kBT)\, V_a} \tag{2-20}$$

Note that in Eqs. (2-19) and (2-20), the plus sign is for the forward-biased case where minority carriers are injected. The minus sign is for the reverse-biased case where minority carriers are extracted.

The concentration of the carriers on the boundaries, for the usual cases, $N_a >> n_i$ and under an external applied voltage $V$ is shown in Fig. 2-5. It should be noted that in the foregoing discussion it is assumed that the applied voltage is small and well below the breakdown voltage. Figure 2-5 is a key diagram, which can be used in deriving various important equations of diodes and transistors.

## 2-5-3 The Junction Capacitance

In this section, we will examine the calculation of the width of the transition region, the expression for the capacitance across the junction, and the variation of the field in the space-charge region, respectively. For this we will consider two types of idealized junctions, which are approximated closely in practice. These are: (i) the abrupt or step, junction, which results from the alloying technique, and (ii) the graded junction, which results from the crystal-growing technique.

### Abrupt junction

Shown in Fig. 2-6(a) is the charge density of a step junction under the assumption that on the $p$-side and the $n$-side of the transition region close to the junction, the charge densities are $eN_a$ and $eN_d$, respectively. For simplicity, the actual charge density is idealized as shown in Fig. 2-6(b).

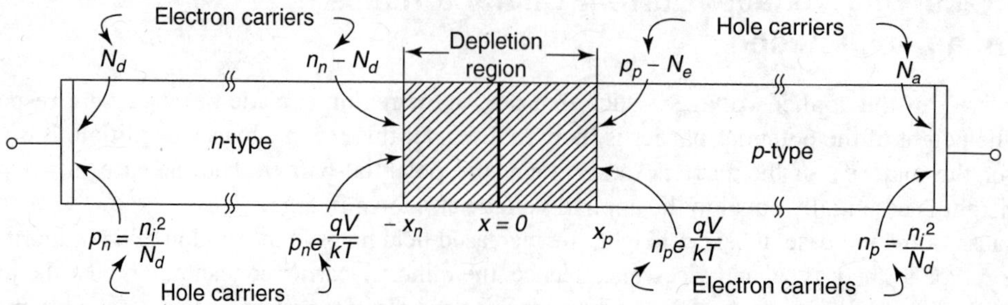


**Figure 2-5** Electron and hole carriers at the boundaries of a $p$–$n$ junction under an externally applied voltage

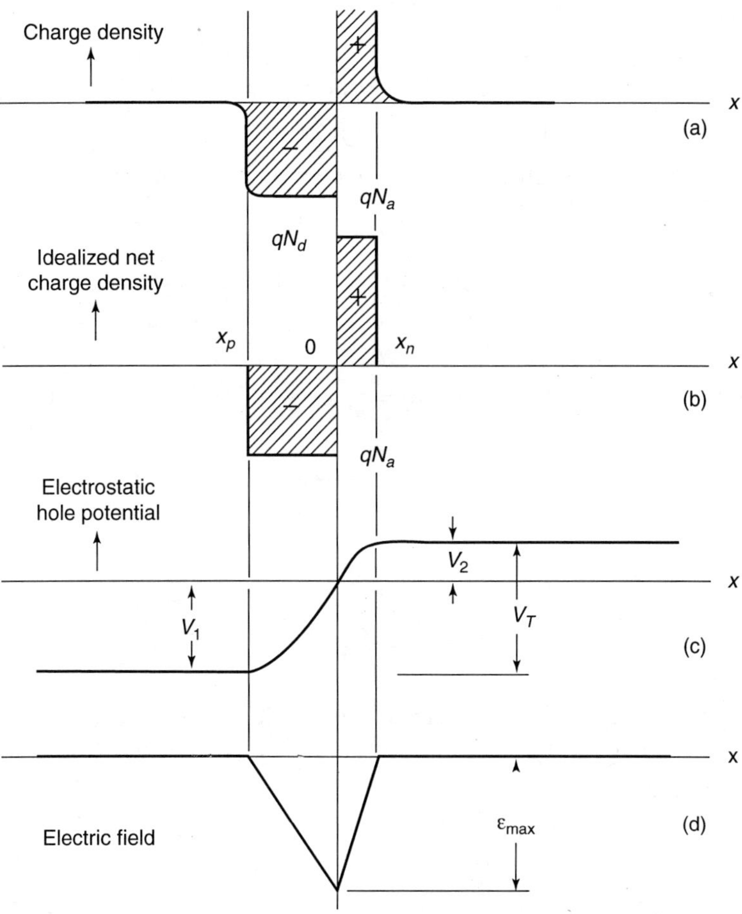


**Figure 2-6** The profiles of charge density, potential, and electric field in an abrupt junction

Assuming one-dimensional geometry, the Poisson equation in the depletion region is given by:

$$\frac{d^2V}{dx^2} = -\frac{\rho}{\varepsilon} = -\frac{e}{\varepsilon}(N_d - N_a) \quad (\text{For } x_p < x < x_n) \tag{2-21}$$

where, $\varepsilon$ and $\rho$ are the semiconductor dielectric permittivity and the volume charge density respectively. For $x_p < x < 0$, Eq. (2-21) can be written as:

$$-\frac{dE_0}{dx} = \frac{d^2V}{dx^2} = \frac{eN_a}{\varepsilon} \tag{2-22}$$

Integrating Eq. (2-22), we get:

$$-E_0 = \frac{eN_a}{\varepsilon}x + C_1 \tag{2-23}$$

The boundary condition restricts us that at $x = -x_p$, $E_0 = 0$ which implies the assumption that the voltage drops in the bulk of the semiconductor are negligible. Thus, substituting $C_1 = x_p eN_a/\varepsilon$ in Eq. (2-23) yields:

$$-E_0 = \frac{dV}{dx} = \frac{eN_a}{\varepsilon}(x + x_p) \tag{2-24}$$

which shows that the electric field versus distance is a linear plot with a negative slope.

Integrating Eq. (2-24) gives:

$$V = \frac{eN_a}{\varepsilon}\left(\frac{x^2}{2} + xx_p\right) + C_2 \tag{2-25}$$

We may choose $V = 0$ at $x = 0$, as shown in Fig. 2-6(e), since the choice of the reference for the potential is arbitrary. Hence, Eq. (2-25) can be written as:

$$V = \frac{eN_a}{\varepsilon}\left(\frac{x^2}{2} + xx_p\right) \tag{2-26}$$

At $x = -x_p$, Eq. (2-26) is:

$$V \equiv V_1 = -\frac{eN_a x_p^{\,2}}{2\varepsilon} \tag{2-27}$$

Similarly, at $x = x_n$:

$$V \equiv V_2 = -\frac{eN_d x_n^{\,2}}{2\varepsilon} \tag{2-28}$$

Thus, the total voltage $V_T$ is given by:

$$V_T = V_2 - V_1 = \frac{e}{2\varepsilon}\left(N_a x_p^2 + N_d x_n^2\right) \tag{2-29}$$

It may be noted that in Eq. (2-29), $V_T$ is the difference between the contact potential and the applied voltage. The total space-charge neutrality requires that the positive and negative charge-density areas must be equal. Mathematically, we can write:

$$|Q| = eAN_d x_n = eAN_a x_p \tag{2-30}$$

where, $A$ is the junction area. Hence, Eq. (2-29) can be expressed in terms of either $x_n$ or $x_p$:

$$V_T = \frac{ex_p^{\,2}N_a}{2\varepsilon}\left(1 + \frac{N_a}{N_d}\right) = \frac{ex_n^{\,2}N_d}{2\varepsilon}\left(1 + \frac{N_d}{N_a}\right) \tag{2-31}$$

Thus, from Eq. (2-31) the depletion region widths in the *p*-type and *n*-type are:

$$x_p = \left(\frac{2\varepsilon}{eN_a}\frac{V_T N_d}{N_d + N_a}\right)^{1/2} \tag{2-32a}$$

$$x_n = \left(\frac{2\varepsilon}{eN_d}\frac{V_T N_a}{N_a + N_d}\right)^{1/2} \tag{2-32b}$$

The maximum field can be determined from the Eq. (2-24):

$$E_{0\max} = -\left.\frac{dV}{dx}\right|_{x=0} = -\frac{eN_a x_p}{\varepsilon} = -\frac{eN_d x_n}{\varepsilon} \tag{2-33}$$

Since there is a voltage-dependent charge associated with the depletion region, this indicates the existence of a junction capacitance, $C_j$. The $C_j$ can be mathematically defined as:

$$C_j = \frac{dQ}{dV_T} = \frac{dQ}{dx_n}\frac{dx_n}{dV_T} = \frac{dQ}{dx_p}\frac{dx_p}{dV_T} \tag{2-34}$$

But from Eq. (2-30):

$$\frac{dQ}{dx_p} = eAN_a \tag{2-35}$$

and from Eq. (2-32a):

$$\frac{dx_p}{dV_T} = \frac{1}{2}\left[\frac{2\varepsilon N_d V_T}{eN_a(N_d+N_a)}\right]^{-1/2}\left[\frac{2\varepsilon N_d}{eN_a(N_d+N_a)}\right]^{-1/2} \tag{2-36}$$

Hence, the junction capacitance is given by:

$$C_j = \left(\frac{Ae\varepsilon}{2}\times\frac{N_aN_d}{N_a+N_d}\right)^{1/2} V_T^{-1/2} \tag{2-37a}$$

or,

$$C_j = K_1 V_T^{-1/2} = K_1(V_d - V_a)^{-1/2} \tag{2-37b}$$

where,

$$K_j \equiv A\left(\frac{e\varepsilon N_aN_d}{N_a+N_d}\right)^{1/2}$$

where, $V_a$ and $V_d$ are the applied and diffusion built-in voltages, respectively. Equation (2-37) can also be expressed in the well known form of the familiar parallel-plate capacitance, namely:

$$C = \frac{\varepsilon A}{|x_p| + |x_n|} = \frac{\varepsilon A}{L} \tag{2-37c}$$

The expression for capacitance derived above is referred to as the *transition capacitance*—also known as junction or space-charge capacitance—and exists primarily at the reverse-biased junction. It should be noted that physics of transition capacitance of a diode is totally different from that of the well known parallel plate capacitor of basic electrical science despite their striking similarity.

## Solved Examples

**Example 2-2** Boron is implanted in an *n*-type Si ($N_d = 10^{16}$ cm$^{-3}$), forming an abrupt *p*–*n* junction with the square-cross sectional area $4 \times 10^{-4}$ cm$^2$ and the accepter concentration ($N_A$) on the *p*-side is $5 \times 10^{18}$ cm$^{-3}$. Find out the width of the depletion zone at 300 K.

**Solution:**

The height of the barrier energy is:

$$V_0 = \frac{k_BT}{e}\ln\frac{N_AN_d}{n_i^2} = 0.0259\ln\frac{5\times10^{34}}{2.25\times10^{20}}$$

$$= 0.0259\ln(2.22\times10^{14}) = 0.86\text{ V}$$

We know that:

$$W = \left[\frac{2\varepsilon V_0}{e}\left(\frac{1}{N_A} + \frac{1}{N_d}\right)\right]^{½}$$

$$= \left[2 \times \frac{(11.8 \times 8.85 \times 10^{-14} \times 0.86)}{1.6 \times 10^{-19}}(0.2 \times 10^{-18} + 10^{-16})\right]^{½}$$

$$= 1.059 \times 10^{-5} \text{ cm}$$

**Example 2-3** An Si *p–n* junction is formed from *p*-material doped with $10^{22}$ acceptors/m³ and *n*-material doped with donors/m³. Find the thermal voltage and barrier voltage at 30°C.

**Solution:**

$$T = (273 + 30) \text{ K} = 303 \text{ K}$$

$$V_T = \frac{k_B T}{e} = \frac{1.38 \times 10^{-23}(303)}{1.6 \times 10^{-19}}$$

$$\doteq 26.1 \text{ mV}$$

$$n_i^2 = (1.5 \times 10^{16})^2$$

$$= 2.25 \times 10^{32}$$

$$V_0 = V_T \ln\left(\frac{N_A N_D}{n_i^2}\right)$$

$$V_0 = V_T \ln\left(\frac{10^{22} \times 1.2 \times 10^{21}}{2.25 \times 10^{32}}\right)$$

$$= 0.635 \text{ V}$$

**Example 2-4** Calculate the barrier potential for a Si junction at: (a) 70°C and (b) 0°C if its value at 25°C is 0.7 V.

**Solution:**

Given: $t_i = 25°C$, $t_2 = 70°C$

(a) We know that $\Delta V = -0.002\Delta t$

$$= -0.002\,(t_2 - t_1)$$

$$= -0.002\,(70 - 25) = -0.69\text{V}$$

Thus the barrier potencial at 100°C, $V_B = 0.7 + (-0.09) = 0.61\text{V}$

(b) $\Delta t = (0 - 25) = -25°C$

$\Delta V = -0.002 \times (-25) = 0.05$ V

The barrier potential at 0°C, $V_B = 0.7 + 0.05 = 0.75$ V

■

## Graded junction

The idealized net charge density for the graded junction may be expressed as:

$$p = eax \qquad |x| < x_p \tag{2-38}$$

where $a$ is the charge-density gradient in the transition region.

Poisson equation for this case is:

$$\frac{d^2V}{dx^2} = -\frac{eAx}{\varepsilon} \qquad |x| < x_p \tag{2-39}$$

Integrating Eq. (2-39), we obtain:

$$-E_0 \equiv \frac{dV}{dx} = -\frac{eax}{2\varepsilon} + C_1 \tag{2-40}$$

From the boundary condition:

$$E_0 \equiv -\frac{dV}{dx} = 0 \quad \text{at } x = x_n \quad \text{and} \quad -x_p \tag{2-41}$$

The charge neutrality condition requires that:

$$x_n = x_p \tag{2-42}$$

Therefore, we shall use only $x_p$ in the following equation. From Eqs. (2-41) and (2-42), we may write:

$$\frac{dV}{dx} = \frac{eA}{2\varepsilon}(-x^2 + x_p^2) \tag{2-43}$$

Integrating Eq. (2-43) again, we obtain:

$$V = \frac{eA}{2\varepsilon}\left(-\frac{x^3}{3} + xx_p^2\right) + C_2 \tag{2-44}$$

Arbitrarily choosing $V = 0$ at $x = 0$, Eq. (2-44) becomes:

$$V = \frac{eA}{2\varepsilon}\left(-\frac{x^3}{3} + xx_p^2\right) \tag{2-45}$$

The total voltage, $V_T$, can be written as:

$$V_T = V(x_p) - V(-x_p) \tag{2-46a}$$

or,

$$V_T = \frac{2eax_p^3}{3\varepsilon} \tag{2-46b}$$

From Eq. (2-46b) the width of the transition region in the *n*- or *p*-type is:

$$|x_n| = |x_p| = \frac{l}{2} = \left(\frac{3\varepsilon}{2ea}\right)^{1/3} V_T^{\,1/3} \tag{2-47}$$

The maximum electric field from Eq. (2-43) is:

$$E_{0\max} = -\frac{eax_p^2}{2\varepsilon} \tag{2-48}$$

To find the junction capacitance, the charge on either side of the depletion region is given by:

$$|Q| = \frac{1}{2}eax_p^2 A \tag{2-49}$$

From Eqs. (2-48) and (2-49):

$$|Q| = \frac{1}{2}eaA\left(\frac{3\varepsilon}{2ae}\right)^{2/3} V_T^{\,2/3} \tag{2-50}$$

Therefore:

$$C_j = \frac{dQ}{dV_T} = A\left(\frac{ea\varepsilon^2}{12}\right)^{1/3} V_T^{1/3} \qquad (2\text{-}51a)$$

or, $$C_j = K_2 V_2^{1/3} = K_2 (V_d - V_a)^{1/3} \qquad (2\text{-}51b)$$

where, $K_2 \equiv A\left(\frac{ea\varepsilon^2}{12}\right)^{1/3}$

Equation (2-51) can also be expressed as:

$$C_j = \frac{\varepsilon A}{2|x_p|} = \frac{\varepsilon A}{L} \qquad (2\text{-}51c)$$

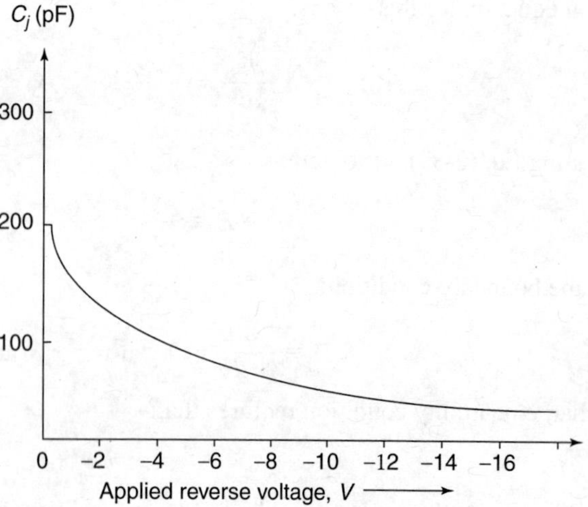


**Figure 2-7** Plot of the junction capacitance as a function of the applied reverse potential for the abrupt *p–n* junction

From Eqs. (2-37) and (2-51) it is seen that this junction capacitance is proportional to $V_T^{1/2}$ in the step-junction case, while it is proportional to $V_T^{1/3}$ in the graded-junction case. The voltage-variable-capacitance property of the *p–n* junction is utilized to maximum advantage in several practical applications. One such application is as a voltage-tuneable element in resonant circuits; other applications are in the field-effect transistor. The non-linear voltage dependence of the junction capacitance also finds application in harmonic generation and in parametric amplification. A *p–n* junction designed for use as a voltage-variable-capacitance is called a *varactor*, or sometimes a *varicap*. The capacitance versus voltage curve of a typical *p–n* junction is shown in Fig. 2-7.

## Solved Examples

**Example 2-5** (a) Prove that for an Si alloy *p–n* junction (with $N_A << N_D$), the depletion layer capacitance in picofarads per square centimetre is given by:

$$C_T = 2.9 \times 10^{-4}\left(\frac{N_A}{V_j}\right)^{1/2}$$

(b) If the resistivity of the *p*-material is 3 Ωcm, the barrier height $V_o$ is 0.4 V, the applied reverse voltage is 4.5 V, and the cross-sectional area is circular of 40 mils diameter, find $C_T$.

**Solution:**

(a) We have, $C_T = \frac{\varepsilon A}{W}$, where $W$ is given by $W = \left(\frac{2\varepsilon V_j}{qN_A}\right)^{1/2}$

or, $$\frac{C_T}{A} = \frac{\varepsilon}{W} = \frac{\varepsilon}{(2\varepsilon V_j)^{1/2}}(qN_A)^{1/2} = \sqrt{\frac{q\varepsilon}{2}} \times \sqrt{\frac{N_A}{V_j}}$$

For Si: $$\varepsilon = \frac{12}{36\prod \times 10^{11}} \text{ F/cm}$$

Thus, $$\left(\frac{q\varepsilon}{2}\right)^{1/2} = \left(\frac{1}{2} \times 1.6 \times 10^{-19} \times \frac{12}{36\pi} \times 10^{-11}\right)^{1/2} = 2.9 \times 10^{-16}$$

and, $$C_T = 2.9 \times 10^{-16} \left(\frac{N_A}{V_j}\right)^{1/2} \text{ F/cm}^2$$

or, $$C_T = 2.9 \times 10^{-4} \left(\frac{N_A}{V_j}\right)^{1/2} \text{ pF/cm}^2$$

(b) $A = \frac{\pi D^2}{4} = \frac{\pi}{4}(40 \times 10^{-3} \text{ in} \times 2.54 \text{ cm/in})^2 = 8.11058 \times 10^{-3} \text{ cm}^2$

$$\frac{1}{N_A \mu_p q} = 3$$

or, $$N_A = \frac{1}{3.5 \times 1.6 \times 10^{-19} \times 500} = 4.18 \times 10^{15}/\text{cm}^3$$

$$W^2 = \frac{V_0 - V_d}{14.13 \times 10^{10}} \text{ m}^2$$

Hence, $$C_T = (2.9 \times 10^{-4}) \times \left[\left(\frac{4.18 \times 10^{15}}{4.9}\right)^{1/2}\right] \times (8.14 \times 10^{-3}) = 68.94 \text{ pF}$$

**Example 2-6** The reverse-biased diodes are frequently employed as electrically controllable variable capacitors. The transition capacitance of an abrupt junction diode is 20 pF at 5 V. Compute the decrease in capacitance for a 1.5 V increase in bias.

**Solution:**
We know that for an abrupt junction diode $C_T = \frac{\lambda}{V^{1/2}}$, where $\lambda$ = constant and $V$ is the reverse-bias voltage across the diode. When, $V = 5$ V, $C_T = 20$ pF.

$\therefore$ $$\lambda = 20 \times \sqrt{5}$$

When, $$V = 6.5 \text{ V}, C_T = \frac{20 \times \sqrt{5}}{\sqrt{6.5}} = 17.57 \text{ pF}$$

This corresponds to a decrease 2.43 pF.

**Example 2-7** Calculate the barrier capacitance of a Ge *p–n* junction whose area is 1.5 mm by 1.5 mm and whose space-charge thickness is $2\times10^{-4}$ cm. The dielectric constant of Ge is 16.

**Solution:**

We have, $$C_T = \frac{\varepsilon A}{W} \text{ where, for Ge } \varepsilon = \frac{16}{36\pi \times 10^{11}} \text{ F/cm}$$

$\therefore$ $$C_T = \left(\frac{16}{36\pi \times 10^{11}}\right) \times \left(\frac{1.5 \times 10^{-1} \times 1.5 \times 10^{-1}}{2 \times 10^{-4}}\right) = 159.3\text{pF}$$

**Example 2-8** The zero-voltage barrier height at a Ge alloy *p–n* junction is 0.2V. The concentration of acceptor atoms in the p side is much smaller than the concentration of donor atoms in the n material, and $N_A = 2.5 \times 10^{20}$ atoms/m$^3$. (a) calculate the width of the depletion layer for an applied reverse voltage of 10 V; (b) calculate the width of the depletion layer for an applied reverse voltage of 0.1 V; (c) calculate the width of the depletion layer for an applied forward bias of 0.1 V; and (d) if the cross-sectional area of the diode is 1, evaluate the space-charge capacitance corresponding to the values of applied voltage in (a) and (b).

**Solution:**
We know that:

$$V_0 - V_d = \frac{eN_A}{2\varepsilon} W^2 = \frac{1.6 \times 10^{-19} \times 2.5 \times 10^{20}}{2 \times \frac{16}{36\pi \times 10^9}} W^2$$

Therefore:

$$V_0 - V_d = \frac{1.6 \times 10^{-19} \times 2.5 \times 10^{20}}{2 \times \frac{16}{36\pi \times 10^9}} W^2$$

or,

$$W^2 = \frac{V_0 - V_d}{14.13 \times 10^{10}} \text{ m}^2$$

Hence,

(a) $W = \left(\frac{10.2}{14.13} \times 10^{-10}\right)^{1/2} = (8.5 \times 10^{-6}) \text{ m} = 8.5\ \mu\text{m}$

(b) $W = \left(\frac{0.3}{14.13} \times 10^{-10}\right)^{1/2} = (1.45 \times 10^{-6}) \text{ m} = 1.45\ \mu\text{m}$

(c) $W = \left(\frac{0.1}{14.15} \times 10^{-10}\right)^{1/2} = (0.84 \times 10^{-6}) \text{ m} = 0.84\ \mu\text{m}$

(d) We know that: $C_T = \frac{\varepsilon A}{W}$

For (a):

$$C_T = \frac{\varepsilon A}{W} = \frac{16 \times 1 \times 10^{-6}}{36\pi \times 10^9 \times 8.5 \times 10^{-6}} = 16.65 \text{ pF}$$

For (b):

$$C_T = \frac{\varepsilon A}{W} = \frac{0.1415 \times 10^{-9}}{1.45} = 97.6\text{pF}$$

■

## 2-5-4 The Varactor Diode

"Varactor" is actually an abbreviated form of "variable reactor". One property of a *p–n* junction is that the width of the junction depletion region (and hence the depletion capacitance) is a function of the applied voltage, which is utilized in this application, as shown in Fig. 2-8. In general, let us consider a $p^+n$ junction where the doping profile in the *n*-region is given by:

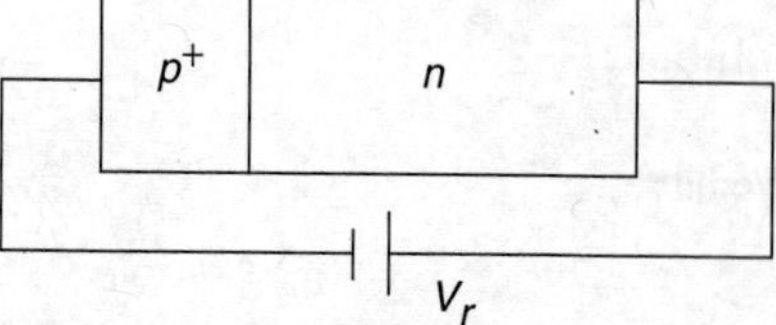


**Figure 2-8** The schematic diagram of the varactor diode

$$N_D(x) = N_{D0}\left(\frac{x}{x_0}\right)^n$$

where, $N_{D0}$ and $n$ are the constants for a particular doping profile, and $x$ is the distance from the *p–n* junction in the *n*-region. Then, following the usual procedure of integrating Poisson equation twice using appropriate boundary conditions, the depletion region width is obtained as a function of the applied reverse bias $V_T$ as:

$$L \,\alpha\, (V_N + V_T)^{\frac{1}{(n+2)}}$$

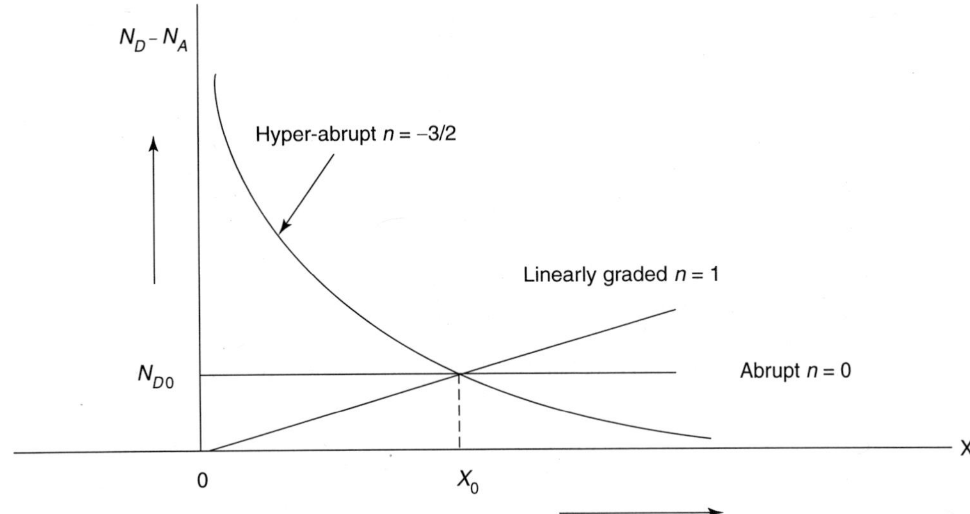


**Figure 2-9** The doping profiles used in varactor diode

Neglecting $V_{bi}$ when a large reverse bias is applied, the depletion capacitance is given by:

$$C_j = \frac{A\varepsilon_5}{L}\,\alpha(V_T)^{\frac{-1}{(n+2)}}$$

For an abrupt junction $n = 0$, and therefore, the depletion capacitance $C_J$ is proportional to $V_T^{-1/2}$. Similarly, it can be shown that for a linearly graded junction ($n = 1$), $C_J$ will be proportional to $V_T^{-1/3}$. These two cases have individually been proved in the previous section. The voltage sensitivity $dC_j/dv$ is found to be greater for an abrupt junction.

Using a hyper-abrupt junction, where the doping profile is tailored specifically to obtain a value of $n < 0$, can further enhance the sensitivity. A special case is seen for $n = -3/2$ as shown in Fig. 2-9. In this case, the depletion capacitance variation can be expressed as $C_J \alpha V_T^{-2}$. If this varactor is used with an inductance $L$ in a resonant circuit, the resonant frequency of the circuit is given by:

$$f_r = \frac{1}{2\pi\sqrt{LC}}\,\alpha V_T$$

In other words, the resonant frequency of the circuit can be varied linearly by changing the applied reverse voltage. This property is widely used for tuning in the radio TV receivers.

It may be noted at this point that there is another type of capacitance, known as the diffusion or storage capacitance, which can be briefly explained. For a forward bias, a capacitance, which is much larger than the transition capacitance $C_T$, comes into play. The origin of this larger capacitance lies in the injected charge stored near the junction outside the transition region. It is convenient to introduce an incremental capacitance, defined as the rate of change of the injected charge with voltage called the diffusion or storage capacitance $C_D$ as mentioned earlier. Thus, we can write:

$$C_D = \frac{dQ}{dV} = \tau\frac{dI}{dV}$$

where, $\tau$ is the lifetime of minority current.

We know that:

$$I = I_s\left(e^{\frac{eV}{\eta k_B T}} - 1\right)$$

So,
$$\frac{dI}{dV} = \frac{I_s\left(e^{\frac{V}{\eta V_T}}\right)}{\eta V_T}$$

Replacing $I_s\left(e^{\frac{V}{\eta V_T}}\right)$ by $(I + I_s)$ we get:
$$\frac{dI}{dV} = \frac{I + I_s}{\eta V_T}$$

Since,
$$I >> I_s$$

Therefore,
$$\frac{dI}{dV} = \frac{I}{\eta V_T}$$

Substituting this value of $dI/dV$ in the basic equation of $C_D$ we get:

$$C_D = \frac{\tau I}{\eta V_T}$$

Therefore, the diffusion capacitance is proportional to the current $I$. We assume that the diode current $I$ is generated due to holes only. Contradicting this assumption will gives us the diffusion capacitance $C_D(p)$ due to holes only and a similar expression can be obtained for $C_D(n)$ due to electrons. The total diffusion capacitance can then be obtained as the sum of $C_D(p) + C_D(n)$.

## 2-6 DERIVATION OF THE I–V CHARACTERISTICS OF A *p–n* JUNCTION DIODE

Let us consider the fact that the drift component of the current is negligible. Then from Eq. (1-122) (see Chapter 1), we can write:

$$D_p \frac{d^2p}{dx^2} = \frac{p - p_n}{\tau_p} \quad (2\text{-}52)$$

Equation (2-52) can be written as:
$$\frac{d^2(p - p_n)}{dx^2} = \frac{p - p_n}{L_p^2} \quad (2\text{-}53)$$

where, $L_p = \sqrt{D_p \tau_p}$ is the diffusion length and $p_n$ is the equilibrium density of holes in the $n$-region far away from the junction.

The solution of the ordinary differential Eq. (2-53) is:

$$p - p_n = C_1 \exp\left(-\frac{x}{L_p}\right) + C_2 \exp\left(\frac{x}{L_p}\right) \quad (2\text{-}54)$$

where, $C_1$ and $C_2$ are two arbitrary constants of integration to be determined from the boundary conditions (see Fig. 2-5). The boundary conditions in this case are:

$$x = x_n: \quad p = p_n \exp\left(\frac{eV}{k_B T}\right) \quad (2\text{-}55)$$

$$x = \infty: \quad p = p_n \quad (2\text{-}56)$$

By using Eqs. (2-54) and (2-56) we get $C_2 = 0$ (2-57a)

Again by using Eqs. (2-54) and (2-55), together with the condition $C_2 = 0$, we get the expression of $C_1$ as:

$$C_1 = p_n\left[\exp\left(\frac{eV}{k_BT}\right) - 1\right]\exp\left(\frac{x_n}{L_p}\right) \tag{2-57b}$$

Substituting the values of constants $C_1$ and $C_2$ in Eq. (2-54) we get:

$$p - p_n = p_n\left[\exp\left(\frac{eV}{k_BT}\right) - 1\right]\exp\left(\frac{x_n - x}{L_p}\right) \tag{2-58a}$$

The current density of holes in $n$-type semiconductors along the $x$ direction by diffusion is given by:

$$J_p(x) = -eD_p\frac{dp}{dx} \tag{2-58b}$$

Using Eqs. (2-58a) and (2.58b) we get:

$$J_p(x) = +e\frac{D_p p_n}{L_p}\left[\exp\left(\frac{eV}{k_BT}\right) - 1\right]\exp\left(\frac{x_n - x}{L_p}\right), \quad x \geq x_n \tag{2-59}$$

Equation (2-59) shows that the minority-carrier flow (hole current) is decreasing exponentially with distance from the junction in the $n$-region. Since the total diode current must be constant, under steady-state condition i.e., the equilibrium condition, the majority carrier flow (electron current) is increasing exponentially with distance from the junction.

The hole current density at the edge of the transition region i.e, at $x = x_n$, from Eq. (2-59) is given as:

$$J_p(x_n) = +e\frac{D_p p_n}{L_p}\left[\exp\left(\frac{eV}{k_BT}\right) - 1\right] \tag{2-60}$$

In the same manner, from Eq. (1-119), and with the boundary conditions as shown in Fig. 2-5, we can have the following equation for electrons in the $p$-type:

$$J_n(x) = +e\frac{D_n n_p}{L_n}\left[\exp\left(\frac{eV}{k_BT}\right) - 1\right]\exp\left(\frac{x + x_p}{L_n}\right), \quad x \leq -x_p \tag{2-61}$$

The hole current density at the edge of the transition region, i.e, $x = -x_p$ can be written from Eq. (2-61) as:

$$J_n(-x_p) = +e\frac{D_n n_p}{L_n}\left[\exp\left(\frac{eV}{k_BT}\right) - 1\right] \tag{2-62}$$

Since the transition region width $L$ is almost always much smaller than the diffusion length, we can assume that the recombination in the transition region can be neglected. This assumption has been found to be valid in most cases. Thus, we can write:

$$J_n = (-x_p) = J_n(x_n) \quad \text{and} \quad J_p(x_n) = J_p(-x_p) \tag{2-63}$$

The total diode-current density is given by:

$$J = J_n(x_n) + J_p(x_n) = J_n(-x_p) + J_p(x_n) \tag{2-64}$$

$$J = e\left(\frac{D_p p_n}{L_p} + \frac{D_n n_p}{L_n}\right)\left[\exp\left(\frac{eV}{k_BT}\right) - 1\right] \tag{2-65}$$

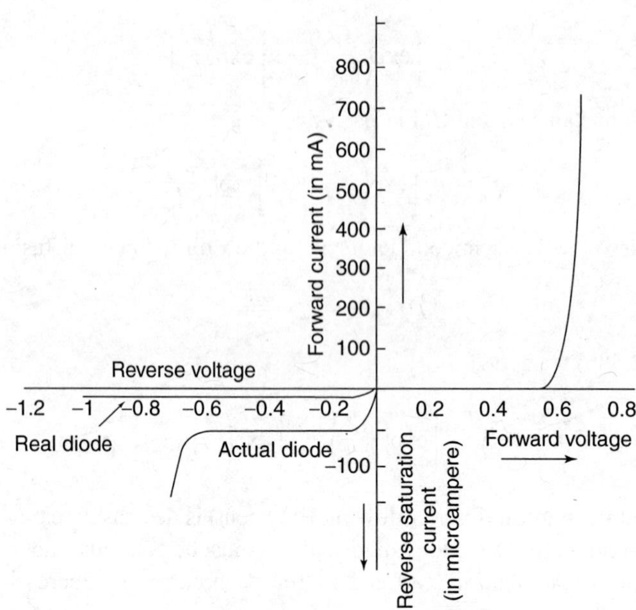


**Figure 2-10** Actual and theoretical I–V characteristics of a typical semiconductor diode

The total direct current of the diode, with a cross-sectional junction area $A$, is:

$$I = AJ_s\left[\exp\left(\frac{eV}{k_BT}\right)-1\right] = I_s\left[\exp\left(\frac{eV}{k_BT}\right)-1\right] \tag{2-66}$$

where,

$$I_s \equiv AJ_s \equiv Ae\left(\frac{D_p p_n}{L_p}+\frac{D_n n_p}{L_n}\right) \tag{2-67}$$

Equation (2-66) is the dc equation of an ideal $p$–$n$ junction diode. The plot of the voltage–current characteristics of the diode, for forward-bias and reverse-bias, is shown in Fig. 2-10.

It should be noted that because of the higher concentration of holes in the $p$-region the hole current is much larger than the electron current.

## Solved Examples

**Example 2-9**

(a) Prove that the reverse saturation current in a $p$–$n$ diode can be written as:

$$I_0 = Ae\left(\frac{D_p}{L_pN_D}+\frac{D_n}{L_nN_A}\right)n_i^2$$

(b) Starting with the expression for $I_0$ found in part (a), verify that the reverse saturation current is given by:

$$I_0 = AV_T\left(\frac{b\sigma_i^2}{(1+b)^2}\right)\left(\frac{1}{L_p\sigma_N}+\frac{1}{L_N\sigma_P}\right)$$

where, $b = \frac{\mu_n}{\mu_p}$ and $\sigma_i = n_i e\,(\mu_n + \mu_p)$

**Solution:**

(a) From Eq. (2-67), we can write that the reverse saturation current $I_0$ is given by:

$$I_0 = Ae\left(\frac{D_p p_n}{L_p}+\frac{D_n n_p}{L_n}\right)$$

We know that:

$$p_n = \frac{n_i^2}{N_D} \quad \text{and} \quad n_p = \frac{n_i^2}{N_A}$$

Substituting these two in the equation for saturation current we get:

$$I_0 = Ae\left(\frac{D_p}{L_p N_D}+\frac{D_n}{L_n N_A}\right)n_i^2 \tag{1}$$

(b) We know from the Einstein relation under the condition of carrier non-degeneracy that:

$$D_n = \mu_n V_T, \quad V_T = \frac{k_B T}{e} \quad \text{and} \quad D_p = \mu_p V_T$$

Substituting the values of $D_n$ and $\mu_n$ in the equation for saturation current, we get

$$I_0 = AeV_T\left(\frac{\mu_p}{L_p N_D}+\frac{\mu_n}{L_n N_A}\right)n_i^2$$

$$= AeV_T\mu_n\mu_p\left(\frac{1}{L_p N_D \mu_n}+\frac{1}{L_n N_A \mu_p}\right)n_i^2$$

$$= Ae^2V_T\mu_n\mu_p\left(\frac{1}{eL_p N_D \mu_n}+\frac{1}{eL_n N_A \mu_p}\right)n_i^2$$

$$= Ae^2V_T\frac{\mu_n\mu_p}{(\mu_n+\mu_p)^2}\left(\frac{1}{\sigma_p L_n}+\frac{1}{\sigma_n L_p}\right)n_i^2(\mu_n+\mu_p)^2$$

$$= Ae^2V_T\frac{(\mu_n/\mu_p)}{1+\left(\frac{\mu_n}{\mu_p}\right)^2}\left(\frac{1}{\sigma_p L_n}+\frac{1}{\sigma_n L_p}\right)n_i^2(\mu_n+\mu_p)^2$$

Since, $\sigma_i = n_i e\,(\mu_n\,\mu_p)$ and by the question $b = \mu_n/\mu_p$, we can write that:

$$I_0 = AV_T\left(\frac{b\sigma_i^2}{(1+b)^2}\right)\left(\frac{1}{L_p\sigma_n}+\frac{1}{L_N\sigma_p}\right)$$

**Example 2-10** Find the reverse saturation point of current for a Si $p$–$n$ junction diode at a room temperature of 27° C. The cross sectional area is 1.5 mm², $\sigma_n = 0.1$ (ohm cm)$^{-1}$, $\sigma_p = 3.0$ (ohm cm)$^{-1}$ and $L_n = L_p = 0.15$ cm. Use the other physical data if required.

**Solution:**

We know that:

$$b = \frac{\mu_n}{\mu_p} = \frac{3800}{1800} = 2.11, \quad A = 1.5 \times 10^{-6}\,\text{m}^2, \quad V_T = 0.026\ \text{eV}$$

$$L_p\sigma_n = 0.15 \times 0.1 = 0.015\ \Omega^{-1}, \quad L_n\sigma_p = 0.15 \times 3.0 = 0.45\ \Omega^{-1}$$

We know that:

$$\sigma_i = (\mu_n + \mu_p)n_i e = (3800 + 1800) \times 2.5 \times 10^{15} \times 1.6 \times 10^{-19} = 2.24\ (\Omega\text{m})^{-1}$$

In this case: $L_n\sigma_p = L_p\sigma_n = 3 \times 10^{-4}$

Therefore, using the results of Example 7(b), we get:

$$I_0 = \frac{1.5 \times 10^{-16}\,\text{m}^2 \times 0.026\ \text{V} \times 2.11 \times (2.24)^2}{(3.11)^2\,(\Omega\text{m})^2}\left(\frac{1}{0.45} + \frac{1}{0.015}\right)\Omega$$

$$= 2.94\ \mu\text{A}$$

**Example 2-11** Consider a $p$–$n$ diode operating under low-level injection. Assuming that the minority current is due entirely to diffusion show that the electric field in the $n$-side is given by:

$$\varepsilon(x) = \frac{I_T + (D_n/D_p - 1)\,I_{pn}(x)}{en_n\mu_n A}$$

where, $I_{pn}(x)$ is the minority diffusion current in the $n$-side of the diode, $I_T$ is the total current which is constant and independent of $x$ and the other notations have the usual meanings.

**Solution:**

The minority (hole) diffusion current in the $n$-side of the diode is $I_{pn}(x) = -AeD_p\,dp/dx$. The majority (electron) current in the $n$-side consists of two parts: majority diffusion current and majority drift current. So total majority (electron) current is:

$$I_{\text{majority}} = Aen_n\mu n\xi(x) - \frac{D_n}{D_p}I_{pn}(x) \quad \text{where } \xi(x) \text{ is the electric field.}$$

Therefore, since we neglect the minority drift current, the total current $I_T$ which is constant and independent of $x$ can be written as

$$I_T = Aen_n\mu_n\xi(x) - \frac{D_n}{D_p}I_{pn}(x) + I_{pn}(x)$$

Solving for $\xi(x)$ we find:

$$\xi(x) = \frac{I_T + \left(\frac{D_n}{D_p} - 1\right)I_{pn}(x)}{Aen_n\mu_n}$$

**Example 2-12** Using the result $I_0 = AV_T\dfrac{b\sigma_i^2}{(1+b)^2}\left(\dfrac{1}{L_p\sigma_n} + \dfrac{1}{L_n\sigma_p}\right)$, find the reverse saturation current for a $p$-Si $p$–$n$ junction diode at room temperature, 300 K. Assume that the cross-sectional area $A = 5.0\ \text{mm}^2$, $L_n = L_p = 0.01$ cm and $\sigma_n = \sigma_p = 0.01\ \Omega\text{cm}^{-1}$.

**Solution:**

$$A = 5 \times 10^{-6}\,\text{m}^2 \quad \text{and} \quad b = \frac{1300}{500} = 2.6$$

$$L_p\sigma_n = 10^{-2} \times 10^{-2}\,(\Omega)^{-1} = L_n\sigma_p$$

$$\sigma_i = 1.5 \times 10^{10}\,(1300 + 500)\,(1.6 \times 10^{-19}) = 4.32 \times 10^{-6}\,(\Omega - \text{cm})^{-1}$$

Then,

$$I_0 = 5 \times 10^{-2}\,\text{cm}^2 \times 0.026 \times \frac{2.6}{(3.6)^2} \times \frac{(4.32 \times 10^{-6})^2}{(\Omega - \text{cm})^2} \times (2 \times 10^4)\,\Omega$$

$$= 97.25\,\text{pA}$$

**Example 2-13** Find the ratio of the reverse saturation current in Ge to that in Si, using the result:

$$I_0 = AV_T\frac{b\sigma_i^2}{(1+b)^2}\left(\frac{1}{L_p\sigma_n} + \frac{1}{L_n\sigma_p}\right)$$

Assume $L_n = L_p = 0.1$ cm and $\sigma_n = \sigma_p = 1.0\,(\Omega\text{cm})^{-1}$ for Ge, whereas the corresponding values are 0.01 cm and $0.01(\Omega\text{cm})^{-1}$ for Si.

**Solution:**

Let:

$$Y = \frac{I_0}{AV_T} = \frac{b\sigma_i^2}{(1+b)^2}\left(\frac{1}{L_p\sigma_n} + \frac{1}{L_n\sigma_p}\right)$$

For Si:

$$b = \frac{\mu_n}{\mu_p} = \frac{1300}{500} = 2.6$$

$$\sigma_i = 1.5 \times 10^{10}\,(1300 + 500)\,(1.6 \times 10^{-19}) = 4.32 \times 10^{-6}\,(\Omega - \text{cm})^{-1}$$

Then,

$$Y_{\text{Si}} = \frac{2.6 \times (4.32 \times 10^{-6})^2}{(3.6)^2}(2 \times 10^4) = 7.49 \times 10^{-8}\,(\Omega - \text{cm}^2)^{-1}$$

For Ge:

$$b = \frac{\mu_n}{\mu_p} = \frac{3800}{1800} = 2.11$$

$$\sigma_i = 2.5 \times 10^{13}\,(3800 + 1800)\,(1.6 \times 10^{-19}) = 2.24 \times 10^{-2}\,(\Omega - \text{cm})^{-1}$$

Then,

$$Y_{\text{Ge}} = \frac{2.11 \times (2.24 \times 10^{-2})^2}{(3.11)^2}(2 \times 10^2) = 2.189 \times 10^{-2}\,(\Omega - \text{cm}^2)^{-1}$$

Therefore,

$$\frac{Y_{\text{Ge}}}{Y_{\text{Si}}} = \frac{2.189 \times 10^{-2}}{7.49 \times 10^{-8}} = 0.29 \times 10$$

■

**Key Points**

**(i)** Equation (2-66) has been derived under the following assumptions:

- Only one-dimensional flow of current has been assumed
- The recombination mechanism has been neglected in the transition region
- The injected minority carrier density is much less than the normally present majority carrier density

- The majority carrier density is much higher than the intrinsic carrier density of both *p*- and *n*-type semiconductor
- Both *p*- and *n*-type semiconductors are essentially non-degenerate
- The carrier dispersion relations of both types of semiconductors are strictly parabolic

**(ii)** Figure 2-10 gives the plot of the current *I* versus the voltage *V*, as described by Eq. (2-69). This plot is termed the current–voltage characteristic or the volt-ampere characteristic of the *p*–*n* junction diode. The practical I–V characteristics of a *p*–*n* junction diode, as obtained by measurements, will deviate from the ideal model characteristics Eq. (2-66) in the manner shown in Fig. 2-10. The deviation in the forward region is due to the voltage drops across the semiconductor bulk. The applied voltage will cause less current than predicted by the theoretical model since the drops were neglected.

It should be noted that in the derivation of the diode equation it was assumed that the forward applied voltage was small, so that the injected minority-carriers were much less than the majority-carrier concentration normally present in the extrinsic semiconductor. In other words, we have assumed low-level injection. If the low-level-injection condition is violated, the electric field outside the depletion region cannot be neglected and the differential equation to be solved is no longer a linear differential equation. Fortunately, in many transistor applications the applied forward voltage is small, so that the low-level-injection assumption is valid. In the presence of non-ideality, Eq. (2-66) assumes the form:

$$I = I_s\left(e^{\frac{eV}{\eta K_B T}} - 1\right) \tag{2-68}$$

where, $\eta$ is a dimensionless number and is known as non-ideality factor, which depends on the band structure of the material. When *V* is positive, the junction is forward-biased, and when *V* is negative, the junction is reverse-biased. It has been observed from Fig. 2-9 that at a particular value of the reverse voltage ($V_B$), the reverse current increases suddenly. When the forward bias *V* is less than a value $V_B$, the current is very small. As *V* exceeds $V_B$, the current increases very sharply. The voltage $V_B$ is known as the break point, offset, threshold or *cutin voltage* of the diode. $V_B = 0.2$ V for Ge diodes and $V_B = 0.6$ V for Si diodes.

## Solved Examples

**Example 2-14** The current flowing in a certain *p*–*n* junction diode at room temperature is $9\times10^{-7}$ A, when the large reverse voltage is applied. Calculate the current flowing, when 0.1 V forward bias is applied.

**Solution:**

Given: $I_0 = 9 \times 10^{-7}$ A, $V_F = 0.1$ V

Current flowing through the diode under forward-bias is given by:

$$I = I_0\,(e^{V/\eta V_T} - 1)$$

$$= I_0\,(e^{40V_F} - 1)$$

$$= 9 \times 10^{-7}\,(e^{40\times 0.1} - 1) = 9 \times 10^{-7}\,(e^{4} - 1)$$

$$= 48.15\ \mu\text{A}$$

**Example 2-15** The saturation current density of a *p*–*n* junction Ge diode is 500 mAm$^{-2}$ at 350 K. Find the voltage that would have to be applied across the junction to cause a forward current density of $10^5$ Am$^{-2}$ to flow.

**Solution:**

Given: $J_0 = 500 \text{ mA/m}^2,\ T = 350 \text{ K},\quad J = 10^5 \text{ Am}^{-2}$

We know that: $I = I_0 (e^{eV/\eta KT} - 1)$

Dividing by the area of the diode, we have an expression for current density:

$$J = J_0 (e^{eV/\eta KT} - 1)$$

or,
$$e^{eV/\eta KT} = \frac{J}{J_0} + 1 = \frac{10^5}{500 \times 10^{-3}} + 1 = 2 \times 10^5$$

$\therefore$
$$\frac{eV}{\eta KT} = \log_e (2 \times 10^5) = 2.303 \log_{10} (2 \times 10^5)$$
$$= 12.20827$$
$$V = \frac{12.208 \times \eta KT}{e}$$
$$V = \frac{12.208 \times 1.38 \times 10^{-23} \times 350}{1.6 \times 10^{-19}} = 0.3685 \text{ V}$$

**Example 2-16** A Si diode has a saturation current of 0.15 pA at 20°C. Find it's current when it is forward-biased by 0.55 V. Find the current in the same diode when the temperature rises to 100°C.

**Solution:**

At $T = 20°\text{C}$:
$$V_T = \frac{k_B T}{e} = \frac{1.38 \times 10^{-23} (273 + 20)}{1.6 \times 10^{-19}}$$
$$= 0.02527 \text{ V}$$

Assuming $\eta = 1$:
$$I = I_s (e^{V/\eta VT} - 1)$$
$$= 1.5 \times 10^{-13} (e^{0.55/0.02527} - 1)$$
$$= 0.4245 \text{ mA}$$

At $T = 100°\text{C}$:
$$V_T = \frac{k_B T}{e} = \frac{1.38 \times 10^{-23} (273 + 100)}{1.6 \times 10^{-19}}$$
$$= 0.03217 \text{ V}$$

Now $100 - 20 = 80$ and $80/10 = 8$.
Therefore, $I_s$ doubles 8 times, i.e., increased by a factor $2^8 = 256$.

So at 100°C:
$$I_s = 1.5 \times 256 \times 10^{-13} \text{ A}$$
$$I = 1.5 \times 256 \times 10^{-13} (e^{0.55/0.032} - 1)$$
$$I = 0.0011 \text{ A}$$

**Example 2-17** The saturation currents of the two diodes are 2 and 4 $\mu$A. The breakdown voltages of the diodes are the same and are equal to 100 V. Calculate the current and voltage for each diode if $V = 90$ V and $V = 110$ V.

**Solution:**

$$I_{01} = 2\ \mu\text{A} \quad \text{and} \quad I_{02} = 4\ \mu\text{A}, \quad V_{z1} = V_{z2}\ 100\text{ V}$$

We have, $I = I_0\,(e^{V/\eta V_T} - 1)$. Therefore, $V = \eta V_T\ \text{In}\left(1 + \frac{I}{I_0}\right)$. We assume that the diodes are Si diodes so that $\eta = 2$.

When $V = 90$ V: None of the diodes will break down and $I$ is determined by the diode with the smallest $I_0$, i.e., $D_1$. Thus, $I = 1\ \mu$A and for $D_2$, $I = -1\ \mu$A.

$$\therefore \qquad V_2 = \eta V_T \ln\left(1 + \frac{I}{I_0}\right) = 2 \times 0.026 \ln\left(1 - \frac{2}{4}\right) = -36\text{ mV}$$

and,
$$V_1 = -89.964\text{ V}$$

When $V = 110$ V: As the applied voltage $V$ is increased to 110 V, $D_1$ will break down, while $D_2$ will be reverse-biased. Thus, $I = I_{02} = 2\ \mu$A, $V_1 = -\ 100$ V and $V_2 = -10$ V

**Example 2-18** (a) For what voltage will the reverse current in a *p–n* junction Ge diode reach 90% of its saturation value at room temperature?

(b) What is the ratio of the current for a forward bias of 0.05 V to the current for the same magnitude of reverse-bias?

(c) If the reverse saturation current is 15 $\mu$A, calculate the forward currents for voltages of 0.1 V, 0.2 V, and 0.3 V, respectively.

**Solution:**

(a) We have, $V_T = \dfrac{T}{11{,}600} = 0.026$ V at room temperature.
Using the formula, $I = I_0\,(e^{V/VT} - 1)$ we have:

$$-0.9\,I_0 = I_0\,(e^{V/0.026} - 1)$$

or,
$$e^{V/0.026} = 0.1$$

or,
$$V = (0.026)\,(-2.3) = -0.060\text{ V}$$

(b) $$\frac{e^{50/26} - 1}{e^{-50/26} - 1} = \frac{e^{1.92} - 1}{e^{-1.92} - 1} = \frac{6.82-1}{0.147-1} = -6.83$$

For $V = 0.1$:

$$I = 15\,(e^{100/26} - 1) = 15\,(e^{3.84} - 1) = 682.5\ \mu\text{A} = 0.682\text{ mA}$$

$$V = 0.2, I = 15\,(e^{200/26} - 1) = 15\,(e^{7.68} - 1) = 32550\ \mu\text{A} = 32.55\text{ mA}$$

$$V = 0.3, I = 15\,(e^{300/26} - 1) = 15\,(e^{11.52} - 1) = 1.515 \times 10^6\ \mu\text{A} = 1.515\text{ A}$$

**Example 2-19** (a) Calculate the anticipated factor by which the reverse saturation current of a Ge diode is multiplied when the temperature is increased from 25 to 70°C.
(b) Repeat part (a) for an Si diode in the range 25–150°C.

**Solution:**

(a) $I_0\,(70°\text{C}) = I_0\,(25°\text{C}) \times 2^{(70-50)/10} = I_0\,(25°\text{C}) \times 2^{4.5} = 45I_0\,(25°\text{C})$

(b) $I_0\,(150°\text{C}) = I_0\,(25°\text{C}) \times 2^{12.5} = 5700\,I_0\,(25°\text{C})$

**Example 2-20** It is predicted that, for Ge, the reverse saturation current should increase by 0.15°C$^{-1}$. It is found experimentally in a particular diode that at a reverse voltage of 10 V, the reverse current is 5 $\mu$A and the temperature dependence is only 0.07°C$^{-1}$. What is the leakage resistance shunting the diode?

**Solution:**

The diode is connected in parallel with a resistance $R$ and the parallel combination is placed in series with the diode with a series with a reverse voltage 10 V. Let the current through diode be $I_0$ and the current through $R$ be $I_R$. Let the main current be $I$ so that the application of Kirchoff's current law gives:

$$I = I_0 + I_R \tag{1}$$

$I_R$ is independent of $T$.

Hence:

$$\frac{dI}{dT} = \frac{dI_0}{dT} \tag{2}$$

It is given that:

$$\frac{1}{I_0}\frac{dI_0}{dT} = 0.15 \quad \text{and} \quad \frac{I}{I}\frac{dI}{dT} = 0.07 \tag{3}$$

Using Eq. (2), we have:

$$\frac{dI_0}{dT} = (0.15)\,I_0 = (0.07)\,I = \frac{dI}{dT}$$

or,

$$(0.07)\,I = (0.15)\,I_0 \tag{4}$$

Multiplying Eq. (1) by 0.15 and subtracting Eq. (4) from Eq. (1), we obtain:

$$0.08\,I = 0.15\,I_R \quad \text{or,} \quad I_R = \frac{0.08}{0.15}\,I \quad \text{or,} \quad I_R = \frac{40}{15}\,\mu\text{A}$$

Hence,

$$R = \frac{10 \times 15}{40} = 3.75\ \text{M}\Omega$$

**Example 2-21** A diode is mounted on a chassis in such a manner that, for each degree of temperature rise above ambient, 0.1 mW is thermally transferred from the diode to its surroundings. (The thermal resistance of the mechanical contact between the diode and its surroundings is 0.15 mW/°C.) The ambient temperature is 25°C. The diode temperature is not to be allowed to increase by more than 10°C above ambient. If the reverse saturation current is 5 $\mu$A at 25°C and increases at the rate 0.07°C$^{-1}$, what is the maximum reverse-bias voltage which may be maintained across the diode?

**Solution:**

Since the maximum permissible diode temperature is 35°C, this must be the temperature at which we have thermal equilibrium. Thus:

$$P_{\text{out}} = 0.15\ \text{mW/°C} \times (35 - 25)\text{°C} = 1.5\ \text{mW}$$

We know that the reverse saturation current approximately doubles for every 10°C rise in temperature. Thus at 35°C, we can write:

$$I_0\,(35) = 2\,I_0\,(25) = 10\ \mu\text{A}$$

$\therefore$

$$P_{\text{in}} = V \times I_0 = 1.5\ \text{mW}$$

or,

$$\text{V} = 150\ \text{V}$$

**Example 2-22** A Si diode operates at a forward voltage of 0.4 V. Calculate the factor by which the current will be multiplied when the temperature is increased from 25°C to150°C.

**Solution:**

At the temperature of 150°C, i.e., 423 K:

$$V_T = \frac{423}{11{,}600} = 0.0364 \text{ V}$$

We know that:

$$I_0(T) = I_{01} \times 2^{\frac{(T-T_1)}{10}}$$

At 150°C:

$$I_0(150)\, I_0(25) \times 2^{12.5} = 5792\, I_0(25)$$

Using $n = 2$ for Si, we obtain form the diode equation:

$$I = 5792(e^{0.4/0.0728} - 1)\, I_0(25)$$

or,

$$I = 5792 \times 242 \times I_0(25)$$

On the other hand, at 25°C:

$$I = I_0(25)(e^{0.4/0.0514} - 1) = 2400 I_0(25)$$

Hence,

$$\frac{I(150)}{I(25)} = 584$$

At 25°C and $V = 0.4$ V, $I(25) = 0.01$ mA. At 150°C and $V = 0.4$ V, $I(150) = 2.42$ mA

or,

$$\frac{I(150)}{I(25)} = 242$$

**Example 2-23** (a) Consider a grown junction for which the uncovered charge density ρ varies linearly with distance. If $\rho = ax$, prove that the barrier voltage $V_j$ is given by:

$$V_j = \frac{aW^3}{12\varepsilon}$$

(b) Verify that the barrier capacitance $C_T$ is given by:

$$C_T = \frac{\varepsilon A}{W}$$

**Solution:**

(a) From Poisson's equation:

$$\frac{d^2V}{dx^2} = -\frac{\rho}{\varepsilon} = -\frac{ax}{\varepsilon}; \quad \frac{dV}{dx} = -\frac{ax^2}{2\varepsilon} + C_1$$

Since, at $x = -\frac{W}{2}$:

$$\frac{dV}{dx} = 0; \quad C_1 = \frac{aW^2}{8\varepsilon}$$

(Assuming electric field does not extend outside the depletion region)

or,

$$\frac{dV}{dx} = -\frac{ax^2}{2\varepsilon} + \frac{aW^2}{8\varepsilon}$$

or,
$$V = -\frac{ax^3}{6\varepsilon} + \frac{aW^2}{8\varepsilon}x + C_2$$

Since, at $x = -\frac{W}{2}$:
$$V = 0;\ C_2 = \frac{ax^3}{6\varepsilon} - \frac{aW^2}{8\varepsilon}x = -\frac{aW^3}{48\varepsilon} + \frac{aW^3}{16\varepsilon} = \frac{aW^3}{24\varepsilon}$$

At $x = +\frac{W}{2}$:
$$V = V_j = -\frac{aW^3}{48\varepsilon} + \frac{aW^3}{16\varepsilon} + \frac{aW^3}{24\varepsilon} = \frac{aW^3}{12\varepsilon}$$

(b) $Q = \int_{0^-}^{W/2} A\rho dx = \int_0^{W/2} Aaxdx = \frac{Aa}{2}\left(\frac{W}{2}\right)^2 = \frac{AaW^2}{8}$

$$C_T = \frac{dQ}{dV} = \frac{AaW}{4}\frac{dW}{dV}$$

From part (a):
$$dV = \frac{aW^2}{4\varepsilon}dW$$

∴
$$C_T = \frac{AaW}{4}\frac{4\varepsilon}{aW^2} = \frac{\varepsilon A}{W}$$

**Example 2-24** Given a forward-bias Si diode with $I = 1$ mA. If the diffusion capacitance is $C_D = 1.5\ \mu$F, what is the diffusion length $L_p$? Assume that the doping of the $p$-side is much greater than that of the $n$-side.

**Solution:**

We know that:
$$C_D = \frac{L_p^2}{D_p}\frac{I}{\eta V_T}$$

or,
$$L_p^2 = \frac{C_D D_p \eta V_T}{I}$$

or,
$$L_p^2 = \frac{1.5 \times 10^{-6} \times 13 \times 2 \times 0.026}{1 \times 10^{-3}} = 1014 \times 10^{-6}$$

∴
$$L_p = 31.84 \times 10^{-3}\ \text{m}$$

■

**(iii)** The $p$–$n$ junction diode when operated in this portion of the characteristic is termed as a breakdown diode. It should be clearly understood that the term breakdown does not mean that the diode has been corrupted or burnt out. It is only a temporary mechanism, which can be restored when the reverse applied voltage is increased in the positive direction. In the forward-bias mode the forward current increases slowly with the increase of applied voltage till the voltage reaches a certain value called saturation voltage. After this, the current rises very sharply but the voltage remains constant.

**(iv)** Substituting $e = 1.6 \times 10^{-19}$ C, $k_B = 1.38 \times 10^{-23}$ J/K, and $T = 300$ K (i.e., at room temperature) in Eq. (2.68), we get:

$$I = I_s\left(e^{\frac{39\,V}{\eta}} - 1\right) \tag{2-69}$$

Neglecting the voltage drops in the bulk $p$-type and $n$-type regions, the voltage $V$ is approximately the voltage applied across the diode terminals.

When $V$ is positive and sufficiently high, the term unity in Eq. (2-69) can be neglected, so that the current $I$ increases exponentially with the voltage $V$. When the diode is reverse biased we have $I = -I_s$. Thus, the reverse current is independent of the applied bias and its magnitude equals the reverse saturation current. The range of forward current for the diode operation is higher than the reverse saturation current. The forward current is in the range of milli-ampere, but the reverse current is in the range of microampere or less.

We have $\eta = 1$ for Si diodes and $\eta = 1$ for Ge diodes, therefore, for a given forward bias (larger than $V_B$) the current for a Si diode is less than that for a Ge diode.

**(v)** A study of the $I$–$V$ characteristics reveals two different regions: a non-linear region for low values of applied voltage and an almost linear region for high values of voltage. This leads us to give two different expressions for diode resistance. The non-linear region can be assumed to be made up of piecewise linear regions of extremely short voltage intervals where Ohm's law is valid.

The slope of the $I$–$V$ characteristic of the $p$–$n$ diode defines the dynamic or the ac resistance $r_{ac}$ of the diode, as shown in Fig. 2-10(a). Thus, for a change of voltage $dV$, the corresponding change in current is $dI$. The corresponding $r_{ac}$ is defined as:

$$r_{ac} = \frac{dV}{dI} = \frac{V_2 - V_1}{I_2 - I_1} \tag{2-71}$$

From Eq. (2-69), we obtain:

$$r_{ac} = \frac{dV}{dI} = \frac{\eta}{39\,(I + I_s)} \tag{2-72}$$

$r_{ac}$ is not a constant and is determined by the operating voltage. The ratio between the voltage $V$ across the junction and the current $I$ flowing through the junction is called the static or the dc resistance $r_{dc}$ of the diode which can be mathematically written from Fig. 2-10(b) as:

$$r_{dc} = \frac{V}{I} \tag{2-70}$$

$r_{dc}$ is not a constant but varies significantly with the applied voltage as shown in Fig. 2-10(b).

For a reverse-bias, such that, $|39\ V/\eta| \approx 1$, the dynamic resistance is extremely large. For a Ge diode at room temperature, the forward ac resistance is $r_{ac} = 26/I$.

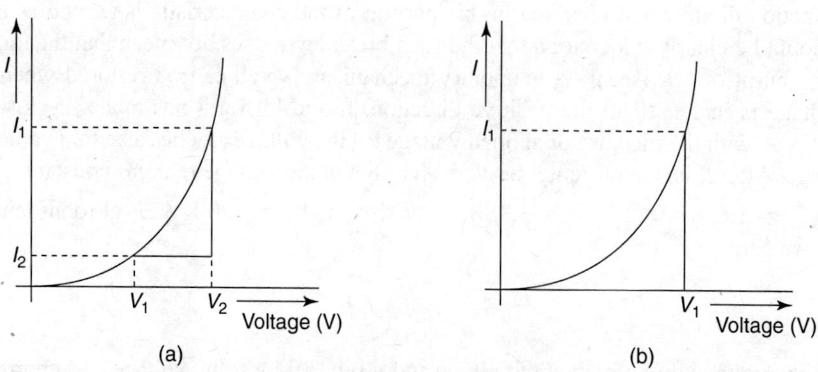


**Figure 2-10** (a) Dynamic resistance calculation from $I$–$V$ characteristics of a $p$–$n$ junction diode (b) Static or dc resistance calculation from $I$–$V$ characteristics of a $p$–$n$ diode

## Solved Examples

**Example 2-25** Find the static resistance of a *p–n* junction germanium diode if the temperature is 27°C and $I_0 = 20\ \mu A$ for an applied voltage of 0.2 V.

**Solution:**
Given:

$$I_0 = 20\ \mu A,\ V_F = 0.2\ V,\ t = 27°C$$

The forward current through the diode: $I = I_0\,[e^{V/\eta V_T} - 1]$

$$= 20 \times 10^{-6}\,(e^{40 V_F} - 1) = 20 \times 10^{-6}\,(e^{40 \times 0.2} - 1)$$

$$= 59.59915\ mA$$

Static resistance: $r_{dc} = \dfrac{0.0343}{80 \times 10^{-6}}\, e^{0.2/0.0343} = 0.14625\ M\Omega$

**Example 2-26** An ideal germanium diode has a reverse saturation current of 80 $\mu$A at a temperature of 125°C. Find the dynamic resistance at that temperature for a 0.2 V bias in: (a) the forward direction and (b) the reverse direction.

**Solution:**
Given:

$$t = 125°C \quad \text{or} \quad T = 398\ K,\ I_0 = 80\ \mu A,\ \eta = 1,\ V_F = 0.2\ V$$

(a) Volt equivalent of the temperature:

$$V_T = \frac{T}{11{,}600} = \frac{398}{11{,}600} = 0.0343\ V$$

Diode current is given by: $I = I_0\,[e^{V_F/\eta V_T} - 1]$

Differentiating with respect to $V$: $\dfrac{dI}{dV} = \dfrac{I_0}{V_T}\, e^{V_F/V_T}$

or, $\dfrac{I}{R_{ac}} = \dfrac{I_0}{V_T}\, e^{V_F/V_T}$

$$R_{ac} = \frac{V_T}{I_0}\, e^{-V_F/V_T}$$

$$= \frac{0.0343}{80 \times 10^{-6}}\, e^{-0.2/0.0343} = 1.258\ \Omega$$

(b) Dynamic resistance in reverse direction:

$$R_{ac} = \frac{V_T}{I_0}\, e^{V_F/V_T}$$

$$= \frac{0.0343}{80 \times 10^{-6}}\, e^{0.2/0.0343} = 0.14625\ M\Omega$$

**Example 2-27** The reverse-bias saturation current for a *p–n* junction diode is 1.5 $\mu$A at 300 K. Determine its ac resistance at 150 mV forward-bias.

**Solution:**

We know that,
$$r_{ac} = dV/dI$$

$$= \frac{1}{\frac{dI}{dV}} = \frac{1}{(I_0/k_BT)\exp(V/V_T)}$$

But at temperature of 300 K:

$$k_BT = 8.62 \times 10^{-5} \times 300$$

$$= 25.86 \times 10^{-3}$$

$$r_{ac} = \frac{25.86 \times 10^{-3}}{1.5 \times 10^{-6} \times \exp(0.15/0.02586)}$$

$$= \frac{25.86 \times 10^{3}}{495.69} = 52.169 \text{ W}$$

**Example 2-28** A diode reaches its maximum power rating of 2.5 W when operating in the forward mode at the forward voltage of 900 mV. Calculate: (a) the maximum allowable forward current $I_{f(\max)}$ (b) the forward diode resistance $R_f$.

**Solution:**

(a)
$$I_{f(\max)} = \frac{P_{\max}}{v_f} = \frac{2.5W}{0.9V} = 2.75 \text{ A}$$

(b)
$$R_j = \frac{P_{\max}}{I^2_{\max}} = \frac{2.5}{2.2^2} = 0.5165\ \Omega$$

**Example 2-29** (a) The resistivities of the two sides of a step-graded germanium diode are 2 Ω − cm (*p*-side) and 1 Ω − cm (*n*-side). Calculate the height of the potential-energy barrier.(b) Repeat the part (a) for silicon *p–n* junction.

**Solution:**

(a)
$$\rho = \frac{1}{\sigma} = \frac{1}{N_A e \mu_p} = 2\ \Omega - \text{cm}$$

or,
$$N_A = \frac{1}{2 \times 1.6 \times 10^{-19} \times 1800} = 1.74 \times 10^{15}\ /\text{cm}^3$$

Similarly,
$$N_D = \frac{1}{1 \times 1.6 \times 10^{-19} \times 3800} = 1.65 \times 10^{15}\ /\text{cm}^3$$

Therefore the height of the potential energy barrier is:

$$V_0 = 0.026 \times \ln \frac{1.65 \times 10^{15} \times 1.74 \times 10^{15}}{(2.5 \times 10^{13})^2} = 0.22 \text{ eV}$$

(b) $$N_A = \frac{1}{2 \times 1.6 \times 10^{-19} \times 500} = 6.25 \times 10^{15}/\text{cm}^3$$

$$N_D = \frac{1}{1 \times 1.6 \times 10^{-19} \times 1300} = 4.8 \times 10^{15}/\text{cm}^3$$

$\therefore$ $$V_0 = 0.026 \times \ln \frac{4.8 \times 10^{15} \times 6.25 \times 10^{15}}{(1.5 \times 10^{10})^2} = 0.667 \text{ eV}$$

**Example 2-30** An ideal Ge *p–n* junction diode has at a temperature of 125°C a reverse saturation current of 35 $\mu$A. At a temperature of 125°C find the dynamic resistance for a 0.2 V bias in: (a) the forward direction (b) the reverse direction.

**Solution:**

(a) At the temperature of 125°C, that is 398 K,

$$V_T = \frac{398}{11{,}600} = 0.0343 \text{ V}$$

For Germanium we know that $\eta = 1$.

Thus, $I = I_0\,(125)\,(e^{V/V_T(125)} - 1) \approx I_0\,(125)\,(e^{V/V_T(125)})$ since $\dfrac{V}{V_T(125)} >> 1$.

Differentiating with respect to $V$:

$$\frac{1}{r} = \frac{dI}{dV} = \frac{I_0\,(125)}{V_T(125)}\, e^{V/V_T(125)} = \frac{35.0 \times 10^{-6}}{34.3 \times 10^{-3}}\, e^{5.83} = 0.348$$

or, $$r = 2.873\ \Omega$$

(b) $$\frac{I}{r} = \frac{dI}{dV} = \frac{I_0\,(125)}{V_T(125)}\, e^{-V/V_T(125)} = 3.185 \times 10^{-6} \text{ mho}$$

or, $$r = 0.314\ \text{M}\Omega$$

■

## 2-7 LINEAR PIECEWISE MODELS

The *p–n* junctions are unilateral in nature, i.e., they conduct current in only one direction. Thus, we can consider an ideal diode as a short circuit when forward-biased and as an open circuit when reverse-biased. Forward-biased diodes exhibit an offset voltage ($V_y$) that can be approximated by the simple equivalent circuit with a battery in series with an ideal diode. The series battery in the model keeps the ideal diode turned off for applied voltage less than $V$; the actual diode characteristic is improved by adding a series resistance ($r$) to the equivalent circuit. The equivalent diode model, as shown in Fig. 2-11, is called the *piecewise linear equivalent model.*

The approximate characteristics are linear over specific voltage and current ranges. This approximated characteristic of the device by a straight-line segment is shown in the Fig. 2-12.

It is obvious from Fig. 2-12 that the straight line segments do not result in an exact duplication of the actual characteristics, especially in the knee region. From a practical point of view, the resulting segments are sufficiently close to the actual curve to establish an equivalent circuit that will provide an excellent first approximation to the actual behaviour of the device.

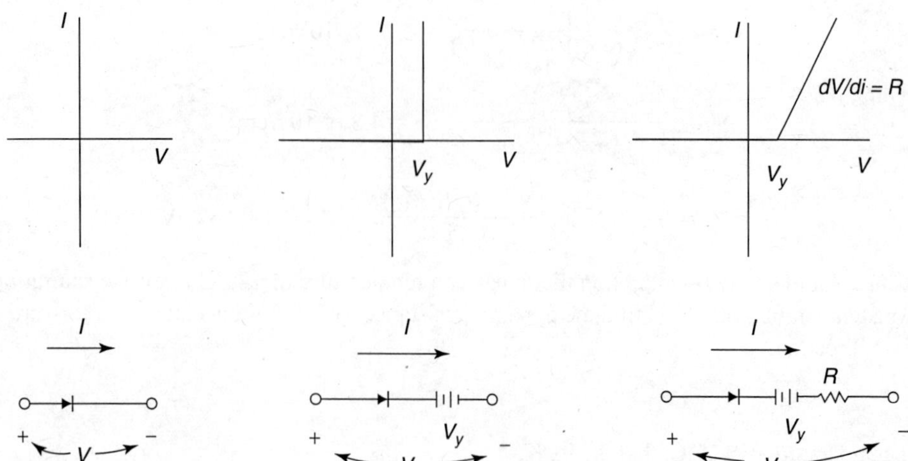


**Figure 2-11** Linear piecewise models of a diode for different order of approximations

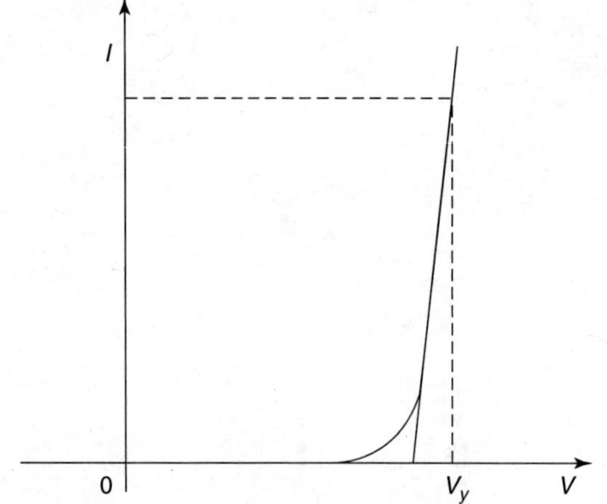


**Figure 2-12** I–V characteristics of *p–n* junction diode

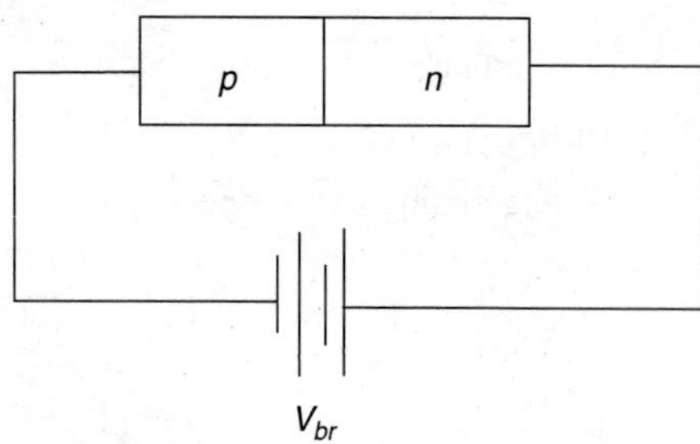


**Figure 2-13** Reverse-biased *p–n* junction

Thus, in the linear piecewise model we neglect the bending of the curve at the knee region, making an idealistic assumption compared to a switch.

## 2-8 BREAKDOWN DIODE

Breakdown diodes are *p–n* junction diodes operated in the reverse-bias mode, as shown in Fig. 2-13.

This breakdown occurs at a critical reverse-bias voltage ($V_{br}$). At this critical voltage the reverse current through the diode increases sharply, and relatively large currents flow with little increase in voltage, as shown in Fig. 2-14. These diodes are designed with sufficient

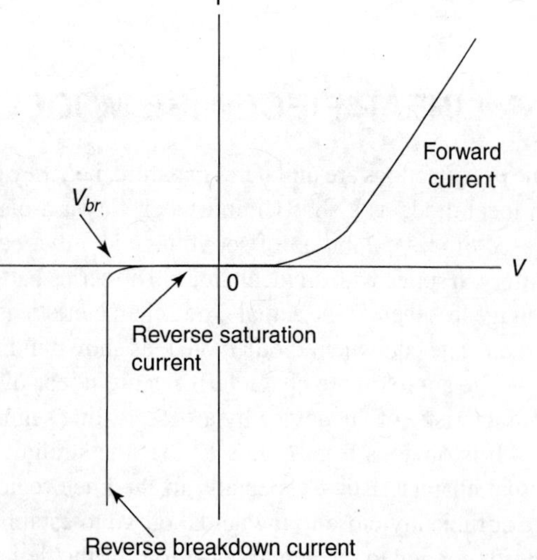


**Figure 2-14** Reverse breakdown in a *p–n* junction

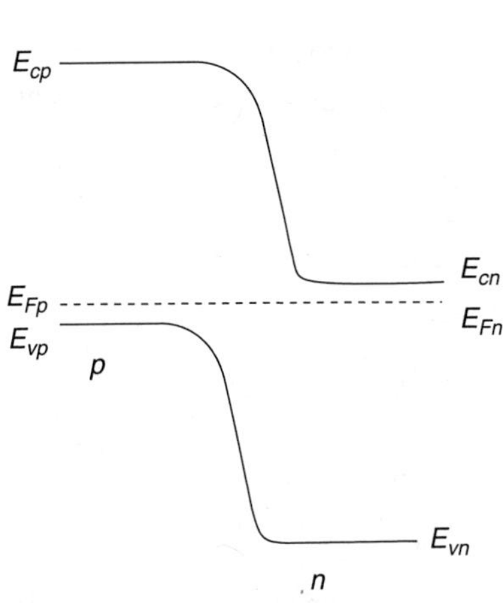


**Figure 2-15(a)** Energy band diagram of a Zener diode

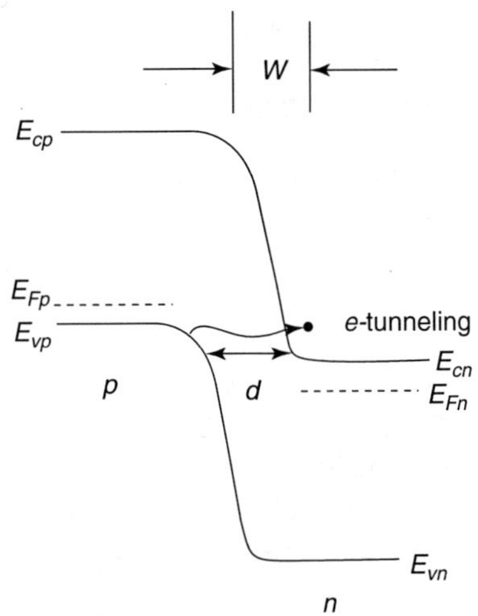


**Figure 2-15(b)** Reverse bias with electron tunneling from *p* to *n* leads to Zener breakdown

power-dissipation capabilities to work in the breakdown region. The following two mechanisms can cause reverse breakdown in a junction diode.

## 2-8-1 Zener Breakdown

Zener breakdown occurs when a sufficiently large reverse-bias is applied across a *p–n* junction diode. The resulting electric field at the junction imparts a very large force on a bound electron, enough to dislodge it from its covalent bond. The breaking of the covalent bonds produces a large number of EHP (electron–hole pairs). Consequently the reverse current becomes very large. This type of breakdown phenomena is known as *Zener breakdown.*

An essential criterion for the Zener mechanism to occur is the existence of a sufficiently thin depletion region, enabling the applied potential to create a high field at the junction, unlike an ordinary diode. This needs high doping and the Zener breakdown voltage tends to zero on increasing the doping. Figure 2-15(a) exhibits the energy band diagram of a Zener diode.

Zener breakdown occurs in a narrow depletion region due to which quantum mechanical tunneling takes place. It is purely a quantum mechanical process, (with no classical analogy) which states that particles with lower energy compared to the barrier energy in front of them can penetrate to the other side of the barrier, thus having a tunneling effect. When the reverse-bias is increased, the valance band of the *p*-side slowly moves up over the conduction band of the *n*-side and the tunneling probability increases as the number of filled states of the *p*-side moves up over *n*-side. Electron tunneling from the *p*-type valence band to the *n*-type conduction band produces a reverse current known as the *Zener effect*. Figure 2-15(b) shows the energy band diagram under reverse-biased condition with electron tunneling from *p* to *n* causing Zener breakdown.

The Zener breakdown voltage decreases with increasing temperature; the temperature coefficient of the Zener breakdown voltage is negative. Figure 2-15(c) exhibits the I–V curve for the Zener diode.

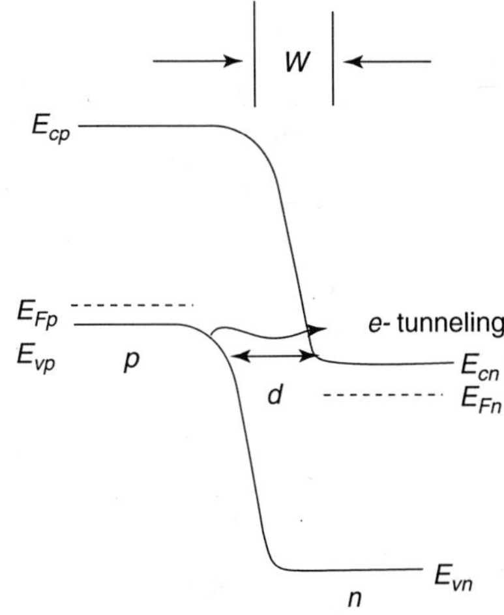


**Figure 2-15(c)** I–V characteristics

## 2-8-2 Avalanche Breakdown

In a reverse-biased junction, the minority-carriers drift across the depletion region. On their way across this region, they occasionally have collisions with atoms in the lattice. With a large enough field, a carrier drifting across the depletion region is accelerated to the point where it has enough energy to knock a valance electron free from its host atom during a collision. The field then separates the electron and hole of this newly created EHP and we now have three mobile carriers instead of one. This process is called avalanche multiplication. The multiplication can become quite large if the carriers generated by this collision also acquire to create more carriers, thereby initiating a chain reaction. Once the process starts, the number of multiplication that can occur from a single collision increases rapidly with further increase in the reverse-bias, so the terminal current grows rapidly, and we say that the junction breaks down. This is called *avalanche breakdown.*

Under the condition of low carrier concentration, the breakdown voltage is larger and avalanche multiplication is a dominant factor since the electric field required for Zener breakdown is much higher and the avalanche breakdown sets in before that can be achieved. In spite of the two different breakdown mechanisms, the breakdown diodes are commonly referred to as Zener diodes.

Under the condition of low doped semiconductors, the electron tunneling is negligible. The breakdown in such cases involves impact ionization of atoms by energetic carriers. For a large electric field $E_o$ inside the depletion region, an electron entering from the *p*-type semiconductor may acquire the energy needed to cause an ionizing collision with the lattice atom creating an EHP. A single such event results in multiplication of carriers; the original electron as well as the secondary electron are swept to the *n*-type semiconductor, while, the generated hole is swept to the *p*-type semiconductors as exhibited in Fig. 2-16. A very high degree of multiplication can be achieved if the carriers generated within the depletion zone also have ionizing collisions with the lattice; that is, an incoming carrier creates a new EHP. Again, each new carrier creates an EHP, and the process continues. This is called *avalanche process*, since each incoming carrier can create a large number of new carriers resulting in an avalanche of carriers.

We shall present a simplified analysis of this avalanche multiplication of carriers. Let us assume that a charge carrier, while being accelerated through the depletion zone of width *L*, as shown in Fig. 2-17 has a probability *P* of creating an EHP by undergoing ionizing collision with the lattice. Then for $n_{in}$ incoming electrons entering the depletion region from the *p*-region, $Pn_{in}$ secondary EHPs will be generated. Now these generated electrons move to the *n*-region while the generated holes travel to the *p*-region under the electric field. As the total distance traversed by this EHP is still W, they in turn generate new EHPs with the same probability. In all, assuming no recombination, the total number of electrons coming out of the depletion region in the *n*-region can be expressed as

$$n_{out} = n_{in}(1 + P + P^2 + \cdots) \tag{2-73}$$

Therefore, by using the summation formula for an infinite geometric progression series, the multiplication factor is given by:

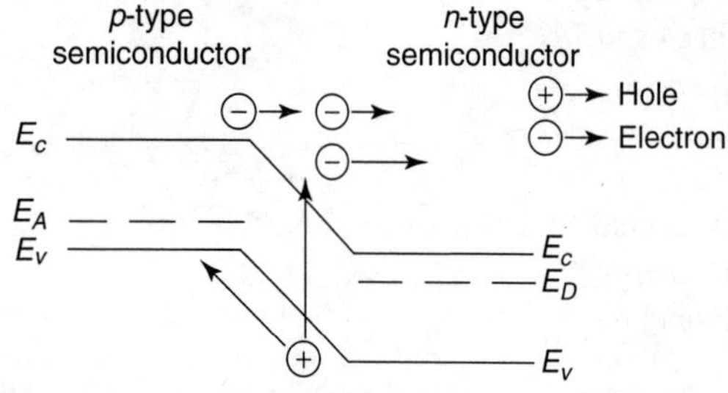


**Figure 2-16** Avalanche breakdown in low doped semiconductor

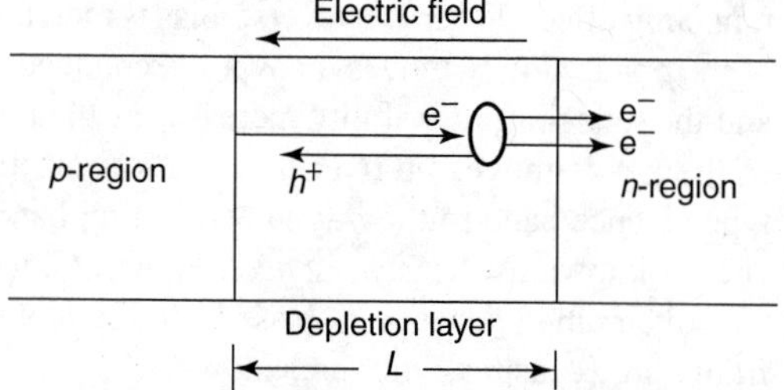


**Figure 2-17** Carrier multiplications in the depletion region due to impact ionization

$$M = \frac{n_{\text{out}}}{n_{\text{in}}} = 1 + P + P^2 + \cdots = \frac{1}{1 - P} \quad (2\text{-}73)$$

The probability of an ionizing collision as a carrier travels through the depletion region can also be expressed as:

$$P = \int_0^L \alpha dx \quad (2\text{-}74)$$

where $\alpha$ is known as the ionization coefficient.

For the avalanche process to be self-sustaining, $M$ should be infinite and thus, we can write:

$$P = \int_0^l \alpha dx = 1 \quad (2\text{-}75)$$

where, the dependence of $\alpha$ on the electric field can, in general, be expressed as

$$\alpha = \alpha_0 \exp\left[-\left(\frac{b}{E_0}\right)^t\right] \quad (2\text{-}76)$$

where, the constants $\alpha_0$, $b$, and $t$ are characteristics of the particular semiconductor.

It must be pointed out that the present analysis is the oversimplified version of the reality and actually the ionization probability is related to the junction parameters in a much more complicated fashion. Qualitatively, we can expect the ionization probability to increase with increasing electric field and therefore, on the application of reverse-bias the ionization probability increases. A widely used empirical relation between the multiplication factor $M$ and the applied reverse voltage near breakdown ($V_{BR}$) is given by:

$$M = \frac{1}{\left[1 - \left(\frac{V}{V_{BR}}\right)^t\right]} \quad (2.77)$$

where, $t$ varies between 3 and 6 depending on the semiconductor material. In general, the critical reverse voltage for breakdown increases with increasing values of band gap, since more energy is required for ionization in the case of larger band gap materials.

From the breakdown conditions described so far and the field dependence of the ionization coefficient, the critical electric field $E_c$ at which the avalanche process and breakdown occurs is given by:

$$V_{BR} = \frac{E_c l}{2} = \frac{\varepsilon_s E_c^2}{2eN_d} \quad (2\text{-}78)$$

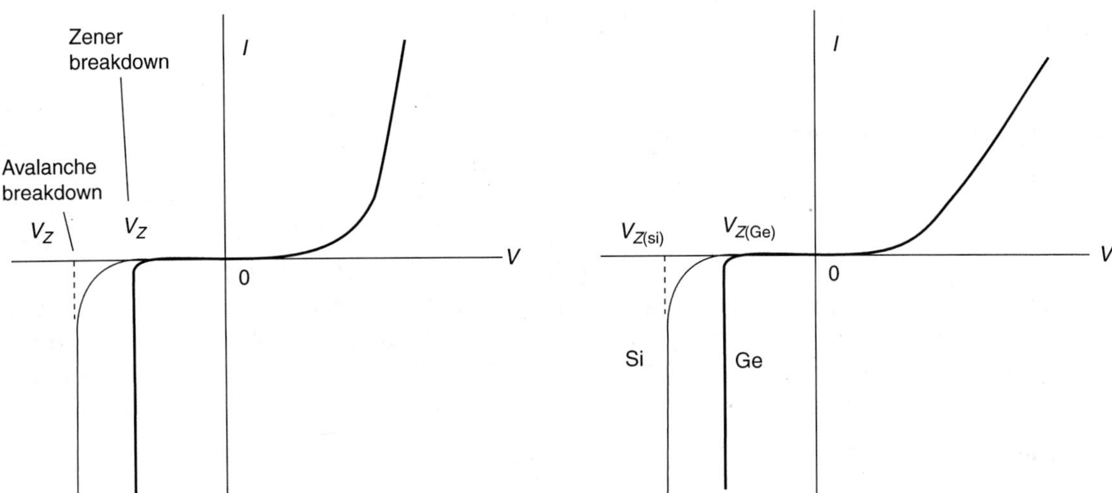


**Figure 2-18** The I–V characteristics comparison between Zener and avalanche breakdown

**Figure 2-19** Comparison of Zener breakdown of Ge and Si semiconductor diodes with respect to I–V curve

**Table 2-1** Comparison between Zener and Avalanche breakdown

| *Zener Breakdown* | *Avalanche Breakdown* |
|---|---|
| 1. Narrow depletion region and quantum mechanical tunneling takes place. | 1. Higher depletion region width and electron tunneling is negligible. |
| 2. Highly doped diode with reverse-bias is required. | 2. Low doped diode with reverse-bias is sufficient. |
| 3. Operates at low voltage up to few volts reverse-bias. | 3. Breakdown occurs at high reverse-bias from a few volts to thousands of volts. |
| 4. Impact ionization does not occur in this case. | 4. This breakdown mechanism involves the impact ionization of host atoms by energetic carriers. |

Thus, we can infer from Eq. (2-78) that for junctions where breakdown takes place due to the avalanche process, the breakdown voltage increases with reduction in the doping concentration. The comparison between Zener and avalanche breakdown with respect to the I–V characteristics has been shown in Figs. 2-18 and 2-19 respectively.

## Solved Examples

**Example 2-31** For a Zener shunt regulator if $V_z = 10$ V, $R_s = 1$ kΩ, $R_L = 10$ kΩ, and the input voltage varies from 25 to 40 V. Find the maximum and minimum values of Zener current.

**Solution:**
Given:

$$V_Z = V_0 = 10\text{ V}, R_s = 1\text{ k}\Omega, R_L = 10\text{ k}\Omega, V_i = 25\text{ to }40\text{ V}.$$

$$V_{i(\min)} = 25\text{ V} \quad \text{and} \quad V_{i(\max)} = 40\text{ V}$$

Maximum value of Zener current:

$$I_z = \frac{V_{I(\max)} - V_z}{R_{L(\min)}}$$

$$= \frac{40 - 10}{1000} - 5 \times 10^{-3} = 25\text{ mA}$$

Minimum value of Zener current is given by:

$$I_{z(\min)} = \frac{V_{i(\min)} - V_z}{R_s} - I_{L(\max)}$$

$$= \frac{25 - 10}{1000} - 5 \times 10^{-3} = 10\text{ mA} \quad (\because I_{L(\min)} = I_{L(\min)} = I_L)$$

**Example 2-32** A 5 V Zener diode has a maximum power dissipation of 250 mW. It maintains a constant voltage when the current through the diode does not fall below 10% of the maximum permissible current. A 15 V supply is given to the Zener through a series resistor *R*. Find the range for *R* so that the Zener maintains its constant voltage. Find the new range when the diode is loaded by 50 W loads.

**Solution:**

$$\text{Maximum permissible current} = \frac{250 \times 10^{-3}}{5} = 50\text{ mA}$$

$$10\%\text{ of }50\text{ mA} = 5\text{ mA}$$

Maximum current through the diode to maintain constant voltage $= 50 - 5 = 45$ mA

**Example 2-33** In a $p^+n^+$ diode, the doping concentration of the $n$-region is $2 \times 10^{15}\ \text{cm}^{-3}$. If the critical field at the avalanche breakdown is $1.5 \times 10^5$ Volt/cm. Find out the breakdown voltage assuming the width of the $n$-region is 10 $\mu$m.

**Solution:**

For this junction diode the depletion region exists only in the $n$-region.
At breakdown the peak electric field is $|E_P| = 1.5 \times 10^5$ V/cm
Therefore the width of the depletion region $W$ is given by:

$$W = \frac{|E_P|\,\varepsilon}{eN_D} = \frac{1.5 \times 10^5 \times 11.9 \times 8.854 \times 10^{-14}}{1.6 \times 10^{-19} \times 2 \times 10^{15}} = 4.939\ \mu\text{m}$$

Therefore, the breakdown voltage:

$$V_{BR} = \frac{|WE_P|}{2} = \frac{1.5 \times 10^5 \times 4.939 \times 10^{-4}}{2} = 37.04\ \text{V}$$

**Example 2-34** (a) Prove that the magnitude of the maximum electric field $E_m$ at a step-graded junction with $N_A >> N_D$ is given by:

$$E_m = \frac{2V_j}{W}$$

(b) It is found that Zener breakdown occurs when $E_m = 2 \times 10^7$ V/m $= E_z$. Prove that Zener voltage $V_Z$ is given by:

$$V_z = \frac{\varepsilon E_z^2}{2eN_D}$$

**Solution:**

(a) We have: $$E(x) = \frac{-eN_D}{\varepsilon}(x - W)$$

Obviously, $E(x)$ is maximum at $x = 0$.

Hence, $$E_{\max} = \frac{eN_D}{\varepsilon} W$$

We have: $$V_j = \frac{eN_D}{2\varepsilon} W^2$$

Thus, $$E_{\max} = \frac{2V_j}{W}$$

(b) We have: $$W = \left(\frac{2\varepsilon}{eN_D} V_j\right)^{1/2}$$

Substituting this value in the result of part (a), we obtain:

$$e_{\max} = \frac{2V_j}{W} = \frac{2V_j}{\left(\frac{2\,\varepsilon}{eN_D} V_j\right)^{1/2}} = \frac{2V_j^{1/2}}{\left(\frac{2\,\varepsilon}{eN_D}\right)^{1/2}}$$

Solving for $V_j$ we obtain: $$V_j = \frac{\varepsilon}{2eN_D} E_{\max}^2 = V_0 - V_d = V_z$$

**Example 2-35** (a) Zener breakdown occurs in Ge at a field intensity of $2\times10^7$ V/m. Prove that the breakdown voltage is $V_Z = 51/\sigma_p$, where $\sigma_p$ is the conductivity of the $p$ material in $(\Omega\text{cm})^{-1}$. Assume that $N_A << N_D$.
(b) If the $p$-material is essentially intrinsic, calculate $V_Z$.
(c) For a doping of 1 part in $10^8$ of $p$-type material, the resistivity drops to 3.9 Ωcm. Calculate $V_Z$.
(d) For what resistivity of the $p$-type material will $V_Z = 1.5$ V?

**Solution:**
(a) From Example 2–34: we can write:

$$V_z = \frac{\varepsilon}{2eN_A} E_z^2.$$

We know, $\sigma_p = N_A e\mu_p$

or,

$$eN_A = \frac{\sigma_p}{\mu_p} \quad \text{Thus,} \quad V_z = \frac{\varepsilon E_z^2 \mu_p}{2\sigma_p}$$

$$V_z = \left(\frac{16}{36\pi \times 10^9}\,\text{F/m}\right) \times \left(\frac{4 \times 10^{14}}{2}\,\text{V}^2/\text{m}^2\right) \times \left(1800\,\frac{\text{cm}^2}{\text{V}-\text{sec}}\right) \times \left(10^{-6}\,\frac{\text{m}^3}{\text{cm}^3}\right) \times \frac{1}{\sigma_p}(\Omega - \text{cm})$$

$$V_z = \frac{51}{\sigma_p} \text{ where } \sigma_p \text{ is in } (\Omega - \text{cm})^{-1}$$

(b)

$$\sigma_i = \frac{1}{45}$$

∴

$$V_z = 51 \times 45 = 2300 \text{ V}$$

(c)

$$\sigma_p = \frac{1}{3.9}(\Omega - \text{cm})^{-1}$$

∴

$$V_z = 51 \times 3.9 = 198.9 \text{ V}$$

(d) From part (a), $I = 6 \times 10^{-6}(e^{100/26} - 1) = 6 \times 10^{-6}(46.5 - 1)$

**Example 2-36** (a) Two $p$–$n$ Ge diodes are connected in series opposing. A 5 V battery is impressed upon this series arrangement. Find the voltage across each junction at room temperature. Assume that the magnitude of the Zener voltage is greater than 5 V.
(b) If the magnitude of the Zener voltage is 4.9 V, what will be the current in the circuit? The reverse saturation current is 6 $\mu$A.

**Solution:**
The current is same in each diode but in one it is in the reverse direction. To obtain a forward current equal to the reverse saturation current requires a very small voltage. We have, $I = I_0\,(e^{V/\eta V_T} - 1) = I_0$. For Ge, $\eta = 1$, and at $T = 300$ K, $V_T = 0.026$ V. Thus, $e^{V/V_T} = 2$ or, $V = 0.693 \times 0.026 = 0.018$ V. Hence, the voltage across the second diode is $5 - 0.018 \approx 4.98$ V. Since this voltage is in the reverse direction, current $I = I_0\,(e^{4.98/0.026} - 1) \approx -I_0$. This confirms the above assumption that saturation current flows.
(b) If the Zener voltage is 4.9 V, then the voltage across the reverse-biased diode is 4.9 V. This leaves 0.1 V across the forward-biased diode. If $I_0 = 5\mu$A, then

$$I = 6 \times 10^{-6}(e^{100/26} - 1) = 6 \times 10^{-6}(46.5 - 1) = 273.6\,\mu\text{A}.$$

**Example 2-37** The diode current of a $p$–$n$ junction diode is 0.5 mA. at 340 mV and again 15 mA at 465 mV. Assuming $k_BT/e = 5$ mV. find out the ideality factor.

**Solution:**
We know that

$$I = I_s\left(e^{\frac{eV}{\eta K_B T}} - 1\right)$$

Since $e^{\frac{eV}{\eta K_B T}} >> 1$, we can write $I = I_s\left(e^{\frac{eV}{\eta K_B T}}\right)$

Therefore, according to the question we get:

$$\frac{1.5\text{mA}}{0.5\text{mA}} = \frac{\exp\left(\frac{465}{25\eta}\right)}{\exp\left(\frac{340}{25\eta}\right)}$$

or,
$$30 = \exp\left(\frac{5}{\eta}\right)$$

or,
$$\frac{5}{\eta} = 2.303\ \log_{10}(30)$$

or,
$$\eta = \frac{5}{3.4} = 1.47$$

**Example 2-38** A series combination of a 12 V avalanche diode and a forward-biased Si diode is to be used to construct a zero-temperature-coefficient voltage reference. The temperature coefficient of the Si diode is −1.7 mV/°C. Express in percent per degree centigrade the required temperature coefficient of the Avalanche diode.

**Solution:**
For the Si diode, the temperature coefficient = −1.7mV/°C. If the series combination is to have a zero-temperature coefficient, then the temperature coefficient of the avalanche diode must be at the biasing current +1.7 mV/°C. In percentage, temperature coefficient:

$$\frac{1.7 \times 10^{-3}\ \text{V}}{12\text{V}}(100\%) = 0.0142\%/^\circ\text{C}$$

**Example 2-39** (a) The avalanche diode regulates at 60 V over a range of diode currents from 5 to 40 mA. The supply voltage $V$ = 200 V. Calculate $R$ to allow voltage regulation from a load current $I_L = 0$ up to $I_{max}$, the maximum possible value of $I_L$. What is $I_{max}$?

(b) If $R$ is set as in part (a) and the load current is set at $I_L$ = 25 mA, what are the limits between which $V$ may vary without loss of regulation in the circuit?

**Solution:**
(a) For $I_L = 0$ and $V_0 = 60$, we have $I_D = I_T = \frac{200 - 60}{\text{R}} \leq 40$ mA

or,
$$R \geq \frac{140}{40} \times 10^3 = 3.5\ \text{K}$$

For $I_L = I_{max}$:
$$I_D \leq 5\ \text{mA},\ R = 3.5\text{K},\ I_T = 40\ \text{mA}$$

Hence,
$$I_{max} = 40 - 5 = 35\ \text{mA}$$

(b) Minimum current $I_D$ for good regulation is 5 mA. Hence, $I_T = 25 + 5 = 30$ mA and $V_{max} = 30 \times 3.5 + 60 = 165$ V. Maximum current $I_D$ for good regulation is 40 mA. Hence, $I_T = 65$ mA and $V_{max} = 65 \times 3.5 + 60 = 287.5$ V.

## 2-9 SPECIAL TYPES OF *p–n* JUNCTION SEMICONDUCTOR DIODES

In this section, we shall discuss the tunnel, light-emitting, photo detector and photovoltaic diodes. These types of diodes find extensive applications in different areas of electronics in general.

### 2-9-1 Tunnel Diode

The tunnel diode is a negative-resistance semiconductor *p–n* junction diode. The negative resistance is created by the tunnel effect of the electrons in the *p–n* junction as already discussed in the section of Zener diode. The doping of both the *p*- and *n*-type regions of the tunnel diode is very high—impurity concentration of $10^{19}$ to $10^{20}$ atoms/cm$^3$ are used (which means both *n*-type and *p*-type semiconductors having parabolic energy bands are highly degenerate)—and the depletion layer barrier at the junction is very thin, in the order of $10^{-6}$ cm. Quantum mechanically, if the barrier is less than 3 Å there is an appreciable probability that particles will tunnel through the potential barrier even though they do not have enough kinetic energy to pass over the same barrier. In addition to the barrier thinness, there must be filled energy states on one side from which the particles will tunnel and allowed energy states on the other side into which the particles will penetrate through at the same energy level. This phenomenon is called the tunnel effect and the diode that operates on the tunneling phenomena is called a tunnel diode. This is also known as *Esaki diode*, named so after Leona Esaki who invented this diode in 1956 and later became a Noble laureate in 1973 for this invention. It may be noted that the Figs. 2-20(a) to 2-20(e) exhibit the energy band diagram of the tunnel diode under five different conditions which are described as follows.

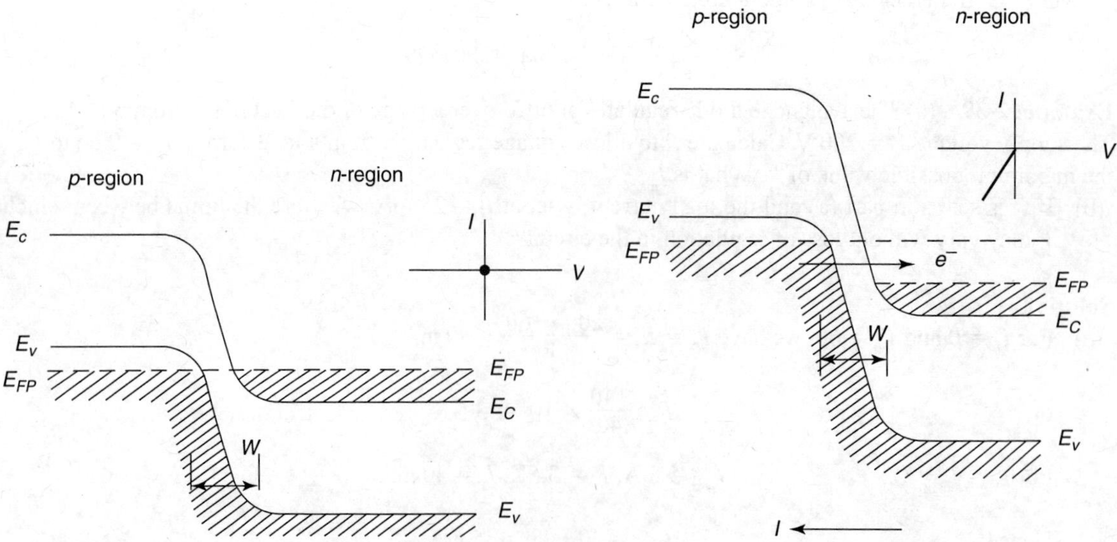


**Figure 2-20(a)** Tunnel diode under zero bias equilibrium **Figure 2-20(b)** Small reverse bias

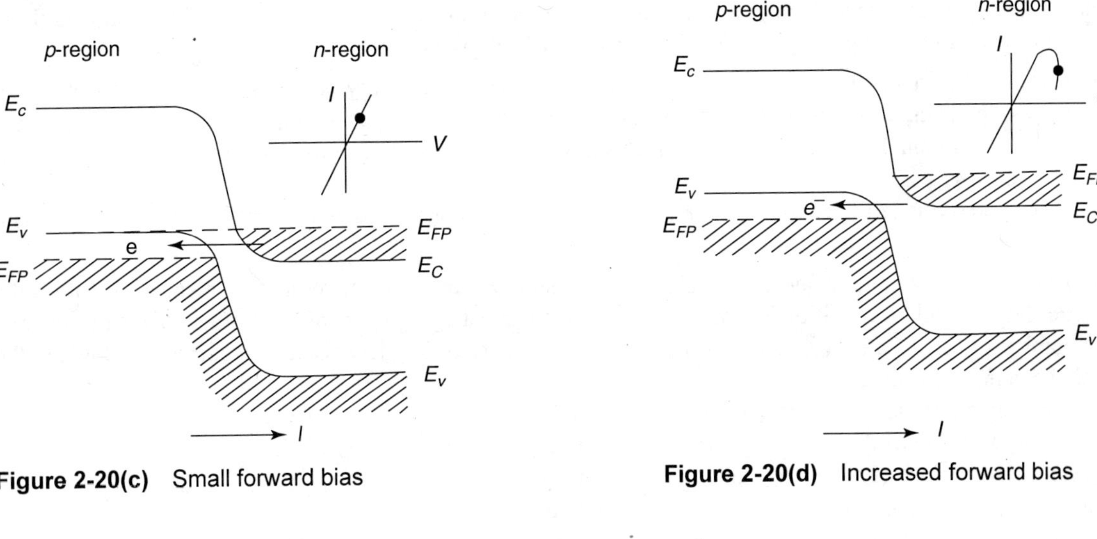


**Figure 2-20(c)** Small forward bias

**Figure 2-20(d)** Increased forward bias

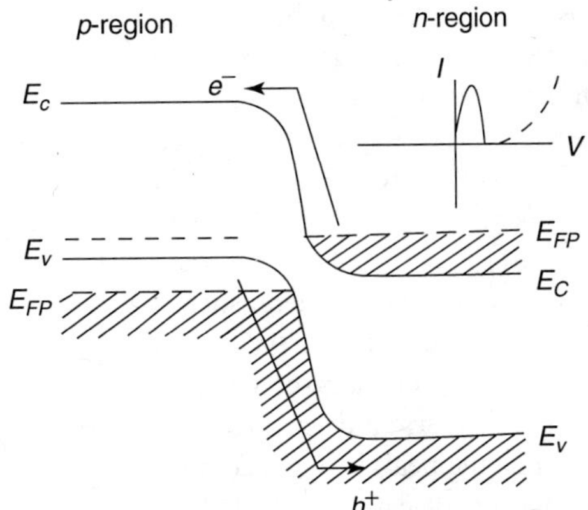


**Figure 2-20(e)** Increased forward bias condition where the current begins to increase again.

**Step I:** This is the equilibrium condition and the Fermi level is constant throughout the junction. The Fermi energy for *p*-type semiconductors ($E_{FP}$) lies below the valence band edge on the *p*-side and lies above the conduction band edge on the *n*-side (both under the conditions of carrier degeneracy). Thus, the bands must overlap on the energy scale in order for $E_F$ to be constant. This overlapping of bands means that with a small reverse-bias or forward-bias, filled states and empty states appear opposite to each other only separated by the width of the depletion region. Under this condition, the upper levels of electron energy of both *p*- and *n*-type are lined up at the same Fermi level. Since there are no filled states on one side of the junction that are at the same energy level as empty allowed states on the other side, there is no net flow of charge in either direction across the junction, and the current is zero as there is equal tunneling from *n* to *p* and from *p* to *n* giving a zero net current.

**Step II:** Since the bands overlap under equilibrium condition, only a small reverse-bias allows electrons to tunnel or move from the filled valence band to the empty conduction band.

This condition is similar to the Zener effect except that no bias is needed to create the condition of overlapping bands. As the reverse-bias is increased, $E_{FP}$ continues to move down on the *n*-region with respect to the *p*-region, placing more filled states on the *p*-side opposite to the empty states on the *n*-side.

The tunneling of electrons from *p* to *n* thus, increases with an increase in the reverse-bias. The resulting conventional current is opposite to that of electron flow from *p* to *n*.

**Step III:** When a small forward bias is applied $E_{FP}$ moves up in energy on the *n*-side with respect to that on the *p*-side. Thus, electrons below $E_{FP}$ on the *n*-side are placed opposite empty states above $E_{FP}$ on the *p* side. Electron tunneling occurs from *n* to *p* as shown with the resulting conventional current flowing from *n* to *p* as shown.

This type of tunneling continues to increase with increased bias as more filled states are placed opposite empty states.

**Step IV:** As the voltage is increased, EFP on the *n*-side continues to move up with respect to that on the *p*-type and gradually a condition is reached at which the bands begin to pass by each other. When this occurs, the number of filled states from empty states decreases. The resulting decrease in the current is shown in the I–V characteristics curve.

This region of the I–V characteristics curve is important as it shows the decrease of the tunneling current with increase in the applied forward voltage and hence depicts the negative resistance region; i.e., the dynamic resistance of *dV*/*dI* is negative.

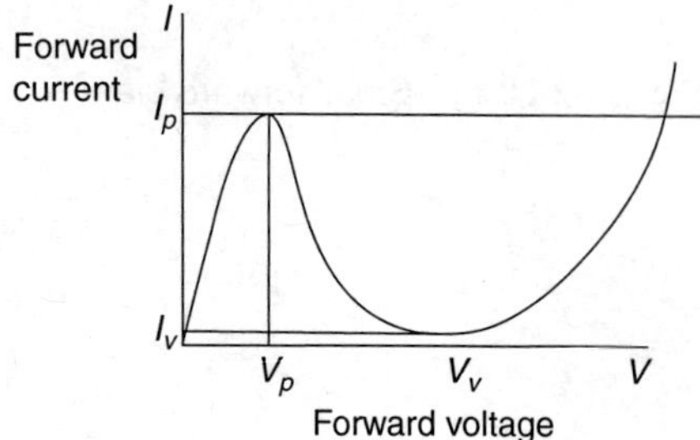


**Figure 2-21 (a)** I–V characteristics of a tunnel diode

**Step V:** When the applied bias is increased beyond the negative resistance region, the current begins to increase again. Once the bands have passed each other the characteristics resemble that of a conventional diode. The forward current is now dominated by the diffusion current. The diffusion current is present in the forward tunneling region, but it is negligible compared to the tunneling current.

The total tunnel diode characteristics have the general shape of a tilted N; therefore it is a common practice to refer to the curve as type N negative resistance. It is also called a voltage-controlled negative resistance, which means that the current decreases rapidly at some critical voltage, known as the peak voltage. As the tunneling process is very fast, tunnel diodes can be operated at microwave frequencies. The I–V curve, the symbol and the linear equivalent circuit of the tunnel diode are shown in Figs. 2.21 (a), (b) and (c) respectively.

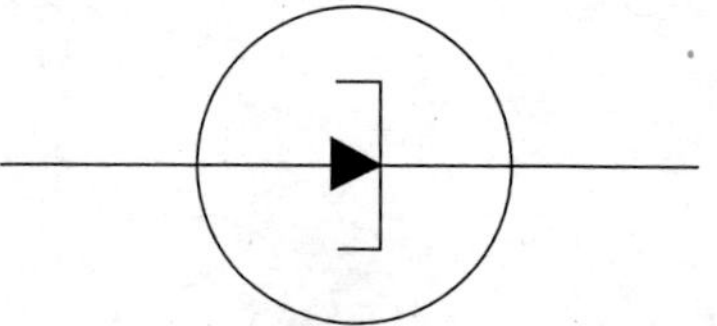

**Figure 2-21(b)** Symbol of tunnel diode.

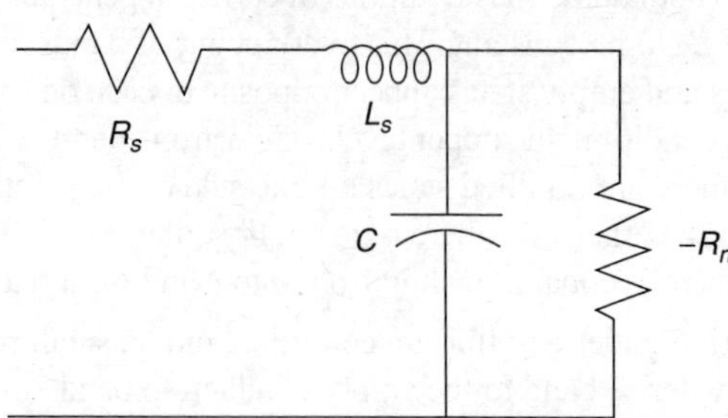


**Figure 2-21(c)** Small-signal model of the tunnel diode. (Typical values for these parameters for a tunnel diode of peak current $I_p = 10$ mA are $-R_n = -30\ \Omega$, $R_s = 1\ \Omega$, $L_s = 5$ nH and capacitance $C = 20$ pF respectively).

## 2-9-2 Light-Emitting Diode

Charge carriers recombination takes place at the *p–n* junction as electron crosses from the *n*-side and recombines with holes on the *p*-side. When the junction is forward-biased the free electron is in the conduction band and is at a higher energy level than the hole located at valence band. The recombination process involves radiation of energy in the form of photons. If the semiconductor material is translucent, the light will be emitted and the junction becomes a light source, *i.e.,* a light-emitting diode (LED). LEDs are *p–n* junctions that can emit spontaneous radiation in *ultraviolet, visible, or infrared regions.*

In other words LED is a diode, which generates visible light from the region of the depletion layer when it is forward-biased. $GaAs_{1-y}P_y$ is the most preferred material for a visible LED. The phosphorus mole fraction in this ternary compound is denoted by *y*, i.e., as *y* is increased, more and more phosphorus atoms replace arsenic in the crystal lattice. For $0 < y < 0.45$, the band gap of the material is direct and increases from 1.424 eV (at $y = 0$) to 1.977 eV (at $y = 0.45$). In Si and Ge, greater percentage of released energy is given up in the form of heat and the emitting light is insignificant. The efficiency of light generation increases with an increase in the injected current and with a decrease in temperature. Since LED is a *p–n* junction device, it has forward current vs forward voltage characteristics similar to that of a diode. LEDs are available in different colours. LEDs have fast response and offer good contrast ratios of visibility. The lifetime of LEDs is high, exceeding 106 hours.

### Advantages of LEDs

1. Low operating voltage, current and power consumption make LEDs compatible with electronic drive circuits.
2. LEDs exhibit high resistance to mechanical shock and vibration and allow them to be used in severe environment conditions.
3. LEDs ensure a longer operating life line, thereby improving the overall reliability and lowering the maintenance costs of equipment.
4. LEDs have low inherent noise levels and also high immunity to externally generated noise.
5. LEDs exhibit linearity of radiant power output with forward current over a wide range.

### Limitations of LEDs

1. Temperature dependence of radiant output power and wavelength.
2. Sensitivity to damages by over voltage or over current.
3. Theoretical overall efficiency is not achieved except in special cooled or pulsed conditions.

### Operation of LEDs

The dominant operating process for LEDs is spontaneous emission. A photon of appropriate energy can be absorbed by a semiconductor, creating an EHP in the process. This is called optical absorption. Let us consider Fig. 2-22 (a) which depicts two energy levels in a semiconductor $E_1$ and $E_2$ where $E_1$ corresponds to the ground state and $E_2$ to the excited state. At room temperature, most of the electrons are in ground state.

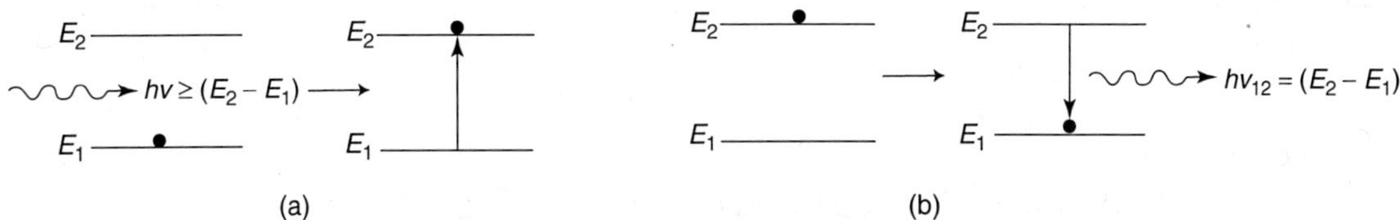


**Figure 2-22** (a) Schematic showing the basic process of absorption (b) emission

When a photon of frequency greater than, or equal to, $\nu_{12} = (E_2 - E_1)/h$ is incident on the system, an electron in the ground state absorbs it and goes to the excited state. However, the excited state is unstable. So, after a short time, without any external stimulus, the electron comes back to the ground state emitting a photon of energy $h\nu_{12}$. The emitted wavelength $\lambda$ is given by:

$$\lambda = \frac{hc}{E_g}$$

where, $E_g$ is the band gap of the semiconductor.

This process is referred to as spontaneous emission and is schematically represented in Fig. 2-22 (b).

Infrared LEDs are used in fibre-optic communication systems where silica fibres are used to guide the optical signal over long distances. An important application of infrared LED is in opto-isolators where an input electrical signal is applied to the LED. Light is generated and subsequently detected by a photodiode and converted back to an electrical signal as a current flowing through a load resistor. Opto-isolators allow signal transmission at the speed of light and are electrically isolated. In this context, it may be noted that the emitted wavelength $\lambda$ is given by:

$$\lambda = \frac{hc}{E_g}$$

where, $E_g$ is the band gap of the semiconductor. The probability of direct (radiative) transition is high in direct band gap semiconductors. Hence, $GaAs_{1-y}P_y$ ($y < 0.45$) is used for light emission in the wavelength range of 627–870 nm. For $y > 0.45$, the material has an indirect band gap. So, special recombination centres have to be introduced to facilitate radiative recombination. Incorporation of nitrogen results in the formation of such a recombination centre. It introduces an electron trap level very close to the bottom of the conduction band and greatly enhances the probability of radiative recombination. In general, red LEDs are fabricated on GaAs substrates while orange, green, and yellow LEDs are fabricated on GaP substrates on which a graded $GaAs_{1-y}P_y$ layer is grown by epitaxy. In optical communications, to take advantage of the 1.3-$\mu$m and 1.55-$\mu$m low-loss windows in optical fibers, InGaAsP substrates are used.

At a low forward voltage, the LED current is dominated by the non-radiative recombination current, mostly due to surface recombination. At higher forward voltages, the radiative diffusion current dominates and light is emitted as the injected minority-carriers recombine with the majority-carriers through a radiative-recombination process. Finally, at very high forward voltages, the series resistance limits the current. Figure 2-23 shows the symbol of an LED.

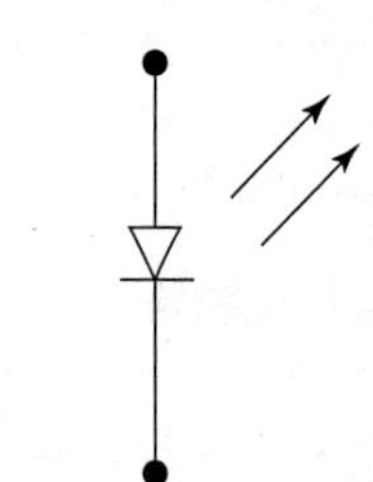

**Figure 2-23** The symbol of an LED

## 2-9-3 Photo Detector Diode

The detector is an essential component of an optical fibre communication system and is one of the crucial elements that dictate the overall system performance. Its function is to convert the received optical signal into an electrical signal, which is then amplified before further processing. The following criteria define important performance and compatibility for the detectors, which are generally similar to the requirements for the sources:

(i) High sensitivity at the operating wavelengths
(ii) High fidelity
(iii) Large electrical response to the received optical signal
(iv) Short response time to obtain a suitable bandwidth

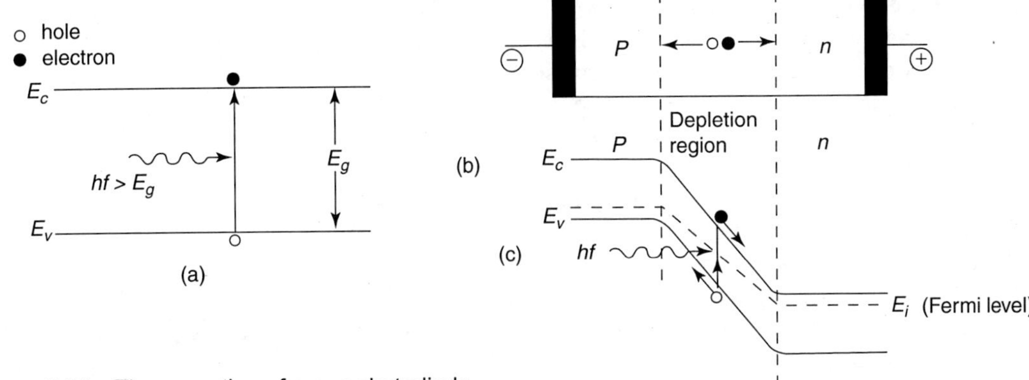


**Figure 2-24** The operation of a *p–n* photodiode

(v) A minimum noise introduced by the detector
(vi) Stability of performance characteristics
(vii) Small size
(viii) Low bias voltages
(ix) High reliability
(x) Low cost

**Detection principles**

The basic detection process in an intrinsic absorber is illustrated in Fig. 2-24. The *p–n* photodiode is reverse-biased and the electric field developed across the *p–n* junction sweeps mobile carriers (holes and electrons) to their respective majority sides (*p*- and *n*-type material). A depletion region or layer is therefore created on either side of the junction. This barrier has the effect of stopping the majority carriers crossing the junction in the opposite direction to the field. However, the field accelerates minority-carriers from both the sides to the opposite side of the junction, forming the reverse leakage current of the diode. Thus, intrinsic conditions are created in the depletion region.

A photon incident in or near the depletion region of this device which has an energy greater than or equal to the band gap energy *Eg*, of the fabricating material will excite an electron from the valence band into the conduction band. This process leaves an empty hole in the valence band and is known as the photo-generation of an electron–hole (carrier) pair. Carrier pairs so generated near the junction are separated and swept (drift) under the influence of the electric field to produce a displacement by current in the external circuit in excess of any reverse leakage current. Figures 2-24 (a), (b) and (c) illustrate the operation of a *p–n* photodiode which can be stated briefly as follows:

(i) Photo-generation of an electron hole pair
(ii) Structure of the reversed biased *p–n* junction illustrating carrier drift in depletion region.
(iii) Energy band diagram of the reversed biased *p–n* junction

## 2-9-4 Photovoltaic Diode

The photovoltaic diode or solar cell is an important technological device for overcoming energy problems. It is also known as solar energy converter; it is basically a *p–n* junction diode which converts solar energy into electrical energy. The energy reaching the earth's surface from the sun is primarily electromagnetic radiation, which covers a spectral range of 0.2 to 0.3 micrometre. The conversion of this energy into electrical energy is called photoelectric effect.

## Construction and working principle

A photovoltaic diode essentially consists of a silicon *p–n* junction diode usually packaged with a glass window on the top. Surface layer of the *p*-material is made extremely thin so that the incident light (photons) can penetrate and reach the *p–n* junction easily, as shown in Fig. 2-25.

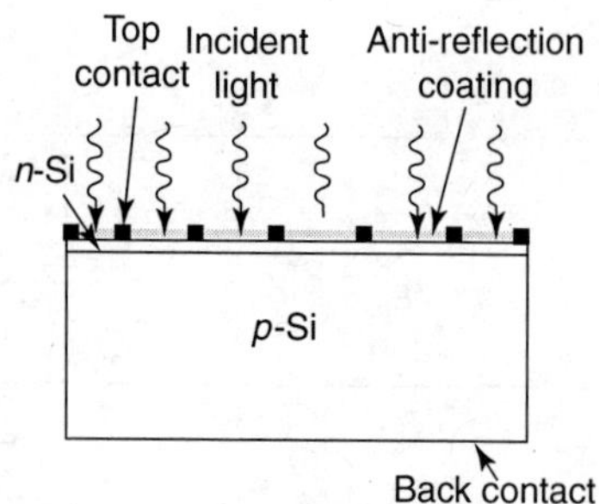


**Figure 2-25** Structure of a solar cell

When these photons collide with the valence electrons, they impart in them sufficient energy so that they gain enough energy to leave the parent atoms. In this way, free electrons and holes are generated on both sides of the junction. Consequently, their flow constitutes a current (minority current). This current is directly proportional to the illumination (lumen/m$^2$ or mW/m$^2$). This, in general depends on the size of the surface being illuminated. The open circuit voltage $V_{oc}$ is a function of illumination. Consequently, power output of a solar cell depends on the level of sun-light illumination. Power cells are also available in the form of a flat strip so as to cover sufficiently large surface areas.

## Current–voltage characteristics

The current voltage characteristic is shown in Fig. 2-26. It is seen that the curve passes through the fourth quadrant and hence the device can deliver power from the curve. We also see that $V_{oc}$ is the maximum voltage obtainable at the load under open-circuit conditions of the diode, and $I_{sc}$ is the maximum current through the load under short-circuit conditions. The power delivered by the device can be maximized by maximizing the area under the curve (see Fig. 2-26) or by maximizing the product ($I_{sc} \times V_{oc}$). By properly choosing the load resistor, output power can be achieved. In the absence of light, thermally generated minority carriers across the junction constitute the reverse saturation current.

To maximize power, we need to maximize both $I_{sc}$ and $V_{oc}$. This can be done by making both the *p*- and *n*-side of the junction heavily doped. The fill factor (*FF*) is defined as $FF = P_m/V_{oc}I_L$ where $P_m$ is the maximum power output of the solar cell. For a well-designed solar cell the *FF* usually lies between 0.7 and 0.8. The efficiency of a

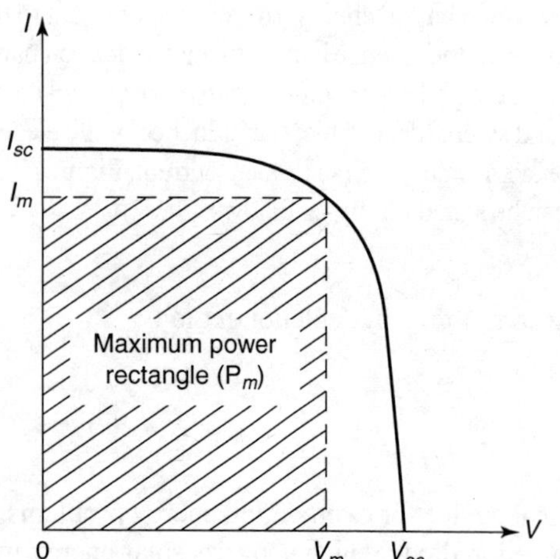


**Figure 2-26** I–V characteristics of an illuminated solar cell showing the point of maximum power

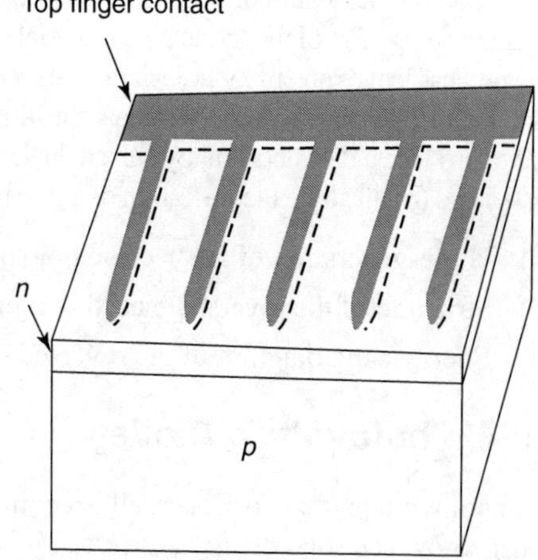


**Figure 2-27** Top finger contact with anti-reflecting coating

solar cell can be written as $\eta = V_m I_m/P_{in} = FF = V_{oc} I_L/P_{in}$ where $V_m$ and $I_m$ are the voltage and the current at the point of maximum power, and $P_{in}$ is the incident optical power.

Thus, to realize a solar cell with high efficiency, it is not only necessary to have high $V_{oc}$ and $I_{sc}$ but also a high *FF*. Solar cells with 15 per cent efficiency are commercially available.

The surface of the solar cell is coated with anti-reflecting coating materials such as $SiO_2$, $TiO_2$ and $Ta_2O_5$ to obtain better conversion efficiency, as shown in Fig. 2-27. Today the solar cell, a non-conventional source of energy of the twenty-first century, has become popular in remote villages and in rural areas. Solar cells are used on board the satellites to recharge their batteries. Since their sizes are small, a large number of cells are required for charging; therefore, series parallel cell combinations are employed for this purpose. Si and Ge are the most widely used semiconductor materials for solar cell but nowadays GaAs is used for better efficiency and better thermal stability.

## Solved Examples

**Example 2-40** For the photovoltaic cell whose characteristics are given in Fig 2-25, find out the power output for different values of the load resistance $R_L$. What is the optimum value of $R_L$?

**Solution:**
The table indicates the Power versus $R_L$ chart.

| $R_L$ | $I$(mA) | $V$(V) | $P(\mu W)$ |
|---|---|---|---|
| 0 | 0.145 | 0 | 0 |
| 800 Ω | 0.14 | 0.12 | 15.4 |
| 3.4 K | 0.10 | 0.35 | 34 |
| 10 K | 0.04 | 0.42 | 16.4 |

The optimum value of $R_L$ = 3.4 K. $P = VI = 0.35\text{ V} \times 0.10\text{ mA} = 35\ \mu\text{W}$. The hyperbola intersects the characteristics at one point. And maximum power is generated at one point only.

■

## 2-10 APPLICATIONS OF DIODE

In the following sections, we shall examine the various applications of the diode. The diode is used in radio demodulation, power conversion, over-voltage protection, logic gates, ionizing radio detectors, charge-coupled devices and measuring temperature,

### 2-10-1 Radio Demodulation

In demodulation of amplitude modulated (AM) radio broadcasts diodes are used. The crystal diodes rectify the AM signal, leaving a signal whose average amplitude is the desired audio signal. The average value is obtained by using a simple filter and the signal is fed into an audio transducer, which generates sound.

## 2-10-2 Power Conversion

In the Cockcroft–Walton voltage multiplier, which converts ac into very high dc voltages, diodes are used. Full-wave rectifiers are made using diodes, to convert alternating current electricity into direct current .

## 2-10-3 Over-Voltage Protection

Diodes are used to conduct damaging high voltages away from sensitive electronic devices by putting them in reverse-biased condition under normal circumstances. When the voltage rises from normal range, the diodes become forward-biased (conducting). In stepper motor, H-bridge motor controller and relay circuit's diodes are used to de-energize coils rapidly without damaging voltage spikes that would otherwise occur. These are called a fly-back diodes. Integrated circuits also use diodes on the pins to protect their sensitive transistors from damaging external voltages. At higher power, specialized diodes are utilized to protect sensitive electronic devices from over-voltages.

## 2-10-4 Logic Gates

AND and OR logic gates are constructed using diodes in combination with other components. This is called diode logic.

## 2-10-5 Ionizing Radiation Detectors

In addition to light, energetic radiation also excites semiconductor diodes. A single particle of radiation, having very high electron volts of energy, generates many charge carrier pairs, as its energy is transmitted in the semiconductor material. If the depletion layer is large enough to catch the whole energy or to stop a heavy particle, an accurate measurement of the particle's energy is possible, simply by measuring the charge conducted and excluding the complexity of using a magnetic spectrometer. These semiconductor radiation detectors require efficient charge collection and low leakage current. They are cooled by liquid nitrogen. For longer range (of the order of a centimetre) particles the requirements are very large depletion depth and large area. Short range particles require any contact or un-depleted semiconductor on at least one surface to be very thin. The back-bias voltages are near breakdown (of the order of a thousand volts per centimetre). Common materials are Ge and Si. Some of these detectors sense both position as well as energy. Due to radiation damage, they have a finite life, especially when detecting heavy particles. Semiconductor detectors for high energy particles are used in large numbers.

## 2-10-6 Temperature Measuring

The forward voltage drop across the diode depends on temperature. A diode can be used as a temperature measuring device. This temperature dependence follows from the Shockley ideal diode equation and is typically around –2.2 mV per degree Celsius.

## 2-10-7 Charge-Coupled Devices

Arrays of photodiode, integrated with readout circuitry are used in digital cameras and similar units.

## POINTS TO REMEMBER

1. When donor impurities are introduced into one side and acceptors into the other side of a single crystal of a semiconductor through various sophisticated microelectronic device-fabricating techniques a *p–n* junction is formed.
2. A *p–n* junction is said to be in thermal equilibrium when it is at a uniform temperature and no external disturbances, such as light or a bias voltage, are acting on it.
3. The thin layer on each side of the *p–n* junction has no free electrons or holes. This thin layer is depleted of mobile carriers and is called depletion layer.
4. The nature of the *p–n* junction so formed may be of two types:
   (a) A step-graded junction: In a step-graded junction, the acceptor or the donor impurity density in the semiconductor is constant up to the junction.
   (b) A linearly-graded: In a linearly-graded junction, the impurity density varies almost linearly with distance away from the junction.
5. The electric field between the acceptor and the donor ions is called a barrier. It is equivalent to a difference of potential called the barrier potential.
6. The width of the depletion region is inversely proportional to the doping strength, as a larger carrier concentration enables the same charge to be achieved over a smaller dimension.
7. The Fermi level is invariant at thermal equilibrium.
8. Connecting the positive terminal of the battery to the *p*-type and the negative terminal to the *n*-type of the *p–n* junction is called forward-biasing.
9. Connecting the negative terminal of the battery to the *p*-type and the positive terminal to the *n*-type of the *p–n* junction is called reverse-biasing.
10. In reverse-biased condition, the current that flows through the diode is called reverse saturation current. At absolute zero, this current is zero.
11. The forward-biased barrier potential is typically 0.7 V for silicon diode and 0.3 V for a germanium diode. These values increase slightly with forward current.
12. There is a very small current called reverse saturation current in a reverse-bias due to the thermally generated minority-carriers.
13. The voltage at which the forward current rises sharply is known as cut-in voltage.
14. The reverse saturation current is dependent on temperature. The reverse saturation current approximately doubles for every 10°C rise in temperature.
15. Reverse breakdown voltage for a diode is typically greater than 50 V.
16. An ideal diode is one which acts as a short circuit (zero resistance) when forward-biased and as an open circuit (infinite resistance) when reverse-biased.
17. The resistance offered by a doped semiconductor is called the bulk resistance.
18. Forward resistance (dc) of a diode is the ratio of the dc voltage across the diode to the resulting dc current flowing through it.
19. Dynamic resistance (ac) is the ratio of small change in forward voltage to the corresponding change in the diode current.
20. The Junction Capacitance is proportional to $V_T^{1/2}$ for step junction and to $V_T^{3/2}$ for graded junction cases respectively.
21. Zener diode is a *p–n* junction diode, which is designed to sustain heavy current at the Zener breakdown region.
22. There are two mechanisms which give rise to the breakdown of a *p–n* junction under reverse bias condition: (i) Avalanche breakdown (ii) Zener breakdown.
23. *I–V* characteristics of a Zener diode is as shown:
24. Breakdown can be made very abruptly at accurately known values ranging from 2.4 V to 200 V with power ratings from 1 W to 50 W.
25. A Zener diode maintains a nearly constant voltage across its terminals over a specified range of Zener current.

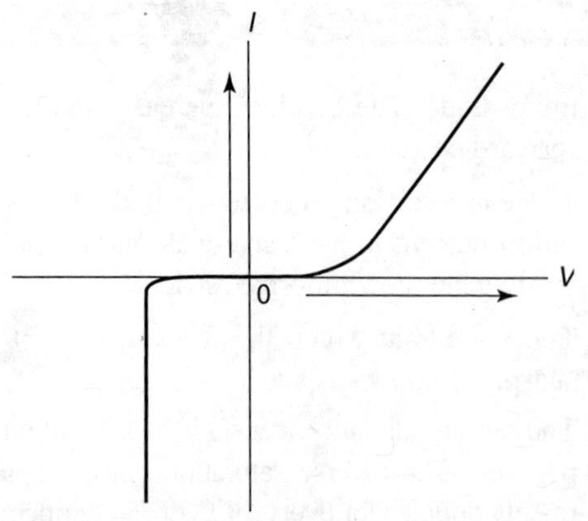


26. Zener diodes are used as voltage regulators and limiters.
27. The tunnel diode is a negative-resistance semiconductor *p*–*n* junction diode. The negative resistance is created by the tunnel effect of the electrons in the *p*–*n* junction.
28. The doping of both the *p*- and *n*-type regions of the tunnel diode is very high—impurity concentration of $10^{19}$ to $10^{20}$ atoms/cm$^3$ are used (which means both *n*-type and *p*-type semiconductors having parabolic energy bands are highly degenerate)—and the depletion layer barrier at the junction is very thin, in the order of $10^{-6}$ cm.
29. Charge carriers recombination takes place at *p*–*n*-junction as electron crosses from the *n*-side and recombines with holes on the *p*-side. When the junction is forward-biased the free electron is in the conduction band and is at a higher energy level than hole located at valence band.
30. The recombination process involves radiation of energy in the form of photons or leaf. If the semiconductor material is translucent, the light will be emitted and the junction becomes a light source i.e., a light emitting diode (LED).
31. Photo detector diodes accept light (optical signal) as input and produces corresponding current as output, which varies proportionally with the intensity of the incident light. The resultant current is amplified before it is passed as output.

## IMPORTANT FORMULAE

1. Fermi level in *n*-side of a *p*–*n* diode is given by:

$$E_{fn} = E_{cn} - k_B T \ln \frac{N_c}{N_d}$$

2. Fermi level in *p*-side of a *p*–*n* diode is given by:

$$E_{fp} = E_{cp} - kT \ln \frac{N_c N_a}{n_i^2}$$

3. Fermi level invariance: $\dfrac{dE_F}{dx} = 0$

4. Electric field in a abrupt junction is given by:

$$-E_0 = \frac{dV}{dx} = \frac{eN_a}{\varepsilon}(x + x_p)$$

5. Total voltage in an abrupt junction is given by:

$$V_T = V_2 - V_1 = \frac{e}{2\varepsilon}(N_a X_p^2 + N_d X_n^2)$$

6. The width of the depletion regions in *p*-type region is given by:

$$X_p = \left(\frac{2\,\varepsilon}{eN_a}\,\frac{V_T N_d}{N_d + N_a}\right)^{\frac{1}{2}}$$

7. The width of the depletion regions in *n*-type region is given by:

$$x_n = \left(\frac{2\,\varepsilon}{eN_d}\,\frac{V_T N_a}{N_a + N_d}\right)^{\frac{1}{2}}$$

8. Maximum electric field is given by:

$$E_{0\max} = -\frac{dV}{dX}\bigg|_{x=0} = -\frac{eN_a\,x_p}{\varepsilon} = -\frac{eN_d x_n}{\varepsilon}$$

9. The junction capacitance is is given by:

$$C_j = K_1\,V_T^{-\frac{1}{2}} = K_1\,(V_d - V_a)^{-\frac{1}{2}}$$

10. The total voltage is given by: $V_T = \dfrac{2eax_p^{\,3}}{3\varepsilon}$

11. The width of the transition region in the *n*- or *p*-type is is given by:

$$|x_n| = |x_p| = \frac{l}{2} = \left(\frac{3\varepsilon}{2ea}\right)^{\frac{1}{3}} V_T^{\frac{1}{3}}$$

12. The maximum electric field is given by:

$$E_{0\max} = -\frac{eax_p^{\,2}}{2\varepsilon}$$

13. The junction capacitance is given by:

$$C_j = \frac{dQ}{dV_T} = A\left(\frac{ea\,\varepsilon^2}{12}\right)^{\frac{1}{3}} V_T^{\frac{1}{3}}$$

14. The depletion capacitance in aVaractor diode is given by:

$$C_j = \frac{A\,\varepsilon_5}{\mathrm{L}}\,\alpha\,(V_{\mathrm{T}})^{-1/(n+2)}$$

15. If varactor is used with an inductance $L$ in a resonant circuit, the resonant frequency of the circuit is given by:

$$f_r = \frac{1}{2\pi\sqrt{LC}}\,\alpha V_{\mathrm{T}}$$

16. The total diode-current density in a semiconductor diode is given by:

$$J = e\left(\frac{D_p p_n}{L_p} + \frac{D_n n_p}{L_n}\right)\left[\exp\left(\frac{eV}{k_B T}\right) - 1\right]$$

17. The total direct current of the diode, with a cross-sectional junction area $A$ is:

$$I = AJ_s\left[\exp\left(\frac{eV}{k_B T}\right) - 1\right]$$

$$= I_s\left[\exp\left(\frac{eV}{k_B T}\right) - 1\right]$$

18. Dynamic diode resistance is given by:

$$r_{ac} = \frac{dV}{dI} = \frac{\eta}{39(I + I_s)}$$

19. Barrier potential is given by:

$$V_B = \frac{kT}{e}\log_e\left(\frac{n}{p}\right)$$

20. Decrease in barrier potential is given by:

$$\Delta V_B = -0.002 \times \Delta t$$

21. Diffusion current density is given by:

$$J_{(\mathrm{diff})} = -\,eD_p\frac{dp}{dx} + eD_n\frac{dn}{dx}$$

22. Drift current density is given by:

$$J_{(\mathrm{diff})} = J_p + J_n$$

23. Forward diode current: $I_f = I_p + I_n$

24. V–I characteristic of a diode:

$$I = I_0\,(e^{V/\eta V_T} - 1) = I_0\,(e^{V/\eta kT} - 1)$$

25. Volt equivalent temperature is given by:

$$V_T = \frac{T}{11{,}600}$$

26. Forward diode current is given by:

$$I_F = I_0\,e^{V/\eta V_T}$$

27. Diode reverse current is given by:

$$I_R = I_0\,[e^{V/\eta V_T} - 1]$$

## OBJECTIVE QUESTIONS

1. In a $p$–$n$ junction diode:
   (a) The depletion capacitance increases with increase in the reverse-bias.
   (b) The depletion capacitance decreases with increase in the reverse-bias.
   (c) The diffusion capacitance increases with increase in the forward-bias.
   (d) The diffusion capacitance is much higher than the depletion capacitance when it is forward-biased.

2. A $p$–$n$ junction in series with a 100 ohms resistor is forward-biased so that a current of 100 mA flows. If the voltage across this combination is instantaneously reversed to 10 volt at $t = 0$ the reverse current that flows through the diode at $t = 0$ is approximately given by:
   (a) 20 mA (b) 100 mA
   (c) 200 mA (d) None of the above

3. The width of the depletion region is:
   (a) Directly proportional to doping
   (b) Inversely proportional to doping
   (c) Independent of doping
   (d) None of the above

4. The Fermi energy in $p$–$n$ junction at thermal equilibrium is:
   (a) Proportional to distance
   (b) Directly increases with the temperature
   (c) Invariant with respect to distance
   (d) None of the above

5. The depletion capacitance, $C_j$ of an abrupt *p*–*n* junction with constant doping on either side varies with reverse bias, $V_R$, as:
   (a) $C_j \alpha V_R^3$
   (b) $C_j \mu V_R^{-2}$
   (c) $C_j \alpha V_R^{-½}$
   (d) None of the above

6. Gold is often diffused into silicon *p*–*n* junction devices to:
   (a) Is proportional to the square of the recombination rate
   (b) Is proportional to the cube of the recombination rate
   (c) Make silicon a direct gap semiconductor
   (d) None of the above

7. In a forward-biased photo diode with increase in incident light intensity, the diode current:
   (a) Increases
   (b) Remains constant
   (c) Decreases
   (d) None of the above

8. The current through a *p*–*n* junction diode with V volts applied to the *p*-region relative to the *n*-region (where $I_O$ is the reverse saturation current of the diode, $\eta$ the ideality factor, K the Boltzmann constant, *T* the absolute temperature and *e* the magnitude of charge on an electron) is:
   (a) $\left(I_0 \frac{-eV}{e^{mKT}-1}\right)$
   (b) $I_0\,(1 + e^{\frac{-eV}{mKT}})$
   (c) $I_0\left(1 - e^{\frac{eV}{mKT}}\right)$
   (d) $I_0\,[\exp\left(\frac{eV}{\eta K_B T}\right) - 1]$

9. The varactor diode is:
   (a) Voltage-dependent resistance
   (b) Voltage-dependent capacitance
   (c) Voltage-dependent inductor
   (d) None of the above

10. The electric field in abrupt *p*–*n* junction is:
   (a) Linear function of distance
   (b) Parabolic function of distance
   (c) Independent of distance
   (d) None of the above

11. In a linearly-graded *p*–*n* junction the doping concentration:
   (a) Changes abruptly at the junction
   (b) Varies linearly with distance from junction
   (c) Has a similar variation in junction capacitance with applied voltage
   (d) (a), (b) and (c)

12. A *p*–*n* junction, which is produced by recrystallisation on a base crystal, from a liquid phase of one or more components and the semiconductor is called:
   (a) Doped junction (b) Alloy junction
   (c) Fused junction (d) None of these

13. In an open circuit *p*–*n* junction, the energy band diagram of *n*-region shifts relative to that of *p*-diagram:
   (a) Downward by $E_0$
   (b) Upward by $E_0$
   (c) Remains invariant
   (d) Upward by $2E_0$

14. The contact potential $V_0$ in a *p*–*n* diode equals:
   (a) $V_T \ln \frac{N_A N_D}{n_i^2}$
   (b) $V_T \exp\left(\frac{n_i^2}{N_A N_D}\right)$
   (c) $V_T \ln \frac{n_i^2}{N_A N_D}$
   (d) $V_T \ln (N_A N_D)$

15. Each diode of full wave centre-tapped rectifier conducts for:
   (a) 45° only
   (b) 180° only
   (c) 360° complete period
   (d) 270° only

16. Bulk resistance of a diode is:
    (a) The sum of resistance values of $n$-material and $p$-material
    (b) The sum of half the resistance value of $n$-material and $p$-material
    (c) Equivalent resistance of the resistance value of $p$- and $n$-material is parallel
    (d) None of the above

17. In unbiased $p$–$n$ junction, thickness of depletion layer is of the order of:
    (a) $10^{-10}$ m (b) 50 $\mu$m
    (c) 0.5 $\mu$m (d) 0.005 $\mu$m

18. In a diode circuit, the point where the diode starts conducting is known as:
    (a) Cut-in point (b) Cut-out point
    (c) Knee point (d) Cut-off point

19. A Zener diode should have:
    (a) Heavily doped $p$- and $n$-regions
    (b) Lightly doped $p$- and $n$-regions
    (c) Narrow depletion region
    (d) Both (a) and (c)

20. When a diode is forward-biased, the recombination of free electron and holes may produce:
    (a) Heat (b) Light
    (c) Radiation (d) All of the above

21. In a linear-graded junction, the width of the depletion layer varies as:
    (a) $V_j$ (b) $V_j^{-2}$
    (c) $3\ \overline{V_j}$ (d) None of the above

22. The transition capacitance in step-graded junction varies as:
    (a) $C_j \alpha V_R^3$ (b) $C_j \alpha V_R^{-1/2}$
    (c) $C_j \alpha V_R^{-1/3}$ (d) None of the above

23. The law of junction gives that in $p$–$n$ diode, the concentration of holes injected to $n$-region at the junction is given by:
    (a) $n_{po}\varepsilon^{V/VT}$ (b) $n_{po}\varepsilon^{-V/VT}$
    (c) $p_{no}\varepsilon^{V/VT}$ (d) $p_{no}\varepsilon^{-V/VT}$

    Where $v$ is voltage applied at $p$ terminal relative to $n$ terminal, $n_{po}$ and $P_{no}$ are the thermal equilibrium concentration of electrons and holes in $p$- and $n$-regions respectively.

24. In a forward-biased diode, with $N_A \gg N_D$, product of the diffusion capacitance $C_D$ and the dynamic diode resistance $r$ equals:
    (a) $1/\tau_p$ (b) $\tau_p$
    (c) $\tau_p^3$ (d) $1/\tau_p^3$

    Where $\tau_p$ is the lifetime of injected minority-carrier holes.

25. In a forward-biased diode, with $N_A \gg N_D$, the following equation relates the diffusion current $I$, the injected excess minority carrier charge $Q$ and the file time of hole $\tau_p$:
    (a) $I = Q\tau_p$ (b) $I = Q^2/\tau_P$
    (c) $I = Q/\tau_P$ (d) $I = \tau_P^2$

26. In a $p$–$n$ diode, for constant value of current at room temperature, $d_v/d_t$ varies approximately at the rate of:
    (a) −2.5 mV/°C (b) −25 mV/°C
    (c) 2.5 mV/°C (d) 25 mV/°C

27. Total space-charge neutrality is given by:
    (a) $|Q| = eAN_d x_n = \frac{e}{A} N_a x_p$
    (b) $|Q| = eAx_n / N_d = eAx_p/N_D$
    (c) $|Q| = eAN_d x_n = eAN_a x_p$
    (d) $|Q| = eAN_d x_n = eAN_a x_p^2$

28. Width of the depletion region in $p$-type is:
    (a) $x_p = \left(\frac{2\varepsilon}{eN_a} \frac{V_T N_d}{N_d - N_a}\right)^{\frac{1}{2}}$
    (b) $x_p = \left(\frac{2\varepsilon}{eN_a} \frac{V_T N_d}{N_d + N_a}\right)^{-\frac{1}{2}}$
    (c) $x_p = \left(\frac{2\varepsilon}{eN_a} \frac{V_T N_d}{N_d - N_a}\right)^{\frac{1}{2}}$
    (d) $x_p = \left(\frac{2\varepsilon}{eN_a} \frac{V_T}{N_d + N_a}\right)^{\frac{1}{2}}$

29. The width of the depletion region in *n*-type is given by:

(a) $x_n = \left(\frac{2\varepsilon}{eN_d}\frac{V_T N_d}{N_d - N_a}\right)^{\frac{1}{2}}$

(b) $x_n = \left(\frac{2\varepsilon}{eN_d}\frac{V_T N_d}{N_a + N_d}\right)^{\frac{1}{2}}$

(c) $x_n = \left(\frac{2\varepsilon}{e}\frac{V_T N_d}{N_a + N_d}\right)^{\frac{1}{2}}$

(d) $x_n = \left(\frac{2\varepsilon}{eN_d}\frac{V_T N_a}{N_a - N_d}\right)$

30. The reverse saturation current $I_s$ is given by:

(a) $I_s \equiv AJ_s \equiv Ae\left(\frac{D_p}{L_p} + \frac{D_n}{L_n}\right)$

(b) $I_s \equiv AJ_s \equiv A\ln\left(\frac{D_p p_n}{L_p} - \frac{D_n n_p}{L_n}\right)$

(c) $I_s \equiv AJ_s \equiv A\left(\frac{D_p p_n}{L_p} + \frac{D_n n_p}{L_n}\right)^2$

(d) $I_s \equiv AJ_s \equiv Ae\left(\frac{D_p p_n}{L_p} + \frac{D_n n_p}{L_n}\right)$

## REVIEW QUESTIONS

1. What is *p–n* junction diode?
2. Explain process of formation *p–n* junction diode.
3. What are the different natures of *p–n* junction can be fabricated?
4. Draw the energy band diagram for an unbiased *p–n* junction.
5. Why band bending occurs during the formation of a *p–n* junction diode?
6. What is space charge? Write the properties of the junction.
7. For an unbiased *p–n* junction sketch the variation of space-charge, electric field electrostatic potential and electron energy as function of distance across the junction.
8. What is the effect of doping on barrier potential?
9. Show that the Fermi level is invariant at thermal equilibrium.
10. Find out the expression of built in potential in terms of donor and accepter concentration.
11. What are basic difference between contact potential and built inpotential?
12. The barrier potential of a *p–n* diode can not be measured with the help of voltmeter. Explain.
13. When a *p–n* junction is said to be forward-biased and reverse-biased? Explain with a figure.
14. Draw the energy band diagram for forward-biased and reverse-biased *p–n* junction diode.
15. What are an abrupt and a linearly-graded *p–n* junction?
16. Draw and explain the I–V characteristics of *p–n* junction.
17. Compare the I–V characteristics of ideal and practical *p–n* junction diode.
18. Compare cut-off voltage ($V_\gamma$) of Ge, Si and GaAs.
19. Derive the equation of I–V characteristics of *p–n* junction diode starting from continuity equation.
20. What is the origin of the reverse saturation current of a *p–n* junction? Does the reverse saturation current change with the temperature and the applied reverse-bias?
21. Define and explain the dynamic or ac and static or dc resistance of *p–n* diode. Do these resistances depend on temperature and biasing voltage?
22. Draw the linear piecewise model of *p–n* junction.
23. What is junction capacitance and how it is formed in *p–n* junction diode?

24. Find out the expressions of junction capacitance of $p$–$n$ junction diode.
25. Explain the working principle of varactor diode.
26. What is breakdown diode? Classify different types of breakdown diode.
27. Explain the working principle of Zener diode.
28. What is the difference between an ordinary semiconductor diode and a Zener diode?
29. Write down the use of Zener diode.
30. What is avalanche breakdown? How does avalanche multiplication occur?
31. What are the basic differences between Zener and avalanche breakdown?
32. Draw the V–I characteristics curves of Zener diode and Avalanche breakdown diode.
33. Compare the Zener breakdown between Si, Ge and GaAs.
34. Explain the working principle of tunnel diode.
35. Explain the basic working of a light emitting diode.
36. Explain the advantages and disadvantages of a light emitting diode.
37. Explain the working principle of a photo detector.
38. What are the advantages and disadvantages of a photo detector?
39. What are the basic criterions to be fulfilled by a photo detector?
40. Explain with diagrams the V–I curve of the following:
    (a) Tunnel diode
    (b) Light-emitting diode
    (c) Photo detector diode
    (d) Solar cell

## PRACTICE PROBLEMS

1. At a certain point in a certain junction at equilibrium, the electric field is +8000 V/cm and $n = 10^{12}/\text{cm}^3$.
    (a) Calculate the density gradient for electrons at manager that point.
    (b) Do the same for holes at that point.
2. A certain symmetric step $p$–$n$ junction has $N_D = 10^{16}/\text{cm}^3$ (donors only) on the left-hand side and $N_A = 10^{16}/\text{cm}^3$ (acceptors only) on the right-hand side. Using the depletion approximation, calculate, for equilibrium conditions,
    (a) Contact potential $\Delta\psi_0$
    (b) Depletion-layer thickness $X_o$ in $\mu$m.
    (c) Field at the junction $E_{OM}$ in kilovolts per centimetre.
3. For the junction of 2, and continuing to use the depletion approximation, sketch dimensioned diagrams of:
    (a) Charge-density profile
    (b) Field profile
    (c) Potential profile
4. Considering the spatial origin to be positioned at the metallurgical junction in the equilibrium sample of 2 and 3:
    (a) Calculate, the four current-density components at $x = 0$, $J_{n\text{'drft}}$, $J_{p\text{,drft}}$, $J_{n\text{,diff}}$ $J_{p\text{,diff}}$ using realistic carrier profiles, but using $E_{OM}$ from the depletion approximation.
    (b) Calculate both density-gradient values at $x = 0$.
5. A certain asymmetric step junction has a doping on the left-hand side of $N_1 = N_D - N_A = 10^{13}/\text{cm}^3$ and on the right-hand side of $N_2 = N_A - N_D = 4 \times 10^{13}/\text{cm}^3$. Using the depletion approximation, calculate, for the junction at equilibrium:
    (a) Contact potential $\Delta\psi_0$
    (b) Depletion-layer thickness $X_o$
    (c) Electric field at the junction, $E_{OM}$
    (d) Thickness of the depletion-layer portion on the $n$-type side, $X_1$
    (e) Thickness of the depletion-layer portion on the $p$-type side, $X_2$

6. For the junction of Problem 5:
   (a) Calculate the ratio $X_1/X_2$
   (b) Explain why $(X_1/X_2) = (N_2/N_1)$
   (c) Sketch a dimensioned diagram of the field profile.
   (d) Calculate the potential drops $\Delta\psi_1$ and $\Delta\psi_2$ on the *n*-type and *p*-types sides, respectively.

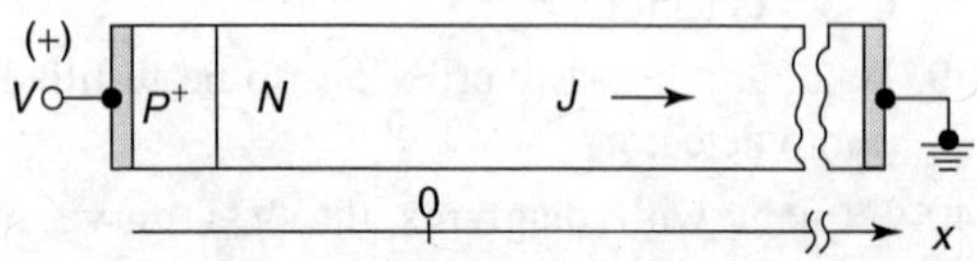


7. Holes are being injected by a forward-biased junction under low level steady-state conditions at the left end of a long extrinsic-*n*-type silicon bar. In this problem you are only concerned with the *N*-region for $x > 0$. The junction is several diffusion lengths to the left of the spatial origin, and $p'_N(0)$ is several times $p_{ON}$. The total current density in the bar is *J*.
   (a) Write an expression for $J(\infty)$, that is, current density where $x >> L_P$. Make reasonable approximations.
   (b) Given that $p'_N(x) = p'_N(0)e^{-x/L}p$, write an expression free of primed variables for the hole-density gradient as a function of *x* for $x > 0$.
   (c) Write an expression for $J_{p,\text{diff}}(X)$ for $x > 0$.
   (d) Write an expression for $J_{n,\text{diff}}(X)$ for $x > 0$.
   (e) Given that $J_{p,\text{diff}}(0) = 0.092\ J(\infty)$, find $E(0)$ in terms of $E(\infty)$.

8. You are given an $N^+P$ forward-biased junction. The minority electron profile on the right-hand side of the sample $n_p(x)$ versus *x* is plotted above with reasonable accuracy using normalized linear coordinates. In the following diagram, the minority-carrier profiles on both sides of the junction are plotted using normalised semi log coordinates, with linear abscissa and logarithmic ordinate.

   Given that $D_n = 19.5$ cm²/s on the right-hand side, find carrier lifetime on the *p*-type side.

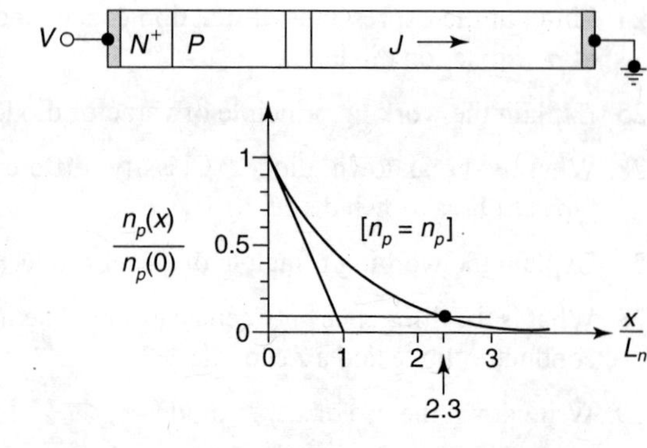


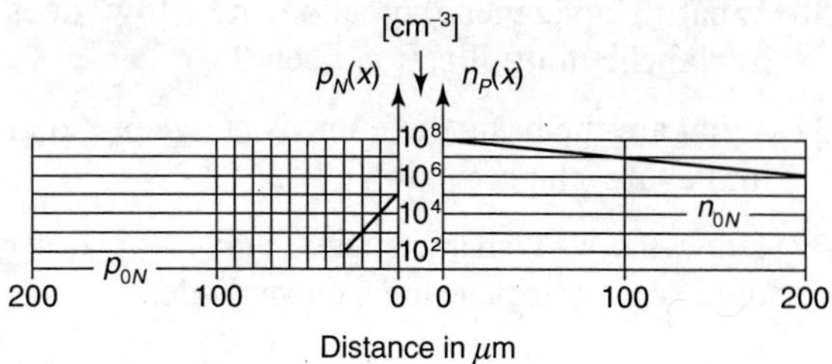


9. Given the $N + P$ junction shown here with $N_1 = 10^{19}/\text{cm}^3$, $N_2 = 10^{15}/\text{cm}^3$, and $\tau_2 = 1\ \mu s$:

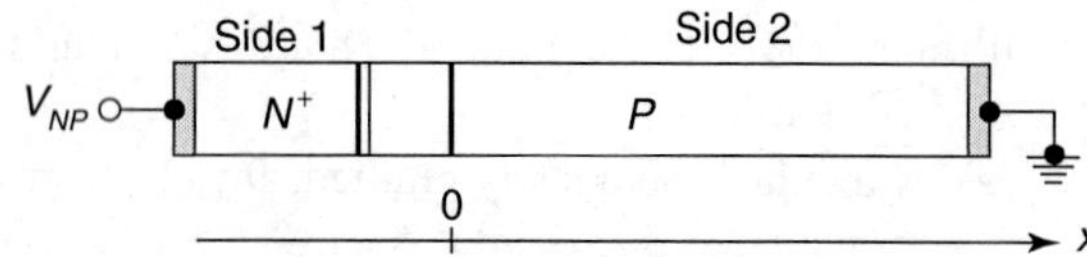


   (a) Find $J_n(0)$ for $V_{Np} = -0.04$ V.
   (b) Find $J_n(0)$ for $V_{Np} = -0.4$ V.
   (c) Repeat (a) and (b) with $\tau_2 = 100\ \mu s$ and all else held constant.
   (d) Repeat (a) and (b) with $N_2 = 10^{15}/\text{cm}^3$ and all else held constant.
   (e) Find $\Delta\psi$ for (a) and (b).
   (f) Find an approximate value for $J_n(0)$ in b as a percentage of total current density *J* in the sample, given $L_p = 1\ \mu$m and $D_p = 1.5$ cm²/s in side 1, and assuming negligible recombination in the space-charge region.
   (g) Assuming the junction to be ideal, find the reverse-current density for:

$$V_{NP} = +100\text{ V}, +10\text{ V}, +1\text{ V},$$

   for the sample of (a) and (b).

10. Following is a reasonably accurate sketch of the numerical solution for an equilibrium symmetric step junction having $N = 10^{13}/\text{cm}^3$.

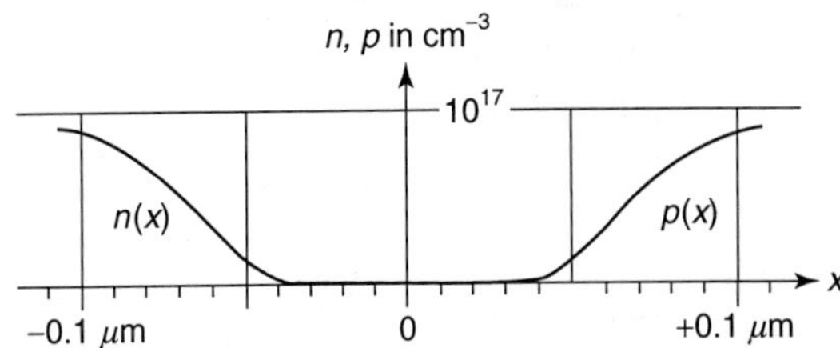


(a) Estimate the maximum value of carrier-density gradient displayed in the diagram without regard for algebraic sign.

(b) Calculate the maximum value of the current densities $|J_{n,\text{diff}}| = |J_{n,\text{drift}}|$ without regard for algebraic sign. Use $\mu_n = 700\ \text{cm}^2/\text{Vs}$.

(c) Given that $X_o = 0.16\ \mu\text{m}$ (as can be confirmed on the sketch), calculate the maximum electric field in the junction, $E_{OM}$

(d) Calculate the value of the current densities $|J_{n,\text{diff}(0)}| = |J_{n,\text{drft}(0)}|$ without regard for algebraic sign.

(e) In what approximate fraction of the space-charge-layer thickness does electric field exceed 50 kV/cm, the value at which the drift velocity of electrons saturates at 107 cm/s?

(f) In view of your answer to part (e), what is the appropriate value of $\mu$m to use in solving part (d)? Explain.

11. A certain one-sided silicon step junction under bias exhibits the carrier profile.

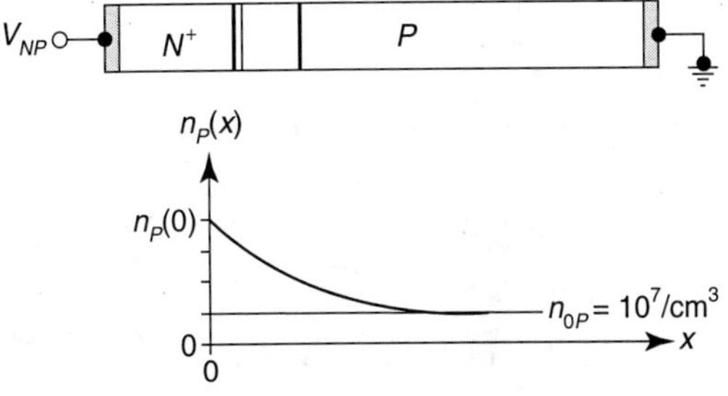


(a) Find the magnitude and sign of the applied voltage $V_{NP}$. Use units in calculation.

(b) $V_{Np}$ is changed so that $n_p(0) = 10^8/\text{cm}^3$. Given that the total current density is $J = -4.19 \times 10^{-8}\ \text{A/cm}^2$, find the minority-carrier diffusion length $L_n$ in $\mu$m for the $p$-type region. Use units in calculation.

(c) The sample of (a) is replaced by another of identical doping that exhibits minority-carrier diffusion length of $L_n = 2$ mm. Find the carrier lifetime $\tau$ in $\mu$s for the $p$-type region of the new sample. Use units in calculation.

12. A silicon sample at equilibrium contains an asymmetric step junction. Shown here is a portion of its field profile that is based upon the depletion approximation.

(a) With the help of Poisson's equation, derive an expression for net doping on the left-hand side. Put your final expression in terms of only symbols given in the diagram and the right-hand side of Poisson's equation.

(b) Determine the potential drop on the $n$-type side of the junction. Calculate numerical value and units.

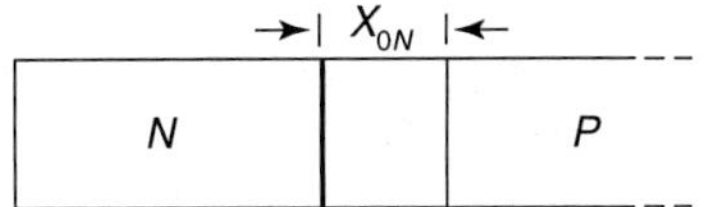


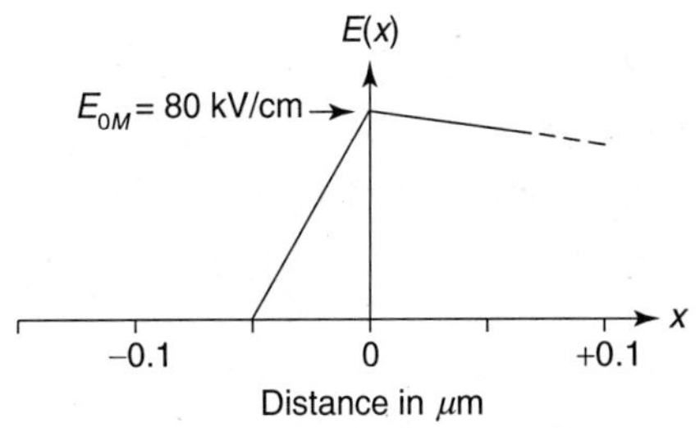


13. This $N+$ $P$ junction is under a bias of $V_{NP} = -kT/q$. In the $p$-type region, it has the values $D_n = 22\ \text{cm}^2/\text{s}$ and $T = 2\ \mu\text{s}$. The law of the junction is not very accurate at such a low bias, but for purposes of this problem, assumes that it is accurate. The value of $n_{op}$ is $10^6/\text{cm}^3$.

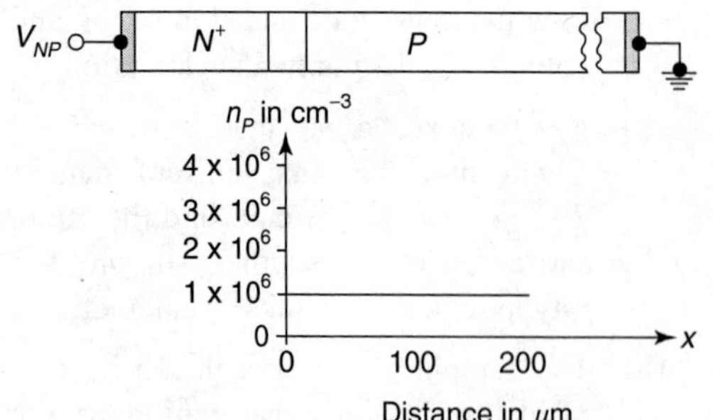


(a) Calculate at the origin the slope of $n_p$. Give the correct units and algebraic sign along with the numerical value.

(b) Using a solid line, sketch $n_p(x)$ for $x > 0$.

(c) Calculate $Q'$, the excess-electron charge stored in the $p$-type region. The cross section of the sample is square, 1 mm × 1 mm. Give units.

14. To analyse this one-sided step junction at equilibrium:

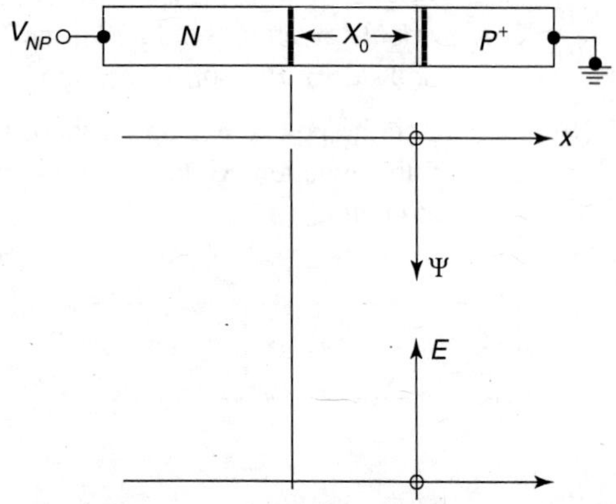


(i) Use the depletion approximation

(ii) Neglect potential drop on the $p$ +– side

(iii) Neglect depletion-layer penetration on the $p$+– side

(iv) Include units in all necessary calculations

The properties of the sample are these: On the left-hand side, $(N_D - N_A) = N_1 = 10^{15}/\text{cm}^3$; on the right-hand side, $(N_A - N_D) = N_2$; the depletion-layer dimension $X_1 = X_0 = 0.5\ \mu$m; $E$ = 12 pF/cm; $(kT/q) = 0.06566$ V; and finally

$$\psi(x) = \frac{qN_1}{\varepsilon}\left(\frac{x^2}{2} + X_0 x\right)$$

(a) Sketch $\psi(x)$ versus $x$ on the upper axis pair. Label the axis.

(b) Derive an expression for $E(x)$ versus $x$

(c) Use your expression to calculate $E(-X_0)$, and $E(0)$ in terms of $X_0$

(d) Plot $E(x)$ in the space provided

(e) Derive an expression for $\Delta\psi_0$

(f) Calculate $\Delta\psi_0$

(g) Derive an expression for $E_{OM}$ in terms of $\Delta\psi_0$

(h) Calculate $E_{OM}$ in kV/cm

(i) Given $\Delta\psi_0 = (kT/q)\ln(N_1N_2/n_i^2)$, find $N_2$

15. Following is a one-sided junction sample having an extensive $p$-type region and a cross-sectional area of $10^{-3}$ cm$^2$. It is subjected to a steady-state forward-bias. Calculate the number of excess minority-carriers stored in the $p$-type region.

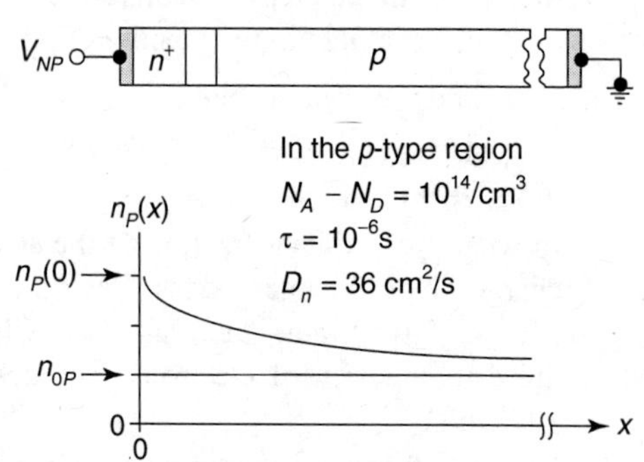


16. Given this symmetric step junction at equilibrium with a net doping on each side of $10^{16}/\text{cm}^3$, perform approximate calculations, estimating where necessary.

(a) Calculate $\Delta\psi_0$

(b) Calculate $X_0$

(c) Calculate $E_{OM}$

(d) Estimate the value of maximum hole gradient.

(e) Compute the approximate value of $J_{p,\text{diff}}$ there.

(f) What is the value of $J_{p,\text{drft}}$ there?

(g) Is $J_{p,\text{drft}}$ higher or lower at $x = 0$?

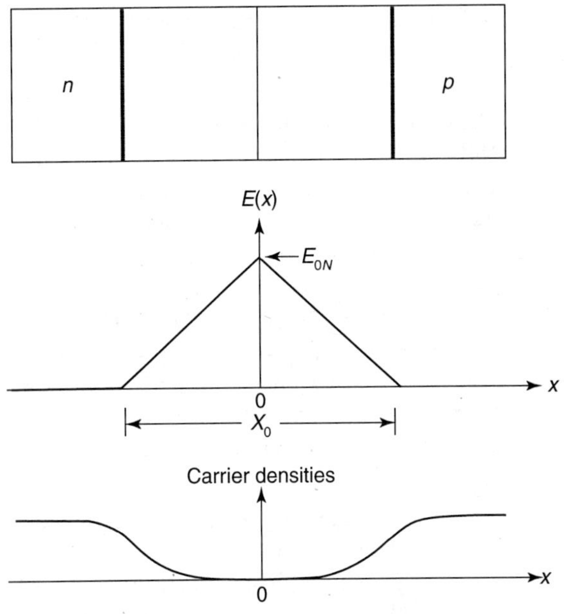


(h) Calculate $E = E_{OM}$ there; why doesn't $J_{p,\mathrm{drft}}$ peak at $x = 0$?

(i) Calculate $J_{p,\mathrm{drft}}$ at $x = 0$

17. A single-crystal silicon sample has a thin ideally intrinsic region flanked by heavily doped $n$-type and $p$-type regions, forming a PIN (or NIP) junction. It is not a step junction.

(a) Considering that there is sufficient space charge in the given sample to launch and terminate four lines of force when the sample is at equilibrium, sketch these lines in the top diagram, using arrowheads to indicate direction.

(b) Sketch and dimension the approximate field profile in the space provided, given $\Delta\psi_0 = 2.0$ V.

(c) Sketch and dimension the equilibrium potential profile in the space provided, taking the potential of the $p$+ region as reference. Your sketch should show potential throughout the entire device, from $x = 0$ to $x = 30\ \mu$m.

(d) Sketch the equilibrium band diagram for this sample, letting $\psi_G = 1.1$ V. Show the Fermi level ø.

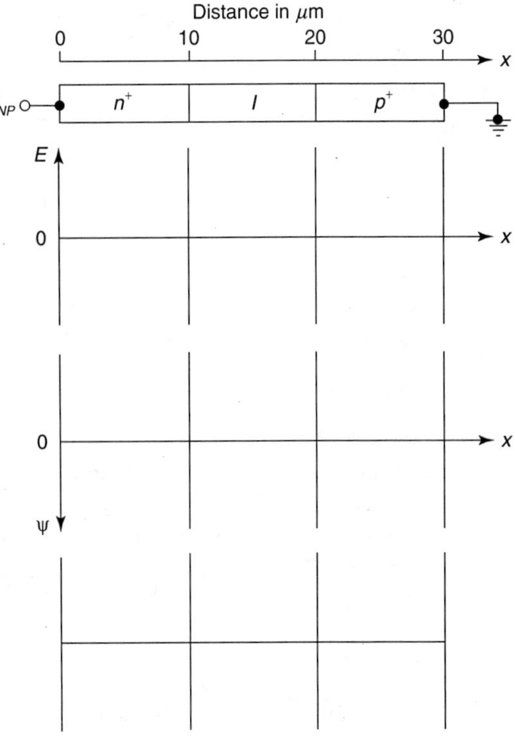


(e) Assuming the two extrinsic regions have equal net-doping magnitudes, calculate net-doping density approximately.

(f) Calculate the magnitude and sign of applied voltage $V_{NP}$ required to produce a maximum field of $E_M = 4$ kV/cm in this junction.

18. Write a completely general expression for volumetric space-charge density $\rho_v(x)$ at equilibrium in a region with arbitrary doping profiles $N_D(x)$ and $N_A(x)$.

19. Given the depletion approximation for an equilibrium step junction of arbitrary doping, one can specify the junction completely by specifying two of its independent variables. With the well known six variables demonstrate the validity of this assertion.

(a) By writing four independent equations in the six variables.

(b) By writing three independent equations in the first five variables.

(c) By writing two independent equations in the first four variables.

(d) By writing one equation in the first three variables.

20. You are given an ideal silicon junction having a saturation current of $I_0 = 10^{-14}$ A, and carrying a forward current of 1 $\mu$A.

    (a) Assuming a reverse resistance of $R = \infty$, compute values for a piecewise-linear model of the junction.

    (b) Sketch and dimension the resulting I–V characteristic.

    (c) Repeat part a for $I = 1.5$ A

    (d) Repeat part b for $I = 1.5$ A

21. A $p^+$ $n$ junction diode with a junction area of 1 mm$^2$ has a charge of $10^{-6}$ C of excess holes stored in it when it carries a current of 49 mA. Compute the minority-carrier diffusion length in the *n*-type region.

22. An abrupt GaAs *p–n* diode has $N_a = 10^{16}$ cm$^{-3}$ and $N_d = 10^{15}$ cm$^{-3}$. Calculate the Fermi level positions at 300 K in the *p*- and *n*-regions.

23. Consider the sample discussed in Problem 22. The diode has a diameter of 25 $\mu$m. Calculate the depletion widths in the *n*- and *p*-regions. Also calculate the charge in the depletion regions and plot the electric field profile in the diode.

24. An abrupt silicon *p–n* diode at 300 K has a doping of $N_a = 10^{18}$ cm$^{-3}$, $N_d = 10^{13}$ cm$^{-3}$. Calculate the built-in potential and the depletion widths in the *n*- and *p*-regions.

25. AGe *p–n* diode has $N_a = 5 \times 10^{17}$ cm$^{-3}$ and $N_d = 10^{17}$ cm$^{-3}$. Calculate the built-in voltage at 300 K. At what temperature does the built in voltage decrease by 2 per cent?

26. The diode of Problem 25 is subjected to bias values of:
    (i) $V_f = 0.1$ V (ii) $V_f = 1.5$ V (iii) $v_f = 1.9$ V
    (iv) $v_f = 5.0$ V. Calculate the depletion widths and the maximum field $F_m$ under these biases.

27. Consider a $p + n$ Si diode with $N_a = 10^{18}$ cm$^{-3}$ and $N_d = 10^{16}$ cm$^{-3}$. The hole diffusion coefficient in the *n*-side is 10 cm$^2$/s and $\tau_p = 10^{-7}$ s. The device area is $10^{-4}$ cm$^2$. Calculate the reverse saturation current and the forward current at a forward bias of 0.4 V at 300 K.

28. Consider a $p + n$ silicon diode with area $10^{-4}$ cm$^2$. The doping is given by $N_a = 10^{18}$ cm$^{-3}$ and $N_d = 10^{17}$ cm$^{-3}$. Plot the 300 K values of the electron and hole currents $I_n$ and $I_p$ at a forward-bias of 0.4 V. Assume $\tau_n = \tau_p = 1\,\mu$s and neglect recombination effects. $D_n = 20$ cm$^2$/s; and $D_p = 20$ cm$^2$/s.

29. A GaAs LED has a doping profile of $N_a = 10^{17}$ cm$^{-3}$, $N_d = 10^{18}$ cm$^{-3}$ at 300 K. The minority carrier time is $\tau_n = 10^{-8}$ s; $\tau_p = 5 \times 10^{-9}$ s. The electron diffusion coefficient is 150 cm$^2$ s$^{-1}$ while that of the holes is 20 cm$^2$ s$^{-1}$. Calculate the ratio of the electron injected current to the total current.

30. The diffusion capacitance of a wide base diode is greater than that of a narrow base diode. Show that the ratio of the diffusion capacitances of such diodes with heavily doped *p*-regions is $3\tau_p/4\tau_B$.

31. Consider a GaAs *p–n* diode with a doping profile of $N_a = 10^{16}$ cm$^{-3}$, $N_d = 10^{16}$ cm$^{-3}$ at 300 K. The minority carrier lifetimes are $\tau_n = 10^{-7}$ s; $\tau_p = 10^{-8}$ s. The electron and hole diffusion coefficients are 150 cm$^2$/s and 24 cm$^2$/s, respectively. Calculate and plot the minority carrier current density in the neutral *n*- and *p*-regions at a forward-bias of 1.0 V.

32. Consider a *p–n* diode made from InAs at 300 K. The doping is $N_a = 10^{17}$ cm$^{-3}$ = $N_d$. Calculate the saturation current density if the electron and hole density of states masses are 0.02 $m_0$ and 0.4 $m_0$, respectively. Compare this value with that of a silicon *p–n* diode doped at the same levels. The diffusion coefficients are $D_n = 800$ cm$^2$/s; $D_p = 30$ cm$^2$/s. The carrier lifetimes are $\tau_n = \tau_p = 10^{-8}$ s for InAs. For the silicon diode use the values $D_n = 30$ cm$^2$/s; $D_p = 20$ cm$^2$/s; $\tau_n = \tau_p = 10^{-7}$ s.

33. Consider a *p–n* diode in which the doping is linearly graded. The doping is given by:

$$N_d - N_a = Gx$$

so that the doping is *p*-type at $x < 0$ and *n*-type at $x > 0$. Show that the electric field profile is given by:

$$F(x) = \frac{e}{2\varepsilon} G\left[x^2 - \left(\frac{W}{2}\right)^2\right]$$

where, $w$ is the depletion width given by:

$$W = \left[\frac{12\ \varepsilon H\,(Vbi - V)}{eG}\right]^{1/3}$$

34. A silicon diode is being used as a thermometer by operating it at a fixed forward bias current. The voltage is then a measure of the temperature. At 300 K, the diode voltage is found to be 0.7 V. How much will the voltage change if the temperature changes by 1 K?

35. Consider a GaAs $p$–$n$ diode with $N_a = 10^{17}$ cm$^{-3}$, $N_d = 10^{17}$ cm$^{-3}$. The diode area is $10^{-3}$ cm$^2$ and the minority carrier mobilities are (at 300 K) $\mu_n = 3000$ cm$^2$/V−s; $\mu_p = 200$ cm$^2$/V−s. The electron–hole recombination times are $10^{-8}$s. Calculate the diode current at a reverse-bias of 5 V.

36. A long base GaAs abrupt $p$–$n$ junction diode has an area of $10^{-3}$ cm$^2$, $N_a = 10^{18}$ cm$^{-3}$, $N_d = 10^{17}$ cm$^{-3}$, $\tau_p = \tau_n = 10^{-8}$ S, $D_p = 6$ cm$^2$ s$^{-1}$ and $D_n = 100$ cm$^2$ s$^{-1}$. Calculate the 300 K diode current at a forward bias of 0.3 V and a reverse bias of 5 V. The electron–hole recombination time in the depletion regions is $10^{-7}$s.

37. The critical field for breakdown of silicon is 4 × 105 V/cm. Calculate the $n$-side doping of an abrupt $p$+$n$ diode that allows one to have a breakdown voltage of 29 V.

38. What is the width of the potential barrier seen by electrons during band to band tunneling in an applied field of 5 × $10^5$ V/cm in GaAs, Si and $In_{0.53}Ga_{0.47}As$ ($E_g = 0.8$ V)?

39. If the electron effective mass is 0.5 $m_0$, and the semiconductor band gap is 1.0 eV, at what applied field does the tunneling probability become $10^{-11}$?

40. Consider a Si $p$–$n$ diode with $N_a = 10^{18}$ cm$^{-3}$ and $N_d = 10^{18}$ cm$^{-3}$. Assume that the diode will break down by Zener tunneling if the peak field reaches $10^6$ V/cm. Calculate the reverse bias at which the diode will break down.

41. A $p$+ − $n$ silicon diode has an area of $10^{-2}$ cm$^2$. The measured junction capacitance (at 300 K) is given by:

$$\frac{1}{C^2} = 5 \times 10^8\,(2.5 - 4\ \mathrm{V})$$

in which C is in units of $\mu$F and $V$ is in volts. Calculate the built-in voltage and the depletion width at zero bias. What are the dopant concentrations of the diode?

42. In a long base $n$+$p$ diode, the slope of the $C_{diff}$ versus $I_F$ plot is 1.6 × $10^{-5}$ F/A. Calculate the electron lifetime, the stored charge, and the value of the diffusion capacitance at $I_F = 2$ mA.

43. Consider a Si $p$+$n$ diode with a long base. The diode is forward-biased (at 300 K) at a current of 2 mA. The hole lifetime in the $n$-region is $10^{-7}$ s. Assume that the depletion capacitance is negligible and calculate the diode impedance at the frequency of 100 KHz, 100 MHz and 500 MHz.

44. Consider a diode with the junction capacitance of 16 pF at zero applied bias and 4 pF at full reverse bias. The minority carrier time is 2 × $10^{-8}$ s. if the diode is switched from a state of forward bias with current of 2.0 mA to a reverse bias voltage of 10 V applied through a 5 Kohm resistance, estimate the response time of the transient.

45. What is the differential resistance of a diode at zero bias?

46. Assume a $p$–$n$ diode with a heavily doped $p$-region. Show that the concentration of excess holes is proportional to the forward current either in a wide or a narrow $n$-region.

47. Derive the built-in potential of a junction for heavy doping, $N_A$ on the $p$-side and near intrinsic doping on the $n$-side.

48. When the junction capacitance is plotted against the applied reverse voltage on log-log graph paper, a straight line results for $V_a >> V_b$. What is the slope of this line?

49. For a $p$+$n$ diode $\mu_p = 450$ cm$^2$/Vs and $\tau_p = 1\ \mu s$ in the $n$-region. Calculate the widths of the $n$-region for which: (a) $W_n \leq 0.1\ L_p$ and (b) $W_n \geq 4\ L_p$.

50. How can you connect two Zener diodes, 6 V and 4 V to obtain a reference of 10 V, if the supply voltage is 15 V and the load resistance is variable? The minimum current for each Zener diode is 1 mA and the wattage of each diode is 0.5 Watts.

## SUGGESTED READINGS

1. Shockley, W. 1949. "The Theory of *p–n* Junctions in Semiconductors and *p–n* Junction Transistors". *Bell System Technical Journal*. Vol. 28: 435–89.
2. Shockley, W. and W. T. Read Jr. 1952. "Statistics of the Recombination of Holes and Electrons". *Physical Review*. Vol. 87: 835–42.
3. Mou, J. N. 1958. "The Evolution of the Theory of the Voltage Current Characteristic of *p–n* Junctions". *Proceedings of the IRE*: 1076–82.

# 3

# Diode Circuits

## Outline



## Objectives

This chapter analyses diode circuits and load line with the *Q*-point concept. The formulation of the diode as a voltage regulator, half-wave rectifier, and full-wave rectifier along with bridge rectification and performance analysis of rectifier circuits will be dealt with in detail. This is followed by a derivation of peak inverse voltage, dc voltage and current, ripple factor, and efficiency. Clipper and clamper circuits, comparators, and additional diode circuits will be analysed at the end of the chapter.

## 3-1 INTRODUCTION

In the field of electronics, the simplest and the most fundamental non-linear circuit element is the diode. The *p–n* junction diode is considered to be a circuit element. For easy and lucid evaluation of the diode element, the concept of load line is extremely important. Among the many applications of diodes, their use in the design of rectifiers, which convert ac to dc, is the most common. The piecewise linear model is used in certain applications of diodes, namely clippers, rectifiers and comparators. Many more such circuits are possible with one or more diodes being implemented in them.

## 3-2 ANALYSIS OF DIODE CIRCUITS

The basic diode circuit consists of a voltage source in series with a resistor and a diode. The circuit might be analysed properly to obtain the instantaneous current and diode voltage. For such an analysis to be done, the concept of load line and its effective use in various circuits has to be thoroughly understood. The concept of load line is absolutely essential.

## 3-3 LOAD LINE AND *Q*-POINT

The applied load will normally have an impact on the region (or point) of operation of a device. If the analysis is performed in a graphical manner, a line can be drawn on the characteristics of the device to represent the applied load. The intersection of the load line with the characteristics will determine the point of operation of the system. Physically, this point of operation mainly determines the conditions under which the device is to be operated in a circuit. This case takes care of the various intriguing attributes of the circuit. This kind of an analysis is known as the *load-line analysis.* We will discuss the concept of load line from all practical points of view. An example has been shown in Fig. 3-1(a) and Fig. 3-1(b).

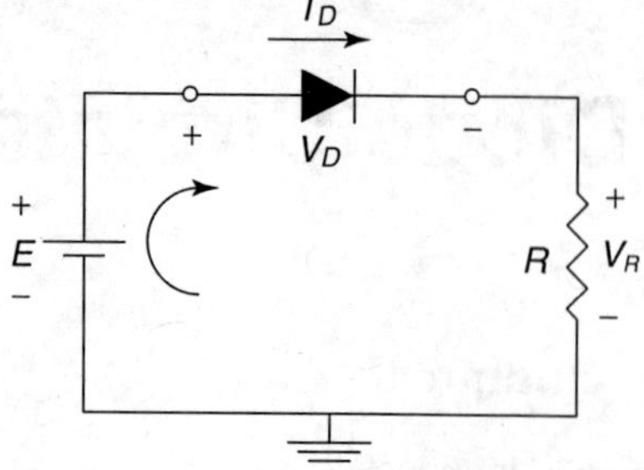


**Figure 3-1(a)** Analysis of a basic diode circuit

Let us consider the network and its characteristics as given in Fig. 3-1(a). The voltage established by the battery *E* is to generate a current through the series resistor *R* of the circuit in the clockwise direction. The fact that this current and the defined direction of conduction of the diode are the same reveals that the diode is in the ON state, i.e., the diode is forward-biased and consequently, the forward resistance of the diode is very low. Under normal conditions, this resistance is approximately 10 Ω. Applying Kirchoff's voltage law (KVL) of circuit theory, to the series circuit of Fig. 3-1(a), we obtain:

$$E - V_D - V_R = 0 \tag{3-1}$$

$$E = V_D + I_D R \tag{3-2}$$

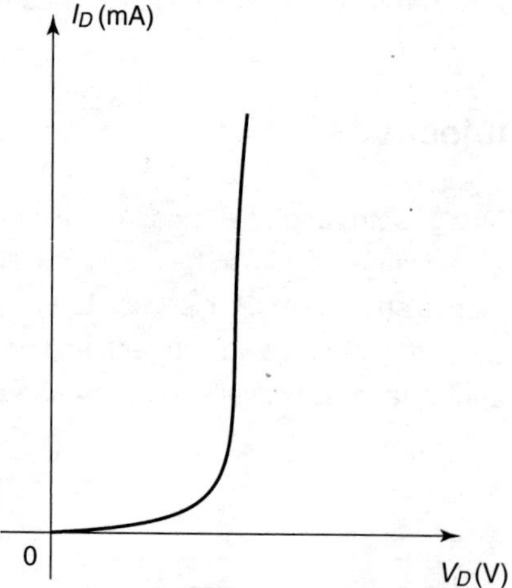


**Figure 3-1(b)** I-V characteristics of the diode

The intersection of the load line with the curve of current-voltage characteristics under forward-biased conditions easily implicates the conditions of operation of the device in the circuit.

If $V_D = 0$ V, we can calculate $I_D$ and plot the magnitude of $I_D$ on the vertical axis.

As $V_D = 0$ V, Eq. (3-2) is modified as:

$$E = V_D + I_D R$$
$$= 0\text{ V} + I_D R$$
$$I_D = \left.\frac{E}{R}\right|_{V_D = 0\text{ V}} \tag{3-3}$$

If $I_D = 0$ A, we can calculate $V_D$ and plot the magnitude of $V_D$ on the horizontal axis.

As $I_D = 0$ A, Eq. (3-2) is modified as:

$$E = V_D + I_D R$$
$$= V_D + (0\text{ A})\,R$$
$$V_D = E\,\big|_{I_D = 0\text{A}} \tag{3-4}$$

A straight line drawn between two points will define the load line, as shown in Fig. 3-2(a).

If the value of $R$ is changed, the intersection on the vertical axis will change. This affects the slope of the load line, and gives a different point of intersection between the load line and the device characteristics. The point of intersection between the device characteristics and the load line ($V_{D_0}$, $I_{D_0}$) is called the point of operation or the quiescent point ($Q$-point) as defined by a dc network.

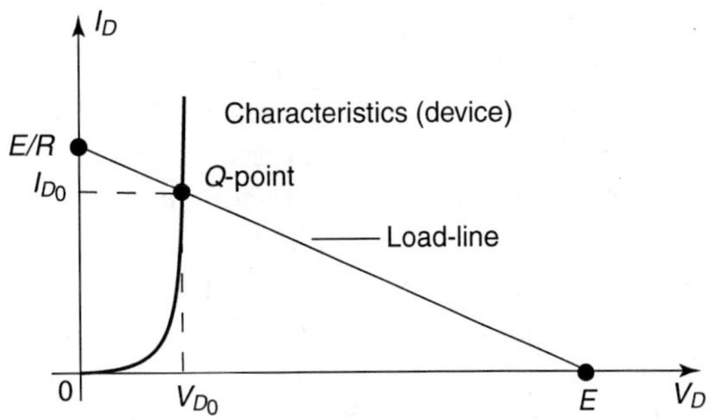


**Figure 3-2(a)** The load line on the characteristics of the diode

From the circuit diagram given in Fig. 3-1(a) it can be seen that the voltage drop across the diode is given by:

$$V_D = E - V_L$$

or,

$$V_D = E - i_D R_L \tag{3-5}$$

where, $V$ is the supply voltage, $V_L$ is the voltage across the load, and $I_a$ is the current flowing through the diode.

Equation (3-5) gives a relation between the voltage across the diode and the current flowing through it. It can be seen that this equation is an equation of a straight line. The load line and the static characteristic curve of the diode intersect at the quiescent point. The co-ordinates of the $Q$-point are $v_Q$, $i_Q$. The point is as shown in Fig. 3-3(a). Again, if the value of the voltage source is changed, another point on the static characteristic of the diode is obtained. The co-ordinates of the new point are $Q'$ ($v_Q'$, $i_Q'$).

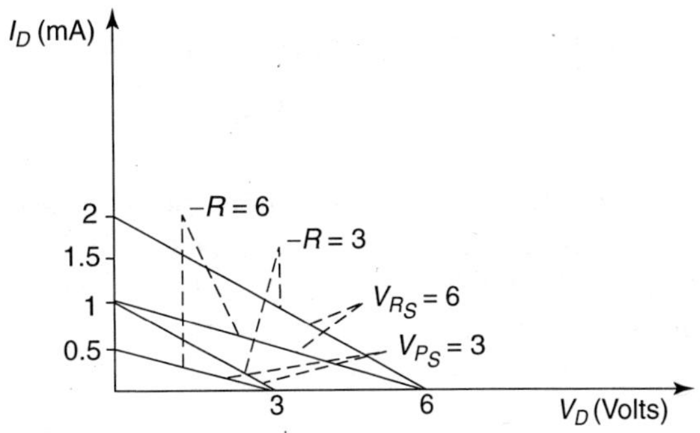


**Figure 3-2(b)** Illustration of a load line when: (1) voltage is constant and resistance varies (2) voltage varies and the resistance is constant

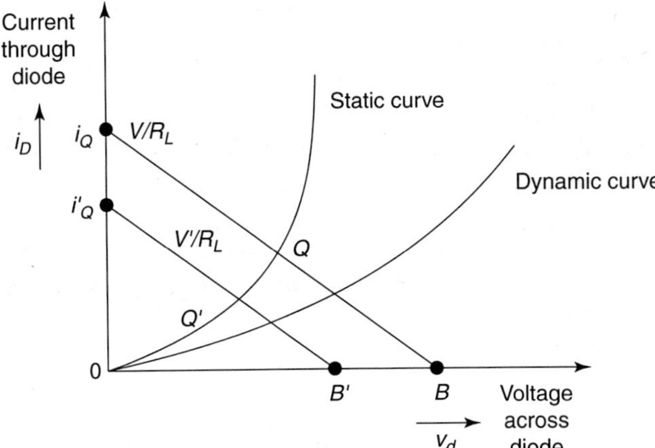


**Figure 3-3(a)** Change of $Q$-point with changes in supply voltage and load

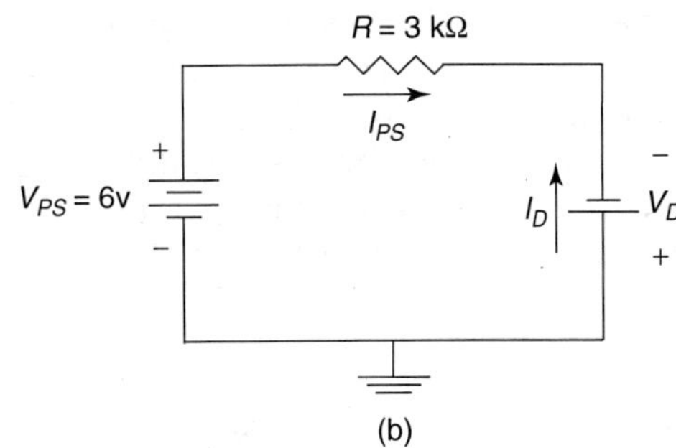


**Figure 3-3(b)** Reverse-biased diode circuit

If the diode currents $i_Q$ and $i_Q'$ are plotted vertically above the corresponding supply voltages, two distinct points are obtained—$B$ and $B'$. The curve passing through these two points is known as the *dynamic load line*. Dynamic load line can be obtained for different load resistances. This dynamic load line is important, because with the help of this, for any given input voltage, the diode current can be obtained directly from the graph. The load-line concept is important in reverse biasing.

Figure 3-3(b) depicts the diode circuit under the reversed condition. The forward-biased parameters over here are $I_D$ (diode current) and $V_D$ (voltage). It can be written with the application of Kirchoff's voltage law, $V_{PS} = I_{PS}R - V_D = -I_DR - V_D$ (where, $I_D = -I_{PS}$).

The equation, $V_{PS} = I_{PS}R - V_D = -I_DR - V_D$, characterizes the load line.

For the first end point, putting $I_D = 0$:

$$V_D = -V_{PS} = -6\text{ V}$$

For the second end point, putting $I_D = 0$:

$$I_D = \frac{-V_{PS}}{R} = -\frac{6}{3} = -2\text{ mA}$$

Figure 3-3(c) depicts the plot of diode characteristics and load line. The reverse-biased condition of the diode is demonstrated by the fact that the load intersects the diode characteristics curve in the third quadrant at $V_D = -6$ V and $I_D = 0$.

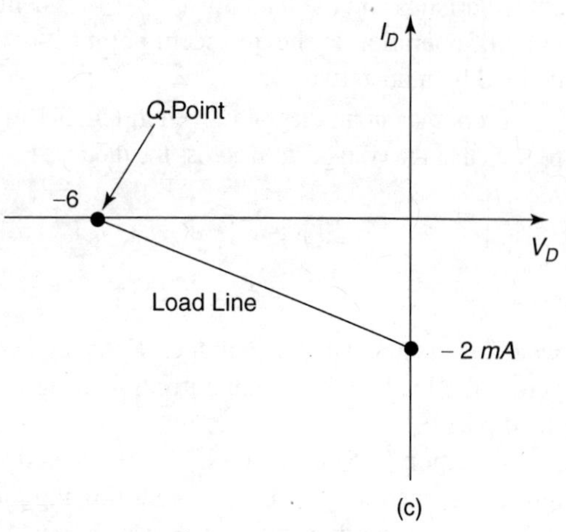


**Figure 3-3(c)** Load line

## 3-4 ZENER DIODE AS VOLTAGE REGULATOR

A Zener diode can be used as a voltage regulator because it maintains a constant output voltage even though the current passing through it changes. It is generally used at the output of an unregulated power supply to provide a constant output voltage free of ripple components.

The circuit diagram of a voltage regulator is shown in Fig. 3-4. The circuit consists of a current limiting resistor $R_S$ and a Zener diode connected in parallel with the load resistance $R_L$. The diode is selected in such a way that its breakdown voltage is equal to the desired regulating output.

For proper voltage regulation, the voltage of an unregulated power supply must be greater than the Zener voltage of the diode selected. The diode does not conduct current when the input voltage is less than the Zener voltage. The value of $R_S$ is chosen to ensure that the diode initially operates in the breakdown region under the Zener voltage across it. The function of the regulator is to keep the output voltage nearly constant with changes in $V_{\text{in}}$ or $I_L$.

The operation is based on the fact that in the Zener breakdown region small changes in the diode voltage are accompanied by large changes in the diode current. The large currents flowing through $R_S$ produce voltages that compensate for the changes in $V_{\text{in}}$ or $I_L$. The relation gives the input current:

$$I = \frac{V_{\text{in}} - V_Z}{R_S} = I_Z + I_L \quad (3\text{-}6)$$

There are two types of regulation:

(i) Regulation with varying input voltage, also known as line regulation
(ii) Regulation with varying load resistance, also known as load regulation

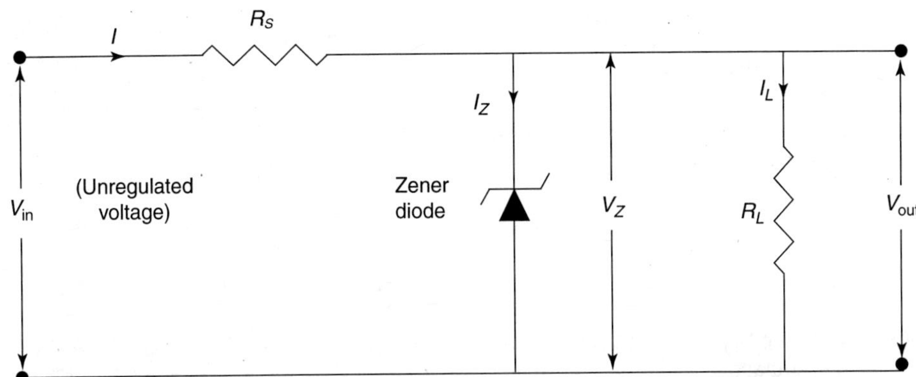


**Figure 3-4** Zener regulation of a variable input voltage

## 3-4-1 Line Regulation

When the input voltage is more than $V_Z$, the Zener diode conducts. With a further increase in $V_{in}$, the input current $I$ will also increase. This increases the current through the Zener diode without affecting the load current $I_L$. The limitations on the input-voltage variations are set by the minimum and maximum current values ($I_{ZK}$ and $I_{ZM}$) within which the Zener diode can operate.

The increase in the input current $I$ will also increase the voltage drop across series resistance $R_S$, thereby keeping the output voltage constant. If the input voltage decreases, the input current through the Zener diode will also decrease. Consequently, the voltage drop across the series resistance will be reduced. Thus, the output voltage and the load current remain constant.

For fixed values of $R_L$ (see Fig. 3-4), the voltage $V_i$ must be sufficiently large to turn the Zener diode on. The turn-on voltage is determined by $V_L$ and $V_{i_{min}}$ as:

$$V_L = V_Z = \frac{R_L V_i}{R_L + R_S} \tag{3-7a}$$

$$V_{i_{min}} = \frac{(R_L + R_S)V_Z}{R_L} \tag{3-7b}$$

The maximum value of $V_i$ is limited by the maximum Zener current $I_{ZM}$.

We have:

$$I_{ZM} = I_R - I_L \tag{3-8}$$

$$I_{R\max} = I_{ZM} + I_L \tag{3-9}$$

As $I_L$ is fixed at $V_Z/R_L$, and $I_{ZM}$ is the maximum value of $I_Z$, the maximum $V_i$ is given by:

$$V_{i_{max}} = V_{R_{S_{max}}} + V_Z \tag{3-10}$$

$$V_{i_{max}} = I_{R_{max}} R_S + V_Z \tag{3-11}$$

Figure 3-5(a) shows a plot of $V_o$ versus $V_i$ for line regulation. It is observed from the graph that the output voltage $V_o$ remains constant when the diode is in the Zener region, i.e., when the stipulated current is flowing. This is called *input* or *line regulation*.

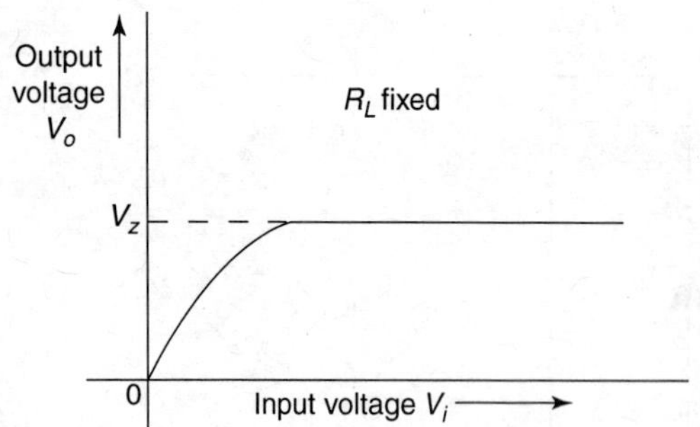


**Figure 3-5(a)** Output voltage vs. input voltage for line regulation

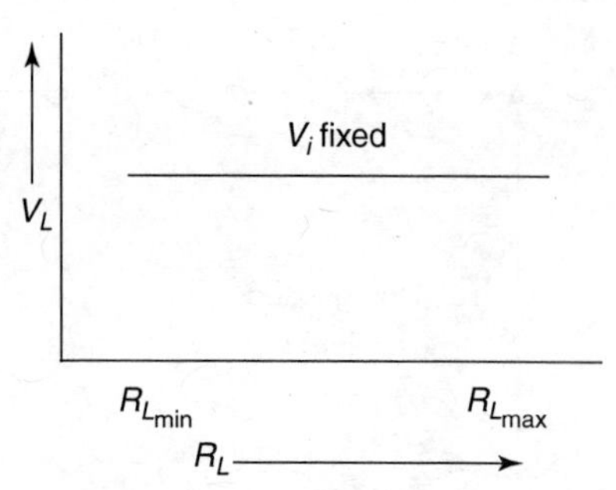


**Figure 3-5(b)** Load regulation showing the variation of load voltage $V_L$ and $R_L$ taking $V_i$ as constant

## 3-4-2 Load Regulation: Regulation with Varying Load Resistance

In this case, the input voltage, $V_{in} > V_z$, is kept fixed and the load resistance, $R_L$, is varied. The variation of $R_L$ changes the current $I_L$ through it, thereby changing the output voltage. When the load resistance decreases, the current through it increases. This ultimately causes a decrease in the Zener current. As a result, the input current and the voltage drop across $R_S$ remains constant. And the output voltage is also kept constant, as shown in Fig. 3-5(b). On the other hand, if the load resistance increases, the load current decreases. As a result, the Zener current $I_Z$ increases. This again keeps the value of input current and voltage drop across the series resistance constant. Thus, the output voltage remains constant. This is called *load regulation*.

Due to the offset voltage $V_Z$, there is a specific range of resistor values, which will ensure that the Zener is in the ON state. Too small a load resistance, $R_L$, will result in a voltage, $V_L$, across the load resistor to be less than $V_Z$, and the Zener device will be in the OFF state, which is usually not required in this kind of operation.

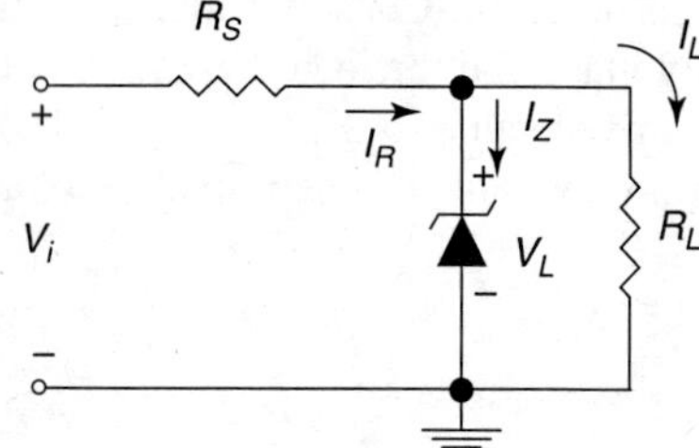


**Figure 3-6** Diode as a voltage regulator

To determine the minimum load resistance which will turn the Zener diode on, simply remove the Zener diode, as shown in Fig. 3-6, and calculate the value of $R_L$ that will result in a load voltage $V_L = V_Z$.

That is:

$$V_L = V_Z = \frac{R_L V_i}{R_L + R_S} \tag{3-13}$$

From the voltage-divider rule and solving for $R_L$, we have:

$$R_{L_{\min}} = \frac{R_S V_Z}{V_i - V_Z} \tag{3-14}$$

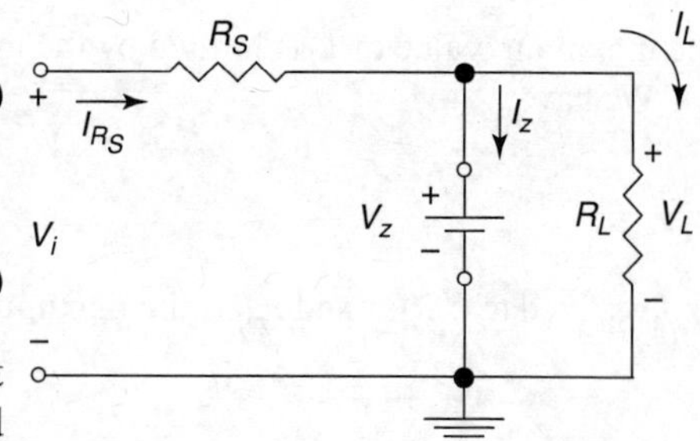


**Figure 3-7** Analytical circuit of Zener diode being used as a regulator

Any resistance with a value greater than the $R_L$ will ensure that the Zener diode is in the ON state and the diode can be replaced by its $V_Z$ source equivalent, as shown in Fig. 3-7.

The condition defined by Eq. (3-14) establishes the minimum $R_L$, but the maximum $I_L$ is:

$$I_{L_{\max}} = \frac{V_L}{R_L} = \frac{V_Z}{R_{L_{\min}}} \tag{3-15}$$

Once the diode is in the ON state, the voltage across $R_S$ will remain fixed at:

$$V_{R_S} = V_i - V_Z \tag{3-16}$$

And $I_R$ remains fixed at:

$$I_{R_S} = \frac{V_{R_S}}{R_S} \tag{3-17}$$

$$I_Z = I_R - I_L \tag{3-18}$$

This will result in a minimum $I_Z$ when $I_L$ is a maximum; and a maximum $I_Z$ when $I_L$ is a minimum value ($I_R$ is constant). Since $I_Z$ is limited to $I_{ZM}$, it does affect the range of $R_L$, and therefore, $I_L$.

Substituting $I_{ZM}$ for $I_Z$ establishes the minimum $I_L$ as:

$$I_{L_{\min}} = I_{R_S} - I_{ZM} \tag{3-19}$$

and the maximum load resistance as:

$$R_{L_{\max}} = \frac{V_Z}{I_{L_{\min}}} \tag{3-20}$$

Thus, load resistance can be calculated by using Eqs. (3-14) and (3-20).

## Solved Examples

**Example 3-1** A *p–n* germanium junction at room temperature has a reverse saturation current of 10 $\mu$A, negligible ohmic resistance, and a Zener breakdown voltage of 100 V. A 1.5 K resistor is in series with this diode, and a 45 V battery is impressed across this combination.

(a) Find the current if the diode is forward-biased.

(b) Find the current if the battery is inserted into the circuit with reverse polarity.

(c) Repeat part (a) and (b) if the Zener breakdown voltage is 10 V with 1 K resistor and 30 V battery.

**Solution:**

(a) The solution can be found graphically by plotting the diode characteristics and drawing the load line. Using the method of successive approximation, we have with us, diode drop equal to zero, in essence neglecting the diode threshold voltage.

$$I = \frac{45}{1.5\text{ K}} = 30\text{ mA}$$

For this current, $V$ is given by:

$$30 \times 10^{-3} = 10 \times 10^{-6}\,(e^{38.4\text{ V}} - 1)$$

or,
$$e^{38.4\text{ V}} = 3000 \quad \text{and} \quad V = 0.208\text{ V}$$

Hence,
$$I = \frac{45 - 0.208}{1.5\text{ K}} \approx 29.8\text{ mA}$$

For this current, $V = 0.2$ V

$\therefore$
$$I = 29.8\text{ mA}$$

(b) The diode drop is $-45$ V and $I = -I_0 = -10\ \mu$A. The voltage drop across the 1.5 K resistors is only 15 mV and may be neglected.

(c) In the forward direction, the answer is the same as in part (a), i.e., $I = 29.8$ mA.

In the reverse direction, we draw a load line from $V = -30$ V to $I = -30$ mA, as shown in the following figure. Then:

$$I' = -30 \times \frac{20}{30} = -20 \text{ mA}.$$

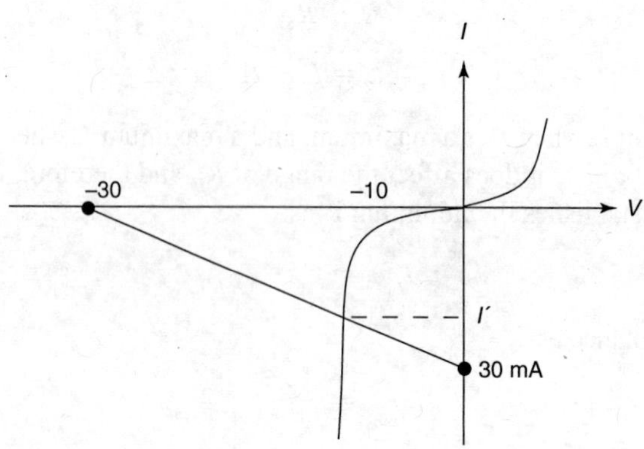


Alternatively, there is a 10 V across the diode, leaving 20 V across the 1 K resistor.

$\therefore$ Current = 20 mA, as there is a 10 V drop

**Example 3-2** Each diode is described by a linearized volt–ampere characteristic with incremental resistance $r$ and offset voltage $V_\gamma$. Diode $D_1$ is germanium with $V_\gamma = 0.2$ V and $r = 20\ \Omega$, whereas $D_2$ is silicon with $V_\gamma = 0.6$ V and $r = 15\ \Omega$. Find the diode currents if: (a) $R = 10$ K, (b) $R = 1.5$ K.

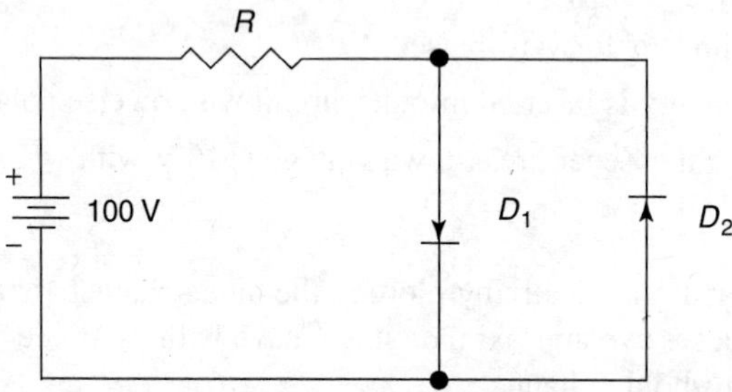


**Solution:**

(a) $R = 10$ K. Assume both diodes are conducting. We have:

$$100 = 10.02I_1 + 10I_2 + 0.2$$
$$100 = 10.015I_2 + 10I_1 + 0.6$$

R

$I_1$ 20 ohm $I_2$ 15 ohm

100V

0.2 V 0.6V

Solving for $I_2$ we find $I_2 < 0$. Thus our assumption that $D$ is ON is not valid. Assume $D_1$ is ON and $D_2$ is OFF. Then:

$$I_1 = \frac{100 - 0.2}{10.02} = 9.97 \text{ mA} \quad \text{and} \quad I_2 = 0$$

(b) $R = 1$ K. Assume both $D_1$ and $D_2$ are ON. We have:

$$100 = 1.52I_1 + 1.5I_2 + 0.2$$
$$100 = 1.515I_2 + 1.5I_1 + 0.6$$

Solving, we find $I_1 = 39.717$ mA and $I_2 = 26.287$ mA. Since both currents are positive, our assumption is valid.

**Example 3-3** Calculate the break region over which the dynamic resistance of a diode is multiplied by a factor of 10,000.

**Solution:**
We have:

$$r = \eta \frac{V_T}{I_0} \varepsilon^{-V/\eta V_T}$$

Hence,
$$\frac{r_1}{r_2} = \varepsilon^{V_2 - V_1/\eta V_T}$$

But,
$$\frac{r_1}{r_2} = 10^4$$

Hence,
$$\frac{\Delta V}{\eta V_T} = \frac{V_2 - V}{\eta V_T} = \ln 10^4$$

For silicon $\eta = 2$, and at room temperature $\Delta V = 2 \times 26 \times 4 \times 2.3 = 478$ mV.
For germanium, $\eta = 1$ and at room temperature $\Delta V = 1 \times 26 \times 4 \times 2.3 = 239$ mV.

■

## 3-5 RECTIFIERS

The process of rectification involves converting the alternating waveforms to the corresponding direct waveforms. Thus, it is one type of converter in which the direct waveforms must be filtered so that the resultant output waveforms become time invariant. Rectifiers can, in general, be classified into two categories.

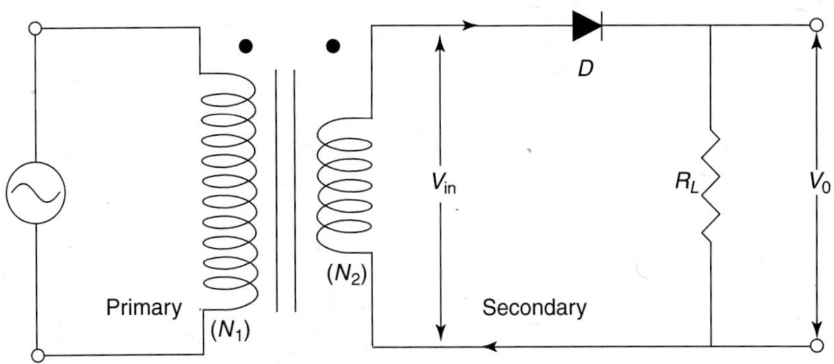


**Figure 3-8** Half-wave rectifier

### 3-5-1 Half-Wave Rectifier

In a half-wave rectifier, the output waveform occurs after each alternate half-cycle of the input sinusoidal signal. Figure 3-8 shows a simple half-wave rectifier circuit.

The half-wave rectifier will generate an output waveform $v_o$. Between the time interval $t$ = 0 to $T/2$, the polarity of the applied voltage $v_i$ is such that it makes the diode forward-biased. As a result the diode is turned on, i.e., the forward voltage is more than the cut-in voltage of the diode. Substituting the short-circuit equivalence of the ideal diode will result in the equivalent circuit of Fig. 3-9 where it is obvious that the output signal is a replica of the applied signal.

For the period $T/2$ to $T$, the polarity of the input voltage ($v_i$) is reversed and the resulting polarity across the diode produces an OFF state with an open circuit equivalent, as shown in Fig. 3-10.

The result is the absence of a path for charge to flow and $v_o = i_R = 0$ V, $R = 0$ V for the period $T/2$ to $T$. The output signal $v_o$ has the average value determined by:

$$V_{dc} = 0.318\ V_m$$

The input $v_i$ and the output $v_o$ are sketched together in Fig. 3-11. It is to be noted and clearly understood that when the diode is in the forward-biased mode, it consumes 0.7 V in the case of a silicon-based diode, and 0.3 V in the case of a germanium-based diode. Thus, in such a case, there must be a drop in the voltage across the diode. Consequently, the voltage across the resistance $R$ in case of a half-wave rectifier is lowered to a value of $E - V_d$. If a sinusoidal source is kept in place of the battery, the sinusoidal voltage will represent the voltage across the resistance. The corresponding figure will then be as shown in Fig. 3-12.

In such a case, the peak of the voltage $V_m$ gets lowered by an amount $V_m - V_d$. Thus, there is a finite time lag between turn-on and turn-off time of the diode in a complete time period.

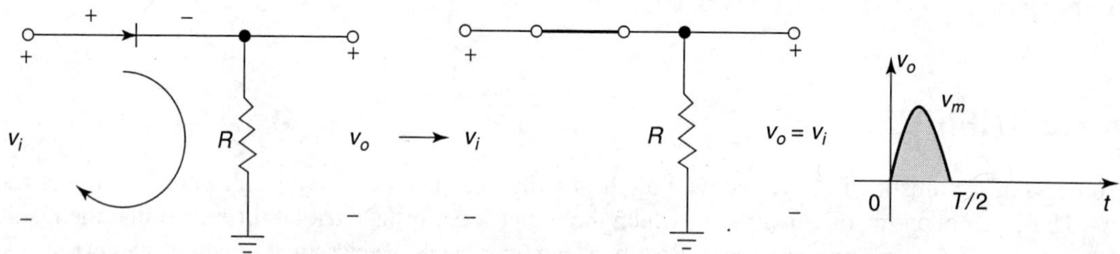


**Figure 3-9** Conduction region (0 to $T/2$)

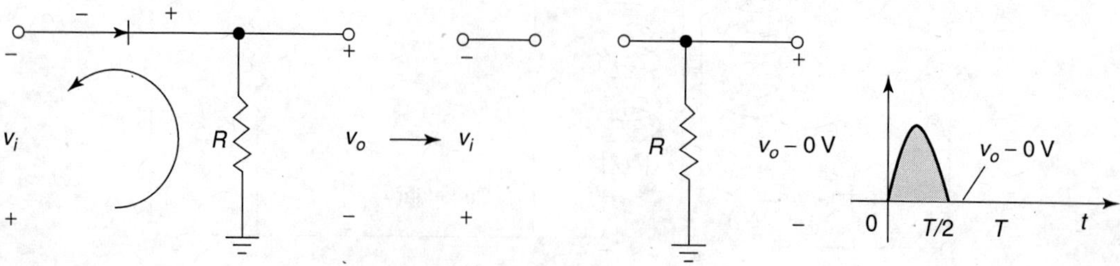


**Figure 3-10** Non-conducting region ($T/2$ to $T$)

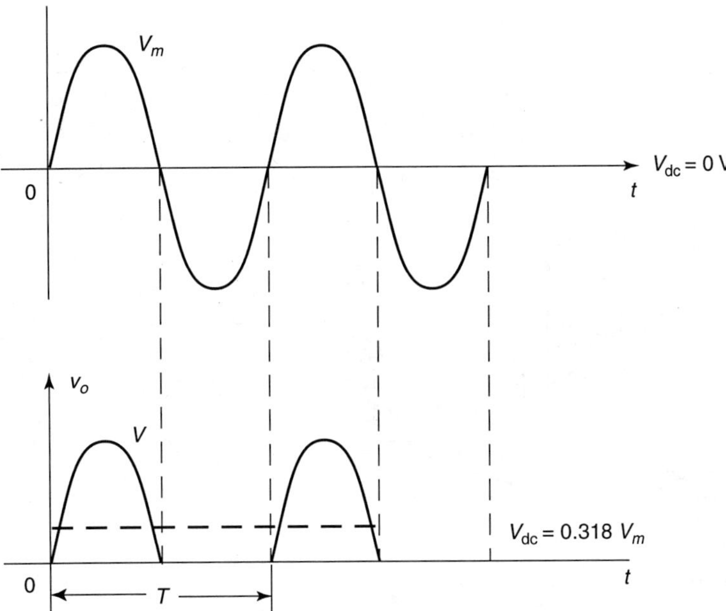


**Figure 3-11** Half-wave rectified signal

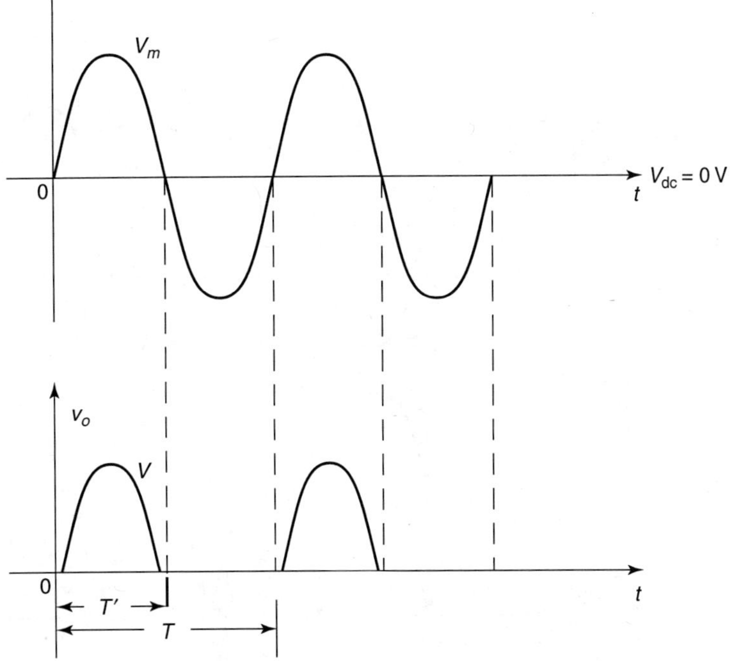


**Figure 3-12** Output signal of the form $V_{in} - V_{diode}$

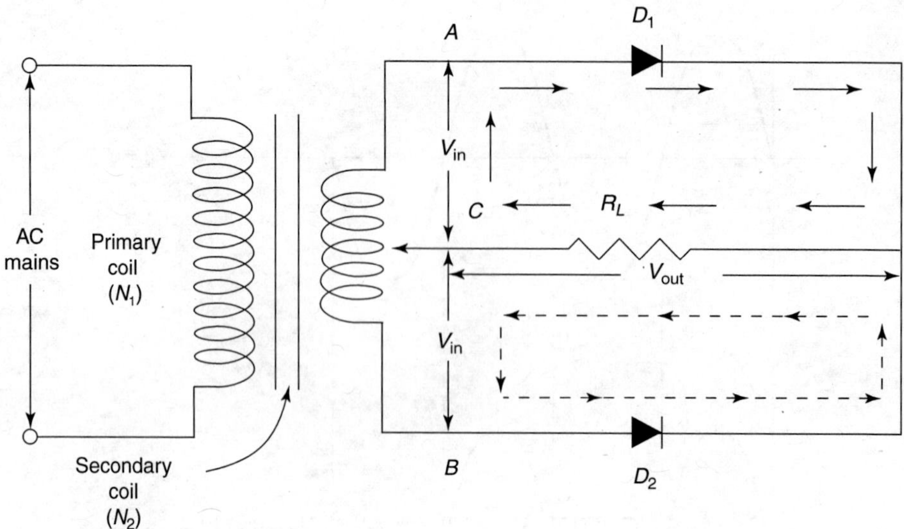


**Figure 3-13** Full-wave rectifier

## 3-5-2 Full-Wave Rectifier

The circuit diagram for full-wave rectifier is shown in Fig. 3-13. The full-wave rectifier can be classified into two distinct types.

(i) Centre-tapped transformer full-wave rectifier

(ii) Bridge type full-wave rectifier

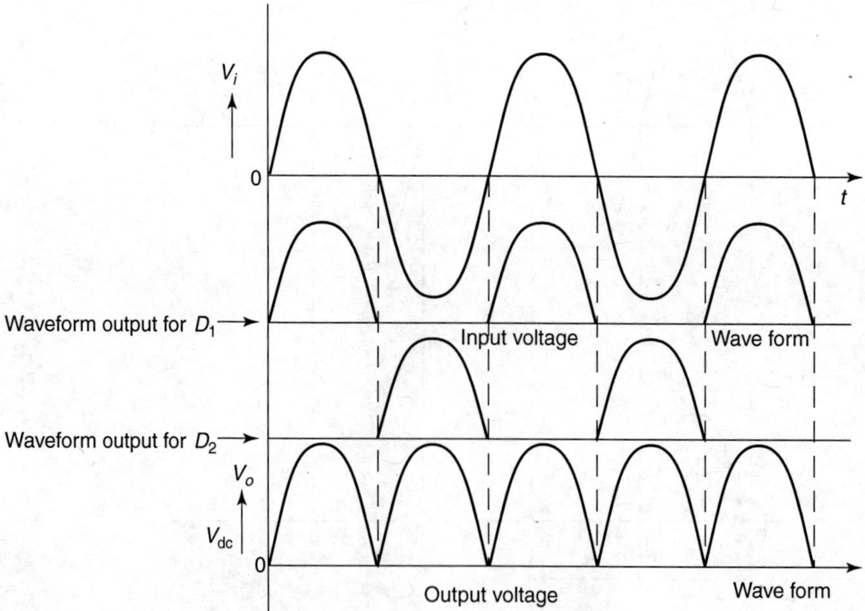


**Figure 3-14** Waveform for full-wave rectifier

### Centre-tapped transformer rectifier

It comprises of two half-wave circuits, connected in such a manner that conduction takes place through one diode during one half of the power cycle and through the other diode during the second half of the cycle.

When the positive half-cycle is applied to the input, i.e., transformer primary, then the top of the transformer secondary is positive with reference to the centre tap, while the bottom of the transformer secondary is negative with reference to the centre tap. As a result, diode $D_1$ is forward-biased and diode $D_2$ is reverse-biased. So the current will flow through $D_1$, but not through $D_2$ during the positive half-cycle. During the negative half-cycle, the condition is reversed. Diode $D_2$ is now forward-biased and diode $D_1$ is reverse-biased. Current will flow through diode $D_2$ and not through $D_1$ for the negative half-cycle. So the load current is shared alternatively by the two diodes and is unidirectional in each half-cycle. As a result, for the full-wave rectifier, we get the output for both the half-cycles. The waveforms for the full-wave rectifier are shown in Fig. 3-14.

### Bridge rectifier

The most important disadvantage of the centre-tapped rectifier is that it brings in the use of a heavy transformer with three terminals at its output, i.e., a centre-tapped transformer. The centre tapping may not be perfect in most cases. This problem can be solved by designing another circuit with four diodes and a simple transformer. This is called a *bridge rectifier*. The circuit of the bridge rectifier is shown in Fig. 3-15(a).

The circuit realizes a full-wave rectifier using four different diodes, connected in such a way that two of these diodes are forward-biased at a time and the other two are kept in OFF state. Consequently, the circuit is completed in both the half-cycles and a rectified output is obtained.

In the positive half-cycle of the input voltage, the current flows through the diode $D_4$, the resistor $R$ and diode $D_3$. Meanwhile the diodes $D_1$ and $D_2$ are reverse-biased. Thus, we observe that two diodes $D_3$ and $D_4$ are in the ON state and two diodes $D_1$ and $D_2$ are in the OFF state. The disadvantage of the bridge rectifier is that it uses four diodes instead of two as used in the full-wave rectifier. This does not matter much because diodes are quite cheap.

In the negative half-cycle, the other two diodes ($D_1$ and $D_2$) are switched ON and the previous two ($D_3$ and $D_4$) are in the OFF state. An important point to be noted is that since the current in both the half-cycle flow in the same direction, the output voltage is positive.

The biggest advantage of such a rectifier is that it does not require the use of a centre-tapped transformer. Again the PIV of the diodes has to be greater than the maximum negative voltage of the input signal.

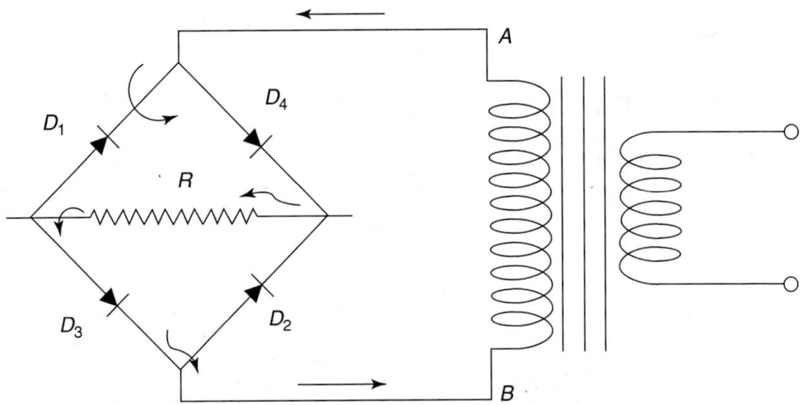


**Figure 3-15(a)** Bridge rectifier

### Advantages of a bridge rectifier

(i) In the bridge circuit a transformer without a centre tap is used.

(ii) The bridge circuit requires a smaller transformer as compared to a full-wave rectifier giving the identical rectified dc output voltage.

(iii) For the same dc output voltage, the PIV rating of a diode in a bridge rectifier is half of that for a full-wave circuit.

(iv) The bridge circuit is more appropriate for high-voltage applications, thus, making the circuit compact.

### Disadvantages of a bridge rectifier

(i) Two or more diodes are required in case of a bridge rectifier, as a full-wave rectifier uses two diodes whereas a bridge rectifier uses four diodes.

(ii) The amount of power dissipated in a bridge circuit is higher as compared to a full-wave rectifier. Hence, the bridge rectifier is not efficient as far as low voltages are concerned.

### Comparison between half-wave and full-wave rectifier

(i) In a half-wave rectifier, a single diode exists and the load current flows through it for only the positive half-cycle. On the other hand, in a full-wave rectifier, the current flows throughout the cycles of the input signals.

(ii) Full-wave rectifiers require a centre-tapped transformer. For a half-wave rectifier, only a simple transformer is required.

(iii) The PIV in a half-wave rectifier is the maximum voltage across the transformer secondary. Whereas, in the case of a full-wave rectifier, the PIV for each diode is two times the maximum voltage between the centre tap and at the either end of the transformer secondary.

(iv) In a half-wave rectifier, the frequency of the load current is the same as that of the input signal and it is twice the frequency of the input supply for the full-wave rectifier.

(v) The dc load current and conversion efficiency for a full-wave rectifier is twice that of a half-wave rectifier. Also, the ripple factor of the full-wave rectifier is less than that of the half-wave circuit. This indicates that the performance of the full-wave rectifier is better than the half-wave rectifier.

(vi) In a full-wave rectifier, two diode currents flow through the two halves of the centre-tapped transformer secondary in opposite directions, so that there is no magnetization of the core. The transformer losses being smaller, a smaller transformer can be used for a full-wave rectifier.

## 3-5-3 Use of Filters in Rectification

A rectifier converts ac to dc Inadvertently, when the output voltage of the rectifier is passed through the load, fluctuating components of currents appear across the load; these are called *ripples*. Filters help in reducing these ripples considerably. The simplest filter in a rectifier can be understood by placing a capacitor across the load, as shown in Fig. 3-15(b).

In the filtered-rectified circuit, diodes $D_1$ and $D_2$ conduct in alternate half cycles at the secondary of the transformer. So, in the absence of the filtering capacitor $C$, the output voltage consists of a series of half sinusoids. This is shown in Fig. 3-16.

During the positive half-cycle the capacitor starts charging and gets charged to the maximum amplitude of the input signal. Beyond this maximum voltage the voltage across the diode $D_1$ is reversed and the diode $D_1$ consequently stops conducting. During this period, the capacitor discharges through the load with a time

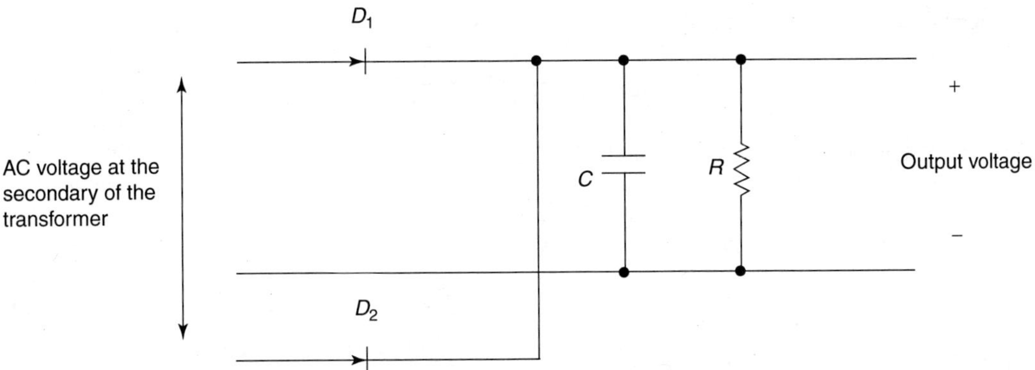


**Figure 3-15(b)** A full-wave capacitor-filtered rectifier

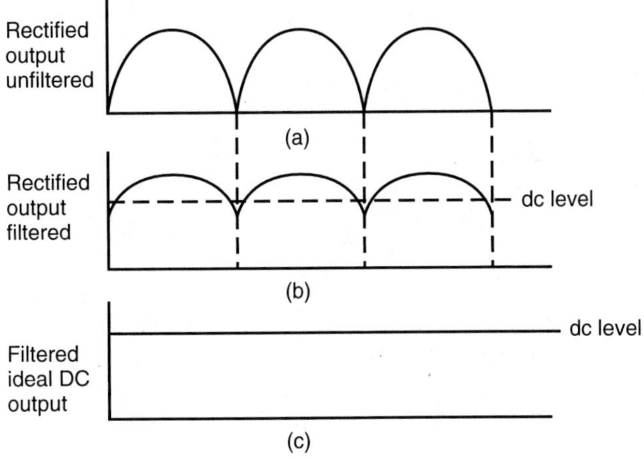


**Figure 3-16** (a) Rectified output without filter
(b) Rectified output with filter
(c) Ideal dc filtered dc output with proper choice of capacitor in filter

constant $\tau = CR$. As the capacitor reactance at the ripple frequency is much smaller than $R$, the time constant $CR$ is much larger than the time period of the alternating voltage. The capacitor, thus, discharges very slowly and the output waveform during this interval is represented by the curve $BD$. In the following half-cycle also, the same thing is repeated with the diode $D_2$.Thus, due to the operation of the filter, it can be seen from Fig. 3-16 that the output curve is smoothened and consequently, the ripple factor also decreases. Also, the regulation increases. The capacitor voltage at any given time is:

$$\frac{V_{NL} - V_{RL}}{V_{RL}} \times 100 = \frac{V_M - I_{dc}R_L}{I_{dc}R_L} \times 100$$

$$= \frac{I_{dc}R_0}{I_{dc}R_L} \times 100 = \frac{100}{4fCR_L}$$

$$\Delta V = \frac{I_{dc}T}{2C} \qquad (3\text{-}21)$$

where, $T$ is the time period of the input signal.

The average voltage across the capacitor is:

$$V_{dc} = V_m - \frac{\Delta V}{2} = V_m - \frac{I_{dc}}{4fC} \tag{3-22}$$

The percentage regulation (discussed in detail in the next section) is:

$$\frac{V_{NL} - V_{RL}}{V_{RL}} \times 100 = \frac{V_M - I_{dc}R_L}{I_{dc}R_L} \times 100$$

$$= \frac{I_{dc}R_0}{I_{dc}R_L} \times 100 = \frac{100}{4fCR_L} \text{ percent} \tag{3-23}$$

The fluctuation $\Delta V$ is a measure of the ripple voltage, we have:

$$V'^2_{rms} = \frac{2}{T}\int_0^{T/2}\left(\frac{\Delta V}{2} - \frac{\Delta V}{T/2}t\right)^2 dt$$

So that,

$$V'_{rms} = \Delta V/2\sqrt{3} = I_{dc}/\left(4\sqrt{3}fC\right) \tag{3-24}$$

noting that $V_{dc} = I_{dc}R_L$, the ripple factor is written as:

$$\gamma = \frac{V'_{rms}}{V_{dc}} = \frac{1}{4\sqrt{3}fCR_L} \tag{3-25}$$

## 3-5-4 Regulation

The average load current can be written as:

$$I_{dc} = KI_m = \frac{KV_s}{R_f + R_L} \tag{3-26}$$

where, $R_f$ is the forward resistance of the diode and $R_L$ is the value of the load resistance.

Again, the value of $K$ is:

$$\frac{V_{NL} - V_{RL}}{V_{RL}} \times 100\% \tag{3-27}$$

The value of $K$ is $1/\pi$ in case of a half-wave rectifier and $2/\pi$ in case of a full-wave rectifier. The dc load voltage is given by:

$$V_{dc} = I_{dc}R_L = KV_s - I_{dc}R_L \tag{3-28}$$

A plot of $V_{dc}$ against $I_{dc}$ gives a linear variation of $V_{dc}$ with $I_{dc}$.

The variation of $V_{dc}$ with $I_{dc}$ is called regulation and the plot of $V_{dc}$ vs. $I_{dc}$ is referred to as the voltage regulation characteristics of the rectifier. In an ideal rectifier, $V_{dc}$ is independent of $I_{dc}$. In practice, the departure of the behaviour of an actual rectifier from that ideal rectifier is expressed by percentage voltage regulation. It is defined as:

Percentage voltage regulation:

$$\frac{V_{NL} - V_{RL}}{V_{RL}} \times 100\% \tag{3-29}$$

In an ideal case, $V_{NL} = V_{RL}$. So we get the percentage regulation of a rectifier as zero.

## Solved Examples

**Example 3-4** A diode, whose internal resistance is 30 Ω, is to supply power to a 990 Ω load from a 110 V (rms) source of supply. Calculate (a) the peak load current, (b) the dc load current, (c) the ac load current, (d) the dc diode voltage, (e) the total input power to the circuit, and (f) the percentage regulation from no load to the given load.

**Solution:**

(a)
$$I_m = \frac{V_m}{R_f + R_L} = \frac{110/2}{1020} = 152.5 \text{ mA}$$

(b)
$$I_{dc} = \frac{I_m}{\pi} = 152.5/\pi = 48.5 \text{ mA}$$

(c)
$$I_{rms} = \frac{1}{2}(152.5) = 76.2 \text{ mA}$$

(d)
$$V_{dc} = \frac{I_m R_L}{\pi} = -48.5 \times .990 = -48 \text{ mA}$$

(e)
$$P_i = I^2_{rms}(R_f + R_L) = (76.2 \times 10^{-3})^2 (1020)$$
$$= 5800 \times 10^{-6} \times 1020 = 5.92 \text{ W}$$

(f)
$$\text{Percentage regulation} = \frac{V_{NL} - V_{FL}}{V_{FL}}(100\%) = \frac{V_m/\pi - I_{dc}R_L}{I_{dc}R_L}(100\%)$$
$$= \frac{49.5 - 48}{48} = 3.125\%$$

■

## 3-5-5 Performance Analysis of Various Rectifier Circuits

### Half-wave rectifier

Peak inverse voltage (PIV). It is the voltage that the diode must withstand, and it is equal to the peak input voltage, $V_m$.

DC voltage. The average output voltage

$$V_{o\,(dc)} = 0.318\, V_m \tag{3-30}$$

And the rms voltage:

$$V_{rms} = 1.21\, V_{o\,(dc)} \tag{3-31}$$

DC value of load current. The load current of a rectifier is unidirectional but fluctuating. The current at the output of the diode is:

$$i_L = I_m \sin \omega t \quad \text{for } 0 \le \omega t \le \pi,$$
$$i_L = 0 \quad \text{for } \pi \le \omega t \le 2\pi.$$

where, $I_m$ is the amplitude of the input signal. If $V_S$ is the amplitude of the transformer secondary voltage, the value of $I_m$ is given by:

$$I_m = \frac{V_S}{R_f + R_L}$$

From the definition of the average value of the load current:

$$I_{\text{dc}} = \frac{1}{2\pi}\int_0^{\pi} I_m \sin \omega t d(\omega t) = \frac{I_m}{\pi} \tag{3-32}$$

The value of rms current can be obtained from the definition, that is:

$$I_{\text{rms}}^2 = \frac{1}{2\pi}\int_0^{\pi} I_m^2 \sin^2 \omega t d(\omega t) = \frac{I_m^2}{4}$$

$$I_{\text{rms}} = \frac{I_m}{2} \tag{3-33}$$

**Ripple factor.** Periodically, fluctuating components—the ripples—are superimposed on $I_{\text{dc}}$ to give the actual load current. Due to these fluctuating components, the conversion from ac to dc by a rectifier is not perfect. The ripple factor gives a measure of this imperfection or the fluctuating components. The ripple factor, $r$, is defined by:

$$r = \frac{\text{rms value of the alternating components of the load current}}{\text{average value of the load current}}$$

The ripple factor $r$ is given by:

$$r = \sqrt{\left(\frac{I_{\text{rms}}}{I_{\text{dc}}}\right)^2 - 1} \tag{3-34}$$

From the expressions of dc and rms current, we can derive the ripple factor. Thus, we obtain:

$$r = \sqrt{\left(\frac{I_m/2}{I_m/\pi}\right)^2 - 1}$$

which, after calculation stands at 1.21.
Therefore, ripple factor = 1.21 or 121%.

**Efficiency.** The effectiveness of a rectifier in delivering the dc output power is generally measured by the rectification efficiency:

$$\eta_R = \frac{P_{\text{o(dc)}}}{P_{\text{o(ac)}}} = \frac{(V_m/\pi)^2/R}{(V_m/2)^2/R} = \frac{4}{\pi^2} = 40.5\% \tag{3-35}$$

Efficiency can also be defined as:

$$\left(\frac{I_{\text{rms}}}{I_{\text{dc}}}\right) = \frac{I_m/2}{I_m/\pi} = \frac{\pi}{2} = 1.571\,\% \tag{3.36}$$

From the above, we see that $(I_{\text{dc}}/I_{\text{rms}}) = 2/\pi$ and putting the value in Eq. (3-35), we get

$$\eta_R = \frac{40.6}{1 + R_f/R_L}\text{percent} \tag{3-37}$$

### Full-wave rectifier

**Peak inverse voltage.** This is the peak voltage that the diode must be able to withstand without any breakdown. In other words, it is the largest reverse voltage that is expected to appear across the diodes.

In this case:

$$\text{PIV} = -2\text{ V}$$

where, $V$ is the voltage peak of the input waveform.

**DC voltage.** The average output voltage:

$$V_{o(\text{dc})} = 0.636\, V_m \tag{3-38}$$

The rms voltage:

$$V_{\text{rms}} = 0.483 V_{o(\text{dc})} \tag{3-39}$$

**Average load current.** From the definition of the load current, we obtain:

$$I_{\text{dc}} = \frac{2}{2\pi}\int_0^{\pi} I_m \sin\omega t d(\omega t) = \frac{2I_m}{\pi} \tag{3-40}$$

And the rms value of the current is given by:

$$I_{\text{rms}} = \left[\frac{2}{2\pi}\int_0^{\pi} I_m{}^2 \sin^2\omega t d(\omega t)\right]^{1/2} = \frac{I_m}{\sqrt{2}} \tag{3-41}$$

Thus, we obtain:

$$\left(\frac{I_{\text{rms}}}{I_{\text{dc}}}\right) = \frac{I_m/\sqrt{2}}{2I_m/\pi} = \frac{\pi}{2\sqrt{2}} = 1.11 \tag{3-42}$$

**Current ripple factor.** Periodically, the fluctuating component—ripples—are superimposed on $I_{\text{dc}}$ to give the actual load current. Due to these fluctuating components, the conversion of ac to dc by a rectifier is not perfect. The ripple factor gives a measure of this imperfection or the fluctuating components. The ripple factor is defined by:

$$r = \frac{\text{rms value of the alternating components of the load current}}{\text{average value of the load current}}$$

The ripple factor $r$ is given by:

$$r = \sqrt{\left(\frac{I_{\text{rms}}}{I_{\text{dc}}}\right)^2 - 1} \tag{3-43}$$

From Eqs. (3-40) and (3-41) we get:

$$I_{\text{rms}} = I_m/\sqrt{2}$$

$$r = \sqrt{\frac{\left(\frac{I_m}{2}\right)^2}{\left(\frac{2I_m}{\pi}\right)} - 1} \tag{3-44}$$

$$r = \sqrt{\left(\frac{\pi}{2\sqrt{2}}\right)^2 - 1}$$

$$= 0.482$$

**Efficiency.** The rectification efficiency is given by:

$$\eta = \frac{P_{dc}}{P_{ac}} \times 100\% \quad (3\text{-}45)$$

For a full-wave rectifier circuit:

$$P_{dc} = (I_{dc})^2 R_L = \left(\frac{2I_m}{\pi}\right)^2 \frac{1}{\left(R_f + R_L\right)^2} = \frac{4V_m{}^2 R_L}{\pi^2 \left(R_f + R_L\right)^2} \quad (3\text{-}46)$$

$$P_{ac} = (I_{rms})^2 \left(R_f + R_L\right) = \frac{V_m{}^2}{2\left(R_f + R_L\right)} \quad (3\text{-}47)$$

$$\eta = \frac{P_{dc}}{P_{ac}} \times 100\% = \frac{0.812}{\left(1 + \frac{R_f}{R_L}\right)} \times 100\% \quad (3\text{-}48)$$

## Solved Examples

**Example 3-5** Show that the maximum dc output power $P_{dc} = V_{dc} I_{dc}$ in a half-wave single-phase circuit occurs when the load resistance equals the diode resistance $R_f$.

**Solution:**
We have:

$$P_{dc} = I_{dc}^2 R_L = \frac{V_m^2 R_L}{\pi^2 \left(R_f + R_L\right)^2}$$

For max,

$$P_{dc} \text{ set } \frac{dP_{dc}}{dR_L} = 0$$

or,

$$\frac{V_m^2}{\pi}\left[\frac{(R_f + R_L)^2 - 2\,(R_f + R_L) R_L}{(R_f + R_L)^4}\right] = 0$$

Then, $R_f + R_L = 2R_L$

$$\therefore R_L = R_f$$

**Example 3-6** The efficiency of the rectification $\eta_r$ is defined as the ratio of the dc output power $P_{dc} = V_{dc} I_{dc}$ to the input power $P_i = (1/2\pi) \int_0^{2\pi} v_i i d\alpha$.

(a) Show that, for the half-wave rectifier circuit:

$$\eta_r = \frac{40.6}{1 + R_f/R_L}\,\%$$

(b) Show that, for the full-wave rectifier, $\eta_r$ has twice the value of the value given in part (a).

**Solution:**

(a)
$$\eta_r = \frac{P_{dc}}{P_i} \times 100\% = \frac{I_{dc}^2 R_L \times 100\%}{I_{rms}^2 (R_f + R_L)} = \left(\frac{I_{dc}}{I_{rms}}\right)^2 \frac{100\%}{1+\dfrac{R_f}{R_L}}$$
$$= \left(\frac{I_m}{\pi/I_m/2}\right)^2 \frac{100\%}{1+\dfrac{R_f}{R_L}} = \frac{40.6}{1+\dfrac{R_f}{R_L}}\%$$

(b)
$$\eta_r = \left(\frac{I_{dc}}{I_{rms}}\right)^2 \frac{100\%}{1+\dfrac{R_f}{R_L}} = \left(\frac{2I_m/\pi}{I_m/2}\right)^2 \frac{100\%}{1+\dfrac{R_f}{R_L}} = \frac{81.2}{1+\dfrac{R_f}{R_L}}\%$$

**Example 3-7** Prove that the regulation of both the half-wave and the full-wave rectifier is given by: Percentage regulation $= R_f/R_L \times 100\%$.

**Solution:**

Consider a half-wave rectifier: $V_{NL} = \dfrac{V_m}{\pi}$ and $V_{FL} = \dfrac{V_m}{\pi}\dfrac{R_L}{R_L + R_f}$

$$\therefore \text{ Percentage regulation } = \frac{V_{NL} - V_{FL}}{V_{FL}} \times 100\% = \frac{\dfrac{V_m}{\pi}\left(1 - \dfrac{R_L}{R_L + R_f}\right)}{\dfrac{V_m}{\pi}\dfrac{R_L}{R_L + R_f}} = \frac{R_f}{R_L} \times 100\%$$

For a full-wave rectifier, $V_{NL}$ and $V_{FL}$ are doubled, and hence, the regulation remains the same as above.

**Example 3-8** A full-wave single phase rectifier consists of a double diode, the internal resistance of each element of which may be considered to be constant and equal to 500 Ω. These feed into a pure resistance load of 2,000 Ω. The secondary transformer voltage to centre tap is 280 V. Calculate (a) the dc load current, (b) the direct current in each tube, (c) the ac voltage across each diode, (d) the dc output power, and (e) the percentage regulation.

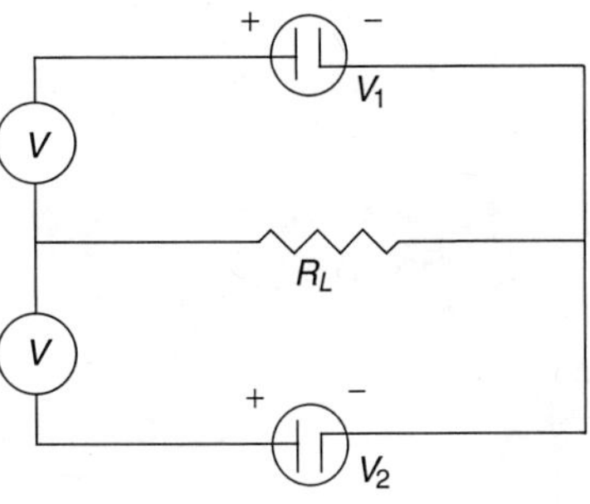


**Solution:**

(a) $$I_{dc} = \frac{2I_m}{\pi} = \frac{2V_m}{\pi(R_L + R_f)} = \frac{2 \times 280}{R2/\pi\,(2500)} = 101 \text{ mA}.$$

(b) $$(I_{dc})_{tube} = \frac{1}{2} I_{dc} = 50.5 \text{ mA}.$$

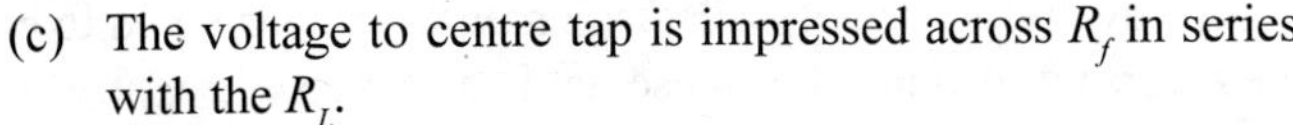

(c) The voltage to centre tap is impressed across $R_f$ in series with the $R_L$.

Hence, the voltage across the conducting is sinusoidal with a peak value:

$$V_m \frac{500}{2500} = 0.2\, V_m$$

During non-conduction of $V_1$, we see by transversing the outside path of the circuit sketched that $v_1 = -2v - v_2$.

Since, $V_2$ is now conducting, its peak value is $0.2V_m$ and that of $v_1$ is $-2V_m + 0.2V_m = -1.8\, V_m$.

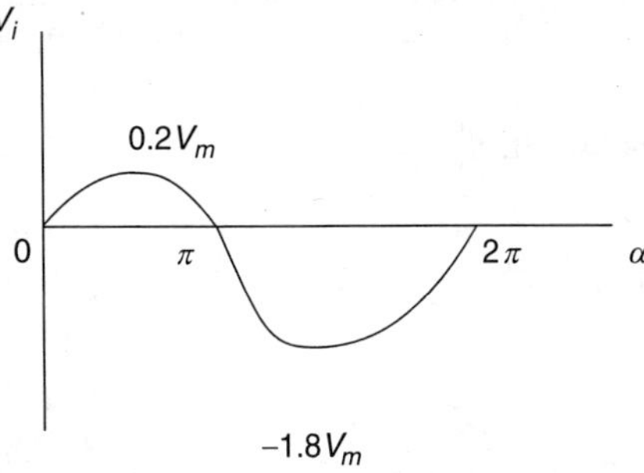

Thus, the voltage $v_1$ across $V_1$ is as shown, and its ac value is:

$$V^2_{\text{rms}} = \frac{1}{2\pi} = \int_0^{\pi} (0.2V_m)^2 \sin^2\alpha\, d\alpha + \frac{1}{2\pi}\int_{\pi}^{2\pi} (-1.8V_m)^2 \sin^2\alpha\, d\alpha = \frac{V^2_m}{4}[(0.2)^2 + (1.8)^2]$$

Hence, $V_{\text{rms}} = 0.905\ V_m = 0.905 \times 280\sqrt{2} = 358$ V.

(d) $$P_{\text{dc}} = I^2_{\text{dc}} R_{\text{L}} = (0.101)^2\,(2000) = 20.4 \text{ W}$$

(e) $$\text{Percentage regulation} = \frac{R_f}{R_L} \times 100\% = \frac{500}{2000} \times 100\% = 25\%$$

**Example 3-9** In the full-wave single phase bridge, can the transformer and the load be interchanged? Explain carefully.

**Solution:**

The load and the transformer cannot be interchanged. If they were, the circuit shown would be the result. Note that if $A$ is positive with respect to $B$, all diodes will be reverse-biased. If $B$ is positive with respect to $A$, then all four diodes conduct, thus, short circuiting the transformer.

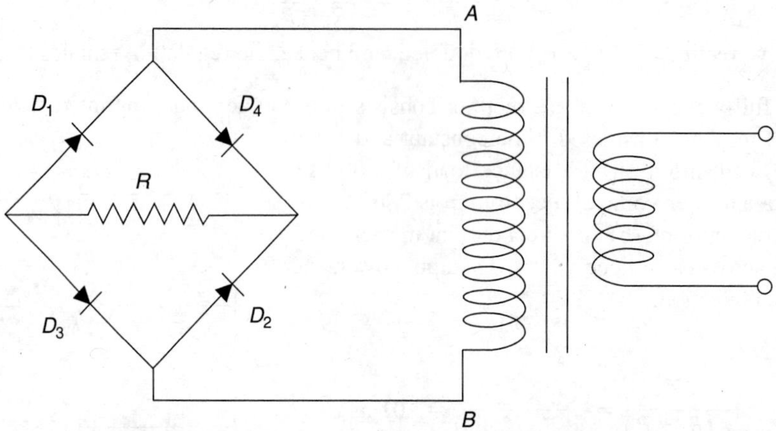


**Example 3-10** A 1 mA dc metre whose resistance is 20 Ω is calibrated to read rms volts when used in a bridge circuit with semiconductor diodes. The effective resistance of each element may be considered to be zero in the forward direction and infinite in the reverse direction. The sinusoidal input voltage is applied in series with a 5 K resistance. What is the full-scale reading of this metre?

**Solution:**

$$I_{\text{dc}} = \frac{2I_m}{\pi} = \frac{2V_m}{\pi R_L} = \frac{2\text{R}2V_{\text{rms}}}{\pi\,(5020)} = 1 \times 10^{-3}$$

Hence, $$v_{0(\text{max})} = 14 \text{ V}$$

**Example 3-11** An ac supply of 220 V is applied to a half-wave rectifier circuit through transformer with a turns ratio 10:1. Find (a) dc output voltage and (b) PIV. Assume the diode to be an ideal one.

**Solution:**
Given $V_i = 220$ V, $N_2/N_1 = 10$.

(a) The secondary voltage:

$$V_2 = V_1 \times \frac{N_2}{N_1} = 220 \times \frac{1}{10} = 22 \text{ V}$$

$$V_m = \sqrt{2}V_2 = \sqrt{2} \times 22 = 31.11 \text{ V}$$

$$V_{dc} = 0.318\ V_m = 0.318 \times 31.11 = 9.89 \text{ V}$$

(b) PIV of a diode is given by PIV $= V_m = 31.11$ V

**Example 3-12** In a half-wave rectifier circuit the input voltage is 230 V and transformer ratio is 3:1. Determine the maximum and the average values of power delivered to the load. Take $R_L$ equal to 200 Ω.

**Solution:**

Given, $V_1 = 230$ V, $\frac{N_2}{N_1} = \frac{1}{3}$, $R_L = 200\ \Omega$.

The rms value of secondary voltage is given by:

$$V_2 = V_1 \times \frac{N_2}{N_1} = 230 \times \frac{1}{3} = 76.67 \text{ V}$$

Maximum value of the secondary voltage is:

$$V_m = \sqrt{2}V_2 = 1.414 \times 76.67 = 108.41 \text{ V}$$

Maximum value of the load current is then given by:

$$I_m = \frac{V_m}{R_L} = \frac{108.41}{200} = 0.542 \text{ A}$$

Maximum load power:

$$P_{max} = I_m^2 \times R_L$$
$$= (0.542)^2 \times 200 = 58.75 \text{ W}$$

Average value of output voltage, $V_{dc} = 0.318V_m = 0.318 \times 108.41 = 34.47$ V

Average value of load current:

$$I_{dc} = \frac{V_{dc}}{R_L} = \frac{34.47}{200} = 0.172 \text{ A}$$

Average value of load power, $P_{dc} = I_{dc}^2 \times R_L = (0.172)^2 \times 200 = 5.92$ W

**Example 3-13** A half-wave rectifier is used to supply 30 V dc to a resistive load of 500 ohms. The diode has a forward resistance of 25 Ω. Find the maximum value of the ac voltage required at the input.

**Solution:**

Given, $V_{dc} = 30$ V, $r_f = 25\ \Omega$, $R_L = 500\ \Omega$.

Average value of the load current:

$$I_{dc} = \frac{V_{dc}}{R_L} = \frac{30}{500} = 0.06 \text{ A}$$

Maximum value of the load current, $I_m = \pi \times I_{dc} = 3.142 \times 0.06 = 0.188$ A

Voltage required at the input is given by:

$$V_{i(max)} = I_m^2 (r_f + R_L)$$

$$= (0.188)^2 (25 + 500) = 18.55 \text{ V}$$

**Example 3-14** A half-wave rectifier is used to supply 100 $V_{dc}$ to a load of 500 Ω. The diode has a resistance of 20 Ω. Calculate (a) the ac voltage required and (b) the efficiency of rectification.

**Solution:**

Given, $V_{dc} = 100$V, $R_L = 500\ \Omega$, $r_f = 20\ \Omega$.

(a) DC voltage:

$$V_{dc} = \frac{V_m}{\pi} = \frac{I_m (R_L + r_f)}{\pi}$$

$$I_{dc} = \frac{V_{dc}}{R_L} = \frac{100}{500} = 0.2 \text{ A}$$

$$I_m = I_{dc} \times \pi = 3.142 \times 0.2 \text{ A} = 0.6284 \text{ A}$$

$$V_m = I_m (R_L + r_f) = 0.6284\,(500 + 20) = 326.77 \text{ V}$$

(b) Rectification efficiency is given by:

$$\eta = \frac{0.406}{1 + \dfrac{r_f}{R_L}} = \frac{0.406}{1 + 20/500} = 0.39 = 39\%$$

**Example 3-15** A half-wave rectifier circuit has a load of 5000 Ω. Find the values of (a) current in the circuit, (b) dc output voltage across $R_L$ and (c) voltage across the load. Given $v = 50 \sin 100\pi t$, $r_f = 20\ \Omega$.

**Solution:**

Given, $v = 50 \sin 100\pi t$.

Comparing the given equation with the standard equation $v = V_m \sin 2\pi f t$, we have:

$$V_m = 50 \text{ V}, f = 50 \text{ Hz}$$

Since the diode conducts only during the positive half of the input voltage, we have:

$$I_m = \frac{V_m}{R_L + r_f} = \frac{50}{5000 + 20} \approx 10 \text{ mA}$$

(a) Hence current $i = 10 \sin 100\pi t$ for $\pi < 100\pi t < 2\pi$

$= 0$ for $0 < 100\pi t < \pi$

(b) DC output voltage, $V_{dc} = I_{dc} \times R_L$

$$= \frac{I_m}{\pi} \times R_L = \frac{10 \times 10^{-3}}{3.142} \times 5000 = 15.9 \text{ V}$$

Output voltage, $V_o = 15.9 \sin 100\pi t$ for $\pi < 100\pi t < 2\pi = 0$

(c) Assuming the diode is an ideal diode, the voltage across it is zero during the forward-biased. When the diode is reverse-biased, the voltage across diode is:

$$v = 15.9 \sin 100\pi t \text{ for } 0 < 100\pi t < \pi = 0 \text{ for } \pi < 100\pi t < 2\pi$$

**Example 3-16** In a half-wave rectifier circuit fed from 230 V, 50 Hz mains, it is desired to have a ripple factor $\gamma << 0.004$. Estimate the value of the capacitance needed. Given, $I_L = 0.5$ A.

**Solution:**

Given $V_{rms} = 230$ V, $f = 50$ Hz, $\gamma \leq 0.005$, $I_L = 0.5$ A. Take $\gamma = 0.003$.

Load current is given by:

$$I_L = \frac{I_m}{\pi} = I_{dc}$$

$$V_m = \sqrt{2} \times V_{rms} = 1.4 \times 230 = 322 \text{ V}$$

$$V_{dc} = \frac{V_m}{\pi} \frac{322}{3.14} = 102.5 \text{ V}$$

Load resistance:

$$R_L = \frac{V_{dc}}{I_{dc}} = \frac{102.5}{0.5} = 205\ \Omega = 200\ \Omega$$

Ripple factor:

$$\gamma = \frac{1}{2\sqrt{3}\, fCR_L}$$

$$= \frac{1}{2\sqrt{3}\, f\gamma R_L} = \frac{1}{2\sqrt{3} \times 50 \times 200 \times 0.003} = 9.62 \text{ mF}$$

**Example 3-17** The voltage across half the secondary winding in a centre-tapped transformer used in a full-wave rectifier is 230sin 314$t$. The forward-bias resistance of each diode is 20 Ω and the load resistance is 3.15 kΩ. Calculate the ripple factor.

**Solution:**

Given, $R_L = 3.15$ k Ω, $r_f = 20$ Ω.

Instantaneous voltage, $v = V_m \sin 2\pi ft$

$= 230 \sin 314t$

Comparing, we have, $V_m = 230$ V, $f = 50$ Hz

The rms value of current:

$$I_{rms} = 0.707\left(\frac{V_m}{R_L + r_f}\right) = 0.707\left(\frac{230}{3150 + 20}\right)$$

$$= 0.707 \times 0.0726 = 0.0513 \text{ A}$$

DC value of current, $I_{dc} = 0.637 I_m = 0.637 \times 0.0726 = 0.0331$ A

Therefore, ripple factor:

$$\gamma = \sqrt{\left(\frac{I_{rms}}{I_{dc}}\right)^2 - 1} = \sqrt{\left(\frac{0.0513}{0.0331}\right)^2 - 1} = 0.48$$

**Example 3-18** Ideal diodes are used in a bridge rectifier with a source of 230 V, 50 Hz. If the load resistance is 150 Ω and turns ratio of transformer is 1:4, find the dc output voltage and pulse frequency of the output.

**Solution:**

Given, $V_p = 230$ V, $f = 50$ Hz, $R_L = 200\ \Omega$, $N_S : N_P = 1 : 4$.

We know that:

$$\frac{V_S}{V_P} = \frac{N_S}{N_P}$$

The rms secondary voltage:

$$V_S = V_P \times \frac{N_S}{N_P} = 230 \times \frac{1}{4} = 57.5 \text{ V}$$

$$V_S = V_{rms} = 57.5 \text{ V}$$

Maximum voltage across secondary, $V_m = V_{rms} \times \sqrt{2} = 57.5 \times \sqrt{2} = 81.3$ V

Average current:

$$I_{dc} = \frac{2V_m}{\pi R_L} = \frac{2 \times 81.3}{3.14 \times 200} = 0.26 \text{ A}$$

DC output voltage, $V_{dc} = I_{dc} \times R_L = 0.26 \times 150 = 39$ V

As there are two output pulses for each complete cycle of the input ac voltage in full-wave rectification, the output frequency is twice that of the ac supply frequency.

$$\therefore \quad f_{out} = 2f_{in} = 2 \times 50 = 100 \text{ Hz}$$

**Example 3-19** Find the maximum dc voltage that can be obtained from the full-wave rectifier circuit.

**Solution:**

Given, $V_p = 220$ V, $f_i = 50$ Hz, $R_L = 1.5$ kΩ, $N_p = 1000$, $N_S = 100$.

We know that:

$$\frac{V_S}{V_P} = \frac{N_S}{N_P}$$

The rms secondry voltage:

$$V_S = V_p \times \frac{N_S}{N_P} = 220 \times \frac{100}{1000} = 22 \text{ V}$$

Maximum secondary voltage, $V_{\text{rms}} = 22 \times \sqrt{2} = 30.8$ V

Maximum voltage across half-secondary winding, $V_m = \dfrac{30.8}{2} = 15.4$ V

Average current: $$I_{\text{dc}} = \frac{2V_m}{\pi R_L} = \frac{2 \times 15.4}{3.14 \times 1500} = 6.53 \times 10^{-3} \text{ A}$$

Dc output voltage, $V_{\text{dc}} = I_{\text{dc}} \times R_L = 6.53 \times 10^{-3} \times 1500 = 9.79$ V

**Example 3-20** A full-wave rectifier using a capacitor filter is to supply 30 V dc to a 1 kΩ load. Assuming the diode and transformer winding resistance to be negligible, calculate the input voltage required and the value of the filter capacitor for a ripple of 0.015.

**Solution:**
Given, $V_{\text{dc}} = 30$ V, $R_L = 1$ kΩ and $\gamma = 0.015$.

DC output current:

$$I_{\text{dc}} = \frac{V_{\text{dc}}}{R_L} = \frac{30 \text{ V}}{1 \text{ k}\Omega} = 30 \text{ mA}$$

DC output voltage: $$V_{\text{dc}} = V_m - \frac{5000\, I_{\text{dc}}}{C}$$

In order to determine $V_m$, we must calculate the value of $C$.
We know that ripple factor:

$$\gamma = \frac{2900}{CR_L} = 0.01$$

Hence: $$C = \frac{2900}{0.015 \times 1000} \,\mu\text{F} = 193 \,\mu\text{F}$$

$$V_m = V_{\text{dc}} + \frac{5000\, I_{\text{dc}}}{C} = 30 + \frac{5000 \times 0.03}{193} = 30.78 \text{ V}$$

Line-to-line secondary voltage: $$\frac{2V_m}{\sqrt{2}} = \frac{2 \times 30.78}{\sqrt{2}} = 43.52 \text{ V}$$

**Example 3-21** The output of a full-wave rectifier is fed from a 40–0–40 V transformer. The load current is 0.1 A. The two 40 $\mu$F capacitors are available. The circuit resistance exclusive of the load is 40 Ω.

(a) Calculate the value of inductance for a two-stage L-section filter. The inductances are to be equal. The ripple factor is to be 0.0001. (b) Calculate the output voltage.

**Solution:**

(a)
$$LC = 1.76\left(\frac{0.472}{r}\right)^{\frac{1}{n}}$$

$$L = \frac{1.76}{40\times10^{-6}}\left(\frac{0.472}{0.0001}\right)^{\frac{1}{2}}$$

$$= 3.02 \text{ H}$$

(b)
$$V_{dc} = \frac{2V_m}{\pi} - I_{dc}R$$

$$= \frac{(2\sqrt{2})\,40}{\pi} - 0.1 \times 40$$

$$= 36 - 4 = 32 \text{ V}$$

**Example 3-22** A full-wave single-phase rectifier employs a $\pi$-section filter consisting of two 4 $\mu$F capacitances and a 20 H choke. The load current is 50 $\mu$A. Calculate the dc output voltage and ripple voltage. The resistance of the choke is 200 Ω.

**Solution:**

$$V_{\text{dc}} = V_m - \frac{4170}{C} I_{\text{dc}} - I_{\text{dc}}R$$

$$= 300\sqrt{2} - \frac{4170}{C}(0.05)\,0 - (0.05) \times 200$$

$$= 362.13 \text{ V}$$

$$r = \frac{3300}{CC_1LR_L}$$

$$= \frac{3300 \times 0.05}{4 \times 4 \times 20 \times 353}$$

$$= 1.46 \times 10^{-3}$$

$$V_{\text{rms}} = rV_{\text{dc}} = 0.015 \text{ V}$$

■

## 3-6 CLIPPER AND CLAMPER CIRCUITS

### 3-6-1 Clipper

A clipper is a type of diode network that has the ability to "clip off" a portion of the input signal without distorting the remaining part of the alternating waveform. The half-wave rectifier is an example of the simplest form of diode clipper—one resistor and a diode. Depending on the orientation of the diode, the positive or negative region of the input signal is "clipped" off.

There are two general categories of clippers: *series* and *parallel*. The series configuration is defined as one where the diode is in series with the load, while the parallel clipper has the diode in a branch parallel to the load.

### Series clipper

A series clipper and its response for two types of alternating waveforms are provided in Figs. 3-17(a) and 3-17(b).

From the Fig. 3-17(b) of the half-wave rectifier, we see that there are no clear cut restrictions on the type of signals that can be applied at the input. The addition of a dc supply, as shown in Fig. 3-18, can have a pronounced effect on the output of a clipper.

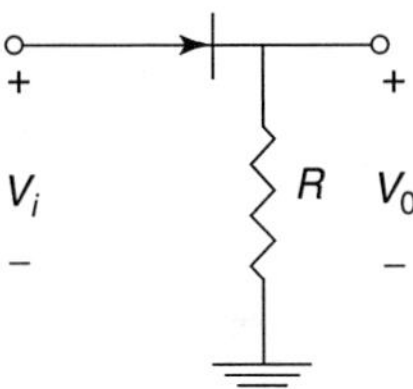


**Figure 3-17(a)** Series clipper circuit

### Key points

(i) The first step is to find out in which interval of the input signal the diode is in forward-bias.

(ii) For Fig. 3-18, the direction of the diode suggests that the signal $v_i$ must be positive to turn it on. The dc supply further requires the voltage $v_i$ to be greater than $v$ volts to turn the diode on. The negative region of the input signal turns the diode into the OFF state. Therefore, in the negative region the diode is an open circuit.

(iii) Determine the applied voltage (transition voltage) that will cause a change in state for the diode. For the ideal diode the transition between states will occur at that point on the characteristics where $v_d = 0$ V and $i_d = 0$ A. Applying this condition to Fig. 3-18 will result in the configuration of Fig. 3-19 and it is recognized that the level of $v_i$ that will cause a transition in state is:

$$v_i = V$$

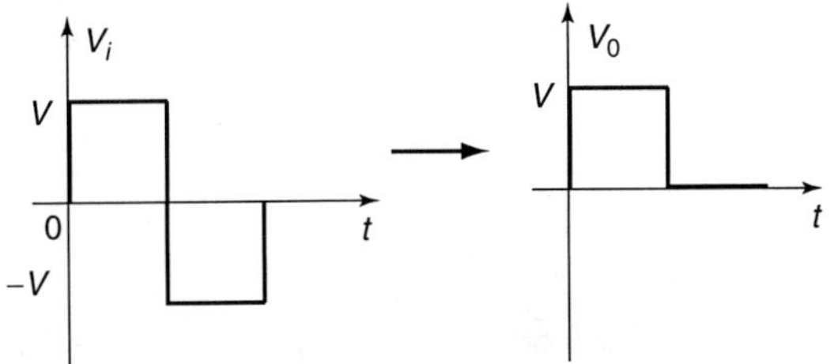


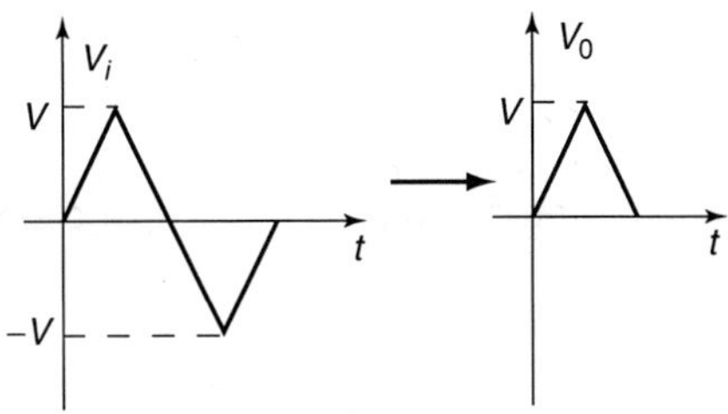


**Figure 3-17(b)** Response of clipper circuit

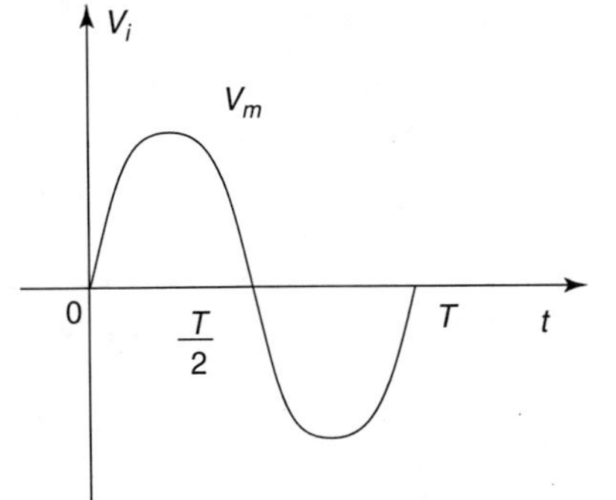


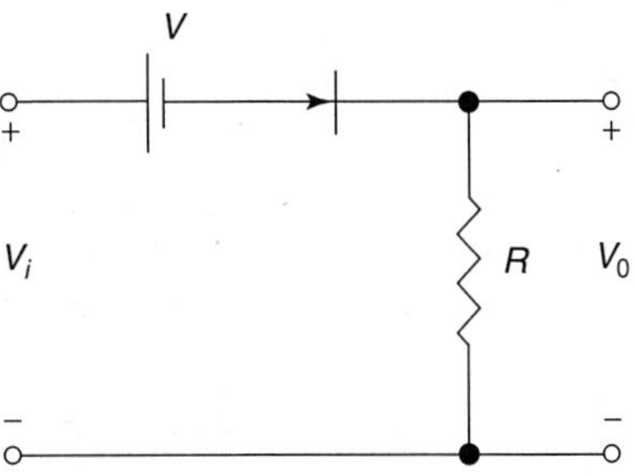


**Figure 3-18** Series clipper with a dc supply

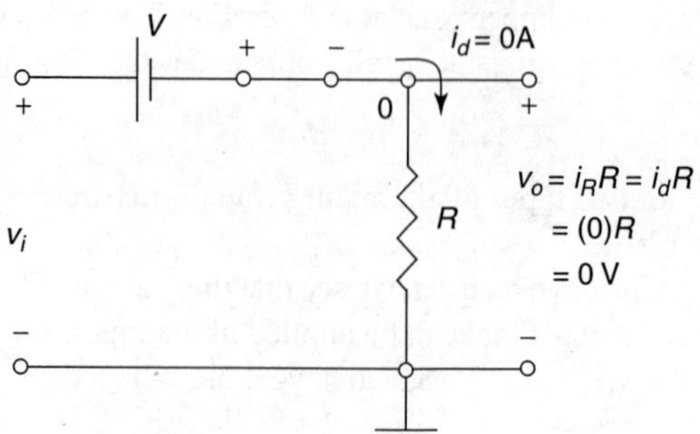


**Figure 3-19** Determining the transition level of the input signal

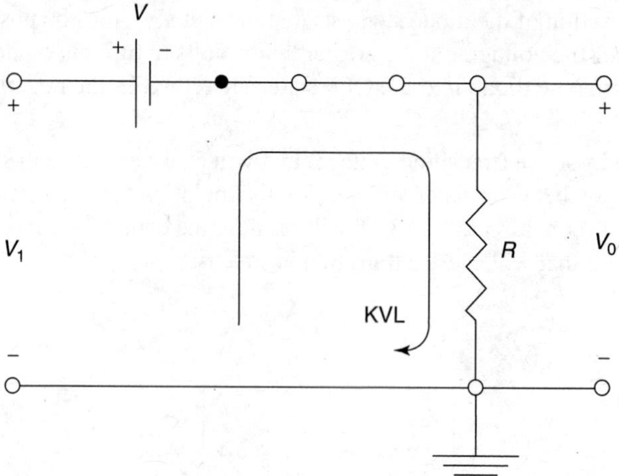


**Figure 3-20** Determining $v_o$ in the clipper circuit

For an input voltage greater than $V$ volts, the diode is in the short-circuit state, while for input voltage less than $V$ volts it is in the open-circuit or OFF state (as it is reverse-biased).

(iv) Be continually aware of the defined terminals and polarity of $v_o$. When the diode is in the short-circuit state, as shown in Fig. 3-20, the output voltage $v_o$ can be determined by applying KVL in the clock-wise direction:

$$v_i - V - v_o = 0$$

or,

$$v_o = v_i - V$$

(v) It can be helpful to sketch the input signal above the output and determine the output at instantaneous values of the input. It is then possible to sketch the output voltage from the resulting data points of $v_o$, as shown in Fig. 3-21.

At $v_i = V_m$, the network to be analysed is shown in Fig. 3-22.

For $V_m > V$, the diode is in the short-circuit state and $v_o = V_m - V$. At $v_i = V$, the diode changes state and $v_i = -V_m$, $v_o = 0$ V. The complete curve for $v_o$ can be sketched, as shown in Fig. 3-23.

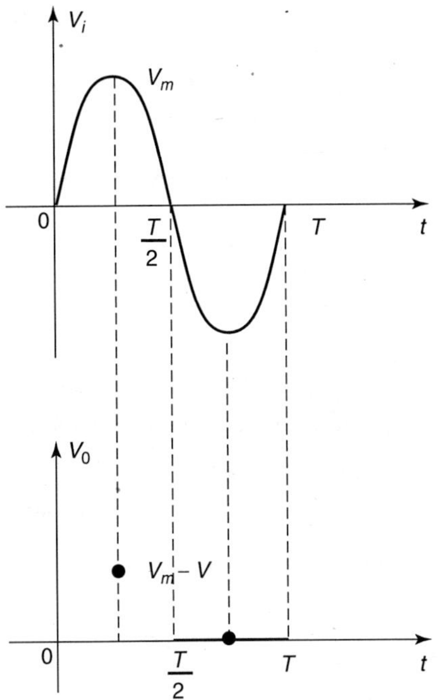


Figure 3-21 Determining levels of $v_o$

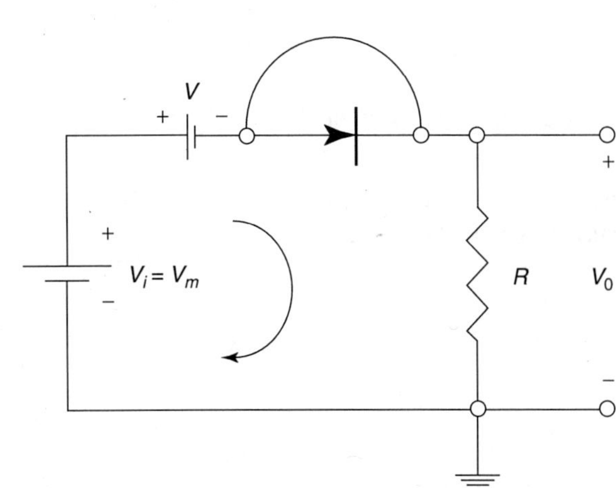


Figure 3-22 Determining $v_o$ when $v_i = V_m$

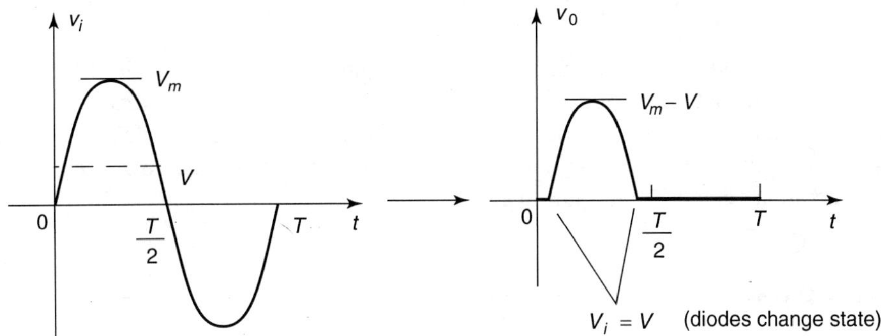


Figure 3-23 Sketch for $v_o$

## Parallel clipper

A simple parallel clipper and its response are shown in Fig. 3-24 and Fig. 3-25 respectively. Input $v_i$ is applied for the output $v_o$. The analysis of parallel configuration is very similar to the series configuration.

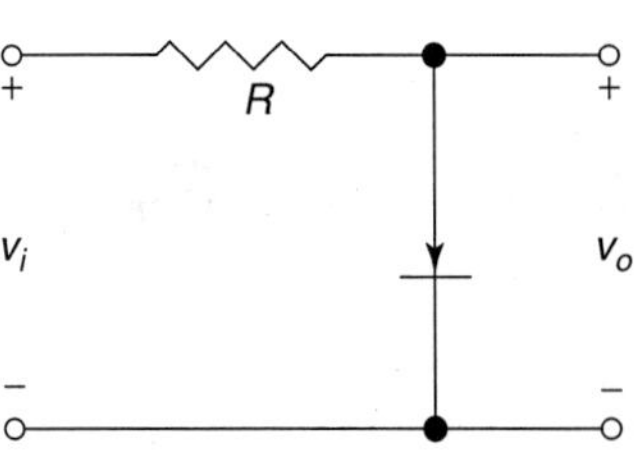


Figure 3-24 Parallel clipper

## Break region

There is a discontinuity at the voltage $V_\gamma$. Actually the transition of a diode state is not exactly abrupt but gradual. Thus, a waveform, which is transmitted through the clipper circuit, will not show an abrupt clipping. Instead, it will show a gradual broken region, exhibiting the regions of

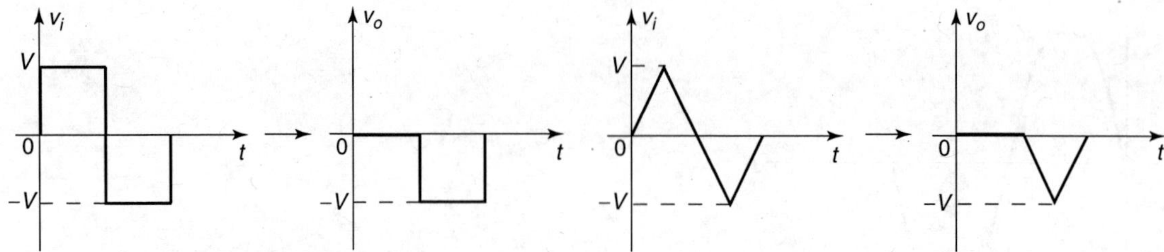


**Figure 3-25** Response of parallel clipper

un-attenuated and attenuated transmission. Now, we will estimate the range of this break region. The output current of a diode is given by:

$$I = I_o\,(e^{V/\eta V_T} - 1) \tag{3-49}$$

Beyond the diode break point, the expression of the current that is large, compared to $I_o$, may be given by:

$$I = I_o\,e^{V/\eta V_T}$$

The incremental diode resistance $r = dv/dI$ and as obtained from the Eq. (3-49) is given by:

$$r = \frac{\eta V_T}{I_o}\,e^{-V/\eta V_T} = \frac{\eta V_T}{I} \tag{3-50}$$

From Eq. (3-50), we note that $r$ varies inversely with the quiescent current, and directly with the absolute temperature. We also note that the break region is independent of the quiescent current.

Again for meaningful clipping to be done, the applied signal must vary from one side of the break point to a point well on the other side. If the signal is only of the order of magnitude of the extent of the break region, the output will not display sharp limiting.

## Solved Examples

**Example 3-23** A symmetrical 5 kHz square wave whose output varies between +10 V and –10 V is impressed upon the clipping circuit shown. Assume, $R_f = 0$, $R_r = 2$ M, and $V_\gamma = 0$. Sketch the steady-state output waveform, indicating numerical values of the maximum, minimum, and constant portions.

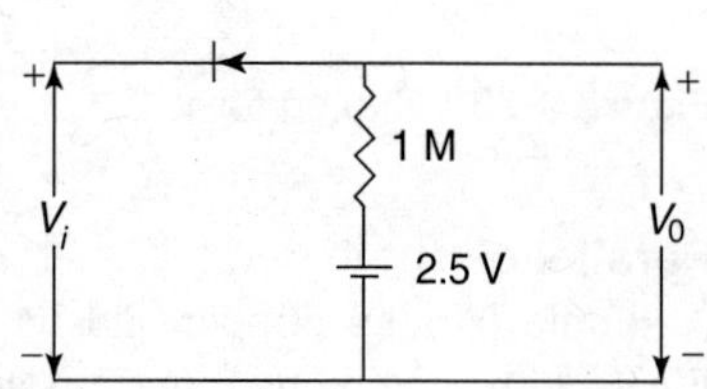


**Solution:**

The diode conducts when, $v_i < 2.5$ V. The diode is open when:

$$v_i > 2.5 \text{ V and } v_o = 2.5 \text{ V} + \frac{v_i - 2.5 \text{ V}}{3}$$

When diode conducts, $v_i = v_0 < 2.5$ V.

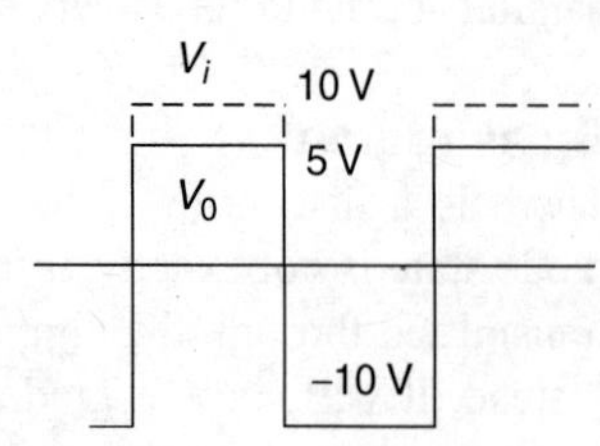

**Example 3-24** (a) The input voltage $v_i$ to the two level clippers shown in Fig. (a) of the figure varies linearly from 0 to 150 V. Sketch the output voltage $v_o$ to the same time scale as the input voltage. Assume ideal diodes.

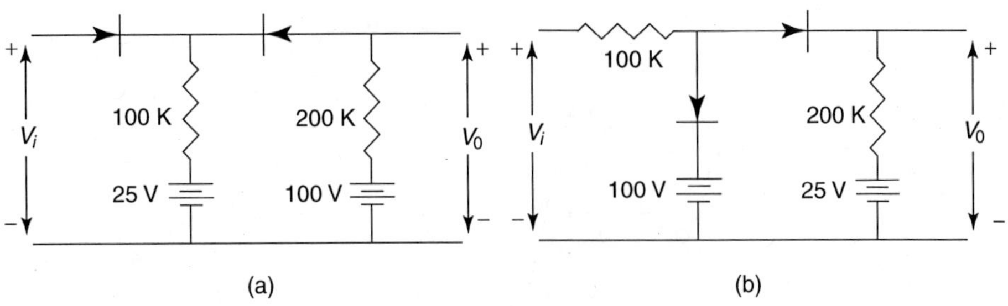


(b) Repeat part (a) for the circuit shown in Fig. (b).

**Solution:**

(a) When $v_i < 50$ V, the first diode is open and second diode conducts, and:

$$v_o = 100 - \frac{2}{3} \times 27 = 50 \text{ V.}$$

When $50 < v_i < 100$, both diodes conduct, and $v_o = v_i$. When $v_i > 100$, the first diode is conducting but the second diode is open, so $v_o = 100$ V.

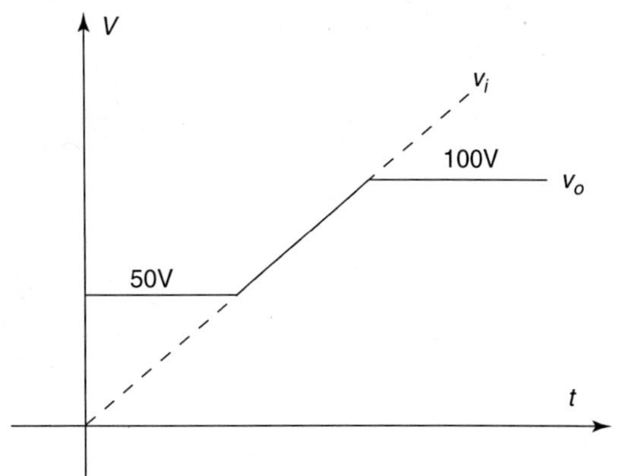


(b) When $v_i < 25$ V, neither diode conducts and $v_o = 25$ V. When $v_i > 25$ V, the upper diode conducts and:

$$v_o = (v_i - 25)\frac{2}{3} + 25$$

When $v_0$ reaches 100 V, $v_i$ rises to 137.5 V. For larger $v_i$, both diodes conduct and $v_o = 100$ V (if $v_i = 40 \sin \omega t$, then $v_0 = 20 \sin \omega t$).

**Example 3-25** The triangular waveform shown is to be converted into a sine wave by using clipping diodes. Consider the dashed waveform sketched as a first approximation to the sinusoid. The dashed waveform is coincident with the sinusoid at 0°, 30°, 60°, etc. Devise a circuit whose output is this broken-line waveform when the input is the triangular waveform. Assume ideal diodes and calculate the values of all supply voltages and resistances used. The peak value of the sinusoid is 50 V.

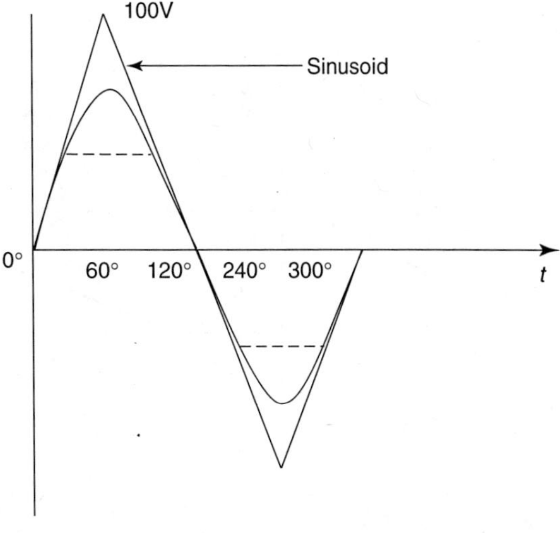

**Solution:**

Desired output:

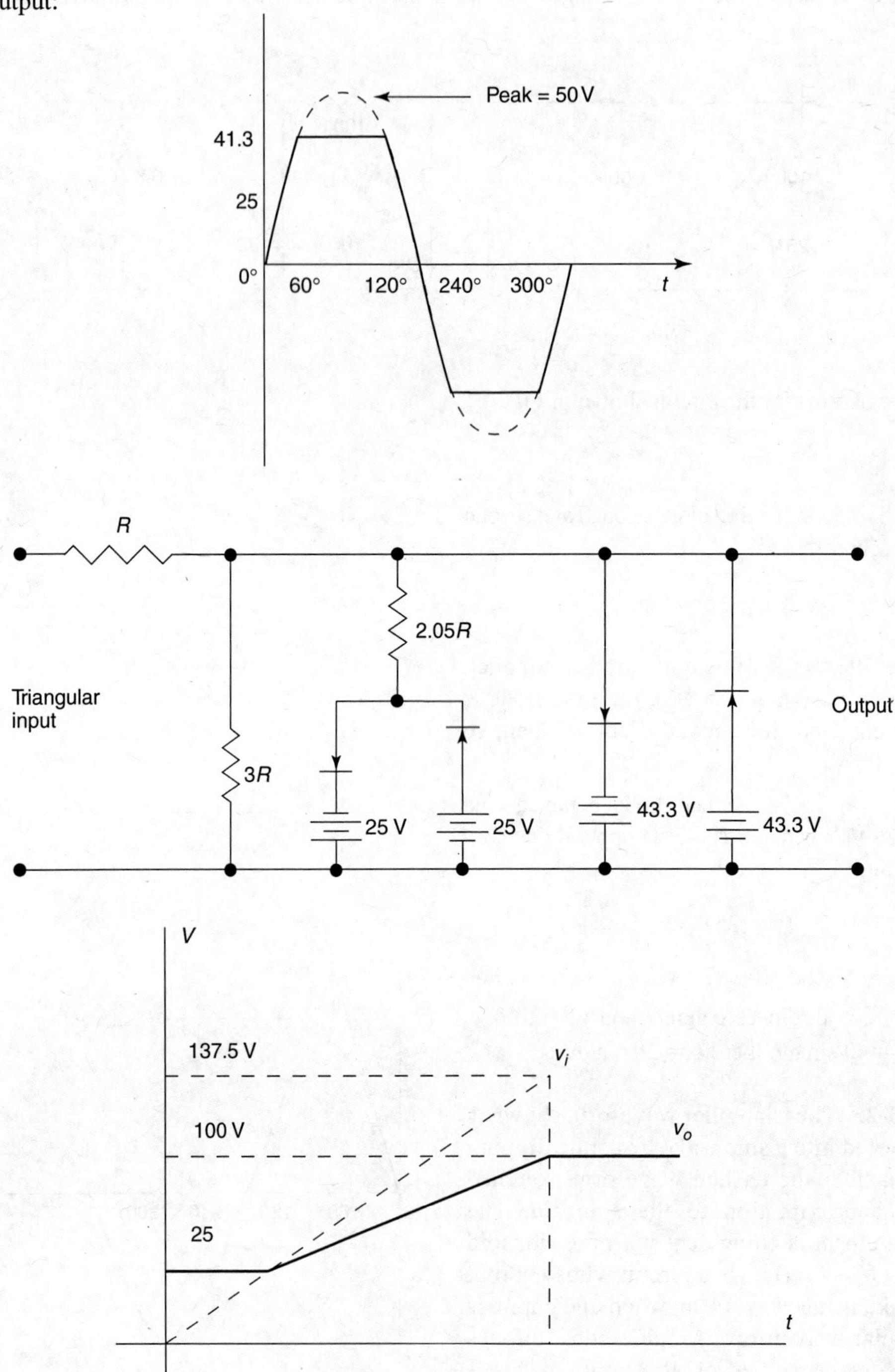


■

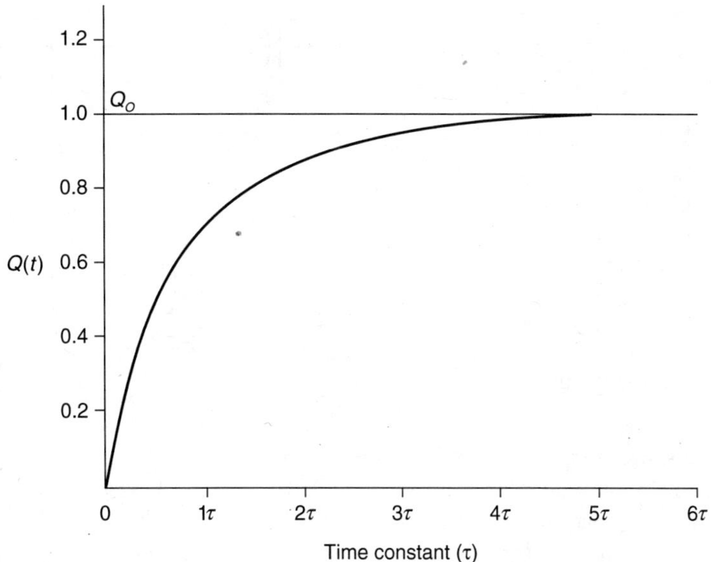


**Figure 3-26** Charging of a *RC* circuit

## 3-6-2 Clamper

A clamping network is one that will "clamp" a signal to a different dc level. The network must have a capacitor, a diode, and a resistive element, but it can also employ an independent dc supply to introduce an additional shift. Before further probing into the clamper circuit one must have a basic understanding of a transient *RC* circuit.

From the basic understanding of a series *RC* transient circuit applied across a dc voltage $E_O$, the instantaneous charge across the capacitor at any time is given by

$$Q(t) = Q_o\left(1 - e^{-t/RC}\right)$$

$Q_o = E_oC$ where, $C$ is the capacitance of the capacitor.

We know that the time constant of the *RC* circuit is given by $\tau = RC$. This was the case when the capacitor was being charged from zero voltage level. From Fig. 3-26 it is at once interpreted that the rise time becomes smaller if we decrease the time constant.

Hence, to reach the maximum charging level quickly, we need to reduce the time constant. In case of discharge through a *RC* circuit, it can again be shown that, $Q(t) = Q_oe^{-t/RC}$ where, $Q_o$ is the initial charge on the capacitor.

Here also the time constant has the same value, and from Fig. 3-27 we can instantly say that the discharge will occur quickly if the time constant of the circuit is decreased. In other words, we can state, that to hold the charge in a capacitor for a sufficiently longer time, we need to increase the time constant of the circuit.

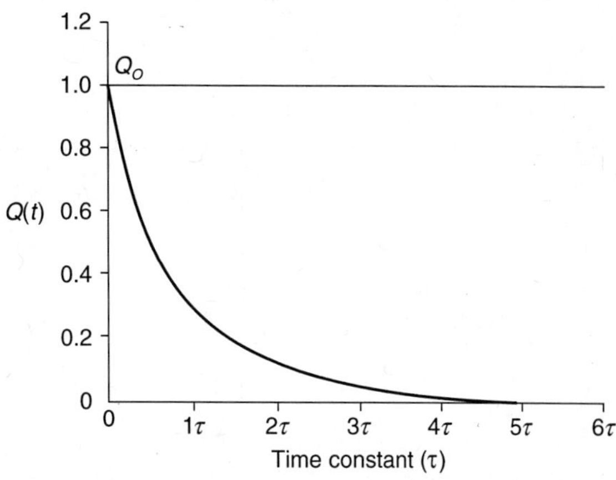


**Figure 3-27** Discharging of an *RC* circuit

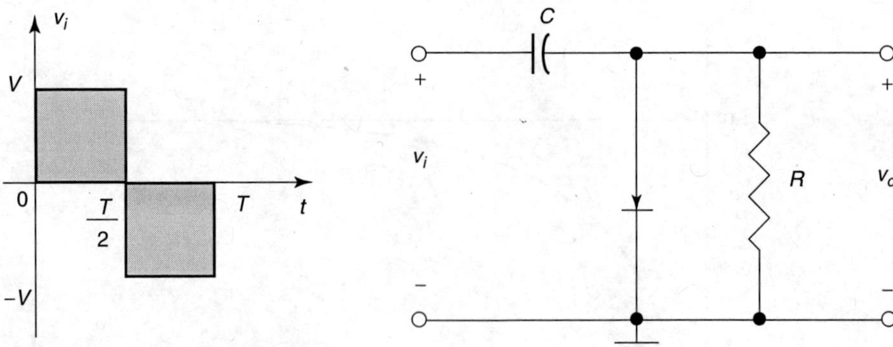


**Figure 3-28** Simple clamper circuit

In electronic circuits, clamping usually refers to holding voltage at its maximum value for a desired period of time. In order to do that, the magnitude of $R$ and $C$ must be so chosen that the time constant, $\tau = RC$, is large enough to ensure that the voltage across the capacitor does not discharge significantly during the interval the diode is non-conducting. Figure 3-28 shows the clamping circuit that will clamp the input signal to the zero level.

During the interval $0–T/2$ the network will appear, as shown in Fig. 3-29, with the diode in the ON state effectively "shorting out" the effect of the resistor $R$.

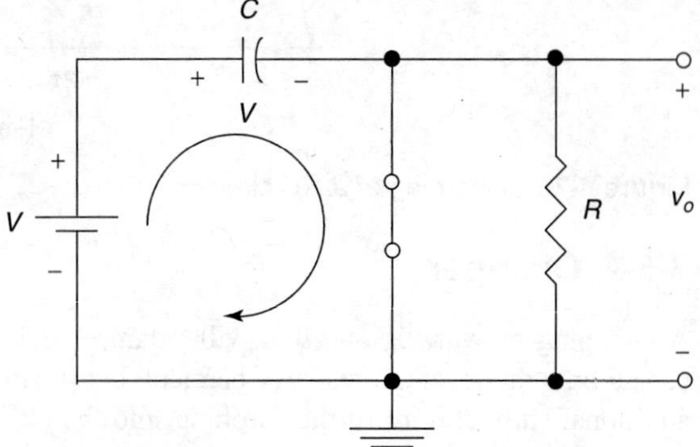


**Figure 3-29** State of the circuit when $v_i$ is more than $v_{diode}$

The resulting $RC$ time constant is so small that the capacitor will charge to $V$ volts very quickly. During this interval the output voltage is directly across the short circuit and $v_o = 0$ V.

When the input switches to $-V$ state, the network will appear as shown in Fig. 3-30.

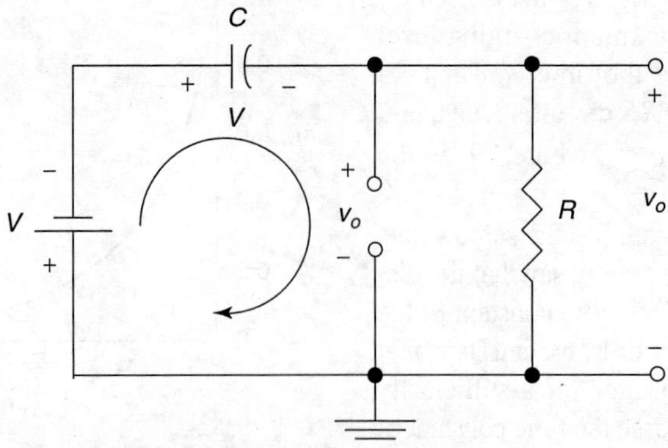


**Figure 3.30** State of the circuit in the negative half-cycle

The diode will now be in the open-state condition. Applying KVL around the input loop of Fig. 3-30 will result in:

$$-V - V - v_o = 0$$

or,

$$v_o = -2\text{ V}$$

The negative sign results from the fact that the polarity of 2 V is opposite to the polarity defined for $v_o$. The resulting output waveform appears with the input signal. The output signal is clamped to 0 V for the interval $0-T/2$ but maintains the same total swing (2 V) as the input. For a clamping network the total swing of the output is equal to the total swing of the input.

### Analysis of clamping networks

In general, the following steps must be kept in mind when analysing the clamping networks.

(i) The first step is to calculate the interval of the input signal in which the diode is in forward bias.

(ii) The second step is to determine the voltage across the capacitor. This is assumed to rise instantaneously.

(iii) Due to the longer time constant of the circuit, the capacitor will hold on to its established voltage level.

(iv) During the whole process, the analysis maintains a continual awareness of the location and reference polarity for $v_o$ to ensure that the proper levels for $v_o$ are obtained.

(v) The general rule that the total swing of the output must match the swing of the input signal, should be kept in mind.

## 3-7 COMPARATORS

The diode circuit which has been used in the design of the clipping circuit can also be used for the purpose of comparison, hence the name comparator. The basic principle on which the comparator works is the switching of the diodes. This action corresponds to the phase when the diode conducts; and when it does, the comparator circuit is used to compare the input arbitrary voltage with the reference voltage. This reference voltage is predefined, and the output attains a sharp slope when the input waveform crosses the predefined level. The basic and foremost difference between a clipper and a comparator is that, in case of a comparator, we are not interested in reproducing the input waveform or any of its parts. In general the comparator output consists of a sharp departure from its quiescent point as the input signal attains the reference level, but otherwise the circuit remains unaffected by the input signal. The basic operation is given in Fig. 3-31. It also gives a comparative study of the input and the output, as can be understood from Fig. 3-32.

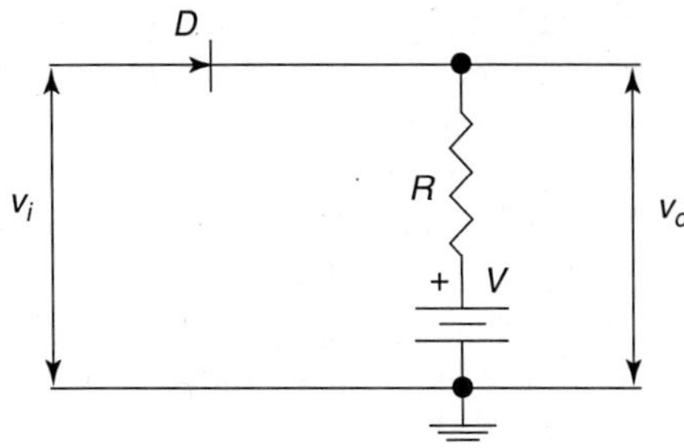


**Figure 3-31** A diode comparator

Figure 3-31 shows that as long as the input signal is below the total threshold voltage, i.e., *cutin voltage* of the diode and that of the voltage source, the output is not affected at all. The output is just the reference voltage of the voltage source. But, as soon as the input voltage exceeds the predefined threshold voltage, the output gives a sharp response and executes the ultimate purpose of the comparator. Thus, the output of

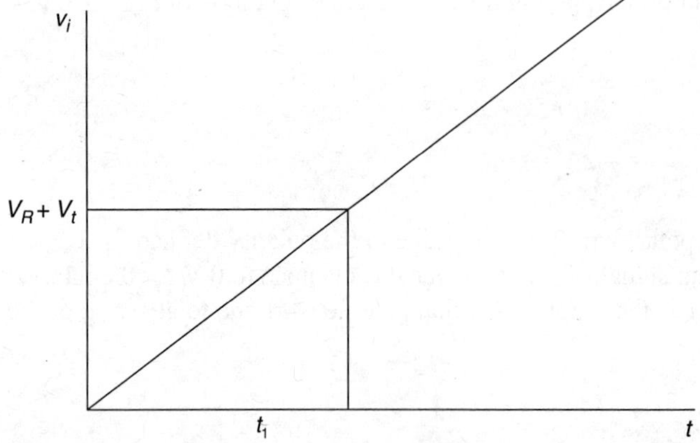


Figure 3-32 Input signal with the threshold voltage

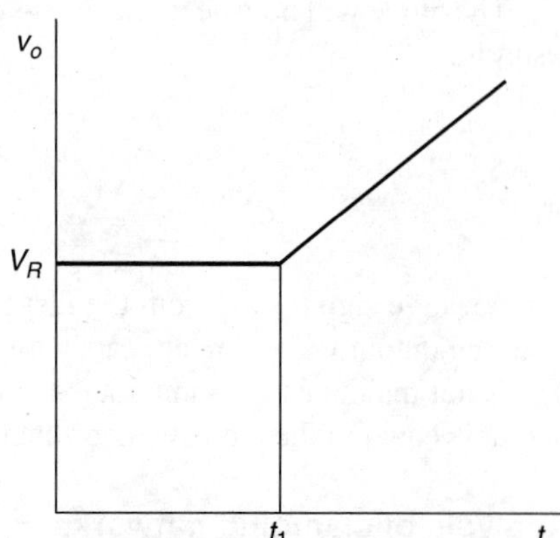


Figure 3-33 Corresponding output waveform

the circuit in Fig. 3-33 shows that the output is very different when the input is below the threshold level than when it is above the threshold value. Consequently, we obtain a device that can make a sharp demarcation in the cases as described earlier.

# 3-8 ADDITIONAL DIODE CIRCUITS

## 3-8-1 Voltage Multiplier

The voltage multiplier is a passive circuit, similar to the rectifier circuit and gives an output which is approximately equal to a certain multiple of the peak value of the peak input voltage. Here it is possible to obtain a dc voltage equal to the peak value of the applied ac voltage.

The circuit of a half-wave doubler is as shown in Fig. 3-34. With terminal $A$ of the ac source assumed positive, diode $D_2$ is forward-biased and diode $D_1$ is reverse-biased, i.e., open. The diode $D_2$ charges the capacitor to the peak supply voltage. In the next half-cycle, the same case is repeated with the second diode, i.e., $D_1$, and consequently, as seen from the circuit, the diode $D_2$ is reverse-biased. Again, in this case, the capacitor $C_1$ charges to its fullest.

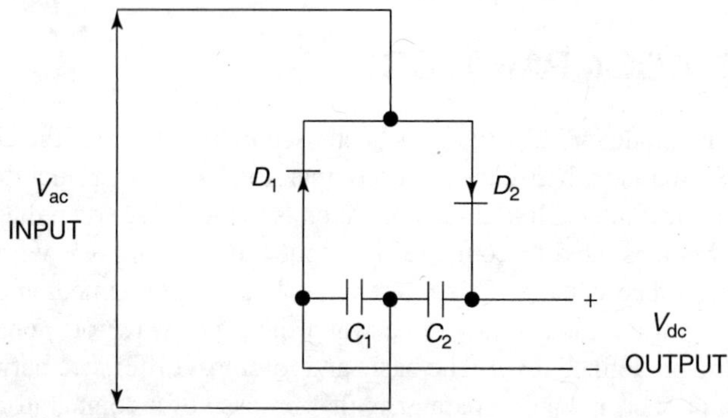


Figure 3-34 Voltage doublers

Now, the capacitors $C_1$ and $C_2$ are in series for the dc output circuit; and at no load, the dc voltage is equal to the sum of the positive and negative peaks of the applied voltage. In other words, the output voltage is equal to twice the peaks of the applied voltage. It should be noted, that for the proper operation of the circuit, the input has to be symmetrical.

Another important requirement is that the values of the capacitors $C_1$ and $C_2$ have to be sufficiently large so that they maintain the voltage level over the intervals in which each diode is conducting.

Similarly, with such designs, various types of multipliers—triplers and quadruplers—can be designed with a proper choice of such components and by suitably connecting them.

## 3-8-2 Peak Detector

The half-wave rectifier circuit can be suitably manipulated to obtain the peak detector circuit. The working principle depends on the charging and discharging of the capacitor, and also on the conducting and non-conducting regions of the diode. Its simple circuit consists of a diode kept in series with a resistor and a load at the end. The circuit of a peak detector is as shown in Fig. 3-35.

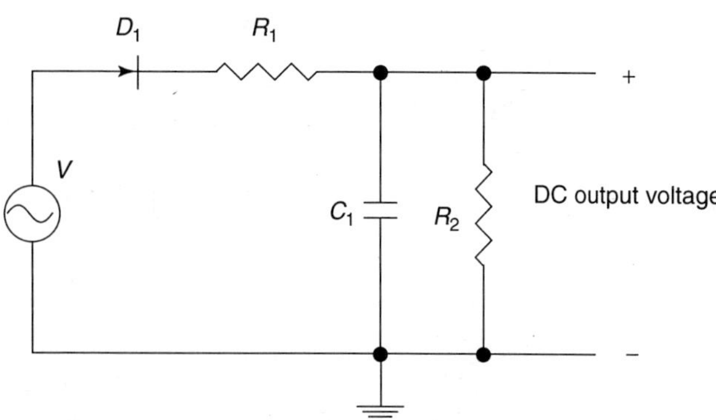

**Figure 3-35** Peak detector

When the input signal forward-biases the diode, the diode conducts and the voltage is obtained at the load. In this process, the output voltage is available across the capacitor, which in this process charges itself to the full.

Again, when the diode is reverse-biased, the diode does not conduct, and the input voltage is not obtainable at the output. During this time, because the capacitor is charged to the maximum, it begins to discharge and an almost steady voltage is obtained at the output. This process continues till the input voltage again forward-biases the diode.

In this sequence of steps, the output becomes the envelope of the input voltage. But proper care has to be taken regarding the value of the capacitor. Its time constant has to be properly set so that proper replication of the envelope is obtained. These circuits are used extensively for communication purposes in various detection devices.

## 3-8-3 Digital Circuits

In various kinds of analog to digital converters, and in many digital circuits, diodes are extensively brought to use. Their primary domain of operation is switching, i.e., to keep a portion of a large circuit in the ON state, selectively for a given interval of time. They, in coherent action with resistors, form many important logic families, which are used in digital electronic circuits.

## 3-8-4 Switching Regulators

Power supplies with switching regulators offer great versatility, as the design of power supplies employing this type of regulator can be lighter and more compact. Another advantage of these power supplies is that the circuit can be designed to give an output voltage that is higher than the unregulated voltage or, has a different polarity.

But there is also a disadvantage. The circuit becomes more complex because of the control circuitry, which is the filtering required to remove the switching noise from the output and the electromagnetic interference that can degrade the performance of the device.

Specifically, a switching regulator has two parts: a converter circuit to perform the switching action and a feedback system to control the switching rate. The structure of the converter can be of three types: *buck, boost* and *buck–boost*. Let us first study the buck-type converter.

#### Buck converter

The basic circuit of the buck converter is as shown in Fig. 3-36.

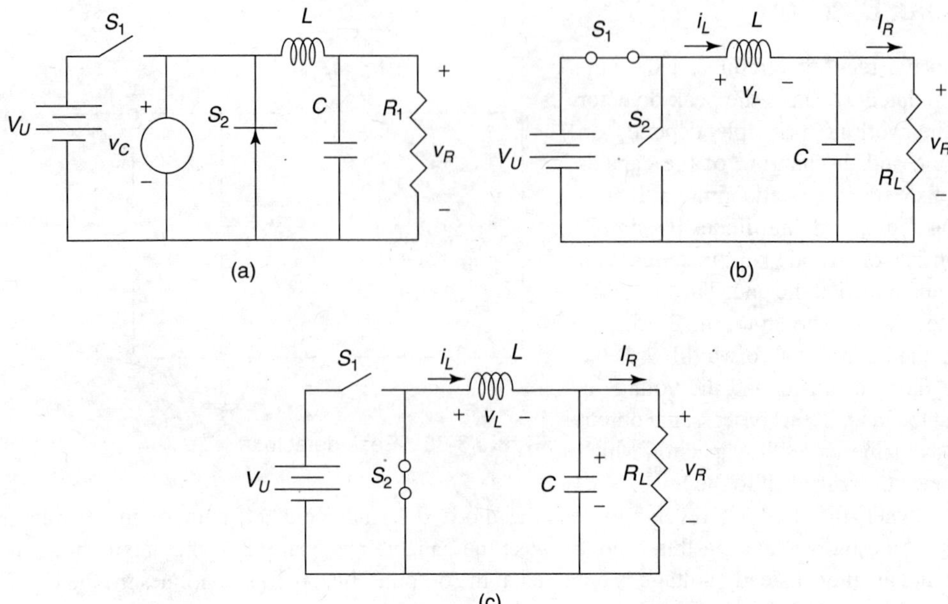


**Figure 3-36** Various stages in the operation of a buck-type switching converter

The total output voltage of the supply is $v_R = V_R + v_r$ where, $V_R$ is the regulated dc output voltage and $v_r$ is a small ac ripple. From Fig. 3-36, we find that we have the switches with two voltage controlled devices that can be switched ON and OFF by critically controlling the control voltage $V_C$. In Fig. 3-36(b), it is seen that the voltage across $L$ is:

$$v_L \approx V_U - V_R \tag{3-51}$$

While forming Eq. (3-51), it is assumed that the capacitor is previously charged to $V_R$, the regulated voltage. At this stage we consider $\delta$ to be the duty cycle of the control voltage. This duty cycle has to be chosen very accurately as it figures out the operational performance of the device. Here, in the period 0–$\delta\, T_S$, the switch $S_1$ is closed and the switch $S_2$ is open. After $\delta\, T_S$, both $S_1$ and $S_2$ are open. Consequently, from Lenz's law—as the inductor must resist the cause of such a change—$v_L$ becomes negative, and the inductor tries to maintain the constant current. At this stage, the voltage $v_L$ is given by:

$$v_L \approx -V_R \tag{3-52}$$

As the cycle expires, the previous state is restored, resulting in the closing of both $S_1$ and $S_2$. Figure 3-37 gives the inductor voltage waveform.

Just as the current through the capacitor is zero, so also must be the average voltage across the inductor. Consequently, it can be written as:

$$\frac{(V_U - V_R)\,\delta T_S + (-V_R)(1-\delta)\,T_S}{T_S} = 0$$

This gives us:

$$V_R = \delta V_U \tag{3-53}$$

To obtain a regulated voltage $V_R$ from an unregulated voltage $V_U$, $v_C$ must provide the required duty cycle. In order to make $V_R$ impervious to the changes in $V_U$, we need a feedback circuit to keep track of the output continuously and adjust the duty cycle as and when required.

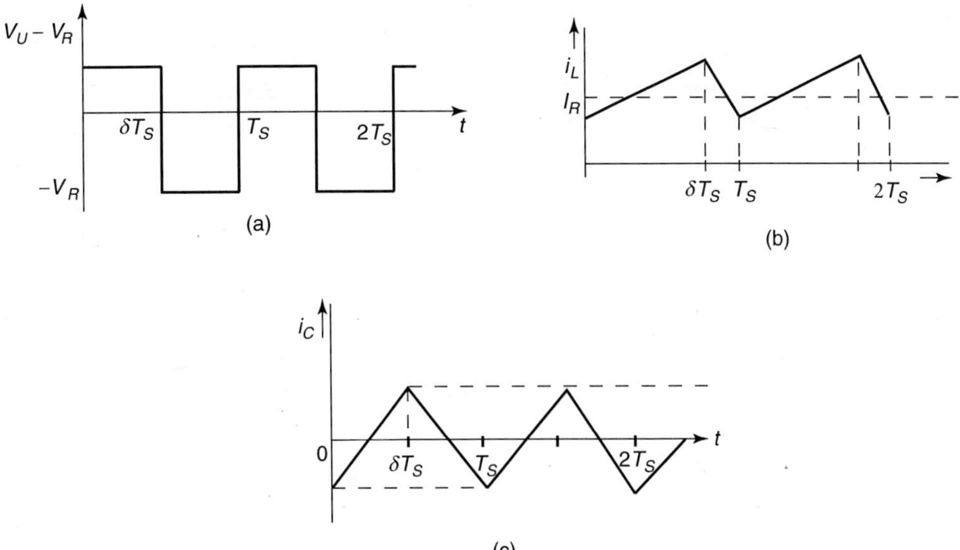


**Figure 3-37** Key waveforms for buck-type converters

Let us now turn to the ripple, $V_r$. The inductor current is given by $i_L = 1/L \int v_L(t)\,dt$, which is represented in Fig. 3-37(b). Also, we note that as the capacitor cannot carry the dc current and consequently the time varying part of $i_l$ must flow through the capacitor, therefore:

$$v_{L,\text{avg}} = \frac{V_U(\delta T_S) + V_R(1-\delta)T_S}{T_S} = 0$$

Subsequently, on solving this:

$$V_R = \frac{-\delta}{1-\delta} V_U \tag{3-54}$$

## Boost converter

The following circuits will also have the previously defined assumptions. The output ripple components can be ignored initially.

$S_1$ closes in the $\delta T_S$ and is open for the subsequent part of the cycle.

$S_2$ closes as soon as $S_l$ opens.

The circuit for the basic boost converter is shown in Fig. 3-38.

Moving ahead, the average voltage across the inductor leads to:

$$v_{L_{\text{avg}}} = \frac{V_U(\delta T_S) + (V_U - V_R)(1-\delta)T_S}{T_S} = 0 \tag{3-55}$$

And upon solving Eq. (3-55), we obtain the constraint $V_R = (1/1-\delta)V_U$. We also find that as $\delta$ is less than 1; and we obtain a boosted output, a higher value at the regulated output.

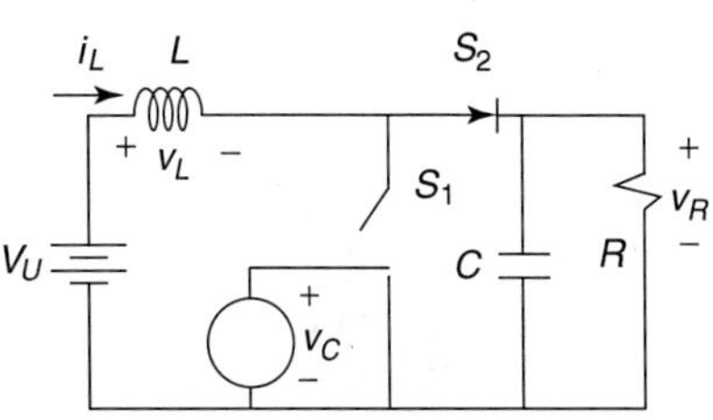


**Figure 3-38** Boost converter

### Buck–boost converter

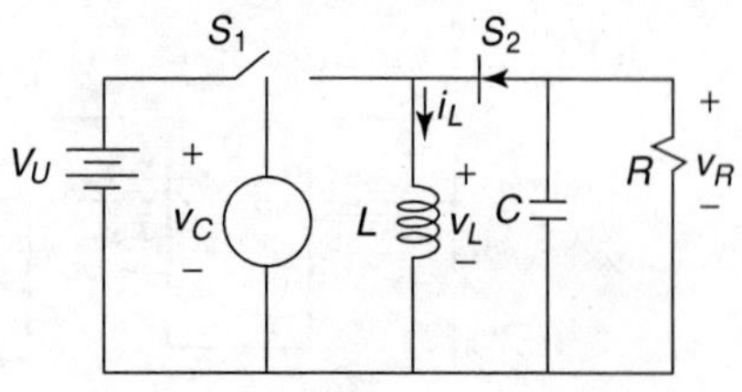


Figure 3-39 Buck–boost converter

The circuit of the buck–boost converter is shown in Fig. 3-39.

Applying the principles discussed earlier, we note that in this circuit also the average voltage across the inductor is zero. Therefore:

$$v_{L,\text{avg}} = \frac{V_U(\delta T_S) + V_R(1-\delta)T_S}{T_S} = 0$$

Subsequently, on solving this:

$$V_R = \frac{-\delta}{1-\delta} V_U \tag{3-56}$$

As we concentrate, we find that in Eq. (3-56), the polarities of $V_R$ and $V_U$ are always opposite.

## Solved Examples

**Example 3-26** By direct integration find the average value of the diode voltage and the load voltage for a diode whose specifications are as given: $V_m = 2.4$ V, $V_\gamma = 0.6$ V, $R_f = 10\ \Omega$ and $R_L = 100\ \Omega$. Note that these two answers are numerically equal and explain why?

**Solution:**

$$\varphi = \arcsin \frac{V_\gamma}{V_m} = \arcsin \frac{1}{4} = 14.5°$$

For $a = \frac{\pi}{2}$:

$$I_{\max} = \frac{1.8}{110} = 16.35 \text{ mA}$$

Hence, $v_{D\max} = 0.6 + 16.35 \times 10^{-3} \times 10 = 0.763$ V

$$\overline{v_L} = \frac{1}{T}\int_0^T iR_L d\alpha = \frac{1}{2\pi}\int_0^{2\pi} iR_L d\alpha = \frac{1}{2\pi}\left(\int_0^{\phi} iR_L d\alpha + \int_{\phi}^{\pi-\phi} iR_L d\alpha + \int_{\pi-\phi}^{2\pi} iR_L d\alpha\right)$$

$$= \frac{1}{2\pi}\left[0 + \int_{\phi}^{\pi-\phi} \frac{R_L}{R_L+R_f}(V_m \sin\alpha - V_\gamma)d\alpha + 0\right]$$

$$= \frac{1}{2\pi} \times \frac{R_L}{(R_L+R_f)}\left[V_m \cos\phi - V_m \cos(\pi-\phi)\right] - \frac{1}{2\pi}\frac{R_L}{R_L+R_f}V_\gamma(\pi-\phi-\phi)$$

$$= \frac{1}{2\pi} \times \frac{R_L}{R_L+R_f}\left[2V_m \cos\phi - V_\gamma(\pi - 2\phi)\right]$$

When the diode is ON:

$$v_D = V_\gamma + iR_f = \frac{R_L}{R_L+R_f}(V_m \sin\alpha + V_\gamma)$$

When the diode OFF:

$$v_D = V_m \sin \alpha - V_\gamma$$

Hence,

$$\overline{v}_D = \frac{1}{2\pi}\left[\int_0^{\phi} (V_m \sin \alpha - V_\gamma)\, d\alpha + \int_{\phi}^{\pi-\phi} \frac{R_L}{R_L + R_f} (V_m \sin \alpha - V_\gamma)\, d\alpha + \int_{\pi-\phi}^{2\pi} (V_m \sin \alpha - V_\gamma) d\alpha\right]$$

$$= -\frac{1}{2\pi} \times \frac{R_L}{R_L + R_f} [2V_m \cos \phi - V_\gamma (\pi - 2\phi)]$$

Since, $v_i = v_D + v_L$ when the diode is ON and $v_i = v_D + v_L = 0$ when the diode is OFF:

$$0 = \frac{1}{T}\int_0^T v_i dt = \frac{1}{T}\int_0^T v_D dt + \frac{1}{T}\int_0^T v_L dt$$

or,

$$\frac{1}{T}\int_0^T v_D dt = \overline{v_D} = -\frac{1}{T}\int_0^T v_L dt = -\overline{v_L}$$

**Example 3-27** For the circuit shown find $I_1$, $V_1$, $I_2$, and $V_2$. Assume ideal diode.

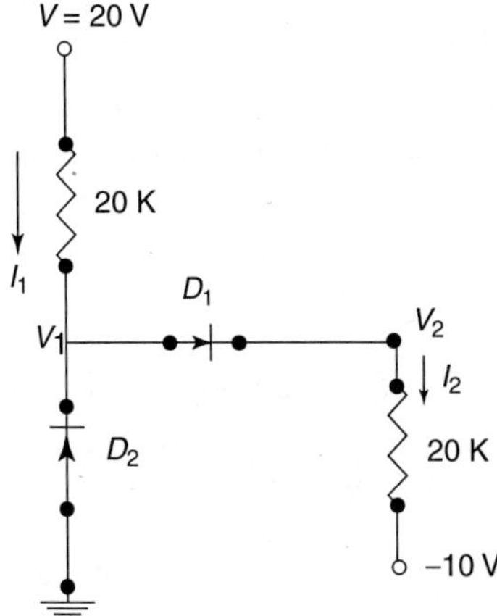


**Solution:**

For ideal diode, $R_f = 0$.

The diode acts as short-circuited. Both the diodes are forward-biased.

$$V_1 = 0 \text{ V}$$

As $D_2$ is short-circuited, therefore:

$$I_1 = \frac{20 - V_i}{20\text{ K}} = \frac{20}{20} \times 10^3$$

$$I_1 = 1 \text{ mA}$$

As $D_1$ is also short-circuited, therefore:

$$V_2 = 0$$

$$I_2 = \frac{V_2 - (-10)}{20\text{ K}} = \frac{10}{20} - 10^3$$

$$I_2 = 0.5 \text{ mA}$$

Therefore,

$$V_1 = 0 \text{ V}$$

$$V_2 = 0 \text{ V}$$

$$I_1 = 1 \text{ mA}$$

$$I_2 = 0.5 \text{ mA}$$

**Example 3-28** Consider the circuit and its $V$-1 characteristics, find voltage across diode $V_D$.

**Solution:**
From the given figure:

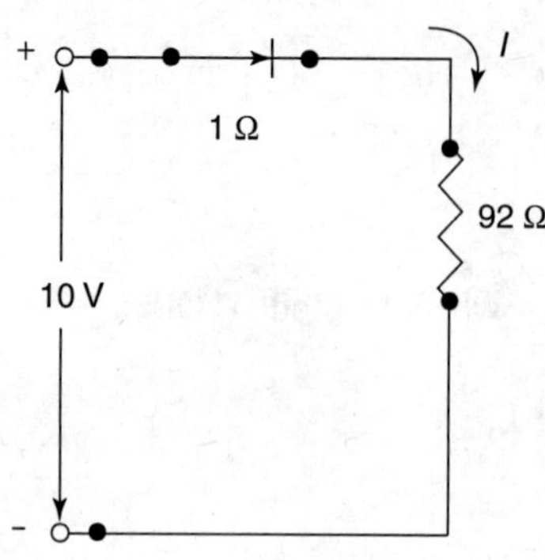


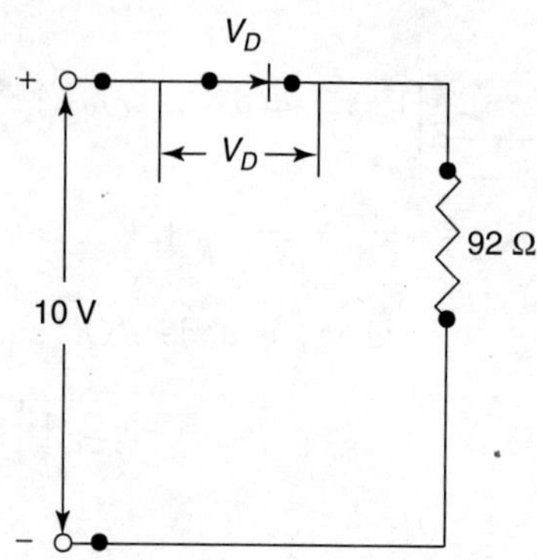


Voltage across diode $V_D$.

$$V_D = IR_D$$
$$= 0.1075 \times 1$$

Therefore, $$V_D = 0.1075 \text{ V}$$

**Example 3-29** A silicon diode is in forward-biased state with constant voltage $V$. Prove that the temperature coefficient of the forward current is given by:

$$\frac{(V_{Go} - V)}{\eta_T V_T}$$

**Solution:**

Forward voltage is given by: $$Vf = IRf$$

Forward resistance: $$Rf = \frac{\eta V_T}{I}$$

The voltage temperature coefficient is given by:

$$\frac{dV}{dT} = \frac{V - (V_{Go} - \eta V_T)}{T}$$

$$Rf\frac{dI}{dT} = \frac{V - (V_{Go} + \eta V_T)}{T} + Rf$$

Therefore: $$\frac{dI}{dT} = \frac{V - (V_{Go} + \eta V_T)}{T.Rf}$$

Putting up the value of $Rf$: $$\frac{dI}{dT} = \frac{V - V_{Go} + \eta V_T}{T} \times \frac{I}{\eta V_T}$$

Therefore: $$\frac{dI}{dT} = \frac{V_{Go} - V}{\eta V_T T}$$

**Example 3-30** The avalanche diode regulates at 50 V. The range of diode current from 5–40 mA. The supply voltage $V_i = 200$ V. Calculate $R$ to allow voltage regulation from a load current of $I_L = 0$ up to $I_{Lmax}$. Hence find $I_{Lmax}$.

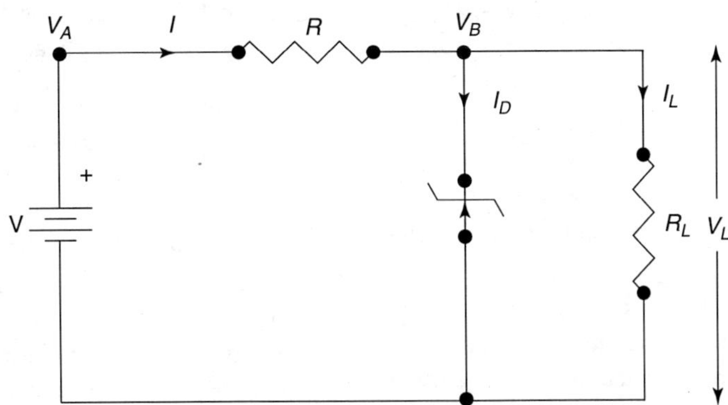


**Solution:**
Given data:

$$V_0 = V_L = 50 \text{ V (Zener Voltage)}$$

$$I_L = 0$$

Diode current ranges from 10 to 40 mA.

Therefore, $$I_D < 40 \text{ mA} = I$$

Therefore, $$R = \frac{V_A - V_B}{I_D} = \frac{200 - 50}{40 \times 10^{-3}}$$

$$R = 3.75 \text{ k}\Omega$$

$\therefore$ $$I_L = I_{max} \text{ when, } I_D = I_{D\,min} = 10 \text{ mA}$$

$\therefore$ $$\text{Maximum load current is, } I_{L\,max} = 40 - 10$$
$$= 30 \text{ mA}$$

■

## POINTS TO REMEMBER

1. A basic diode circuit consists of a diode in series with a voltage source. It mainly corresponds to the inclusion of a diode in the concerned circuit or a diode as a circuit element.

2. The circuit analysis of a diode in a circuit is made simpler by analysing it using the concept of load line.

3. A load line corresponds to the basic equation of the circuit concerned. It mainly deals with the relation of the voltage across the diode and the current flowing through it.

4. A load line actually has an impact on the region of operation of the device. It is an analysis performed in a graphical manner—a line drawn

on the characteristics of the device that represents the applied load.

5. The intersection of the load line with the characteristic curve of the device determines the point of operation.
6. Dynamic load line can be obtained from different load resistances. This dynamic load line is important because with the help of this, we can directly obtain the diode current for any given input voltage, as the corresponding current can be obtained from the graph.
7. The principle behind the use of the Zener diode as a voltage regulator is the fact that it maintains a constant output voltage even though the current through it changes. For proper operation, the voltage of an unregulated power supply must be greater than the Zener voltage of the diode selected. If the input voltage is less, the diode does not conduct.
8. A rectifier mainly works on the principle of conduction through the diode, i.e., it conducts when forward-biased, and not when reverse-biased. Depending on the design of the circuit, two types of rectifiers result—half-wave rectifiers and full-wave rectifiers.
9. In case of half-wave rectifiers, diode conducts for only one half-cycle of the input signal till the diode is forward-biased. Also, the rectified voltage appearing at the load is 0.7 V less than the input signal, the loss owing to the cut in the voltage of the diode concerned.
10. Depending upon the design implemented, two different types of rectifiers result:
    (a) Centre-tapped transformer rectifiers that use a bulky centre tapped transformer with three terminals and two diodes.
    (b) Bridge-rectifiers use a lightweight, simple transformer with four diodes in the circuit.
11. Advantages of a bridge rectifier:
    (a) A transformer without a centre tap can be used in a bridge circuit.
    (b) Since both the primary and the secondary currents in the transformer are sinusoidal, the bridge circuit requires a smaller transformer than that needed by a full-wave rectifier giving the same dc output voltage.
    (c) The peak inverse voltage rating of a diode in a bridge rectifier is half of that of a full-wave circuit yielding the same dc output voltage. The bridge circuit is therefore suitable for high voltage applications.
    (d) Crystal diodes are usually used to construct the bridge rectifier and the assembly of four such diodes is available in the market in a block form. The bridge circuit is therefore more compact and cheaper.
12. Disadvantages of the bridge rectifier:
    (a) A full-wave rectifier uses two diodes whereas a bridge rectifier uses four diodes.
    (b) Since the current flows through two diodes in a series in a bridge circuit, a large power is dissipated in the diodes. Hence, the bridge is not efficient for the low voltages.
    (c) Vacuum diodes with directly heated cathodes are not convenient for a bridge circuits because the cathodes for the diodes do not have the same potential.

    Hence, they cannot be connected in parallel across a transformer secondary meant for a filament supply. Bridge circuits generally employ Se and copper oxide rectifiers.
13. Comparison between half- and full-wave rectifiers:
    (a) In a half-wave rectifier, a single diode exists and the load current flows through it for only the positive half-cycle. On the other hand, in a full-wave rectifier, the current flows throughout the cycles of the input signals.
    (b) In a full-wave rectifier, we usually require a centre-tapped transformer. For a half-wave rectifier, only a simple transformer is required.
    (c) The peak inverse voltage in a half-wave rectifier is the maximum voltage across the transformer secondary. Whereas in the case of full-wave rectifier, the PIV for each diode is two times the maximum voltage between the centre tap and at the either end of the transformer secondary.

(d) In the case of a half-wave rectifier, the frequency of the load current is the same as that of the input signal and it is twice the frequency of the input supply for a full-wave rectifier.

(e) The dc load current and conversion efficiency for a full-wave rectifier is twice that of a half-wave rectifier. Also, we see the ripple factor of the full-wave rectifier is less than that of the half-wave circuit. This indicates that the performance of the full-wave rectifier is better than the half-wave rectifier.

(f) In a full-wave rectifier two diode currents flow through the two halves of the centre-tapped transformer secondary in opposite directions, so that there is no direct current magnetization of the core. The transformer losses being smaller, a smaller transformer can be used for a full-wave rectifier. This is an important advantage of the half-wave rectifier over a full-wave rectifier.

14. Filters form an integral part of a rectifier for obtaining a regulated, steady dc output at the end of a rectifier.

15. Clipper is a type of diode circuit where the diode network is able to clip off a portion of the input signal without disturbing the remaining portion of the input signal.

16. A clamper circuit clamps a signal to a different dc level. The network must have a capacitor, a diode, a resistive network and should also employ an independent voltage source which introduces an additional voltage shift.

17. Comparators are circuits that employ diode for comparisons between two different input voltages. The basic principle on which the comparator works is the switching of the diodes i.e., the action corresponding to the portion when the diode conducts and when it does not.

## IMPORTANT FORMULAE

1. For a simple diode circuit with a diode connected in series with a resistor and a voltage source, the equation governing the behaviour of the circuit is given by:

$$V_a = V - iR_L$$

2. Average value of load current is given by:

$$I_{dc} = \frac{1}{2\pi}\int_0^{2\pi} i_L d(\omega t)$$

3. Ripple factor is given by:

$$\gamma = \frac{(I_{rms}^2 - I_{dc}^2)^{1/2}}{I} = \frac{(V_{rms}^2 - V_{dc}^2)^{1/2}}{V_{dc}}$$

4. Rectification efficiency is given by:

$$\eta = \frac{P_{dc}}{P_i} \times 100\% = \left(\frac{I_{dc}^2}{I_{rms}^2}\right)\frac{1}{1+\dfrac{R_f}{R_L}} \times 100\%$$

5. For a half-wave rectifier:

$$I_{dc} = \frac{I_m}{\pi}$$

$$I_{rms} = \frac{I_m}{\sqrt{2}}$$

$$\gamma = \left[\left(\frac{I_{rms}}{I_{dc}}\right)^2 - 1\right]^{1/2} = 1.21$$

$$\eta = \frac{40.6}{1 + R_f/R_L}\%$$

6. For a full-wave rectifier:

$$I_{dc} = \frac{2I_m}{\pi}$$

$$I_{rms} = \frac{I_m}{\sqrt{2}}$$

$$\gamma = 0.482$$

$$\eta = \frac{81.2}{1 + R_f/R_L}\%$$

7. Percentage voltage regulation:

$$\frac{V_{NL} - V_{RL}}{V_{RL}} \times 100\%$$

## OBJECTIVE QUESTIONS

1. A More Zener diode is based on the principle of:
   (a) Thermionic emission
   (b) Tunneling of charge carriers across the junction
   (c) Diffusion of charge carriers across the junction
   (d) None of the above
2. Silicon diode is less suited for low voltage rectifier operation because:
   (a) Its breakdown voltage is high
   (b) Its reverse saturation current is low
   (c) Its cut-in voltage is high
   (d) None of the above
3. Silicon is not suitable for fabrication of light-emitting diodes because it is:
   (a) An indirect band gap semiconductor
   (b) A direct band gap semiconductor
   (c) A wide band gap semiconductor
   (d) None of the above
4. For an abrupt junction Varactor diode, the dependence of the device capacitance ($C$), an applied reverse-bias ($V$) is given by:
   (a) $C \propto VI/3$ (b) $C \propto VI/2$
   (c) $C \propto V\text{-}I/3$ (d) None of the above
5. A Zener diode:
   (a) Has a high forward voltage rating
   (b) Has a sharp breakdown at low reverse voltage
   (c) Is useful as an amplifier
   (d) None of the above
6. Which of these is a best description of a Zener diode?
   (a) It operates in the reverse region
   (b) It is a constant voltage device
   (c) It is a constant current device
   (d) None of the above
7. When two Zener diodes each of 10 V and 15 V are connected in series, then the overall voltage between them when they are in conduction is
   (a) 10 V (b) 25 V
   (c) 15 V (d) Zero
8. In a standard regulator circuit that uses Zener diode 10 V, the input voltage varies from 25 to 40 V, load current varies from 10 to 20 mA, and the minimum Zener current is 5 mA, the value of the series resistance in ohms will be:
   (a) 1500 (b) 1200
   (c) 600 (d) None of the above
9. The LED is usually made of materials like:
   (a) GaAs (b) C and Si
   (c) GeAs (d) None of the above
10. Varactor diodes are used in FM receivers to obtain:
    (a) Automatic frequency control
    (b) Automatic gain control
    (c) Automatic volume control
    (d) None of the above
11. No-load voltage of power supply is 100 V and full-load voltage is 80 V, the percentage of regulation is:
    (a) 0 (b) 25
    (c) 15.75 (d) None of the above
12. Zener diodes are used as:
    (a) Reference voltage elements
    (b) Reference current elements
    (c) Reference resistance
13. Zener diodes are:
    (a) Specially doped $p$–$n$ junction
    (b) Normally doped $p$–$n$ junction
    (c) Lightly doped $p$–$n$ junction
    (d) None of the above
14. Silicon diode is less suited for low voltage rectifier operation because:
    (a) It can withstand high temperatures
    (b) Its reverse saturation current is low
    (c) Its breakdown voltage is high
    (d) None of the above

15. Silicon is not suitable for fabrication of light emitting diodes because it is:
    (a) An indirect band gap semiconductor
    (b) A direct band gap semiconductor
    (c) A wide band gap semiconductor
    (d) None of the above

16. In an abrupt junction Varactor diode, the dependence of the device capacitance ($C$) and applied reverse-bias ($V$) is given by:
    (a) $C \propto VI/3$ (b) $C \propto VI/2$
    (c) $C \propto V\text{-}I/3$ (d) None of the above

17. A general propose diode is more likely to suffer an avalanche breakdown rather than a Zener breakdown because:
    (a) It is lightly doped
    (b) It is heavily doped
    (c) It has weak covalent bonds
    (d) None of the above

18. A Zener diode:
    (a) Has a high forward voltage rating
    (b) Is useful as an amplifier
    (c) Has a sharp breakdown at low reverse voltage
    (d) None of the above.

19. If the junction temperature of LED is increased the radiant output power:
    (a) Decreases (b) Increases
    (c) Remains the same (d) None of the above

20. The Zener effect is valid approximately:
    (a) Below 5 V (b) Above 5 V
    (c) Equal to 5 V (d) None of these

21. Each diode of full-wave centre-tapped rectifier conducts for:
    (a) 360° (b) 270°
    (c) 90° (d) 180°

22. When a capacitor filter is used, the PIV for a half-wave rectifier:
    (a) Increases (b) Decreases
    (c) Remains unaltered (d) None of the above

23. The transfer characteristics of a diode relates to:
    (a) The diode current and the input voltage
    (b) The diode current and the output voltage
    (c) The output voltage and the input voltage
    (d) None of the above

24. The clipping action of a diode requires that its forward resistance:
    (a) Be zero (b) Have a finite value
    (c) Be infinite (d) None of the above

25. For a low voltage rectification:
    (a) Two diode full-wave rectifier is suitable
    (b) Both bridge and full-wave rectifier are suitable
    (c) Bridge rectifier is suitable
    (d) None of the above

26. If $V_m$ is the peak value of an applied voltage in a half-wave rectifier with a large capacitor across the load, then PIV is:
    (a) $V_m$ (b) $V_m/2$
    (c) $2V_m$ (d) None of the above

27. The induction filter is mostly used for rectifiers with:
    (a) Half-wave rectifiers
    (b) Light loads
    (c) High loads
    (d) None of the above

28. The most significant component of ripple voltage in a half-wave rectifier is contained in:
    (a) Fundamental frequency
    (b) Second harmonic
    (c) DC component
    (d) None of the above

29. The disadvantages of capacitor input $LC$ filter are:
    (a) High cost, more weight and external field produced by a series inductor
    (b) High cost, less weight
    (c) Low cost, more weight
    (d) None of the above

30. Larger the value of the capacitor filter:
   (a) Smaller the dc voltage across the load
   (b) Longer the time that current pulse flows through the diode
   (c) Larger the peak current in the rectifying diode
   (d) None of the above

31. Which rectifier requires four diodes?
   (a) Half-wave voltage doublers
   (b) Full-wave voltage doublers
   (c) Full-wave bridge circuit
   (d) None of the above

## REVIEW QUESTIONS

1. What is the dynamic characteristic of a diode? How can you obtain it from the dynamic characteristics?
2. What is a load line in connection with a diode connected to a supply voltage through a series load resistance? How does the load line change with the change of the supply voltage, the load resistance, and the type of diode?
3. What do you mean by rectification? How can you study the performance of the diode rectifier with the help of its dynamic characteristic?
4. Draw the circuit diagram of a half-wave rectifier and explain the operation of the circuit.
5. Draw the circuit of a full-wave rectifier and explain the operation of the circuit.
6. Distinguish between the following:
   (a) Full-wave rectifier and half-wave rectifier
   (b) Full-wave rectifier and bridge rectifier.
7. Draw the waveforms of the diode current and the load voltage for a sinusoidal input voltage applied to
   (a) Half-wave rectifier
   (b) Full-wave rectifier

   Is it necessary for the two diodes of the rectifier to be identical?
8. Draw the circuit diagram of a full-wave rectifier using junction diodes and explain clearly its action.
9. Discuss how a semiconductor diode can be used as a rectifier. Do you prefer a valve diode or a junction diode for rectification?
10. Explain the phenomena of a bridge rectifier with the help of a circuit diagram. Mention its advantages and disadvantages when compared with a full-wave rectifier with a centre-tapped transformer.
11. Explain the term peak inverse voltage in connection with a diode rectifier. Is it different for half and full-wave rectifiers with a centre tap? Does it change if we use a capacitor filter?
12. Define the following terms:
   (a) dc load current
   (b) Ripple factor
   (c) Conversion efficiency
13. For a half-wave rectifier, calculate
   (a) The dc load current
   (b) The peak load current
   (c) The rms load current
   (d) The rms value of ripple current
   (e) The ripple factor
   (f) The dc power output
   (g) The efficiency of the rectifier
14. What do you mean by the regulation characteristics of a rectifier? Figure out the main quantity determining the regulation characteristics. Define percentage voltage regulation.
15. Explain the significance of percentage voltage regulation. Find the percentage voltage regulation for both the half and full-wave rectifiers.
16. Why is a filter used in a rectifier? Enumerate the different types of filters used at the output of the rectifiers.
17. Explain how the dc voltage of a full-wave rectifier is improved when a capacitor filter is used. Draw waveforms of the load voltage and the diode current.

18. Derive expressions of the percentage regulation for a half-wave, full-wave and a bridge rectifier circuit each employing the same capacitor and the same load resistance.
19. What is a voltage multiplier? Draw the circuit diagram of a half-wave voltage doubler and explain its operation. What is the advantage of the circuit?
20. What is the function of the clipping circuit? Draw the circuit diagram of a diode clipper that limits the positive peak of the input voltage. Explain how the circuit works.
21. Draw the neat diagram of a full-wave voltage doubler and explain its operation. How can you construct a voltage tripler?
22. What is the transfer characteristic of a diode? What is the utility of this characteristic?
23. Explain the working of a diode clipper that limits the lower portion of the input voltage.
24. What is a double diode clipper? Draw the circuit diagram of a double diode clipper and explain.
25. What do you understand by a clamping circuit? Draw the circuit diagram of a dc restorer. How does the circuit function?

## PRACTICE PROBLEMS

1. A 1 mA diode (i.e., one that has $v_D = 0.7$ V at $I_D = 1$ mA) is connected in series with a 200 Ω resistor to a 1 V supply. Provide a rough estimate of the diode current.

   If the diode is characterized by $n = 2$, estimate the diode current closely using iterative analysis.
2. Assuming the availability of the diodes for which $v_D = 0.7$ V and $I_D$ =1-mA and $n = 1$, design a circuit that utilizes four diodes in series with a resistor $R$ connected to ac 15 V power supply. The voltage across the string of diodes is to be 3.0 V.
3. Find the parameters of a piecewise-linear model of a diode, for which $v_D = 0.7$ V at $I_D$ = 1-mA and $n = 2$. The model is to fit exactly at 1-mA and 10 mA. Calculate the error—in millivolts—in predicting $v_D$, using the linear piecewise-linear model at $I_D$ at 0.5, 5 and 14 mA.
4. A junction diode is operated in a circuit in which it is supplied with a constant current $I$. What is the effect on the forward voltage of the diode if an identical diode is connected in parallel? Assume $n = 1$.
5. A diode measured at two operating currents, 0.2 mA and 10 mA, is found to have corresponding voltages 0.650 and 0.750. Find the values of $n$ and $I_S$.
6. A diode for which the forward voltage drop is 0.7 V at 1.0 mA, and for which $n = 1$, is operated at 0.5 V. What is the value of the current?
7. When a 10-A current is applied to a particular diode it is found that the junction voltage immediately becomes 700 mV. However, as the power being dissipated in the diode raises its temperature, it is found that the voltage decreases and eventually reaches 600 mV. What is the apparent rise in junction temperature? What is the power dissipated in the diode in its final state? What is the temperature rise per watt of power dissipation?
8. The small-signal model is said to be valid for voltage variations of about 10 mV. To what percent current change does this correspond for:

   (i) $n = 1$

   (ii) $n = 2$
9. What is the incremental resistance of ten 1 mA diodes connected in parallel and fed with a dc current 10 mA. Let $n = 2$.
10. In the circuit given, $I$ is the dc current and $v_S$ is the sinusoidal signal. Capacitor $C$ is very large; its function is to couple the signal to the diode but block the dc current from the source. Use the diode small signal model to show that the signal component of the output voltage is:

$$v_o = v_S \frac{nV_T}{nV_T + IR_S}$$

   If $v_S = 10$ mV, find $v_o$ for $I = 1$ mA, 0.1 mA and 1 $\mu$A. Let $R_S = 1$ kΩ and $n = 2$. At what value of $I$ does $v_o$ become one half of $v_s$?

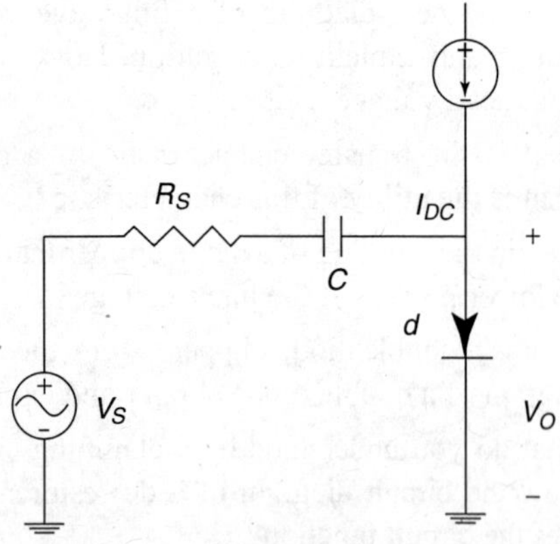


11. Construct circuits which exhibit terminal characteristics as shown in parts (a) and (b) of the given figure.

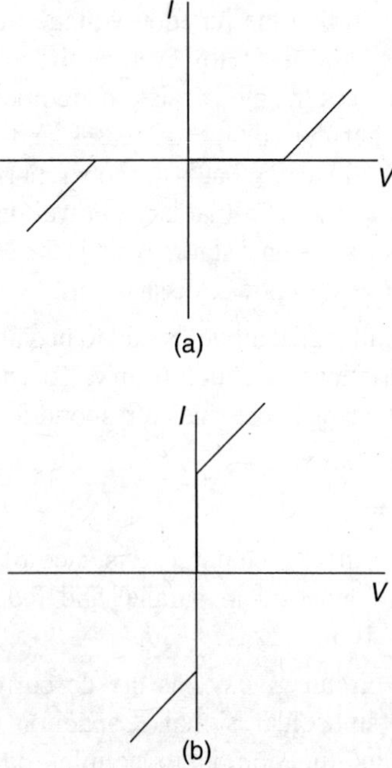


12. At what forward voltage does a diode for which $n = 2$ conduct a current equal to $1000I_S$? In terms of $I_S$, what current flows in the same diode when its forward voltage is 0.7 V?

13. A diode modeled by the 0.1 V/decade approximately operates in a series circuit with $R$ and $V$. A designer, considering using a constant voltage model, is uncertain whether to use 0.7 V or 0.5 V for $V_D$. For what value of $V$ is the difference only 1%? For $V = 2$ V and $R = 1$ K, what two currents would result from the use of the two values of $V_D$?

14. The circuit in the given figure utilizes three identical diodes having $n = 1$ and $I_S = 10^{-14}$ A. Find the value of the current $I$ required to obtain an output voltage $V_o = 2$ V. If a current of 1 mA is drawn away from the output terminal by a load, what is the change in output voltage?

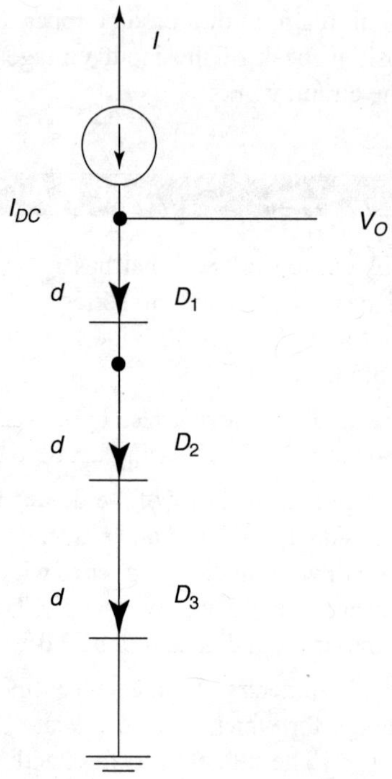


15. Assuming that the diodes in the circuits of the given figure are ideal, utilize Thevenin's theorem to simplify the circuits, and thus, find the values of the labeled currents and voltages.

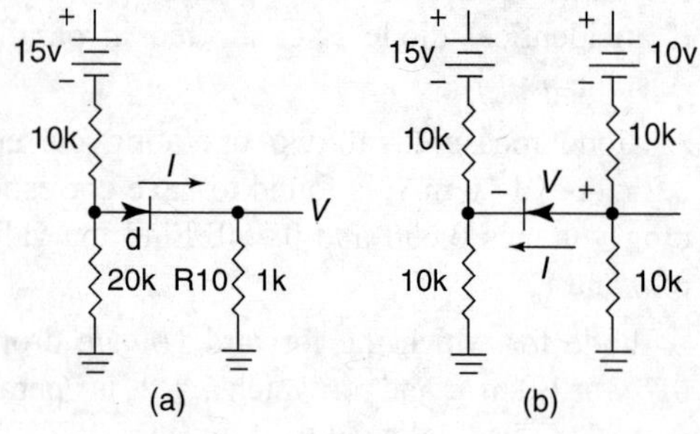

16. For the rectifier circuit as given, let the input sine wave have 120 V rms value, and assume the diode to be ideal. Select a suitable value for $R$ so that the peak diode current does not exceed 0.1 A. What is the greatest reverse voltage that will appear across the diode?

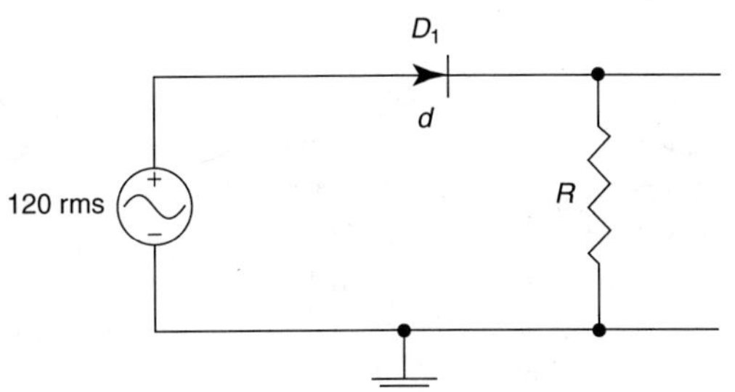


17. For the logic gate as shown in the following figure, assume ideal diodes and input voltage levels of 0 and +5 V. Find a suitable value for $R$ so that the current required from each of the input signal sources does not exceed 0.2 mA.

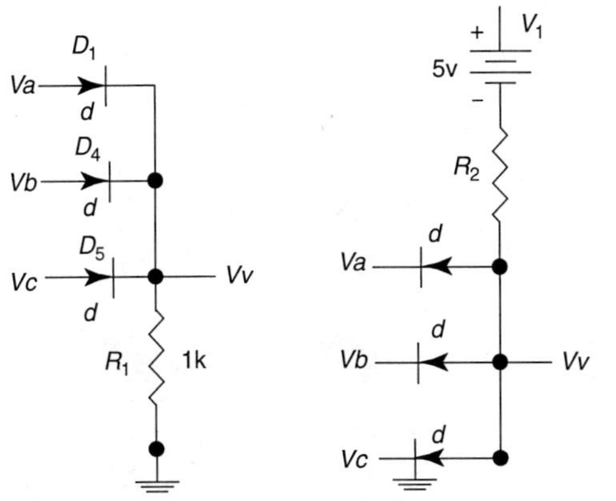


18. Consider the voltage regulator circuit as shown in the following figure, under the condition that a load current $I_L$ is drawn from the output terminal. Denote the output voltage across the diode by $V_o$. If the value of $I_L$ is sufficiency small so that the corresponding change in the regulator output voltage $\Delta V_o$ is small enough to justify using the diode small signal model, show that:

$$\pm \frac{\Delta V_0}{I_L} = -(r_d // R)$$

(Note: This quantity is known as the load regulation and is usually expressed in mV/mA.)

19. In the above problem, if the value of $R$ is selected such that at no load the voltage across the diode is 0.7 V and the diode current is $I_D$, show that the expression derived becomes:

$$\frac{\Delta V_0}{I_L} = \frac{nV_T}{I_D} \frac{V^+ - 0.7}{V^+ - 0.7 + nV_T}$$

Select the lowest possible value for $I_D$ that results in a load regulation $<=$ 5 mV/mA. Assume $n = 2$. If $V^+$ is nominally 10 V, what value of $R$ is required?

20. With reference to the Problem 18 and 19, generalize the expression derived in Problem 19 for the case of M diodes connected in series and $R$ adjusted to obtain $V_o = 0.7$ mV, at no load.

21. A voltage regulator consisting of a 6.8 V Zener diode, a 100 ohm resistor, and intended for operation with a 9 V supply is accidentally connected to 15 V supply instead. Assuming the $r_Z$ is very small; calculate the expected values of Zener current and the power dissipated in both the Zener diode and the resistor, for both the normal as well as aberrant situations. Also compare the ratios.

22. A shunt regulator utilizing a Zener with an incremental resistance of 6 ohms is fed through an 82 ohms resistor. If the raw supply changes by 1.4 V, what is the corresponding change in the regulated output voltage?

23. A 9.1 V Zener diode exhibits its nominal voltage at a current of 28 mA. At this current the incremental resistance is specified as 5 ohms. Find $V_{ZO}$ of the Zener model. Find the Zener voltage at a current of 10 mA and at 100 mA.

24. Consider a half-wave peak rectifier fed with a voltage $v_S$ having a triangular waveform with 20 V peak to peak amplitude, zero average and 1 Khz frequency. Assume that the diode has a 0.7 V drop when conducting. Let the load resistance $R = 100\ \Omega$ and the filter capacitor $C = 100\ \mu$F. Find the average dc output voltage, the time interval during which the diode conducts—the average diode current during conduction, and the maximum diode current.

25. Sketch the transfer characteristics $V_o$, $V_s$, $V_I$ for the limiter circuits as shown in the following figures. All diodes start conducting at a forward voltage drop of 0.5 V and display voltage drops of 0.7 V when fully conducting.

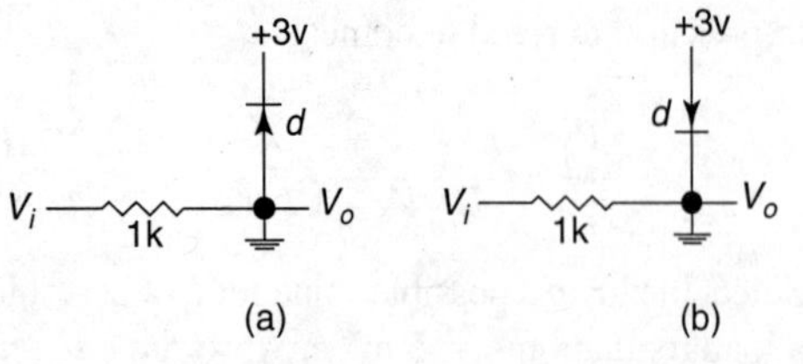


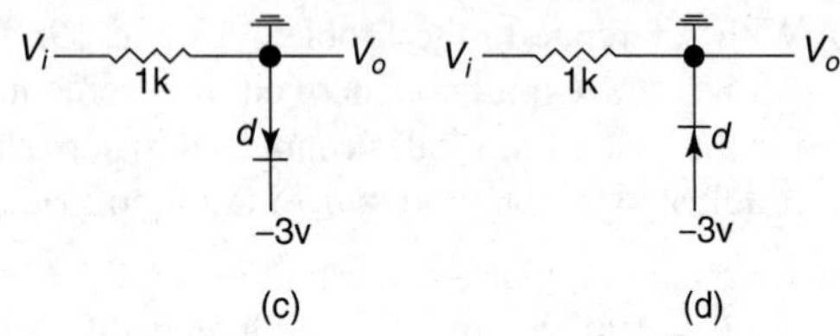


26. In the figure provided, (a) and (b) are connected as follows: The two input terminals are tied together, and the output terminals are tied together. Sketch the transfer characteristic of the resulting circuit, assuming that the cut in voltage of the diodes is 0.5 V and their voltage drop when fully conducting is 0.7 V.

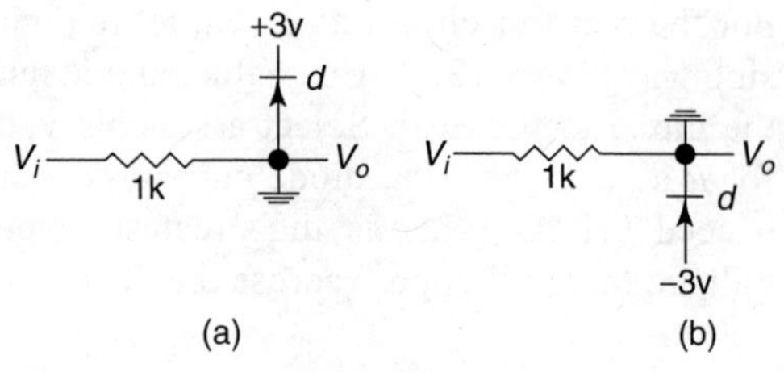


27. Plot the transfer characteristics of the circuit as shown in the following figure by evaluating $V_I$ corresponding to $V_o$ = 0.5 V, 0.6 V, 0.7 V, 0.8 V, 0 V, −0.5 V, −0.6 V, -0.8 V. Assume that the diodes are 1 mA units having a 0.1 V/decade logarithmic characteristi(c) Characterize the circuit as a hard or a soft limiter. What is the vale of *K*? Estimate $L_+$ and *L*.

28. A clamped capacitor using an ideal diode is supplied with a sine wave of 10 − V rms. What is the average dc value of the resulting output?

29. Design limiter circuits using only diodes and 10 kΩ resistors to provide an output signal limited to the range:

    (a) −0.7 V and above

    (b) −2.1 V and above

    (c) ± 4.1 V

    Assume that each diode has a 0.7 V drop when conducting.

## SUGGESTED READINGS


1. Millman, J. and H. Tau(b) 1965. *Pulse, Digital, and Switching Waveforms*. New York: McGraw-Hill Book Company.
2. Boylsted, R. and L. Nashelsky. 2007. *Electronic Devices and Circuit Theory*. New Delhi: Pearson Education.

# 4

# BJT Fundamentals

## Outline



## Objectives

This chapter introduces one of the most important semiconductor devices, the bipolar junction transistor. This is followed by a discussion on the formation of the $p$–$n$–$p$ and $n$–$p$–$n$ junctions. Transistor mechanism and energy band diagrams are examined in detail. Transistor current components for $p$–$n$–$p$ and $n$–$p$–$n$ transistors are provided, and subsequently the CE, CB, CC configurations of transistors are explained. The chapter ends with a discussion on the Ebers–Moll model of a transistor, the transistor characteristics, the concept of load line, and Early effect.

## 4-1 INTRODUCTION

With the advent of junction transistors, leading to the first Nobel Prize in electronics to William Shockley, Walter Brattin and John Bardeen in 1956, the modern electronic era begins in the real sense of the term. The junction transistors are listed at the top among all the amplifying semiconductor devices. They form the key elements in computers, space vehicles and satellites, and in all modern communications and power systems.

A bipolar junction transistor (BJT) is a three-layer active device that consists of two $p$–$n$ junctions connected back-to-back. Although two $p$–$n$ junctions in a series is not a transistor since a transistor is an active device whereas a $p$–$n$ junction is a passive device. Besides, their designs are also different. A BJT is actually a current-amplifying device. In a BJT, the operation depends on the active participation of both the majority carrier, and the minority carrier; hence, the name "bipolar" is rightly justified.

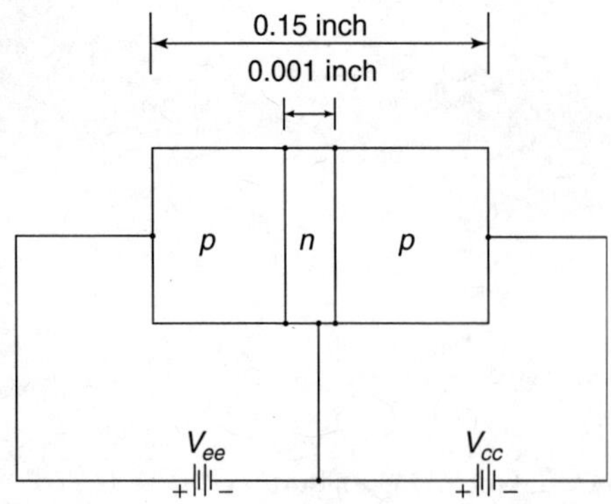


Figure 4-1(a) *p–n–p* transistor

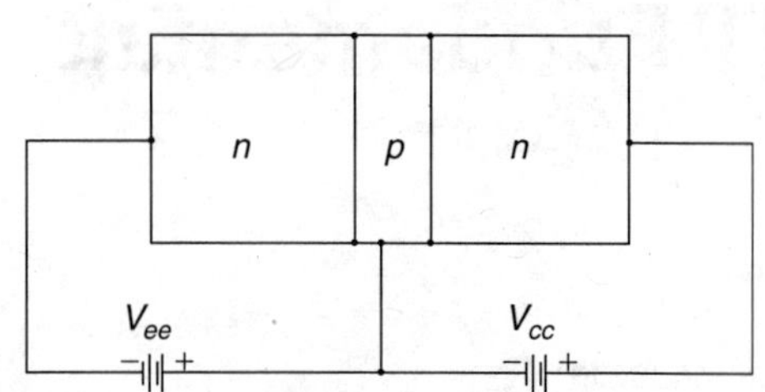


Figure 4-1(b) *n–p–n* transistor

## 4-2 FORMATION OF *p–n–p* AND *n–p–n* JUNCTIONS

When an *n*-type thin semiconductor layer is placed between two *p*-type semiconductors, the resulting structure is known as the *p–n–p* transistor. The fabrication steps are complicated, and demand stringent conditions and measurements. When a *p*-type semiconductor is placed between two *n*-type semiconductors, the device is known as the *n–p–n* transistor. Both these types of transistors are shown in Fig. 4-1(a) and Fig. 4-1(b) respectively.

## 4-3 TRANSISTOR MECHANISM

The basic operation of the transistor is described using the *p–n–p* transistor. The *p–n* junction of the transistor is forward-biased whereas the base-to-collector is without a bias, as shown in the Fig. 4-2.

The middle portion is termed as the base (*B*) while the two end portions are known as the emitter (*E*) and the collector (*C*). The junction between the emitter and the base is called the emitter–base junction, or the emitter junction ($J_E$). The junction between the collector and the base is called the collector-base junction, or briefly the collector junction ($J_C$).

The depletion region gets reduced in width due to the applied bias, resulting in a heavy flow of majority carriers from the *p*-type to the *n*-type material gushing down the depletion region and reaching the base. The forward-bias on the emitter–base junction will cause current to flow. This flow of current consists of two components: (i) holes injected from emitter to base and (ii) electrons injected from base to emitter. The design of the transistor ensures that the base region is fabricated very lightly compared to the emitter or the collector regions. Consequently, we have the holes injected from the emitter into the base in large numbers, and we neglect the injection of electrons from the base region.

For easy analysis, let us now remove the base-to-emitter bias of the *p–n–p* transistor, as shown in the Fig. 4-3.

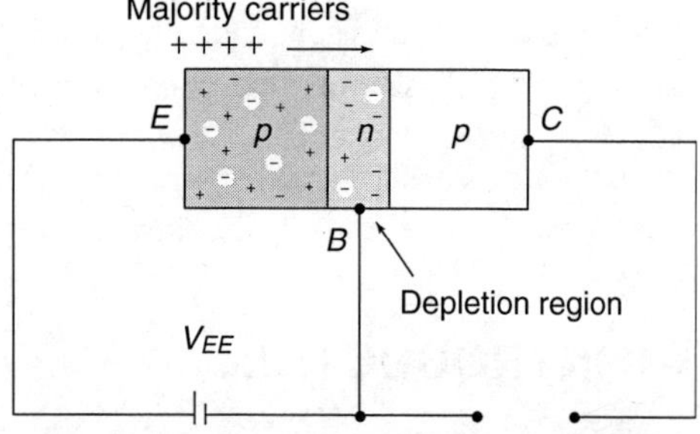


Figure 4-2 Forward-biased junction of a *p–n–p* transistor

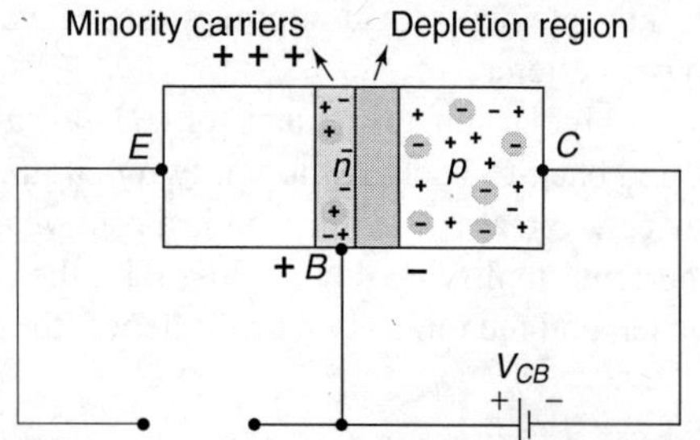


Figure 4-3 Reverse-biased junction of a *p–n–p* transistor

The flow of majority carriers is zero, resulting in a minority-carrier flow. Thus, one *p–n* junction of a transistor is reverse-biased, while the other is kept open. The operation of this device becomes much easier when they are considered as separate blocks. In this discussion, the drift currents due to thermally generated minority carriers have been neglected, since they are very small.

## 4-4 ENERGY BAND DIAGRAMS

Since a transistor can be seen as two *p–n* diodes connected back-to-back, the bending of the energy levels will take place—as is evident from the concepts of diodes—under both forward- and reverse-biased conditions. Under equilibrium conditions, the bending will be such that the Fermi level will remain at par for both the emitter and the base regions. Similarly, for the collector and the base regions, the energy levels will bend sufficiently for the alignment of the Fermi level. If further bias is applied, respective changes will take place at both the junctions.

If the emitter–base junction is forward-biased, the barrier potential is decreased and can be expressed as $V_{bi} - V_f$ where, $V_{bi}$ is the barrier potential difference and $V_f$ is the applied voltage across the junction,

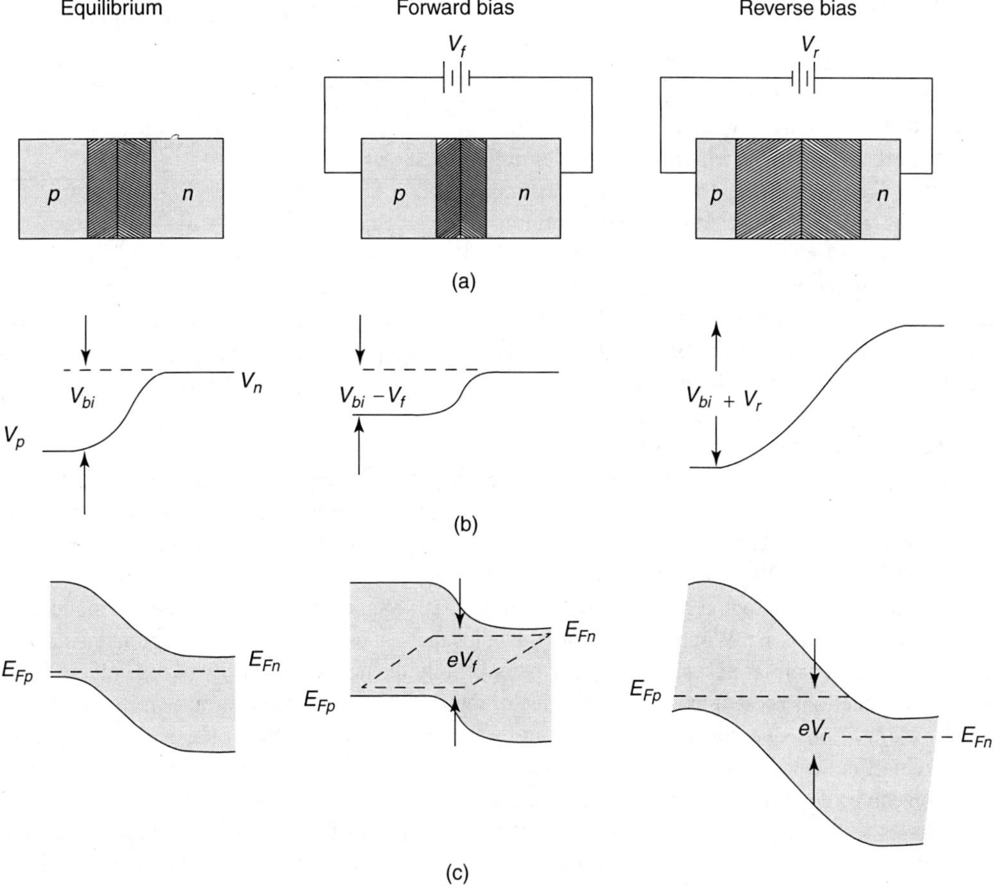


**Figure 4-4** State of energy bands under (a) no bias (b) forward-biased state (c) reverse-biased state

due to reverse-bias of the collector-base junction there is an increase in the barrier potential in the collector–base junction. The changes in the barrier potential and energy levels for both unbiased and biased transistors have been shown in Fig. 4-4.

The energy band diagram of an *n–p–n* bipolar transistor under zero bias and forward-mode bias is shown in the Fig. 4-5.

Thus, the basic transistor operation can be described as the state when the transistor is biased in the forward-active mode of operation. The current at one terminal of the transistor (collector-current) is controlled by the voltage applied across the other two terminals of the transistor (base–emitter voltage).

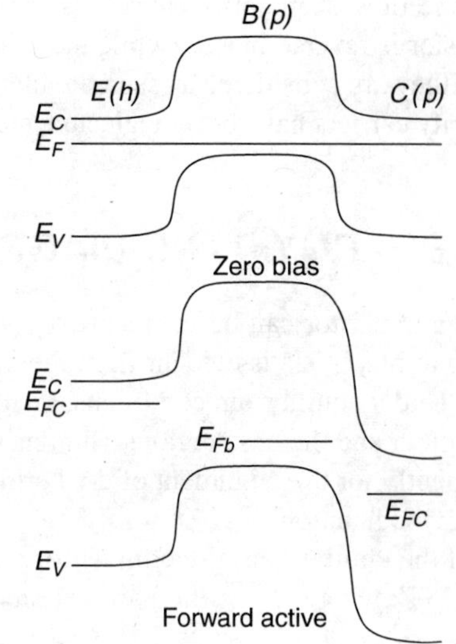


**Figure 4-5** Bending of the energy states under no bias and forward-bias

## 4-5 TRANSISTOR CURRENT COMPONENTS

The transistor current components in a non-degenerate *p–n–p* transistor can be formulated from Fig. 4-6.

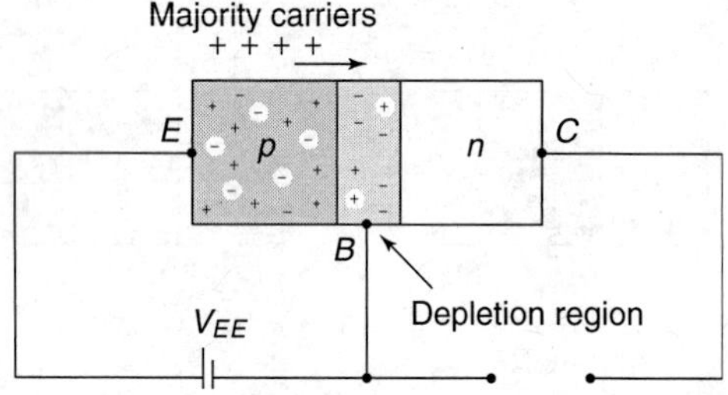


**Figure 4-6** Transistor with forward-biased emitter junction and open-collector junction

Since the emitter junction is connected to the positive pole of the battery $V_{EE}$, which makes the emitter base region forward-biased, the majority carriers (holes) from the *p*-side diffuse into the base region (*n*-type). From Fig. 4-6 we can state that for a forward-biased *p–n* junction, with collector–base open circuited, a forward current flows in the hole direction, i.e., from *p*-side (emitter) to *n*-side (base), and hence is termed as the emitter current $I_E$. In the base region, the holes coming from the *p*-side act as minority carriers, which have a large probability of meeting an electron in the base. In such a case, both the electron and hole disappear, forming a covalent bond. This act is highly dependent on the doping levels as well as on the temperature in the base region. This whole process of the hole meeting an electron is known as *recombination.* The electrons and holes that have been lost in the base region are supplied externally by the $V_{EE}$ through the base lead.

With the collector–base region connected through the $V_{CC}$ voltage source in reverse-bias mode, the whole situation takes a different twist. When this is done, the holes that have entered the base region and act as minority carriers "see" a reverse-biased system and thus, get swept by the negative polarity of the voltage $V_{CC}$. This is also valid for the electrons present in the *p*-side collector region. These electrons migrate from *p*-side (collector) to *n*-side (base), and consequently are also swept out by the positive pole of the voltage source $V_{CC}$. These currents that take part in the reverse-biased collector–base region are the reverse saturation currents. The effect of recombination can be dealt with by using the following example. Let us suppose that the emitter region (*p*-side) transmits 8 holes to the base region (*n*-side). Now, the doping level of the base region is such that for every 8 holes 2 electrons supplied by the base recombine 2 holes. In such a case $8 - 2 = 6$ holes reach the collector region. These two holes contribute to the reverse collector saturation current. Evidently, to increase the amount of current in the collector region, a lower amount of recombination requires to be done in the base. This is achieved by lowering the doping level of the base region, and narrowing the base width. In such a case most of the holes get transmitted to the collector region, thus, increasing the reverse collector saturation current.

Now consider two cases:

(i) When the collector side is open-circuited: In such a case only the emitter current $I_E$ flows from emitter to base and to the voltage source $V_{EE}$.

(ii) When the collector side is closed: In such a case recombination occurs in the base creating the recombination current $I_{E\text{ minority}}$ plus $I_{E\text{ majority}}$. Thus:

$$I_E = I_{E\text{ majority}} + I_{E\text{ minority}} \tag{4-1}$$

Now, $I_{E\text{ majority}}$ when transferred to *p*-region from the base gets converted to $I_{C\text{ majority}}$ and the minority carriers due to the open-circuited emitter–base region flow from *n*-side (base) to *p*-side (collector). This minority carrier is of a very low magnitude—in the order of microamperes for silicon and nanoamperes for germanium—and is designated as $I_{CO}$ where "O" implies open-circuited emitter-base terminal. Thus, we can write:

$$I_{E\text{ majority}} = I_{C\text{ majority}}$$

Hence the current coming out of the collector region:

$$I_C = I_{C\text{ majority}} + I_{C\text{ minority}} \tag{4-2}$$

Meanwhile the recombination current in the close-circuited emitter–base region, which was termed as $I_{E\text{ minority}}$, is nothing but the base current $I_B$.

Thus, applying Kirchoff's current rule in the collector terminal:

$$I_E = I_B + I_C \tag{4-3}$$

## 4-5-1 Current Components in *p–n–p* Transistor

Now, as we can see in the Fig. 4-7(a), both biasing potentials have been applied to a *p–n–p* transistor, with the resulting majority and minority carrier flow indicated. The width of the depletion region clearly indicates

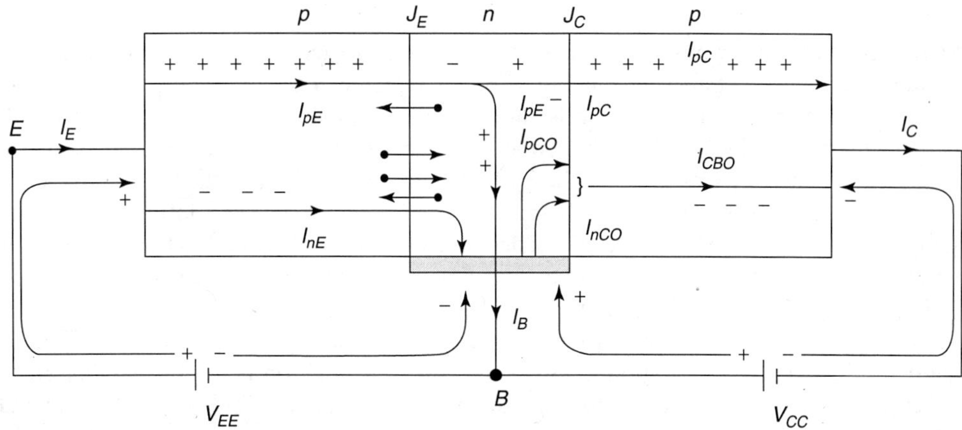


**Figure 4-7(a)** Direction of flow of current in *p–n–p* transistor with the base–emitter junction forward-biased and the collector–base junction reverse-biased

which junction is forward-biased and which is reverse-biased. A large number of majority carriers will diffuse across the forward-biased *p–n* junction into the *n*-type material. Since the sandwiched *n*-type material is very thin and has a low conductivity, a very small number of these carriers will take this path of high resistance to the base terminal. The magnitude of the base current is typically in the order of microamperes as compared to milliamperes for the emitter and collector currents. The large number of these majority carriers will diffuse across the reverse-biased junction into the *p*-type material connected to the collector terminal, as indicated in the Fig. 4-7(a).

The majority carriers can easily cross the reverse-biased junction, because for the reverse-biased diode the injected majority carriers will appear as minority carriers in the *n*-type material. These minority carriers assist in the reverse collector saturation current, and hence cross the reverse-biased junction.

Now, applying Kirchoff's current law to the transistor of the Fig. 4-7(a) as if it was a single node, we get:

$$I_E = I_C + I_B \tag{4-4}$$

and we find that the emitter current is the sum of the collector and base currents. The collector current, however, is comprised of two components: the majority and the minority carriers. The minority carrier components are termed as *leakage current*, and given the symbol $I_{CO}$. The collector current, therefore, is determined in total as:

$$I_C = I_{C\text{ majority}} + I_{CO\text{ minority}} \tag{4-5}$$

For a general-purpose transistor, $I_C$ is measured in milliamperes, while $I_{CO}$ is measured in microamperes or nanoamperes. The basic operation of a *p–n–p* transistor with its emitter–base junction ($J_E$) forward-biased and collector–base junction ($J_C$) reverse-biased, as shown in Fig. 4-7(a), can be visualized as follows: when there is no external bias, the current through the transistor should be zero because of the development of an intrinsic potential barrier across the *p–n* junction. Now the forward-biasing voltage ($V_{EE}$) decreases the emitter–base potential barrier whereas the reverse-biasing voltage ($V_{CC}$) increases the potential barrier across the collector junction ($J_C$). The lowering of emitter-junction barrier allows injection of holes from emitter to base and injection of electrons from base to emitter. These two flows produce the emitter current:

$$I_E = I_E(p) + I_E(n) \tag{4-6}$$

where, $I_E(p)$ is due to the holes moving from emitter to base, and $I_E(n)$ is due to the electrons going from base to emitter. In commercial transistors doping of emitter region is made much higher than the base. This makes $I_E(p) \gg I_E(n)$, and the emitter current is almost due to the holes only. With only 2–3% of the total carriers entering the base region, it is bound to be so. This is desirable, because the current component $I_E(n)$ does not contribute to the collector current. The injected holes diffuse through the base region towards the collector region. While diffusing through the base, a few of the injected holes are lost due to recombination with majority electrons. The holes that reach the collector junction cross the potential barrier and are immediately collected by the collector region. This gives one of the two current components—$I_C(p)$ of the collector current $I_C$. $I_C(p)$ is slightly smaller than $I_E(p)$.

The base current constitutes of electrons flowing from the battery to the base in order to maintain the charge neutrality of the base region. Despite the collector junction being reverse-biased, a small amount of current flows from the collector due to the minority carriers. This current has two components. $I_{CO}(n)$, due to the minority electrons flowing from collector (*p*-side) to base (*n*-side), and $I_{CO}(p)$, due to the minority holes flowing from base to collector region across the collector–base junction ($J_C$). The resultant $I_{CO}(n) + I_{CO}(p)$ is denoted by $I_{CBO}$. This is called the *leakage current* or reverse collector saturation current. This component is very much temperature sensitive; for every 10° rise in temperature the reverse saturation current nearly doubles. Hence, the current component relations of a *p–n–p* transistor are as follows:

$$I_C = I_C(p) + I_{CBO} \tag{4-7}$$

$$I_E = I_B + I_C \tag{4-8}$$

## 4-5-2 Current Components in *n–p–n* Transistor

The operation of an *n–p–n* transistor is the same as that of a *p–n–p* transistor, but with the roles played by the electrons and holes interchanged. The polarities of the batteries and also the directions of various currents are to be reversed, as shown in Fig. 4-7(b). Here the majority electrons from the emitter are injected into the base and the majority holes from the base are injected into the emitter region. These two constitute the emitter current.

$$I_E = I_E(n) + I_E(p) \tag{4-9}$$

Since the doping of the emitter region is much higher than that of the base, i.e., $I_E(n) \gg I_E(p)$, we have $I_E \approx I_E(n)$. Thus, the emitter current is almost entirely due to the electrons moving from emitter to base. Since electrons are negatively charged, the direction of conventional current is opposite to the movement of the electrons. The injected electrons diffuse through the base towards the collector junction. A few of the injected electrons are lost due to recombination with the majority carrier holes in the base region.

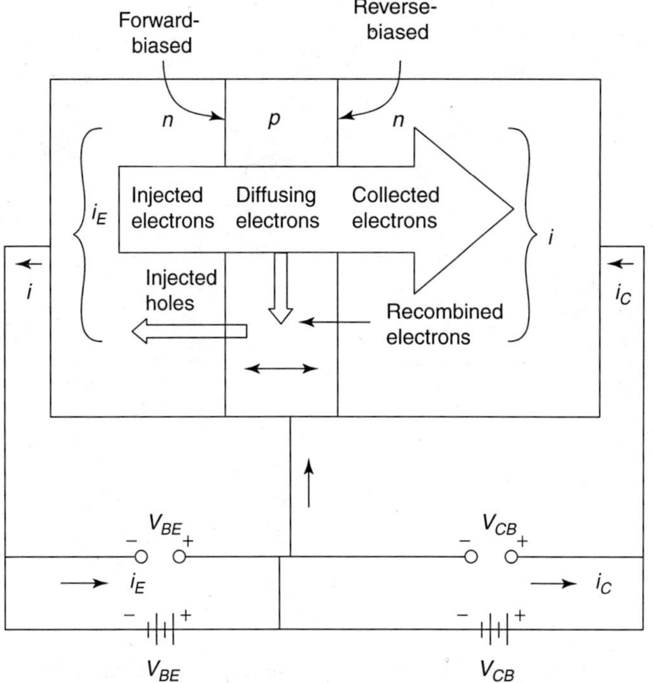


**Figure 4-7(b)** The majority and the minority carrier current flow in a forward-biased *n–p–n* transistor

The electrons moving from the base to the collector junction are collected by the collector region and result in the current $I_C(n)$. The difference, $I_E(n) - I_C(n)$, constitutes a part of the base current $I_B$. Since the collector junction is reverse-biased, there is a reverse saturation current ($I_{CBO}$) through the junction. $I_{CBO}$ consists of two parts, $I_{CO}(n)$ and $I_{CO}(p)$. $I_{CO}(n)$ is due to the minority electrons from the base to the collector and $I_{CO}(p)$ is due to the minority holes from collector to base. The relations for the current components can be written as:

$$I_C = I_C(n) + I_{CBO} \tag{4-10}$$

$$I_E = I_C + I_B \tag{4-11}$$

## 4-6 CB, CE AND CC CONFIGURATIONS

Depending on the common terminal between the input and the output circuits of a transistor, it may be operated in the common-base mode, or the common-emitter mode, or the common-collector mode keeping any one of the three terminals common to both halves of the circuit.

### 4-6-1 Common-Base (CB) Mode

In this mode, the base terminal is common to both the input and the output circuits. This mode is also referred to as the ground–base configuration. Figure 4-8 shows a *p–n–p* transistor connected in the common-base (CB) mode and Fig. 4-9 shows an *n–p–n* transistor connected in the common–base mode.

### 4-6-2 Common-Emitter (CE) Mode

When the emitter terminal is common to both the input and the output circuits, the mode of operation is called the common-emitter (CE) mode or the ground–emitter configuration of the transistor. Figure 4-10 shows this type of configuration.

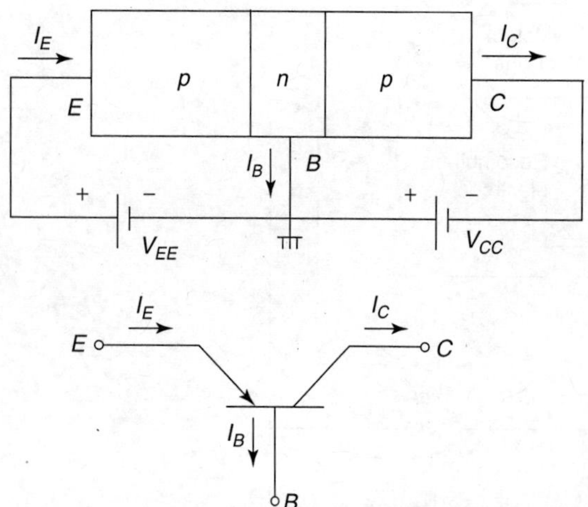


**Figure 4-8** Notation and symbols used for the common-base configuration of a *p–n–p* transistor

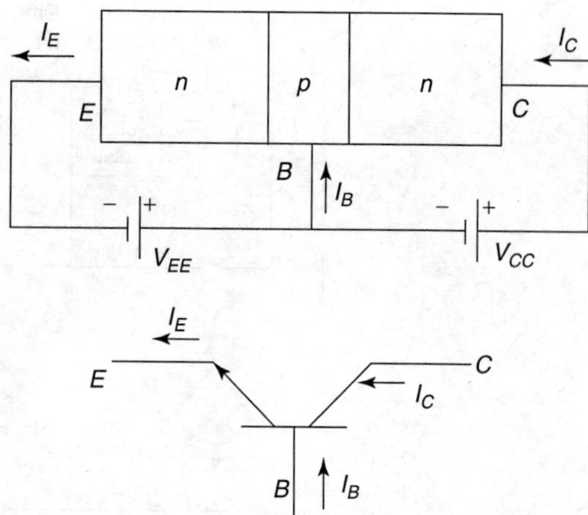


**Figure 4-9** Common-base configuration of an *n–p–n* transistor

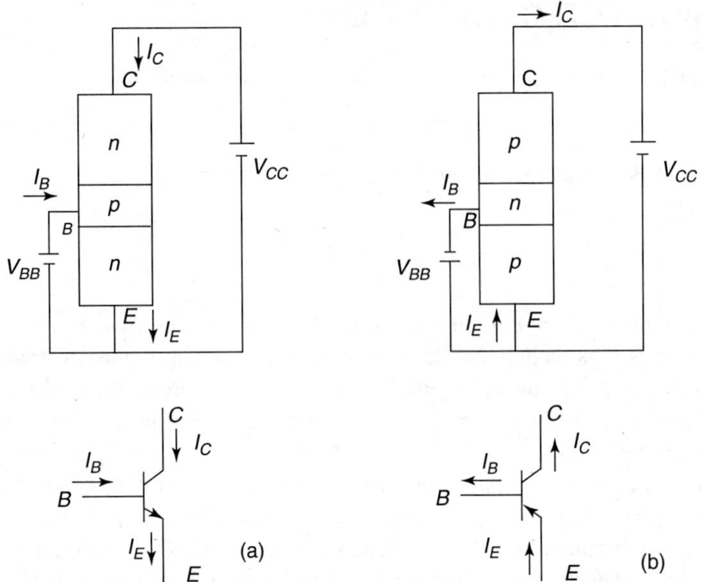


**Figure 4-10** Notation and symbols for common-emitter configuration (a) *n–p–n* transistor (b) *p–n–p* transistor

## 4-6-3 Common-Collector (CC) Mode

When the collector terminal of the transistor is common to both the input and the output terminals, the mode of operation is known as the common-collector (CC) mode or the ground–collector configuration. Figure 4-11 shows the common-collector configuration.

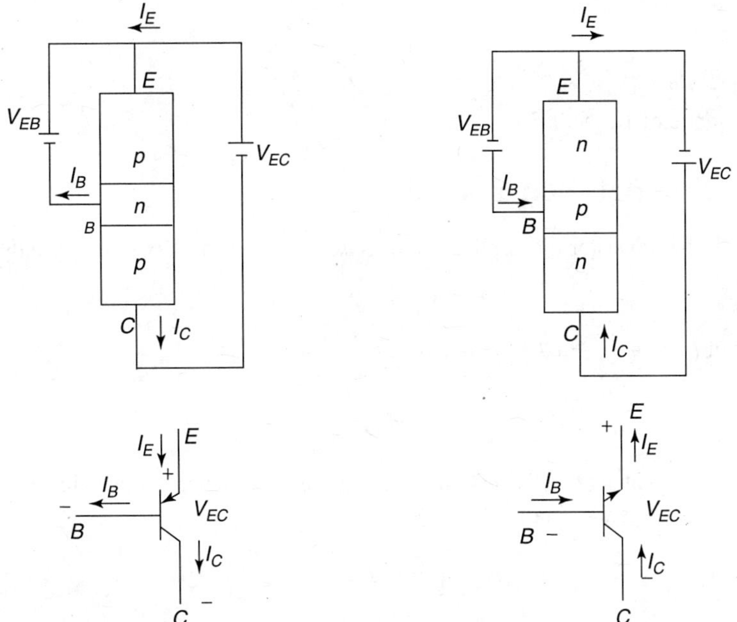


**Figure 4-11** Common-collector configuration

## 4-7 EXPRESSION FOR CURRENT GAIN

The collector current, when the emitter junction is forward-biased is given by:

$$I_C = I_{CO} - \alpha I_E \tag{4-12}$$

where, $I_{CO}$ is the reverse saturation current, and $I_E$ is the emitter current.

Thus, $\alpha$ is given by:

$$\alpha = -\frac{I_C - I_{CO}}{I_E} \tag{4-13}$$

$\alpha$, represents the total fraction of the emitter current contributed by the carriers injected into the base and reaching the collector. $\alpha$ is thus, called the dc current gain of the common-base transistor. $I_E$ and $I_C$ are opposites as far as their signs are concerned, therefore, $\alpha$ is always positive. Generally $\alpha$ lies within the 0.95–0.995 range. It is not a constant but varies with respect to the emitter current $I_E$, the collector to base voltage $V_{CB}$, and temperature.

As we know, the value of the reverse saturation current is in the order of nanoamperes. This can be neglected as compared to the collector current. Thus, the expression for $\alpha$ reduces to $-I_C/I_E$.

The small-signal short-circuit current transfer ratio or the current gain for a common-base configuration is denoted by $\alpha'$. It is defined as the ratio of the change in the collector current to the change in the base current at a constant collector to base voltage. Consequently, it is given by:

$$\alpha' = -\frac{\Delta I_C}{\Delta I_B} \mid V_{CB} = 0 \tag{4-14}$$

Here $\Delta I_C$ and $\Delta I_B$ represent the change of collector and base current.

The maximum current gain of a transistor operated in the common-emitter mode is denoted by the parameter $\beta$ (a detailed discussion on this has been included in Chapter 5). It is defined as the ratio of the collector current to the base current.

$$\beta = \frac{I_C}{I_B}$$

Its value lies in the range of 10–500.

### 4-7-1 Relationship Between $\alpha$ and $\beta$

In the general model of a transistor the application of Kirchoff's current law (KCL) yields:

$$I_E = -(I_C + I_B) \tag{4-16}$$

Replacing the value of $I_E$ ($I_C = I_{CO} - \alpha I_E$), we obtain:

$$I_C = \frac{\alpha}{1-\alpha} I_B + \frac{I_{CO}}{1-\alpha} \tag{4-17}$$

Again we know that as the value of $I_{CO}$ is very small, therefore, we can neglect its value in comparison with $I_B$.

Upon neglecting its value we obtain:

$$I_C = \frac{\alpha}{1-\alpha} I_B \tag{4-18}$$

or,

$$\frac{I_C}{I_B} = \frac{\alpha}{1-\alpha} = \beta$$

or,
$$\beta = \frac{\alpha}{1-\alpha} \tag{4-19}$$

The relation between $\alpha$ and $\beta$ is expressed in terms of Eq. (4-19).

## FOR ADVANCED READERS

### Ebers–Moll Model of Transistor

Investigators Ebers and Moll introduced a general model for transistors, now known as the *Ebers–Moll model*. This generalization enables us to understand the limits of forward-active operation; it operates the transistor in the reverse-active mode and shows how to operate the transistor as a switch. The Ebers–Moll model for a *p–n–p* transistor is shown in the Fig. 4-12.

The model involves a pair of ideal diodes connected back-to-back with two independent current sources.

Now, by applying KCL at the collector we get:

$$i_C = \alpha_F i_{DE} - i_{DC} \tag{4-20}$$

where, $i_C$ is the ordinary diode current in the collector-base junction and represents the special transistor action. Now, upon substituting the values of $i_{DE}$ and $i_{DC}$ in the Eq. (4-20) we obtain:

$$i_C = \alpha_F I_{ES}(e^{V_{BE}/V_T} - 1) - I_{CS}(e^{V_{BC}/V_T} - 1) \tag{4-21}$$

where, $I_{ES}$ and $I_{CS}$ are the reverse saturation currents of the respective junctions.

The Ebers–Moll model also describes the operation of the transistor in the reverse mode, i.e., forward-biasing the collector–base junction and reverse-biasing the emitter–base junction. In the reverse-active mode of operation, the electrons injected from the collector diffuse across the base and are collected at the emitter. Simply put, the collector now acts as the emitter and the emitter as the collector. $\alpha_R$ is the reverse current gain of the transistor.

Now, applying KCL at the emitter, we obtain:

$$i_E = I_{ES}(e^{V_{BE}/V_T} - 1) - \alpha_R I_{CS}(e^{V_{BC}/V_T} - 1) \tag{4-22}$$

Solid-state electronics also describes a reciprocity law:

$$\alpha_F I_{ES} = \alpha_R I_{CS} = I_S \tag{4-23}$$

This relates the two reverse saturation currents and defines the quantity $I_S$. Using reciprocity and on simplifying the equations for reverse saturation currents we get the final form of Ebers–Moll equations:

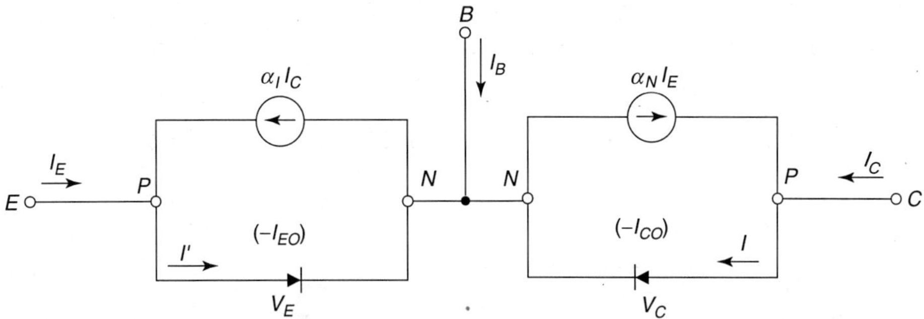


**Figure 4-12** Ebers-Moll model of a *p–n–p* transistor

$$i_C = I_S(e^{V_{BE}/V_T} - 1) - \frac{I_S}{\alpha_R}(e^{V_{BC}/V_T} - 1) \tag{4-24}$$

$$i_E = \frac{I_S}{\alpha_F}(e^{V_{BE}/V_T} - 1) - I_S(e^{V_{BC}/V_T} - 1) \tag{4-25}$$

## 4-8 TRANSISTOR CHARACTERISTICS

The graphical forms of the relations between the various current and voltage variables (components) of a transistor are called *transistor static characteristics*. By considering any two of the variables as independent variables it is possible to draw different families of characteristic curves. However, two sets of characteristic curves known as the input and the output characteristics for common-base and common-emitter modes are of practical use and importance.

### 4-8-1 Input Characteristics

The plot of the input current against the input voltage of the transistor in a particular configuration with the output voltage as a parameter for a particular mode of operation gives the input characteristics for that mode.

#### Common-emitter mode

The CE input characteristics constitute the plot of the input current $I_B$ against the input voltage $V_{BE}$, with the output voltage $V_{CE}$ as the parameter since the emitter is common to both the input and output sections of the device. The characteristics, as shown in Fig. 4-13, are similar to that of the forward-biased *p–n* diode, as is expected from the basic device study. For a constant $V_{BE}$, the effective base width decreases with an increasing $|V_{CE}|$.

#### Common-base mode

In this mode the base is common to both the input and output sections of the transistor. The input characteristic for the CB mode, as shown in Fig. 4-14, constitutes the plot of the input current $I_E$ against the input voltage $V_{EB}$ with the output voltage $V_{CB}$ as the parameter.

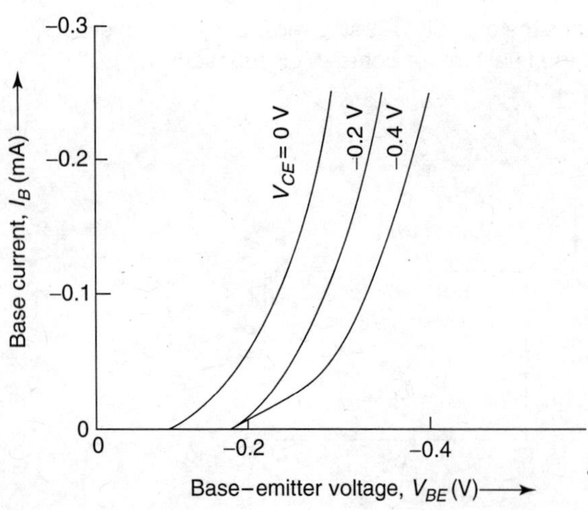


**Figure 4-13** Input characteristics in the CE mode

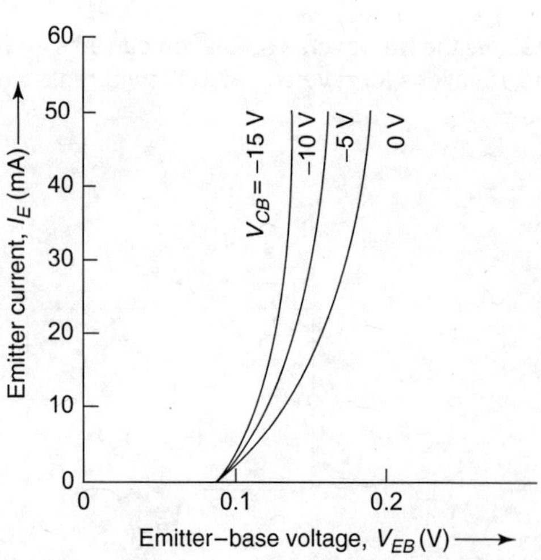


**Figure 4-14** Input characteristics in the CB mode

Since, the emitter–base junction is forward-biased in normal operation, the input characteristics are similar to that of a forward-biased *p–n* diode. For a fixed $V_{EB}$, $I_E$ increases with an increase in $|V_{CB}|$. When $|V_{CB}|$ increases, the width of the depletion region at the collector–base junction increases, and as a result the effective base width decreases. The change of the effective base width by the collector voltage is known as *Early effect.* $I_E$ increases with an increasing reverse collector voltage.

## 4-8-2 Output Characteristics

Similarly a plot for the output current against the output voltage with the input current as a parameter gives the output characteristics.

The output characteristics can be divided into four distinct regions:

(i) The active region
(ii) The saturation region
(iii) The inverse active region
(iv) The cut-off region

Table 4-1 provides the definitions for the four transistor states. These four BJT states or operating modes correspond to the four possible ways in which we can bias the transistor junctions. Figure 4-15 shows these four transistor states and Fig. 4-16 shows the various regions of operation as defined by junction biasing.

**Table 4-1** Definitions of transistor states

| *Transistor State* | *Base–Emitter Junction* | *Base–Collector Junction* |
|---|---|---|
| Forward active | Forward ($V_{BE} \geq V_\gamma$) | Reverse ($V_{BC} \leq V_\gamma$) |
| Reverse active | Reverse ($V_{BE} \leq V_\gamma$) | Forward ($V_{BC} \geq V_\gamma$) |
| Cut-off | Reverse ($V_{BE} \leq V_\gamma$) | Reverse ($V_{BC} \leq V_\gamma$) |
| Saturation | Forward ($V_{BE} \geq V_\gamma$) | Forward ($V_{BC} \geq V_\gamma$) |

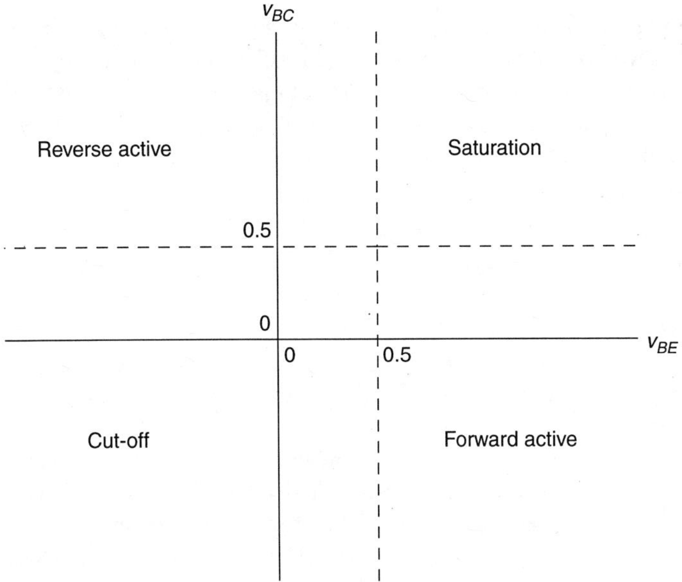


**Figure 4-15** Transistor states defined by junction biasing

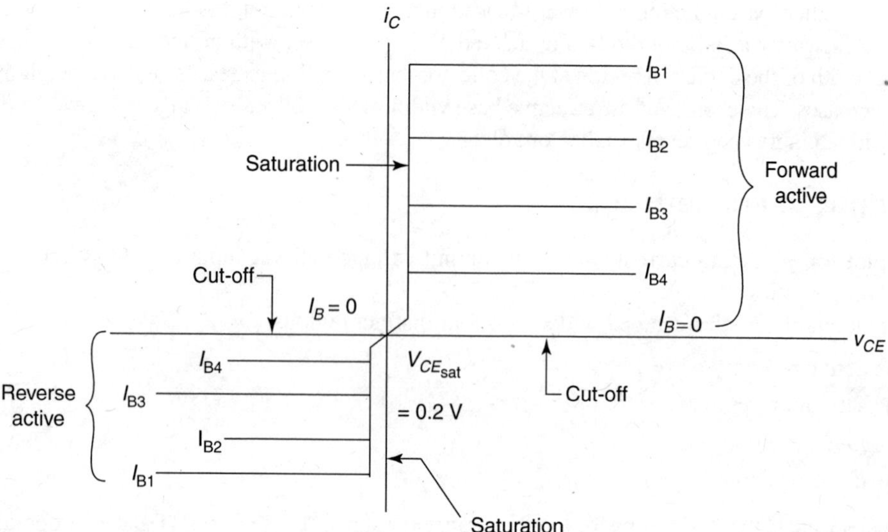


**Figure 4-16** Regions of operation for the four transistor states in terms of the output characteristic curves

The active region is the region normally employed for linear (undistorted) amplifiers. In the active region particularly, the collector–base junction is reverse-biased, while the base–emitter junction is forward-biased. In the active region, as the emitter current increases above zero and thé collector current increases to a magnitude essentially equal to that of the emitter current. Therefore:

$$I_E \approx I_C$$

In the reverse-active region the base-emitter junction is reverse-biased and the base-collector region is forward-biased. In other words, in reverse-active mode, the roles of the emitter and the collector are reversed compared to forward-active mode. However, this mode of operation is hardly used. Figure 4-17 shows the model for the BJT in reverse-active mode with a 0.7 V voltage source connected between the base and the collector, and the dependent source directed from the emitter to the collector.

The emitter current is $\beta_R i_B$.

Here, $$\beta_R = \frac{\alpha_R}{1-\alpha_R} \tag{4-26}$$

This defines the reverse $\beta$ of the transistor. Now, the cut-off region is defined as that region where the collector current is 0 ampere (A). Moreover in the cut-off region the collector–base and base–emitter junctions of a transistor are both reverse-biased. Figure 4-18 shows the cut-off model for the transistor.

This circuit is consistent with the operation in the cut-off region, $I_C = I_B = 0$. At high temperatures, a temperature sensitive dc current, $I_{CBO}$ flows from the collector to the base. Open-circuiting the emitter lead and calculating the collector-to-base current that results from reverse-biasing the collector–base junction measures this $I_{CBO}$ as quite small and is of the order of picoamperes.

The saturation region is defined as the region of the characteristics that lies to the left of $V_{CB} = 0$ V. The horizontal scale in this region was expanded to clearly show the dramatic change in the characteristics in this region. The collector current increases exponentially as the voltage $V_{CB}$ increases toward 0 V. In the saturation

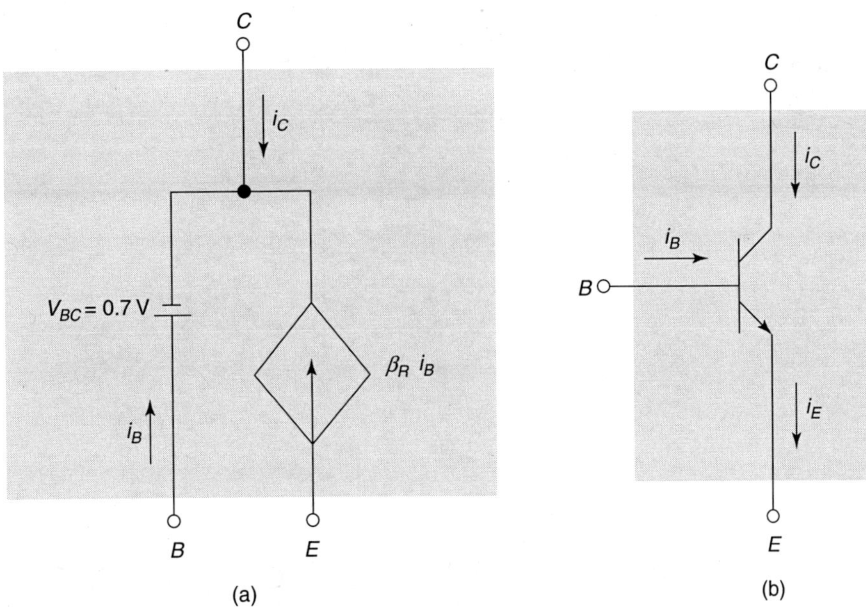


**Figure 4-17** Model for BJT in reverse-active mode of operation

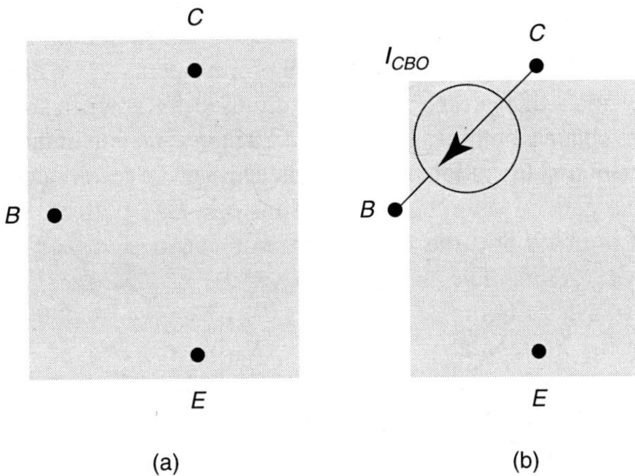


**Figure 4-18** Large signal cut-off model: (a) simple model (b) model for high temperatures

region the collector–base and base–emitter junctions are forward-biased. A necessary condition for a transistor to be saturated is:

$$\beta i_B \geq i_C \tag{4-27}$$

Figure 4-19 shows the circuit model for a saturated transistor.

Since the curves merge into a constant $V_{CE}$ line in the saturation region, a battery of $V_{CE\,\text{sat}}H \approx 0.2$ V is connected between the collector and the emitter in the saturation model. Since the base–emitter junction is forward-biased, the model also employs a dc input source of $V_{BE}$ = 0.7 V between the base and the emitter.

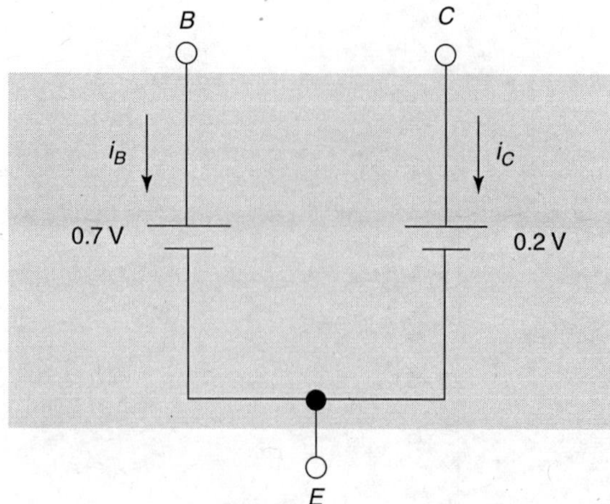


**Figure 4-19** Circuit model for a saturated transistor

# 4-9 OPERATING POINT AND THE CONCEPT OF LOAD LINE

In the case of transistor amplifiers, the operating point refers to the particular condition of the circuit where, with some definite values of voltage and current, we can define the region or the point of operation of the circuit. Lexically, quiescent means stationary, i.e., the voltage or the current should be such that it sets the stage for proper operation of the circuit when a signal is applied. By proper operation we mean that the circuit performs the desired operation. Since most of the time transistors are used for amplification, the region should be so selected that at the output we obtain a faithful and an amplified representation of the input signal. One important attribute of this case is the region of operation of the circuit. This can be readily understood from Fig. 4-20.

Figure 4-20 depicts the general output characteristics of a BJT with three operating points as indicated. These three points indicate that the biasing circuits can be realized for all the indicated operating points in the active region. The maximum ratings for the transistor being operated are indicated by the horizontal line for maximum collector current, and the horizontal line for maximum collector-to-emitter voltage. Keeping the operation confined within this active region, one can select many operating points, but their selection depends on the intended operation.

If no bias is given, the device would be initially in the OFF state. But, if under such a circumstance a signal is applied at the input, the device would not be able to trace the entire signal. The judicious selection for the operating point should be such that the device can vary in current and voltage from the operating point, and allowing the device to react to both the positive and the negative excursions of the input signal.

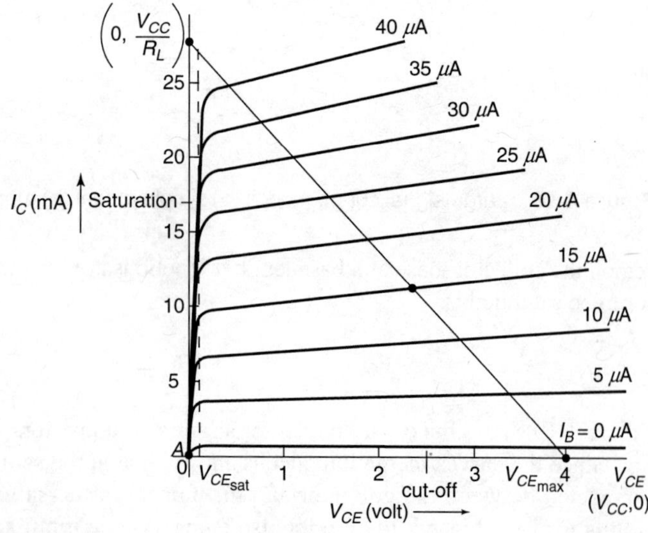


**Figure 4-20** Region of operation of a BJT

Also, it should be kept in mind that the selection of the input signal should be such that it does not drive the device in the cut-off and the saturation regions. In general, the operating point should be such that its operation is in that region where the gain is fairly constant.

Having selected the operating point and the input signal, ample measures should be taken for nullifying the effect of temperature while shifting the operating region. Higher temperatures cause a rapid change in the operating conditions of the device. As a result, a factor called the stability factor is introduced (we will study this in much detail in the later chapters).

The load line is a graphical function used to find the device currents and voltages when the device is described by its characteristic curves. Even when the characteristic curves of the device are not available, the load line solves the purpose as it gives the locus of all such points on the curve where the device can be operated and a corresponding output can be obtained. Let us first consider the circuit as shown in Fig. 4-21.

Here, $R_C$ is the resistor at the collector and $R_B$ is the resistance at the base terminal. The respective currents are shown by $i_C$ and $i_B$ with two voltage sources for proper biasing.

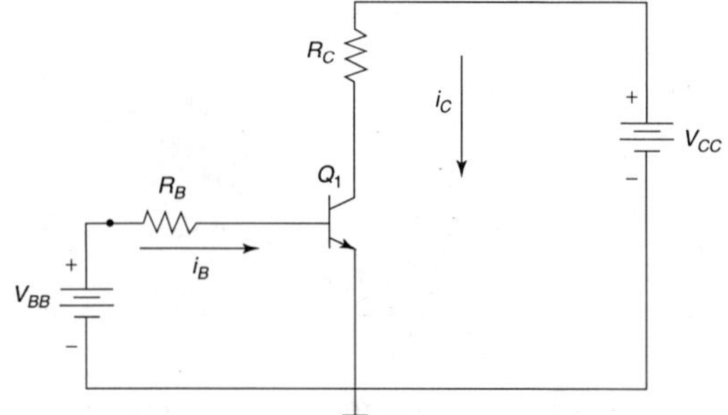


**Figure 4-21** A common-emitter amplifier

From Fig. 4-21, we find that for this common-emitter circuit, the equation for the output voltage, i.e., the collector-to-emitter voltage, ($V_{CE}$) can be obtained by applying KVL.

$$v_{CE} = V_{CC} - i_C R_L \tag{4-28}$$

And this can be written as:

$$i_C = \frac{V_{CC}}{R_L} - \frac{1}{R_L} v_{CE} \tag{4-29}$$

From Eq. (4-29) we find that it represents the output of the circuit. If this equation is plotted on the output characteristic, we obtain a straight line. The intercept on the $y$-axis is $V_{CC}/R_L$ and that on the $x$-axis is $V_{CC}$ as shown in Fig. 4-20. For any such given system, a relation between the output voltage and the output current can always be found and can be superimposed on the output characteristics to judge its conditions for operation.

On the $i_B$ versus $v_{BE}$ coordinate system, we obtain a straight line for Eq. (4-29). The straight line has a slope of $-1/R_B$, which is called the input load line. It is the locus of the points in which the device can be operated. The intersection of this load line with the characteristic curve gives the quiescent operating point $Q$, as shown in Fig. 4-20. We can change the operating point by suitably changing various parameters such as $V_{BB}$ and $R_B$.

An analysis of Eq. (4-29) indicates that any variation in the dc biasing voltage and the circuit elements can have a heavy impact on the operation of the device. But, a judicious selection would be required to place it in the active region, i.e., where the device characteristics also change according to the changes in the input signal.

## 4-10 EARLY EFFECT

In the operating region of a transistor or for a normal operation of the transistor, the emitter–base junction is forward-biased. So the emitter current variation with the emitter-to-base voltage will be similar to the forward characteristic of a $p$–$n$ junction diode. An increase in the magnitude of the collector-to-base voltage ($V_{CB}$) causes the emitter current to increase for a fixed $V_{EB}$. When $|V_{CB}|$ increases, the depletion region in the collector–base junction widens and reduces the base width. This is known as the Early effect. As a result of this effect, the

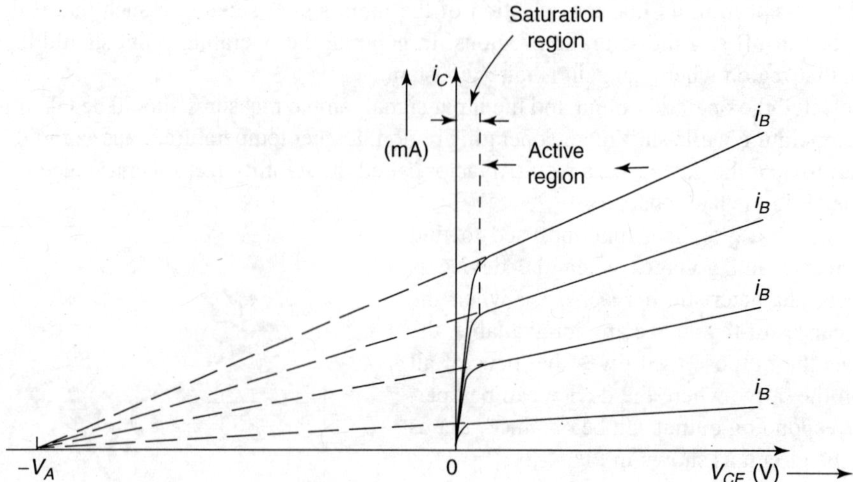


**Figure 4-22** Graphical representation of early voltage

gradient of the injected hole in the base region increases. The injected hole current across the emitter junction is proportional to the gradient of the hole concentration. So, an increase in the emitter current with an increase in $|V_{CB}|$ is obtained. Again, since the collector is also dependent on the emitter current, the collector current also sees a sharp increase. The graphical representation of Early effect is shown in Fig. 4-22.

By including a resistance $r_o$ in parallel with the controlled source, we can represent the linear dependence of $I_C$ on $V_{CE}$ in a condition where there is no current flow since the channel is completely void of electrons. This condition is known as pinch-off. If the early voltage is greater than the pinch-off voltage, then:

$$r_o \approx \frac{V_A}{I_C}$$

## Solved Examples

**Example 4-1** An $n$–$p$–$n$ transistor having current gain $\alpha = 0.90$ is connected in the CB mode and gives a reverse saturation current $I_{CO} = 15\ \mu$A. Calculate the base and the collector currents for an emitter current of 4 mA.

**Solution:**

From Eq. (4-12), we have:

$$I_C = I_{CO} - \alpha I_E$$

$I_E$ is negative for an $n$–$p$–$n$ transistor, therefore:

$$I_C = I_{CO} + \alpha I_E$$

By substituting the required values, we get:

$$I_C = 0.90 \times 4 \times 10^{-3} + 15 \times 10^{-6}$$

$$= 3.615 \text{ mA}$$

From Eq. (4-11), we have:

$$-I_E = -(I_C + I_B)$$

or,

$$I_B = I_E - I_C = (4 - 3.615) \text{ mA} = 385\ \mu\text{A}$$

**Example 4-2** What is the value of $\alpha$ for a BJT that has a $\beta$ of 90? Find the base and the emitter current if the collector current is 4 mA.

**Solution:**
From Eq. (4-19), the common-emitter current gain $\beta$ is given by:

$$\beta = \frac{\alpha}{1-\alpha}$$

We find $\alpha = 0.989$ by putting $\beta = 90$:

$$I_B = I_C/\beta = 44.44\ \mu\text{A}$$
$$I_E = I_C + I_B = 4.04\ \text{mA}$$

**Example 4-3** When used in the common-base configuration mode, a transistor with $\alpha = 0.90$ gives a reverse saturation current $I_{CO} = 15\ \mu$A. Calculate the collector current when the transistor is connected in the common-emitter mode with a base current of 0.5 mA.

**Solution:**
From Eq. (4-19), the common-emitter current gain $\beta$ is given by:

$$\beta = \frac{\alpha}{1-\alpha} = \frac{0.90}{1-0.90} = 9$$

From Eq. (4-17), it follows that:

$$I_C = \beta I_B + (\beta + 1) I_{CO}$$

By substituting the adequate values, we have:

$$I_C = [9 \times 0.5 + (9 + 1) \times 0.015]\ \text{mA}$$
$$= 4.65\ \text{mA}$$

**Example 4-4** For a particular transistor having a very thin base, a base current of 20 $\mu$A and a corresponding collector current of 5 mA are measured. What is $\beta$ for this device?

**Solution:**

We know that: $$\beta = \frac{I_C}{I_B}$$

Therefore: $$\beta = \frac{5 \times 10^{-3}}{20 \times 10^{-6}}$$
$$= 250$$

**Example 4-5** For a BJT having a base current of 50 $\mu$A and a collector current of 5 mA, what is the emitter current? What is $\beta$ for this transistor? Using your computed value of $I_E$. With the given value of $I_C$, find the value of $I_E$. With the given value of $I_C$, find the value of the common-base current gain $\alpha$. Verify that $\alpha = \beta/(\beta + 1)$ and that $\beta = \alpha/(1 - \alpha)$.

**Solution:**
From Eq. (4-4), we have:

$$I_E = I_C + I_B = 5 + .05 = 5.05\ \text{mA}$$

$$\beta = \frac{I_C}{I_B} = (5 \times 10^{-3})/(50 \times 10^{-6}) = 100$$

$$\alpha = \frac{I_C}{I_E}$$

$$= (5/5.05) = 0.990$$

Verification:

$$\alpha = \frac{\beta}{\beta + 1}$$

$$= \frac{100}{100 + 1}$$

$$\alpha = \frac{100}{101} = 0.990099$$

$$\beta = \frac{\alpha}{1 - \alpha}$$

$$= \frac{\frac{100}{101}}{1 - \frac{100}{101}}$$

$$\beta = \frac{0.990099}{1 - 0.990099} = 100$$

**Example 4-6** From the following figure we note that the transistor is not a symmetrical device; therefore, interchanging the collector and the emitter terminals will result in a device with different values of $\alpha$ and $\beta$, known as the inverse or reverse values and denoted as $\alpha_R$ and $\beta_R$ respectively. An *n–p–n* transistor is accidentally connected with the collector and emitter leads interchanged. The resulting emitter and base currents are 10 mA and 5 mA, respectively. What are the values of $\alpha_R$ and $\beta_R$?

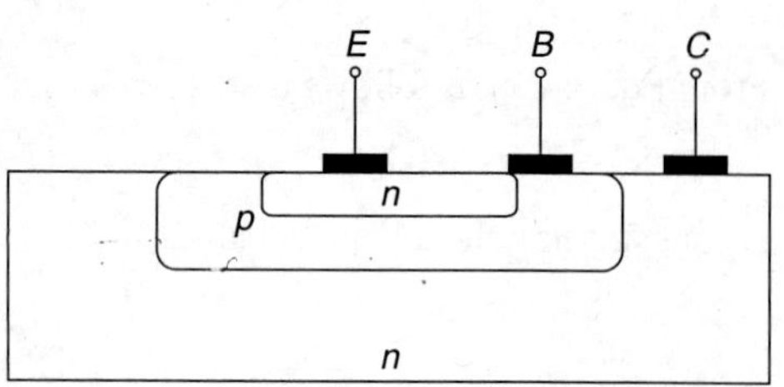


**Solution:**

$$I_E = 10 \text{ mA}, I_B = 5 \text{ mA}$$

So,

$$I_C = I_E - I_B = 5 \text{ mA}$$

$$\beta_R = \frac{I_C}{I_B} = 1$$

$$\alpha_R = \frac{I_C}{I_E} = 2$$

**Example 4-7** Find the labeled currents and voltages for the circuits, as shown in the following diagrams. Let $\beta$ = 100 and $|V_{BE}|$ = 0.7 V.

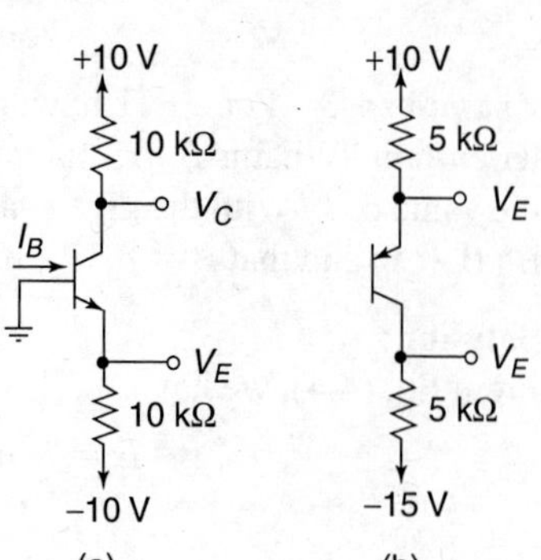


**Solution:**

(a) $V_E = -0.700$ V

$$I_E = (10 - 0.7)/10 \text{ K} = 0.93 \text{ mA}$$

$$I_B = \frac{I_E}{\beta} + 1 = \frac{0.93}{101} = 9.2\ \mu\text{A}$$

$$V_C = 10 - 10\,(0.930 - 0.0092) = 0.792 \text{ V}$$

(b) $V_E = -0.700$ V

$$I_E = \frac{10 - 0.7}{5\text{ K}} = 1.86 \text{ mA}$$

$$I_C = \frac{\beta}{\beta + 1} = \frac{100}{101}(1.86) = 1.842 \text{ mA}$$

$$V_C = -15 + 5\,(1.842) = -5.79 \text{ V}$$

**Example 4-8** With reference to the following diagrams, we may assume that base currents are negligibly small, since the transistors shown in the circuits have very large values of $\beta$. Find the values of the labeled voltages if they are determined by the measurement $|V_{BE}| = 0.7$ V.

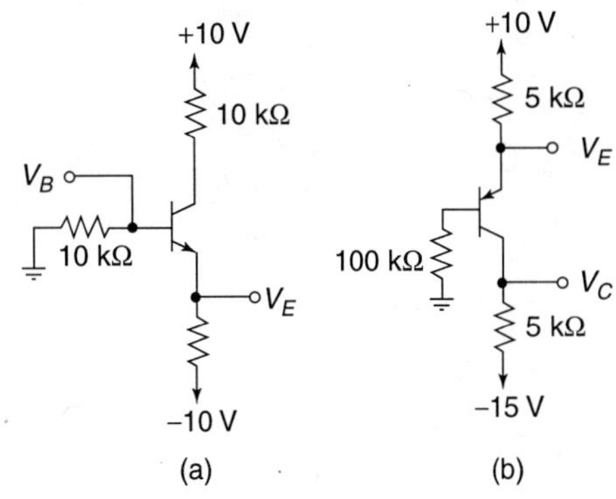


**Solution:**

$$\beta = \infty, |V_{BE}| = 0.7$$

(a) $V_B = 0.0$ V, $V_E = -0.7$ V

(b) $V_E = +0.7$ V, $V_C = -15 + 5\left(\frac{10 - 0.7}{5}\right) = -5.7$ V

**Example 4-9** The emitter voltage of the transistor in the circuit as shown in the following diagram is 1.0 V. Find $V_B$, $I_B$, $I_E$, $I_C$, $\beta$ and $\alpha$ under the assumption that $V_{BE} = 0.7$ V.

**Solution:**

$$V_E = 1.0 \text{ V}$$

$$V_B = 1.0 - 0.7 = 0.3 \text{ V}$$

$$I_B = \frac{0.3}{20\text{ K}} = 0.015 \text{ mA}$$

$$I_E = \frac{5 - 1}{5\text{ K}} = 0.80 \text{ mA}$$

$$I_C = 0.80 - 0.015 = 0.785 \text{ mA}$$

$$V_C = -5 + 5(0.785) = -1.075 \text{ V}$$

$$\beta = \frac{0.785}{0.015} = 52.3$$

$$\alpha = \frac{0.785}{0.800} = 0.98$$

+5 V
5 kΩ
$V_E$
$V_B$
20 kΩ
$V_C$
5 kΩ
−5 V

**Example 4-10** An *n–p–n* transistor has its base connected to −5 V, its collector is connected to ground via a 1 kΩ resistor and its emitter is connected to a 2 mA constant current source that pulls current out of the emitter terminal. If the base voltage is raised by 0.4 V, what voltage changes are measured at the emitter and at the collector?

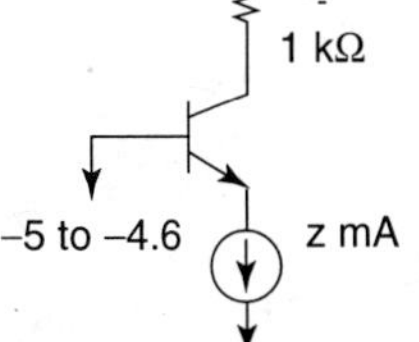

**Solution:**

$$\Delta V_B = +0.4\text{ V}$$

$$\Delta V_E = +0.4\text{ V}$$

$$\Delta V_C = 0.0\text{ V}$$

**Example 4-11** With reference to the circuits, as shown in the following diagrams, identify whether they operate in the active mode or saturation mode. What is the emitter voltage in each case? If active, what is the collector voltage? Given $|V_{BE}| = 0.7$ V, $\beta = 100$.

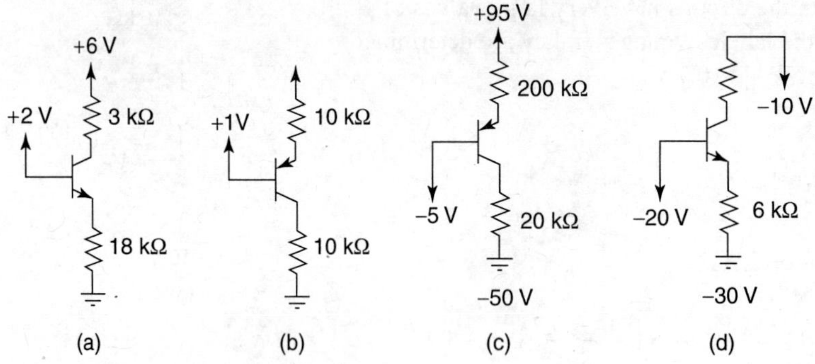


**Solution:**

$$|V_{BE}| = 0.7, \beta = 100$$

(a) Assume active: $V_E = 2 - 0.7 = 1.3$ V

$$I_E = 1\text{ mA}; I_C = 1 \times \frac{100}{101} = 0.99\text{ mA}$$

$$V_E = 6 - 3(0.99) = 3.03\text{ V}$$

Thus, the circuit operates in an active mode.

(b) Assume active: $V_E = 1.0 + 0.7 = 1.7$ V

$$I_E = \frac{6 - 1.7}{10\text{ K}} = 0.43\text{ mA} \approx I_C$$

$$V_C = 0 + 10(0.43) >> V_B$$

Thus, the circuit operates in a saturated mode.

(c) Assume active: $V_E = -5 + 0.7 = -4.3$ V

$$I_E = \frac{9.5 + 4.3}{200\text{ K}} = 0.4965\text{ mA}$$

$$I_C = I_E \frac{100}{101} = 0.492\text{ mA}$$

$$V_C = -50 + 0.492(20\text{ K}) = -40.2\text{ V}$$

Therefore, the circuit operates in an active mode.

(d) Assume active: $V_E = -20.7$ V

$$I_E = \frac{30-20.7}{5\text{ K}} = 1.86\text{ mA}$$

$$V_C = -1.86\left(\frac{100}{101}\right)(2\text{ K}) - 10 = -13.68\text{ V}$$

Therefore, the circuit operates in an active mode.

■

## POINTS TO REMEMBER

1. A bipolar junction transistor (BJT) is a three-terminal active device and can be considered to be made up of two *p–n* junctions connected back-to-back.
2. The operation of a BJT depends mainly on the active participation of both the majority and minority carriers. Thus, the significance of the term bipolar.
3. For normal operation, the emitter–base junction is forward-biased and the collector–base junction is reverse-biased.
4. Consequently, the width of the depletion region of the emitter–base junction is less and that of the collector–base junction is high.
5. In the common-base mode, the base terminal is common to both the input and the output terminals.
6. In the common-emitter mode, the emitter terminal is common to both the input and the output terminals.
7. When the collector terminal of the transistor is made common to the input and output terminals, the mode of operation is called common-collector mode.
8. The factor or the current gain in the common-base mode gives the fraction of the total emitter current injected into the base and reaching the collector. Its value ranges between 0.9–0.995.
9. The factor gives the current gain when the transistor is in the common-emitter mode. Typically, its value ranges from 20–200.
10. The Ebers–Moll model is a generalized model of a transistor. The model involves two diodes connected back-to-back with two independent current sources.
11. The Ebers–Moll model provides a model for the forward-active and reverse-active modes.
12. In the active mode, the emitter–base junction is forward-biased and the collector–base junction is reverse-biased.
13. In the saturation mode, both the emitter–base and the collector–base junctions are forward-biased.
14. In cut-off mode, both the emitter–base and the collector-base junctions are reverse-biased.
15. Active region is mainly employed for linear operation of the device and when it is so required that the device parameters change with changes in the input.
16. The junction temperature of a transistor rises due to self heating and ambient temperature. Due to junction temperature, the collector current may rise, which in turn increases the power dissipation. This is called thermal runaway.

## IMPORTANT FORMULAE

1. Components of the emitter current are:

$$I_E = I_{\text{minority}} + I_{\text{majority}}$$

2. Total emitter current:

$$I_E = I_C + I_B$$

3. Total collector current:

$$I_C = I_{CO} - \alpha I_E$$

4. Forward current gain in the common base configuration is given by:

$$\alpha = \frac{-I_C}{I_B}$$

5. Forward current gain in the common emitter configuration is given by:

$$\beta = \frac{I_C}{I_B}$$

6. The relationship between $\alpha$ and $\beta$ is given by:

$$\beta = \frac{\alpha}{1 - \alpha}$$

## OBJECTIVE QUESTIONS

1. BJT is a:
   (a) Current-controlled device
   (b) Voltage-controlled device
   (c) Power-controlled device
   (d) None of the above
2. A BJT is in the saturation region if:
   (a) Base–emitter junction is reverse-biased and base–collector junction is forward-biased
   (b) Both the junctions are reverse-biased
   (c) Both the junctions are forward-biased
   (d) Base–emitter junction is forward-biased and base-collector junction is reverse-biased
3. Doping concentration of BJT is high in the:
   (a) Emitter region
   (b) Base region
   (c) Collector region
   (d) None of the above
4. The Ebers–Moll model is valid for:
   (a) Bipolar junction transistors
   (b) MOS transistors
   (c) Unipolar junction transistors
   (d) Junction field-effect transistors
5. If a transistor is operating with both of its junctions forward-biased, but with the collector–base forward-bias greater than emitter–base forward-bias, then it is operating in the:
   (a) Forward-active mode
   (b) Reverse-saturation mode
   (c) Reverse-active mode
   (d) Forward-saturation mode
6. In a bipolar transistor at room temperature, if the emitter current is doubled, the voltage across its base–emitter junction:
   (a) Doubles
   (b) Halves
   (c) Increases by 1/3 Volt
   (d) No change occurs
7. The Early effect in a bipolar transistor is caused by:
   (a) Base width modulation
   (b) Large collector–base reverse-bias
   (c) Large emitter–base forward-bias
   (d) Increase in junction temperature
8. $\beta$ is the symbol of current gain for:
   (a) Common-base mode
   (b) Common-emitter mode
   (c) Common-collector mode
   (d) None of the above
9. $\alpha$ is the symbol of current gain for:
   (a) Common-base mode
   (b) Common-emitter mode
   (c) Common-collector mode
   (d) None of the above
10. Magnitude of $\alpha$ is:
    (a) $<1$ (b) $>1$
    (c) $<0$ (d) None of the above
11. Magnitude of $\beta$ is:
    (a) $<1$ (b) $>0$ and $<1$
    (c) $\gg 1$ (high value) (d) None of the above

12. For the BJT, the impurity concentration in the emitter ($E$), base ($B$) and collector ($C$) are such that:

(a) $E > B > C$ (b) $B > C > E$

(c) $C = E = B$ (d) $C > E > B$

13. When a junction transistor is operated under saturated conditions:

(a) Both the CB and EB junction are forward-biased

(b) The CB junction is forward-biased but the EB junction is reverse-biased

(c) The CB junction is forward-biased but the EB junction is forward-biased

14. The modulation of effective base width by collector voltage is known as Early effect. Hence reverse collector voltage:

(a) Increases both $\alpha$ and $\beta$

(b) Decreases both $\alpha$ and $\beta$

(c) Increases $\alpha$ but decreases $\beta$

(d) Decreases $\beta$ but increases $\alpha$

15. If $\propto = 0.995$, $I_E = 10$ mA and $I_{CO} = 0.5$ mA, then $I_{CEO}$ will be:

(a) 100 $\mu$A (b) 25 $\mu$A

(c) 10.1 mA (d) 10.5 mA

16. Match list A (transistor parameter) with list B (typical value):

| List-A | List-B |
|---|---|
| (a) $R_B$ | 1. 20 k |
| (b) $R_C$ | 2. 1 M |
| (c) $\alpha$ | 3. 479 |
| (d) $\beta$ | 4. 0.98 |

17. In a transistor as an amplifier, the reverse saturation current:

(a) Doubles for every $C$ rise in temperature

(b) Doubles for every 10°C rise in temperature

(c) Decreases linearly with temperature

(d) Increase linearly with temperature

18. Which configuration of bipolar transistors of similar geometry has the highest current gain, bandwidth product?

(a) *n–p–n* Ge transistor

(b) *p–n–p* Si transistor

(c) *p–n–p* Ge transistor

(d) *n–p–n* Si transistor

19. Base-to-emitter voltage in forward-biased transistor decreases with the increase of temperature at the rate of:

(a) 2.5 mv/°C (b) 25 mv/°C

(c) 0.25 mv/°C (d) 0.6 mv/°C

## REVIEW QUESTIONS

1. What is a transistor?
2. The metal lead of the *p*-side of a *p–n* diode is soldered to the metal lead of the *p*-side of another *p–n* junction diode. Will the structure form an *n–p–n* transistor? If not, why?
3. Indicate the reference current directions and voltage polarities of a transistor. Give the signs of the actual current directions for an *n–p–n* and *p–n–p* transistor operating normally?
4. Why are junction transistors called bipolar devices?
5. The emitter region of the transistor is more heavily doped compared to the base region. Why?
6. Mention briefly the different techniques used in the manufacture of transistors.
7. Show the doping profile in the emitter, base and collector regions of a transistor. Why is the width of the base thin?
8. Discuss the mechanism of amplification obtained in a transistor. What is the origin of the name “transistor”?

9. Give the physical arrangement of a *p–n–p* junction transistor and discuss how it provides current amplification.
10. Discuss how a transistor is to be used as a current amplifier.
11. Explain how voltage amplification is obtained in a transistor CB amplifier although the current gain is less than unity.
12. Show the different current components of a *p–n–p* transistor when the emitter junction is forward-biased and the collector junction is reverse-biased.
13. Give the minority carrier concentration profile in a *p–n–p* transistor operating normally.
14. Draw the energy variation curve in the conduction band for an open circuited *n–p–n* transistor. How is the curve modified when the transistor is operating in the active region?
15. What do you mean by the static characteristics of a transistor? Draw the circuit diagram of a transistor operating in the common-base configuration and sketch the output characteristics.
16. Draw the common-base input characteristics of the transistor. What is an Early effect and how can it account for the CB input characteristics?
17. With respect to CB output characteristics of a transistor, explain the active, saturation and cut-off regions.
18. Draw the common-emitter circuit of a junction transistor. Sketch its output characteristics and indicate the active, saturation and cut-off regions.
19. Define base-spreading resistance and saturation resistance in connection with a transistor.
20. Explain the phenomenon of punch-through in a transistor.
21. Explain the current amplification factors for CB and CE configurations of a *p–n–p* transistor. Obtain a relation between them.
22. How can you find the CE output characteristics of a transistor?
23. Explain the various switching times when a transistor makes a transition from the cut-off state to the saturation state and back.
24. The value of $\alpha$ increases with the increasing reverse-bias voltage of the collector junction. Why?

## PRACTICE PROBLEMS

1. An *n–p–n* transistor has an emitter area of 10 $\mu$m × 10 $\mu$m. The doping concentrations are as follows: In the emitter $N_D = 10^{19}/\text{cm}^3$, in the base $N_A = 10^{17}/\text{cm}^3$, and in the collector $N_D = 10^{15}/\text{cm}^3$. The transistor is operating at T = 300 K where, $n_i = 1.5 \times 10^{10}/\text{ cm}^3$. For electrons diffusing in the emitter, $L_n = 19$ $\mu$m and $D_n = 21.3$ cm²/s. For holes diffusing in the emitter, $L_p = 0.6$ $\mu$m and $D_p = 1.7$ cm²/s. Calculate $I_s$ and $\beta$ assuming the base width *W* is:

   (a) 1 $\mu$m

   (b) 2 $\mu$m

   (c) 5 $\mu$m
2. Two transistors, fabricated with the same, technique, but having different junction areas, when operated at a base-emitter voltage of 0.69 V, have collector currents of 0.13 and 10.9 mA. Find $I_S$ for each device. What are the relative junction areas?
3. In a particular BJT, the base current is 7.5 $\mu$A and the collector current is 940 $\mu$A. Find $\beta$ and $\alpha$ for the device.
4. For a properly biased *n–p–n* transistor, the collector current is measured to be 1 mA and 10 mA for base-to-emitter voltages of 0.63 V and 0.7 V respectively. Find the corresponding values for *n* and $I_S$ for this transistor. If two such devices are connected in parallel, and 0.65 V is applied between the combined base and the emitter in the conducting direction, what total current do you expect?

5. Show that in a transistor with $\alpha$ close to unity, if $\alpha$ changes by a small per unit amount $\Delta\alpha/\alpha$, the corresponding per unit change in $\beta$ is given by:

$$\frac{\Delta\beta}{\beta} = \beta\left(\frac{\Delta\alpha}{\alpha}\right)$$

6. From Problem 5, find $\Delta\beta/\beta$ when $\beta = 100$ and $\alpha$ changes by 0.1 percent.

7. Consider the following diagrams.

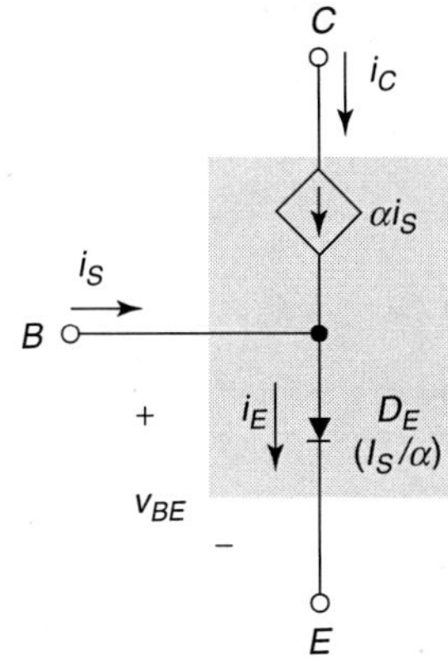


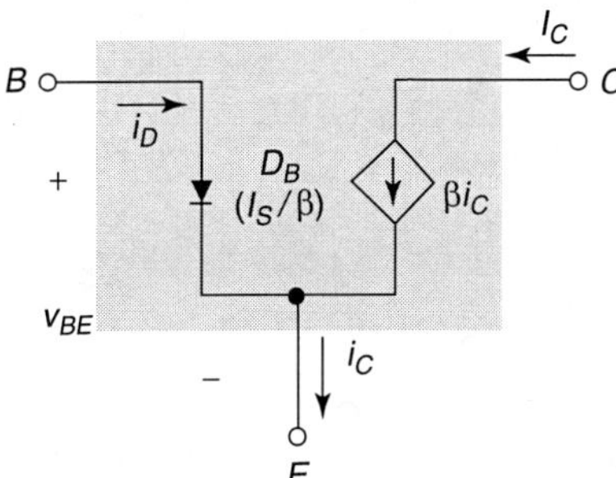


What are the relative sizes of the diodes $D_E$ and $D_B$ for the transistors for which $\beta = 10$ and $\beta = 100$?

8. A particular BJT when conducting a collector current of 10 mA is known to have $v_{BE} = 0.7$ V and $i_B = 100\ \mu$A. Use this data to create a transistor model along the same lines as shown in the Problem 7.

9. The current $I_{CBO}$ of a small transistor is measured to be 15 mA at 25°C. If the temperature of the device is raised to 75°C, what do you expect $I_{CBO}$ to become?

10. Using the model of the transistor, as shown in the following diagram, consider the case of a transistor for which the base is connected to the ground, the collector is connected to a 10 V dc source through a 1 kΩ resistor and a 5 mA current source is connected to the emitter so that the current is drawn out of the emitter terminal. If $\beta = 100$ and $I_S = 10^{-14}$ A, find the voltages at the emitter and the collector, and calculate the base current.

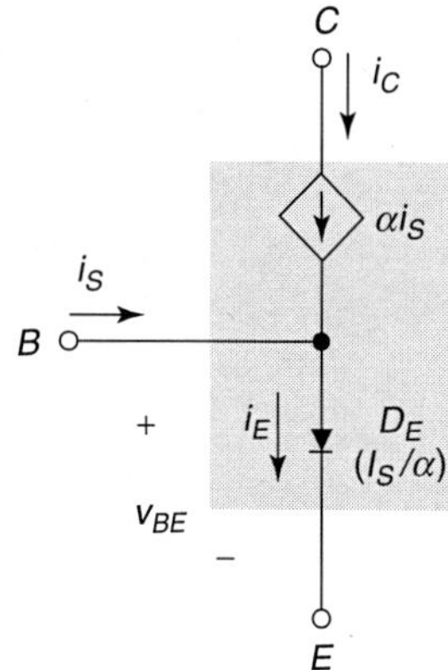


11. A *p–n–p* transistor has $V_{EB} = 0.8$ V at collector current 1 A. What do you expect the voltage to become at collector current 10 mA and 5 mA?

12. A *p–n–p* transistor has a common-emitter current gain of 50. What is the common-base current gain?

13. Augment the model of the *n–p–n* BJT, as shown in the following diagram, by a base current source representing $I_{CBO}$. In terms of this addition what do the terminal currents $i_B$, $i_C$ and $i_E$ become?

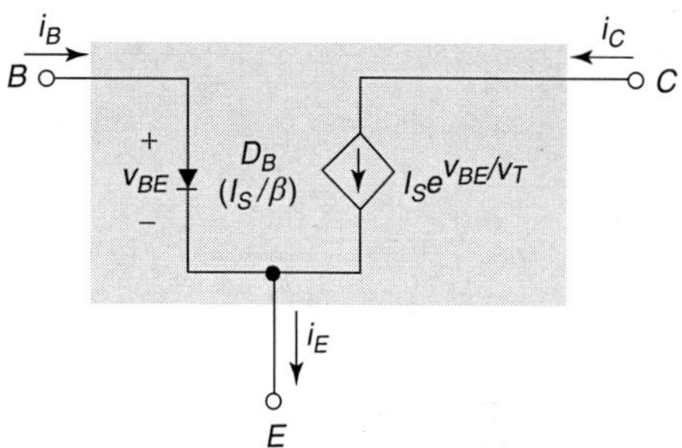


14. In Problem 13, if the base lead is open circuited while the emitter is connected to ground, and the collector is connected to a positive supply, find the emitter and collector currents.

15. The following diagram shows two large signal models of *p–n–p* transistors operating in the active region. And this large signal model is applied to the transistor having $I_S = 10^{-13}$A and $\beta = 40$. If the emitter is connected to ground, the base is connected to a current source that pulls out of the base terminal a current of 10 $\mu$A, and the collector is connected to a negative supply of $-10$ V via a 10 kΩ resistor, find the base voltage, the collector voltage and the emitter current.

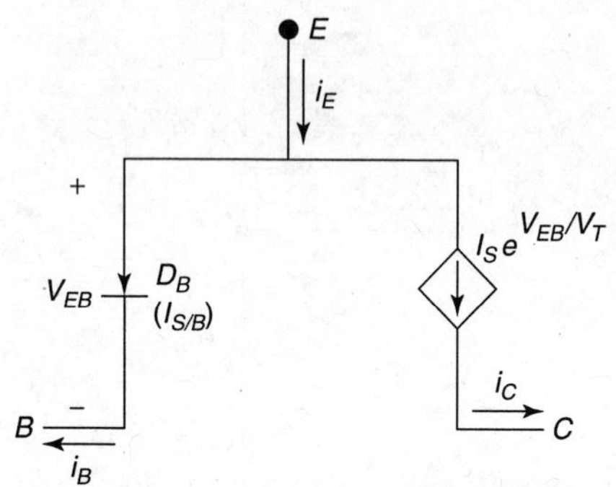


16. In the large signal models, as shown in the following diagrams, contrast the sizes of the two diodes for the situation in which $\beta = 99$.

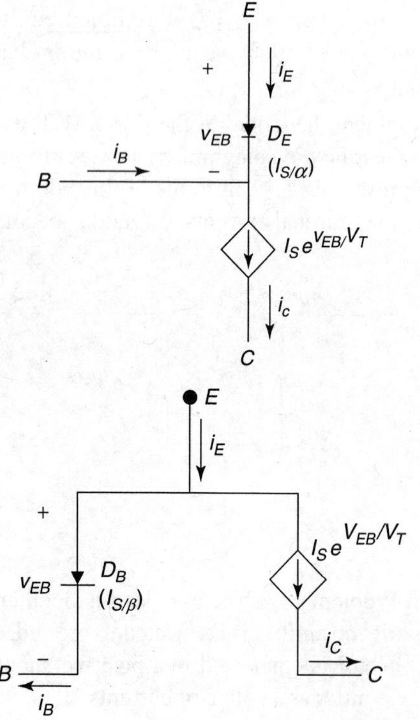


17. For the circuits, as shown in the following diagrams, assume that the transistors have very large $\beta$. Some measurements have been made on the circuits, and the corresponding readings are indicated in the diagrams. Find the values of the other labeled parameters.

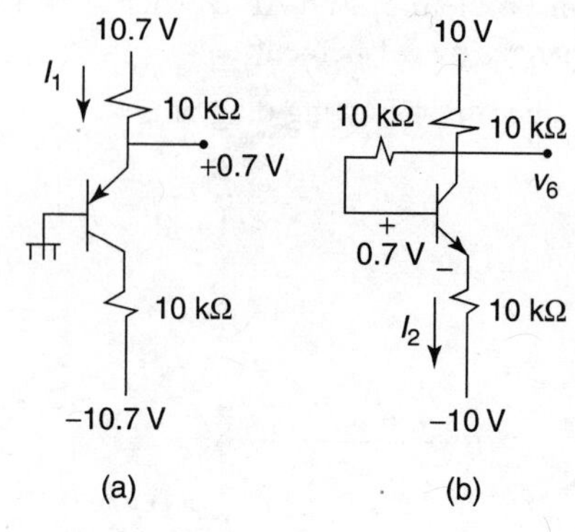


12 V
2.8 kΩ
$V_2$
−5 V
−2.7 V
1 kΩ
−10 V
(c)

+1 V
10 kΩ
$I_3$
$V_3$
−8 V
2 kΩ
−10 V
(d)

18. Measurements on the circuits, as shown in the following diagrams produce labeled voltages as indicated. Find the value of $\beta$ for each transistor.

+5 V
+4.3 V
430 kΩ
+2 V
2 kΩ
(a)

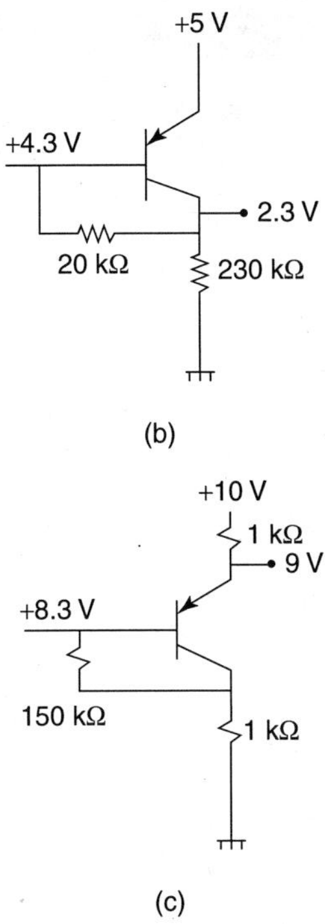


(b)

(c)

19. A particular $p$–$n$–$p$ transistor operating at an emitter current of 0.5 mA at 20 °C has an emitter–base voltage of 692 mV.
    (a) What does $v_{EB}$ become if the junction temperature rises to 50 °C?
    (b) If the transistor has $n = 1$ and is operated at a fixed emitter–base voltage of 700 mV, what emitter current flows at 20 °C and at 50 °C?
20. For a particular $p$–$n$–$p$ transistor operating at a $v_{BE}$ of 670 mV and $I_C = 3$ mA, th e $i_C - v_{CE}$ characteristics has a slope of $3 \times 10^{-5}$ mho. To what value of output resistance does this correspond? What is the value of the Early voltage for this transistor? For operation at 30mA, what should be the output resistance?
21. For a BJT having an early voltage of 200 V, what is the output resistance at 1 mA and at 100 A?
22. For a BJT having an output resistance of 10 MΩ at 10 $\mu$A, what will the early voltage be? If the current is raised to 10 mA, what does the output resistance become?
23. A BJT whose emitter current is fixed at 1 mA has the base–emitter voltage of 0.67 V at 25 °C. What base emitter voltage do you expect at 0 °C and 100 °C?
24. The $p$–$n$–$p$ transistor, as shown in the following diagram, has $\beta = 90$. Find the value of $R_C$ to obtain $V_C = 5$ V. What happens when the transistor is replaced by another having $\beta = 50$.

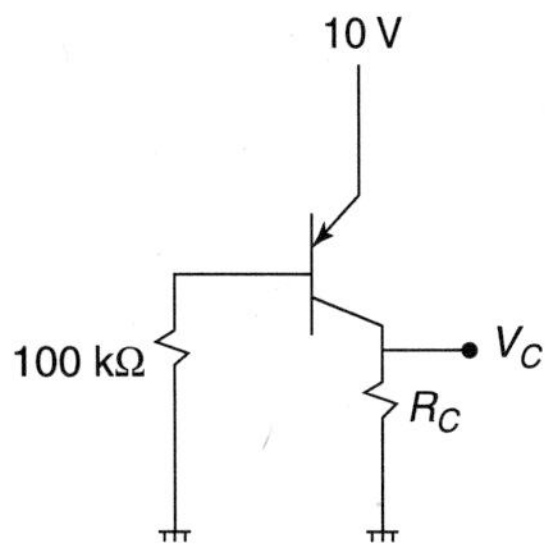


25. For the circuit, as shown in the following diagram, find the labeled node voltages for the following situations where:
    (a) $\beta$ = infinity
    (b) $\beta = 89$
    (c) $\beta = 10$

26. Find the labeled voltages for the given values of $\beta$= infinity and for $\beta = 99$.

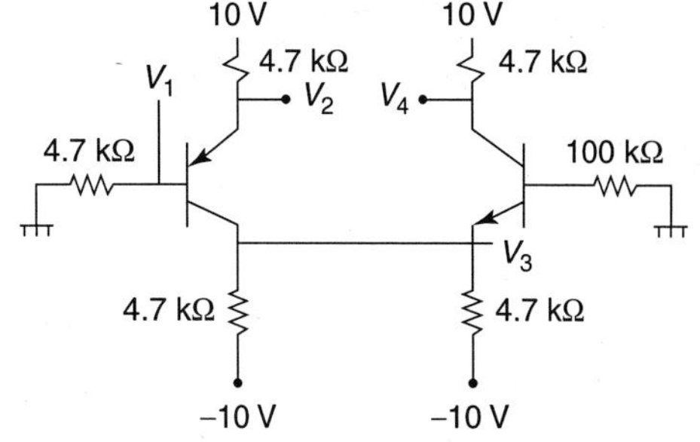

## SUGGESTED READINGS

1. Singh, J. 1994. *Semiconductor Devices: An Introduction*. New York: McGraw-Hill.
2. Streetman, B. G. and S. Banerjee. 2000. *Solid State Electronic Devices*. New Delhi: Pearson Education.
3. Millman, Jacob and Christos C. Halkias. 1986. *Integrated Electronics: Analog and Digital Circuits and Systems*. New Delhi: McGraw Hill Book Company.
4. Pierret R. F. and G.W. Neudeck. 1989. *Modular Series on Solid State Devices*. Boston, M.A.: Addison Wesley.

# 5

# BJT Circuits

## Outline



## Objectives

This chapter introduces biasing of BJT and bias stability. Biasing ensures that the transistor has proper gain and input impedance with undistorted output voltage when used as an amplifier. After the *Q*-point is established, maintaining the stability of the *Q*-point with respect to variations in temperature leads us to the concept of stability.

Next, the small-signal low-frequency operation of the transistor is analysed. Here the circuit operates in the linear region and the calculations can be done using Kirchoff's voltage law and Kirchoff's circuit law. Transistor circuit models are described and designed using hybrid (*h*) parameters.

Finally the frequency response for *CE* amplifier (with and without source impedance) for finding the bandwidth of the amplifier and few applications of transistors like emitter follower and Darlington pair are discussed in detail.

## 5-1 INTRODUCTION

The BJT as a circuit element operates various circuits with many major and minor modifications. For the analysis of such circuits, we obtain the various conditions for proper operation of the device, and also determine the projected range of operation of the device. A detailed study of the device in a two-port mode simplifies the circuit analysis of the device to a large extent. Thus, we calculate the various parameters of the devices' performance, namely voltage gain, current gain, input impedance, and output impedance. The frequency response of the device is dealt with in detail, and a study of the various regions of operation in the frequency scale is also explained.

Finally, we will discuss the various configurations of the device and take a look into the high-frequency operation of the device and its performance in those regions.

## 5-2 BIASING AND BIAS STABILITY

Biasing refers to the establishment of suitable dc values of different currents and voltages of a transistor. Through proper biasing, a desired quiescent operating point of the transistor amplifier in the active region (linear region) of the characteristics is obtained. The selection of a proper quiescent point generally depends on the following factors:

(a) The amplitude of the signal to be handled by the amplifier and distortion level in signal

(b) The load to which the amplifier is to work for a corresponding supply voltage

It is desired that the quiescent point be stable irrespective of changes in temperature or transistor characteristics. With the change in the input signal, the characteristics of the device should also change keeping the output of the device linear, i.e., undistorted.

The operating point of a transistor amplifier shifts mainly with changes in temperature, since the transistor parameters—$\beta$, $I_{CO}$ and $V_{BE}$ (where the symbols carry their usual meaning)—are functions of temperature.

The stability of a system is a measure of the sensitivity of a network or the transistor circuit to variations in parameters (mainly due to change in temperature). In any amplifier employing a transistor, the collector current $I_C$ is sensitive to each of the following parameters:

(a) $\beta$ increases with increase in temperature.

(b) $I_{CO}$ doubles in value for every 10 degree Celsius increase in temperature. This is the intrinsic current flowing, which is a strong function of temperature. The variation of silicon transistor parameters are shown in Table 5-1.

**Table 5-1** Variation of silicon transistor parameters with temperature

| $T$ | $I_{CO}$ (nA) | $B$ | $V_{BE}$ (V) |
|---|---|---|---|
| −65 | 0.0002 | 20 | 0.85 |
| 25 | 0.1 | 50 | 0.65 |
| 100 | 20 | 80 | 0.48 |
| 175 | 0.0033 | 120 | 0.3 |

Any or all of these factors can cause the bias point to drift from the designed or desired point of operation, thus, limiting the operation of the device in the linear region. This shift in the operating point may greatly vary the operation of the device and the output rather than changing the undistorted voltage swing to a distorted voltage swing.

### 5-2-1 Circuit Configurations

Here we study some of the circuit configurations where we design and analyse these particular configurations and their stability of operation.

#### Fixed-bias circuit

The configuration, as shown in Fig. 5-1(a), provides a relatively simple and easy introduction to the concept of biasing and their analysis. We consider the *n–p–n* transistor and this method equally applies to the analysis

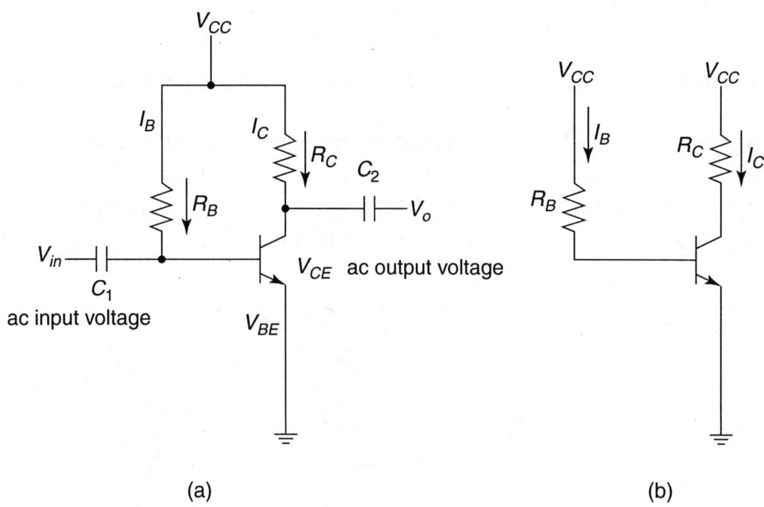


**Figure 5-1** (a) Representation of fixed-bias circuit (b) Equivalent circuit

of *p–n–p* transistor. In the CE mode, for dc analysis, the circuit is made devoid of the ac source and the capacitor, as indicated in the equivalent circuit in Fig. 5-1(b). We simply isolate the input and output, and analyse both of them independently.

$R_B$ indicates the resistance connected to the base lead, $R_C$ indicates the resistance connected to the collector leg of the transistor, and the respective currents are indicated as $I_B$ and $I_C$.

**Base–emitter loop.** From Fig. 5-1, applying KVL to the input circuit, we obtain the following equation:

$$V_{CC} = I_B R_B + V_{BE} \tag{5-1}$$

or,

$$I_B = \frac{V_{CC} - V_{BE}}{R_B} \tag{5-2}$$

Solving Eq. (5-2) will result in obtaining the base current for this particular circuit configuration. Also, we see that the value of $I_B$ changes with the change in the value of the base resistor and the value of the dc voltage source.

**Collector–emitter loop.** In the collector–emitter loop, i.e., the output loop, applying KVL we obtain the following equation:

$$V_{CE} = V_{CC} - I_C R_C \tag{5-3}$$

and,

$$I_C = \beta I_B \tag{5-4}$$

Also,

$$V_{CE} = V_C - V_E \tag{5-5}$$

In this case we have:

$$V_{CE} = V_C$$

Similarly, for $V_{BE}$ we obtain:

$$V_{BE} = V_B - V_E \tag{5-6}$$

which, is equal to $V_B$, since $V_E$ is zero because in Fig. 5-1 there is no resistance between the emitter terminal and ground.

It is worth mentioning that the voltage is measured with respect to ground. $V_E$ is the voltage between emitter and ground. $V_B$ is the voltage between base and ground. $V_C$ is the voltage between collector and ground.

The transistor saturation corresponds to the operation of the transistor in the saturation region. This stage corresponds to the operation when the collector is shorted to the emitter region and $V_{CE}$ = 0 V. Thus, from Eq. (5-3) we get:

$$I_C = \frac{V_{CC} - V_{CE}}{R_C}$$

Substituting $V_{CE} = 0$ and $I_C \rightarrow I_{Csat}$, we get:

$$I_{Csat} = \frac{V_{CC} - 0}{R_C} = \frac{V_{CC}}{R_C} \tag{5-7}$$

### Fixed bias with emitter resistance

The only modification in the circuit is that an emitter resistor is kept at the emitter lead of the transistor. The analysis is very simple; only the resistor is kept at the emitter terminal. The circuit is as shown in Fig. 5-2.

Figure 5-2 shows only a minor change from the previous configuration, i.e., a resistor $R_E$ is connected to the emitter leg of the transistor, and the current flowing through this resistor is given by $I_E$. All the other circuit components carry their usual meaning.

**Base–emitter loop.** For the base–emitter loop, we obtain the following equation:

$$V_{CC} = I_B R_B + V_{BE} + I_E R_E \tag{5-8}$$

and the emitter current can be written as $I_E = (\beta + 1) I_B$

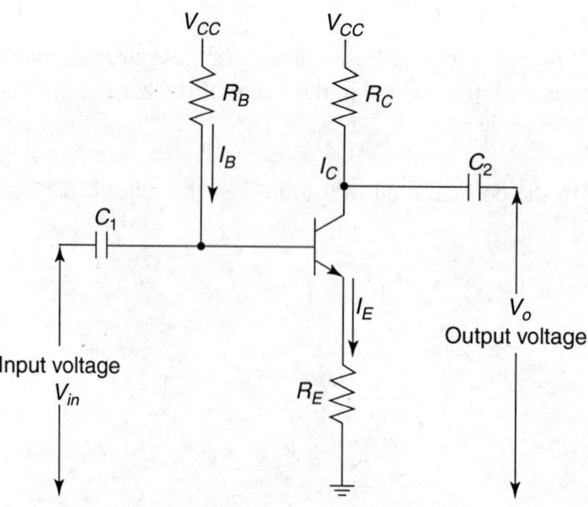


**Figure 5-2** Fixed-bias circuit with emitter resistance

Putting this value of $I_E$ in Eq. (5-8) yields:

$$I_B = \frac{V_{CC} - V_{BE}}{R_B + (\beta + 1)R_E} \tag{5-9}$$

This circuit is very similar to the previous bias circuit; however, we have the extra term $(\beta + 1)R_E$ as the only difference.

**Collector–emitter loop.** Applying KVL to the output loop, we obtain:

$$V_{CE} = V_{CC} - I_C(R_C + R_E) \tag{5-10}$$

with the base current known, $I_C$ can be easily calculated by the relation $I_C = \beta I_B$. With this value of $I_C$, we can obtain the value of the collector current and collector voltage that gives a location on the output characteristics, and also provides the operating point. This location gives an indication whether the biasing of the BJT is proper at that point or not.

### Voltage-divider bias

The study of the previous circuits shows that the quiescent current and the quiescent voltage are dependent on the current gain $\beta$ of the transistor. However, knowing that $\beta$ is a temperature dependent factor, the design of such a circuit with these factors independent of $\beta$ is required. Such a design is provided in Fig. 5-3.

This circuit is also called the self-bias circuit. In this circuit, we find that the voltage at the base is provided with the help of a voltage divider where, the resistors are indicated by $R_1$ and $R_2$.

The analysis of the input side is made exclusively by representing it through the Thevenins equivalent circuit. Thus, the Thevenins equivalent voltage for the input side is given by:

$$V_{TH} = V_{CC}\frac{R_2}{R_1 + R_2} \tag{5-11}$$

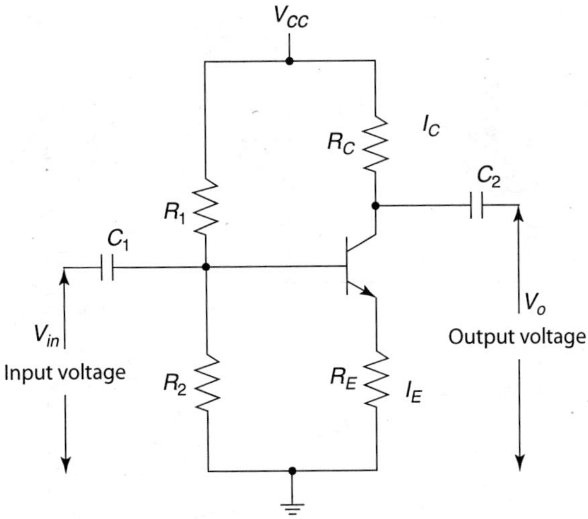


**Figure 5-3** Voltage-divider bias circuit

The Thevenin resistance is given by:

$$R_{TH} = \frac{R_1 R_2}{R_1 + R_2} \tag{5-12}$$

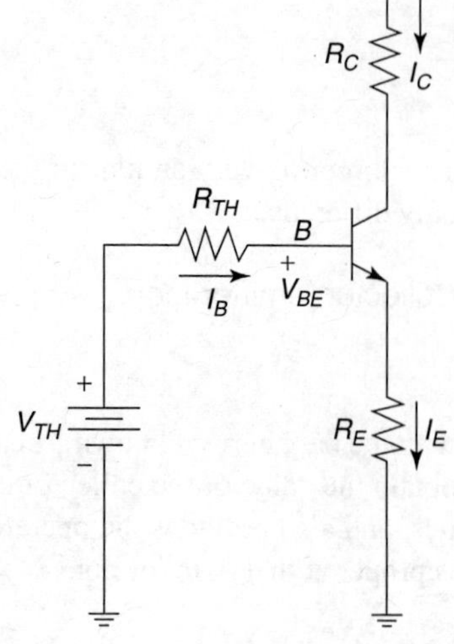


Figure 5-4 Simplified voltage-divider circuit

The equivalent circuit is shown in Fig. 5-4.

The KVL equation for the input circuit [see Eq. (5-9)] is given as:

$$I_B = \frac{V_{TH} - V_{BE}}{R_{TH} + (\beta + 1)R_E} \tag{5-13}$$

The KVL equation for the output circuit gives:

$$I_C(R_C + R_E) + I_B R_E + V_{CE} = V_{CC}$$

The output equation results in determining the collector-to-emitter voltage which is given by:

$$V_{CE} = V_{CC} - I_C(R_C + R_E) \tag{5-14}$$

This gives us the quiescent operating point.

### Voltage-feedback biasing

The objective behind discussing these different topologies is to obtain a better stability. In our desire to obtain a more stable circuit, we end up with the circuit, which has a resistance $R_B$, included in the path connecting the base and the collector. The circuit for such a case is represented by Fig. 5-5.

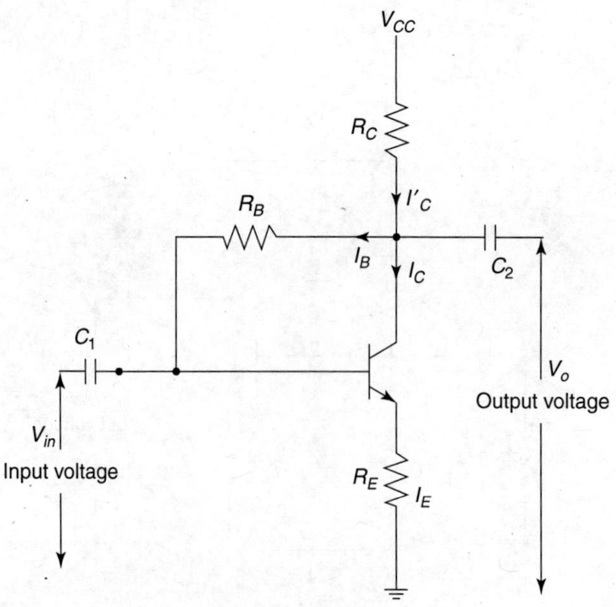


Figure 5-5 Representation of Voltage-feedback biased circuit

**Base–emitter loop.** The analysis for the base–emitter loop is as follows.
Applying KVL for this part, we get:

$$V_{CC} = I'_C R_C + I_B R_B + V_{BE} + I_E R_E = (I_C + I_B) R_C + I_B R_B + V_{BE} + I_E R_E \tag{5-15}$$

Neglecting $I_B$ for the reasons stated, we obtain:

$$V_{CC} = I_C R_C + I_B R_B + V_{BE} + I_E R_E$$

Here, it is important to note that the current $I'_C = I_C + I_B$ is the sum of the current entering the base and the collector terminal of the BJT. But due to the design constraint of the transistor, the value of $I_C$ and $I'_C$ far exceed the value of the base current. For a simpler analysis, we equate $I_C$ and $I'_C$. Putting the values of $I_C$ and $I'_C$ in Eq. (5-15) results in:

$$V_{CC} = \beta I_B R_C + I_B R_B + V_{BE} + \beta I_B R_E \tag{5-16}$$

Rearranging the terms of Eq. (5-16) will result into:

$$V_{CC} - V_{BE} = I_B [R_B + \beta(R_C + R_E)] \tag{5-17}$$

Thus, the base current can be obtained as:

$$I_B = \frac{V_{CC} - V_{BE}}{R_B + \beta(R_C + R_E)} \tag{5-18}$$

## 5-2-2 Stabilization Against Variations in $I_{CO}$, $V_{BE}$ and $\beta$

Our focus should always be on the operation of the device under varying temperature conditions. In this regard, the following temperature dependent sources cause instability of the collector current $I_C$. These include:

(i) the reverse saturation current $I_{CO}$, which doubles itself for every 10°C increase in temperature;

(ii) the base-to-emitter voltage $V_{BE}$, which decreases at the rate of 2.5 mV/°C for both Ge and Si transistors;

(iii) $\beta$, which is a strong function of temperature, and increases with an increase in temperature.

We shall neglect the effect of the change of $V_{CE}$ with temperature because this variation is very small, and we can assume that the transistor operates in the active region where $I_C$ is approximately independent of $V_{CE}$. Under normal operation this value is near about 0.3 V when the transistor is operating in the saturation region.

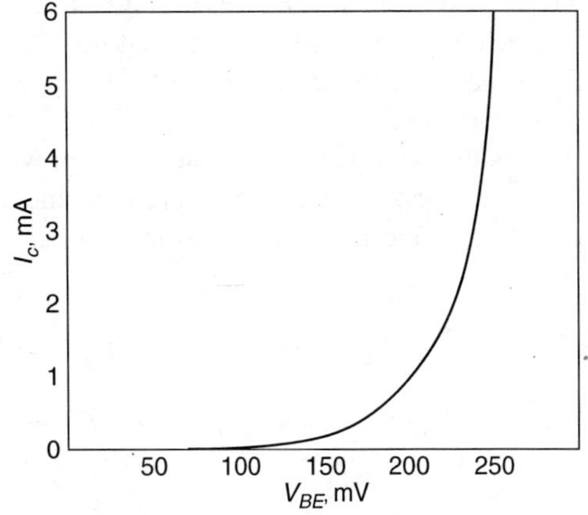


**Figure 5-6** Transfer characteristics for germanium *p–n–p* alloy type transistor

### Transfer characteristic

In this particular characteristic, the output current $I_C$ is plotted (see Fig. 5-6) as a function of input voltage for the germanium transistor. Thus, the word "transfer" is used for this characteristic.

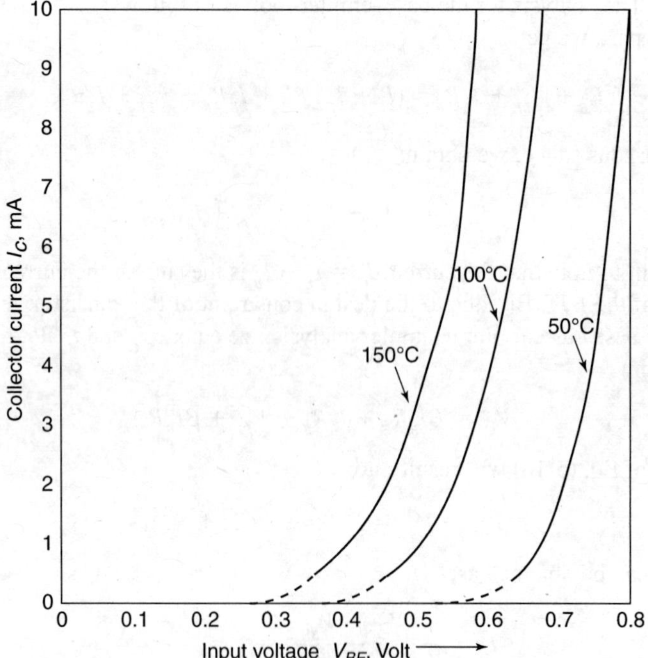


**Figure 5-7** Collector current vs. base-to-emitter voltage for a silicon transistor

Transfer characteristic in the case of a silicon transistor is shown in Fig. 5-7.

In Fig. 5-7 we can see that there is a shift of the curves towards the left at a rate of 2.5 mV/°C with constant $I_C$ and with an increase in temperature. We now examine in detail the effect of the shift in transfer characteristics and the variation of $\beta$ and $I_{CO}$ with temperature. This analysis makes it simpler for us to analyse the variation of the operating point with all of these factors taken into consideration. For easy analysis and evaluation of various stability factors, we consider a self-bias circuit.

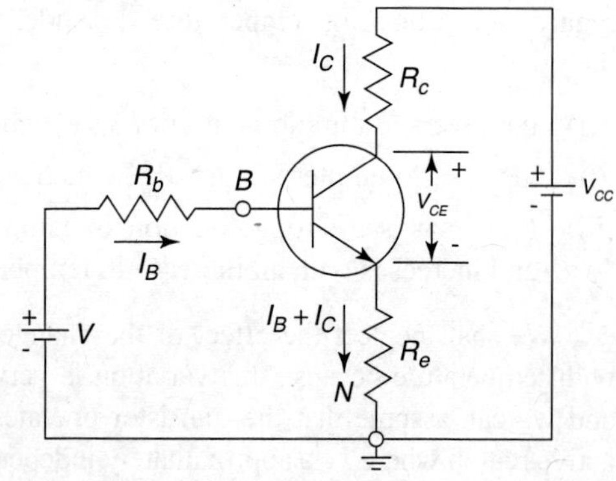


**Figure 5-8** Self-bias circuit

If the equation obtained from Fig. 5-8—by applying KVL around the base–emitter circuit of the self-bias circuit—is combined with the equation obtained by applying KVL around the collector circuit that represents the collector characteristics in the active region, the resultant equation will be:

$$V_{BE} = V + (R_b + R_e)\frac{\beta + 1}{\beta} I_{CO} - \frac{R_b + R_e(1 + \beta)}{\beta} I_C \tag{5-19}$$

Equation (5-19) represents a load line in the $I_C - V_{BE}$ plane, and is indicated in Fig. 5-9. The intercept on the $V_{BE}$ axis is $V + V_1'$, where:

$$V_1' = (R_b + R_e)\frac{\beta + 1}{\beta} I_{CO} \approx (R_b + R_e) I_{CO} \tag{5-20}$$

$\beta \gg 1$, therefore, $\frac{\beta + 1}{\beta} = 1$.

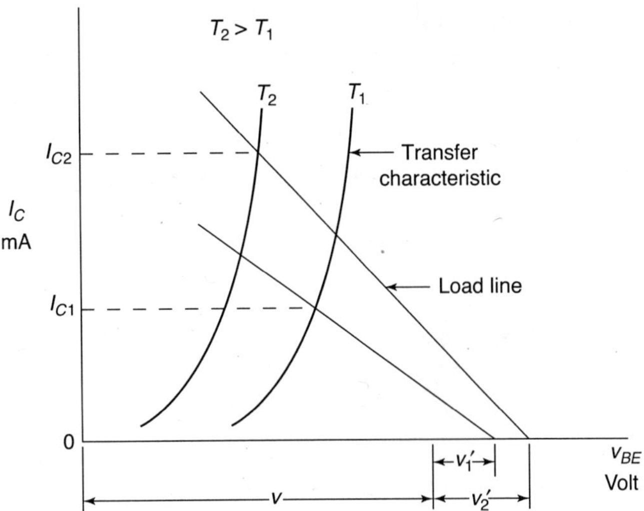


**Figure 5-9** Variation of the collector current with temperature because of $V_{BE}$, $I_{CO}$ and $\beta$

If we put $T = T_1(T_2)$, $I_{CO} = I_{CO_1}(I_{CO_2})$ and $\beta = \beta_1(\beta_2)$, then we have:

$$V_1' \approx (R_b + R_e)I_{CO_1} \tag{5-21}$$

And consequently we obtain:

$$V_2' \approx (R_b + R_e)I_{CO_2} \tag{5-22}$$

Thus, the intercept of the load line on the $V_{BE}$ axis is found to be dependent on temperature because $I_{CO}$ increases with $T$. The slope of the load line as obtained is:

$$\sigma = \frac{-\beta}{R_b + R_e(1 + \beta)} \tag{5-23}$$

and, as is evident, $|\sigma|$ increases with $T$ since $\beta$ increases with $T$. The transfer characteristic for $T = T_2 > T_1$ shifts to the left of the corresponding curve for $T = T_1$ as $V_{BE}$ (at constant $I_C$) varies with $T$, which is indicated above. The intersection of the load line with the transfer characteristic gives us the collector current $I_C$. We also see that $I_{C2} > I_{C1}$ because $I_{CO}$, $\beta$ and $V_{BE}$ all vary with temperature.

From Eq. (5-19) it is seen that with $I_C$ being a function of $I_{CO}$, $V_{BE}$ and $\beta$, it is possible to introduce the three partial derivatives of $I_C$ with respect to these variables in order to indicate the variation of $I_C$ with respect to each of the independent parameters, taken one at a time. These derivatives are called the stability factors $S$, $S'$ and $S''$.

## 5-3 CALCULATION OF STABILITY FACTORS

### 5-3-1 Stability Factor *S*

We here define the stability factor $S$, as the change of collector current with respect to the reverse saturation current, keeping $\beta$ and $V_{BE}$ constant. This can be written as:

$$S \equiv \frac{\partial I_C}{\partial I_{CO}} \tag{5-24}$$

The larger the value of $S$, the more likely it is for the circuit to be thermally unstable, i.e., the operation of the device will be marked by the change in operating temperature. Using the above definition and Eq. (5-19), we find:

$$S = (1 + \beta)\frac{1 + R_b/R_e}{1 + \beta + R_b/R_e} \tag{5-25}$$

If $R_b/R_e \simeq 1$, $S \approx 1$, $R_b/R_e \to \infty$ and $S \approx 1 + \beta$, $S$ remains between 1 and $1 + \beta$.
If $(\beta + 1) \gg R_b/R_e$, then Eq. (5-25) reduces to:

$$S = 1 + \frac{R_b}{R_e} \tag{5-26}$$

Thus, for constant $\beta$, $V_{BE}$ and small $S$, we obtain:

$$\frac{\Delta I_C}{I_c} \approx S\frac{\Delta I_{CO}}{I_c} \approx \frac{\Delta I_{CO}}{I_C} + \frac{R_b}{R_e}\frac{\Delta I_{CO}}{I_C} \tag{5-27}$$

For the typical design $R_b/R_e > 1$, we make the second term in Eq. (5-28) larger than the first term. The denominator of the second term is the dc voltage drop across $R_e$ (since $|I_C| \approx |I_e|$) and should always be under the circuit designer's supervision for its operation to be controlled.

### 5-3-2 Stability Factor *S*′

The variation of $I_C$ with $V_{BE}$ is given by the stability factor $S'$ defined by the partial derivative:

$$S' \equiv \frac{\partial I_C}{\partial V_{BE}} \approx \frac{\Delta I_C}{\Delta V_{BE}} \tag{5-28}$$

where, both $I_{CO}$ and $\beta$ are considered to be constant.
From Eq. (5-19) we find:

$$S' = \frac{-\beta}{R_b + R_e(1 + \beta)} = \frac{-\beta/R_e}{1 + \beta + R_b/R_e} \tag{5-29}$$

Again, if we assume that $\beta + 1 \gg R_b/R_e$ and also that $\beta \gg 1$, then from Eq. (5-29) we obtain:

$$S' \approx \frac{\Delta I_C}{\Delta V_{BE}} \approx -\frac{1}{R_e} \tag{5-30}$$

or,

$$\frac{\Delta I_C}{I_C} \approx \frac{S'\Delta V_{BE}}{I_C} \approx -\frac{\Delta V_{BE}}{I_C R_e} \tag{5-31}$$

From Eq. (5-31) and Eq. (5-27) we find that the dominant factor is stabilising against $I_{CO}$ and $V_{BE}$ is the quiescent voltage drop across the emitter resistance $R_e$. The larger the drop, the smaller is the percentage change in collector current because of the changes in $I_{CO}$ and $V_{BE}$.

### 5-3-3 Stability Factor *S*″

The variation of $I_C$ with respect to $\beta$ is represented by the stability factor, $S''$, given as:

$$S'' \equiv \frac{\partial I_C}{\partial \beta} \approx \frac{\Delta I_C}{\Delta \beta} \tag{5-32}$$

where, both $I_{CO}$ and $V_{BE}$ need to be considered constant. From Eq (5-19):

$$I_C = \frac{\beta(V + V' - V_{BE})}{R_b + R_e(1 + \beta)} \tag{5-33}$$

From Eq. (5-20), $V'$ may be taken to be independent of $\beta$. After differentiation and algebraic manipulation we obtain:

$$S'' = \frac{\partial I_C}{\partial \beta} = \frac{I_C S}{\beta(1+\beta)} \tag{5-34}$$

The change in the collector current due to a change in $\beta$ is given by:

$$\Delta I_C \approx S''\,\Delta\beta = \frac{I_C S}{\beta(1+\beta)}\Delta\beta \tag{5-35}$$

where, $\Delta\beta = \beta_2 - \beta_1$ may represent a large change in $\beta$. Thus, it is not clear whether to use $\beta_1$, $\beta_2$ or some average value of $\beta$ in the expressions for $S''$. This difficulty is avoided if $S''$ is obtained by taking finite differences rather than by evaluating a derivative. Thus:

$$S'' \approx \frac{I_{C2} - I_{C1}}{\beta_2 - \beta_1} = \frac{\Delta I_C}{\Delta\beta} \tag{5-36}$$

From Eq. (5-33) we have:

$$\frac{I_{C2}}{I_{C1}} = \frac{\beta_2}{\beta_1}\,\frac{R_b + R_e(1+\beta_1)}{R_b + R_e(1+\beta_2)} \tag{5-37}$$

Subtracting 1 from both sides of Eq. (5-33) yields:

$$\frac{I_{C2}}{I_{C1}} - 1 = \left(\frac{\beta_2}{\beta_2} - 1\right)\frac{R_b + R_e}{R_b + R_e(1+\beta_2)} \tag{5-38}$$

or,

$$S'' = \frac{\Delta I_C}{\Delta\beta} = \frac{I_{C1} S_2}{\beta_1(1+\beta_2)} \tag{5-39}$$

where, $S_2$ is the value of the stabilizing factor $S$ when $\beta = \beta_2$, as given by Eq. (5-24). This equation gets reduced to Eq. (5-34) as $\Delta\beta = \beta_2 - \beta_1 \rightarrow 0$.

If we assume $S_2$ to be small so that the approximate value given in Eq. (5-25) is valid, then from Eq. (5-39) with $\beta \gg 1$, we find:

$$\frac{\Delta I_C}{I_{C1}} \approx \left(1 + \frac{R_b}{R_e}\right)\frac{\Delta\beta}{\beta_1\beta_2} = \left(1 + \frac{R_b}{R_e}\right)\frac{\dfrac{\beta_2}{\beta_1} - 1}{\beta_2} \tag{5-40}$$

$$\frac{\Delta I_C}{I_{C1}} \approx \left(1 + \frac{R_b}{R_e}\right)\frac{\Delta\beta}{\beta_1\beta_2} = \left(1 + \frac{R_b}{R_e}\right)\frac{\dfrac{\beta_2}{\beta_1} - 1}{\beta_2} \tag{5-41}$$

It is clear that $R_b/R_e$ should be kept small. Also, for a given spread in the value of $\beta$, a high-$\beta$ circuit will be more stable than one using a lower-$\beta$ transistor.

## 5-3-4 General Remarks on Collector Current Stability

The stability factors have been defined earlier keeping in mind the change in collector current with respect to changes in $I_{CO}$, $V_{BE}$ and $\beta$. These stability factors are repeated here for simplicity.

$$S = \frac{\Delta I_C}{\Delta I_{CO}},\quad S' = \frac{\Delta I_C}{\Delta V_{BE}},\quad S'' = \frac{\Delta I_C}{\Delta\beta} \tag{5-42}$$

Each differential quotient is calculated with all other parameters kept constant. If we desire to obtain the total change in collector current over a specified temperature range, we can do so by expressing this change as the sum of the individual changes due to the stability factors calculated from the variation of $I_{CO}$, $V_{BE}$ and $\beta$. Thus, by taking the total differential of $I_C = f(I_{CO}, V_{BE}, \beta)$, we obtain:

$$\Delta I_C = \frac{\partial I_C}{\partial I_{CO}} \Delta I_{CO} + \frac{\partial I_C}{\partial V_{BE}} \Delta V_{BE} + \frac{\partial I_C}{\partial \beta} \Delta \beta \tag{5-43}$$

$$= S\Delta I_{CO} + S' \Delta V_{BE} + S'' \Delta \beta \tag{5-44}$$

Equation (5-44) gives the summary of all the explanations given earlier regarding change in collector current or in other words, the essence of stability factors. The stability factors may be expressed in terms of the parameter $M$ defined by:

$$\mathrm{M} \equiv \frac{1}{1 + R_b/[R_e(1+\beta)]} \approx \frac{1}{1 + R_b/\beta R_e} \tag{5-45}$$

In this case we assume $\beta \gg 1$. Note that if, $\beta R_e \gg R_b$, then $M \approx 1$. Substituting Eqs. (5-22), (5-26) and (5-36) in Eq. (5-39) for the fractional (minute) change in collector current we get:

$$\frac{\Delta I_C}{\Delta_{C1}} = \left(1 + \frac{R_b}{R_e}\right)\frac{M_1 \Delta I_{CO}}{I_{C1}} - \frac{M_1 \Delta V_{BE}}{I_{C1} R_e} + \left(1 + \frac{R_b}{R_e}\right)\frac{M_2 \Delta \beta}{\beta_1 \beta_2} \tag{5-46}$$

where $M_1(M_2)$ corresponds to $\beta_1(\beta_2)$. As $T$ increases, $\Delta I_{CO}/I_{C1}$ and $\Delta\beta$ increase, and $\Delta V_{BE}/I_{C1}$ decreases. Hence all the terms in Eq. (5-42) are positive for an increase in $T$ and negative for a decrease in $T$.

## 5-4 CE, CB MODES AND THEIR PROPERTIES

### 5-4-1 Common-Emitter (CE) Mode

#### Input characteristics

The input characteristiscs are as shown in Fig. 5-10. Here $I_B$ is the input current, $V_{BE}$ is the input voltage and $V_{CE}$ is the output voltage. The variation of the base current $I_B$ with respect to the base-to-emitter voltage $V_{BE}$, considering $V_{CE}$ as a constant is shown in Fig. 5-10. This set of curves represents the CE input characteristics. The characteristics are similar to that of a forward-biased diode. However, for a constant $V_{BE}$, the magnitude of the base current decreases with an increase in $V_{CE}$. This is because, with an increasing $V_{CE}$ the effective base width decreases and thus, the recombination base current also decreases.

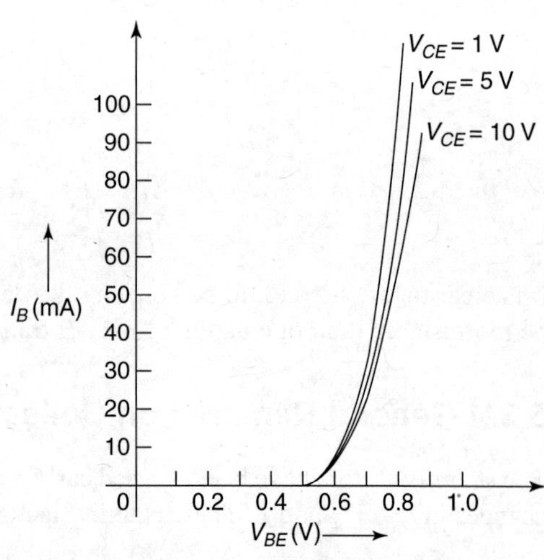


**Figure 5-10** Input characteristic for common-emitter configuration

### Output characteristics

The output characteristics are as shown in Fig. 5-11(a).

Here the collector current $I_C$ is plotted against the collector-to-emitter voltage $V_{CE}$ with base current $I_B$ as parameter. The characteristics can be divided into three regions:

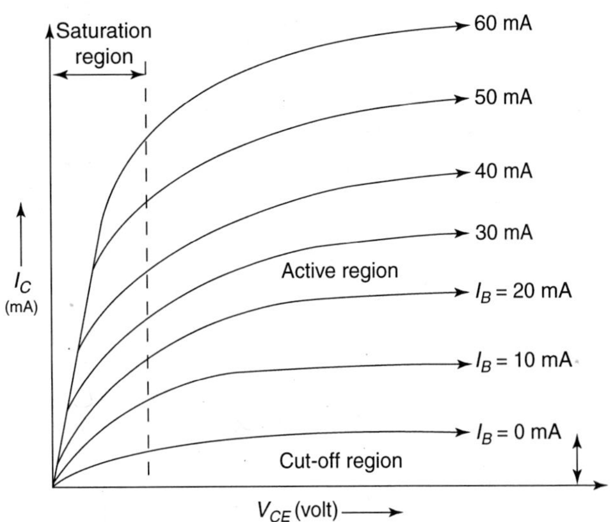


**Figure 5-11(a)** Plot of the collector current against the collector-to-emitter voltage

**Active region.** In this region the transistor works in a mode where the emitter–base junction is forward-biased and the collector–base junction reverse-biased. A transistor when operated in the active region can be used to amplify signals almost faithfully, as this region corresponds to the linear region or rather the device bears a linear relationship between the input and the output signals. In this region the characteristics of the device change according to the changes in the input signal, thus keeping the output a faithful replication of the input signal. The output characteristics in the active region are not horizontal lines. This is because, for a fixed value of the base current $I_B$ the magnitude of the collector current increases with $V_{CE}$ (due to *Early effect*).

**Cut-off region.** For operation in the CE mode, only by making the base current $I_B = 0$, the cut-off collector current or $I_{CO}$ is not obtained. It is also necessary to reverse-bias the emitter–base junction only for it to be zero; the collector current can be equal to $I_{CO}$ (applicable only for germanium transistors). Thus, this region of operation corresponds to the operation of the device in which the both the emitter–base and the collector–base junctions are reverse-biased, and the current is solely in the reverse saturation region. Its magnitude is very small.

**Saturation region.** A transistor operating in this particular region has both the collector–base and the emitter–base junctions forward-biased by at least the cutin ($V\gamma$) voltage. Under this condition the collector current becomes approximately independent of the base current, and the current at the collector attains an overall saturation value.

Table 5-2 provides the four different transistor states under four different combinations of biasing, and Fig.5 -11(b) illustrates the operating modes.

**Table 5-2** Definitions of transistor states

| *Transistor State or Operating Mode* | *Junction Biasing* | |
|---|---|---|
| | *Base–Emitter* | *Base–Collector* |
| Forward active | Forward | Reverse |
| Reverse active | Reverse | Forward |
| Cut-off | Reverse | Reverse |
| Saturation | Forward | Forward |

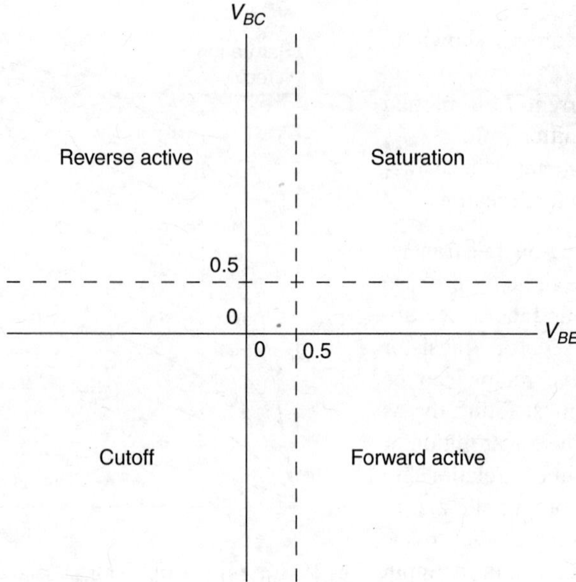


**Figure 5-11(b)** Transistor states defined by junction biasing

## 5-4-2 Common-Base Mode

This configuration corresponds to the arrangement in which the base is common to both the input and the output of the circuit concerned.

### Input characteristics

The input characteristics are as shown in Fig. 5-12. In this case, the emitter current is the input current, and the emitter–base voltage is the input voltage. The collector–base voltage is the output voltage. The variation of the emitter current with respect to the emitter-to-base voltage is equivalent to the situation in a forward-biased *p–n* junction. But due to *Early effect*, an increase in the magnitude of the collector voltage $V_{CB}$ causes the emitter current to increase for a definite $V_{EB}$.

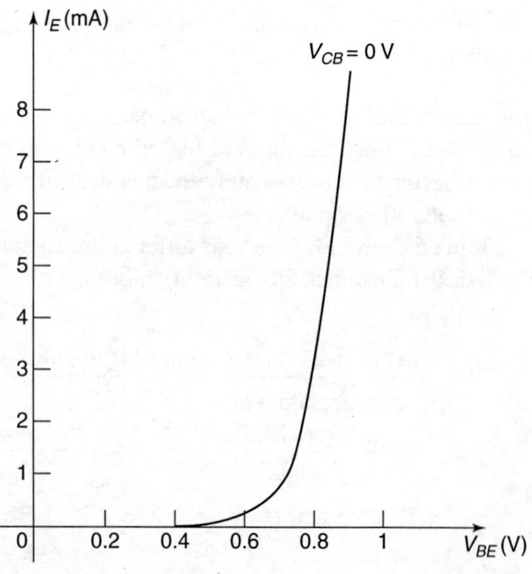


**Figure 5-12** Input characteristics

### Output characteristics

The output characteristics are as shown in Fig. 5-13. Here the current flowing in the collector is the output current, and the collector-to-base voltage is the output voltage. The emitter current is the input current in this case. Again there are three regions in the characteristic curve that can be clearly distinguished.

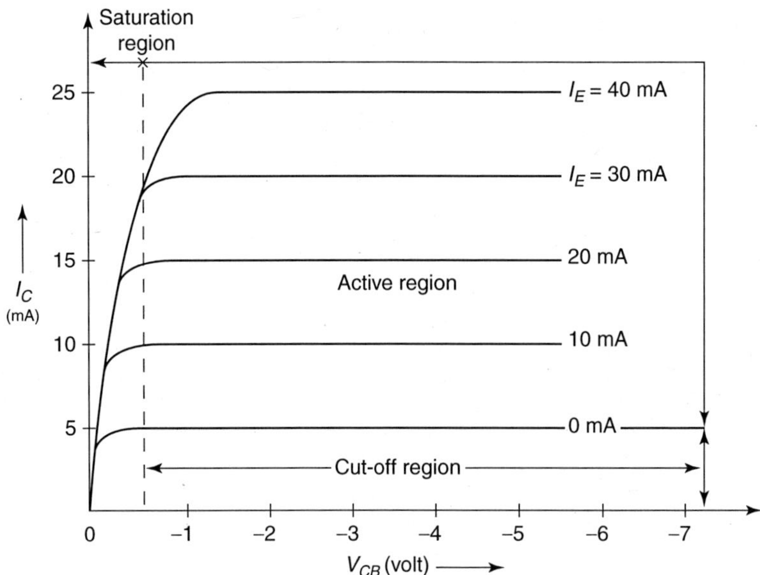


**Figure 5-13** Output characteristics

**Active region.** In this region the collector junction is biased in the reverse direction and the emitter junction in the forward direction. Let us first assume that the emitter current is zero. Then, as is evident, the collector current is small and equals the reverse saturation current $I_{CO}$ (microamperes for Ge and nanoamperes for Si) of the collector junction, which might be considered as a diode. Now, suppose that the forward emitter current $I_E$ is made to flow in the emitter circuit. Then a fraction of $-\alpha I_E$ of this current will reach the collector, and the actual value of the collector current $I_C$ is thus given by:

$$I_C = I_{CO} - \alpha I_E \tag{5-47}$$

In the active region the collector current is mostly independent of the collector voltage and depends on the emitter current. However, because of *Early effect* there is actually a change and thus, an increase in $|I_C|$ with respect to $|V_{CB}|$. Because $\alpha$ is less than but almost equal to 1, the magnitude of the collector current is slightly less than the emitter current. Physically, it can be clearly realised that a part of the emitter current goes on to recombine with the carriers in the base region.

**Saturation region.** This region is located to the left of the line $V_{CB} = 0$ and above the output characteristic of $I_E = 0$. This region is realised by forward-biasing of both the emitter–base and the collector–base junctions. Forward-biasing of the collector–base junction results in an exponential variation in the collector current, and thus, accounts for a large change in the collector current with $V_{CB}$ in the saturation region.

**Cut-off region.** The region to the right of the line $V_{CB} = 0$ and below the characteristic for $I_E = 0$ is the cut-off region of the transistor. In this region the operation of the transistor is realised by reverse-biasing both the emitter–base and the collector–base junctions.

## Solved Examples

**Example 5-1** For the circuit shown in the diagram:

(a) Calculate $I_B$, $I_C$, and $V_{CE}$ if a silicon transistor is used with $\beta = 100$.

(b) Specify a value for $R_b$ so that $V_{CE} = 7$ V.

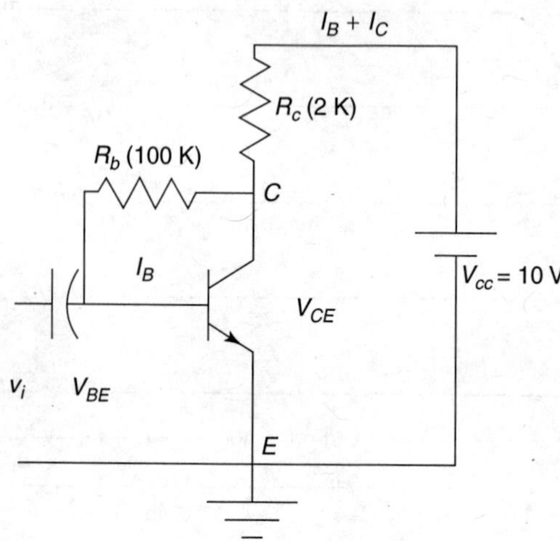


**Solution:**

(a) Applying KVL around the loop $V_{CC} - C - B - E$, we have:

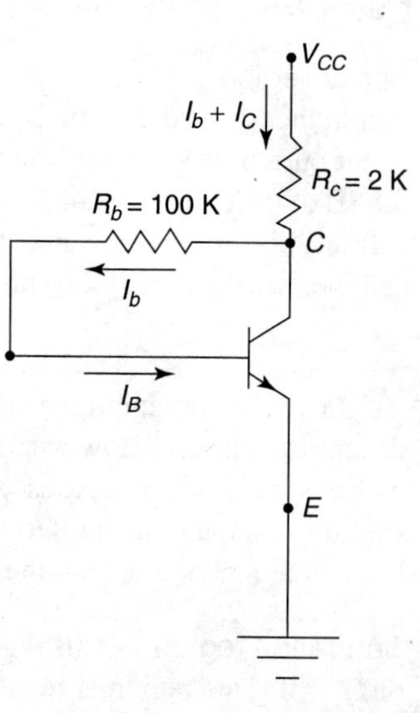


$$V_{CC} = R_C(I_B + I_C) + I_B R_b + V_{BE}$$

or, $$I_B = \frac{V_{CC} - V_{BE}}{\beta R_C + R_C + R_B}$$

But the transistor is in the active region, hence, $I_C = 100 I_B$ and $V_{BE} = 0.70$.

Substituting these values yields:

$$I_B = \frac{10\text{ V} - 0.7\text{ V}}{100 \times 2\text{ k}\Omega + 2\text{ k}\Omega + 100\text{k}\Omega}$$

or, $$I_B = \frac{9.30\text{ V}}{302\text{ k}\Omega} = 0.031\text{ mA}$$

and $$I_C = 100 \times 0.031\text{ mA} = 3.1\text{ mA}$$

Then $$V_{CE} = V_{CC} - (I_B + I_C)R_C = 3.74\text{V}$$

(b) Given $V_{CE} = 7$ V

We apply KVL in the collector–emitter circuit and we have:

$$I_C + I_B = \frac{V_{CC} - V_{CE}}{R_C} = \frac{3\text{ V}}{2\text{ k}\Omega} = 1.5\text{ mA}$$

or, $$I_B = \frac{1.5\text{ mA}}{101} = 0.01485\text{ mA}$$

and $$I_C = 100\, I_B = 1.485\text{ mA}$$

Applying KVL around the loop $V_{CC} - C - B - E$, we have:

$$V_{CC} = R_C(I_B + I_C) + I_B R_b + V_{BE}$$

$$R_b = \frac{V_{CC} - V_{BE} - R_B(I_C + I_B)}{I_B}$$

or,
$$R_b = \frac{10\text{ V} - 0.7\text{ V} - 2\text{ k}\Omega \times 1.5\text{ mA}}{0.01485\text{ mA}} = 424.24\ \Omega$$

**Example 5-2** (a) In Fig. 5-8 verify the stability factor $S = (1+\beta)\dfrac{1 + \dfrac{R_b}{R_e}}{1 + \beta + \dfrac{R_b}{R_e}}$

(b) Show that for the circuit of Example 5-1, $S$ is given by:

$$S = \frac{\beta + 1}{1 + \beta R_c/(R_c + R_b)}$$

**Solution:**

(a) From Fig. 5-8, solving for $I_B$ we get: $I_B = \dfrac{V - V_{BE} - I_C R_e}{R_e + R_b}$

From Eq. (5-19) we have:

$$I_C = (1+\beta)I_{CO} + \beta I_B = (1+\beta)I_{CO} + \frac{\beta}{R_e + R_b}(V - V_{BE} - I_C R_e)$$

or,
$$I_C\left(1 + \frac{\beta R_e}{R_e + R_b}\right) = (1+\beta)I_{CO} + \frac{\beta}{R_e + R_b}(V - V_{BE})$$

Then,
$$s = \frac{\partial I_C}{\partial I_{CO}} = \frac{1+\beta}{1 + \dfrac{\beta R_e}{R_e + R_b}} = (1+\beta)\frac{1 + \dfrac{R_b}{R_e}}{1 + \beta + \dfrac{R_b}{R_e}}$$

(b) Applying KVL around $V_{CC} - C - B - E$, we have:

$$V_{CC} = I_C R_c + (R_b + R_c)I_B + V_{BE}$$

But,
$$I_C = \beta I_B + (1+\beta)I_{CO}$$

Thus, we have:

$$V_{CC} = I_C R_c + \frac{R_b + R_c}{\beta} I_C - (R_b + R_c)\frac{(1+\beta)}{\beta} I_{CO} + V_{BE}$$

Differentiating the equation for $V_{CC}$ with respect to $I_{CO}$, we obtain:

$$\left[R_c + \frac{R_b + R_c}{\beta}\right]\frac{\partial I_C}{\partial I_{CO}} = (R_b + R_c)\frac{(1+\beta)}{\beta}$$

or,
$$S = \frac{\partial I_C}{\partial I_{CO}} = \frac{(R_b + R_c)(1+\beta)}{R_b + R_c + \beta R_c} = \frac{(1+\beta)}{1 + \dfrac{\beta R_c}{(R_b + R_c)}}$$

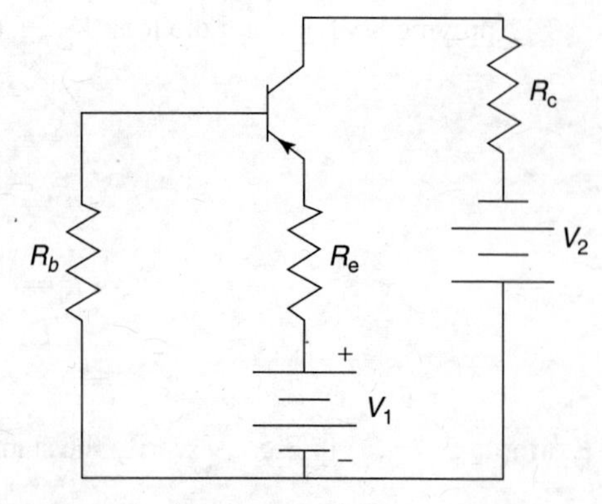


**Example 5-3** For the two-battery transistor circuit as shown in the diagram, prove that the stabilization factor $S$ is given by:

$$S = \frac{1+\beta}{\dfrac{1+\beta R_e}{R_e + R_b}}$$

**Solution:**

Neglecting $V_{BE}$ we obtain from the base circuit:

$$V_1 = I_E R_e - I_B R_b$$

But $I_E = -(I_B + I_C)$; thus, $V_1 = -(I_B + I_C)R_e - I_B R_b$

or, $$I_B = -\frac{I_C R_e + V_1}{R_e + R_b}$$

$$I_C = (1+\beta)I_{CO} + \beta I_B$$

$$= (1+\beta)I_{CO} - \frac{\beta}{R_e + R_b}(I_C R_e + V_1)$$

or, $$I_C\left(1 + \frac{\beta R_e}{R_e + R_b}\right) = (1+\beta)I_{CO} - \frac{\beta V_1}{R_e + R_b}$$

$$S = \frac{\partial I_C}{\partial I_{CO}} = \frac{(1+\beta)}{1 + \dfrac{\beta R_e}{R_e + R_b}}$$

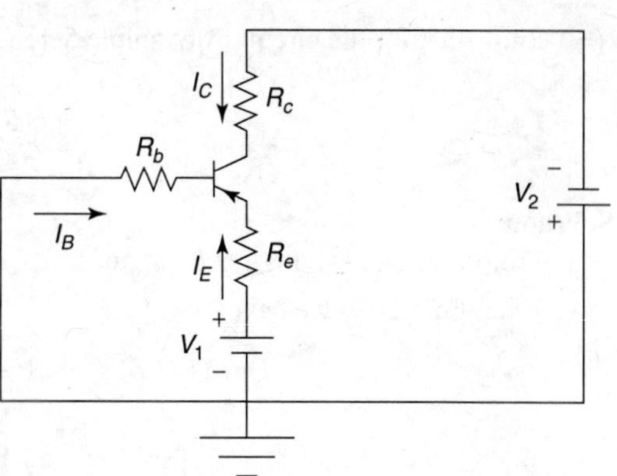


**Example 5-4** Assume that a silicon transistor with $\beta = 50$, $V_{BE(\text{active})} = 0.7$ V, $V_{CC} = 22.5$ V and $R_c = 5.6$ K is used in the given diagram. It is desired to establish a $Q$-point at $V_{CE} = 12$ V, $I_C = 1.5$ mA, and stability factor $S \le 3$. Find $R_e$, $R_1$ and $R_2$.

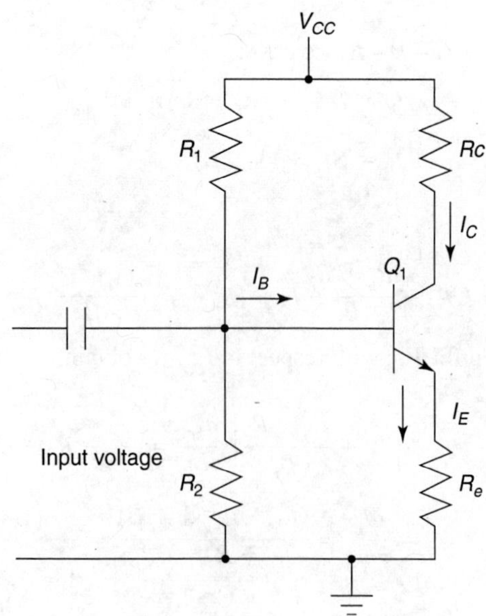

**Solution:**

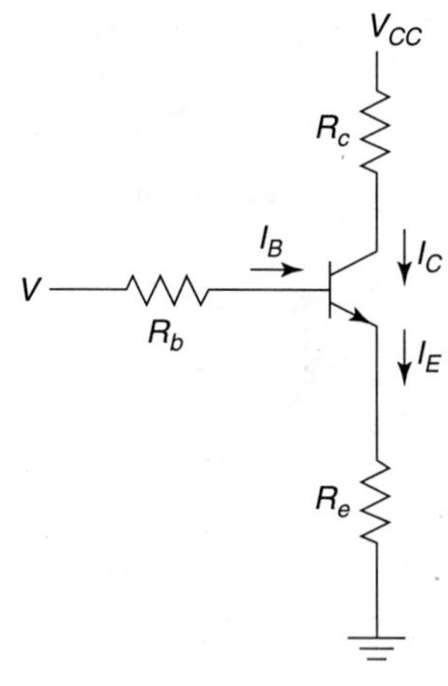


The dc equivalent of the given diagram is as shown.

The current in $R_e$ is $I_C + I_B \approx I_C$. Hence, from the collector circuit, we obtain:

$$R_e + R_c = \frac{V_{CC} - V_{CE}}{I_C} = \frac{22.5\text{ V} - 12\text{ V}}{1.5\text{ mA}} = 7.0\text{ k}\Omega$$

or, $\quad R_e = (7.0 - 5.6)\text{ k}\Omega = 1.4\text{ k}\Omega$. Solving for $R_b/R_e$, we get:

$$3 = 51 \times \frac{1 + R_b R_e}{51 + R_b/R_e} \qquad \text{(Considering } S = 3\text{)}$$

or,

$$R_b/R_e = 2.12$$

Hence,

$$R_b = (2.12 \times 1.4)\text{ k}\Omega = 2.96\text{ k}\Omega$$

If $R_b < 2.96$ then $S < 3$.

The base current is:

$$I_B = \frac{I_C}{\beta} = \frac{1.5}{50} = 30\ \mu\text{A}$$

Solving for $R_1$ and $R_2$, we find:

$$R_1 = R_b \times \frac{V_{CC}}{V} \text{ and } R_2 = \frac{R_1 V}{V_{CC} - V}$$

But we have:

$$V = 0.030 \times 2.96 + 0.7 + (0.03 + 1.5) \times 1.4 = 2.93\text{ V}$$

Thus,

$$R_1 = \frac{2.96 \times 22.5}{2.93} = 22.8\text{ K and } R_2 = \frac{22.8 \times 2.93}{22.5 - 2.93} = 3.4\text{ K}$$

**Example 5-5** (a) A germanium transistor is used in the self-biasing arrangement, as shown in the following figure, with $V_{CC} = 20$ V and $R_c = 1.5$ K. The quiescent point is chosen to be $V_{CE} = 8$ V and $I_C = 4$ mA. A stability factor $S = 12$ is desired. If $\beta = 50$, find $R_1$, $R_2$ and $R_e$.

(b) Repeat part (a) for $S = 3$.

**Solution:**

(a) $I_B = \dfrac{I_C}{\beta} = \dfrac{4}{50} = 80\ \mu\text{A}$

Hence,

$$R_e = \frac{V_{CC} - V_{CE} - I_C R_c}{I_B + I_C}$$

$$= \frac{20\text{ V} - 8\text{ V} - 4\text{ mA} \times 1.5\text{ k}\Omega}{4.08\text{ mA}} = 1.47\text{ k}\Omega$$

$$S = 12 = 51 \times \frac{1 + \dfrac{R_b}{R_e}}{51 + \dfrac{R_b}{R_e}}$$

or,

$$\frac{R_b}{R_e} = 14.4 \text{ and } R_b = 21.17\text{ k}\Omega$$

The base-to-ground voltage:

$$V_{BN} = V_{BE} - I_E R_e = 0.2\text{ V} + 4.08\text{ mA} \times 1.47\text{ k}\Omega = 5.91\text{ V}$$

The Thevenin voltage is:

$$V = V_{BN} + I_B R_b = 5.91\text{ V} + 0.08\text{ mA} \times 21.17\text{ k}\Omega = 7.60\text{ V}$$

We have:
$$\frac{V}{R_b} = \frac{V_{CC}}{R_1} = \frac{20}{R_1}$$

$\therefore$
$$R_1 = \frac{20\text{ V} \times R_b}{V}$$

$$= \frac{20\text{ V} \times 21.17\text{ k}\Omega}{7.60\text{ V}} = 55.71\text{ k}\Omega$$

Then,
$$I_{R1} = \frac{V_{CC} - V_{BN}}{R_1} = \frac{20\text{ V} - 5.91\text{ V}}{55.71\text{ k}\Omega} = 0.253\text{ mA}$$

and
$$I_{R2} = I_{R1} - I_B = 0.173\text{ mA}$$

Hence,
$$R_2 = \frac{V_{BN}}{I_{R2}} = \frac{5.91\text{ V}}{0.173\text{ mA}} = 34.16\text{ k}\Omega$$

(b) $R_e = 1.47\text{ k}\Omega$

$$S = 3 = 51 \times \frac{1 + \dfrac{R_b}{R_e}}{51 + \dfrac{R_b}{R_e}}$$

or,
$$\frac{R_b}{R_e} = 2.13 \quad \text{and} \quad R_b = 3.13\text{ k}\Omega$$

Then,
$$V = 5.91\text{ V} + 0.08\text{ mA} \times 3.13\text{ k}\Omega + 6.16\text{ V}$$

Thus,
$$\frac{V}{R_b} = \frac{V_{CC}}{R_1}$$

$\therefore$
$$R_1 = \frac{V_{CC} R_b}{V} = \frac{20\text{ V} \times 3.13\text{ k}\Omega}{6.16\text{ V}} = 10.16\text{ k}\Omega$$

$$I_{R1} = \frac{20\text{ V} - 5.91\text{ V}}{10.16\text{ k}\Omega} = 1.387\text{ mA}$$

$$I_{R2} = 1.387\text{ mA} - 0.08\text{ mA} = 1.307\text{ mA}$$

$\therefore$
$$R_2 = \frac{5.91\text{ V}}{1.307\text{ mA}} = 4.52\text{ k}\Omega$$

**Example 5-6** In the transformer coupled amplifier stage, as shown, $V_{BE} = 0.7$ V, $\beta = 50$, and the quiescent voltage is $V_{CE} = 4$ V. Determine (a) $R_e$ (b) the stability factor $S$.

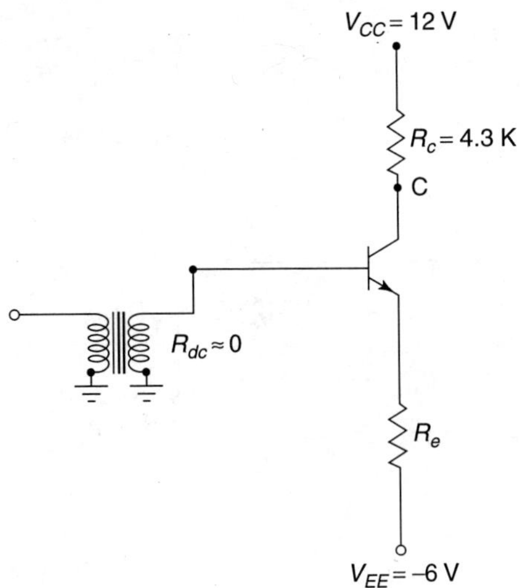


**Solution:**

(a) Collector–emitter circuit:

$$V_{CC} - V_{EE} = R_c I_C + R_e(I_C + I_B) + V_{CE}$$

or,
$$18\text{ V} = 4.3\text{ k}\Omega \times I_C + R_e\left(1 + \frac{1}{50}\right) I_C + 4\text{ V} \quad (1)$$

Base–emitter circuit:

$$-V_{EE} = (I_C + I_B)R_e + V_{BE}$$

or,
$$6 = \left(1 + \frac{1}{50}\right) I_C R_e + 0.7 \quad (2)$$

Solving (1) and (2) simultaneously, we get $I_C = 2$ mA and $R_e = 2.74$ K

(b) Comparing the circuit given with that in the following diagram:

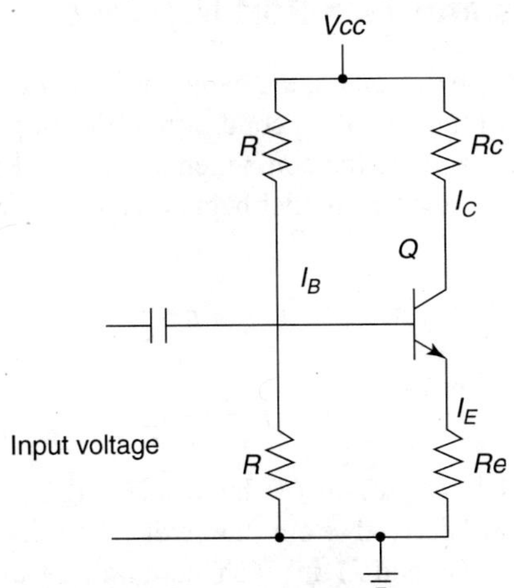

we see that they are similar except here $R_b = 0$. The equation for $S$, therefore, is valid with $R_b = 0$. Under this condition $S = 1$.

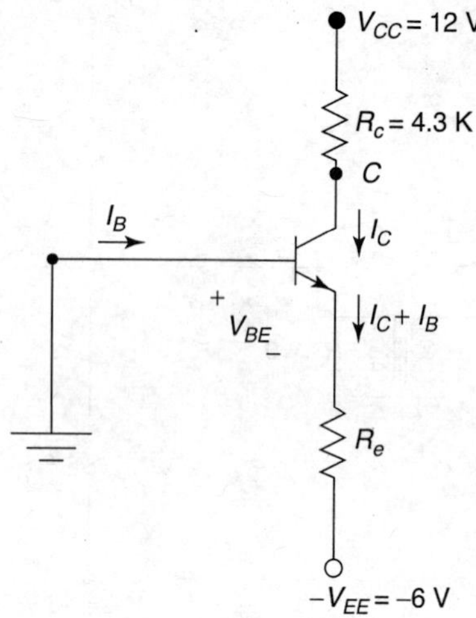


■

## 5-5 SMALL-SIGNAL LOW-FREQUENCY OPERATION OF TRANSISTORS

The small signal model of a transistor operates with reasonable linearity and consequently, requires a small-signal linear model that can represent the operation of the transistor in the active region. This small-signal model is required for the analysis of the system where the input signal has amplitude, which is small in relation to the proximity of the region in the output characteristics where the device operates linearly. One such important model is the $h$-parameter model that makes the analysis of these particular types of transistors excessively easy and lucid.

### 5-5-1 Hybrid Parameters and Two-Port Network

For the hybrid equivalent model to be described, the parameters are defined at an operating point that may or may not give an actual picture of the operating condition of the amplifier. The quantities $h_{ie}$, $h_{re}$, $h_{fe}$ and $h_{oe}$ are called the hybrid parameters and are the components of a small-signal equivalent circuit.

The description of the hybrid equivalent model begins with the general two-port system as shown in Fig. 5-14.

$$V_i = h_{11}I_i + h_{12}V_o$$

$$I_o = h_{21}I_i + h_{22}V_o \tag{5-48}$$

Figure 5-14 is a black box model realization of the whole circuit. The parameters relating the four variables are called $h$-parameters, derived from the word "hybrid". The term hybrid was chosen because the mixture of variables ($V$ and $I$) in each equation results in a "hybrid" set of units of measurement for these $h$-parameters.

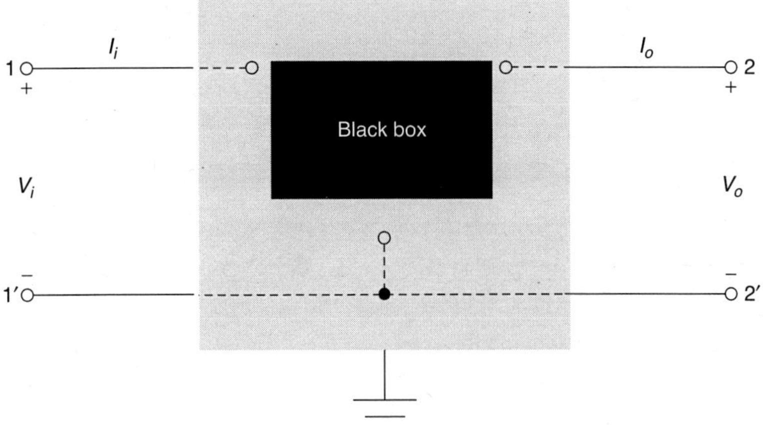


**Figure 5-14** Two-port system representation (Black model realisation)

Now, setting $V_o = 0$ (short circuit the output terminals) and solving for $h_{11}$ we obtain:

$$h_{11} = \left.\frac{V_i}{I_i}\right|_{V_o = 0} \text{ ohms} \tag{5-49}$$

Since $h_{11}$ is the ratio of the input voltage to the input current with the output terminals shorted, it is termed as the short-circuit input-impedance parameter. The subscript 11 of $h_{11}$ defines the fact that the parameter is determined by a ratio of quantities measured at the input terminals.

If $I_i$ is set to zero, i.e., by opening the input leads and replacing this condition in Eq. (5-49), we get the value of $h_{12}$ as:

$$h_{12} = \left.\frac{V_i}{V_o}\right|_{I_i = 0} \tag{5-50}$$

It has no units since it is a ratio of voltage levels and is called the open-circuit reverse transfer voltage ratio parameter. The subscript 12 of $h_{12}$ reveals the fact that the parameter is a transfer quantity determined by the ratio of input to output measurements.

Now, we set $V_o = 0$ by shorting the output terminals. The following will result for $h_{21}$:

$$h_{21} = \left.\frac{I_o}{I_i}\right|_{V_o = 0} \tag{5-51}$$

It is also a unit less since it is the ratio of the current levels. It is formally called the short-circuit forward transfer current ratio parameter. The subscript 21 indicates that it is a transfer parameter with the ratio of output to input quantity.

The last parameter $h_{22}$, which can be found by again opening the input leads to set $I_i = 0$. Thus, by replacing this condition in the basic sets of equations, we obtain:

$$h_{22} = \left.\frac{I_o}{V_o}\right|_{I_i = 0} \text{ siemens} \tag{5-52}$$

Since it is the ratio of output current to the output voltage, it is the output conductance parameter, and is measured in siemens (S). It is called the open-circuit output admittance parameter. The subscript 22 reveals that it is determined by a ratio of output quantities. Therefore, we have:

(i) $h_{11} \rightarrow$ input impedance $\rightarrow h_i$

(ii) $h_{12} \rightarrow$ reverse transfer voltage ratio $\rightarrow h_r$

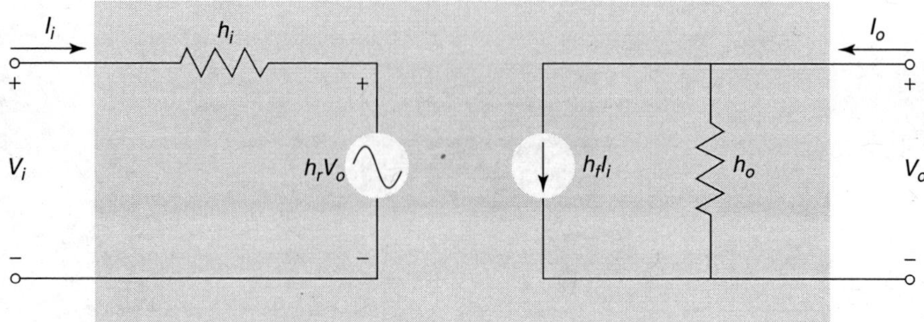


**Figure 5-15** Complete hybrid equivalent model

(iii) $h_{21}\rightarrow$ forward transfer current gain $\rightarrow h_f$

(iv) $h_{22}\rightarrow$ output admittance $\rightarrow h_o$

Figure 5-15 shows the complete hybrid equivalent circuit.

## 5-6 EQUIVALENT CIRCUITS THROUGH HYBRID PARAMETERS AS A TWO-PORT NETWORK

For the transistor, even though it has three basic configurations, they are all four-terminal configurations, and thus, the resulting equivalent circuit will have the same format as shown in Fig. 5-15. The *h*-parameter will however change with each configuration. To distinguish which parameter has been used or which is available, a second subscript has been added to the *h*-parameter notation.

(i) For the common-base configuration: the lower case letter *b*

(ii) For the common-emitter configuration: the lower case letter *e*

(iii) For the common-collector configuration: the lower case letter *c*

The hybrid equivalent network for the common-emitter and common-base is shown in Fig. 5-16.
For common-emitter $I_i = I_b$, $I_o = I_c$, and through an application of Kirchoff's current law we have:

$$I_e = I_b + I_c \tag{5-53}$$

For the common-base configuration, $I_i = I_e$, $I_o= I_c$ with $V_{eb} = V_i$ and $V_{cb} = V_o$

## 5-7 TRANSISTOR AS AMPLIFIER

A load resistor $R_L$ is in series with the collector supply voltage $V_{cc}$, as shown in Fig. 5-17.

A small voltage change, $\Delta V_i$, between the emitter and the base causes quite a large emitter-current change, given by $\Delta I_E$. We define by the symbol $\alpha$ as the fraction of this current change or the effective current which is collected at the collector and passes through $R_L$ or $\Delta I_C = \alpha' \Delta I_E$, i.e., that which reaches the collector region. Consequently the change in output voltage across the load resistor can be given by:

$$\Delta V_L = -R_L \Delta I_C = -\alpha' R_L \Delta I_E \tag{5-54}$$

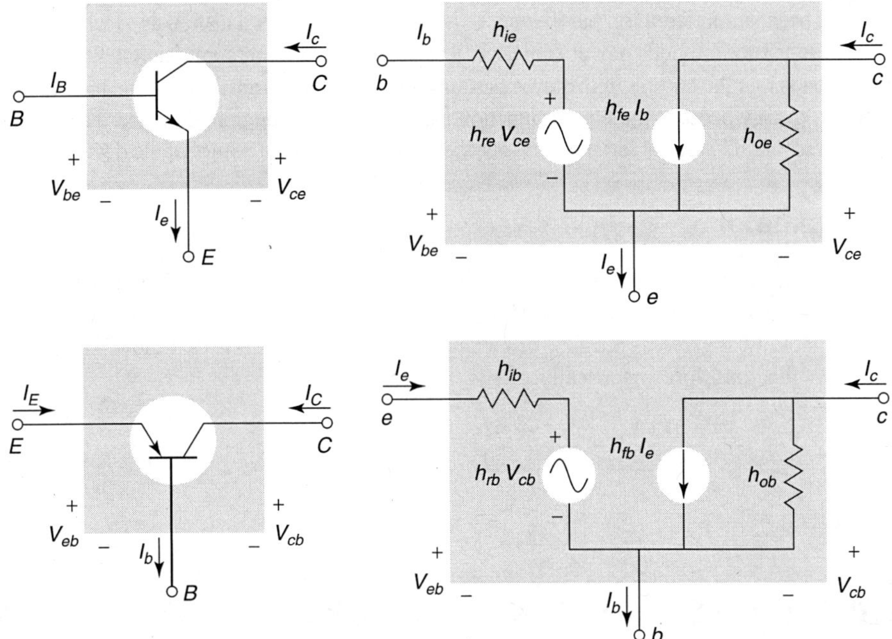


**Figure 5-16** Common-emitter and common-base configuration: (a) graphic symbol (b) hybrid equivalent circuit

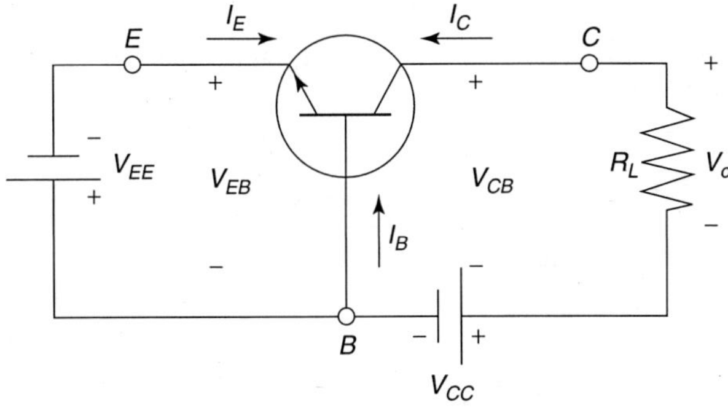


**Figure 5-17** An *n–p–n* transistor in the common-base bias mode

and this may be many times the change in input voltage $V_i$. Under these circumstances, the voltage amplification $A \equiv \Delta V_L/\Delta V_i$ will be greater than unity and thus, the transistor acts as an amplifier. If the dynamic resistance of the emitter junction is given by $r_e$, then corresponding change in the input voltage $\Delta V_i = r_e \Delta I_E$, and thus, the gain is given by:

$$A \equiv \frac{\alpha' R_L \Delta I_E}{r_e \Delta I_E} = -\frac{\alpha' R_L}{r_e} \tag{5-55}$$

We know that the resistance of the emitter is the forward resistance of a diode, i.e, $r_e = 26/I_E$ where, $I_E$ is the quiescent emitter current (in milliamperes). For example, if $r_e = 40\ \Omega$, $\alpha' = -1$ and $R_L = 3{,}000\ \Omega$, we have $A = +75$.

This calculation has been simplified a lot, but in essence it gives a physical explanation as to why the transistor acts as an amplifier. The transistor provides power gain as well as voltage or current amplification. From this explanation we reach the conclusion that the current in the low-resistance input circuit is transferred to the high-resistance output circuit which occurs mainly because the emitter junction has a very low dynamic resistance. The word "transistor" originated as a contraction of "transfer resistor" and is based upon the physical nature of the device outlined here.

### 5-7-1 The Parameter $\alpha'$

The parameter $\alpha'$ is defined as the ratio of the change in the collector current to the change in the emitter current at constant collector-to-base voltage and is called the negative of the small-signal short-circuit current transfer ratio or gain. Qualitatively, it gives us a figure of the amount of current carriers that move from the emitter and are collected at the collector. More specifically:

$$\alpha' = \left.\frac{\Delta I_C}{\Delta I_E}\right|_{V_{CB}} \tag{5-56}$$

where, $V_{CB}$ = constant.

## 5-8 EXPRESSIONS OF CURRENT GAIN, INPUT RESISTANCE, VOLTAGE GAIN AND OUTPUT RESISTANCE

Figure 5-18 shows the $h$-parameter equivalent circuit of a transistor amplifier having a voltage source $V_g$, with its input resistance $R_g$ connected to the input terminals and a load resistance $R_L$ connected to the output terminals.

Now we will derive expressions for the current gain ($A_I$), the input resistance ($R_I$), the voltage gain ($A_V$) and the output resistance ($R_O$).

### 5-8-1 Current Gain ($A_I$)

Current gain is defined as the ratio of the output current to input current. If the load current is $I_L$ and it is assumed to flow from the top to the bottom through $R_L$, as shown in Fig. 5-18, then:

$$A_I = \frac{I_L}{I_1} = -\frac{I_2}{I_1} \tag{5-57}$$

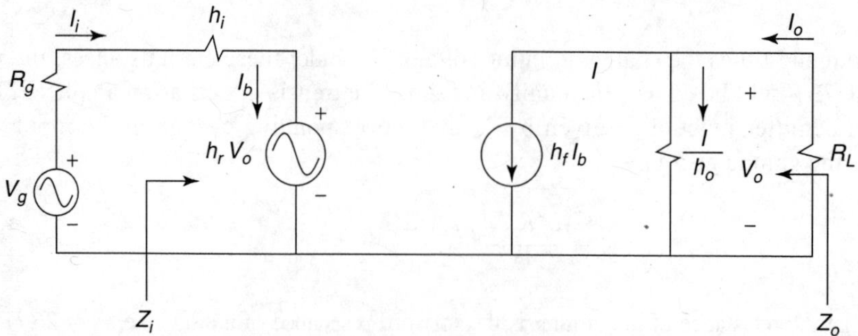


**Figure 5-18** $h$-Parameter equivalent circuit of a transistor

Again, applying Kirchoff's current rule for the output circuit we get:

$$I_2 = h_o V_2 + h_f I_1 \tag{5-58}$$

where, $$V_2 = I_L R_L = -I_2 R_L \tag{5-59}$$

From Eqs. (5-58) and (5-59) we get:

$$I_2 = -h_o I_2 R_L + h_f I_1$$

or, $$I_2(1 + h_o R_L) = h_f I_1$$

or, $$\frac{I_2}{I_1} = \frac{h_f}{(1 + h_o R_L)}$$

or, $$A_I = -\frac{I_2}{I_1} = -\frac{h_f}{(1 + h_o R_L)} \tag{5-60}$$

$$A_I = \frac{-h_f}{(1 + h_o R_L)}$$

## 5-8-2 Input Resistance ($R_I$)

Input resistance is defined as the ratio of the input voltage across the input terminals of the amplifier to the current $I_1$. Therefore:

$$R_I = \frac{V_1}{I_1} \tag{5-61a}$$

From the input circuit we get by applying KVL:

$$V_1 = h_i I_1 + h_r V_2 \tag{5-61b}$$

or, $$\frac{V_1}{I_1} = h_i + h_r \frac{V_2}{I_1} = h_i - h_r \frac{I_2}{I_1} R_L \qquad (\because V_2 = -I_2 R_L) \tag{5-62}$$

Hence, $$R_I = h_i + A_I h_r R_L \qquad \left(\because A_I = -\frac{I_2}{I_1}\right) \tag{5-63}$$

Substituting the value of $A_I$ from Eq. (5-60) we get:

$$R_I = h_i - \frac{h_f h_r R_L}{1 + h_o R_L} \tag{5-64}$$

## 5-8-3 Voltage Gain ($A_V$)

Voltage gain or voltage amplification is defined as the ratio of the output voltage $V_2$ to the input voltage $V_1$.

$$A_V = \frac{V_2}{V_1} \tag{5-65}$$

Since $$V_2 = -I_2 R_L = A_I I_1 R_l \tag{5-66}$$

we get, $$A_V = A_I R_L \frac{I_1}{V_1} = \frac{A_I R_L}{R_I} \tag{5-67}$$

or, $$A_V = -\frac{h_f R_L}{h_i + \Delta h R_L} \tag{5-68}$$

where, $$\Delta h = h_i h_o - h_f h_r$$

### 5-8-4 Output Resistance ($R_o$)

The output resistance is defined as the ratio of $V_2$ and $I_2$, where $I_2$ is the current delivered by the generator $V_2$. In order to find this, we set the source voltage $V_g$ to zero and replace the load resistance $R_L$ by a voltage generator $V_2$. Therefore:

$$R_o = \frac{V_2}{I_2} \tag{5-69}$$

Considering the output circuit of Fig. 5-18 we obtain:

$$I_2 = h_o V_2 + h_f I_1 \tag{5-70}$$

From the input mesh, we can write: $$(R_g + h_i)I_1 + h_r V_2 = 0 \tag{5-71}$$

or, $$I_1 = -\frac{h_r V_2}{R_g + h_i} \tag{5-72}$$

Hence, $$I_2 = h_o V_2 - \frac{h_f h_r V_2}{R_g + h_i} \tag{5-73}$$

or, $$I_2 = V_2\left(h_o - \frac{h_f h_r}{R_g + h_i}\right)$$

Thus, the expansion for $R_o$ finally becomes:

$$R_o = \frac{V_2}{I_2} = \frac{R_g + h_i}{R_g h_o + h_i h_o - h_f h_r} \tag{5-74}$$

## 5-9 FREQUENCY RESPONSE FOR CE AMPLIFIER WITH AND WITHOUT SOURCE IMPEDANCE

At different frequencies of the input signal, the performance of the device is different. The analysis till now has been limited to the mid-frequency spectrum. Frequency response of an amplifier refers to the variation of the magnitude and phase of the amplifier with frequency. A plot of gain vs. frequency for a CE amplifier is shown in Fig. 5-19(a) and phase angle vs. frequency for a CE amplifier is shown in Fig. 5-19(b).

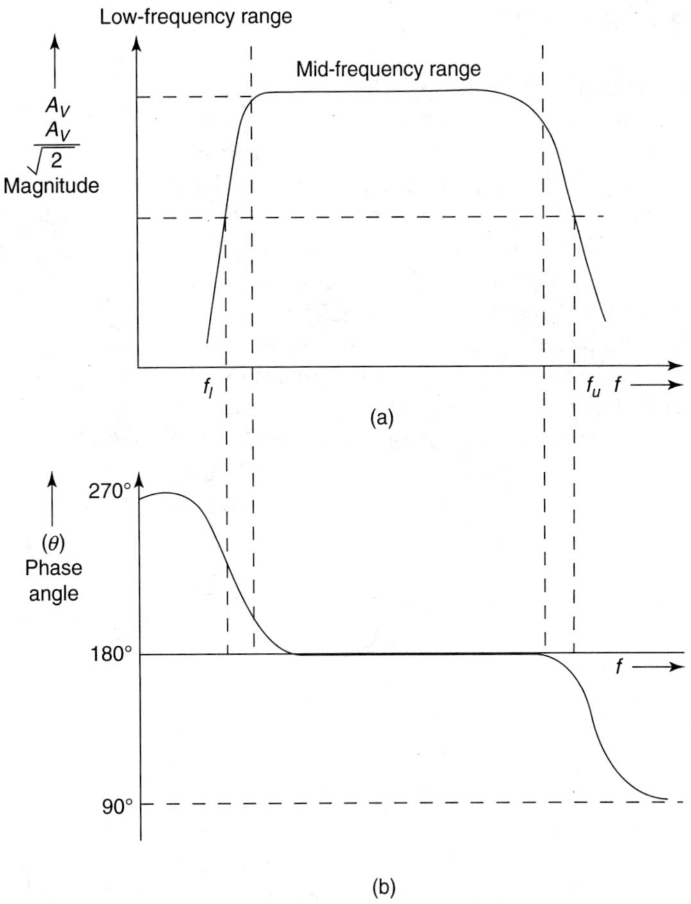


**Figure 5-19** (a) Gain vs. frequency for a CE amplifier (b) Phase angle vs. frequency for a CE amplifier

## 5-9-1 Conclusions

I. The frequency response can be divided into three ranges.

  (i) High-frequency range: The magnitude of voltage gain decreases slowly with the increase in frequency and the phase angle decreases below 180 degrees.

  (ii) Mid-frequency range: Both gain and phase remain *nearly* constant.

  (iii) Low-frequency range: The magnitude of voltage gain increases slowly with the increase in frequency and the phase angle increases up to 180 degrees but below 270 degrees.

II. Bandwidth of the amplifier is the difference between upper and lower cut-off frequency range:

$$BW = f_u - f_l \tag{5-75}$$

III. Upper cut-off frequency: For which voltage gain is 0.707 time of its maximum gain at upper frequency range, as shown in Fig. 5-19(a). It is also known as half-power frequency.

IV. Lower cut-off frequency: For which voltage gain is 0.707 time of its maximum gain at lower frequency range, as shown in Fig. 5-19(a). It is also known as half-power frequency.

## 5-10 EMITTER FOLLOWER

The emitter follower transistor is a design which is basically a CC amplifier. As seen from Fig. 5-20, the output signal is taken from the emitter with respect to ground and the collector is connected directly to $V_{cc}$. Since $V_{cc}$ is at signal ground in the ac equivalent circuit, we have the name common-collector. This is further illustrated in Fig. 5-20.

From Fig. 5-20, we find that the direction of the emitter current for the transistor is opposite to the reference (stipulated) direction. Again, we see that with an increase in the base voltage, the emitter–base junction becomes more forward-biased, giving rise to an increase in the emitter current. The output emitter voltage, follows the input base voltage, with zero phase shift; thus, the significance of the name emitter follower. The expressions for current gain, input resistance, voltage gain and output resistance of the CC amplifier are as follows. The detailed derivation is left for the reader to perform as they can be obtained by a simple $h$-parameter model of the respective circuit.

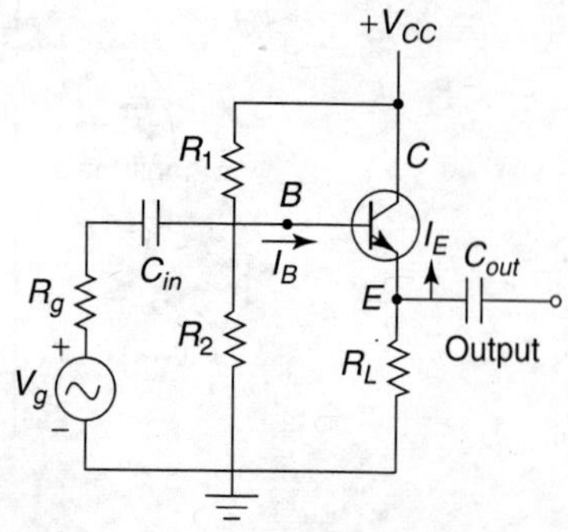


**Figure 5-20** An emitter follower configuration with biasing

(i) Current gain:

$$A_i = -\frac{I_e}{I_b} = 1 + \frac{h_{fe}}{1 + h_{oe}R_L} \tag{5-76}$$

(ii) Input resistance:

$$R_i = \frac{V_i}{I_b} = h_{ie} + A_i R_L \tag{5-77}$$

(iii) Voltage gain:

$$A_V = \frac{V_L}{V_i} = \frac{A_i R_L}{R_i} \tag{5-78}$$

(iv) Output resistance:

$$R_o = \frac{R_G + h_{ie}}{1 + R_G h_{oe} + h_{ie}h_{oe} + h_{fe}} \tag{5-79}$$

Also, in Fig. 5-20, we find that the there are two capacitors which solve the purposes of coupling. The emitter follower is used for impedance matching.

### Solved Examples

**Example 5-7** Consider an emitter follower. Neglect $h_{re}$ and show that as Re $\rightarrow \infty$

(a) $R_i = h_{ie} + \dfrac{1 + h_{fe}}{h_{oe}} = \dfrac{1}{h_{ob}}$

Explain the result physically.

(b) $1 - A_V \approx \dfrac{h_{ie}h_{oe}}{1 + h_{fe}}$

Evaluate $A_V$ using $h$-parameter values given in the following table.

| $h$-parameters | Values |
|---|---|
| $h_{ie}$ | 1.1 kΩ |
| $h_{re}$ | $2.5 \times 10^{-4}$ |
| $h_{fe}$ | 50 |
| $1/h_{oe}$ | 40 kΩ |

**Solution:**

(a) From the equation for current amplification for a emitter follower connection,

$$A_I = \frac{1 + h_{fe}}{1 + h_{oe}R_e} = \frac{1 + h_{fe}}{h_{oe}R_e} \quad \text{(for very large } R_e\text{)}$$

From equation of input resistance:

$$R_i = h_{ie} + A_I R_e = h_{ie} + \frac{1 + h_{fe}}{h_{oe}}$$

But,

$$h_{ob} = \frac{h_{oe}}{1 + h_{fe}}$$

Hence,

$$R_i = \frac{1}{h_{ob}} \approx 2\text{M} > h_{ie} = 1\text{ K}$$

(b) $A_V = 1 - \dfrac{h_{ie}}{R_i} = 1 - \dfrac{h_{ie}h_{oe}}{1 + h_{ie}}$ where, use of its voltage gain and result in part (a) has been made.

$$A_V = 1 - \frac{\frac{1.1}{40}}{1+50} = 1 - \frac{1.1}{2040} = 0.99946$$

**Example 5-8** For the emitter follower with $R_s = 0.5$ K and $R_L = 5$ K, calculate $A_I$, $R_i$, $A_V$, $A_{VS}$, and $R_o$. Assume $h_{fe} = 50$, $h_{ie} = 1$ K, $h_{oe} = 25\ \mu$A/V.

**Solution:**

From the equation for current gain:

$$A_I = \frac{1 + h_{fe}}{1 + h_{oe}R_L} = \frac{51}{1 + 0.185} = 45.3$$

From the equation for input resistance:

$$R_i = h_{ie} + A_I R_L = 1 + 226 = 227\text{ K}$$

From the equation for voltage gain:

$$A_V = 1 - \frac{h_{ie}}{R_i} = 1 - \frac{1}{227} = 0.9956$$

$$A_{VS} = A_V \frac{R_i}{R_i + R_S} = A_V \frac{227}{227 + 0.5} \approx 0.9956\left(1 - \frac{0.5}{227.5}\right) = 0.9934$$

**Example 5-9** (a) Design an emitter follower having $R_i = 500$ K, and $R_o = 20\ \Omega$. Assume $h_{fe} = 50$, $h_{ie} = 1$ K, $h_{oe} = 25\ \mu$A/V.

(b) Find $A_I$ and $A_V$ for the emitter follower of part (a).

(c) Find $R_i$ and the necessary $R_L$ so that $A_V = 0.999$.

**Solution:**

(a) From input resistance, $R_i = h_{ie} + A_I R_L$

or, $$A_I R_L = 499 \text{ K}$$

From the equation of current gain $$A_I = \frac{1 + h_{fe}}{1 + h_{oe} R_L}$$

or, $$A_I + A_I h_{oe} R_L = 1 + h_{fe} \qquad (1)$$

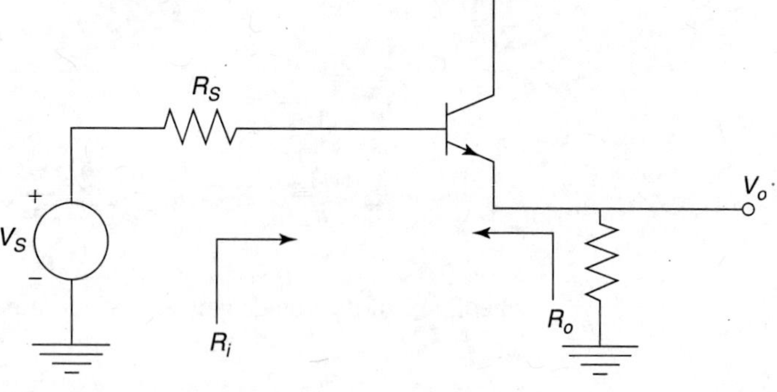


Substituting the given values in Eq. (1) we obtain:

$$A_I + 499 \times 10^3 \times 25 \times 10^{-6} = 1 + 50$$

or, $$A_I = 38.5.$$

Hence $$R_L = \frac{499 \text{ K}}{38.5} = 13 \text{ K}$$

Using of the equation for output conductance we have:

$$Y_0 = 0.05 = 25 \times 10^{-6} + \frac{51}{1000 + R_S}$$

or, $$R_S = 20\ \Omega$$

(b) We found $A_I$ in part (a), $A_I = 38.5$. From the equation for voltage gain we obtain:

$$A_V = 1 - \frac{1 \times 10^3}{0.5 \times 10^6} = 0.998$$

(c) From the equation for voltage gain with $A_V = 0.999$:

$$\frac{h_{ie}}{R_i} = 1 - A_V = 0.001,\ R_i = \frac{1000}{0.001} = 10^6 = 1\ \text{M}\Omega$$

From the equation for input resistance $R_i = h_{ie} + A_I R_L = 10^6$

For $h_{ie} = 10^3, A_I R_L \approx 10^6 = \dfrac{1 + h_{fe}}{1 + h_{oe} R_L}$

$10^6 + 25R_L = 51\ R_L;\ R_L = 38.4\ \text{K}$

**Example 5-10** For the transistor circuit of a transistor in hybrid mode show that:

(a) $A_{IS(\max)} = -h_f\ (\text{if } R_L = 0 \text{ and } R_s = \infty)$

(b) $R_i = h_i\ (\text{if } R_L = 0)$

(c) $R_i = \dfrac{h_i h_o - h_r h_f}{h_o},\ (\text{if } R_L = \infty)$

(d) $(A_{VS})_{\max} = \dfrac{-h_f}{h_i h_o - h_r h_f}\ (\text{if } R_L = \infty \text{ and } R_s = 0)$

(e) $R_0 = \dfrac{h_i}{h_i h_o - h_r h_f}\ (\text{if } R_s = 0)$

(f) $Y_0 = h_o\ (\text{if } R_S = \infty)$

**Solution:**

(a) For current gain, taking into account the source resistance:

$$A_{IS} = A_I \frac{R_S}{R_i + R_S} = -\frac{h_f}{1 + h_o R_L} \times \frac{R_S}{R_S + h_i + h_r A_I R_L} = -\frac{h_f R_S}{R_S + h_i}$$

Since $R_L = 0,\ \lim\limits_{x \to \infty} A_{IS} = -h_f = A_{IS(\max)}$

(b) For input impedance, $R_i = h_i + h_r A_I R_L = h_i$ for $(R_L = 0)$

(c) $R_i = h_i - \dfrac{h_f h_r}{\dfrac{1}{R_L} + h_o} = h_i - \dfrac{h_f h_r}{h_o} = \dfrac{h_i h_o - h_f h_r}{h_o}$

(d) $A_{VS} = A_I \dfrac{R_S}{R_i + R_S} \dfrac{R_L}{R_S}$

So, $A_{VS} = \dfrac{-h_f}{\dfrac{1}{R_L} + h_o} \times \dfrac{1}{h_i - \dfrac{h_r h_f}{\dfrac{1}{R_L} + h_o}}$ since $R_S = 0$ but for $R_L = \infty$

Hence $A_{VS} = \dfrac{-h_f}{h_i h_o - h_r h_f}$

(e) Also $Y_0 = h_o - \dfrac{h_f h_r}{h_i} + R_S = \dfrac{h_i h_o - h_r h_f}{h_i}$ since $R_S = 0$ or $R_0 = \dfrac{h_f}{h_i h_o - h_r h_f}$

(f) Using previous result, if $R_S = \infty$, $Y_0 = h_o$ ■

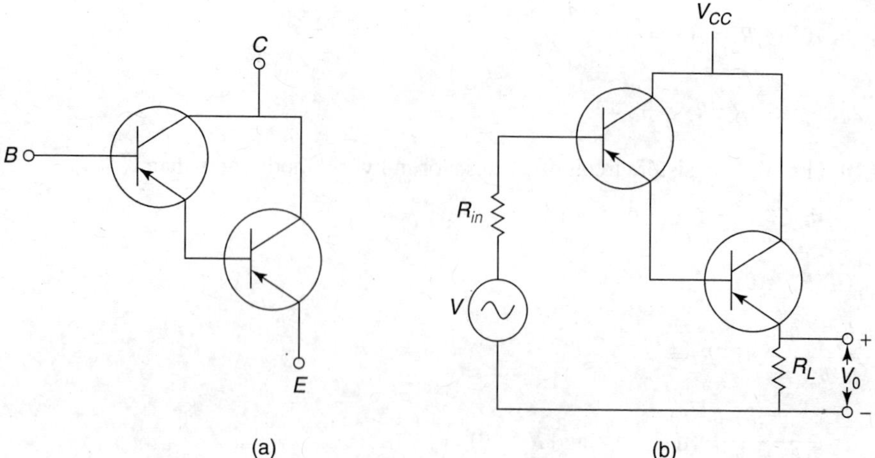


**Figure 5-21** Darlington pair: (a) basic circuit (b) darlington amplifier

## 5-11 DARLINGTON PAIR

The Darlington pair is achieved by connecting the collectors of the transistors where, the emitter of one transistor is connected to the base of the other transistor. Figure 5-21 gives the essence of the Darlington connection.

Input resistance of the second transistor is $(h_{ie} + h_{fe}R_L)$. This is the load resistance of the first stage. Overall current gain is the product of the current gains of the two transistors. As a result, the overall current gain is very high. This is the prime advantage of this type of connection. The effective input resistance of the Darlington connection is higher and the output value is lower. In digital circuits such as the "totem-pole" connection, often the diode following the emitter of a transistor at the output stage can be replaced by an arrangement of a Darlington pair. This helps in increasing the current gain and also the input impedance—the two desired features.

## 5-12 TRANSISTOR AT HIGH FREQUENCIES

Till now we have considered the fact that the carriers responded at once to changes in the input signal. Also, we had neglected the junction capacitance of the transistor for simplicity. These were the approximations that we made for the transistor at low frequencies. But at higher frequencies these capacitances have profound effect and must be included in the analysis of the circuits. Also, at higher frequencies, the time delay between the emitter and the collector currents introduced by the movement of the charges is important and has to be taken into account. The time delay that occurs causes a phase shift between the emitter and the collector currents. Further, the transit time is different in many cases which, in turn, increases such distortion.

At high frequencies a significant amount of current carriers injected into the base fail to reach the collector as the polarity of the input signal reverses by the time the carriers reach. Consequently, a number of carriers can get trapped in this process and take an enormously long time to reach and get cleared from the base region. As the carriers remain in the base region for such a long time, a considerable fraction of current is lost. The current gains of the transistors decrease. This puts an upper limit to the proper operation of the device and results in it behaving in an unexpected manner. Also, there is an increase in the noise which has to be well taken care of.

## 5-13 REAL-LIFE APPLICATIONS OF THE TRANSISTOR

Transistor circuits are the basic component of any convenient electronics circuit. They are widely used in the case of signal amplification, switching, microprocessor and microcontroller designing, etc. These are the most common applications of transistors that go a long way in building powerful microprocessors and other appliances. Microprocessors are the real brains of most electronic appliances around us and a vital role is played by their grey cells—the transistors. The full potential of these transistors is realized through the various appliances they revolutionized, from the simplest of digital watches to the most complex computer hardware.

The first devices that were invented as a direct result of the transistor were phone receivers and broadcasters. Without phones, phone lines would never have come into existence. This, in turn, would have rendered impossible the invention of the Internet or the e-mail or the fax. The lack of signal receivers in phones meant that there would have never been any cell phones that today play a large role in the working world. The ability of a transistor to act as a semiconductor or as an insulator helped solve many problems that inventors had to face while working with the various crystal elements.

Of all the military appliances, the handheld radio was probably the most frequently used when troops went out to the battlefield. It was necessary for them to remain in contact with their commanders, with the base and with the other members of their troop. The transistor was instrumental in the development of this radio. Because of its broadcasting and receiving qualities, the transistor was used in the same manner as it is now used in telephones. In the hi-tech world that we live in today, the global positioning system, better known as GPS, would not have existed were it not for the transistor. The GPS is used extensively in the military to track their naval and air forces, and provides assistance in air-traffic control. In addition to this, the GPS also assists in mapping better routes for the different types of road transport. This technology is now also used in expensive luxury cars. Speaking of cars, car ignition would not have been possible were it not for the invention of the transistor. It's all well and good to build a car, but the most important thing is that it works, and to do that, you have to start it!

The world of computers is probably the largest and most expansive area that the invention of the transistor has graced. There are so many different sub-sections to this one, but we will name a few. A computer is basically made up of only transistors. Computers use small devices like microprocessors and integrated circuits. These hold millions of transistors, and in fact, an entry-level Intel chip (microprocessor) consists of about 10 million transistors, so you can just imagine how many there are in the entire computer. From graphics cards that run the latest games to 800 W speakers, anything that has a circuit has transistors.

Another huge form of entertainment is the radio. Hi-fi's, radios and car radios would definitely not be in existence were it not for the transistor. In fact, one of the first transistorized inventions—the fully transistorized radio—was released in 1954, just a year after the first transistor device—the hearing aid—was invented. We all realise how much the radio is used, and not just for entertainment. Undoubtedly this was an invention much needed for the progress of mankind.

The press, the most widely used media format, would also not be in existence, because the mechanical devices that are used today (computers, printing machines, etc.) would not have been invented. We would be getting a weekly or monthly newspaper telling us what happened two weeks earlier, and even this would not be possible due to the lack of machinery to print this. And finally, one of the most widely used tools today, the Internet, would not have existed for obvious reasons.

What all this tells you, and we are sure you are aware, is that without the invention of the transistor basically everything that contains a circuit would either not be in existence, or they would definitely not be as developed as they are today. Computers, probably the most important device used in the world today, would not be around and it could have been very detrimental to the development of our world. As you can see, behind every electronic device that we use is a tiny, microscopic semiconductor device—the transistor.

## POINTS TO REMEMBER

1. Biasing is the establishment of suitable dc values of different current and voltages of a transistor by connecting it to an external voltage source through suitable circuits.
2. Proper biasing renders a desired operating point in the operating region of the device; either in the active or in the saturation or in the cut-off region.
3. The selection of the $Q$-point depends on (a) the amplitude of the signal to be handled by the amplifier, (b) the load to which the amplifier is to work, (c) the available supply potentials, and (d) the allowable distortion in the signal.
4. The $Q$-point of the transistor should be stable irrespective of changes in temperature or transistor characteristics.
5. The operating point mainly shifts due to changes in temperature. The temperature depends on the $\beta$, $I_{CO}$ and $V$. $\beta$ increases with increase in temperature. $|V_{BE}|$ decreases about 7.5 mV per degree Celsius increase in temperature. $I_{CO}$, which is the reverse saturation current, doubles in value for every 10 degree Celsius increase in temperature.
6. Stability factor determines the stability of the collector current due to the changes in $\beta$, $I_{CO}$ and $V_{BE}$
7. The $h$-parameter gives a two-port model of a transistor.
8. Gain decreases in the low-frequency region because of parasitic capacitances of the network and frequency dependence of the gain of the transistor.
9. In an emitter follower circuit, the output at the emitter follows the input signal and thus, the name.
10. Darlington pair is a special type of connection where the collectors of the transistors are connected together and the emitter of one is connected to the base of the other.
11. Input resistance of Darlington pair circuit increases and the overall gain of the system is high.
12. Typical $h$-parameters of a transistor (at $I_E$= 1.3 mA).

| **Parameters** | **CE** | **CC** | **CB** |
|---|---|---|---|
| $H_{11} = h_i$ | 1,100 Ω | 1,100 Ω | 21.6 Ω |
| $H_{12} = h_r$ | 2.5 × 10⁻⁴ | ~1 | 2.9 × 10⁻⁴ |
| $H_{21} = h_f$ | 50 | −51 | −0.98 |
| $H_{22} = h_o$ | 24 μA/V | 25 μA/V | 0.49 μA/V |
| $1/h_o$ | 40 kΩ | 40 kΩ | 2.04 MΩ |

## IMPORTANT FORMULAE

1. The value of the stability factor when the variation is taken w.r.t $I_{CO}$:

$$S \equiv \frac{\delta I_C}{\delta I_{CO}}$$

2. $S = (1+\beta)\dfrac{1 + R_b/R_e}{1 + \beta + R_b/R_e}$

3. For constant $\beta$, $V_{BE}$, and small $S$, we have:

$$\frac{\Delta I_c}{I_c} \approx S\frac{\Delta I_{co}}{I_c} \approx \frac{\Delta I_{co}}{I_c} + \frac{R_b}{R_e}\frac{\Delta I_{co}}{I_c}.$$

4. $S' \equiv \frac{\delta I_c}{\delta V_{BE}} \approx \frac{\Delta I_c}{\Delta V_{BE}}$

5. The collector current is given by:

$$I_c = \frac{\beta(V + V' - V_{BE})}{R_b + R_e(1+\beta)}$$

6. The variation of the collector current w.r.t $\beta$ is given by:

$$I_c = \frac{\beta(V + V' - V_{BE})}{R_b + R_e(1+\beta)}$$

7. By taking the total differential of $I_C = f(I_{CO}, V_{BE}, \beta)$, we obtain:

$$\Delta I_c = \frac{\delta I_c}{\delta I_{co}}\Delta I_{co} + \frac{\delta I_c}{\delta V_{BE}}\Delta V_{BE} + \frac{\delta I_c}{\delta \beta}\Delta\beta$$

8. The current gain of a transistor amplifier using $h$-parameters is given by:

$$A_I = -\frac{I_2}{I_1} = -\frac{h_f}{1 + h_o R_L}$$

9. The voltage gain is given by:

$$A_V = -\frac{h_f R_L}{h_i + \Delta h R_L}$$

where, $\Delta h = h_i h_o - h_f h_r$

10. The input resistance of the amplifier is given by:

$$R_I = h_i - \frac{h_f h_r R_L}{1 + h_o R_L}$$

11. The output resistance is given by:

$$R_o = \frac{V_2}{I_2} = \frac{R_g + h_i}{R_g h_o + h_i h_o - h_f h_r}$$

12. For emitter follower:

(a) Current gain, $A_i = -\frac{I_e}{I_b} = \frac{1 + h_{fe}}{1 + h_{oe} R_L}$

(b) Input resistance:

$$R_i = \frac{V_i}{I_b} = h_{ie} + A_i R_L$$

(c) Voltage gain:

$$A_V = \frac{V_L}{V_i} = A_i R_L / R_i$$

From these expressions of current gain, and also by making approximations that $h_{fe} \gg 1$ and $h_{oe} \approx 0$, $A_v = h_{fe} R_L / h_{ie} + h_{fe} R_L$

(d) Output resistance:

$$R_o = R_G + h_{ie}/(1 + R_G h_{oe} + h_{ie} h_{oe} + h_{fe})$$

## OBJECTIVE QUESTIONS

1. For good stabilized biasing of the transistor of the CE amplifier of figure. We should have:
   (a) $\frac{R_E}{R_B} \ll 1$ (b) $\frac{R_E}{R_B} \gg 1$
   (c) $\frac{R_E}{R_B} \ll h_{rb}$ (d) $\frac{R_E}{R_B} \gg h_{rb}$

2. Current stability of a CC amplifier can be increased by:
   (a) Reducing both emitter and base resistance
   (b) Increasing both emitter and base resistance
   (c) Reducing emitter resistance and increasing base resistance
   (d) Increasing emitter resistance and decreasing base resistance

3. Which of the following statements are correct for basic transistor amplifier configuration?
   (a) CB amplifier has low input impedance and a low current gain
   (b) CC amplifier has low output impedance and a low current gain
   (c) CE amplifier has very poor voltage gain but very high input impedance
   (d) The current gain of CB amplifier is higher than the current gain OF CC amplifier

4. Which of the following configuration ifs normally used in cascading?
   (a) Common-emitter configuration
   (b) Common-base configuration
   (c) Common-collector configuration
   (d) None of the above

5. A transistor has $h_{fe} = 27$, then its $h_{fe}$ will be:

   (a) $-0.96$ (b) $0.96$

   (c) $-27$ (d) $-28$

6. In the BJT amplifier, the transistor is biased in the forward active region putting a capacitor across $R_E$ will:

   (a) Decrease the voltage gain and decrease the input impedance

   (b) Increase the voltage gain and decrease the input impedance

   (c) Decrease the voltage gain and increase the input impedance

   (d) Increase the voltage gain and increase the input impedance

7. In a common emitter BJT amplifier, the maximum usable supply voltage is limited by:

   (a) Avalanche breakdown of base–emitter junction

   (b) Collector–base breakdown voltage with emitter open (BVCBO)

   (c) Collector–emitter breakdown voltage with base open (BVCEO).

   (d) Zener breakdown voltage of the emitter–base junction

8. The transconductance $g_m$ is defined as $g_m = \partial i_o / \partial V_{BE}$. Its value in terms of $h$-parameters is:

   (a) $\dfrac{h_{ie}}{h_{fe}}$ (b) $\dfrac{h_{fe}}{25}$

   (c) $\dfrac{h_{te}}{h_{ie}}$ (d) None of the above

9. Introducing a resistor in the emitter of a common emitter amplifier stabilizes the dc operating point against variations in:

   (a) Only the temperature

   (b) Only the $\beta$ of the transistor

   (c) Both temperature and $\beta$

   (d) None of the above

10. In a CE transistor amplifier, if collector–emitter voltage increases the instantaneous operating point:

    (a) Moves up the load line

    (b) Moves down the load line

    (c) Moves at right angle to the load line

    (d) Remains stationary

11. The $h$-parameter equivalent circuit of a junction transistor is valid for:

    (a) High-frequency, large-signal operation

    (b) High-frequency, small-signal operation

    (c) Low-frequency, small-signal operation

    (d) Low-frequency, large-signal operation

12. In a transistor $h_{fe} = 50$, $h_{ie} = 830$ ohms, $h_{oe} = 10^{-4}$ mho. When used as in the CB mode, then its output resistance will be:

    (a) 2 mohms (b) 500 K

    (c) 2.5 mohms (d) 780 K

13. The small-signal input impedance of a transistor when the output is shorted for the measuring signal, is: (where the symbols have then usual measuring)

    (a) $h_{11} = \left.\dfrac{V_1}{I_1}\right|_{V_2=0}$ (b) $h_{12} = \left.\dfrac{V_1}{i_1}\right|_{V=0}$

    (c) $h_{21} = \left.\dfrac{i_2}{i_1}\right|_{V_2=0}$ (d) $h_{22} = \left.\dfrac{i_2}{i_1}\right|_{V_1=0}$

14. For obtaining hybrid parameters of a transistor:

    (a) Variable $V_{be}$ and $I_e$ are taken as independent variables

    (b) The two independent variables are the ones that are most easily measurable for a CE configuration

    (c) Variable $i_b$ and $v_{ce}$ ie are taken as dependent variables

    (d) Variable $v_{be}$ and $i_e$ are taken as dependent variables

15. The condition necessary to calculate $h_{oe}$ of a transistor:
    (a) DC base current is to be zero
    (b) Base to emitter voltage is to be constant
    (c) Collector current is to be constant
    (d) Base current is to be constant

16. The approximate value of input impedance of a common-emitter amplifier with emitter resistance $R_e$ is given by:
    (a) $h_{ie} + A_1 R_e$ (b) $h_{ie} + (1 + h_{fe}) R_e$
    (c) $h_{ie}$ (d) $(1 + h_{fe}) R_e$

## REVIEW QUESTIONS

1. What do you mean by the quiescent point of a transistor?
2. Draw the circuit diagram of a common emitter transistor amplifier and explain its operation graphically.
3. What do you mean by the term load line? Explain its significance.
4. What do you mean by distortion in amplifiers? Discuss the origin of the input and outputnon-linearities in transistor amplifiers. What is the desired position of the $Q$-point for minimum distortion and why?
5. Explain the term transistor biasing. What are the factors affecting the position of $Q$-points?
6. Is the operating point of the transistor fixed? If not, what are the factors responsible for its change or shift?
7. What are the factors responsible for affecting the bias stability of a transistor?
8. Define the stability factors with respect to the changes in $I_{CO}$, $V_{BE}$ and $\beta$.
9. Why is stability with respect to changes in $V_{CE}$ not considered?
10. What is thermal runaway?
11. Draw the circuit for fixed bias by considering an $n$–$p$–$n$ transistor in the CE mode. Derive the expressions for stability factors. What are the functions of the coupling capacitors?
12. Draw the circuit diagram for the collector to base arrangement for an $n$–$p$–$n$ transistor. Obtain the stability factors and mention the demerits of the circuit.
13. What is self bias? Draw the circuit showing self bias of an $n$–$p$–$n$ transistor in the CE mode. Explain physically how the self bias improves stability.
14. What is bias curve? How is the $Q$-point in a self bias circuit determined with and without the help of bias curve?
15. Derive the expressions for stability factors in the case of self bias of a CE mode transistor.
16. How is the transistor represented as a two-port device?
17. Define the hybrid parameters for a basic transistor circuit in any configuration and give its hybrid model.
18. Draw the low-frequency $h$-equivalent circuit of a transistor amplifier operating in the CE mode. Why is this circuit not valid for high frequencies?
19. Draw and label the circuit diagram of a small-signal single stage low-frequency transistor amplifier in the CE mode. Using $h$-parameters obtain expression for current gain, input impedance, voltage gain and output impedance.
20. Show that the power gain in the above problem is the product of the current gain and the voltage gain.
21. Draw the circuit diagram of an emitter follower. Why is the circuit called so? Obtain the expressions for current gain, voltage gain, input impedance and output impedance.
22. Draw the approximate hybrid model for any transistor configuration at low frequencies. Show that only $h_{ie}$ and $h_{fe}$ are important in the model. Is the approximation justified?

23. What is Darlington pair? Compare between Darlington pair and emitter follower.

24. Explain how the model of transistor changes in high frequency. What are the prime reasons for such a change?

## PRACTICE PROBLEMS

1. The transistor, as shown in the following diagram has a very high $\beta$.

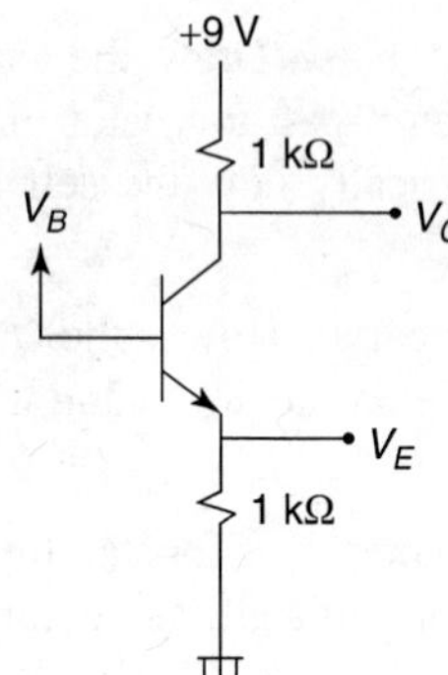


Find $V_E$ and $V_C$ for $V_B$ equal to:

(a) 3 V

(b) 1 V

(c) 0 V

2. The transistor, as shown in Problem 1, has a very high $\beta$. Find the highest value of $V_B$ for which the transistor still operates in the active mode.

3. Consider the circuit, as shown in the following diagram with the base voltage $V_B$ obtained using a voltage divider across the 9 V supply.

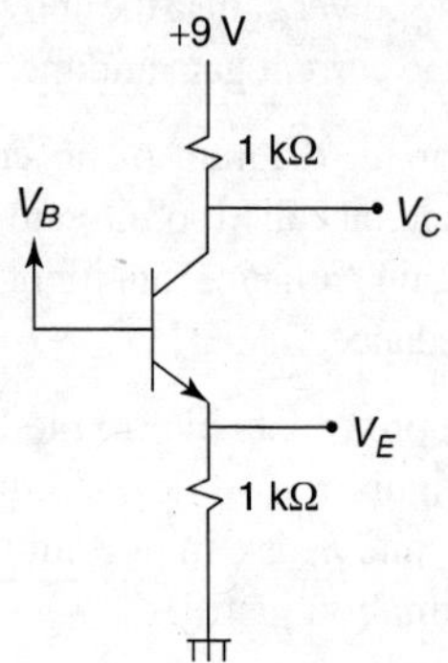


Assuming that the transistor has a very high value of $\beta$, design the voltage divider to obtain $V_B = 3$ V. Design for a 0.2 mA current in the voltage divider.

4. For the circuits, as shown in the following diagrams, find the values of the labeled node voltages and branch currents.

(a)

(b)

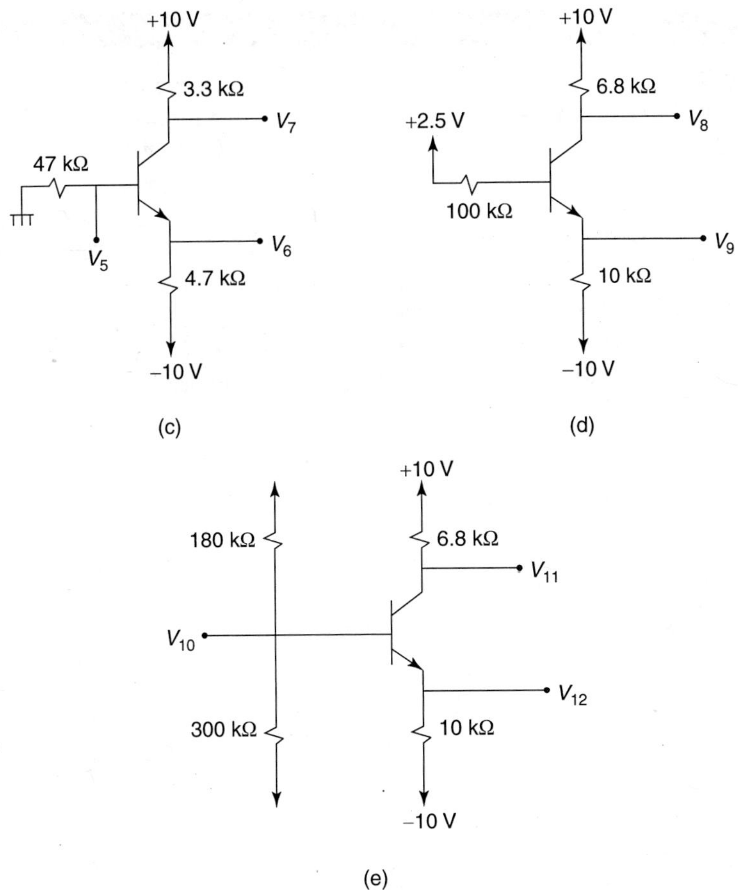


5. Repeat the analysis of Problem 4, considering $\beta = 100$. Assume $V_{BE} = 0.7$ V.

6. The *n–p–n* transistor in the circuit of the diagram has $\beta = 50$. Find the value of $R_C$ required to obtain $V_C = 5$ V. What happens if the transistor is replaced with another having $\beta = 100$?

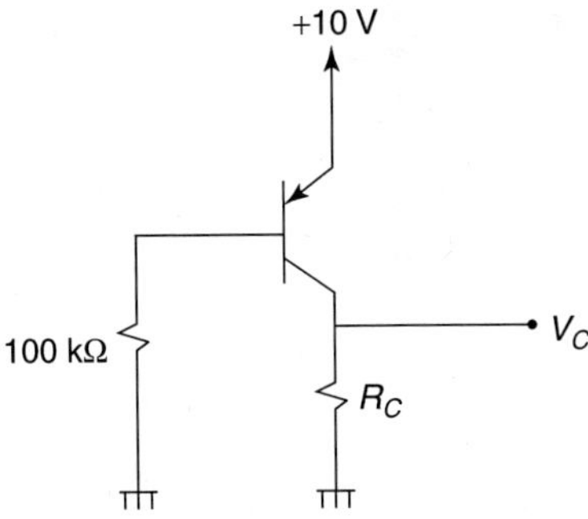

## SUGGESTED READINGS

1. Grove, A. S. 1967. *Physics and Technology of Semiconductor Devices*. New York: Wiley.
2. Sze, S. M. 1981. *Physics of Semiconductor Devices*. New York: Wiley.
3. Singh, J. 1994. *Semiconductor Devices: An Introduction*. New York, NY: McGraw-Hill.
4. Streetman, B. G. and S. Banerjee. 2000. *Solid State Electronic Devices*. New Delhi: Pearson Education.
5. Millman, Jacob and Christos C. Halkias. 1986. *Integrated Electronics: Analog and Digital Circuits and Systems*. New Delhi: McGraw Hill Book Company.
6. Pierret, R. F. and G.W. Neudeck. 1989. *Modular Series on Solid State Devices* Boston, MA: Addison Wesley.
7. Singh, B. P. and Rekha Singh. 2006. *Electronic Devices and Integrated Circuits*. New Delhi: Pearson Education.

# 6

# Field-Effect Transistor

## Outline



## Objectives

In this chapter the junction field-effect transistor (JFET) is introduced and its practical uses, construction, biasing and characteristics are emphasized in an elaborate manner. In addition to this construction of the metal-oxide semiconductor (MOS), the structural characteristics and biasing of MOS field-effect transistor (MOSFET) are also discussed in detail. The last portion of the chapter deals with the various complimentary metal-oxide semiconductor (CMOS) circuits.

## 6-1 INTRODUCTION

The invention of the BJT has brought a great twist in the modern era of semiconductor technology. This device, along with its field-effect counterpart, known as the field-effect transistor (FET), has had a huge impact on virtually every area of modern life. Practical field-effect transistors were first made in the form of JFET in 1953 and MOSFET in 1963.

The field-effect transistor has taken various forms like that of the junction field-effect transistor (JFET), in which the gate voltage controls the depletion width of a reverse-biased *p–n* junction; the metal-semiconductor field-effect transistor (MESFET), in which the junction is replaced by a Schottky barrier; the metal-insulator-semiconductor field-effect transistor (MISFET), where the metal gate electrode is separated from the semiconductor by an insulator; and the metal-oxide-semiconductor field-effect transistor (MOSFET), which is the most common field-effect transistor in both digital and analog circuits. In this chapter, we intend to study and examine the operation of these devices, and also look into some derived circuit configurations obtained from a combination of these devices.

## 6-2 THE FIELD-EFFECT TRANSISTOR

The FET is a single carrier device and is often called the unipolar transistor because the carriers involved in the operation are either electrons or holes. The FET is also a semiconductor device in which the output quantity is controlled by an electric field, which is often the input quantity, i.e., these devices work on the principle that a voltage, when applied to the metallic plate, modulates the conductance of the underlying semiconductor which, in turn, controls the current flowing through the device. The phenomenon where the conductivity of the semiconductor is modulated by an electric field applied normally to the surface of the semiconductor is called *field effect*, and this principle is brought into operation by extending the depletion region deep into the bulk of the semiconductor.

### 6-2-1 Junction Field-Effect Transistor (JFET)

In a junction FET, the control voltage modulates the depletion width of a reverse-biased *p–n* junction which, in turn, varies with the various parameters of the device.

### 6-2-2 Insulated Gate Field-Effect Transistor (IGFET)

The IGFET is also called the metal-oxide-semiconductor field-effect transistor (MOSFET). In this, the metal gate electrode is separated from the semiconductor by an insulator.

### 6-2-3 Metal-Semiconductor Field-Effect Transistor (MESFET)

If the MOS junction is replaced by a direct metal-semiconductor contact, i.e., a Schottky barrier, it is called metal-semiconductor FET (MESFET). This construction enjoys a faster operation compared to the previous structures. The first successful MESFET unit was built by Carver Mead in 1966. A MESFET is similar to a JFET except for the following differences:

(i) It has a single gate
(ii) The gate is formed by a metal-semiconductor junction

In the early day's of the semiconductors, the BJT was of prime interest to the scientists and devices designers, but gradually, its position has been taken up by field-effect devices. The primary reasons for this are discussed in the comparison shown in Table 6-1.

**Table 6-1** Comparison between the BJT and the FET

| *BJT* | *FET* |
|---|---|
| 1. Two types of carriers (electrons and holes) are required. | 1. Only one type of carrier (electron or hole) is required. |
| 2. Carriers move through the base by diffusion process. | 2. Carriers move through the channel by drift process. |
| 3. The BJT has a comparatively lower switching speed due to the diffusion process. | 3. The FET has a higher switching speed due to the drift process; the drift of the carrier is faster than diffusion. |

| | |
|---|---|
| 4. The BJT is not a thermally stable device. | 4. The FET has a negative temperature coefficient at high-current operations, i.e., the current decreases as temperature increases Due to this particular feature, a uniform temperature distribution and protection against breakdown can be achieved. |
| 5. In case of IC fabrication, the BJT requires more space than the FET. | 5. In case of IC fabrication the FET requires lesser space than the BJT. |
| 6. At audio frequencies the BJT offers less power gain. | 6. At audio frequencies the FET offers greater power gain. |
| 7. The BJT is a current-controlled device. | 7. The FET is a voltage-controlled device. |
| 8. The BJT offers low input impedance. | 8. The FET offers high input impedance, therefore, it can be used as a buffer. |
| 9. BJT is much noisier than FET. | 9. FET is less noisy. |
| 10. The BJT has offset voltage. | 10. The FET has no offset voltage. |
| 11. The BJT can also be used as a switch; it is taken to be in the OFF state when operating in the cut off region, and in the ON state when it is operating in the saturation region. | 11. The FET is particularly useful for its operation as a controlled switch, operating in both the conducting and the non-conducting zones. |

## 6-3 CONSTRUCTION OF THE JFET

The JFET is a three-terminal device whose one terminal is capable of controlling the current between the other two. In JFETs, the width of a junction is used to control the effective cross-sectional area of the channel that conducts current. JFETs are basically of two types: *n*-channel and *p*-channel. Figure 6-1 shows the basic construction of the *n*-channel JFET.

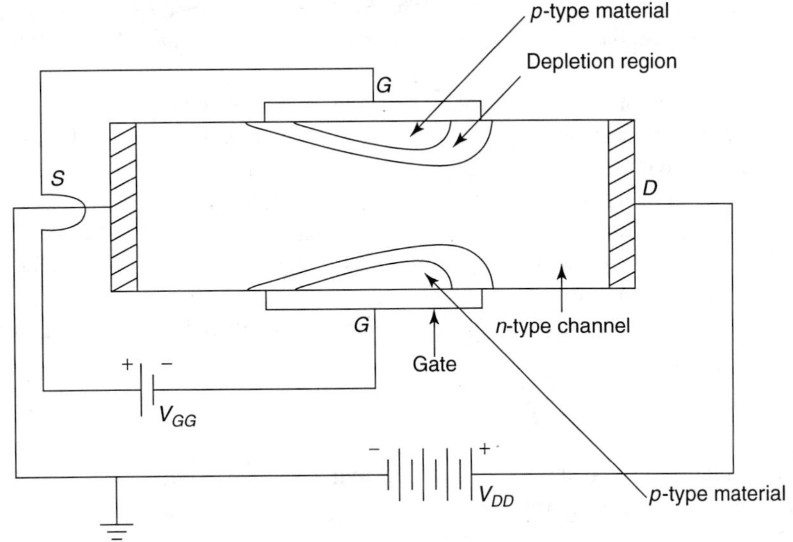


**Figure 6-1** Construction of *n*-channel JFET

The $n$-channel JFET consists of a uniformly doped $n$-type silicon semiconductor bar with ohmic contacts at both ends and semiconductor junctions made on either sides of the bar. The top portion of the $n$-type channel is connected through the ohmic contact to a terminal called the drain ($D$) while the lower end is connected to the terminal referred to as the source ($S$). The two $p$-type materials, fabricated on the two sides, are connected together and then to the third terminal called gate ($G$). The source terminal gets its name from the fact that the carriers contributing to the current flow move out from the external circuit into the semiconductor at this electrode. The carriers travel through the bulk of the semiconductor and are subsequently collected at the drain electrode. The gate is called so because it controls the flow of charges though the bulk.

With proper biasing of the device, current is allowed to flow from the source and gets collected at the drain terminal of the bar. As the drain current ($I_D$) flows through the channel, a reverse-bias between the two $p$-regions and the channel reduces the effective width of the channel. This reduction is mainly an effect of the intrusion of the depletion region deep into the semiconductor channel. As the effective width of the conducting channel has a striking effect on the resistance of the channel, the current flowing through it also varies correspondingly.

The electrons in the $n$-region move from the source to the drain region; they are carried by majority carriers that drift through the channel. The majority carriers enter the channel region through the source terminal and leave the channel through the drain terminal, in agreement with the naming of the drain and the source terminals.

In the absence of any applied potentials at the two $p$–$n$ junctions, the JFET is under no-bias condition. The result is a depletion region at each junction that resembles the same region of a junction diode under no-bias conditions. A minor current starts to flow through the sandwiched layer of the channel between the slightly extending depletion regions of the two $p$–$n$ junctions by the sides of the $n$-type semiconductor bar. Also, under bias the depletion region attains a particular shape, i.e., its width is more towards the drain terminal and keeps on decreasing at a steady rate as we move towards the source terminal. The reason for this is that the channel can be considered to be a series of resistors. For normal operation, the current enters the channel through the drain and as it flows through the channel the voltage drop constantly decreases. This can be a linear variation for low values of current.

The device is basically a voltage-controlled resistor and its resistance can be changed by varying the width of the depletion layer extending into the channel region, i.e., by constricting the cross-sectional area of the channel for the carriers to flow through it. The FETs have considerably higher input impedance as compared to BJTs. This allows the input of an FET to be more suitably matched to the standard microwave systems.

It is evident that the control voltage is applied to a reverse-biased junction. So, the FET has a negative temperature coefficient at high current levels, i.e., the current decreases with an increase in temperature. This characteristic leads to a more uniform temperature distribution over the device area and prevents the FET from thermal runaway, which is a major point of concern in bipolar transistors. The device is thermally stable, even when the active area is large and when many such devices are connected in parallel. Because FETs are unipolar devices, they hardly suffer from any minority-shortage effects, and thus, have higher switching speeds and higher cut off frequency. This makes their operation rather smooth. In addition to this they are square-law devices, i.e., inter-modulation and cross-modulation products are much smaller than those of a bipolar transistors.

## 6-4 BIASING OF THE JFET

Let us now study the operation of the device with respect to the characteristics of an $n$-channel JFET. We shall consider two different cases in this regard.

**When $V_{GS}$ = 0 V, $V_{DS}$ = some positive value:** In Fig. 6-2, a positive voltage $V_{DS}$ has been applied across the channel and the gate has been connected directly to the source to establish the condition $V_{GS} = 0$ V.

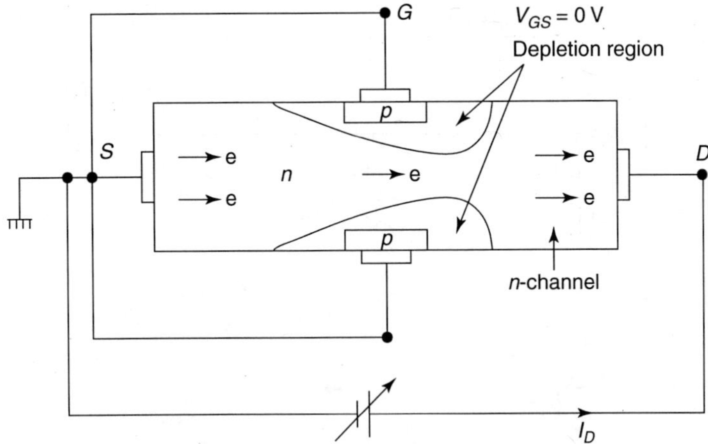


**Figure 6-2** State of the device with zero source–gate voltage and positive drain–source voltage

As a result, the gate and source terminals are at the same potential, and a depletion region is formed in the low end of each *p*-type material, i.e., at the sides of the channel. Again, when $V_{DS} = 0$, the only thing one observes is that the device is in a thermal equilibrium. Above all, the small depletion region of the $n$–$p^+$ region extends into the lightly doped *n*-type semiconductor similar to the depletion region extending in an unbiased *p*–*n* junction diode. The moment, the voltage $V_{DD}$ (= $V_{DS}$) is applied, the electrons will be drawn from the source to the drain terminal establishing the conventional current $I_D$ with the direction as shown in Fig. 6-2. The path of the charge flow and thus, the un-depleted channel clearly indicates that the drain and the source currents are almost equal ($I_D = I_S$). The flow of charge is relatively uninhibited, and limited totally by the resistance of the *n*-channel between drain and source. This resistance, in the case of JFETs, corresponds to the intrusion of the depletion region in the channel. The depletion region is wider near the top of both *p*-type materials.

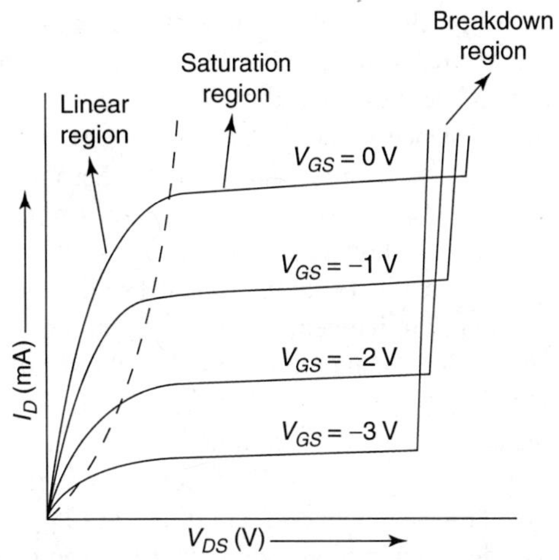


**Figure 6-3** Drain current vs. drain–source voltage: Static characteristics of an *n*-channel JFET

As this condition persists, the top and bottom of the depletion region widens going down the channel from the source to the drain. Now, as the voltage $V_{DS}$ is increased from 0 to a few volts, the current will increase—as determined by Ohm's law—and the plot of $I_D$ versus $V_{DS}$ will appear as shown in Fig. 6-3.

The device in this stage goes into a completely new phase of operation. The relative straightness of the plot indicates that for the region of low values of $V_{DS}$, the resistance of the channel is almost kept constant. The $p+$ sides of the $p+$–$n$ junctions are nearly at zero voltage. Consequently, the bias applied to the drain leads to a reverse-biasing of the gate junctions and increase in the depletion widths. As $V_{DS}$ increases and approaches a level—referred to as $V_P$—the depletion regions will cause a considerable reduction in the effective channel width. The reduced path of conduction causes the resistance to increase and the curve takes the shape as shown in Fig. 6-3. The more horizontal the curve, the higher is the resistance, suggesting that the resistance is approaching "infinite" ohms in the

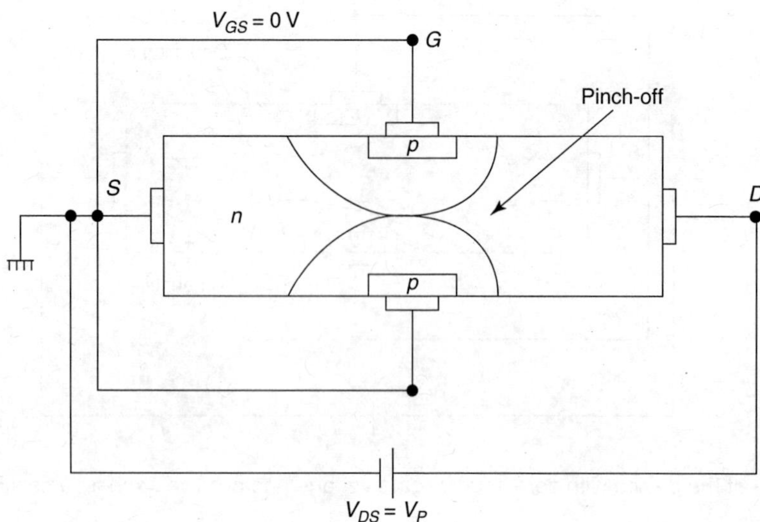


**Figure 6-4** FET in Pinch-off condition

horizontal region. If $V_{DS}$ is increased to a level where the two depletion regions touch each other, a condition called pinch-off will result. At the complete depletion of the channel, first the top and the bottom depletion regions touch in the near vicinity of the drain, and as the reverse-bias is further increased, this depletion region keeps on extending throughout the channel. An immediate observation from such a situation is that the resistance of the channel also keeps on increasing. When the channel pinches-off, the slope of the $I_D$ and the $V_D$ curve approximately goes to zero, indicating that the resistance reaches a very high value. The level of $V_{DS}$ that establishes this condition is referred to as pinch-off voltage, and is denoted by $V_P$. The pinch-off condition is shown in Fig. 6-4.

In actuality the term pinch-off suggests that the current $I_D$ is pinched-off and should drop to 0 A. However, this does not happen; rather $I_D$ maintains a saturation level current defined as $I_{DSS}$. The explanation for the above phenomena can be given by the fact that if $I_D$ were zero, there would be no current in the channel, i.e., same as at $V_D = 0$. If the channel potential is zero everywhere, the *p*–*n* junctions would be zero biased, and the channel would be completely open from the source to the drain truly contradicting the initial assumption of the pinched-off condition for a channel. In other words, a current must flow in the JFET to induce and maintain the pinched-off condition. As $V_{DS}$ is increased beyond $V_P$, the region of close encounter between the two depletion regions will increase in length along the channel and the level of $I_D$ will remain the same.

If $V_{DS} > V_P$, the JFET has more or less the characteristic of a current source. The current is fixed at $I_D = I_{DSS}$ but the voltage $V_{DS}$ (for levels $> V_P$) is determined by the applied load. $I_{DSS}$ is the maximum drain current for a JFET and is given by the conditions:

$$V_{GS} = 0 \text{ V} \quad \text{and} \quad V_{DS} > V_P$$

**When $V_{GS} < 0$ V:** The voltage from gate to source, denoted as $V_{GS}$, is the controlling voltage parameter of the JFET, as it modulates the channel width for the carriers to flow through. Now, for the *n*-channel, it should be taken into notice that the device-controlling voltage $V_{GS}$, is made more and more negative from its $V_{GS} = 0$ V level, i.e., the unbiased level. In other words, for an *n*-channel FET, the gate terminal has to be set at lower potential levels compared to the source. In Fig. 6-5 a negative voltage of –1 V has been applied between the gate and the source terminal for a low level of $V_{DS}$.

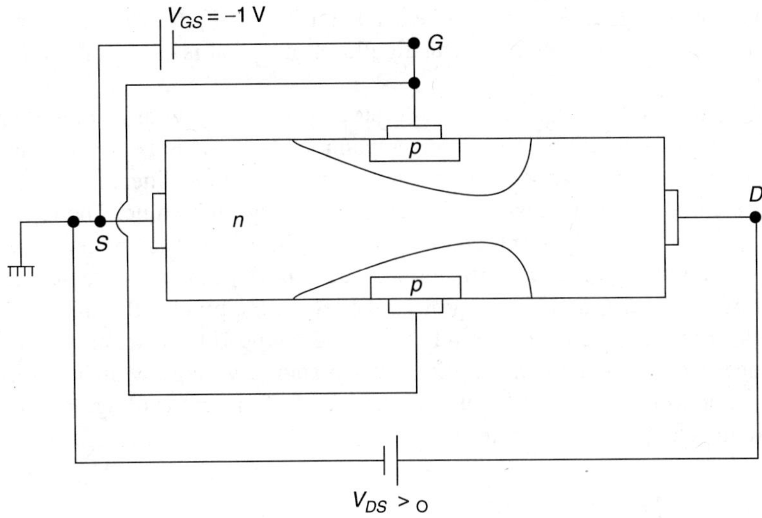


**Figure 6-5** Application of a negative voltage to the gate of a JFET

The effect of the applied negative bias $V_{GS}$ is to form the depletion regions and to intrude into the effective channel width, similar to those obtained with $V_{GS} = 0$ V, but obviously at lower levels of $V_{DS}$. Therefore, the result of applying a negative bias to the gate is to reach the saturation level at a lower level of $V_{DS}$ rather more quickly, as shown in Fig. 6-6.

As is evident, the resulting saturation level for $I_D$ gets reduced and continues to decrease as $V_{GS}$ is made more and more negative. Also, it is noticeable that the fall of the pinch-off voltage follows a parabolic

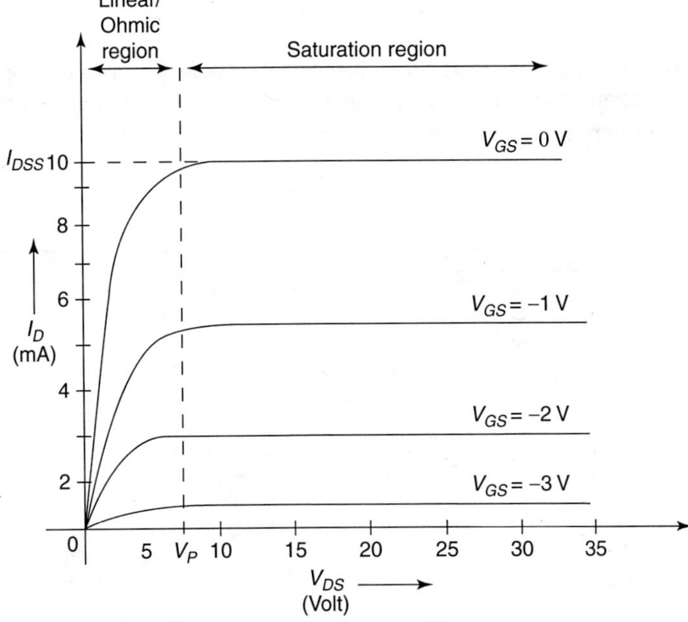


**Figure 6-6** *n*-channel JFET characteristics

nature as $V_{GS}$ becomes more and more negative. Eventually for $V_{GS} = -V_P$, the stage is reached where the gate voltage is sufficiently negative to establish a saturation level of $I_D$ that is essentially 0 mA, and the device will be turned OFF.

The level of $V_{GS}$ that results in $I_D = 0$ mA is defined by $V_{GS} = V_P$, with $V_P$ being a negative voltage for *n*-channel devices and a positive voltage for *p*-channel JFETs. This is the situation where for any further increase of the drain voltage $V_D$, the channel current $I_D$ cannot increase, i.e., there is an ideal pinch-off. The differential channel resistance increases to a very high value. The region to the right of the pinch-off is the region typically employed in linear amplifiers, and is commonly referred to as the constant-current saturation or linear-amplification region, as the device characteristics vary linearly with the variation of the input signal. The region to the left of the pinch-off is referred to as the ohmic or voltage-controlled region or the triode region. In this region the JFET is employed as a variable resistor whose resistance is controlled by the applied gate-to-source voltage, which in turn modulates the channel width. We can now quantitatively analyse the effect of the gate voltage on the intrusion of the depletion region in the channel of the device.

## 6-4-1 Effect of the Gate Voltage

The effect of the negative gate potentials is to increase the resistance of the channel and bring in "pinch-off" for a lower value of channel current. Since the depletion width is higher in the case of negative gate bias, the pinch-off is brought in more quickly. Beyond pinch-off, the drain current is controlled only by the gate voltage, and by suitably varying it we can obtain amplification of an ac signal. The calculation of the pinch-off voltage becomes rather simple if the analysis is made from Fig. 6-7.

The expression for the width of the depletion region ($W$) under a biased condition is given by Eq. (6-1).

$$W = \left[\frac{2\varepsilon(V_0 - V)}{q}\left(\frac{N_a + N_d}{N_a N_d}\right)\right]^{1/2} \tag{6-1}$$

where, $\varepsilon$ is the dielectric constant of the material, $V_0$ is the contact potential, $V$ is the applied voltage, $q$ is the magnitude of electronic charge, $N_a$ is the concentration of the acceptor ions in the *p*-type gate, and $N_d$ is the concentration of the donor ions in the *n*-type bar.

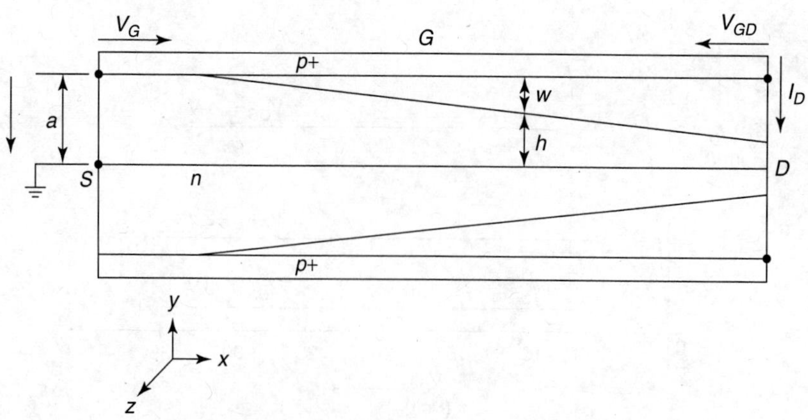


**Figure 6-7** Geometry of a JFET under a biased condition

In Fig. 6-7, our attention is restricted to the channel half-width measured from the centre line. Here, $a$ represents the actual width of the channel under no bias, $h$ is the width of the channel at a distance $x$ from the source terminal and is a function of $x$, i.e., $h(x)$ and $W(x)$ represent the corresponding width of the depletion region at the sides of the channel.

$$W(x) = a - h(x) \tag{6-1a}$$

From this relation giving the width of the depletion region, its value at a distance $L$ away from the source (at the drain terminal) can be calculated as:

$$W(x = L) = \left[\frac{2\varepsilon(-V_{GD})}{qN_D}\right]^{1/2} \tag{6-2}$$

As the width of the depletion region can not be negative, so:

$$W(x = L) = \left[\frac{2\varepsilon|-V_{GD}|}{qN_D}\right]^{1/2} \tag{6-2a}$$

The condition of the pinch-off is: $\quad W(x = L) = a - h(x = L) = 0$ (6-3)

Poisson equation for the depletion region in the $n$-type bar is:

$$\frac{d^2V}{dy^2} = -\frac{q}{\varepsilon} N_D \tag{6-4}$$

The barrier electric field is: $\quad F = -\frac{dV}{dy}$

Integrating the Poisson equation of Eq. (6-4) we get:

or, $$\frac{dV}{dy} = -\frac{qN_D}{\varepsilon}(y - W)$$

or, $$-F = \frac{dV}{dy} = -\frac{qN_D}{\varepsilon}(y - W)$$

or, $$\int_0^V dV = -\frac{qN_D}{\varepsilon}\int_0^y (y - w)dy$$

or, $$V(y) = -\frac{qN_d}{2\varepsilon}(y^2 - 2wy)$$

At, $y = w$, $\quad V(y) = V_p$ = pinch-off voltage; Eq. (6-4) is reduced to:

$$V(w) = -\frac{qN_d}{2\varepsilon}(w^2 - 2ww)$$

$$V_p = -\frac{qN_d}{2\varepsilon}(w^2 - 2w^2)$$

$$V_p = -\frac{qN_d}{2\varepsilon}(-w^2)$$

$$V_p = -\frac{qN_d}{2\varepsilon}(w^2) \tag{6-5}$$

At pinch-off condition $h(x) = 0$; therefore, from Eq. (6-1a), $w = a$. Substituting the value of $w$, Eq. (6-5) is modified as:

$$V_P = \frac{qa^2N_D}{2\varepsilon} \qquad \text{(6-5a)}$$

The pinch-off voltage is a positive number and its relation with the drain and the gate voltage is given by:

$$V_P = -V_{GD}(\text{pinch-off}) = -V_G + V_D \qquad \text{(6-5b)}$$

It is important to note that in the above analysis we assumed that the equilibrium contact voltage is zero.

## 6-5 CURRENT–VOLTAGE CHARACTERISTICS

In this approach to determine the current–voltage relationship, we find the current corresponding to the pinched-off condition of the device. We extend our analysis from Fig. 6-7, where the source end is taken to be the origin and the total length of the channel is $L$. Also, the depth of the channel is taken to be $Z$ along the $z$-axis.

The differential channel volume can be expressed as $2h(x)Zdx$ and the corresponding differential resistance is given by:

$$dR = \frac{\rho dx}{A(x)}$$

where, $A$ is the cross-sectional area.

The formulation of area $A$ can be expressed as:

$$A = \frac{2Zh(x)dx}{dx} = 2Zh(x)$$

The incremental resistance of the channel is controlled in the cross-sectional area $A$, therefore:

$$dR = \frac{\rho dx}{2Zh(x)} \qquad \text{(6-6)}$$

where, $\rho$ is the resistivity of the channel. As the current through the channel does not change with distance, the current can be expressed in the following form:

$$I_D = -\frac{dV_x}{dR}$$

$$I_D = -\frac{dV_x}{\frac{\rho dx}{2Zh(x)}}$$

$$I_D = -\frac{2Zh(x)dV_x}{\rho dx} \qquad \text{(6-7)}$$

where, $-dV_x$ is the differential voltage drop along the channel.

The width of the channel depends on the reverse-bias being applied at that particular point, and thus, the local reverse-bias is given by $-V_{Gx}$ ($x$ is the distance from the source from where the width is calculated).

The effective width is then obtained in the following form:

$$h(x) = a - \left[\frac{2\varepsilon|-V_{Gx}|}{qN_D}\right]^{1/2} \qquad \text{(6-8a)}$$

As inferred from Eq. (6-8a), we finally reach an expression for the effective channel width for a distance in the channel. Using Eq. (6-5a), we get:

$$\frac{2\varepsilon}{qN_D} = \frac{a^2}{V_P} \tag{6-8b}$$

$$V_{Gx} = V_G - V_x \tag{6-8c}$$

Substituting the value of $2\varepsilon/qN_D$ and $V_{Gx}$ from the Eqs. (6-8b) and (6-8c) in Eq. (6-8a) we get a simplified equation of effective width $h(x)$, as expressed by Eq. (6-9).

$$h(x) = \left\{ a - \left[ a^2 \left( \frac{V_x - V_G}{V_P} \right) \right]^{1/2} \right\}$$

$$h(x) = a\left[1 - \left(\frac{V_x - V_G}{V_P}\right)^{1/2}\right] \tag{6-9}$$

Having obtained this expression for $h(x)$ in Eq. (6-9), we can replace $h(x)$ in Eq. (6-7) to find out the current–voltage relationship as follows:

$$\frac{2Za}{\rho}\left[1 - \left(\frac{V_x - V_G}{V_P}\right)^{1/2}\right] dV_x - I_D dx \tag{6-10}$$

Solving Eq. (6-10) yields:

$$I_D = \frac{2aZ}{L\rho}\left[\frac{V_D}{V_P} + \frac{2}{3}\left(\frac{-V_G}{V_P}\right)^{2/3} - \frac{2}{3}\left(\frac{V_D - V_G}{V_P}\right)^{3/2}\right] \tag{6-11}$$

Equation (6-11) illustrates the phenomena up to pinch-off where, $V_D - V_G = V_P$. And, an extension of this gives us the expression for current in the saturation region which is given by:

$$I_D = \frac{2aZ}{L\rho}V_P\left[\frac{V_D}{V_P} + \frac{2}{3}\left(\frac{-V_G}{V_P}\right)^{3/2} - \frac{2}{3}\right] \tag{6-12}$$

where, $V_D/V_P = 1 + (V_G/V_P)$.

Equation (6-12) expresses the characteristic curves, and we observe that the saturation current increases when $V_G$ tends to zero and decreases when the gate voltage becomes negative.

## 6-6 TRANSFER CHARACTERISTICS OF THE JFET

In the BJT, a linear relationship between the output current $I_C$ and the input controlling-current exists. In the JFET, unlike the BJT, there is a non-linear relationship existing between the output and the input quantities. The relationship between $I_D$ and $V_{GS}$ is known as Shockley's equation [Eq. (6-12)].

Equation (6-12) indicates that the relationship between $I_D$ and $V_{GS}$ is non-linear, producing a curve that grows exponentially with decreasing magnitudes of $V_{GS}$, as shown in Fig. 6-8.

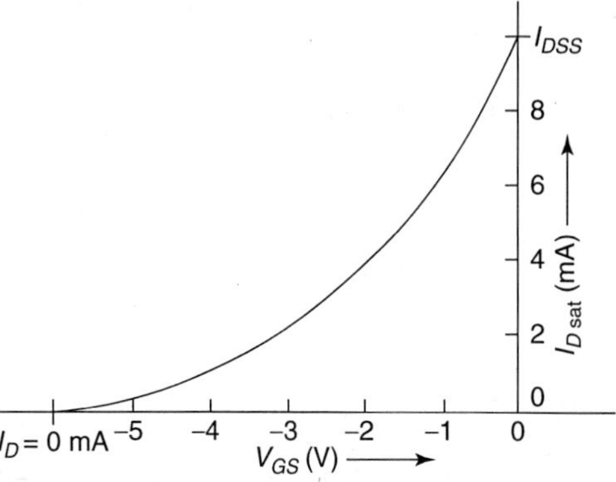


**Figure 6-8** Transfer characteristics for an *n*-channel JFET

The transfer characteristics—as defined by the Shockley's equation—are independent of the network in which the device is employed. In other words, Eq. (6-12) takes care of all the intrinsic device parameters. The transfer curve is obtained either by using Shockley's equation, or by carefully studying and observing the output characteristics of the JFET. In Fig. 6-8, a graph is provided with the vertical scaling in milliamperes. It shows a plot of $I_D$ versus $V_{GS}$. When $V_{GS} = 0$ V, $I_D = I_{DSS}$. When $V_{GS} = V_P$, the drain current $I_D$ is 0 mA.

The transfer characteristics are a plot of an output current versus an input controlling quantity which gives an assessment of the performance of the device along with the various regions of operation. There is, therefore, a direct "transfer" from input to output variables.

## Solved Examples

**Example 6-1** An *n*-channel JFET has $I_{DSS} = 10$ mA and $V_P = -2$ V. For the values of $v_{DS}$, determine the approximate drain–source resistance when:

(a) $v_{GS} = 0$ V and (b) $v_{GS} = -0.5$ V.

**Solution:**

$$I_{DS} = 10 \text{ mA}, V_P = -2 \text{ V}, r = \frac{V_P^2}{2I_{DSS}(V_{GS} - V_P)}$$

(a) $V_{GS} = 0, r_{DS} = \dfrac{(-2 \text{ V})^2}{2(10 \text{ mA})(0 \text{ V} + 2 \text{ V})} = \dfrac{1}{10} \text{ k}\Omega = 100\ \Omega$

(b) $V_{GS} = 0.5 \text{ V}, r_{DS} = \dfrac{(-2 \text{ V})^2}{2(10 \text{ mA})(-0.5 \text{ V} + 2 \text{ V})} = 133.33\ \Omega$

**Example 6-2** For the circuit, as shown in the given figure, the JFET has $I_{DSS} = 12$ mA and $V_P = -4$ V. Given that $R_D = 3$ kΩ, $R_S = 0$ Ω, $V_{DD} = 15$ V and $V_{GG} = -2$ V; find (a) $i_D$ and (b) $v_{DS}$.

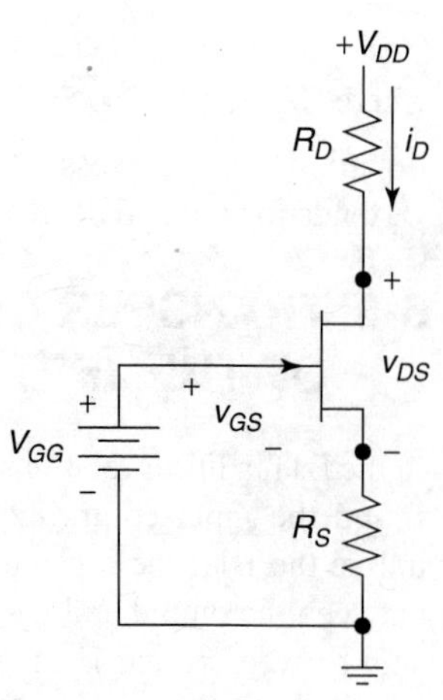


**Solution:**

$$I_{DSS} = 12 \text{ mA}, V_P = -4 \text{ V}, v_{GS} = -2 \text{ V}$$

Assuming the saturation region operation:

(a) $i_D = I_{DSS}\left(1 - \dfrac{v_{GS}}{V_P}\right)^2$

$= 12 \times 10^{-3}\left(1 - \dfrac{2}{4}\right)^2 = 3 \text{ mA}$

(b) $v_{DS} = -i_D R_D + V_{CC}$

$= -3 \times 10^3 \times 3 \times 10^{-3} + 15 = -9 \text{ V} + 15 \text{ V} = 6 \text{ V}.$

Since $V_{DS} = 6 > (v_{GS} - V_P) = 2$ V, the device is operating in the saturation region.

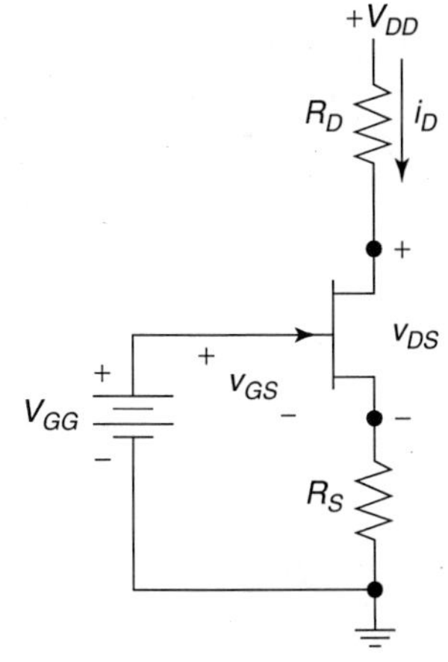


**Example 6-3** For the circuit, as shown in the given diagram, the JFET has $I_{DSS} = 12$ mA and $V_P = -4$ V. Given that $R_S = 0\ \Omega$ and $V_{DD} = 15$ V, find the value of $R_D$ for which $v_{DS} = 0.1$ V and $V_{GG} = 0$ V.

**Solution:**

$$I_{DSS} = 12 \text{ mA}, V_P = -4 \text{ V}$$

$$v_{DS} = 0.1 < v_{GS} - V_P = 4 \text{ V}$$

$\therefore$ ohmic region operation is confirmed.

$$i_D = I_{DSS}\left[2\left(1 - \frac{v_{GS}}{V_P}\right)\frac{v_{DS}}{-V_P} - \left(\frac{v_{DS}}{V_P}\right)^2\right]$$

$$i_D = 12 \times 10^{-3}\,(50 \text{ m} - 625\ \mu) = 592.5\ \mu\text{A}$$

$$\therefore \qquad R_D = \frac{15 - v_{DS}}{i_D} = 25.15 \text{ k}\Omega$$

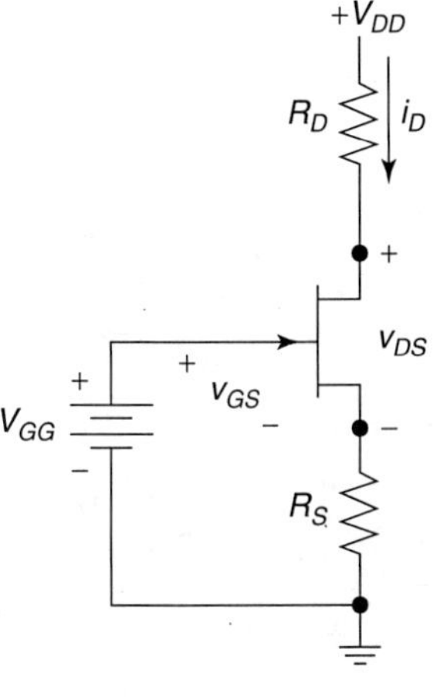


**Example 6-4** For the circuit, as shown in the given figure, the JFET has $I_{DSS} = 12$ mA and $V_P = -4$ V. Given that $R_D = 1$ kΩ, $R_S = 0\ \Omega$, $V_{DD} = 15$ V and $V_{GG} = 0$ V, find (a) $i_D$ (b) $v_{DS}$ and (c) slope of the operation of JFET.

**Solution:**
Assume saturation region operation. Then:

$$i_D = I_{DSS}\left(1 - \frac{V_{GS}}{V_P}\right)^2 = 12 \text{ mA}$$

$$\therefore \quad V_{DS} = 1 \text{ k}\Omega \times i_D + 15 \text{ V}$$
$$= -12 \text{ V} + 15 \text{ V} = 3 \text{ V}$$

Also, $$V_{GS} - V_P = 0 \text{ V} - (-4 \text{ V}) = 4 \text{ V}$$

Since, $v_{DS} < v_{GS} - V_P$, the JFET is not operating in the saturation region.

Let us consider it to be operating in the ohmic region. Then:

$$i_D = I_{DSS}\left[2\left(1 - \frac{v_{GS}}{V_P}\right)\frac{v_{DS}}{-V_P} - \left(\frac{v_{DS}}{V_P}\right)^2\right] = 6 \text{ m}v_{DS} - \frac{3}{4}\text{ m}v_{DS}^2$$

But, $$i_D = \frac{15 - v_{DS}}{1 \text{ k}} = 15 \text{ m} - 1 \text{ m}v_{DS}$$

Thus, $$\frac{3}{4}v_{DS}^2 - 7v_{DS} + 15 = 0$$

$$v_{DS} = \frac{10 \text{ V}}{3}, 6 \text{ V}$$

$\because \quad 6\text{ V} < -V_P$

$\therefore \quad v_{DS}$ is neglected

(a) $i_D = \dfrac{15 - v_{DS}}{1\text{ k}} = 11.67\text{ mA}$

(b) $v_{DS} = \dfrac{10}{3} = 3.33\text{ V}$

(c) Since, $v_{DS} < v_{GS} - V_P = 4$ V, the operation is confirmed in the ohmic region.

**Example 6-5** For the circuit, as shown in the given figure, the JFET is in active region with $I_{DSS} = 10$ mA and $V_p = -4$ V. Suppose that $V_{DD} = 12$ V and $V_{GG} = 0$ V. Find:

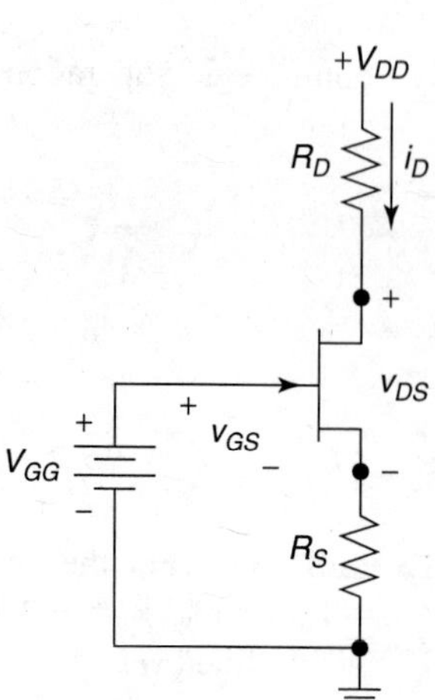


(a) $i_D$ when $v_{GS} = -2$ V
(b) $v_{GS}$ when $i_D = 9$ mA
(c) $R_D$ when $v_{GS} = -3$ V and $v_{DS} = 4.5$ V
(d) $v_{DS}$ when $R_D = 3\text{ k}\Omega$ and $v_{GS} = -2$ V

**Solution:**

(a) $i_D = I_{DSS}\left(1 - \dfrac{v_{GS}}{V_P}\right)^2 = 10 \times 10^{-3}\left(1 - \dfrac{-2}{-4}\right)^2 = 2.5\text{ mA}$

(b) $9 \times 10^{-3} = 10 \times 10^{-3}\left(1 + \dfrac{v_{GS}}{4}\right)^2$

or, $\sqrt{\dfrac{9}{10}} = 1 + \dfrac{v_{GS}}{4}$

or, $v_{GS} = -0.205\text{ V}$

(c) $v_{GS} = -3\text{ V}, v_{DS} = 4.5\text{ V}$

From part (c), $i_D = 0.625$ mA

$\therefore \; 12 = R_D i_D + v_{DS} - v_{GS}$

or, $12 = R_D \times 0.625 \times 10^{-3} + 4.5 + 3$

$\therefore \; R_D = 7.2\text{ k}\Omega$

(d) $12 = R_D i_D + v_{DS} - v_{GS}$

or, $12 = 3(0.625) + v_{DS} + 2$

$\therefore \; v_{DS} = 8.125\text{ V}$

**Example 6-6** For the circuit, as shown in the given figure, the JFET has $I_{DSS} = 16$ mA and $V_P = -4$ V. Given that $V_{DD} = 18$ V, $V_{GG} = 0$ V, and $R_D$ and $R_S = 500\ \Omega$, determine (a) $v_{GS}$, (b) $i_D$, (c) $v_{DS}$ and (d) the region of operation of JFET.

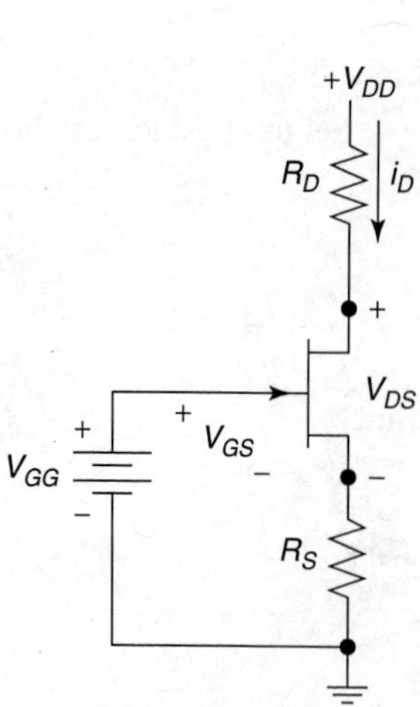


**Solution:**
Assuming the saturation region operation.

$$\therefore \quad i_D = I_{DSS}\left(1 - \frac{v_{GS}}{V_P}\right)^2$$

(a) $i_D = \dfrac{-v_{GS}}{500} = 16\text{ mA}\left(1+\dfrac{v_{GS}}{4}\right)^2$

or, $v_{GS}^2 + 10v_{GS} + 16 = 0$

or, $v_{GS} = -2\text{ V}, -8\text{ V}$

Since, $0 \geq v_{GS} \geq -4\text{ V}, v_{GS} = -2\text{ V}$

(b) $i_D = \dfrac{-v_{GS}}{500} = \dfrac{2}{1/2\text{ k}} = 4\text{ mA}$

(c) $v_{DS} = -500i_D + 18 - 500i_D = 18 - 1\text{ k}(4\text{m}) = 18 - 4 = 4\text{ V}$

(d) Since, $v_{DS} = 14\text{ V} \geq v_{GS} - V_P = -2 + 4 = 2\text{ V}$, the saturation region operation is confirmed.

**Example 6-7** For the circuit as shown in the given figure, the JFET has $I_{DSS} = 8$ mA and $V_P = -4$ V. Given that $V_{DD} = 18$ V, $V_{GG} = 0$ V, $R_D = 8$ kΩ, $R_S = 1$ kΩ and the JFET operates in the ohmic region, determine (a) $v_{GS}$ (b) $i_D$ and (c) $v_{DS}$.

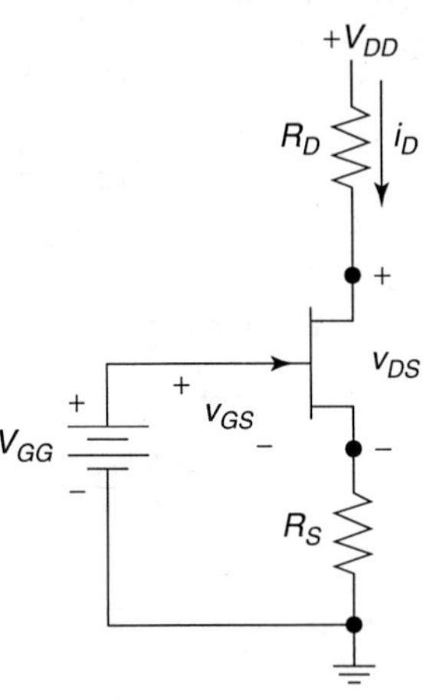


**Solution:**
For ohmic region operation:

$$i_D = I_{DSS}\left[2\left(1-\frac{v_{GS}}{V_P}\right)\frac{v_{DS}}{-V_P} - \left(\frac{v_{DS}}{V_P}\right)^2\right]$$

or,
$$i_D = 8\text{ mA}\left[2\left(1+\frac{v_{GS}}{4}\right)\frac{v_{DS}}{4} - \frac{v_{DS}^2}{16}\right]$$

Also,
$$i_D = \frac{-v_{GS}}{1\text{ k}}$$

∴
$$v_{DS} = 18 + 9\,v_{GS}$$

(a) $-\dfrac{-v_{GS}}{1\text{ k}} = 8\text{ mA}\left[2\left(1+\dfrac{v_{GS}}{4}\right)\left(\dfrac{18 + 9v_{GS}}{4}\right) - \left(\dfrac{18 + 9v_{GS}}{4}\right)^2\right]$

or $63v_{GS}^2 + 214v_{GS} + 180 = 0$

or $v_{GS} = -1.53\text{ V}$ or $-1.86\text{ V}$

(b) $i_D = \dfrac{-v_{GS}}{1\text{ k}} = 1.53\text{ mA}, 1.36\text{ mA}$

(c) $v_{DS} = -8\text{ k}i_D + 18 - 1\text{ k}i_D = -9 \times 10^3\, i_D + 18 = 4.23\text{ V}, 5.76\text{ V}$

**Example 6-8** For the circuit given in the figure, suppose that $R_D = R_S = R$, $V_{DD} = 15$ V, and the JFET has $V_P = -3$ V. (a) Assuming active region operation, $V_{GG} = 5$ V and $I_{DSS} = 10$ mA, determine the value of $R$ for which $v_{GS} = 0$ V. (b) Assuming active region operation, find $I_{DSS}$, given that $R = 400$ Ω, $V_{GG} = 5$ V and $v_{GS} = 0$ V. (c) If $I_{DSS}$, $R$ and the region of operation are known, what is $v_{GS}$ when $V_{GG} = 5$ V and $v_{DS} = 1$ V?

**Solution:**

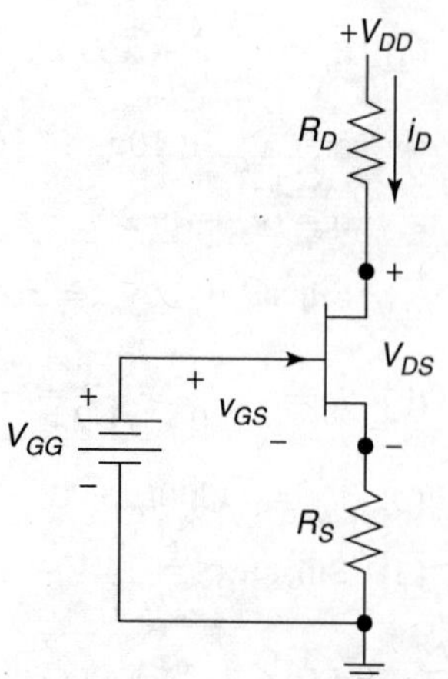


(a) Active region:

$$i_D = I_{DSS}\left(1 - \frac{v_{GS}}{V_P}\right)^2 = 10 \text{ mA}$$

$$v_{GG} = v_{GS} = R_{i_D}$$

or, $5 = 10 \text{ mA} \times R$

or, $R = 500\ \Omega$

$$v_{DS} = 2i_D R + 15 = 5 \text{ V} > v_{GS} - V_P = 3 \text{ V}$$

This confirms active region operation.

(b) Active region: $i_D = I_{DSS}$

$V_{GG} = v_{GS} + R_{i_D}$

or, $I_{DSS} = \dfrac{5}{400} = 12.5 \text{ mA}$

$v_{DS} = 2i_D R + 15 = -300(12.5) + 15 = 5 \text{ V} > v_{GS} - V_P = 3 \text{ V}$

This confirms active region operation.

(c) $15 = 2Ri_D + v_{DS} + 2R_{i_D} + 1$

or, $14 = 2R_{i_D}$

or, $R_{i_D} = 7$

But, $V_{GG} = v_{GS} + R_{i_D}$

or, $v_{GS} = -2 \text{ V}$

(d) $v_{DS} = 2 > v_{GS} - V_P = -1.5 + 3 = 1.5 \Rightarrow$ active region.

**Example 6-9** Both the JFETs in the given circuit have $I_{DSS} = 4$ mA and $V_P = -2$ V. Given that both the JFETs are in the active region, $R_S = 0\ \Omega$ and $V_{DD} = V_{SS} = 10$ V. Find (a) $i_D$ (b) $v_{GS}$ (c) $v_{DS}$; and confirm the region of operation of the JFETs.

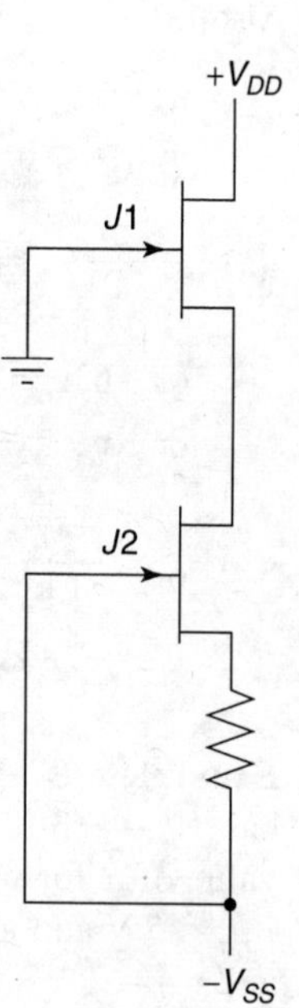


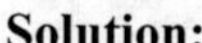

**Solution:**

Assume active region:

(a) Since, $v_{GS2} = 0$, then:

$$i_{D2} = I_{DSS}\left(1 - \frac{v_{GS}^{\ 2}}{V_P}\right) = I_{DSS} = 4 \text{ mA } (\because v_{GS2} = 0)$$

(b) Since, $i_D = I_{DSS}$, then $v_{GS} = 0$ V

(c) $10 = v_{DS} - v_{GS} = v_{DS} - 0 = 10$ V

Since, $v_{DS} = 10 \text{ V} > v_{GS} - V_P = -(-2) = 2$ V, then active region operation of the upper JFET is confirmed.

**Example 6-10** (a) A silicon JFET has a donor concentration of $3 \times 10^{21}/m^3$ and a channel width of 4 $\mu$m. If the dielectric constant of silicon is 12, find the pinch-off voltage. (b) If the JFET operates with a gate–source voltage of $-2$ V, what is the saturation voltage?

**Solution:**

(a) The expression for required pinch-off voltage is:

$$V = \frac{eN_D}{2\varepsilon} a^2$$

According to the question:

$N_D = 3 \times 10^{21}/m^3$, $2a = 4 \times 10^{-6}$ m, $\varepsilon = 12$, $\varepsilon_0 = 12 \times 8.85 \times 10^{-12}$ F/m and $q = 1.6 \times 10^{19}$ C

Therefore, $$V_P = \frac{1.6 \times 10^{19} \times 3 \times 10^{21} \times (2 \times 10^{-6})^2}{2 \times 12 \times 8.854 \times 10^{-12}} = 9.035 \text{ V}$$

(b) We know that:

$$V_P = V_{DS} - V_{GS}$$

Therefore, the saturation voltage is given by:

$$V_{DS} = V_P + V_{GS}$$

where, $$V_{GS} = -2 \text{ V}$$

$$V_{DS} = 9.035 - 2$$

$$= 7.035 \text{ V}$$

■

## 6-7 CONSTRUCTION AND CHARACTERISTICS OF THE MOSFET

The MOSFET is the fundamental building block of metal-oxide-semiconductor (MOS) and complementary-metal-oxide semiconductor (CMOS) integrated circuits. It has great commercial importance and is also an important power device. The MOS devices are extremely useful and suitable for large scale integration, and are the building blocks of large scale integrated circuits with low power consumption. The technological importance of the MOS along with the simplicity of its operation has made it the most sought after circuit element in modern technology.

The basic structure of MOS is shown in Fig. 6-9. The structure consists of three layers:

(i) The metal gate electrode

(ii) The insulating oxide layer

(iii) The bulk semiconductor (whichever is the choice, *n*-type or *p*-type)

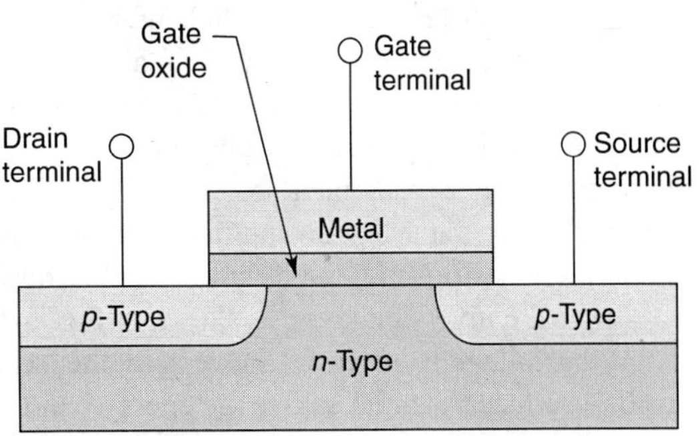


**Figure 6-9** Geometry of a basic MOS device

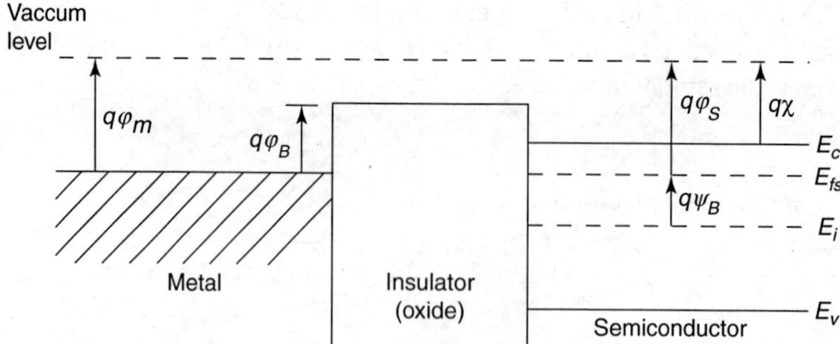


**Figure 6-10** Band diagram for a MOS device under no bias

The MOS structure forms a capacitor along with the gate and the substrate semiconductor as the two terminals, and the oxide layer as the dielectric. The thickness of the oxide layer typically varies from 10 nm to 50 nm. The local distribution and the carrier contribution within the substrate can be affected by manipulating the gate, and the voltages at the gate and the substrate.

The analysis of the MOS structure is incomplete without the band diagram of the metal semiconductor interface. The band structure indicating different levels of energy bands is shown in Fig. 6-10.

Here, the Fermi potential is given by:

$$\phi_F = \frac{E_F - E_i}{q} \tag{6-13}$$

where, the local Fermi level is indicated by $E_F$, the intrinsic Fermi level is denoted by $E_i$ and the magnitude of electronic charge is indicated by $q$. The electron affinity of the silicon bar is taken into consideration. This is the potential difference between the conduction band level and the vacuum and is indicated by $\chi$. Thus, the energy required for moving an electron from the Fermi level up to the vacuum level is called the work function, and is given by:

$$q\phi_S = q\chi + (E_C - E_F) \tag{6-14}$$

The insulating $SiO_2$ layer has a large band gap of about 8 eV and an electron affinity of 0.9 eV approximately.

Let us consider a case where the three components of an ideal MOS system are actually brought together, and thus, there is an interaction between them which causes the energy bands to bend. These are mainly because of the difference in potentials between the adjoining layers in this system.

The energy band diagram takes an abrupt turn when the system is brought under an external bias. This results in changes in the energy levels in varying proportions, as shown in Fig. 6-11.

If the voltage applied to the gate terminal is negative, the holes in the *p*-type substrate are attracted towards the semiconductor-oxide interface. The majority-carrier concentration near the surface becomes larger than what is there in the substrate and this phenomenon is called *accumulation*. The hole density in the interface region increases pushing the electrons further down the substrate. There is a variation in the band structure that forms the inversion layer, as shown in Fig. 6-12.

If the voltage is increased above zero, the oxide electric field will be directed towards the substrate. The positive surface potential causes the energy bands to bend downwards as shown in Fig. 6-12. The holes will be repelled back into the substrate as a result of this positive gate bias. These holes will leave the negatively charged field acceptor ions behind. Thus, a depletion region is created near the surface.

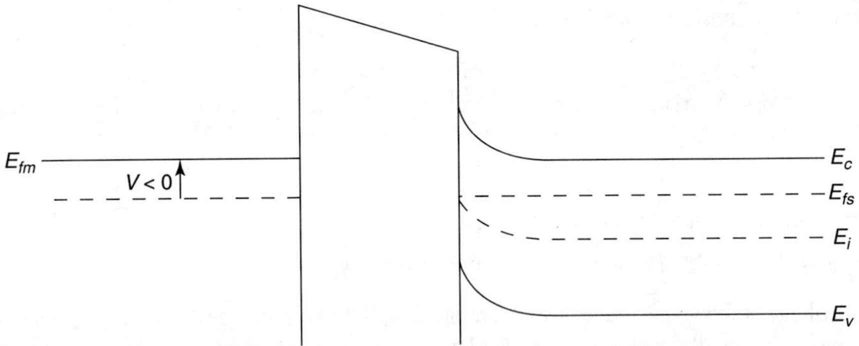


**Figure 6-11** Band structure where the device operates in depletion mode

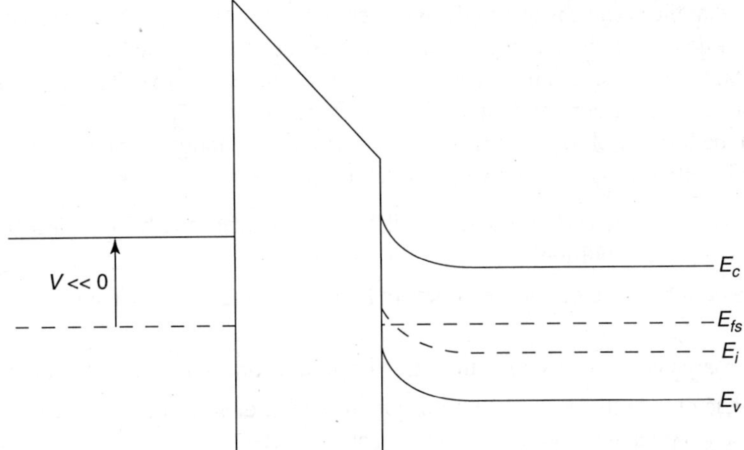


**Figure 6-12** Condition of the bands when the device is operating in the inversion mode

For quantitative analysis, we assume that a thin layer of mobile charge at the surface is created. This is given by:

$$dQ = -qN_A dx \tag{6-14a}$$

The change required in the surface potential to displace this sheet of charge through a distance $x_d$ away from the surface can be found by Poisson equation:

$$d\phi_s = -x\frac{dQ}{\varepsilon_{Si}} \tag{6-15}$$

Substituting the value of $dQ$ from Eq. (6-14a), we get:

$$d\phi_s = \frac{xqN_A}{\varepsilon_{Si}} \tag{6-16}$$

Integrating with proper limits gives:

$$\int_0^{\phi_s} d\phi_s = \int_0^{x_d} \frac{qN_A}{\varepsilon_{Si}} xdx \tag{6-17}$$

Here, $x_d$ is the depletion region width. Thus:

$$V_G = V_{ox} + \phi_S \tag{6-18}$$

Here, the voltage across the oxide is related to the charge on either side divided by the capacitance:

$$V_{ox} = \frac{-Q_s}{C_{ox}} \tag{6-19}$$

where, $C_{ox}$ = gate oxide capacitance per unit area
$Q_s$ = negative charge per unit area in the semiconductor

When the voltage at the gate is increased above the threshold voltage, the downward bending of the energy bands will increase. Eventually the intrinsic energy level $E_i$ becomes smaller than Fermi level $E_{FS}$ on the surface, which means that the substrate semiconductor in this region becomes *n*-type. The *n*-type region created near the surface by the positive gate bias is called the inversion layer, and this condition is called *surface inversion*.

While it is true that the surface is inverted whenever $\phi_S$ is larger than $\phi_F$, a practical criterion is required to tell us whether a true *n*-type conducting channel exists at the surface. The best criterion for strong inversion is that the surface should be as strongly *n*-type as the substrate is *p*-type. That is to say, $E_i$ should lie as far below the Fermi level at the surface as it is above $E_F$ from the surface.

MOSFETs can be broadly classified into two types: *n*-channel and *p*-channel types.

The ideal MOS structure has the following explicit properties:

(i) The metallic gate is sufficiently thick so that it can be considered to be an equipotential region under both ac and dc biasing conditions.

(ii) The oxide is a perfect insulator with zero current flowing through the oxide layer under all static biasing conditions.

(iii) There are no charge centers located in the oxide or in the oxide-semiconductor interface.

(iv) The semiconductor is sufficiently thick to ensure that regardless of the applied gate potential, a field-free region is encountered before reaching the back contact.

(v) The semiconductor is uniformly doped.

(vi) An ohmic contact has been established between the semiconductor and the metal on the back side of the device.

Figure 6-13 shows the basic construction of an *n*-channel MOSFET.

The *n*-channel MOSFET consists of a lightly doped *p*-type substrate in which two heavily doped *n*-regions are formed at the two ends by the process of diffusion. These two $n^+$ regions act as the source and the drain respectively. A thin layer of insulating silicon dioxide ($SiO_2$) is formed on the surface. The metallic gate electrode is then developed on the silicon dioxide surface. Generally, for ease in the fabrication process, poly silicon is used in place of metals. The gate-metal, the semiconductor substrate, and the oxide layer in between them, together form a parallel plate capacitor with $SiO_2$ as the dielectric. Since the gate is insulated from the semiconductor channel by the $SiO_2$

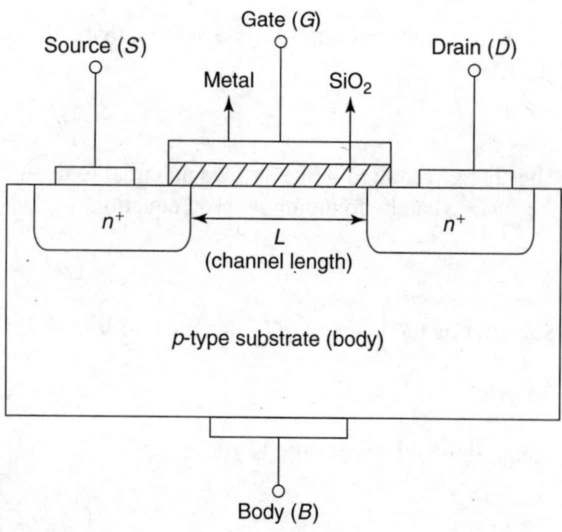


**Figure 6-13** *n*-channel MOSFET

layer grown in between, the device is called an insulated-gate FET. The presence of the oxide layer gives very high input impedance for this device, which is a much desired feature in the case of such devices.

The *n*-channel devices are faster in switching applications since electron mobility is greater than hole mobility. Their fabrications also take less space and thus, these are smaller in size than the *p*-channel devices. The disadvantage of *n*-channel devices is that their fabrication is difficult and hence, they are more expensive than *p*-channel devices. MOSFETs are mainly classified into two types:

(i) Depletion-type MOSFET
(ii) Enhancement-type MOSFET

## 6-7-1 Depletion-Type MOSFET

An *n*-channel MOSFET with a negative threshold voltage is called depletion-type MOSFET.

### Basic construction

Figure 6-14 depicts the basic construction of an *n*-channel depletion-type MOSFET.

A slab of *p*-type material formed from a silicon base is used as the *substrate.* It is the base upon which the device will be constructed. In some cases the substrate is often internally connected to the source terminal and in some discrete devices an additional terminal for the substrate, *SS*, is provided, resulting in a four-terminal device. Unless otherwise stated, it is assumed that the source and the bulk terminals are grounded. The source and drain terminals are connected through metallic contacts to *n*-doped regions in the *p*-type substrate linked by an *n*-channel, as shown in Fig. 6-14. The gate is also connected to a metal contact surface but remains insulated from the *n*-channel by a very thin silicon dioxide ($SiO_2$) layer, which forms the dielectric of the MOS capacitor. Since $SiO_2$ is used as an insulating layer, there is no direct electrical connection between the gate terminal and the channel of a MOSFET. The insulating layer of $SiO_2$ in the MOSFET construction leads to a very desirable high input impedance of the device, and also acts as the basis for the operation of the MOS device.

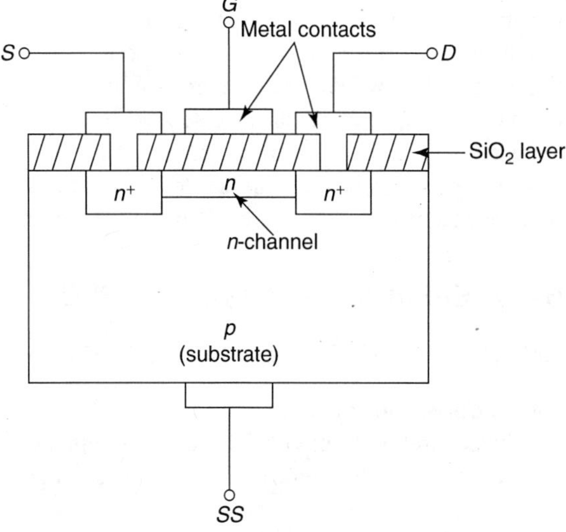


**Figure 6-14** *n*-channel depletion-type MOSFET

The name MOS or metal-oxide semiconductor FET comes from the metal for the drain, source and gate connections; the oxide for the silicon dioxide insulating layer; and the semiconductor for the basic structure on which the *n*- and *p*-type regions are diffused. Due to the insulating layer between the gate and the channel, it gets the name insulated-gate FET or IGFET.

### Basic operation and characteristics

For the operation of an *n*-channel depletion-type MOSFET, the gate-to-source voltage is taken to be zero volts, as shown in Fig. 6-15, by a direct connection between the two terminals, and a voltage $V_{DS}$ is applied across the drain-to-source terminals. The result is an attraction for the positive potential at the drain by the free electrons of the *n*-channel, and a current ($I_{DSS}$) is established through the channel of the JFET. This refers to such a condition of the device where electrons enter the structure through the source terminal and leave via the drain terminal.

To a large extent the movement of the charge carriers is subject to the voltage applied at the gate terminal as it acts as the controlling agent.

Now $V_{GS}$ has been set at a negative voltage of −1 V. The negative potential at the gate will tend to pressurize electrons towards the *p*-type substrate, as shown in Fig. 6-16.

Based on the magnitude of the negative bias by $V_{GS}$, a recombination between electrons and holes takes place, which reduces the number of free electrons in the *n*-channel available for conduction. The larger the negative bias, the higher will be the rate of recombination. Thus, the resulting level of drain current is reduced with increasing negative bias for $V_{GS}$ as shown in Fig. 6-17.

Now for positive values of $V_{GS}$ the positive gate will draw additional electrons from the *p*-type substrate due to the reverse leakage current and establish new carriers through the collisions occurring between the accelerating particles. So, as the gate-to-source voltage keeps on increasing in the positive direction, the drain current also increases at a rapid rate.

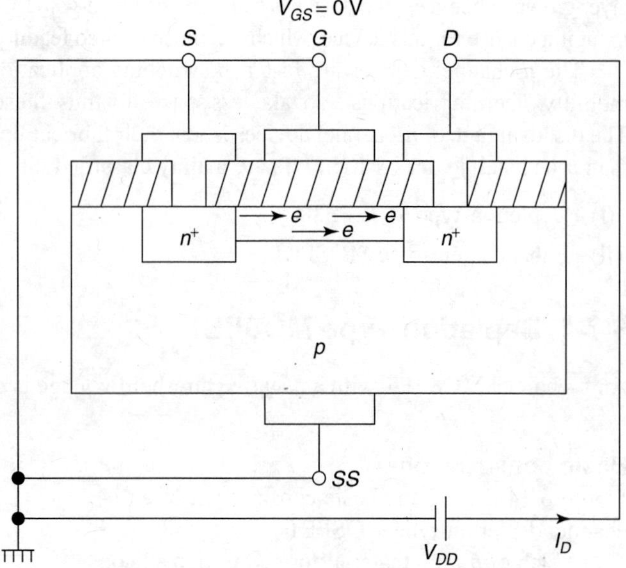


**Figure 6-15** *n*-channel depletion-type MOSFET with $V_{GS}$ = 0 V and an applied voltage $V_{DD}$

## 6-7-2 Enhancement-Type MOSFET

An *n*-channel MOSFET with a positive threshold voltage is called the enhancement-type MOSFET.

### Basic construction

The basic guiding figure of the *n*-channel enhancement-type MOSFET is provided in Fig. 6-18. A slab of *p*-type material is formed from a silicon base; this is referred to as the substrate or the base. The source and

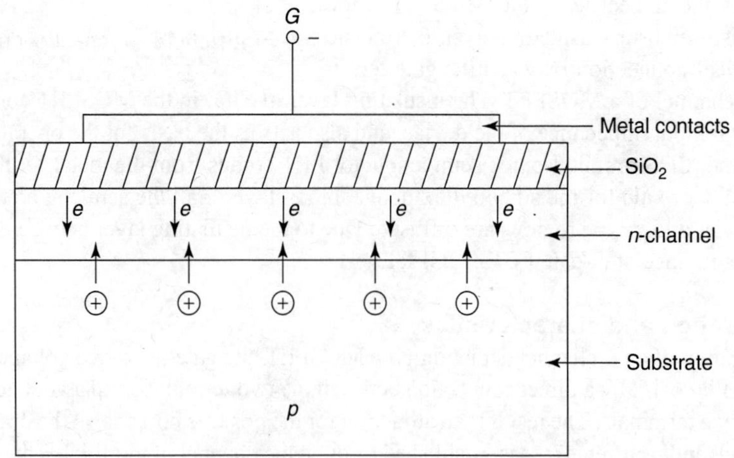


**Figure 6-16** Reduction of free carriers in the channel due to a negative potential at the gate terminal

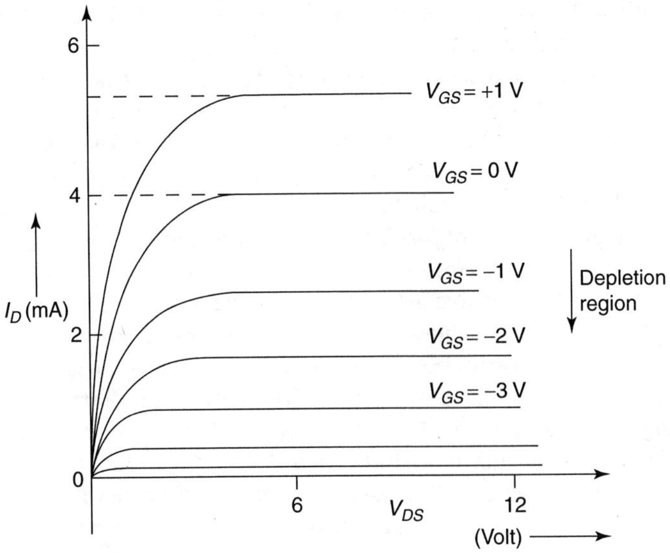


**Figure 6-17** Drain characteristics of an *n*-MOS

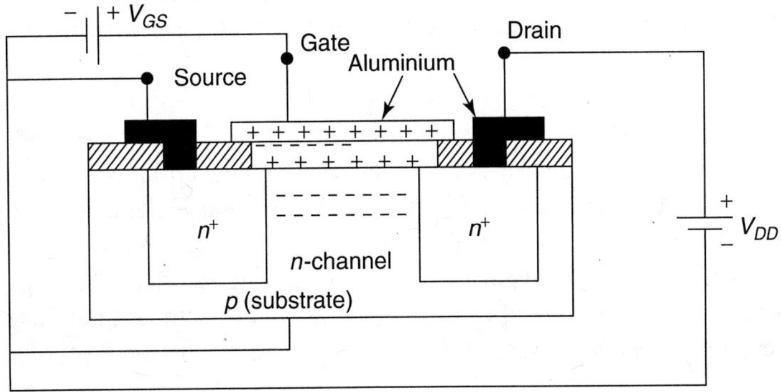


**Figure 6-18** Enhancement-type (*n*-channel) MOSFET

drain regions are diffused on the two sides of the semiconductor bar and are connected through metallic contacts to the *n*-doped regions, but in this case the channel is absent between the two *n*-doped regions. This is the primary and the most striking difference between the construction of a depletion-type MOSFET and an enhancement-type MOSFET. The $SiO_2$ layer is still present to isolate the gate metallic plate from the region between the drain and source. However, now it is simply separated from a section of the *p*-type material. In other words, the construction of the enhancement-type MOSFET is quite similar to that of the depletion-type MOSFET except for the absence of a channel between the drain and source terminals in the former. The enhancement-type MOSFET requires a proper bias voltage for current to flow through it.

## Basic operation and characteristics

If $V_{GS}$ is set at 0 V and a sufficient voltage is applied between the drain and source of the device given in Fig. 6-18, the absence of an *n*-channel will result in a current of effectively zero amperes—quite different from the depletion-type MOSFET and the JFET, where the drain current will be the saturation current through the channel, i.e., $I_D = I_{DSS}$.

Naturally, it is not sufficient to have a large accumulation of carriers (electrons) at the drain and source (due to the *n*-doped regions) if a path fails to exist between the two. With $V_{DS}$ at some positive voltage, $V_{GS}$ at 0 V and with the terminal *S* directly connected to the source, there are in fact two reverse-biased *p–n* junctions between the *n*-doped regions and the *p*-substrate to oppose any significant flow between the drain and the source.

In Fig 6-18, both $V_{DS}$ and $V_{GS}$ have been set at some positive voltage greater than zero volts establishing the drain and gate at a positive potential with respect to the source. The positive potential at the gate will pressurize the holes (since like charges repel) in the *p*-substrate along the edge of the $SiO_2$ layer so as to leave the area and move deeper into the region of *p*-substrate (see Fig. 6-18). The result is a depletion region near the $SiO_2$ insulating layer void of holes. However, the electrons in the *p*-substrate (the minority carriers of the material) will be attracted to the positive gate and accumulate in the region near the surface of the $SiO_2$ layer. The $SiO_2$ layer and its insulating qualities will prevent the negative carriers from being absorbed at the gate terminal. As $V_{GS}$ increases in magnitude, the concentration of electrons near the $SiO_2$ surface increases until the induced *n*-type region can support a measurable flow between the drain and the source. The level of $V_{GS}$ that results in the significant increase in drain current is called the *threshold voltage*, and is given by the symbol $V_T$. Since the channel is non-existent when $V_{GS}$ = 0 V and "enhanced" by the application of a positive gate-to-source voltage, this type of MOSFET is called an enhancement-type MOSFET.

As $V_{GS}$ is increased beyond the threshold level, the density of free carriers in the induced channel will increase resulting in an increased level of drain current. However, if we hold $V_{GS}$ constant and increase the level of $V_{DS}$, the drain current will eventually reach a saturation level similar to the JFET and the depletion-type MOSFET. The leveling-off of $I_D$ is due to a pinching-off process depicted by the narrower channel at the drain end of the induced channel. Applying Kirchoff's voltage law to the terminal voltages of the MOSFET as shown in Fig. 6-18, we have:

$$V_{DG} = V_{DS} - V_{GS} \tag{6-20}$$

If $V_{GS}$ is held fixed at some value and $V_{DS}$ is increased, the voltage $V_{DG}$ will drop and the gate will become less and less positive with respect to the drain. This reduction in gate-to-drain voltage in turn will reduce the attractive forces for free carriers (electrons) in this region of the induced channel causing a reduction in the effective channel width. Eventually, the channel will be reduced to the point of pinch-off and a saturation condition will be established as described earlier for the JFET and the depletion-type MOSFET. In other words, any further increase in $V_{DS}$ at the fixed value of $V_{GS}$ will not affect the saturation level of $I_D$ until breakdown conditions are encountered. The saturation level for $V_{DS}$ is related to the level of applied $V_{GS}$ by:

$$V_{DS_{sat}} = V_{GS} - V_T \tag{6-21}$$

For values of $V_{GS}$ less than the threshold level, the drain current of an enhancement-type MOSFET is 0 mA. Figure 6-19(a) illustrates the transfer characteristics of the *n*-channel MOSFET and Fig. 6-19(b) reveals that as the level of $V_{GS}$ is increased from $V_T$, the resulting saturation level for $I_D$ also increases from a level of 0 mA to that of several mA. In addition, it is quite noticeable that the spacing between the levels of $V_{GS}$ increases as the magnitude of $V_{GS}$ is increased resulting in increments in the drain current.

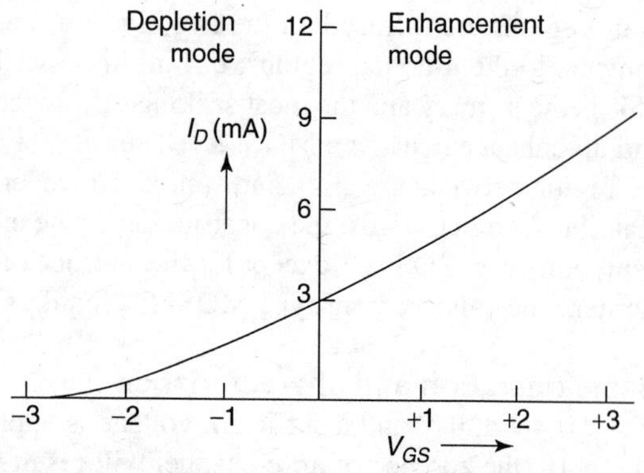

**Figure 6-19(a)** Transfer characteristics for *n*-channel MOSFET

For levels of $V_{GS} > V_T$ the drain current is related to the applied gate-to-source voltage by the following non-linear relationship:

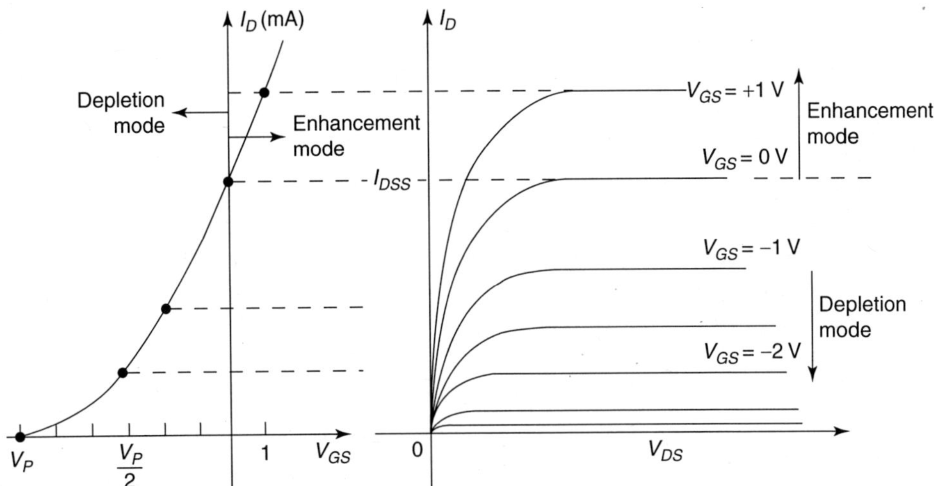


**Figure 6-19(b)** Comparison between enhancement-type and depletion-type MOSFET

$$I_D = k(V_{GS} - V_T)^2 \tag{6-22}$$

It is the squared term that results in the non-linear relationship between $I_D$ and $V_{GS}$. The $k$ term is a constant that is a function of the construction of the device. The value of $k$ can be determined from Eq. (6-23) where $I_{D(\text{on})}$ and $V_{GS(\text{on})}$ are the values for each at a particular point on the characteristics of the device.

$$k = \frac{I_{D(\text{on})}}{(V_{GS(\text{on})} - V_T)^2} \tag{6-23}$$

The transfer curve for $n$-channel enhancement-type MOSFET, and the comparison between an enhancement-type and a depletion-type MOSFET are shown in Fig. 6-19(a) and Fig. 6-19(b) respectively.

The drain characteristics and transfer characteristics have been set side by side to describe the transfer process from one to the other. The curve is certainly quite different from those obtained earlier. For an $n$-channel device it is now totally in the positive $V_{GS}$ region and does not rise until $V_{GS} = V_T$.

## Solved Examples

**Example 6-11** An $n$-channel depletion region MOSFET has $I_{DSS}$ = 10 mA and $V_P$ = −2 V. For small values of $v_{DS}$ determine the actual value of drain-to-source resistance $r_{DS}$ when $v_{GS}$ is (a) 1 V and (b) 2 V.

**Solution:**

$$I_{DSS} = 10 \text{ mA}, V_P = -2 \text{ V}$$

$$r_{DS} = \frac{V_P^2}{2 I_{DSS}(v_{GS} - V_P)}$$

(a) $v_{GS} = 1 \text{ V}, r_{DS} = \dfrac{(-2)^2}{2\,(10 \text{ mA})\,(1 + 2)} = 66.7\ \Omega$

(b) $v_{GS} = 2 \text{ V}, r_{DS} = 50\ \Omega$

**Example 6-12** An $n$-channel enhancement-type MOSFET has $K$ = 0.25 mA/V² and $V_t$ = 2 V. For small values of $v_{DS}$, determine the approximate drain source resistance $r_{DS}$ when (a) $v_{GS}$ = 4 V (b) 6 V and (c) 10 V.

**Solution:**

$$r_{DS} = \frac{1}{2K(v_{GS} - V_T)}$$

(a) $v_{GS} = 4\text{ V} \Rightarrow r_{DS} = 1\text{ k}\Omega$
(b) $v_{GS} = 6\text{ V} \Rightarrow r_{DS} = 500\ \Omega$
(c) $v_{GS} = 10\text{ V} \Rightarrow r_{DS} = 250\ \Omega$

**Example 6-13** For the circuit, as shown in the diagram, the enhancement NMOS transistor has $K = 0.2\text{ mA/V}^2$ and $V_t = 1$ V. Given that $V_{DD} = 10$ V and $v_{DS} = 6$ V. Find (a) $v_{GS}$ and (b) the region of operation (c) $i_D$ and (d) $R_D$.

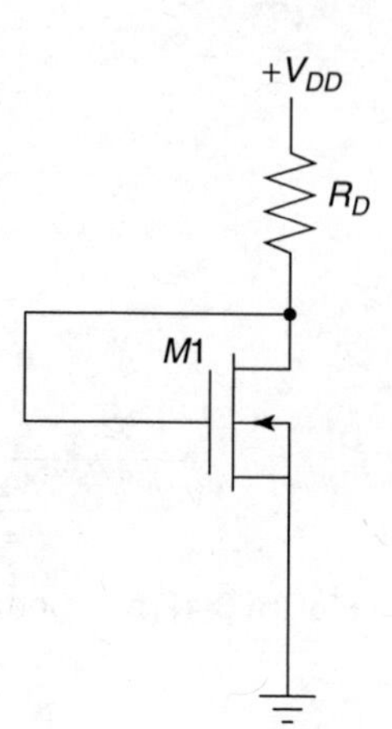

**Solution:**

(a) $v_{GS} = v_{DS} = 6\text{ V}$

(b) $v_{DS} = 6\text{ V} > v_{GS} - V_T = 6 - 1 = 5\text{ V}$

(c) $i_D = K(v_{GS} - V_T)^2 = 0.2\text{ m}\,(6 - 1)^2 = 5\text{ mA}$

(d) $R_D = \dfrac{V_{DD} - v_{DS}}{i_D} = \dfrac{10 - 6}{5\text{ m}} = 800\ \Omega$

**Example 6-14** For the MOSFET circuit as shown in the diagram, the enhancement NMOS has $K = 0.2\text{ mA/V}^2$ and $V_t = 1$ V. Given $V_{DD} = 10$ V and $i_D = 3.2$ mA. Find (a) the region of operation (b) $v_{GS}$ (c) $v_{DS}$ and (d) $R_D$.

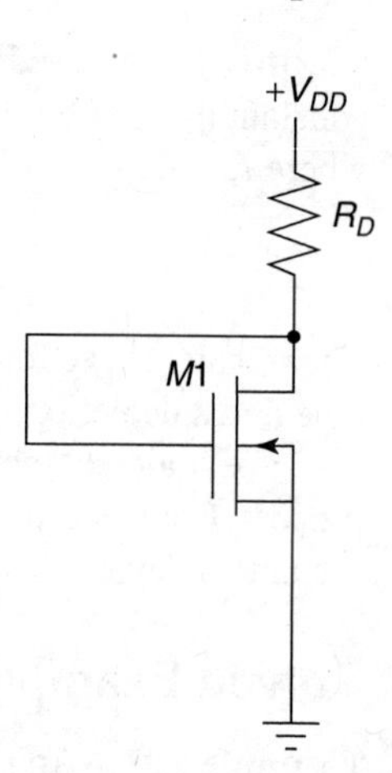


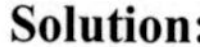

**Solution:**

(a) $v_{DS} = v_{GS} > v_{GS} - V_T = v_{GS} - 1 \Rightarrow$ active region operation

(b) $i_D = K(v_{GS} - V_T)^2 \Rightarrow v_{GS} = V_T + \sqrt{\dfrac{i_D}{K}} = 5\text{ V}$

(c) $v_{DS} = v_{GS} = 5\text{ V}$

(d) $R_D = \dfrac{V_{DD} - v_{DS}}{i_D} = \dfrac{10 - 5}{3.2\text{ m}} = 1.56\text{ k}\Omega$

**Example 6-15** For the circuit as shown in the diagram, $K = 0.15\text{ mA/V}^2$, $V_T = 2$ V and the circuit operates in the active region. Suppose that $V_{DD} = 12$ V. (a) Find $V_{GG}$ when $R_S = 0$ and $i_D = 5.4$ mA (b) Find $i_D$ when $v_{GS} = 4$ V.

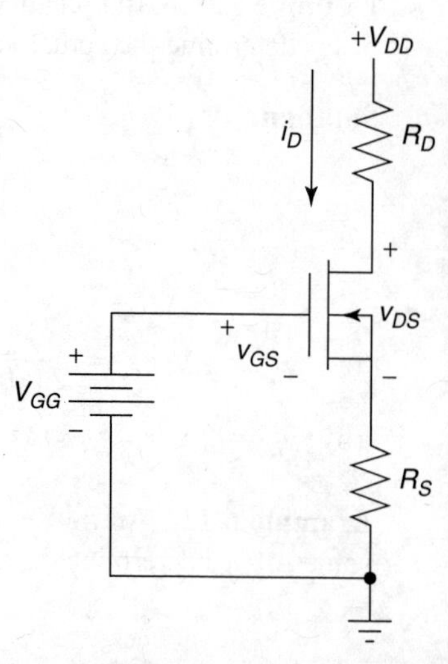


**Solution:**

Active region consideration:

(a) $R_S = 0\ \Omega$, $i_D = 5.4$ mA

$i_D = K(v_{GS} - V_T)^2$

or $5.4\text{ mA} = 0.15\text{ m}\,(V_{GG} - 2)^2$

or $v_{GG} = 8\text{ V}$

(b) $v_{GS} = 4\text{ V}$

$i_D = K(v_{GS} - V_T) = 0.15\text{ m}\,(4 - 2)^2 = 0.6\text{ mA}$

**Example 6-16** For the circuit, as shown in the diagram, $K = 0.25$ mA/V$^2$ and $V_T = 2$ V. Given that $R_S = 0\ \Omega$ and $V_{DD} = 16$ V. Determine the value of $R_D$ for which the MOSFET will operate on the border between the active and the ohmic region when $V_{GG}$ is (a) 4 V and (b) 10 V.

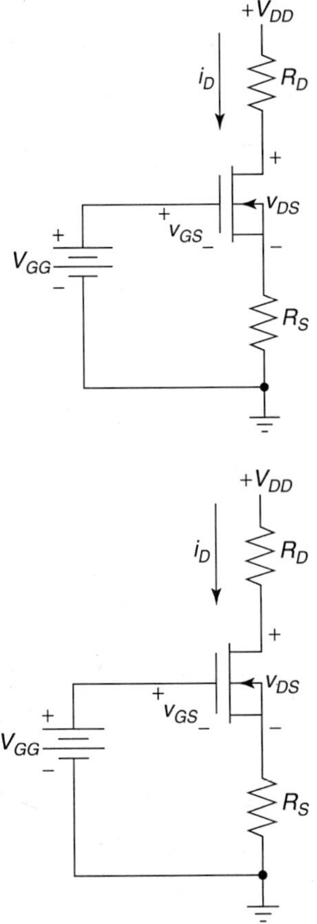


**Solution:**

When operating on the border of active and ohmic region, we have:

$$V_{DS} = v_{GS} - V_T = V_{GG} - 2$$

and,

$$i_D = K(v_{GS} - V_T)^2 = 0.25\text{ m}\,(v_{GG} - 2)^2$$

(a) $V_{GG} = 4\text{ V} \Rightarrow i_D = 0.25\text{ m}\,(4 - 2)^2 = 1\text{ mA}$

$16 = R_D i_D + v_{DS} = R_D i_D + (4 - 2)$

$R_D = 14\text{ k}\Omega$

(b) $V_{GG} = 10\text{ V} \Rightarrow i_D = 0.25\text{ m}\,(10 - 2)^2 = 16\text{ mA}$

$16 = R_D i_D + (v_{GG} - 2) \Rightarrow R_D = 500\ \Omega$

**Example 6-17** For the circuit, as shown in the diagram, the enhancement MOSFET has $K = 0.25$ mA/V$^2$ and $V_T = 2$ V. Given that $R_D = 1$ kΩ, $R_S = 0\ \Omega$ and $V_{DD} = 16$ V and $V_{GG} = 4$ V. Find (a) $i_D$ (b) $v_{DS}$ and (c) the region of operation.

**Soluton:**

Assume active region operation.

(a) $i_D = K(v_{GS} - V_T)^2 = 0.25\text{ m}\,(4 - 2)^2 = 1\text{ mA}$

(b) $v_D = -1ki_D + 16 = 15\text{ V}$

(c) Since, $v_{DS} = 15 > v_{GS} - V_T = 4 - 2 = 2$ V, the active region operation is confirmed.

**Example 6-18** For the MOSFET as shown in the diagram, $M_1$ has $I_{DSS} = 8$ mA and $V_P = -4$ V, whereas $M_2$ has $I_{DSS} = 16$ mA and $V_P = -4$ V. When $V_{DD} = 11$ V and $V_{GG} = 10$ V, then $M_1$ is in the ohmic region and $M_2$ is in the active region. Find (a) $i_D$ (b) $v_{DS1}$ and (c) $v_{DS2}$.

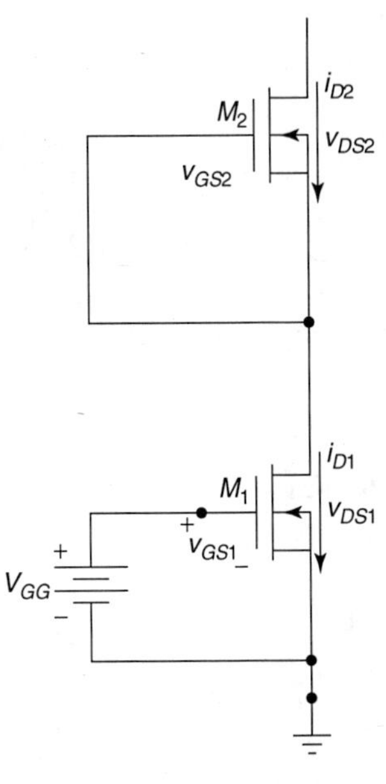


**Solution:**

(a) $$i_D = I_{DSS2}\left(1 - \frac{v_{GS2}}{V_T}\right)^2 = I_{DSS2} = 16\text{ mA}$$

(b) $$i_D = I_{DSS1}\left[2\left(1 - \frac{v_{GS1}}{V_T}\right)\frac{v_{DS1}}{-V_T} - \left(\frac{v_{DS1}}{V_T}\right)^2\right]$$

or $$16\text{ m} - 8\text{ m}\left[2\left(1 + \frac{10}{4}\right)\frac{v_{DS1}}{4} - \frac{v_{DS1}^2}{16}\right]$$

or $$v_{DS1}^2 - 28v_{DS1} + 32 = 0 \Rightarrow v_{DS1} = 1.19,\ 26.8$$

For ohmic region operation, $v_{DS1} < v_{GS1} - V_T = 10 + 414$ V.

Thus, $v_{DS1} = 1.19$ V

(c) $v_{DS2} = V_{DD} - v_{DS1} = 11 - 1.19 = 9.81$ V

**Example 6-19** For the MOSFET, as shown in the diagram, the depletion NMOS transistor has $I_{DSS} = 4$ mA and $V_P = -4$ V. Given that $V_{DD} = 10$ V and $R_D = 1$ kΩ. Find (a) the region of operation, (b) $v_{GS}$ (c) $v_{DS}$ and (d) $i_D$.

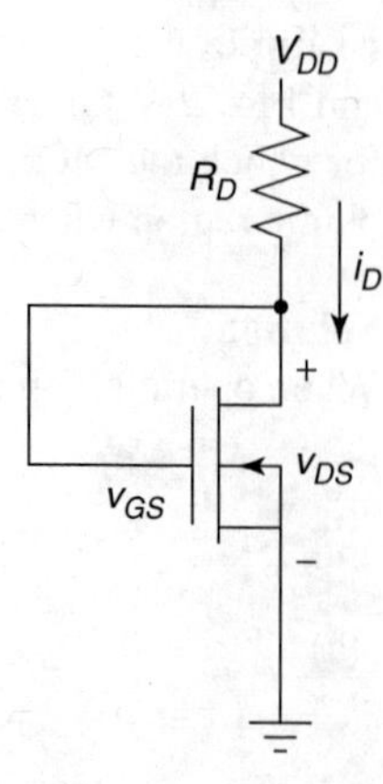


**Solution:**

(a) $v_D = v_{GS} < v_{GS} - V_P = v_{GS} + 4 \Rightarrow$ ohmic region operation.

(b) $i_D = I_{DSS}\left[2\left(1 - \frac{-V_{GS}}{V_P}\right)\frac{v_{DS}}{-V_P} - \left(\frac{v_{DS}}{V_P}\right)^2\right],\ i_D = \frac{10 - v_{DS}}{1k}$

Also, we know: $v_{DS} = v_{GS}$

$$4 \times 10^{-3}\left[2\left(1 + \frac{v_{GS}}{4}\right)\frac{v_{GS}}{4}\left(\frac{v_{GS}^2}{16}\right)\right] = \frac{10 - v_{DS}}{1k}$$

or,
$$v_{GS}^2 + 12v_{GS} - 40 = 0$$

$$v_{GS} = -14.72 \ \&\ 2.72 \text{ V}$$

$\because$
$$0 \le v_{GS} \le 10 \text{ V}$$

$$v_{GS} = 2.72 \text{ V}$$

(c) $v_{DS} = v_{GS} = 2.72$ V

(d) $i_D = \frac{10 - v_{DS}}{1k} = 7.28$ mA

**Example 6-20** For the MOSFET circuit, as shown in the diagram, the depletion NMOS transistor has $I_{DSS} = 4$ mA and $V_P = -4$ V. Given that $V_{DD} = 10$ V and $v_{DS} = 6$ V. Find (a) $v_{GS}$ and (b) the region of operation and (c) $i_D$.

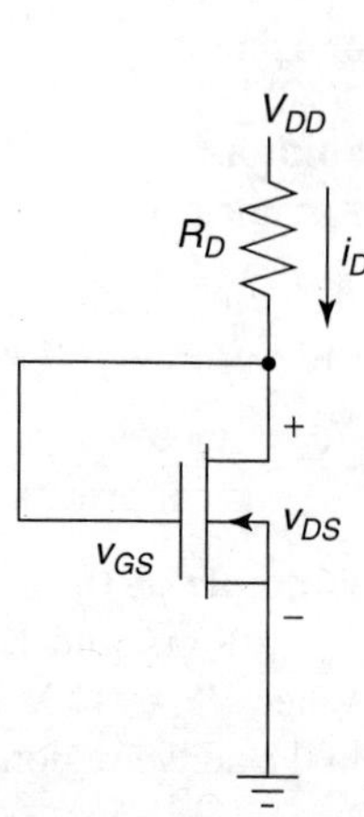


**Solution:**

(a) $v_{GS} = v_{DS} = 6$ V

(b) $v_{DS} = v_{GS} < v_{GS} - V_P = v_{GS} + 4 \Rightarrow$ ohmic region operation.

(c) $i_D = I_{DSS}\left[2\left(1 - \frac{V_{GS}}{V_P}\right)\frac{v_{DS}}{-V_P} - \left(\frac{v_{DS}}{V_P}\right)^2\right]$

$$i_D = 4 \times 10^{-3}\left[2\left(\frac{5}{2}\right)\left(\frac{3}{2}\right) - \frac{9}{4}\right] = 21 \text{ mA}$$

**Example 6-21** For the MOSFET as shown in the diagram, $K = 0.2$ mA/V² and $V_T = 1$ V. Given that $V_{DD} = 10$ V and $R_D = 1$ kΩ. Find (a) the region operation (b) $v_{GS}$ (c) $v_{DS}$ and (d) $i_D$.

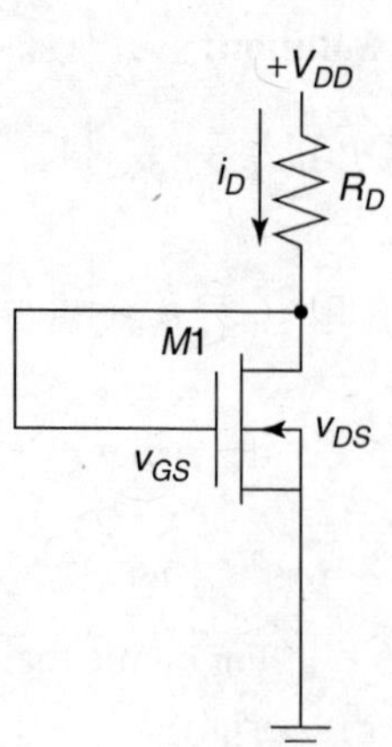


**Solution:**

(a) $v_{DS} = v_{GS} > v_{GS} - V_T = v_{GS} - 1 \Rightarrow$ active region operation.

(b) $i_D = K(v_{GS} - V_T)^2 = 0.2 \times 10^{-3}(v_{GS} - 1)^2 ;\ i_D = \frac{V_{DD} - v_{DS}}{1k} = \frac{10 - v_{GS}}{1k}$

$$0.2 \times 10^{-3}(v_{GS} - 1)^2 = \frac{10 - v_{GS}}{1k}$$

$$v_{GS} = 5.66, -8.66 \text{ V}$$

Since, $0 \le v_{GS} \le 10$ V, $v_{GS} = 5.66$ V

(c) $v_{DS} = v_{GS} = 5.66$ V

(d) $i_D = \dfrac{V_{DD} - v_{DS}}{R_D} = \dfrac{10 - 5.66}{1k} = 4.34$ mA

■

## 6-8 COMPLEMENTARY MOS

The complementary metal-oxide semiconductor (CMOS) belongs to a logic family and it consists of an *n*-MOS coupled with a *p*-MOS. With the upsurge of LSI and VLSI integrated circuits, and with the availability of *n*-MOS, early VLSI integrated circuits were available with the employment of the *n*-MOS technology. Such circuits utilized the enhancement-load amplifier and later the depletion-load amplifier as the basic inverter configuration. But with the advancement in technology, CMOS virtually replaced NMOS technology at all levels of integration.

In general the CMOS circuit looks complex compared to the circuit designed with only *n*-MOS or *p*-MOS. Still CMOS circuits have become popular because of the following reasons:

(i) The CMOS circuit draws negligible current in low- or high-output states.

(ii) The CMOS circuit allows the full power supply voltage $V_{DD}$ to be available at the output in the logic-high state.

(iii) The power supply voltage can be changed to any value between 3 V to 15 V without changing the logic operation of the circuit.

(iv) Such a configuration has a high fan-out.

(v) The output impedance of CMOS in both states lies between 100 Ω and 400 Ω.

The basic operation of a CMOS amplifier can be understood from the study of the basic inverter circuit. To construct an effective logic circuit we implant a *p*-channel and an *n*-channel MOSFET on the same substrate. Thus, a CMOS is formed. This is the key element in the development of the VLSI system. It is widely used in computer logic design due the following reasons:

(i) It has high input impedance

(ii) It has a faster switching speed

(iii) It has lower operating power levels

(iv) It behaves like an inverter

### 6-8-1 Construction of the CMOS

#### The *p*-tub process of fabrication of the CMOS

In this process of constructing the *p*-tub CMOS, as shown in Fig. 6-20, we use an *n*-type substrate in which a *p*-type dopant is diffused to form the tub. Thus, the *p*-MOS is formed in *n*-substrate as a *p*-tub. And naturally, the *n*-channel MOSFET is produced in the *p*-tub. On the other side the *p*-channel is already present in the main *n*-type substrate.

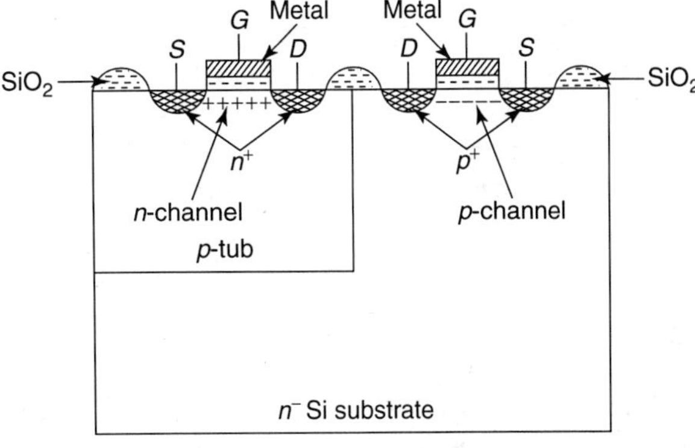


**Figure 6-20** Design of the CMOS using an *p*-tub structure

## The *n*-tub process of fabrication of the CMOS

In this process of constructing the *n*-tub CMOS, as shown in Fig. 6-21, we use a *p*-type substrate in which an *n*-type dopant is diffused to form the tub. Thus, the *n*-MOS is formed in the *p*-substrate as an *n*-tub. And naturally, the *p*-channel MOSFET is produced in the *n*-tub. On the other side the *n*-channel is already present in the main *p*-type substrate. The *n*-MOS and the *p*-MOS are electrically isolated by reversing the *p*–*n* isolation island.

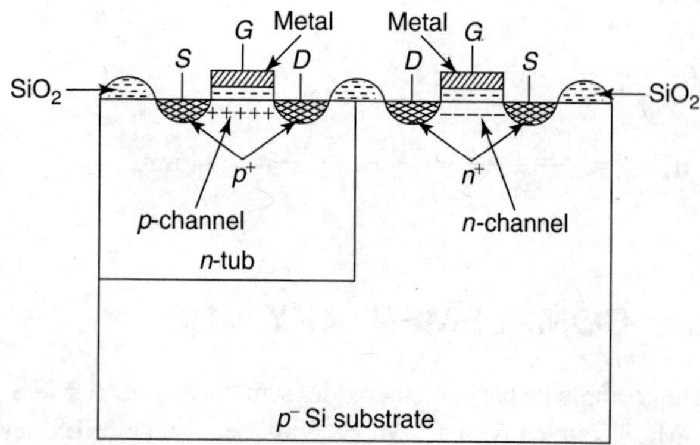


**Figure 6-21** Design of the CMOS using an *n*-tub structure

## The twin-tub process of fabrication of the CMOS

The process of constructing twin-tub CMOS is illustrated in Fig. 6-22(a). We use a heavily doped silicon substrate (either *p*-type or *n*-type) into which a lightly doped epitaxial layer of silicon is grown. Then we form a composite layer of $SiO_2$ and $Si_3N_4$ over the epitaxial layer, but the Si over the *n*-tub region is exposed. Now to form the *n*-tub the exposed Si is region-doped with phosphorus (P) which is masked from the adjacent region by the $Si_3N_4$ layer. After this the wafer is oxidized over the region of the *n*-tub selectively. To construct the *p*-tub, boron (B) is implanted (after stripping the nitrite). Boron enters the silicon through the thin oxide layer but is prevented from entering further by the

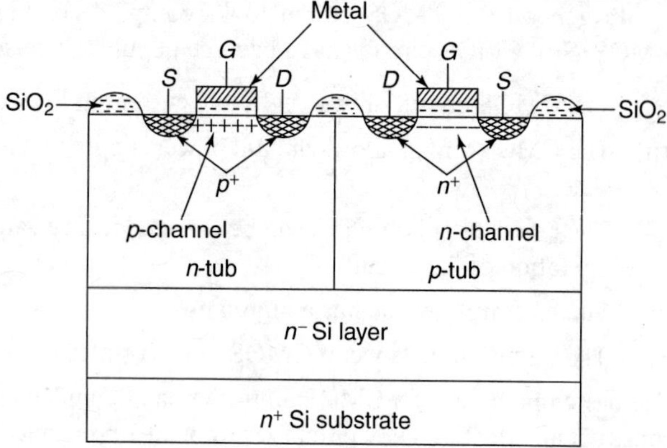


**Figure 6-22(a)** Design of CMOS using twin-tub structure

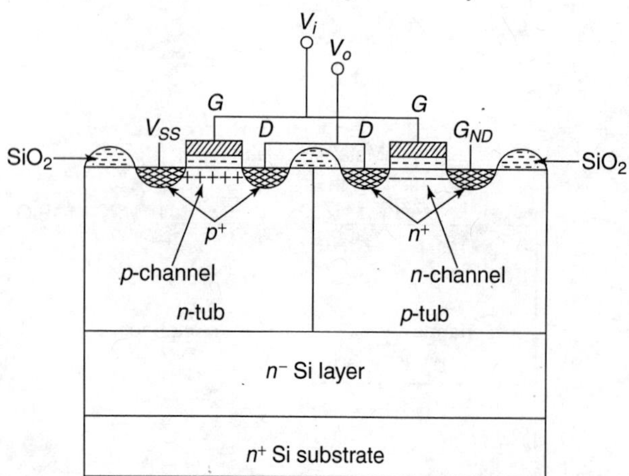


**Figure 6-22(b)** Schematic diagram of the CMOS inverter with electrical connection: source connected to supply, common input gate and common drain output

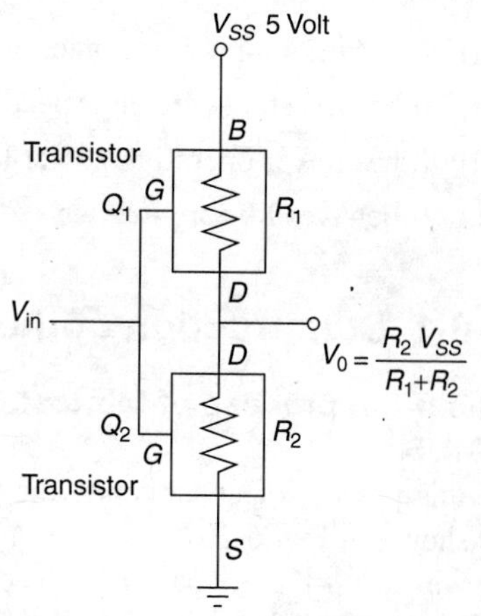


**Figure 6-22 (c)** Load diagram of the CMOS inverter

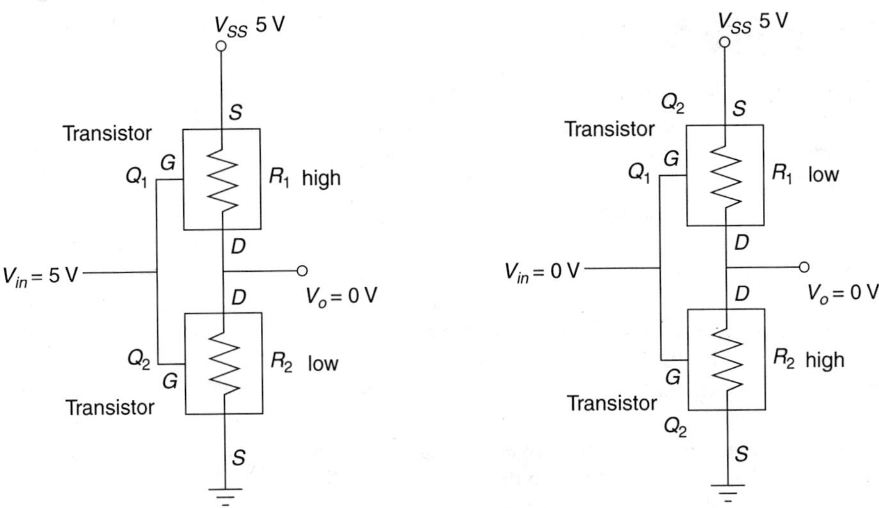


**Figure 6-23** Equivalent load diagrams of CMOS inverter in different voltage conditions showing inverter operations

thicker silicon-di-oxide layer over the *n*-tub. In the end all the thick oxides are stripped off and we get the twin-tub CMOS.

The twin-tub arrangement is widely used in modern semiconductor industry for USLI design. There are other uses of CMOS configuration for processor and controller design. It is also possible to fabricate devices below 45 nm. Some of these basic structures are discussed in the following sections.

## 6-8-2 CMOS Inverter

CMOS is mostly used as an inverter circuit when the source *S* of the *p*-channel of the *n*-tub is connected to the supply voltage $V_{ss}$ having a common input gate-connection and common drain output. The source of the *n*-channel of the *p*-tub is grounded. Typical logic-high is 5 V and logic-low is 0 V. When we apply low (logic '0') 0 V input voltage to the input gate-terminal—shown in Fig. 6-22(b)—the output of common drain yields logic high (logic '1') 5 V.

Figure 6-22(c) provides the load diagram of the CMOS inverter. In order to analyse the inverter operation, equivalent load diagrams of the CMOS inverter in different voltage conditions is shown in detail in Fig. 6-23.

Equivalent channel resistance $R_1$ offered by transistor $Q_1$ and channel resistance $R_2$ offered by transistor *Q*.

If $V_{in}$ = 5 V, $R_1$ is high and $R_2$ is low; so $V_0$ = 0 V

If $V_{in}$ = 0 V, $R_1$ is low and $R_2$ is high; so $V_0$ = 5 V

The working principle of the CMOS is more or less the same as the enhancement-type MOSFET yet it consists of two MOSFETs in a same bath; one *n*-type and the other *p*-type. The interesting thing is that it is a complementary arrangement. This means that the low-level input voltage gives out a high-level output voltage. The schematic CMOS circuit is as shown in Fig. 6-24.

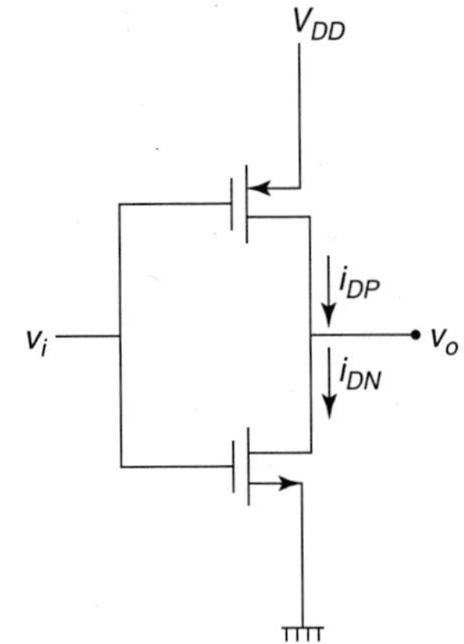


**Figure 6-24** A basic CMOS inverter circuit

## FOR ADVANCED READERS

### Examination of Mosfets Under Two Extremes

We have here two types of MOSFETs—an *n*-MOS and a *p*-MOS. Consider the two extremes: when $V_i$ is at logic 1 and when $V_i$ is at at logic 0. Figure 6-25 shows the basic characteristics of the circuit when the input voltage is at logic 1, i.e., $V_{DD}$.

Under such circumstances, the *p*-MOS remains in the OFF state and the *n*-MOS is in the ON state. The operating point, as shown in Fig. 6-25, is at the intersection of the curve when the source gate potential for the *p*-MOS is zero and the gate-to-source voltage for the *n*-MOS is $V_{DD}$. From Fig. 6-25 we also find that the operating point is at a location where the output voltage is nearly zero, and thus, the power dissipation is really less. However, the *n*-MOS is operating at nearly zero current and the operating point is nearly on a steep segment of the $i_D - v_{DS}$ characteristics. The *n*-MOS under this circumstance provides a low resistance path and the output is pulled down. This explains the operation of the CMOS circuit under logic input 1. The resistance in this case is given by:

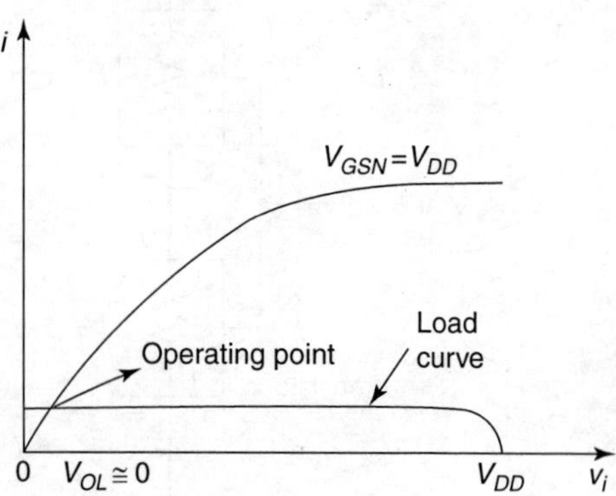


**Figure 6-25** Graphical representation of the operating point

$$r_{DS} = 1/k'_n\left(\frac{W}{L}\right)_n (V_{DD}-V_{tn}) \tag{6-24}$$

A similar description also prevails for the case when the input voltage is low. In that case, the output is pulled up and the output of the circuit is at logic 1. The resistance under such a circumstance is given by:

$$r_{DS} = \frac{1}{k'_p\left(\frac{W}{L}\right)_p (V_{DD}2V_{tp})} \tag{6-25}$$

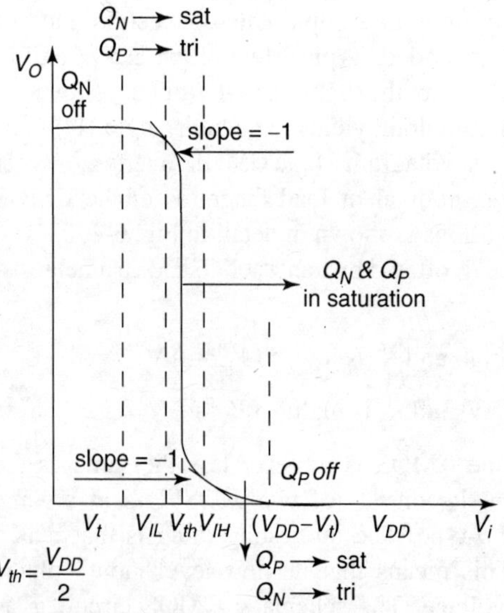


**Figure 6-26** The voltage transfer characteristics of the CMOS circuit

Now, we investigate the voltage transfer characteristics for the case as mentioned earlier. For this analysis, it is necessary to study the critical points of operation. For the *n*-MOS, the drain current is given by:

$$i_{DN} = k_n'\left(\frac{W}{L}\right)_n\left[(V_i - V_{tn})V_o - \frac{V_0^2}{2}\right] \quad (\text{for } V_0 \leq V_i - V_{tn}) \tag{6-26}$$

and,

$$i_{DN} = \frac{1}{2}k_n'\left(\frac{W}{L}\right)_n (V_i 2V_{tm})^2 \quad (\text{for } V_0 \geq V_i - V_{tn}) \tag{6-27}$$

Similarly for the p-MOS we have the following set of equations:

$$i_{DP} = k_p'\left(\frac{W}{L}\right)_P\left[(V_{DD} - V_i - |V_{tp}|)(V_{DD} - V_o) - \frac{1}{2}(V_{DD} - V_o)^2\right] \quad (\text{for } V_o \geq V_i + |V_{tp}|) \tag{6-28}$$

and,

$$i_{DP} = k_p'\left(\frac{W}{L}\right)_P\left[(V_{DD} - V_i - |V_{tp}|)(V_{DD} - V_o) - \frac{1}{2}(V_{DD} - V_o)^2\right] \quad (\text{for } V_o \geq V_i + |V_{tp}|) \tag{6-29}$$

The CMOS inverter is usually built with $V_{tn} = V_{tp}$, and $k_n'(W/L)_n\, k_p'(W/L)_p$

This design is basically done to render symmetric transfer characteristics. The transfer characteristics, as obtained, are shown in Fig. 6-26.

We shall now proceed with an illustration of the inverter circuit from the digital point of view and if the operation and the characteristics are understood, the results can be extended for complex circuits. The basic circuit of the CMOS inverter is as shown in Fig. 6-27.

We first consider the two extreme cases: when $v_i$ is at logic 1 level, which is nearly equal to $V_{DD}$ volts, and when $v_i$ is at logic 0 level, which is approximately equal to 0 V.

**When $v_i = V_{DD}$:** In the circuit we find that as the input voltage $V_{SGP} = 0$; the PMOS is switched off. Consequently we find that for the NMOS, the condition is just the opposite and the NMOS provides a path of low resistance given by the mathematical formula:

$$r_{DS} = \frac{1}{k_n'\left(\frac{W}{L}\right)_n (V_{DD} - V_{tn})} \tag{6-30}$$

This means that the power dissipation in the circuit is zero. The output point in the circuit gets an easy and a low resistance path for the current to flow, and thus, the output is zero. Figure 6-28 shows the state of the circuit when input is a logic 1.

**When $v_i = 0$ V:** The state of the circuit when the input voltage is 0 V is as shown in Fig. 6-29.

From the above circuit, we find that the PMOS is conducting and the NMOS does not conduct. As a result the output point is set at a voltage $V_{DD}$ which is logic 1. In this case the resistance offered by the PMOS is given by:

$$r_{DSP} = \frac{1}{\left[k_p'\left(\frac{W}{L}\right)_P (V_{DD} - |V_{tp}|)\right]} \tag{6-31}$$

From Fig. 6-29 we can conclude that the basic CMOS logic inverter behaves as an ideal inverter.

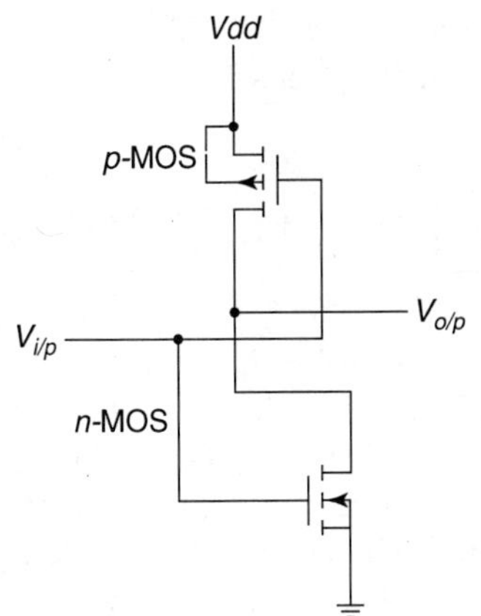


**Figure 6-27** The CMOS inverter

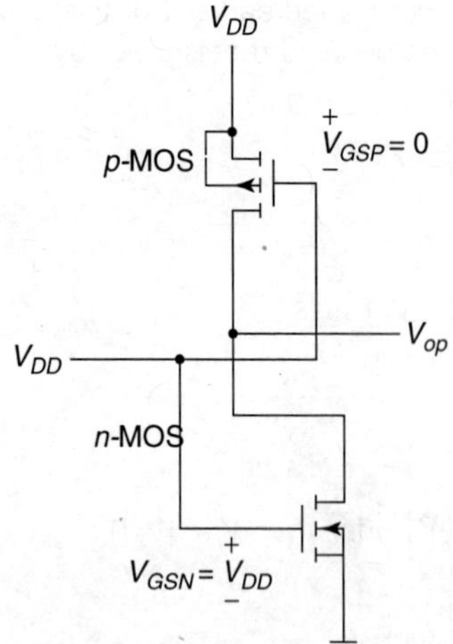


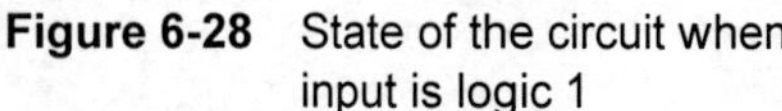
**Figure 6-28** State of the circuit when input is logic 1

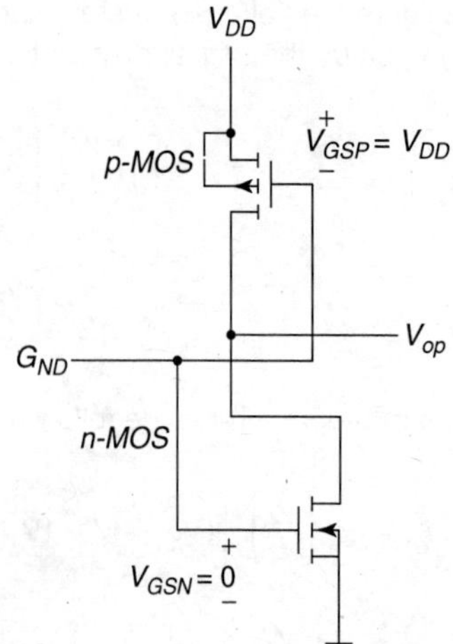


**Figure 6-29** State of the circuit when input logic is 0

These concepts enable us to use the MOS devices as switches in order to obtain a logic inverter. We can make various such connections and obtain many such logic circuits devised only by *n*-MOS and *p*-MOS, i.e., CMOS. Let us now make a comparison between CMOS and BJT (see Table 6-2).

**Table 6-2** Comparison between CMOS and BJT

| *CMOS* | *BJT* |
|---|---|
| 1. CMOS devices occupy less space, i.e., higher packing density | BJT devices requires more space |
| 2. Dissipates lower power | Dissipates comparatively higher power |
| 3. Higher input impedance | Lower input impedance |
| 4. Low driving current | High driving current |
| 5. Output current is low | Output current is high |
| 6. Low-noise device | High-noise device |
| 7. Source and drain interchangeable, i.e., bidirectional device | Unidirectional device |
| 8. Lower fan-out loading | Higher fan-out loading |
| 9. High delay sensitivity | Low delay sensitivity |

## 6-9 REAL-LIFE APPLICATIONS OF THE FET

The main advantage of the FET amplifier over the BJT amplifier is that the former has very high input impedance. This makes it compatible with devices of high output impedance. An example of such a case is the multi-channel audio mixer. Like the BJT, the FET is massively used in electronic circuits for amplification and switching. In IC fabrication, the FET requires much lesser space than the BJT. It consumes low power and the fabrication technique is also easier. So, gradually it occupies the position of the BJT. The CMOS-FET has today become the basic building block of digital ICs.

## POINTS TO REMEMBER

1. The field-effect transistor is a semiconductor device in which the output quantity is controlled by an electric field which is also the input quantity.
2. The field-effect transistor is a majority-carrier device and it is also called unipolar transistor.
3. The various field-effect transistors are:
   (a) Junction field-effect transistor
   (b) Insulated gate field-effect transistor
   (c) Metal insulator field-effect transistor
   (d) Metal-oxide-semiconductor field-effect transistor, which is a derived class of IGFET.
4. Field-effect transistor happens to steal the interest of the scientists from BJT because unlike the BJTs, the various types of field-effect devices have high input impedance, which is quite a desired feature for their operation. These devices are particularly suited for their operation as a controlled switch operating in the conducting and the non-conducting regions.
5. In JFET, the width of a junction is used to control the effective cross-sectional area of the conducting channel.
6. With proper biasing of the device current is allowed to flow from the source. In case of proper biasing in JFETs, the depletion region attains a particular shape where its width is more towards the drain terminal and narrows towards the source terminal.
7. As a consequence of the previous fact, the device can be operated as a voltage-controlled resistor.
8. The FET has a considerably higher resistance than the BJT.
9. The FET has a negative temperature coefficient at high-current levels.
10. The FET device is thermally stable even when the active area is large, and when many such devices are connected in parallel.
11. As FETs are unipolar devices they hardly suffer from any minority carrier shortage.
12. The term pinch-off suggests a level where the drain current reaches zero but in actuality at pinch-off the drain current maintains a constant drain current $I_{DSS}$.
13. Unlike a BJT, a non-linear relationship exists between the output and input quantities of a JFET.
14. Being the fundamental building block of CMOS, the MOSFET has great commercial importance and is suitable for very large-scale integration.
15. The MOS structure mainly forms a capacitor along with the gate and the substrate semiconductor with two terminals, and with the oxide layer as the dielectric.
16. The operation of the MOS has three stages that depend on the applied bias voltage.
17. An ideal MOS structure has the following explicit properties:
    (a) The metallic gate is sufficiently thick so that it can be considered to be an equipotential region under ac as well as dc biasing conditions.
    (b) The oxide is a perfect insulator with zero current flowing through the oxide layer under all static biasing conditions.
    (c) There are no charge centers located in the oxide or in the oxide–semiconductor interface.
    (d) The semiconductor is sufficiently thick so that, regardless of the applied gate potential, a field-free region is encountered before reaching the back contact.
    (e) The semiconductor is uniformly doped.
    (f) An ohmic contact is established between the semiconductor and the metal on the back side of the device.
    (g) The presence of the oxide layer gives very high input impedance for this device.
18. The *n*-channel devices are faster in switching applications, since electron mobility is greater than hole mobility.
19. In terms of fabrication, MOSFETs are classified into:
    (a) Depletion-type MOSFET
    (b) Enhancement-type MOSFET

20. The circuits with CMOS have become more popular because:
    (a) The CMOS circuit draws negligible current either in low or in high output states.
    (b) The CMOS circuit allows the full power supply voltage $V_{DD}$ to be available at the output in the logic high state.
    (c) The power supply voltage can be changed to any value between 3 − 15 V without changing the logic operation of the circuit.
    (d) This kind of a configuration has a high fan-out.
    (e) The output impedance of CMOS in both the states lies between 100 − 400 Ω.

## IMPORTANT FORMULAE

1. $I_{DSS}$ (saturation drain current) is the maximum drain current for a JFET and is given by the conditions:

$$V_{GS} = 0 \text{ V and } V_{DS} > |V_P|$$

2. The width of the depletion region at a distance $L$ from the source is given by:

$$W(x = L) = \left[\frac{2\epsilon(-V_{GD})}{qN_D}\right]^{1/2}$$

3. The condition at pinch-off becomes $h(x = L) = a - W(x = L) = 0$, and we obtain the expression for pinch-off as:

$$V_P = \frac{qa^2N_D}{2\epsilon}$$

where, $a$ = width of the channel under no bias, $h$ = width of the channel at a distance $x$ from the source terminal and is a function of $x$.

4. The pinch-off voltage is a positive number and its relation with the drain and gate voltage is given by:

$$V_P = -V_{GD}(\text{pinch-off}) = -V_G + V_D$$

5. Let $Z$ be the depth of the channel in the $z$-axis, and the current through the channel is given by:

$$I_D = \frac{2aZ}{L\rho}\left[\frac{V_D}{V_P}\frac{2}{3}\left(\frac{-V_G}{V_P}\right)^{3/2} - \frac{2}{3}\left(\frac{V_D - V_G}{V_P}\right)^{3/2}\right]$$

6. Considering the case of the MOSFET, the work function is given by:

$$q\phi_S = q\chi + (E_C - E_F)$$

7. For a given MOSFET, when $V_{GS} > V_T$, the drain current is given by:

$$I_D = k(V_{GS} - V_T)^2$$

## OBJECTIVE QUESTIONS

1. FET is:
   (a) A unipolar device
   (b) A bipolar device
   (c) A tripolar device
   (d) Not a device
2. FET is:
   (a) A current-controlled device
   (b) A voltage-controlled device
   (c) A power-controlled device
   (d) None of the above
3. The pinch-off voltage of JFET is 5.0 volts. Its cut off voltage is:
   (a) $(5.0)^{1/2}$ V (b) 5.0 V
   (c) 2.5 V (d) $(5.0)^{3/2}$ V
4. An $n$-channel JFET has a pinch-off voltage of $V_p = -5$ V, $V_{DS\,(\max)} = 20$ V, and $g_m = 2$ mA/V. The minimum ON resistance is achieved in the JFET for:
   (a) $V_{GS} = -7$ V and $V_{DS} = 0$ V
   (b) $V_{GS} = 0$ V and $V_{DS} = 0$ V
   (c) $V_{GS} = 0$ V and $V_{DS} = 20$ V
   (d) $V_{GS} = -7$ V and $V_{DS} = 20$ V
5. The threshold voltage of an $n$-channel MOSFET can be increased by:
   (a) Increasing the channel dopant concentration
   (b) Reducing the channel length
   (c) Reducing the gate oxide thickness
   (d) Reducing the channel dopant concentration

6. Two identical FET's, each characterized by the parameters $g_m$ and $r_d$, are connected in parallel. The composite FET is then characterized by the parameters:
   (a) $g_m/2$ and $2\,r_d$ (b) $g_m/2$ and $r_d/2$
   (c) $2g_m$ and $r_d/2$ (d) $2g_m$ and $2r_d$
7. An $n$-channel JEET has $I_{DSS} = 2$ mA and $V_p = -4$ V. Its transconductance $g_m$ (in milli mho) for an applied gate-to-source voltage $V_{Gs}$ of $-2$ V is:
   (a) 0.25 (b) 1.0
   (c) 0.75 (d) 0.5
8. The V-I characteristic of an $n$-channel depletion FET drain–source output has:
   (a) $I_{DS} = 0$ at $V_{GS} = 0$ V
   (b) $I_{DS}$ is independent of $V_{GS}$
   (c) $I_{DS}$ = positive maximum at $V_{GS} = 0$ V
   (d) $I_{DS}$ = negative maximum at $V_{GS} = 0$ V
9. For an enhancement-type MOSFET the output V–I characteristic of has:
   (a) Only an ohmic region
   (b) An ohmic region at low voltage value followed by a saturation region at higher voltages
   (c) Only a saturation region
   (d) An ohmic region at large voltage values preceded by a saturation region lower voltage
10. For an $n$-channel JFET, having drain–source voltage constant, if the gate–source voltage is increased to more negative pinch-off would occur for:
   (a) Saturation value of drain current
   (b) High values of drain current
   (c) Zero drain current
   (d) Source current equal to the drain current
11. For a junction FET in the pinch-off region, as the drain voltage is increased, the drain current:
   (a) Becomes zero (b) Abruptly decreases
   (c) Abruptly increases (d) Remains constant
12. Now a days in a MOSFET the material used for the gate is:
   (a) Heavily doped polycrystalline silicon
   (b) Pure silicon
   (c) High purity silica oxide
   (d) Epitaxial grown silicon
13. An $n$-channel JFET has $I_{DS}$ whose value is:
   (a) Maximum for $V_{GS} = 0$, and minimum for $V_{GS}$ = negative and large
   (b) Minimum for $V_{GS} = 0$, and maximum for $V_{GS}$ = negative and large
   (c) Maximum for $V_{GS} = 0$, and minimum for $V_{GS}$ = positive and large
   (d) Minimum for $V_{GS} = 0$, and maximum for $V_{GS}$= positive and large
14. The threshold voltage of an $n$-channel enhancement mode MOSFET is 0.5 V. When the device is biased at a gate voltage of 3 V, pinch-off would occur at a drain voltage of:
   (a) 1 V (b) 0.5 V
   (c) 3.5 V (d) 2.5 V
15. Bipolar transistors are ____________ than field effect transistor.
   (a) Less sensitive and slower
   (b) More sensitive and faster
   (c) More sensitive and slower
   (d) Less sensitive and faster
16. The main factor, which differentiates a D-MOSFET from an E-MOSFET, is the absence of:
   (a) $p$–$n$ junction (b) Insulated gate
   (c) Electrons (d) Channel
17. In order to protect a MOSFET against damage from any stray voltage at the gate:
   (a) Grounding rings are provided
   (b) Source terminal is earthen
   (c) Terminals are shorted
   (d) None
18. $n$-channel FETs are superior to $p$-channel FETs because:
   (a) Mobility of electrons is smaller than that of holes
   (b) Mobility of electrons is greater than that of holes
   (c) They have high switching time
   (d) They consume less power

19. Pinch-off voltage $V_p$ for an FET is the drain voltage at which:
    (a) Drain current becomes zero
    (b) All free charges get removed from the channel
    (c) Significant drain current starts flowing
    (d) Avalanche breakdown takes place
20. The saturation drain current $I_{DS}$ in an FET equals:
    (a) $I_{DSS}\left(1-\frac{V_{GS}}{V_P}\right)^2$ (b) $I_{DSS}\left(1-\frac{V_{GS}}{V_P}\right)$
    (c) $I_{DSS}\sqrt{\frac{V_{GS}}{V_P}}$ (d) $I^2_{DSS}\left(\frac{V_{GS}}{V_P}\right)$
21. The transconductance $g_m$ of a diffused JFET is the order of:
    (a) 1 mS (b) 1 S
    (c) 100 S (d) 1000 S
22. The gate-to-source resistance of an FET is of the order of:
    (a) 100 MΩ (b) 10 MΩ
    (c) 1 MΩ (d) 0.1 MΩ
23. Inter-electrode capacitances in an FET are in the order of:
    (a) 1 pF (b) 100 pF
    (c) 0.1 $\mu$F (d) 1 $\mu$F
24. FET has offset voltage of about:
    (a) 0.2 V (b) 0.6 V
    (c) 1.5 V (d) Zero
25. A JFET has a potential divider bias arrangement if the resistor between the gate and power supply terminal is removed. The JFET will:
    (a) Continue to work as an amplifier
    (b) Have a forward-bias gate w. r. t source
    (c) Not work as an amplifier but as a switch
    (d) Immediately burn out
26. Which of the following transistors is symmetrical in there structure?
    (a) *n–p–n* transistor (b) JFET
    (c) *p–n–p* transistor (d) MOSFET
27. The drain current in JFET is controlled by:
    (a) Voltage drop across channel
    (b) Depletion regions
    (c) Channel resistance
    (d) Reverse-bias at the gate

## REVIEW QUESTIONS

1. Why is the field-effect transistor called a unipolar transistor? Draw the diagram of an *n*-channel transistor and explain the terms drain, gate and source.
2. What is the significance of the term "field-effect" in the name JFET? Provide the circuit symbol of the JFET.
3. With the aid of a neat sketch of an *n*-channel transistor explain the principle of its operation.
4. Explain the nature of the typical common-source drain characteristics of a JFET. What is its transfer characteristic?
5. Explain the terms saturation current and saturation voltage in connection with the common-source drain characteristic. How do they vary when the reverse gate-bias increases?
6. How can you obtain the static characteristics of a JFET?
7. When is the channel of a JFET said to be pinched-off?
8. Define pinch-off voltage.
9. Give the relation between the pinch-off voltage, the saturation voltage and the gate–source voltage. What is the pinch-off current?
10. Sketch the depletion region before and after pinch-off condition, and obtain the significance of the term pinch-off.
11. What are the significant differences between enhancement-type and depletion-type MOSFETs? What type of gate voltage is required to induce the flow of charges through the channel of an *n*-channel MOSFET?

12. Which is a faster device as far as the operation is concerned: a BJT or an FET? Justify your answer.
13. Why is the FET called a voltage amplifier and the BJT a current amplifier?
14. Explain the total operation of the MOSFET device citing the regions of operation using band diagrams.
15. Explain the cause of change in the band pattern under external bias and give the band diagrams for various regions of operation.
16. Describe the operation of a CMOS circuit.
17. Design a NAND gate using NMOS and PMOS.
18. Design a NAND gate using CMOS.
19. Discuss the design of CMOS as a combination of PMOS and NMOS sub-circuits.
20. Draw the transfer characteristics of a CMOS amplifier and describe its operation from the transfer curve.
21. In what way does the rudimentary analysis of MOSFETs yield a square law model? Name the two most important operating regimes.
22. How does MOSFET transconductance depend on input voltage?
23. Does the load device really affect the turn-on or turn-off properties of the inverter and why?
24. Which factors primarily cause the switching delays and why?
25. What is meant by band bending and how does it happen?
26. Describe the field profile in an ideal MOS capacitor in depletion but well below threshold. Explain why its major features exist.
27. Describe the oxide-silicon boundary conditions for an ideal MOS capacitor.
28. Explain vacuum level and work function.
29. Explain electron affinity.

## PRACTICE PROBLEMS

1. With 2 $\mu$m, what must be the relative width of $n$-channel and $p$-channel devices if they are to have equal conductance parameters?
2. An $n$-channel device has $k_n' = k_n'$. The device is to operate as a switch for small $v_{DS}$ utilizing a control voltage of $v_{GS}$ in the range of 0−5 V. Find the switch closure resistance and the closure voltage obtained when $v_{GS} = 5$ V and $i_D = 1$ mA.
3. In Problem 2, if $\mu_p = 0.4\ \mu_n$, what must be the $W/L$ of the $p$-channel device that provides the same performance as the $n$-channel device?
4. An $n$-channel MOSFET with oxide thickness 20 nm, minimum gate length 1 $\mu$m and $V_t =$ 0.8 V, operates in the triode region with the gate–source voltage in the range 0−5 V. What device width is needed to ensure that the minimum available resistance is 1 k$\Omega$?
5. Consider an $n$-channel MOSFET with $t_{ox} = 20$ nm, $v_t = 0.8$ V and $W/L = 10$. Find the drain current for the following cases:
   (a) $v_{GS} = 5$ V and $V_{DS} = 1$ V
   (b) $V_{GS} = 2$ V and $v_{DS} = 1.5$ V
6. The design of a $n$-MOS power transistor capable of saturation mode operation with $i_D = 1$ A for $v_{DS}$ as 2.5 V is being considered. For the available process technology, $V_t = 0.9$ V and the gate oxide thickness is 2 $\mu$m. For this process, what are the values of $C_{ox}$ and $k_n'$?
7. For Problem 6, what is the channel width required?
8. A particular enhancement-type MOSFET for which $V_t = 1$ V and $k_n'\ (W/L) = 0.1$ mA/V$^2$ is to be operated in the saturation region. If $i_D$ is to be 0.2 mA, find the required $v_{GS}$.
9. In Problem 8, find the minimum required voltage $v_{DS}$.
10. A particular $n$-channel enhancement-type MOSFET is measured to have a drain current of 4 mA at $V_{GS} = V_{DS} = 9$ V and 1 mA at $V_{GS} = V_{DS} = 5$ V. What are the values of $k_n'(W/L)$ and $V_t$ for the device?
11. For a transistor, the transconductance parameter $k_n' = 50\ \mu$A/V$^2$ and $V_t = 1$ V. In an application where $v_{GS} = v_{DS} = 5$ V and the drain current is 0.8 mA with the device minimum length of 2 $\mu$m, what must be the width that the designer should use?

12. Consider an enhancement-type MOSFET operated at constant gate-to-source voltage. Show that the drain current decreases to a fraction $\alpha$ of the value at the onset of saturation $\alpha$. For $V_t = 1$ V and $V_{GS} = 2$ V, find $V_{DS}$ for $\alpha = 0.5$ and 0.1.
13. For an $n$-MOS transistor, with $V_t = 0.8$ V, operating with $v_{GS}$ in the range of 1.5−4 V, what is the largest value of $v_{DS}$ for which the channel remains continuous?
14. An $n$-MOS transistor fabricated with $W/L = 100/5$, for which $k_n' = \lambda = 0.02\ V^{-1}$ and $V_t = 1$ V, is to be operated at very low values as a linear resistor. For $v_{GS}$ varying from 1.1−11 V, what range of resistor values can be obtained?
15. In Problem 14, what is the range of resistors available if the device width is halved, and when both the device width and length are halved?
16. In a device, for which the gate oxide thickness is 20 nm, find the value of $N_A$ for which $\gamma = 0.5\ V^{1/2}$. If the doping level is maintained but the gate thickness is increased to 100 nm, what is the resulting value of $\gamma = 0.5\ V^{1/2}$?
17. An enhancement PMOS transistor has $k_P\ (W/L) = 80\ \mu A/V^2$ and $v_t = -1.5$ V. The gate is connected to ground and the source to 6 V. Find the drain current for $v_D = 1.6$ V, 0 V and −3 V.
18. A PMOS operates in saturation with $V_t = 1$ V, $|v_{GS}| = 3$ V, $|v_{DS}| = 4$ V and $i_D = 3$ mA. Find the corresponding values for $V_{GS}$, $V_{DS}$ and $V_t$.
19. A depletion $n$-type transistor with $k_P\ (W/L) = 2\ \mu A/V^2$ and $V_t = -3$ V has its source and gate grounded. Find the region of operation and the drain current for $V_D = 0.1$ V, 3 V, 5V and 0 V.
20. For a particular depletion-type $n$-MOS $V_t = -2$ V, $k_n\ (W/L) = 200\ \mu A/V^2$ and $\gamma = 0.02 V^{-1}$. When operated with $v_{GS} = 0$, what is the drain current that flows through for $v_{DS} = 1$, 2 and 10 V?
21. In the Problem 20, what do the currents become if: (a) the width is doubled the width is halved? (b) both the width and length are doubled?
22. A depletion-type n-MOS operating in the triode region with $v_{DS} = 0.1$ V, conducts a drain current of 1 mA at $v_{GS} = -1$ V and 3 mA for $v_{GS} = +1$ V. Find $I_{DSS}$ and $V_t$.
23. A depletion-type $n$-MOS transistor operating in the saturation region with $v_{DS} = 5$ V conducts a drain current of 1 mA for a $v_{GS} = -1$ V and 9 mA for $v_{GS} = 1$ V. Find $I_{DSS}$ and $V_t$.

## SUGGESTED READINGS


1. Mead, C.A. "Schottky Barrier Gate Field Effect Transistor". February 1966. *Proceedings of the IEEE* 2: 307−8.
2. Grove, A. S. 1967. *Physics and Technology of Semiconductor Devices*. New York: Wiley.
3. Kahng, D. July 1976. "A Historical Perspective on the Development of MOS Transistors and Related Devices". *IEEE Transactions on Electron Devices* 7: 655−7.
4. Hofstein, S. R. and F. P. Heiman. September 1963. "The Silicon Insulated-Gate Field-Effect Transistor". *Proceedings of the IEEE* 9: 1190−1202.
5. Sze, S. M. 1981. *Physics of Semiconductor Devices*. New York: Wiley.
6. Singh, J. 1994. *Semiconductor Devices: An Introduction*. New York, NY: McGraw-Hill.
7. Streetman, B. G. and S. Banerjee. 2000. *Solid State Electronic Devices*. New Delhi: Pearson Education.
8. Millman, Jacob and Christos C. Halkias. 1986. *Integrated Electronics: Analog and Digital Circuits and Systems*. New Delhi: McGraw Hill Book Company.
9. Pierret, R. F. and G.W. Neudeck. 1989. *Modular Series on Solid State Devices*. Boston, MA: Addison Wesley.
10. Singh, B. P. and Rekha Singh. 2006. *Electronic Devices and Integrated Circuits*. New Delhi: Pearson Education.

# 7

# FET Circuits

## Outline



## Objectives

This chapter begins with a discussion on the FET biasing technique and proceeds to formulate the working of the FET as an amplifier. AC equivalent circuit is provided for Small-signal analysis, and all possible kinds of problems are solved. After this, the high-frequency MOSFET model has been explained, followed by a discussion on a number of additional FET circuits supplemented with detailed theoretical investigations. Lastly, a comparison between the FET and the BJT is provided with a mapping of their respective advantages and disadvantages.

## 7-1 INTRODUCTION

The first modern day field-effect device was proposed and analysed by W. Shockley in 1952. The operational JFET was subsequently designed by Ross and Dancy in 1953. In the previous chapter we studied the basics of the FET as a three-terminal unipolar voltage-controlled device. Now let us move forward toward and understand the FET as an active circuit element. The various configurations of the FET, their corresponding features, and related operations will be examined.

Before we begin, a glance into the relative comparison between the BJT and the FET is required. One of the most important characteristics of the FET is its high input impedance. Contrary to this, the BJT has a much higher sensitivity to changes in the applied signal. In other words, the variation in the output current is typically a great deal more for the BJT than for the FET for the same change in the input voltage. Generally, the FET is more thermally stable than the BJT, and is often of a smaller size than the BJT. Thus, the FET, especially the GaAs FET is used in core devices meant for very high-scale monolithic integrated circuits, such as the microwave.

## 7-2 FET BIASING

In the FET, the relationship between the input and output quantities is governed by the non-linear Shockley's equation. The current–voltage relationship can be written as:

$$I_D = I_{DSS}\left(1 - \frac{V_{GS}}{V_P}\right)^2 \tag{7-1}$$

In the case of the JFET and MOSFET, and also for the enhancement-type MOSFET, Eq. (7-1) gets modified as:

$$I_D = k(V_{GS} - V_T)^2 \tag{7-2}$$

It is essential to note that these equations do not change with each network configuration as long as the device is being operated in the active region. The biasing assembly simply defines the level of current and voltage associated with the operating point through its own set of equations.

### 7-2-1 Fixed-Bias Arrangement

The FET fixed-bias arrangement is the simplest of all the biasing arrangements. The fixed-bias arrangement and its corresponding network for dc analysis are shown in Fig 7-1.

In Fig. 7-1(a), $v_i$ and $v_o$ are the input and output ac voltages, and $C_1$ and $C_2$ are the coupling capacitors. The coupling capacitors become open-circuited for the dc analysis, and have low impedance for the ac analysis. The resistor $R_G$ ensures that for the ac analysis, the voltage $V_i$ appears across the input source of the circuit.

For dc analysis, $I_g = 0$ A and $V_{RG} = I_g R_G = 0$ V. This zero voltage drop across the resistor enables us to replace the resistor $R_G$ by a short, and hence, Fig. 7-1(b) is obtained. Thus, applying Kirchoff's law across the input loop, $V_{GS} = V_{GG}$. As the fixed supply is the dc voltage, and the voltage does not get divided anywhere, a fixed-bias configuration is obtained. Consequently, the drain current is given by:

$$I_D = I_{DSS}\left(1 - \frac{V_{GS}}{V_P}\right)^2 \tag{7-3}$$

which is the same as Eq. (7-1).

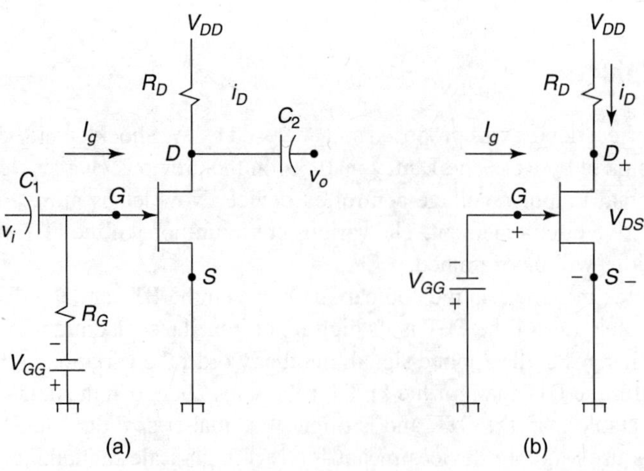


**Figure 7-1** (a) Fixed-bias circuit (b) Corresponding network for dc analysis

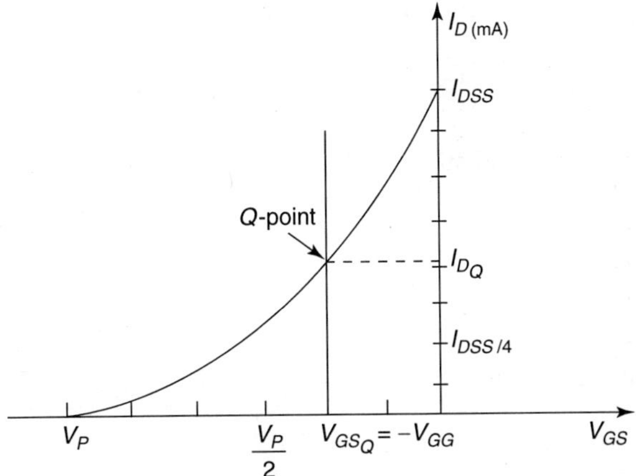


**Figure 7-2** Shockley's equation and calculation of the operating point

Since the gate-to-source voltage in this case is a fixed quantity, the corresponding sign can be replaced in the above Shockley's equation, and the value of $I_D$ can be calculated. If the drain current $I_D$ [from Eq. (7-3)] is plotted against $V_{GS}$, we obtain the following curve as shown in Fig. 7-2.

Figure 7-2 shows that the operating $Q$-point can also obtained by simply superimposing the vertical line at $V_{GS} = -V_{GG}$. This intersection also gives the drain current $I_{DQ}$ in the given situation.

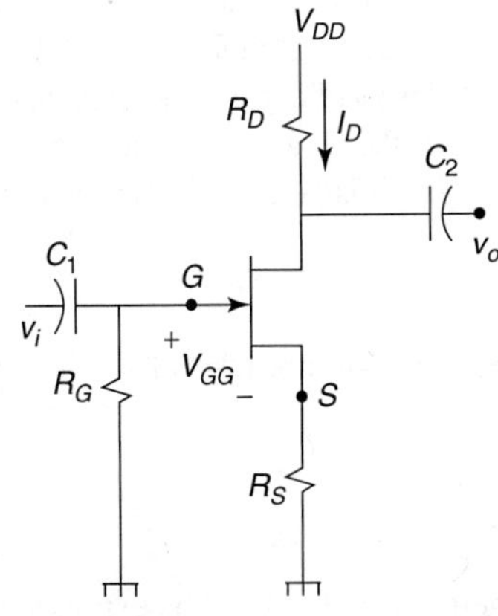


**Figure 7-3** Self-bias arrangement

## 7-2-2 Self-Bias Arrangement

The self-bias arrangement is a better biasing arrangement as it eliminates the need for two dc supplies. The controlling gate-to-source voltage is being determined by the voltage across the resistor $R_s$ introduced in the series with the source of Fig. 7-3.

In case of dc analysis, the capacitors are replaced by open circuits and the circuit is modified, as shown in Fig. 7-4.

In Fig. 7-4, the current through the resistor at the source leg, $R_S$, is the drain current, for which the voltage drop across $R_S$ is:

$$V_{R_S} = I_S R_S = I_D R_S \tag{7-4}$$

Again, considering the loop including the gate and the drain in Fig. 7-4, we can write:

$$V_{GS} = -I_D R_S \tag{7-5}$$

It is essential to note that the gate-to-source voltage $V_{GS}$ is a function of the output current, and this is remarkably different in comparison to the fixed bias arrangement.

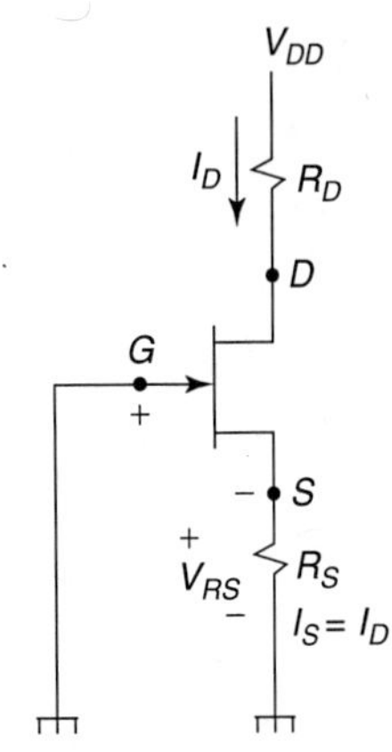


**Figure 7-4** Simplified circuit for dc analysis

Applying Eq. (7-5) in Eq. (7-1), we obtain:

$$I_D = I_{DSS}\left(1 - \frac{V_{GS}}{V_P}\right)^2 = I_{DSS}\left(1 + \frac{I_D R_S}{V_p}\right)^2 \quad (7\text{-}6)$$

The solution of Eq. (7-6) leads to a quadratic equation which can be obtained for a solution of $I_D$.The operating point can be obtained by superimposing the characteristic curve and the expression for the gate-to-source voltage as obtained from Eq. (7-5). Thus, we obtain the curve as shown in Fig. 7-5. The $Q$-point is shown at the intersection point on the characteristic curve.

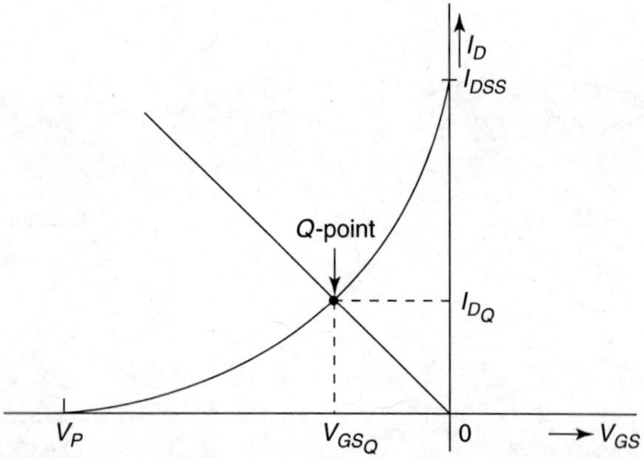


**Figure 7-5** Sketch of the self-bias circuit along with the Q-point.

## 7-2-3 Voltage Divider Biasing Arrangement

The basic configuration in this case is the same as that applied for the BJT, although the dc analysis is quite different. The gate current for the FET is zero, but the magnitude of $I_B$ for the common-emitter BJT amplifiers can affect the dc levels of current and voltage in both the input and output circuits. The circuit is as shown in Fig. 7-6.

For the dc analysis, the bypass capacitors included in the circuit should be replaced by an open circuit. The circuit is simplified for easy analysis. This enables us to find the Thevenin voltage $V_G$ as:

$$V_G = \frac{R_2 V_{DD}}{R_1 + R_2} \tag{7-7}$$

Applying Kirchhoff's voltage law in the clockwise direction to the gate-source circuit, we obtain:

$$V_G - V_{GS} - V_{R_S} = 0 \tag{7-8}$$

We can also write:

$$V_{GS} = V_G - V_{R_S} \tag{7-9}$$

where, the value of $V_{R_S}$ is expressed as:

$$V_{R_S} = I_S R_S = I_D R_S \tag{7-10}$$

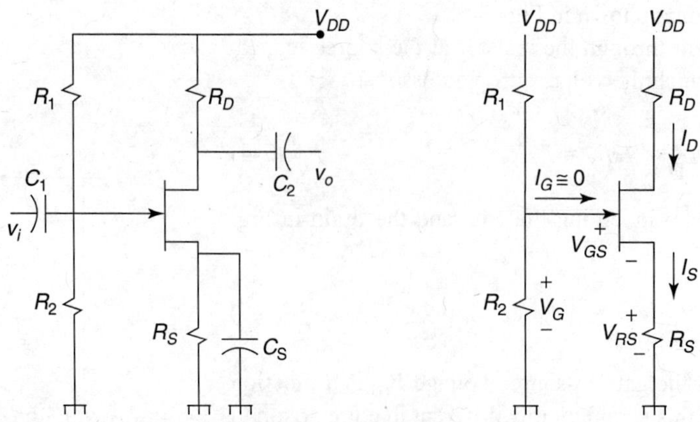


**Figure 7-6** (a) Voltage divider bias circuits (b) Circuit for dc analysis

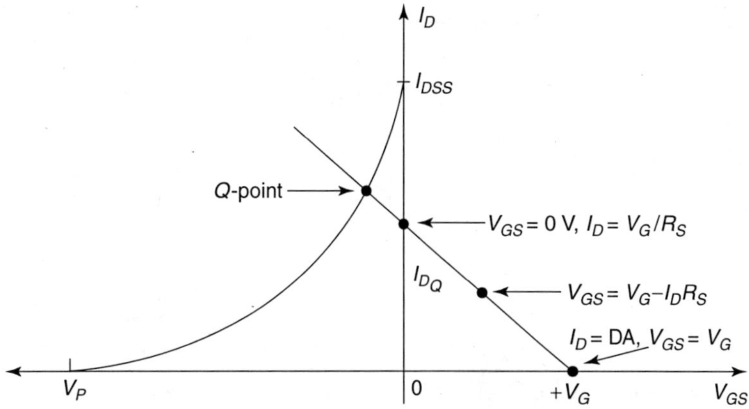


**Figure 7-7** Determination of the Q-point for the voltage divider bias configurations

Therefore, by substituting the value of $V_{R_S}$ from Eq. (7-10) in Eq. (7-9) we can write:

$$V_{GS} = V_G - I_D R_S \tag{7-11}$$

An analysis of Eq. (7-11) reveals the fact that the equation includes the same variables that appear in the Shockley's equation. This also includes $V_G$ and $R_S$ which are determined by the network in consideration. Equation (7-11) is a relation of a straight line, and their points of intersection with the axis are found as follows:

For $I_D = 0$,
$$V_{GS} = V_G - 0 = V_G \tag{7-12a}$$

and for, $V_{GS} = 0$,
$$I_D = \frac{V_G}{R_S} \tag{7-12b}$$

Figure 7-7 shows the points of intersection and the subsequent determination of the Q-point from the graphical analysis for the voltage divider bias configurations.

## 7-3 FET AS AN AMPLIFIER

In this section, we shall study the operation of the FET as a common-source amplifier. This observation is made by considering an enhancement-type MOSFET biased by a dc voltage $V_{GS}$, with the time-varying input signal $v_{gs}$ superimposed on it. The investigation begins with the calculation of the dc bias point.

### 7-3-1 DC Bias Point

To operate the MOSFET as an amplifier, it must be biased in the saturation region where we obtain a linear variation between the input voltage and the output voltage. The circuit for the study of MOSFET as an amplifier is shown in Fig. 7-8.

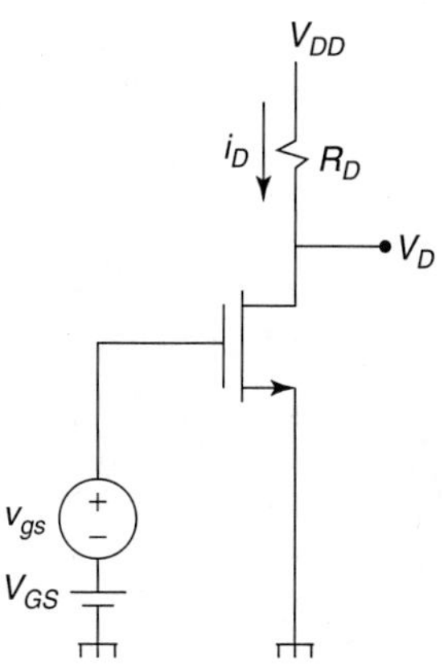


**Figure 7-8** Circuit for the study of the MOSFET as an amplifier

With the input signal set to zero, the drain current can be expressed as:

$$I_D = \frac{1}{2} k_n' \frac{W}{L} (V_{GS} - V_T)^2 \tag{7-13}$$

(neglecting the channel-width modulation).

Also from Fig. 7-8, we obtain the KVL for the output part of the circuit as:

$$V_D = V_{DD} - R_D I_D \tag{7-14}$$

The condition that has to be maintained in order to keep the device operating in the saturation region is:

$$V_D > V_{GS} - V_T \tag{7-15}$$

After considering the calculation of the dc bias point, the instantaneous drain current is given by:

$$i_D = \frac{1}{2} k_n' \frac{W}{L} (V_{GS} + v_{gs} - V_t)^2$$

$$= \frac{1}{2} k_n' \frac{W}{L} (V_{GS} - V_t)^2 - k_n' \frac{W}{L} (V_{GS} - V_t) v_{gs} + \frac{1}{2} k_n' \frac{W}{L} v_{gs}^2 \tag{7-16}$$

In Eq. (7-16), we observe that the first term on the right hand side results in the dc bias current $I_D$. Now, comparing the second and the last terms, it is seen that the last term introduces an unwanted distortion, and thus, Eq. (7-16) can be equivalently written as:

$$i_D = I_D + i_d \tag{7-17}$$

where, $i_d = k_n' \frac{W}{L} (V_{GS} - V_t) v_{gs}$.

This is obtained only on the condition that:

$$\frac{1}{2} k_n' \frac{W}{L} v_{gs}^2 \ll k_n' \frac{W}{L} (V_{GS} - V_t) v_{gs}$$

This results in:

$$v_{gs} \ll 2(V_{GS} - V_t) \tag{7-18}$$

Thus, from the definition of transconductance, we obtain:

$$g_m \equiv \frac{i_d}{v_{gs}} = k_n' \frac{W}{L} (V_{GS} - V_t) \tag{7-19}$$

## 7-3-2 Voltage Gain of the FET

Voltage gain of the FET is defined as the ratio of drain voltage ($v_d$) and gate-to-source voltage ($v_{gs}$), and is given by:

$$A_v = \frac{V_D}{V_{gs}} \tag{7-20a}$$

The expression for the total instantaneous output voltage, obtained by considering the circuit as shown in Fig. 7-8, is given by:

$$v_D = V_{DD} - i_D R_D \tag{7-20b}$$

Under the condition of small-signal analysis, substituting the value of $i_d$ from Eq. (7-17), we get:

$$v_D = V_{DD} - (I_D + i_d)R_D \tag{7-20c}$$

Equation (7-20c) can be approximated as:

$$v_D = V_D - R_D i_d \tag{7-21}$$

Thus, the signal component of the drain voltage is:

$$v_d = -i_d R_D = -g_m R_D v_{gs} \tag{7-22}$$

The negative sign indicates that the output drain voltage is exactly 180° out of phase with respect to the input signal. Thus, the voltage gain is given by:

$$A_v = \frac{V_D}{V_{gs}} = -g_m R_D$$

or,

$$A_v = -g_m R_D \tag{7-23}$$

## 7-4 ELECTRICAL PARAMETERS OF THE FET

In the FET, the drain current is a function of the drain-to-source voltage and gate-to-source voltage. For small-signal analysis, the ac component of the drain current is expressed as a combination of the ac component drain-to-source voltage $v_{ds}$, and gate-to-source voltage $v_{gs}$, in the following way:

$$i_d = g_m v_{gs} + \frac{v_{ds}}{r_d} \tag{7-24}$$

In Eq. (7-24), $g_m$ represents the mutual conductance or transconductance, and the unit of $g_m$ is mho or siemens or A/V. The $g_m$ is expressed as a ratio of drain current to the corresponding gate-to-source voltage for constant drain-to-source voltage.

$$g_m = \frac{i_d}{v_{gs}}, \quad \text{for } v_{ds} = 0 \tag{7-25}$$

In Eq. (7-24), $r_d$ represents the ac resistance or channel resistance, and the unit is in ohms. It is the ratio of the drain-to-source voltage to the drain current for a constant gate-to-source voltage, and can be written as:

$$r_d = \frac{v_{ds}}{i_d}, \quad \text{for } v_{gs} = 0 \tag{7-26}$$

The amplification factor is expressed as the ratio of $v_{ds}$ to $v_{gs}$ for a constant drain current, and is given by:

$$\mu = -\frac{v_{ds}}{v_{gs}}, \quad \text{for } i_d = 0 \tag{7-27}$$

$\mu$ is a positive quantity. Substituting $i_d = 0$ in Eq. (7-24), we can write:

$$-\frac{v_{ds}}{v_{gs}} = r_d g_m \tag{7-28a}$$

$$\mu = r_d g_m \tag{7-28b}$$

Therefore, the amplification factor of the FET is the product of channel resistance and transconductance.

## 7-5 AC EQUIVALENT CIRCUIT FOR SMALL-SIGNAL ANALYSIS

From the previous analysis it is observed that that the signal quantities are superimposed on the dc quantities. For instance, the total drain current $i_D$ is given by the superimposition of two currents—the dc drain current $I_D$, and the small-signal current $i_d$. Similar is the case for the drain voltage $v_D$.

Upon examination, it is observed that the analysis becomes exceptionally simplified if the ac analysis is segregated from the dc analysis. In order to obtain some stable circuit configurations for their lucid analysis, the following circuit forms are studied.

### 7-5-1 Small-Signal Model for the MOSFET

If the circuit of an FET is considered from the signal point of view, it is found that the FET is basically a voltage-controlled current source. It provides a current of $g_m v_{gs}$ upon accepting a voltage of $v_{gs}$ applied between the gate and the source terminal. The input impedance of this controlled source is very high. Also, the output resistance is high, and is ideally assumed to be infinite. Using all these factors together Fig. 7-9 is obtained.

In the analysis of the FET amplifier circuits, the FET is replaced by the ac equivalent circuit. The rest of the circuit, i.e., the circuit configuration is left unaltered, and the dc voltage sources are replaced with a short circuit. The reason behind this replacement is that the voltage across an ideal dc voltage source does not change, and thus, there will always be a zero voltage change at the output. Similar statements should be used for the dc current sources, but with the only difference that their equivalent replacement is an open circuit.

The severe deficiency of the above depicted small-signal representation is that it has been assumed that the drain current in saturation is independent of the drain voltage. Also, from previous study, it should be noted that the drain current depends on $v_{DS}$ in a linear manner. Such dependence is easily depicted using the resistance $r_o$, whose value is given approximately by:

$$r_o \cong \frac{|V_A|}{I_D} \tag{7-29}$$

where, $V_A$ is the applied voltage.

Besides, the small-signal model parameters $g_m$ and $r_o$ depend on the dc bias point. As done previously, we can write:

$$\frac{v_d}{v_{gs}} = -g_m(R_D \| r_o) \tag{7-30}$$

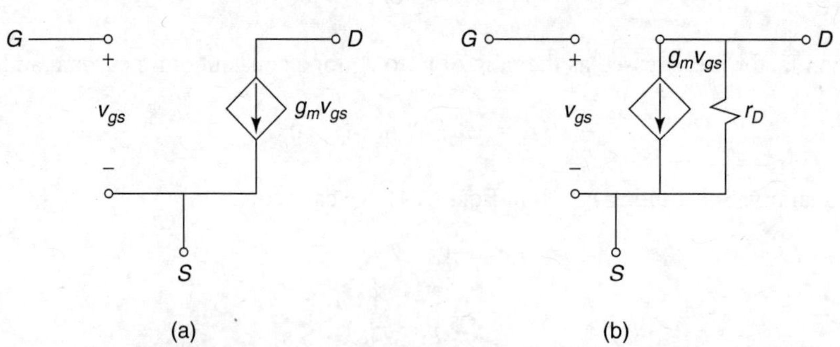

**Figure 7-9** Small-signal models of MOSFET: (a) Neglecting channel length modulation (b) Including channel length modulation

## FOR ADVANCED READERS

### T Equivalent-Circuit Model

A simple circuit transformation of the equivalent-circuit model for the MOSFET is called the T model. The circuit is depicted in Fig 7-10.

At this stage it is advisable to look at the transformation of the circuit into an equivalent T circuit. This is explained in the following steps:

(i) A second current source is added, i.e., $g_m v_{gs}$, in addition to the current source in series with the previous one. In doing so, we also make sure that the terminal currents are not changed, which ensures that we get an allowable transformation.

(ii) The newly created circuit node is attached to the gate terminal as shown Fig. 7-10.

(iii) It is notable that a controlled current source $g_m v_{gs}$ is connected across its control voltage $v_{gs}$.

(iv) The controlled source is replaced by a resistance as long as an equal current flows through this resistance. Therefore, the value of the resistance is:

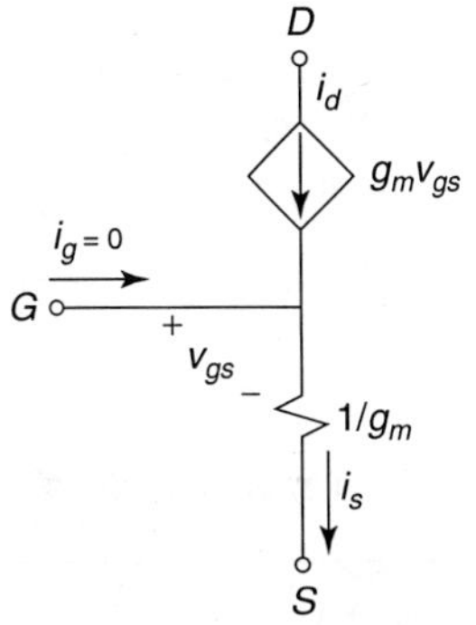

**Figure 7-10** T equivalent-circuit model

$$\frac{v_{gs}}{g_m v_{gs}} = \frac{1}{g_m}$$

The circuit with this replaced resistance is given in Fig. 7-11.

It is to be noted that in the development of the T model, we did not consider the inclusion of the resistance $r_o$ in the circuit. If such a model is required, then the resistance $r_o$ can be inserted between the drain and the source. Such a circuit is depicted in Fig. 7-12.

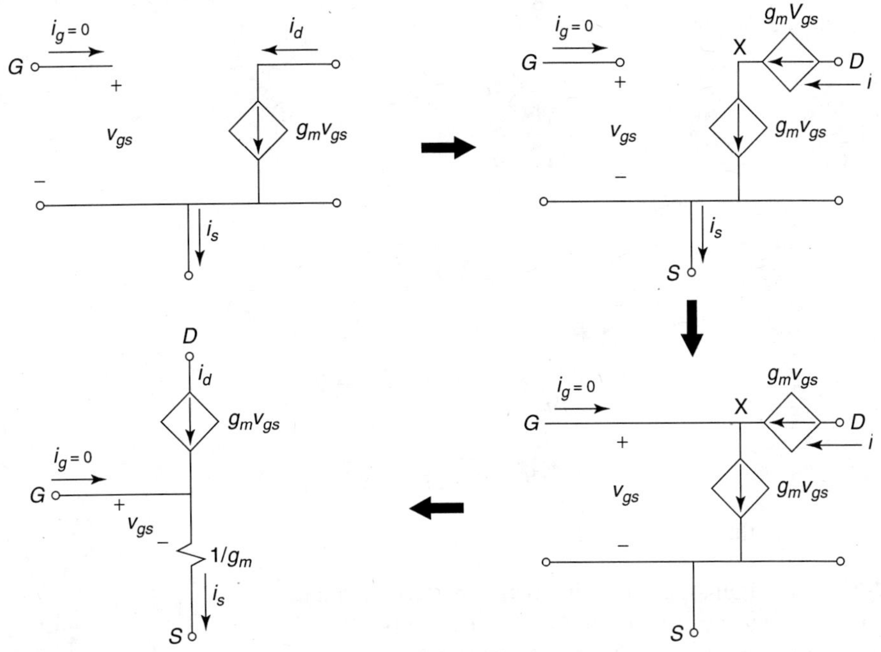

**Figure 7-11** Conversion of small-signal model into T equivalent-circuit model

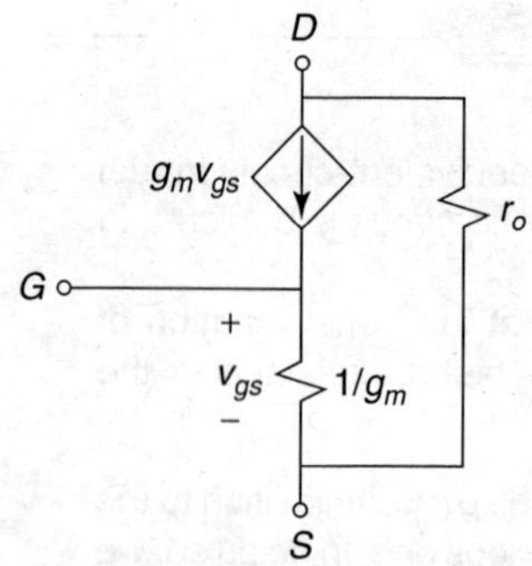


**Figure 7-12** T model circuit along with the drain-to-source resistance $r_o$

## Solved Examples

**Example 7-1** Derive an expression for voltage gain for the FET amplifier, as shown in the diagram.

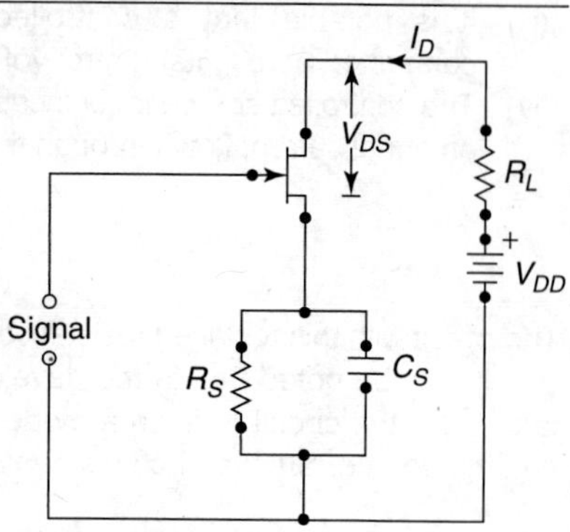


**Solution:**

The value of $R_S$ can be determined from the following relation:

$$R_S = \frac{V_{GS}}{I_D}$$

where, $V_{GS}$ = voltage drop across $R_S$

$I_D$ = current through $R_S$

Voltage gain of the FET amplifier is:

$$A_V = \frac{\mu R_L}{r_D + R_L}$$

As,

$$\mu = r_d \times g_m$$

Substituting the value of $\mu$, we get:

$$A_V = \frac{r_d g_m R_L}{r_d + R_L} = \frac{g_m R_L}{1 + \frac{R_L}{r_d}}$$

If $r_d >> R_L$:

$$\frac{r_d}{R_L} >> 1 \quad \text{and} \quad \frac{R_L}{r_d} \rightarrow 0$$

∴

$$A_V = \frac{g_m R_L}{1 + 0} = g_m R_L$$

or, $A = g_{fs} \times R_L$, i.e., the gain of the FET.

**Example 7-2** If an input signal $V_i$ is impressed between the gate and the ground, find the amplification $A_V = V_o/V_i$. Apply Miller's theorem to the 50 K resistor. The FET parameters are $\mu$ = 30 and $r_d$ = 5 K. Neglect capacitances.

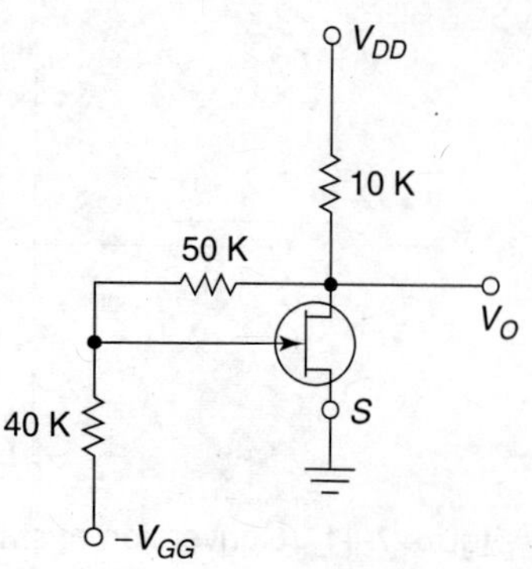

**Solution:**
Applying Miller's theorem to the 50 kΩ resistor, we obtain the following equivalent circuit.

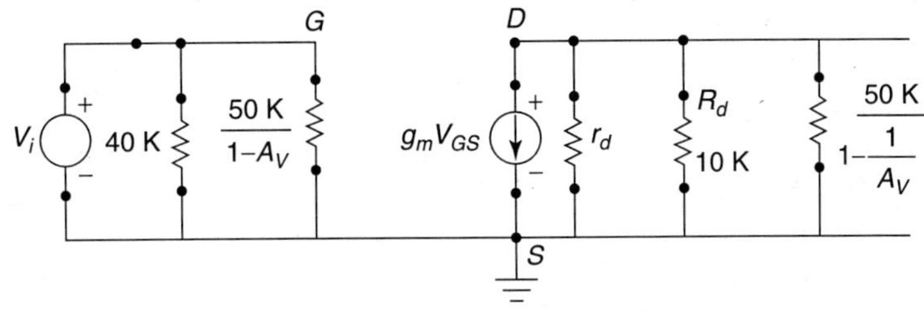


The voltage gain of the amplifier, as calculated in Example 7-1, is given by:

$$A_V = \frac{-g_m}{\dfrac{1}{r_d} + \dfrac{1}{R_d} + \dfrac{1 - \dfrac{1}{A_V}}{R}} = -g_m$$

or,
$$A_V\left(\frac{1}{r_d} + \frac{1}{R_d} + \frac{1}{R} - \frac{1}{RA_V}\right) = -g_m$$

or,
$$A_V\left(\frac{1}{r_d} + \frac{1}{R_d} + \frac{1}{R}\right) = -g_m + \frac{1}{R}$$

or,
$$A_V\left(\frac{1}{5} + \frac{1}{10} + \frac{1}{50}\right) = -\frac{30}{5} + \frac{1}{50} = -\frac{299}{50}$$

$$A_V\left(\frac{16}{50}\right) = -\frac{299}{50} \quad \therefore \quad A_V = -\frac{299}{16} = -18.7$$

**Example 7-3** If in Example 7-2, the signal $V_i$ is impressed in series with the 40 kΩ resistor (instead of from gate to ground), find $A_V = V_o/V_i$.

**Solution:**
From Example 7-2, $A_V = -18.7$.
If the signal is impressed in series with the 40 kΩ resistor:

$$A_{VS} = \frac{A_V R_1}{R_1 + 40\text{ K}}$$

where, $R_1 = \dfrac{50\text{ K}}{1 + 18.7} = 2.54\text{ K}$

Hence,
$$A_{Vs} = (-18.7)\,\frac{2.54}{1.54 + 40} = -1.11$$

**Example 7-4** The circuit, as shown in the diagram, is called the common-gate amplifier. For this circuit, find (a) the voltage gain (b) the input impedance and (c) the output impedance. Power supplies are omitted for simplicity. Neglect capacitances.

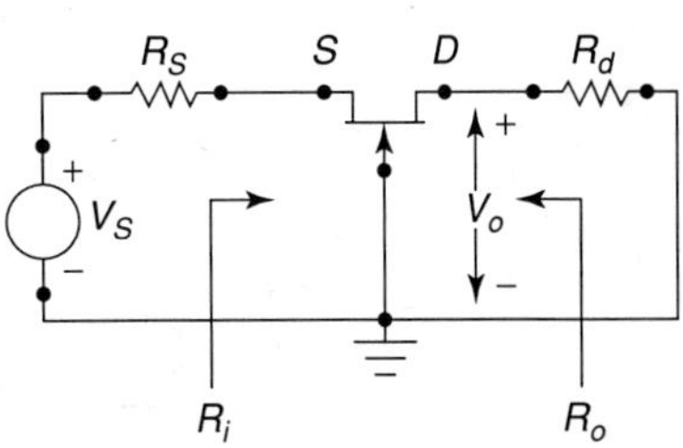


**Solution:**

(a) The small-signal equivalent circuit is shown in the following diagram.

Then, $\quad V_{gs} = -V_s + IR_s \qquad (1)$

Applying KVL around the loop we have:

$$V_s - \mu V_{gs} = I(R_s + r_d + R_d)$$

Now, substituting $V_{gs}$ from Eq. (1), we have:

$$V_s(\mu + 1) = I\,[r_d + R_d + (\mu + 1)R_s]$$

or, $$\frac{I}{V_s} = \frac{(\mu + 1)}{[r_d + R_d + (\mu + 1)R_s]} \qquad (2)$$

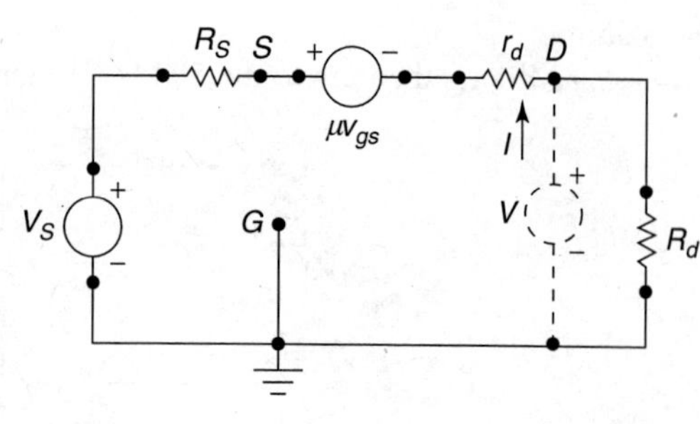


but, $\quad V_o = IR_d$

hence, $$A_V = \frac{V_o}{V_s} = \frac{(\mu + 1)R_d}{[r_d + R_d + (\mu + 1)R_s]}$$

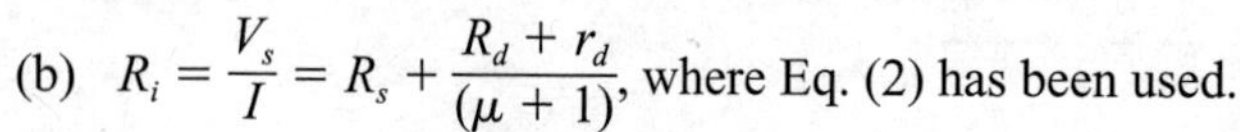

(b) $R_i = \dfrac{V_s}{I} = R_s + \dfrac{R_d + r_d}{(\mu + 1)}$, where Eq. (2) has been used.

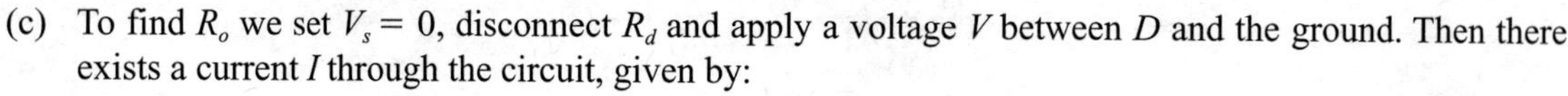

(c) To find $R_o$ we set $V_s = 0$, disconnect $R_d$ and apply a voltage $V$ between $D$ and the ground. Then there exists a current $I$ through the circuit, given by:

$$V + \mu V_{gs} = (r_d + R_s)I$$

but, $$V_{gs} = -IR_s$$

$\therefore$ $$V - \mu R_s I = (r_d + R_s)I$$

or, $$V = \mu R_s I + r_d I + R_s I$$

or, $$V = I\,[R_s(\mu + 1) + r_d]$$

or, $$R_o = \frac{V}{I} = r_d + (\mu + 1)R_s$$

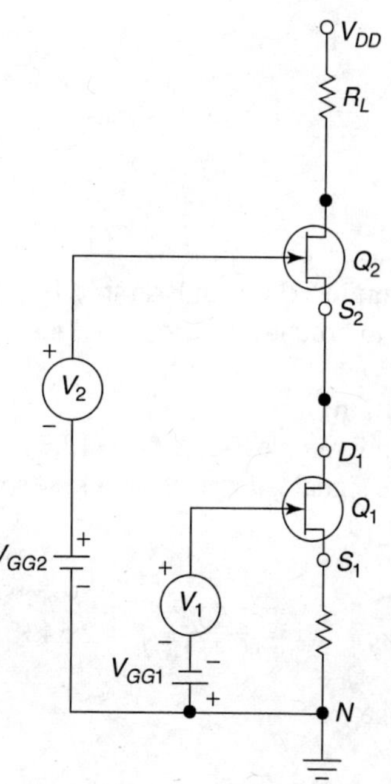


**Example 7-5** Find an expression for the signal voltage across $R_L$. The two FETs are identical, with parameters $\mu$, $r_d$ and $g_m$.
**Hint:** Use the equivalent circuits for the generalized amplifier of a common drain arrangement at $S_2$ and $D_1$.

**Solution:**
Using small-signal equivalent circuit for FET $Q_2$ and for $Q_1$, we have the following figure.

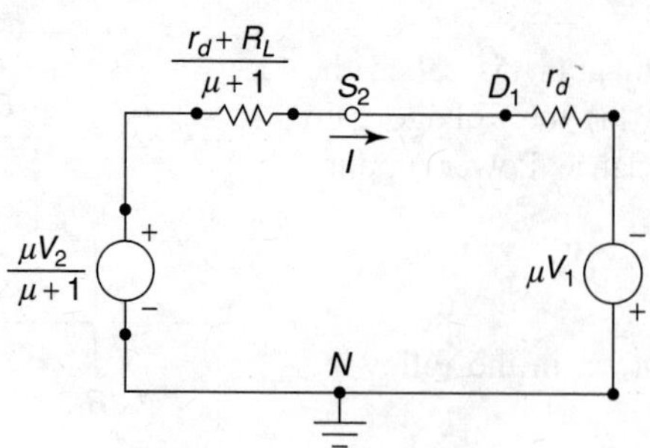

Applying KVL around the loop, we obtain:

$$\frac{\mu V_2}{\mu + 1} + \mu V_1 = I\left[r_d + \frac{r_d + R_L}{\mu + 1}\right]$$

Solving for $I$, we obtain:

$$I = \frac{\mu}{(\mu + 2)r_d + R_L} V_2 + \frac{\mu(\mu + 1)}{(\mu + 2)r_d + R_L} V_1$$

but

$$V_L = -IR_L$$

$\therefore$

$$V_L = -\frac{\mu R_L}{(\mu + 2)r_d + R_L}[V_2 + (\mu + 1)V_1]$$

**Example 7-6** Each FET, as shown in the diagram has parameters $r_d = 15$ K and $g_m = 2$mA/V. Using the equivalent circuits at $S_2$ and $D_1$, find the gain $v_o/v_i$ under the conditions:

(a) If $v_2 = 0$

(b) If $v_1 = 0$

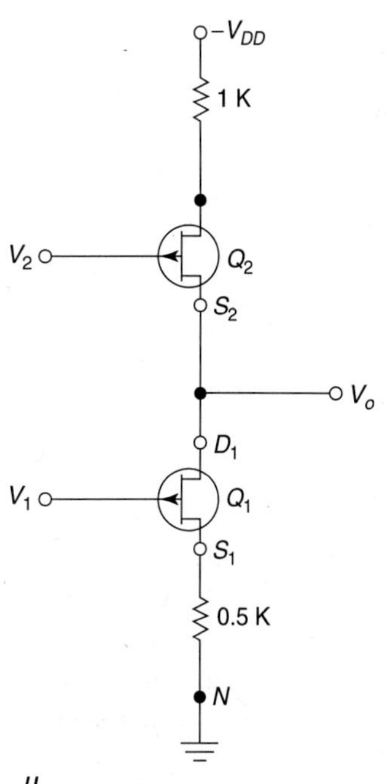


**Solution:**

Using super position we obtain:

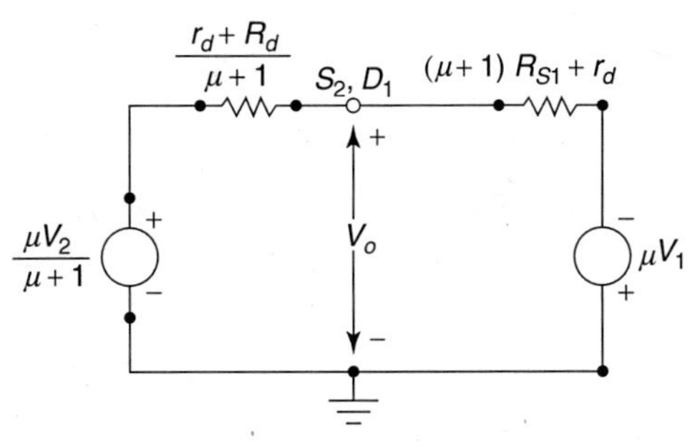


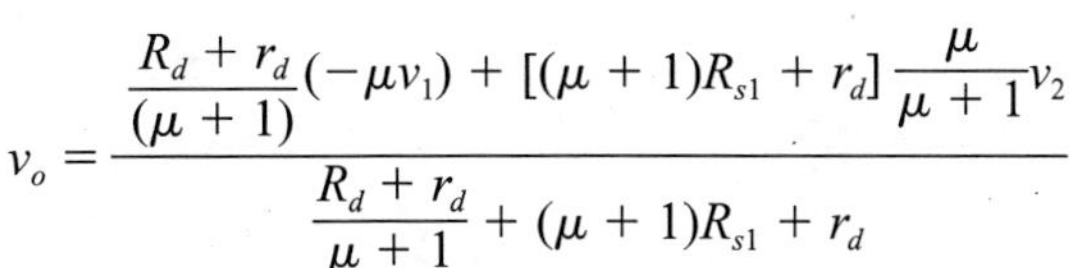

$$v_o = \frac{\frac{R_d + r_d}{(\mu + 1)}(-\mu v_1) + [(\mu + 1)R_{s1} + r_d]\frac{\mu}{\mu + 1}v_2}{\frac{R_d + r_d}{\mu + 1} + (\mu + 1)R_{s1} + r_d}$$

(a) If $v_2 = 0$:

$$\frac{v_o}{v_1} = \frac{-\mu(R_d + r_d)}{R_d + (\mu + 1)^2 R_{s1} + (\mu + 2)r_d}$$

$$= \frac{-30(16\text{ K})}{(1 + 31^2 \times 0.5 + 32 \times 15)\text{K}}$$

$$= -\frac{480}{961.5} = -0.499$$

(b) If $v_1 = 0$:

$$\frac{v_o}{v_2} = \frac{\frac{\mu}{\mu+1}[(\mu+1)R_{s1} + r_d]}{\frac{R_d + r_d}{\mu+1} + (\mu+1)R_{s1} + r_d}$$

$$= \frac{30 \times 0.5 + \frac{30 \times 15}{31}}{\frac{16}{31} + 31(0.5) + 15} = \frac{29.52}{31.02} = 0.952$$

**Example 7-7** (a) Prove that, $r = 1/g_m + 2R_L/\mu$, provided that the magnitude of the signal current is the same in both the FETs. Neglect the reactance of the capacitors. (b) If $r$ is chosen, as in part (a), prove that the voltage gain is given by:

$$A = \frac{-\mu^2}{\mu+1} \frac{R_L}{R_L + \frac{r_d}{2}}$$

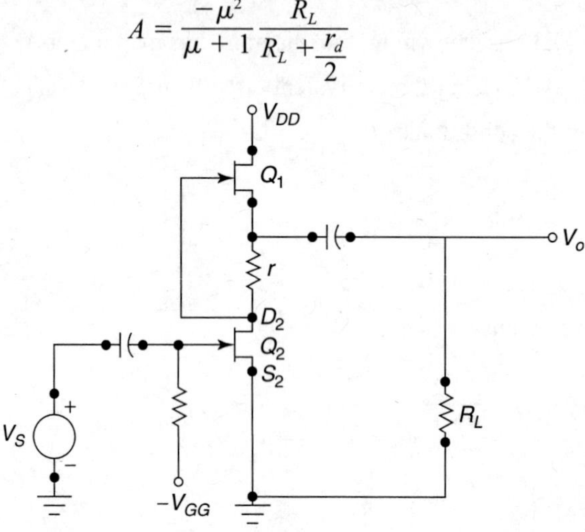


**Solution:**

(a) If the equivalent circuit is under the condition that the signal current is the same in both FETs as indicated, KVL in loop 1 gives:

$$2IR_L + \mu V_{gs1} + Ir_d = 0$$

where, $$V_{gs1} = -Ir$$

Thus, $$I(2R_L + r_d) = I\mu r$$

or, $$r = \frac{2R_L}{\mu} + \frac{1}{g_m}$$

(b) KVL in loop 2 gives $(2R_L + r + r_d)I = \mu V_s$

Hence, $$A = \frac{V_o}{V_s} = \frac{2IR_L}{V_s} = -2R_L \frac{\mu}{2R_L + r + r_d}$$

Using the result of part (a), we obtain:

$$A = \frac{-2R_L\mu^2}{(\mu+1)(2R_L + r_d)} = \frac{-\mu^2}{\mu+1} \times \frac{R_L}{R_L + \frac{r_d}{2}}$$

**Example 7-8** (a) If $R_1 = R_2 = R$ and the FETs have identical parameters, verify that the voltage amplification is $V_o/V_s = -\mu/2$, and the output impedance is $1/2\ [r_d + (\mu + 1)R]$. (b) Given $r_d = 62$ K, $\mu = 10$, $R_1 = 2$ K and $R_2 = 1$ K. Find the voltage gain and the output impedance.

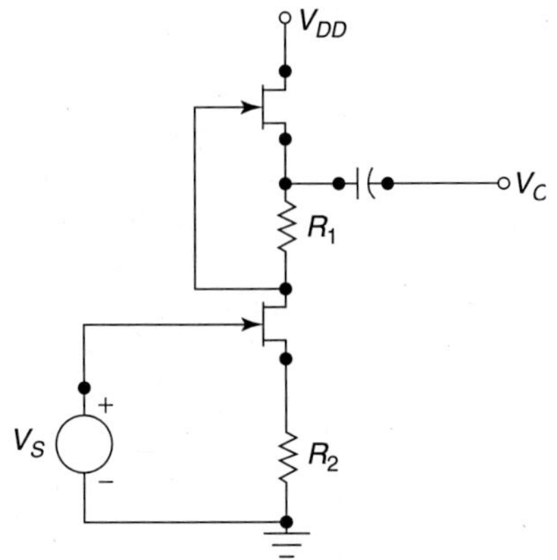


**Solution:**

(a) The small-signal equivalent circuit is shown in the following diagram.

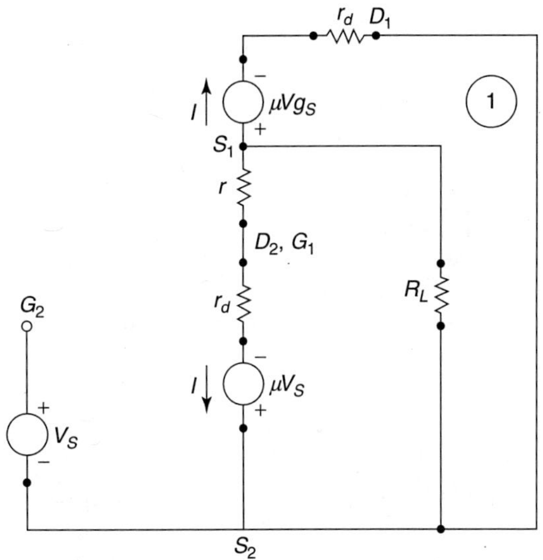


KVL in the loop gives:

$$\mu V_{gs1} + \mu V_s = [2r_d + (\mu + 2)R]I$$

But, $\quad V_{gs1} = -IR$

Hence, $\quad -\mu RI + \mu V_s = [2r_d + (\mu + 2)R]I$

or, $\quad I = \dfrac{\mu}{2[r_d + (\mu + 1)R]} V_s$

We notice that, $V_o = -I(r_d + R) + \mu V_{gs1} = -I[r_d + (\mu + 1)R]$

Hence, $$\frac{V_o}{V_s} = -\frac{\mu}{2}$$

To find the output impedance we ground $D_1$ and we have:

$$I_s = \frac{\mu V_s}{r_d + (\mu + 1)R}$$

For $D_2$ open circuit, we have exactly the $V_o$ that we had found previously:

$\therefore$ $$V_o = -\frac{\mu}{2} V_s$$

Hence, $$R_o = \frac{V_o}{I_s} = \frac{1}{2}[r_d + (\mu + 1)R]$$

(b) In this case KVL around the loop gives:

$$\mu V_{gs1} + \mu V_s = I[2r_d + (\mu + 1)R_2 + R_1]$$

or, $$-\mu R_1 I + \mu V_s = I[2r_d + (\mu + 1)R_2 + R_1]$$

or, $$I = \frac{+\mu}{[2r_d + (\mu + 1)(R_2 + R_1)]} V_s$$

and, $$V_o = -I(r_d + R_1) + \mu V_{gs1}$$

or, $$V_o = -I[r_d + (\mu + 1)R_1]$$

Hence, $$A = \frac{V_o}{V_s} = \frac{-\mu[r_d + (\mu + 1)R_1]}{[2r_d + (\mu + 1)(R_2 + R_1)]}$$

$$= \frac{-10(62 + 22)}{124 + 33} = \frac{-84}{157} \times 10 = -5.35$$

In this case: $$I_s = \frac{-\mu V_s}{r_d + (\mu + 1)R_2}$$

or, $$R_o = \frac{V_o}{I_s} = \frac{V_o}{V_s} \times \frac{V_s}{I_s} = A \times \left[\frac{r_d + (\mu + 1)R_2}{-\mu}\right]$$

$$= 5.3 \times 7.3 \text{ K} = 39 \text{ K}$$

**Example 7-9** (a) If in the amplifier stage, as shown in the diagram, the positive supply voltage $V_{DD}$ changes by $\Delta V_{DD} = v_a$, how much does the drain-to-ground voltage change? (b) How much does the source-to-ground voltage change under the conditions as given in (a)?

(c) Repeat (a) and (b) if $V_{DD}$ is constant, but $V_{SS}$ changes by $\Delta V_{SS} = v_s$.

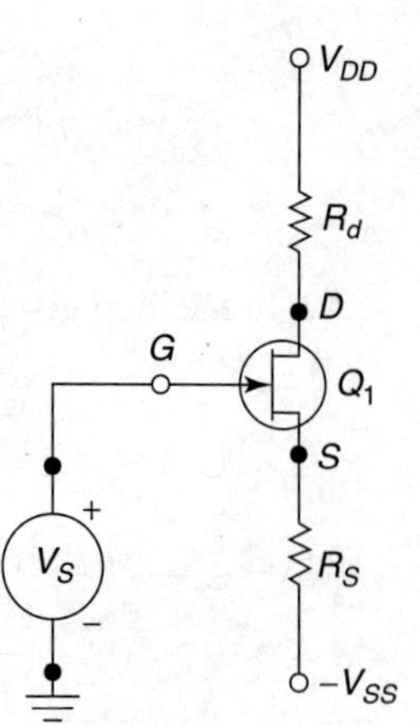


**Solution:**

(a) The equivalent circuit of this amplifier is shown in the following diagram:

A voltage source is in series with $R_d$ and $V_1 = 0$ (see diagram). KVL around the loop gives:

$$I = \frac{V_a}{R_d + r_d + (\mu + 1)R_s}$$

Notice, $\quad V_{dn} = I[r_d + (\mu + 1)R_s]$

or, $$V_{dn} = \frac{r_d + (\mu + 1)R_s}{R_d + r_d + (\mu + 1)R_s} \times V_a$$

(b) Also, $\quad V_{sn} = I(\mu + 1)R_s$

Hence, $$V_{sn} = \frac{(\mu + 1)R_s}{R_d + r_d + (\mu + 1)R_s} \times V_a$$

(c) The equivalent circuit is given in with a source in series with $R_s$ and $v_1 = 0$:

Hence, $$I = \frac{(\mu + 1)V_s}{R_d + r_d + (\mu + 1)R_s}$$

Now, $\quad V_{dn} = R_d \times I$

or, $$V_{dn} = \frac{(\mu + 1)R_d}{R_d + r_d + (\mu + 1)R_s}V_s$$

Similarly, $$V_{sn} = (r_d + R_d)I = \frac{(\mu + 1)\,(r_d + R_d)}{R_d + r_d + (\mu + 1)R_s}V_s$$

**Example 7-10** If in the circuit, as shown in the diagram, $V_2 = 0$, then this circuit becomes a source-coupled phase inverter, since $V_{o1} = -V_{o2}$. Solve for the current $I_2$ by drawing the equivalent circuit, looking into the source of $Q_1$ hen replace $Q_2$ by the equivalent circuit, looking into its drain. The source resistance $R_s$ may be taken to be arbitrarily large.

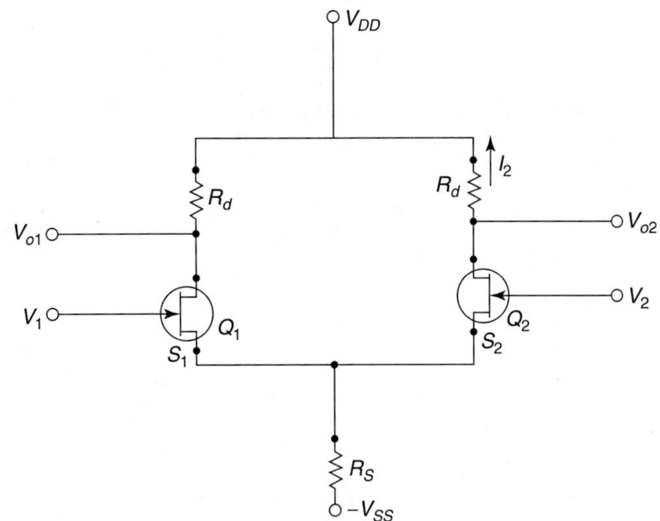

**Solution:**

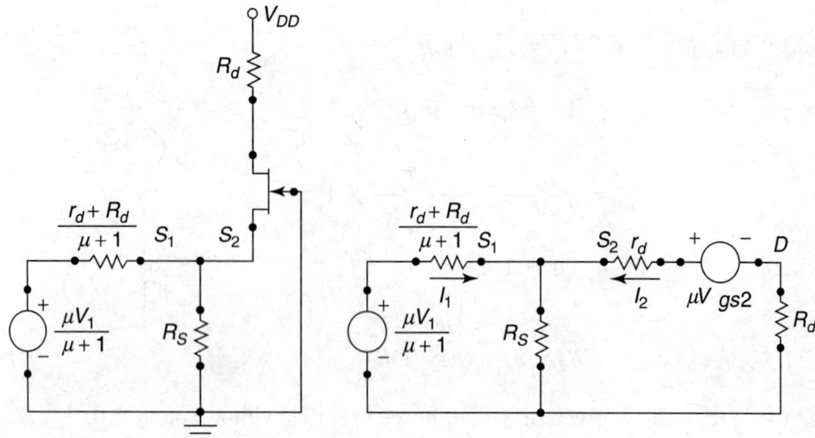


We note that:

$$V_{gs2} = -\frac{\mu}{\mu+1}V_1 + \frac{r_d + R_d}{\mu+1}I_1$$

Also, since $R_s$ is arbitrarily large, $I_1 = -I_2$.
Applying KVL around the loop we have:

$$I_1\left(\frac{r_d + R_d}{\mu+1} + r_d + R_d\right) = \frac{\mu}{\mu+1}V_1 + \frac{\mu^2}{\mu+1}V_1 - \frac{\mu}{\mu+1}(r_d + R_d)I_1$$

or,
$$I_1 \times 2(r_d + R_d) = \mu V_1$$

Hence,
$$I_1 = \frac{\mu V_1}{2(r_d + R_d)} = -I_2$$

but,
$$V_{o1} = -I_1 \times R_d = \frac{-\mu R_d}{2(r_d + R_d)}V_1$$

and,
$$V_{o2} = -I_2R_d = \frac{\mu R_d}{2(r_d + R_d)}V_1$$

**Example 7-11** In the circuit of Example 7-10, assume that $V_2 = 0$, $R_d = r_d = 10$ K, $R_s = 1$ K and $\mu = 19$. If the output is taken from the drain of $Q_2$, find (a) the voltage gain, (b) the output impedance.

**Solution:**

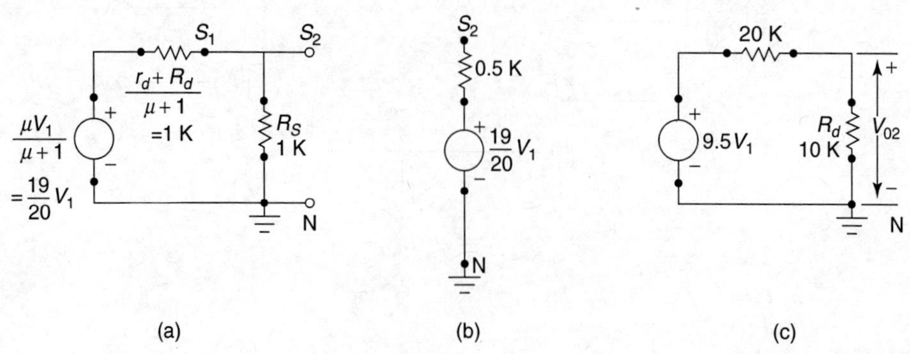

(a) Looking into the source $S_1$ we see the equivalent circuit, as shown in Fig. (a). The Thevenin equivalent of $S_2$ is indicated in Fig. (b). Looking into $D_2$ we see the circuit in Fig. (c), from which:

$$A = \frac{V_{o2}}{V_1} = \frac{9.5 \times 10}{10 + 20} = 3.17$$

(b) From Fig. (c), $R_o' = 10\text{ K} \| 20\text{ K} = \dfrac{20 \times 10}{30} = 6.67\text{ K}$

**Example 7-12** In the circuit of Example10, $V_2 \neq V_1$, $R_d = 30$ K, $R_s = 2$ K, $\mu = 19$, and $r_d = 10$ K. (a) Find the voltage gain $A_1$ and $A_2$ defined by $V_{o2} = A_1V_1 + A_2V_2$. (b) If $R_s$ is arbitrarily large, show that $A_2 = -A_1$. Note that the circuit now behaves as a different amplifier.

**Solution:**

(a) The equivalent circuit as seen upon looking into the sources of the FETs is as shown in the following diagram.

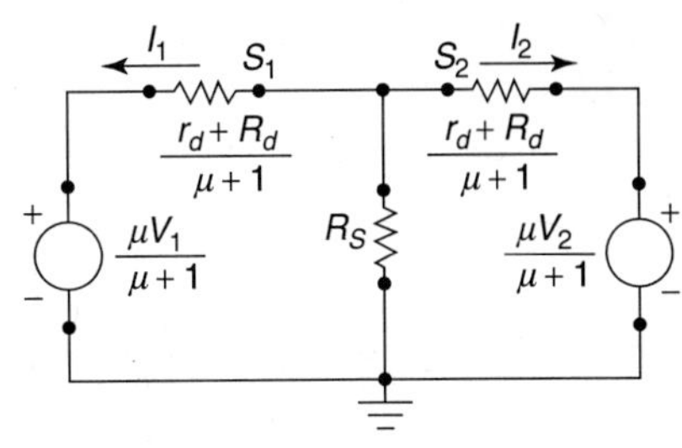


KVL in loop 1 and 2 respectively, gives:

$$-\frac{\mu}{\mu+1}V_1 = \left(\frac{r_d+R_d}{\mu+1} + R_s\right)I_1 + R_sI_2$$

$$-\frac{\mu}{\mu+1}V_2 = I_1R_s + \left(\frac{r_d+R_d}{\mu+1} + R_s\right)I_2$$

Solving for $I_2$ we obtain:

$$I_2 = \frac{\dfrac{-\mu}{\mu+1}\left(\dfrac{r_d+R_d}{\mu+1} + R_s\right)}{\left(\dfrac{r_d+R_d}{\mu+1}\right)\left(\dfrac{r_d+R_d}{\mu+1} + 2R_s\right)}V_1 + \frac{R_s\dfrac{\mu}{\mu+1}}{\left(\dfrac{r_d+R_d}{\mu+1}\right)\left(\dfrac{r_d+R_d}{\mu+1} + 2R_s\right)}V_2$$

But, $V_{o2} = R_dI_2$

Hence, $$A_1 = \frac{-\mu[r_d + R_d + (\mu+1)R_s]R_d}{(r_d+R_d)[r_d + R_d + 2(\mu+1)R_s]}$$

and, $$A_2 = \frac{\mu(\mu+1)R_sR_d}{(R_d+r_d)[R_d + r_d + 2(\mu+1)R_s]}$$

For the given values of $R_d$, $R_s$, $r_d$ and $\mu$ we have:

$$A_1 = \frac{-19 \times 80 \times 30}{40 \times 120} = -9.5$$

and, $$A_2 = \frac{19 \times 20 \times 30 \times 2}{40 \times 120} = 4.75$$

(b) If $R_s \to \infty$, then $$A_1 = \frac{-\mu(\mu+1)R_d}{2(\mu+1)(r_d+R_d)}$$

$$= \frac{-\mu R_d}{2(r_d+R_d)} = -A_2$$

or, $$A_1 = -A_2 = -7.14$$

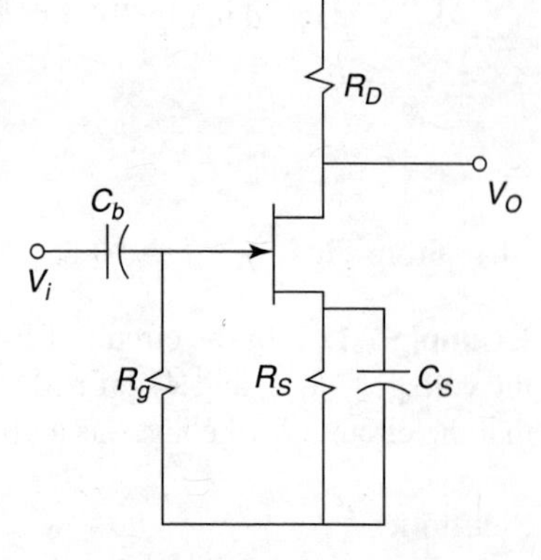


**Example 7-13** The common source mode amplifier stage shown in the diagram has the following parameters: $R_d = 12\text{K}$, $R_g = 1\text{M}$, $R_s = 470\,\Omega$, $V_{DD} = 30\,\text{V}$, $C_s$ is arbitrarily large, $I_{DDS} = 3\,\text{mA}$, $V_p = -2.4\,\text{V}$ and $r_d >> R_d$. Determine (a) the gate-to-source bias voltage $V_{GS}$, (b) the drain current $I_D$, (c) the quiescent voltage $V_{DS}$ and (d) the small-signal voltage gain $A_V$.

**Solution:**

(a) From the basic equation of the JFET:

$$I_D = I_{DSS}\left(1 - \frac{V_{GS}}{V_P}\right)^2$$

where, from the given diagram:

$$V_{GS} = -I_D R_s$$

or,

$$V_{GS} = -I_{DSS}R_s\left(1 - \frac{V_{GS}}{V_P}\right)^2 = -3 \times 0.47\left(1 + \frac{V_{GS}}{2.4}\right)^2$$

$$= -1.41\left(1 + \frac{2V_{GS}}{2.4} + \frac{V_{GS}^2}{2.4^2}\right)$$

Upon solving, we get $V_{GS} = -0.7\,\text{V}$

(b) $I_D = 3\left(1 - \frac{0.7}{2.4}\right)^2 = 3\left(\frac{1.7}{2.4}\right)^2 = 1.5\,\text{mA}$

(c) $V_{DS} = V_{DD} - I_D(R_d + R_s) = 30 - 1.5 \times 12.47 = 11.25\,\text{V}$

(d) Since $r_d >> R_d$, $A_V = -g_m R_d$ where, we have:

$$g_m = -\frac{2I_{DSS}}{V_P}\left(1 - \frac{V_{GS}}{V_P}\right)$$

$$= \frac{2 \times 3 \times 10^{-3}}{2.4}\left(1 - \frac{0.7}{2.4}\right)$$

$$= 1770\,\mu\text{mho} = 1.77\,\text{mA/V}$$

Hence,

$$A_V = -1.77 \times 12 = 21.2$$

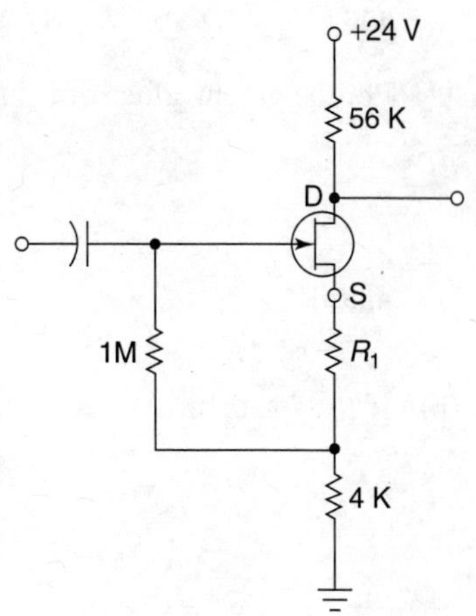


**Example 7-14** The amplifier stage, as shown in the diagram, uses an $n$-channel FET having $I_{DSS} = 2\,\text{mA}$ and $V_P = -1\,\text{V}$. If the quiescent drain-to-ground voltage is 10 V, find $R_1$.

**Solution:**

From the circuit given in the diagram:

$$I_D = \frac{V_{DD} - V_{DN}}{R_d} = \frac{24 - 10}{56} = 0.25\,\text{mA}$$

$$= I_{DDS}\left(1 - \frac{V_{GS}}{V_P}\right)^2 = 2(1 + V_{GS})^2$$

or,

$$1 + V_{GS} = \left(\frac{1}{8}\right)^{\frac{1}{2}} = 0.35$$

$$\therefore \qquad V_{GS} = -0.65 \text{ V}$$

but, $$V_{GS} = -I_D R_1$$

or, $$R_1 = -\frac{V_{GS}}{I_D} = \frac{0.65}{0.25 \times 10^{-3}} = 2.6 \text{ K}$$

**Example 7-15** The FET, as shown in the diagram, has the following parameters: $I_{DSS} = 5.6\text{mA}$ and $V_P = -4\text{V}$.

(a) Find $v_o$, if $v_i = 0$

(b) Find $v_o$, if $v_i = 10\text{V}$

(c) Find $v_i$, if $v_o = 0$

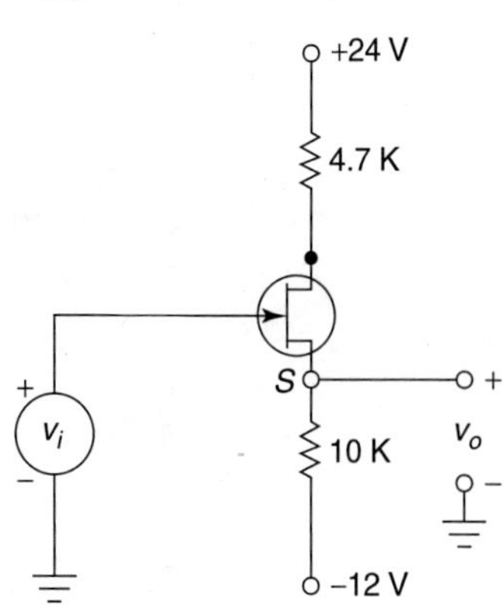


**NOTE:** $v_i$ and $v_o$ are constant voltages (and not small-signal voltages)

**Solution:**
KVL in the G–S loop gives:

$$v_i = V_{GS} + 10i_d - 12$$

or, $$i_d = \frac{v_i + 12 - V_{GS}}{10}$$

(a) If $v_i = 0$:

$$i_d = \frac{12 - V_G}{10} = I_{DSS}\left(1 - \frac{V_{GS}}{V_P}\right)^2$$

$$= 5.6\left(1 + \frac{V_{GS}}{4}\right)^2 = 5.6\left(1 + \frac{V_{GS}}{2} + \frac{V_{GS}^2}{16}\right)$$

Solving, we obtain $V_{GS} = -2$ V and $i_d = 1.4$ mA.

$\therefore v_o = 10i_d - 12 = 14 - 12 = 4$ V

(b) If $v_i = 10\text{V}$:

$$i_d = \frac{v_i + 12 - V_{GS}}{10} = 2.2 - \frac{V_{GS}}{10}$$

$$= 5.6\left(1 + \frac{V_{GS}}{2} + \frac{V_{GS}^2}{16}\right)$$

or, $V_{GS} \approx -1.4$ V, $\quad i_d = 2.43\,\text{mA}$

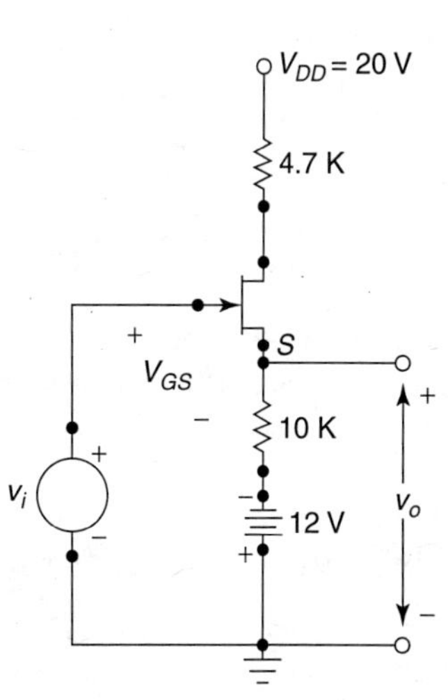


(c) If $v_o = 0$, $i_d = \dfrac{12\text{ V}}{10\text{ K}} = 1.2\,\text{mA} = 5.6\left(1 + \dfrac{V_{GS}}{4}\right)^2$

or, $$\left(1 + \frac{V_{GS}}{4}\right)^2 = \frac{1.2}{5.6} = 0.214,\ V_{GS} = -2.15\text{V}$$

Then, $v_i = V_{GS} = -2.15\text{V}$

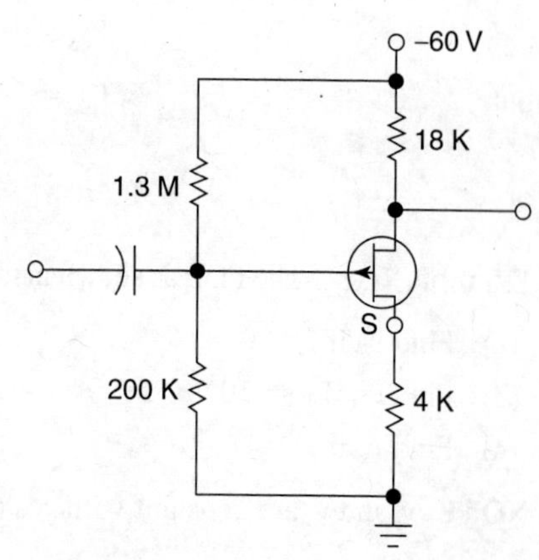


**Example 7-16** If $|I_{DSS}| = 4\,\text{mA}$, $V_P = 4\,\text{V}$, calculate the quiescent values of $I_D$, $V_{GS}$ and $V_{DS}$.

**Solution:**
If we find Thevenin's equivalent to the left of the gate of the given circuit, we obtain the circuit as shown in the following diagram:

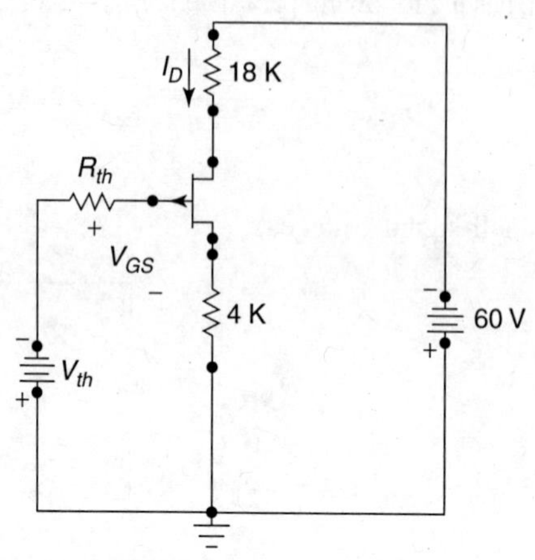


where, $R_{TH} = 200\,\text{K} \| 1.3\,\text{M} = 173.5\,\text{K}$

and,
$$-V_{Th} = \frac{200\,\text{K}}{1500\,\text{K}}(-60) = -8\ \text{V}$$

KVL in the G–S loop gives:

$$V_{GS} = -4\text{I}_\text{D} - 8 \text{ or } I_D = -\frac{8 + V_{GS}}{4} = -2 - \frac{V_{GS}}{4}$$

$$= I_{DSS}\left(1 - \frac{V_{GS}}{V_P}\right)^2 = -4\left(1 - \frac{V_{GS}}{4}\right)^2$$

Solving, we get: $V_{GS} = 1\text{V}$ and $I_D = -2.25$ mA

Hence,
$$V_{DS} = -60 + (18 + 4)(2.25)$$
$$= -60 + 49.5$$
$$= -10.5\ \text{V}$$

**Example 7-17** In the given diagram, two extreme transfer characteristics are indicated. The values of $V_{P(\max)}$ and $V_{P(\min)}$ are difficult to determine accurately. Hence, these values are calculated from the experimental values of $I_{DSS(\max)}$, $I_{DSS(\min)}$, $g_{m(\max)}$, and $g_{m(\min)}$ measured at a drain current corresponding to $I_{DSS(\min)}$. Verify that:

(a) $V_{P(\max)} = -\dfrac{2}{g_{m(\min)}}(I_{DSS(\max)} I_{DSS(\min)})^{\frac{1}{2}}$

(b) $V_{P(\min)} = -\dfrac{2I_{DSS,\,(\min)}}{g_{m(\max)}}$

(c) If for a given FET, $I_{DSS(\min)} = 1.5\,\text{mA}$, $I_{DSS(\max)} = 8\,\text{mA}$, $g_{m(\min)} = 1.4\,\text{mA/V}$ and $g_{m(\max)} = 3.5\,\text{mA/V}$, evaluate $V_{P(\max)}$ and $V_{P(\min)}$.

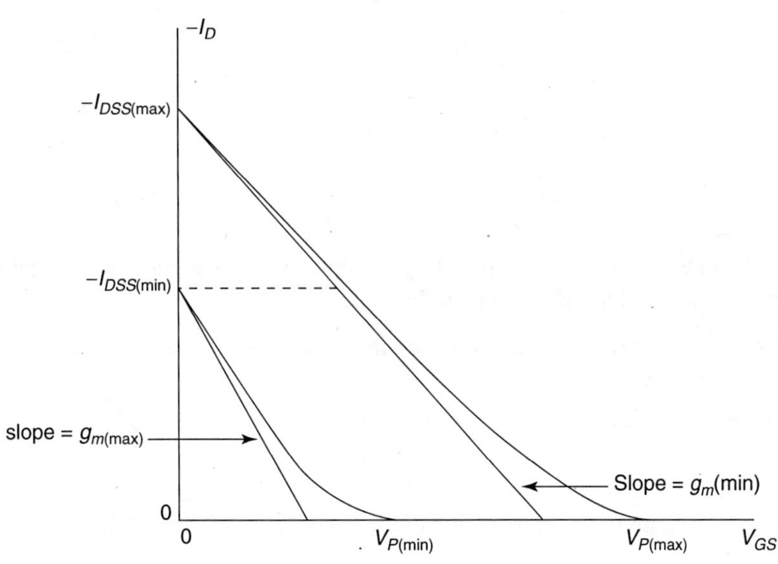


**Solution:**

(a) The basic equation of transconductance is given by:

$$\left[g_m = g_{m0}\left(1-\frac{V_{GS}}{V_P}\right) = -\frac{2}{V_P}(I_{DSS}I_{DS})^{1|2}\right]$$

Applying this to the upper curve we have:

$$g_{m(\min)} = -\frac{2I_{DSS(\max)}}{V_{P(\max)}}\left(1-\frac{V_{GS}}{V_{P(\max)}}\right)$$

where, $V_{GS}$ corresponds to $I_D = I_{DSS(\min)}$ or $I_D = I_{DSS(\min)} = I_{DSS(\max)}$

From, $\left(1-\dfrac{V_{GS}}{V_{P(\max)}}\right)$: $\quad V_{GS} = V_{P(\max)}\left(1-\sqrt{\dfrac{I_{DSS(\min)}}{I_{DSS(\max)}}}\right)$

Substituting, we have:

$$g_{m(\min)} = \frac{-2I_{DSS(\max)}}{V_{P(\max)}}\left(1-1+\sqrt{\frac{I_{DSS(\min)}}{I_{DSS(\max)}}}\right)$$

or,

$$V_{P(\max)} = \frac{-2I_{DSS(\max)}}{g_{m(\min)}}\left(\sqrt{\frac{I_{DSS(\min)}}{I_{DSS(\max)}}}\right)$$

$$= \frac{-2}{g_{m(\min)}}\left(I_{DSS(\max)} \times I_{DSS(\min)}\right)^{\frac{1}{2}}$$

(b) For the lower curve $g_{m0} = \dfrac{-2I_{DSS}}{V_P}$ becomes:

$$g_{m(\max)} = \frac{-2I_{DSS(\min)}}{V_{P(\min)}}$$

or, $$V_{P(\min)} = \frac{-2I_{DSS(\min)}}{g_{m(\max)}}$$

(c) $V_{P(\max)} = \dfrac{-2}{1.4}(8 \times 1.5)^{\frac{1}{2}} = -4.95\,\text{V}$

$V_{P(\min)} = \dfrac{-2 \times 1.5}{3.5} = -0.86\,\text{V}$

**Example 7-18** In the circuit, as shown in the diagram, the FET is used as an adjustable impedance element by varying the dc bias, and thereby the $g_m$ of the FET.

(a) Assume that there is a generator $V$ between the terminals $A$ and $B$. Draw the equivalent circuit. Neglect inter-electrode capacitances.

(b) Show that the input admittance between $A$ and $B$ is:

$$Y_i = Y_d + (1 + g_m R)Y_{CR}$$

where, $Y_d$ is the admittance corresponding to $r_d$, and $Y_{CR}$ is the admittance corresponding to $R$ and $C$ in series.

(c) If $g_m R >> 1$, show that the effective input capacitance is:

$$C_i = \frac{g_m \alpha}{\omega(1 + \alpha^2)}$$

and the effective input resistance is:

$$R_i = \frac{(1 + \alpha^2)r_d}{1 + \alpha^2(1 + \mu)}$$

where, $\alpha \equiv \omega CR$.

(d) At a given frequency, show that the maximum value of $C_i$ (with either $C$ or $R$ is varied) is obtained when $\alpha = 1$ and $(C_i)_{\max} = g_m/2\omega$. Also show that the value of $R_i$ corresponding to this $C_i$ is:

$$(R_i)_{\max} = \frac{2r_d}{2 + \mu}$$

which, for $\mu >> 2$, is reduced to $(R_i)_{\max} = 2/g_m$

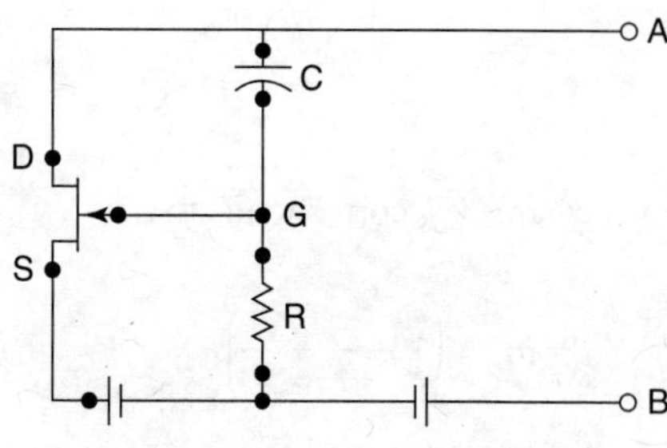


**Solution:**

(a) Using the small-signal equivalent circuit we have:

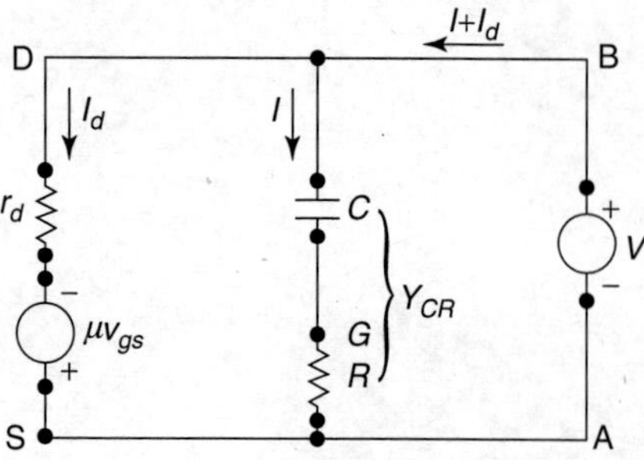


(b) $I_d = \dfrac{V + \mu V_{gs}}{r_d}$, but $V_{gs} = I_R$ and $I = V \times Y_{CR}$

Hence, $V_{gs} = VY_{CR} \times R$

$$I_d = \left(\frac{1}{r_d} + g_m R Y_{CR}\right)V$$

but, $$Y_1 V = I + I_d = \left( Y_{CR} + \frac{1}{r_d} + g_m R Y_{CR} \right)$$

or, $$Y_1 = Y_d + (1 + g_m R) Y_{CR}$$

(c) Since $g_m R >> 1$, we have:

$$Y_1 = Y_d + g_m R \frac{1}{R - j\left(\frac{1}{\omega C}\right)}$$

$$= \frac{1}{rd} + \frac{g_m R^2}{R^2 + \left(\frac{1}{\omega^2 C^2}\right)} + j\frac{g_m\left(\frac{R}{\omega C}\right)}{R^2 + \left(\frac{1}{\omega^2 C^2}\right)} = G_1 + j\omega C_1$$

hence, $$\omega C_1 = \frac{g_m R C \omega}{1 + (\omega R C)^2} = \frac{g_m \alpha}{1 + \alpha^2} \text{ or } C_1 = \frac{g_m a^2}{(1 + \alpha^2)\omega}$$

also, $$G_1 = \frac{1}{r_d} + \frac{g_m \omega^2 C^2 R^2}{\omega^2 C^2 R^2 + 1} = \frac{1}{r_d} + \frac{a^2}{1 + a^2} = \frac{1 + (1 + r_d g_m) a^2}{r_d (1 + a^2)}$$

hence, $$R_1 = \frac{1}{G_1} = \frac{(1 + a^2) r_d}{1 + a^2 (1 + \mu)}$$

(d) For $\omega$ = constant, $C_1$ is a function of $\alpha$, hence:

$$\frac{dC_1}{d\alpha} = \frac{g_m \omega(\alpha^2 + 1) - 2\alpha^2 g_m \omega}{\omega^2 (1 + \alpha^2)^2} = 0$$

or, $\alpha^2 = 1$ or $\alpha = 1$, since $\alpha$ is always greater than zero.

also, $$\frac{d^2 C_1}{d\alpha^2} = \frac{-2 g_m \omega^2 \alpha(\alpha^2 + 1)^2 - g_m \omega^2 (1 - a^2) 2 (1 + \alpha^2) \times 2\alpha}{\omega^2 (1 + \alpha^2)^2} \quad (< 0 \text{ for } \alpha < \sqrt{3})$$

hence, for $\alpha = 1$: $$(C_1)_{\max} = \frac{g_m}{2\omega}$$

and, follows that: $$(R_1)_{\max} = \frac{2 r_d}{2 + \mu}$$

**Example 7-19** Solve Example 7-18 if the capacitance $C$ is replaced by an inductance $L$.

**Solution:**
If we replace $C$ with $L$ in Example 7-18 (a), we obtain:

$$Y_1 = Y_d + (1 + g_m R) Y_{LR}$$

where, $$Y_{LR} = \frac{1}{R + j\omega L}$$

hence, $$Y_1 = G_1 + j\omega C_1 = \frac{1}{r_d} + \frac{g_m R^2}{R^2 + \omega^2 L^2} - j\frac{g_m R L}{R^2 + \omega^2 L^2}$$

which gives, $$j\omega C_1 = -j\frac{\omega g_m R L}{R^2 + \omega^2 L^2}$$

But then $Y_1$ is a parallel combination of a conductance $G_1$ and inductance $L_1$ and:

$$Y_1 = G_1 - j\,\frac{1}{\omega L_1}$$

or,

$$\frac{1}{\omega L_1} = \frac{\omega g_m RL}{R^2 + \omega^2 L^2}$$

or,

$$L_1 = \frac{R^2 + \omega^2 L^2}{\omega^2 g_m RL} = \frac{1 + \omega^2 \frac{L^2}{R^2}}{\omega g_m \left(\frac{\omega L}{R}\right)}$$

If we say:

$$\frac{\omega L}{R} = \alpha$$

then,

$$L_1 = \frac{1 + \alpha^2}{\omega g_m \alpha}$$

$$G_1 = \frac{r_d g_m R^2 + R^2 + \omega^2 L^2}{(R^2 + \omega^2 L^2) r_d} = \frac{\mu + 1 + \alpha^2}{(1 + \alpha^2) r_d}$$

$$R_1 = \frac{(1 + \alpha^2) r_d}{\mu + 1 + \alpha^2}.$$

For $\omega$ constant, $L_1$ is a function of $\alpha$ hence [for $(L_1)_{max}$] we find:

$$\frac{dL_1}{d\alpha} = \frac{2\omega g_m \alpha^2 - \omega g_m (1 + \alpha^2)}{\omega^2 g_m \alpha^2} = 0$$

or,

$$\alpha = 1 \text{ and } (L_1)_{max} = \frac{2}{\omega g_m}$$

For $\alpha = 1{:}1$

$$(R_1)_{max} = \frac{2r_d}{\mu + 2}$$

**Example 7-20** For a constant drain-to-source voltage, if the gate-to-source voltage is changed from 0 to −2 V, the corresponding change in the drain current becomes 2 mA. Calculate the transconductance of the FET if the ac drain resistance is 200 K. Also calculate the amplification factor of the FET.

**Solution:**
Transconductance is given by:

$$g_m = \left.\frac{i_d}{v_{gs}}\right|_{v_{ds} = 0}$$

hence, $g_m = \frac{2}{2} = 1$ mA/V

again, $\mu = g_m r_d = 1 \times 10^{-3} \times 200 \times 10^3 = 200$, which is the amplification factor.

**Example 7-21** Calculate the dynamic resistance of a JFET having an amplification factor of 80 and transconductance 400 $\mu$mho.

**Solution:**
Using the formula $\mu = g_m r_d$, we have:

$$80 = 400 \times 10^{-6} \times r_d \Rightarrow r_d = 0.2 \times 10^6 \text{ ohm}$$

**Example 7-22** The following data were obtained in an experiment with an FET:

| | | | |
|---|---|---|---|
| $V_{GS}$ (Volt) | 0 | 0 | 0.3 |
| $V_{DS}$ (Volt) | 6 | 16 | 16 |
| $I_D$ (mA) | 12 | 12.3 | 12 |

Calculate: (a) ac drain resistance, (b) transconductance and (c) amplification factor.

**Solution:**

(a) AC drain resistance:

$$r_d = \left.\frac{v_{ds}}{i_d}\right|_{v_{ds}=0}$$

$$\therefore \qquad r_d = \left.\frac{16-6}{12.3-12}\right|_{v_{gs}=0} = \frac{10}{0.3} = 3.33\ \text{k}\Omega$$

(b) Transconductance is given by:

$$g_m = \left.\frac{i_d}{v_{gs}}\right|_{r_{ds}=0}$$

$$\therefore \qquad g_m = \left.\frac{i_d}{v_{gs}}\right|_{r_{ds}=0} = \frac{12.3-12}{0-0.3} = \frac{0.3}{-0.3} = 1 \text{ (neglecting the sign)}$$

(c) Amplification factor is given by:

$$\mu = g_m r_d$$

$$\mu = 3.33 \times 1 = 3.33$$

**Example 7-23** The *Q*-point of a JFET in a source self-bias arrangement is chosen at $V_{GS} = -1.5$ V and $I_{D\text{sat}} = 2$ mA. Find the value of the resistance $R_S$.

**Solution:**

We have:

$$R_S = \frac{|V_{GS}|}{|I_{D\text{sat}}|} = \frac{1.5}{2 \times 10^{-3}}\ \text{Ohm} = 750\ \text{Ohm}$$

**Example 7-24** An FET amplifier in the common-source configuration uses a load resistance of 250 kΩ and the transconductance is 0.5 mA/V. What is the voltage gain of the amplifier? Given $r_d = 200$ kΩ.

**Solution:**

The voltage gain is:

$$A_V = -\frac{\mu R_L}{r_d + R_L}$$

Here, $r_d = 200$ kΩ, $g_m = 0.5$ mA/V and $R_L = 250$ kΩ.

We have:

$$\mu = r_d g_m = 200 \times 0.5 = 100$$

Hence,

$$A_V = -\frac{100 \times 250}{100 + 250} = -71.42$$

**Example 7-25** An *n*-channel JFET has $I_{DSS} = 10$ mA and pinch-off voltage $V_p = -4$ V. Find the drain current for $V_{GS} = -2$ V. If the transconductance $g_m$ of the JFET with the same $I_{DSS}$ at $V_{GS} = 0$ is 4 millimho, find the pinch-off voltage.

**Solution:**

We know that:

$$I_{DS} = I_{DSS}\left(1 - \frac{V_{GS}}{V_p}\right)^2$$

The transconductance is:

$$g_m = \left.\frac{\partial I_{DS}}{\partial V_{GS}}\right|_{V_{DS} = \text{const.}} = -\frac{2I_{DSS}}{V_P}\left(1 - \frac{V_{GS}}{V_P}\right) = g_{mo}\left(1 - \frac{V_{GS}}{V_P}\right)$$

Clearly, $g_{mo}\left(1 - \frac{V_{GS}}{V_P}\right)$ is the value of $g_m$ when $V_{GS} = 0$

Here $I_{DSS} = 10$ mA, $V_P = -4$ V, $V_{GS} = -2$ V, and $g_{mo} = 4$ mS.
Substituting these values, we obtain:

$$I_{DS} = 10\left(1 - \frac{2}{4}\right)^2 = 2.5 \text{ mA}$$

$$V_P = -\frac{2I_{DS}}{g_{mo}} = -\frac{2 \times 10}{4} = -5 \text{ V}$$

**Example 7-26** The drain current of a JFET, as shown in the diagram, is given by:

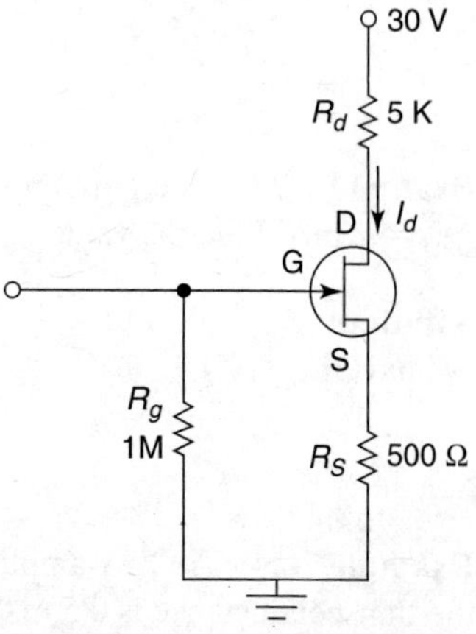

$$I_D = 20\left(1 + \frac{V_{GS}}{4}\right)^2 \text{ mA}$$

Calculate the quiescent values of $I_D$, $V_{DS}$ and $V_{GS}$.

30 V
$R_d$ 5 K
D $I_d$
G
S
$R_g$ 1M
$R_S$ 500 Ω

**Solution:**

We know that:

$$I_D = I_{DS}\left(1 - \frac{V_{GS}}{V_P}\right)^2$$

Therefore, by comparison, we get $I_{DS} = 20$ mA and $V_p = -4$ V.
The gate-source voltage $V_{GS}$ is equal to voltage drop across $R_S$:

$$V_{GS} = -I_D \times 500\ \Omega = -I_D \times 0.5 \text{ K}$$

Substituting the values of $I_{DS}$, $V_p$ and $V_{GS}$ we get:

$$I_D = 20\left(1 + \frac{-0.5 \text{ K } I_D}{4}\right)^2$$

$$= 20\,(1 - 0.125\, I_D)^2$$

$$= 20\,[(1) - (2 \times 1 \times 0.125)\, I_D + (0.125)^2\, I_D^2]$$

or,

$$I_D = 20 - 5\, I_D + 0.3125\, I_D^2$$

or,

$$0.3125\, I_D^2 - 6\, I_D + 20 = 0$$

$$\therefore \quad I_D = \frac{-(-6) \pm \sqrt{(-6)^2 - 4 \times 0.3125 \times 20}}{2 \times 0.3125}$$

$$= \frac{6 \pm \sqrt{36 - 25}}{0.625} = \frac{6 \pm 3.31}{0.625}$$

$$\therefore \quad I_D \text{ is either, } \frac{6 + 3.31}{0.625} = 14.9 \text{ mA}$$

or,
$$\frac{6 - 3.31}{0.625} = 4.3 \text{ mA}$$

Out of these two values, $I_D = 14.9$ mA is physically absurd because $I_D (R_d + R_s)$, i.e., 14.9 mA (5 K + 0.5 K) = 81.95 volts.
This is more than the supply voltage. We shall only consider $I_D = 4.3$ mA.

$$\therefore \quad V_{GS} = -I_D \times 0.5 \text{ K} = -4.3 \times 0.5 = -2.15 \text{ V}$$

and,
$$V_{DS} = V_{DD} - I_D (R_d + R_s)$$
$$= 30 - 4.3 \text{ mA} (5 \text{ K} + 0.5 \text{ K})$$
$$= 30 - 23.65 \text{ V}$$
$$= 6.35 \text{ V}$$

**Example 7-27** The diagram shows an FET amplifier circuit. If the FET has $I_{DSS} = 3$ mA, $V_p = -2.4$ V and $r_d >> R_d$, calculate:

(a) Quiescent values of $I_D$, $V_{DS}$ and $V_{GS}$

(b) Voltage gain $A_v$

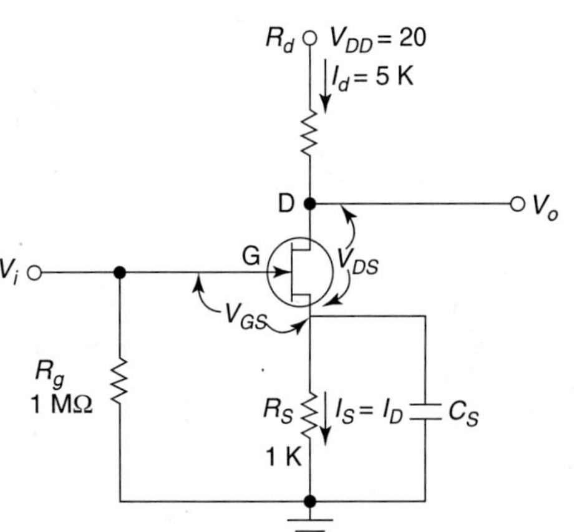


**Solution:**

$$0.52\, I_D^2 - 3.5\, I_D + 3 = 0$$

or,
$$I_D^2 - 6.73\, I_D + 5.76 = 0$$

$$\therefore \quad I_D = \frac{6.73 \pm \sqrt{(6.73)^2 - 4 \times 1 \times 5.77}}{2}$$

$$= 5.72 \text{ mA or } 1.01 \text{ mA}$$

We have,
$$V_{GS} = -I_D \times R_s$$
$$= -I_D \times 1 \text{ K}$$

We know that the drain current:

$$I_D = I_{DSS}\left(1 - \frac{V_{GS}}{V_P}\right)^2$$

Substituting the given values of $I_{DSS}$ and $V_P$, we get:

$$I_D = 3 \times \left[1 - \frac{V_{GS}}{(-2.4)}\right]^2 \text{ mA}$$

$$= 3 \times \left(1 + \frac{V_{GS}}{2.4}\right)^2$$

Substituting the equation of $I_D$ we get:

$$I_D = 3 \times \left(1 - \frac{I_D \times 1\text{K}}{2.4}\right)^2$$

$$= 3 \times \left[1 - \frac{2}{2.4} I_D + \left(\frac{1}{2.4}\right)^2 I_D^2\right]$$

$$= 3 - 2.5\, I_D + 0.52\, I_D^2$$

(a) The possible value is $I_D = 1.01$ mA

$\therefore$ $V_{GS} = -I_D R_S = -1.01 \times 1\text{ K} = -1.01\text{ V}$

and, $V_{DS} = V_{DD} - I_D R_S - I_D R_d$

$= 20 - I_D (R_S + R_d)$

$= 20 - 1.01 \times (1\text{ K} + 10\text{ K}) = 8.89\text{ V}.$

(b) The voltage gain $A_v = -g_m (R_d \| r_d)$

Since it is given that $r_d >> R_d$, therefore, $r_d$ can be ignored for $(R_d \| r_d)$ value. Taking magnitude only $|A_V| = |g_m \,.\, R_d|$ the value of $g_m$ is:

$$g_m = -\frac{2}{V_p}\sqrt{I_D I_{DSS}}$$

$$= \frac{-2}{-2.4}\sqrt{1.01\text{ mA} \times 3\text{ mA}}$$

$$= \frac{2}{2.4}\sqrt{3.03}\text{ mA/V}$$

$$= 1.45\text{ mA/V} = 1.45\text{ mA/V}$$

$\therefore$ Voltage gain $|A_v| = 1.45 \times 10 = 14.5$

**Example 7-28** A common source FET amplifier with unbypassed $R_s$ has the following circuit parameters: $R_d = 15$ K, $R_g = 1$ M, $r_d = 5$ k, $g_m = 5$m and $V_{DD} = 20$ V. Calculate $A_V$, $Z_o$ and $Z_i$.

**Solution:**

Using the given values we can find:

$$\mu = r_d\, g_m = 5\text{ K} \times 5\text{ mho} = 25$$

The expression for $A_v$ and $R_o$:

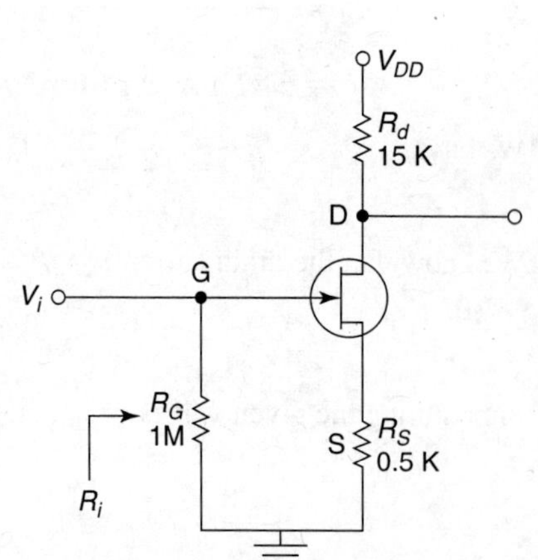


$$A_v = \frac{-\mu\, R_d}{r_d + R_d}$$

$$= \frac{-25 \times 15\text{ K}}{5\text{ K} + 15\text{ K}}$$

$$= \frac{-375}{20} = 18.75$$

$$Z_o = r_d$$

$$= 5\text{ K}$$

$Z_i$ = Input resistance of the FET amplifier

$\cong R_g$ (resistance of gate-to-source − infinite) = 1 MΩ

**Example 7-29** The circuit, as shown in the diagram, uses a JFET with $I_{DSS}$ = 5 mA and $V_p$ = −4.5 V. If $V_i$ and $V_o$ represent voltages and not the small-signal values, calculate the following:

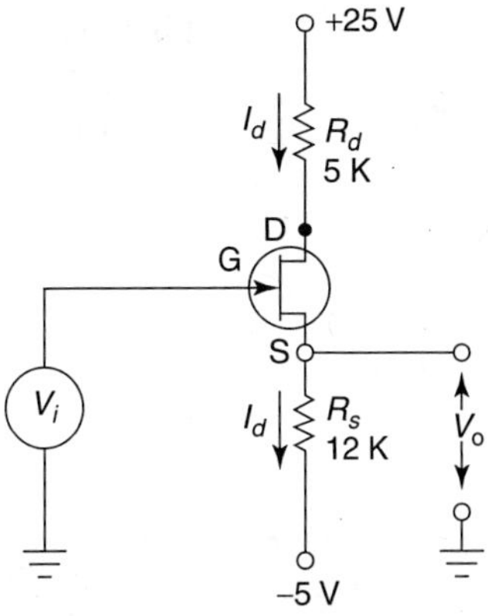

(a) $V_o$ for $V_i = 0$

(b) $V_i$ for $V_o = 0$



**Solution:**

We know that:

$$I_D = I_{DSS}\left(1 - \frac{V_{GS}}{V_p}\right)^2$$

Substituting the given value of $I_{DSS}$ and $V_p$ we get:

$$I_D = 5\left(1 + \frac{V_{GS}}{4.5}\right)^2 \text{ mA} \qquad (1)$$

(a) When $V_i = 0$

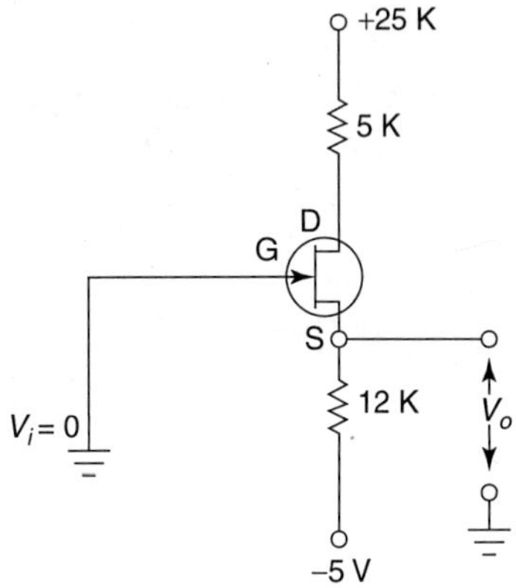

For this condition, we first redraw the circuit.



Taking KVL around the input loop, we get:

$$V_{GS} = I_D R_S - 5 \text{ V}$$

$$= I_D \times 12 - 5 \text{ V} \quad (I_D \text{ in mA})$$

or,
$$I_D = \frac{V_{GS} + 5}{12} \qquad (2)$$

Substituting Eq. (2) in Eq. (1) we get:

$$\frac{V_{GS} + 5}{12} = 5\left(1 + \frac{V_{GS}}{4.5}\right)^2$$

or,
$$V_{GS} + 5 = 60\left(1 + \frac{2V_{GS}}{4.5} + \frac{V_{GS}^2}{4.5^2}\right)$$

$$= 60 + 26.67\, V_{GS} + 2.963 V_{GS}^2$$

or,
$$2.963\, V_{GS}^2 + 25.67\, V_{GS} + 55 = 0$$

∴
$$V = \frac{-25.67 \pm \sqrt{(25.67)^2 - 4 \times 2.963 \times 55}}{2 \times 2.963}$$

$$= -4.78 \text{ V} \quad \text{or} \quad -3.88 \text{ V}$$

The output voltage $V_o$ is voltage of the source w.r.t the ground. Since the gate is connected to the ground, $V_o = V_{SG} = -V_{GS}$

∴
$$V_o = +4.78 \text{ V} \quad \text{or} \quad +3.88 \text{ V}$$

(b) When $V_o = 0$

The output point (the source $S$) is grounded, as shown in the diagram.

∴
$$I_D R_s = 5 \text{ V}$$

or,
$$I_D = \frac{5 \text{ V}}{12 \text{ K}} = 0.415 \text{ mA}$$

Substituting this value of $I_D$ in Eq. 1:

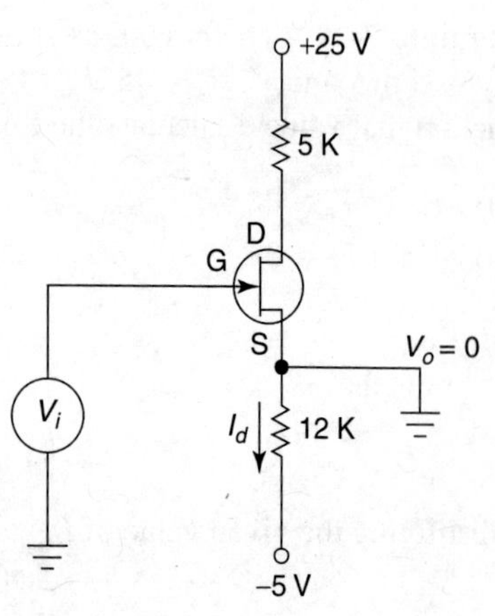


$$0.415 = 5\left(1 + \frac{V_{GS}}{4.5}\right)^2$$

or,
$$\left(1 + \frac{V_{GS}}{4.5}\right)^2 = \frac{0.415}{5} = 0.083$$

or,
$$1 + \frac{V_{GS}}{4.5} = \sqrt{0.083} = 0.288$$

or,
$$V_{GS} = 4.5(0.288 - 1) = -3.2 \text{ V}$$

Taking KVL around the input loop:

$$V_i = V_{GS}$$

$\therefore$
$$V_i = -3.2 \text{ V}$$

**Example 7-30** For the circuit, as shown in the following diagram, calculate $A_v = V_o/V_s$ and $Z_o$ if $g_m$ = 5 mA/V and $r_d$ = 10 KΩ.

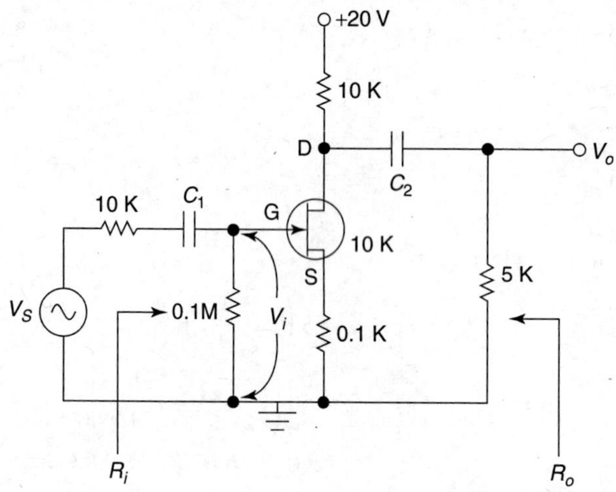


**Solution:**

$$\mu = g_m\, r_d = 5 \text{ mA/v} \times 10 \text{ k}\Omega = 50$$

At signal frequency, both $C_1$ and $C_2$ are assumed to be short-circuited. Therefore, the 10 K load resistance effectively comes in parallel with $R_d$ (10 K).

Therefore, effective $R_d$ = 10 K || 10 K = 5 kΩ.

The voltage gain ($V_o$ / $V_i$) is:

$$A_v = \frac{-\mu R_d}{r_d + R_d + (\mu + 1)R_s}$$

$$= \frac{-50 \times 5 \text{ K}}{10 \text{ K} + 5 \text{ K} + (50 + 1) \times 0.1 \text{ K}}$$

$$= -12.44$$

The right voltage $V_i$ appearing across the 0.1 M resistance is:

$$V_i = \frac{0.1\text{ M}}{0.1\text{ M} + 10\text{ K}} \times V_s$$

$$\therefore \quad \frac{V_i}{V_s} = \frac{100\text{ K}}{110\text{ K}} = 0.909$$

The overall voltage gain:

$$A_{vs} = \frac{V_o}{V_s} = \frac{V_o}{V_i} \times \frac{V_i}{V_s}$$

$$= A_v \frac{V_i}{V_s}$$

$$= -12.44 \times 0.909 = -11.3$$

The output impedance:

$$Z_o = r_d + (\mu + 1) R_s$$

$$= 10\text{ K} + (50 + 1)\, 0.1\text{ K}$$

$$= 15.1\text{ K}$$

The effective output impedance:

$$Z_o' = Z_o \,//\, 10\text{ K} \,//\, 10\text{ K}$$

$$= 15.1\text{ K} \,//\, 5\text{ K} = 3.75\text{ k}\Omega$$

**Example 7-31** The $n$-channel JEFT, as shown in the following diagram, has $V_p = -4$ V and $I_{DSS} = 1.65$ mA. The operating point desired is $I_{DQ} = 0.8$ mA. Assume $r_d >> R_L$. Calculate (a) $V_{GSQ}$, (b) $g_m$, (c) $R_s$, (d) $V_{DSQ}$ and (e) $R_L$, such that a voltage gain of 20 dB is obtained. $R_s$ is passed by the large capacitor $C_s$ and reactance of $C_c$ is negligible at signal frequency.

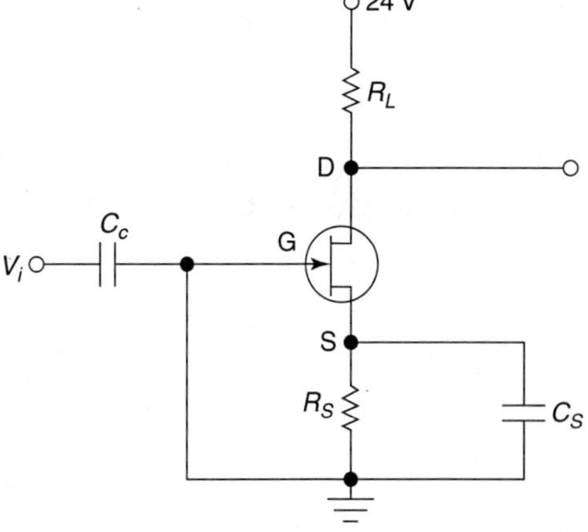


**Solution:**

(a) We know that

$$I_D = I_{DSS}\left(1 - \frac{V_{GS}}{V_p}\right)^2$$

Substituting $I_{DSS} = 1.65$ mA, $V_p = -4$ V and $I_{DQ} = 0.8$ mA:

$$0.8 = 1.65\left(1 + \frac{V_{GS}}{4}\right)^2$$

or,

$$1\left(1 + \frac{V_{GS}}{4}\right)^2 = \frac{0.8}{1.65} = 0.485$$

$$1 + \frac{V_{GS}}{4} = \sqrt{0.485} = 0.696$$

$$V_{GS} = 4 \times (0.696 - 1)$$

$$= -1.214 \text{ V}$$

$\therefore$ $$V_{GSQ} = -1.214 \text{ V}$$

(b) $$g_m = g_{mo}\left(1 - \frac{V_{GS}}{V_p}\right)$$

where, $$g_{mo} = \frac{-2\, I_{DSS}}{V_p}$$

$\therefore$ $$g_{mo} = \frac{-2 \times 1.65}{-4} = 0.825 \text{ mA/V}$$

$\therefore$ $$g_m = 0.825\left(1 - \frac{(-1.214)}{(-4)}\right)$$

$$= 0.825 \times 0.6965 = 0.575 \text{ mA/V}$$

(c) Looking into the input loop:

$$V_{GS} = I_{DQ}\, R_s$$

$\therefore$ $$-1.214 \text{ V} = -0.8 \text{ mA} \times R_s$$

$\therefore$ $$R_s = \frac{1.214}{0.8} = 1.518 \text{ k}\Omega$$

(d) It is given that the voltage gain has to be 20 dB. Now voltage gain in dB = 20 log (voltage gain)

$\therefore$ $$20 \text{ dB} = 20 \log_{10} A_v$$

$\therefore$ $$A_v = 10$$

Since it is given that $r_d \gg R_L$, the voltage gain of the CS amplifier:

$$A_v = -g_m\, R_L$$

$$|A_v| = g_m\, R_L$$

$\therefore$ $$10 = 0.575 \text{ mA/V} \times R_L$$

$\therefore$ $$R_L = \frac{10}{0.575 \text{ mA/V}} = 17.4 \text{ k}\Omega$$

(e) Taking the KVL for the output loop:

$$24 \text{ V} = I_D\, R_L + V_{DS} + I_D\, R_s$$

or, $$V_{DS} = 24 \text{ V} - I_D\, R_L - I_D\, R_s$$

$$= 24 \text{ V} - 0.8 \text{ mA} \times 17.4 \text{ K} - 0.8 \text{ mA} \times 1.518 \text{ k}\Omega$$

$$= 8.86 \text{ V}$$

$\therefore$ $$V_{DSQ} = 16.48 \text{ V}$$

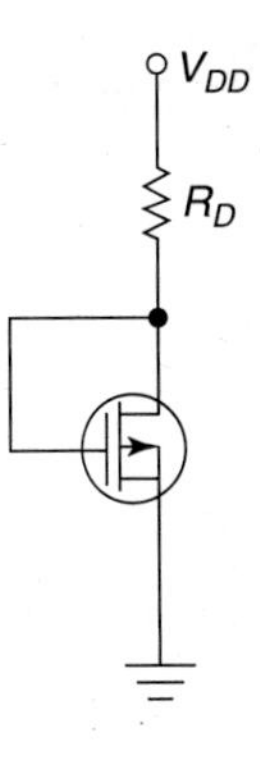


**Example 7-32** (a) Draw the low-frequency small-signal model of the circuit shown in the diagram (b) Determine $Z_o$. (c) Evaluate $Z_o$ for $r_d = 50$ kΩ and $R_D = 20$ kΩ.

**Solution:**

(a) The equivalent circuit at low-frequency small-signal model is:

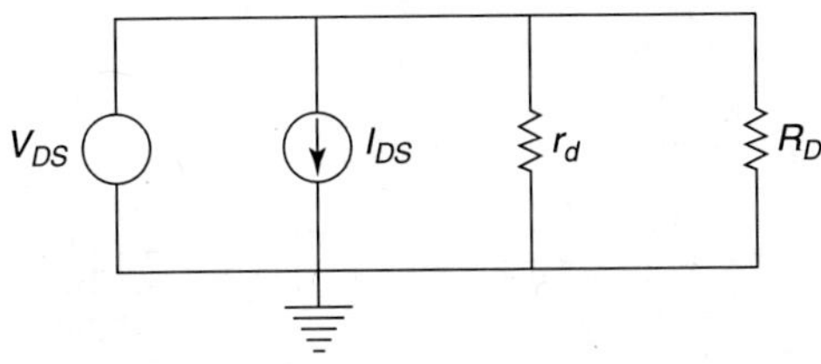


(b) Output resistance:

$$Z_o = r_d \,||\, R_D$$

(c) Hence, for $R_D = 20$ kΩ and $r_d = 50$ kΩ

$$Z_o = \frac{r_d R_D}{r_d + R_D} = \frac{20 \times 10^3 \times 50 \times 10^3}{20 \times 10^3 + 50 \times 10^3}$$

$$= \frac{1000}{70} \times 10^3$$

$$= 14.28 \text{ k}\Omega$$

**Example 7-33** In the CS amplifier, as shown in the diagram, let $R_D = 5$ kΩ, $R_G = 500$ kΩ, $\mu = 60$, and $r_{ds} = 30$ kΩ. Find the value of the voltage-gain ratio $A_v = v_o/v_i$ and current gain ratio $A_i = i_d/i_i$.

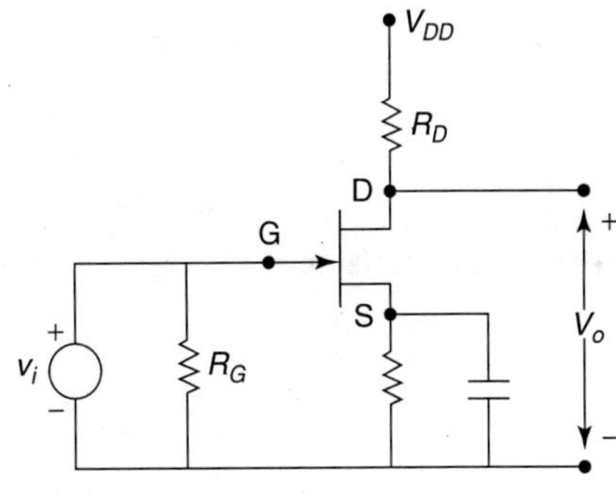


**Solution:**

Voltage-gain ratio:

By voltage division at the output network of the applicable equivalent circuit, as shown in the following diagram, we have:

$$v_o = \frac{R_D}{R_D + r_{ds}} \mu v_{gs}$$

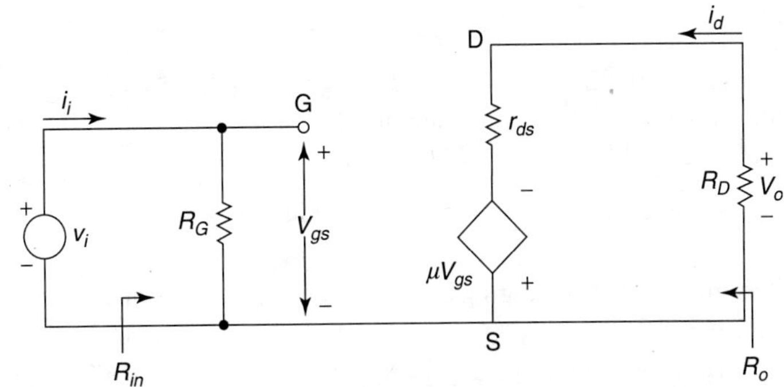

Substitution of $v_{gs} = v_i$ and rearrangement gives:

$$A_v = \frac{v_o}{v_i} = \frac{\mu R_D}{R_D + r_{ds}} = \frac{60 \times 5}{5 + 30} = 8.57$$

Current-gain ratio:
KVL around the output network leads to:

$$i_d = \frac{\mu v_{gs}}{r_{ds} + R_D}$$

But, Ohm's law requires that $v_{gs} = i_i R_G$ which, when substituted gives:

$$A_i = \frac{i_d}{i_i} = \frac{\mu R_G}{r_{ds} + R_D} = \frac{60 \times 500}{30 + 5} = 857.14$$

**Example 7-34**

(a) Draw the low frequency small signal model of the circuit as shown in the diagram.

(b) Determine $Z_O$.

(c) Evaluate $Z_O$ for $g_m = 1.0$ m-mho, $r_d = 50\ \text{k}\Omega$, $R_D = 5\ \text{k}\Omega$.

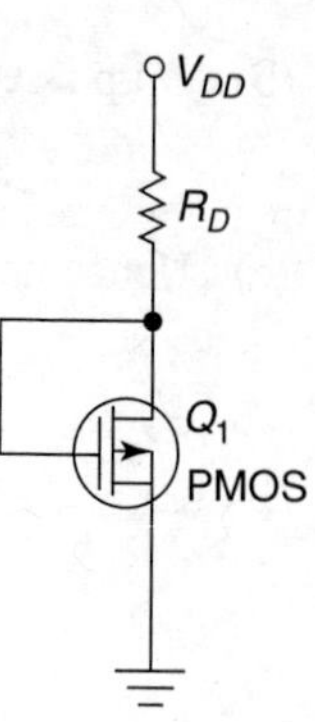


**Solution:**

(a) The equivalent circuit at low-frequency small-signal model is:

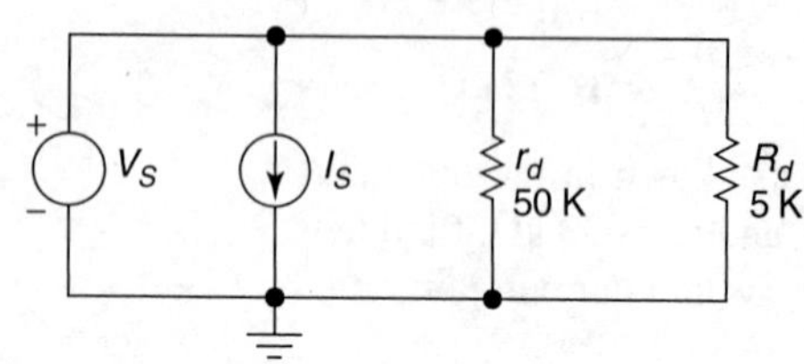


(b) Output resistance, $Z_O = R_{DS} \parallel R_D$

(c) Hence, for $R_D = 5\ \text{k}\Omega$ and $R_{DS} = 50\ \text{k}\Omega$:

$$Z_O = \frac{50 \times 5}{50 + 5}$$

$$= 4.54\ \text{k}\Omega$$

**Example 7-35** An *n*-channel JFET has $V_P = -5$ V, and $I_{DSS} = 12$ mA and is used in the circuit shown. The parameter values are $V_{DD} = 18$ V, $R_S = 2\ \text{k}\Omega$, $R_D = 2\ \text{k}\Omega$, $R_1 = 400\ \text{k}\Omega$ and $R_2 = 90\ \text{k}\Omega$.

(a) If the resistance $R_2$ is changed, what must be the new value of $R_2$ if $I_D = 8$ mA?

(b) Using the values as given, but changing $V_{DD}$, find the new values of $V_{DD}$ for which $I_D = 8$ mA.

(c) For the condition in (b), what is the new value of $V_{DS}$?

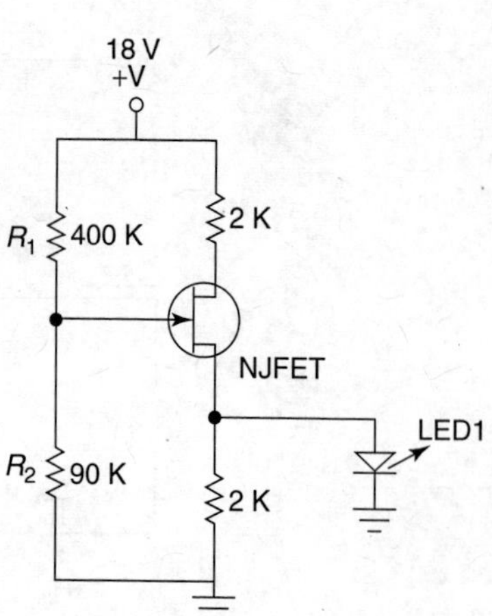

**Solution:**

Let us consider the voltage across the $R_2$ to be $V_2$.

Now, applying the KVL in the lower loop of the circuit:

$$V_2 = I_D R_S + V_{GS}$$

where, $I_D$ is the drain current and $V_{gs}$ is the gate-to-source voltage.

$\therefore$ Voltage drop across $R_2$ is, $V_2 = \dfrac{R_2}{R_1 + R_2} V_{DD}$

Putting the value of $V_2$, we get:

$$\frac{R_2}{R_1 + R_2} V_{DD} = I_D R_S + V_{GS}$$

or, $$\frac{90}{90 + 400} 18 = 2\, I_D + V_{GS}$$

or, $$2I_D + V_{GS} = 3.3$$

In case of the JFET, the drain current is:

$$I_D = I_{DSS}\left(1 - \frac{V_{GS}}{V_P}\right)^2 = 12\left(1 + \frac{V_{GS}}{5}\right)^2$$

or, $$1.65 - \frac{V_{GS}}{2} = 12 + \frac{24}{5} V_{GS} + 0.48\, V_{GS}^2$$

or, $$0.48\, V_{GS}^2 + 5.3\, V_{GS} + 10.35 = 0$$

or, $$V_{GS} = -8.5 \text{ V} \quad \text{or} \quad -2.53 \text{ V}$$

$\therefore$ $$V_{GS} = -2.53 \text{ V}$$

From equation for $I_D$:

$$I_D = (3.306 - V_{GS})/2 = 2.92 \text{ mA}$$

Applying KVL in the outer loop of the circuit:

$$V_{DD} = I_D R_D + V_{DS} + I_D R_S$$

or, $$18 = 2.92 \times 2 + V_{DS} + 2.92 \times 2$$

$\therefore$ $$V_{DS} = 6.32 \text{ V}$$

(a) Again we have:

$$\frac{R_2}{R_1 + R_2} V_{DD} = I_D R_S + V_{GS}$$

or, $$\frac{R_2}{400 + R_2} 18 = 8 \times 2 - 2.53$$

or, $$R_2 = 1189.4 \text{ k}\Omega$$

The new value of $R_2$ is 1189.4 K, if $I_D = 8$ mA

(b) We have:

$$\frac{R_2}{R_1 + R_2} V_{DD} = I_D R_S + V_{GS}$$

or,
$$V_{DD} \frac{90}{400 + 90} = 8 \times 2 - 2.53$$

or,
$$V_{DD} = 73.34 \text{ V}$$

The new value of $V_{DD}$ is 73.34 V when $I_D$ = 8 mA.

(c) We know that:

$$V_{DD} = I_D \times R_D + V_{DS} + I_D \times R_S$$

or,
$$73.34 = 8 \times 2 + V_{DS} + 8 \times 2$$

∴
$$V_{DS} = 41.34 \text{ V}$$

The new value of $V_{DS}$ is 41.34 V.

**Example 7-36** The given diagram shows the transfer characteristics of an FET. Write an equation for drain current.

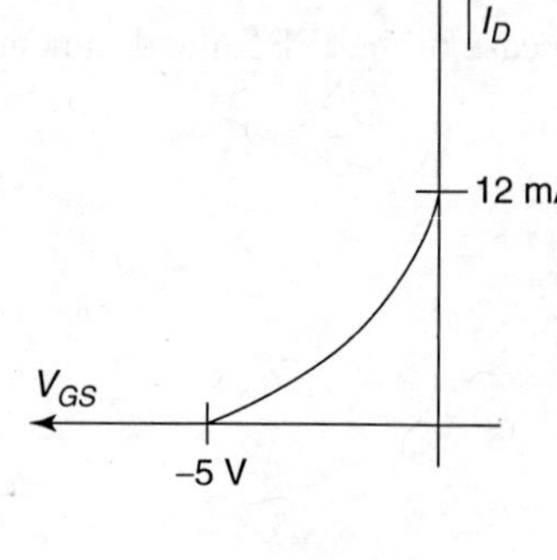


**Solution:**
From the figure:

$$I_{DSS} = 12 \text{ mA}$$

$$V_{GS(\text{off})} = -5 \text{ V}$$

We know that:

$$I_D = I_{DSS}\left(1 - \frac{V_{GS}}{V_{GS(\text{off})}}\right)^2 = 12\left(1 - \frac{V_{GS}}{-5}\right)$$

∴
$$I_D = 12\left(1 + \frac{V_{GS}}{5}\right)^2 \text{mA}$$

**Example 7-37** An FET has a drain current of 5 mA. If $I_{DSS}$ = 12 mA and $V_{GS(\text{off})} = -6$ V, find the value of (a) $V_{GS}$ and (b) $V_P$.

**Solution:**

(a) ∵
$$I_D = I_{DSS}\left(1 - \frac{V_{GS}}{V_{GS(\text{off})}}\right)^2$$

⇒
$$5 = 12\left(1 + \frac{V_{GS}}{6}\right)^2$$

$$\frac{5}{12} = \left(1 + \frac{V_{GS}}{6}\right)^2$$

$$\sqrt{\frac{5}{12}} = 1 + \frac{V_{GS}}{6}$$

$$-0.355 = \frac{V_{GS}}{6}$$

∴
$$V_{GS} = -2.13 \text{ V}$$

(b)
$$V_{GS(\text{off})} = V_P = 6 \text{ V}$$

**Example 7-38** In a self-bias $n$-channel FET, the operating point is to be set at $I_D = 1.5$ mA and $V_{DS} = 10$ V. The FET parameters are $I_{DSS} = 5$ mA and $V_P = -2$ V. Find the values of $R_S$ and $R_D$. Given that $V_{DD} = 20$ V.

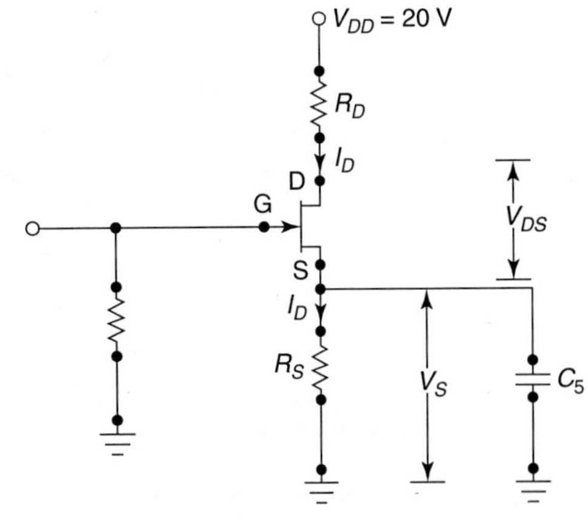


**Solution:**

$$\because \quad I_D = I_{DSS}\left(1 - \frac{V_{GS}}{V_P}\right)^2$$

$$\therefore \quad 1.5 = 5\left(1 + \frac{V_{GS}}{2}\right)^2$$

$$V_{GS} = -0.9 \text{ V}$$

Now, $$V_{GS} = V_G - V_S$$

$$\therefore \quad V_S = V_G - V_{GS} = 0 - (-0.9)$$

$$\therefore \quad V_S = 0.9 \text{ V}$$

$$R_S = \frac{V_S}{I_D} = \frac{0.9}{1.5} \text{ mA} = 0.6 \text{ k}\Omega$$

Now, apply KVL to the outer loop:

$$\therefore \quad V_{DD} = I_D R_D + V_{DS} + I_D R_S$$

$$\therefore \quad 20 = 1.5 \text{ mA} \times R_D + 10 + 0.9 \quad (\because \; I_D R_S = V_S)$$

$$\therefore \quad R_D = \frac{20 - 10.9}{1.5 \times 10^{-3}}$$

$$R_D = 6.06 \text{ k}\Omega$$

**Example 7-39** In an $n$-channel FET biased by potential divider method, it is desired to set the operating point at $I_D = 2.5$ mA and $V_{DS} = 8$ V. If $V_{DD} = 30$ V, $R_1 = 1$ M$\Omega$ and $R_2 = 500$ k$\Omega$, find the value of $R_S$. The parameters of the FET are $I_{DSS} = 15$ mA and $V_P = -5$ V.

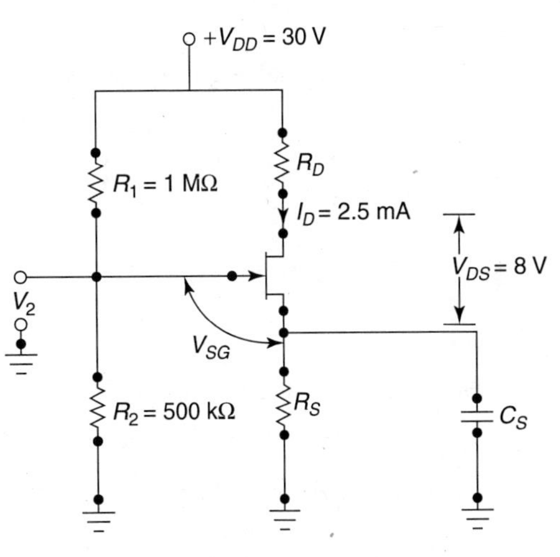


**Solution:**

$$I_D = I_{DSS}\left(1 - \frac{V_{GS}}{V_P}\right)^2$$

or, $$2.5 = 15\left(1 + \frac{V_{GS}}{5}\right)^2$$

or, $$\frac{5}{2 \times 15} = \left(1 + \frac{V_{GS}}{5}\right)^2$$

or, $$V_{GS} = -2.96 \text{ V}$$

Now,
$$V_2 = \frac{V_{DD} R_2}{R_1 + R_2} = \frac{30 \times 500}{1000 + 500}$$

$$V_2 = 10 \text{ V}$$

or,
$$V_2 = V_{GS} + I_D R_S$$

or,
$$10 = -2.96 + 2.5 \times 10^{-3} R_S$$

or,
$$R_S = \frac{12.5}{2.5} \times 10^3$$

or,
$$R_S = 5.18 \text{ k}\Omega$$

Applying KVL to outer loop:

$$V_{DD} = I_D R_D + V_{DS} + I_D R_S$$

or,
$$30 = 2.5 \times 10^{-3} R_D + 8 + 2.5 \times 10^{-3} \times 5.18 \times 10^3$$

or,
$$R_D = 3.62 \text{ k}\Omega$$

**Example 7-40** (a) Calculate the voltage gain $A_V = V_o/V_i$ at 1 KHz for the circuit, as shown in the diagram. The FET parameters are $g_m = 2\text{mA/V}$ and $r_d = 10 \text{ k}\Omega$. Neglect capacitance. (b) Repeat part (a) if the capacitance 0.005 $\mu$F is taken underconsideration.

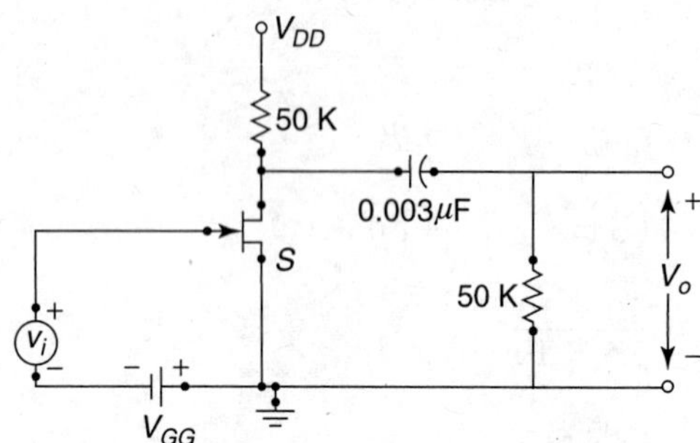


**Solution:**

(a) The equivalent small-signal circuit is shown as follows.

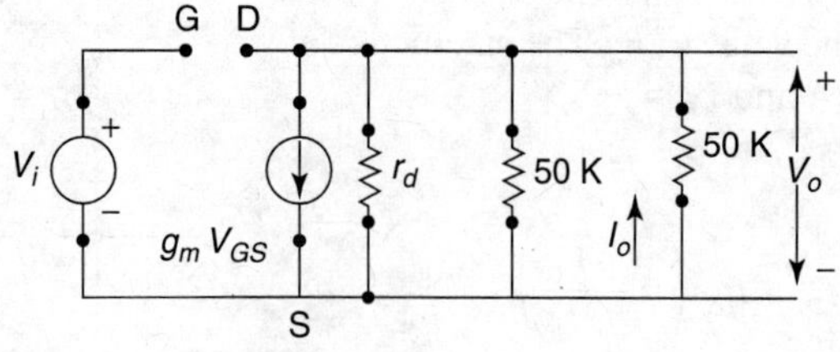


We neglect the 0.005$\mu$F capacitance and we have $V_{GS} = V_i$

$R = r_d \| 50 \text{ K} \| 50 \text{ K} = r_d \| 25 \text{ K} = 7.14 \text{ K}$

The voltage gain of the amplifier, as calculated in Example 7-1, is:

$$A_V = -g_m R$$

Hence,
$$A_V = -2 \times 7.14 = -14.28$$

(b) The small-signal equivalent circuit seen at the drain is as shown in the following diagram.

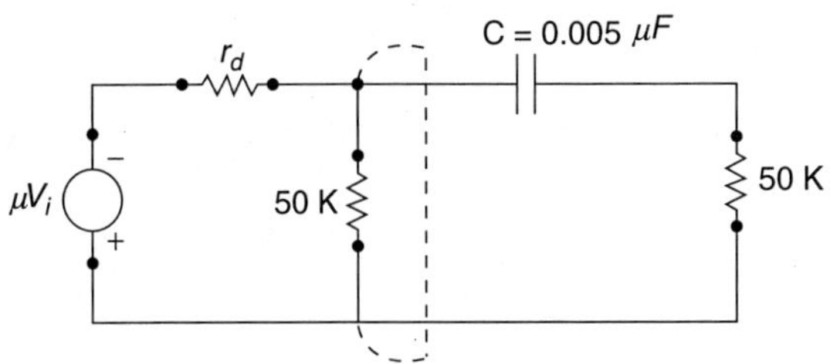


Then the Thevenin's equivalent consists of:

$$R_{Th} = r_d \| 50\text{ K} = 10\text{ K} \| 50\text{ K} = 8.33\text{ K}$$

and,

$$V_{Th} = \mu V_i \frac{50\text{ K}}{50\text{ K} + 10\text{ K}} = \frac{20 \times 50}{60} V_i$$

$$= 16.67\ V_i \text{ at 1 KHz, } Z_C = -j\,31.83\text{ K}$$

Then,

$$\frac{V_o}{V_i} = \frac{-50\text{ K}(16.67)}{50\text{ K} + 8.33\text{ K} - j31.83\text{ K}} = \frac{-833}{58.33 - j31.83}$$

$$= \frac{-833}{66.45} \angle\ 28.62^\circ = -12.54 < 28.62^\circ$$

**Example 7-41** The transconductance of an FET used in a voltage-amplifier circuit is 2 mS, and the load resistance is 10 kΩ. Calculate the voltage amplification of the circuit. Assume that $r_d \gg R_L$.

**Solution:**
Given,

$$g_{fs} = 2\text{ mS}$$

$$R_L = 10\text{ k}\Omega$$

As,

$$r_d \gg R_L$$

The voltage gain is given by:

$$A_V = g_{fs} R_L = 2\text{ mS} \times 10\text{ k}\Omega = 20$$

**Example 7-42** Show that, if two identified FETs are connected in parallel, then $g_m$ is double and $r_d$ is half of that of the individual. If two FET's are not identical, show that:

$$\frac{1}{r_d} = \frac{1}{r_{d1}} + \frac{1}{r_{d2}} \quad \text{and} \quad \mu = \frac{\mu_1 r_{d2} + \mu_2 r_{d1}}{r_{d1} + r_{d2}}$$

**Solution:**
The relationship among the FET parameters can be written as:

$$\mu = \frac{\partial V_{DS}}{\partial V_{GS}}$$

This can be further modified and written as:

$$\mu = \frac{\partial V_{DS}}{\partial I_D} \times \frac{\partial I_D}{\partial V_{GS}}$$

$\therefore$
$$\mu = \frac{\partial V_{DS}}{\partial I_D} \times \frac{\partial I_D}{\partial V_{GS}} \quad (1)$$

where, $r_d = \dfrac{\partial V_{DS}}{\partial I_D}$ and $g_{fs} = \dfrac{\partial I_D}{\partial V_{GS}}$

$\mu$ is the amplification factor.

If two FETs are in parallel then the current change is the double of that of a single FET for a given change in gate voltage.

Now,
$$g_m = \frac{I_{dS1}(\text{total})}{V_{gS}}$$

$\therefore$
$$g_m = g_{m1} + g_{m2} \quad (2)$$

Also,
$$g_d = g_{d1} + g_{d2} \quad (3)$$

$\because$
$$g_d = \frac{1}{r_d} \text{ when } \mu = 1$$

$\therefore$
$$\frac{1}{r_d} = \frac{1}{r_{d1}} + \frac{1}{r_{d2}}$$

Also,
$$\mu = g_m r_d$$

$$= (g_{m1} + g_{m2}) \frac{r_{d1} r_{d2}}{r_{d1} + r_{d2}}$$

$$= \left(\frac{\mu_1}{r_{d1}} + \frac{\mu_2}{r_{d2}}\right) \frac{r_{d1} \times r_{d2}}{r_{d1} + r_{d2}}$$

$$= \frac{\mu_1 r_{d2} + \mu_2 r_{d1}}{r_{d1} + r_{d2}} \times \frac{r_{d1} r_{d2}}{r_{d1} + r_{d2}}$$

$\therefore$
$$\mu = \frac{\mu_1 r_{d2} + \mu_2 r_{d1}}{r_{d1} + r_{d2}}$$

**Example 7-43** The JFET, as shown in the diagram, has an $I_{dss} = 10$ mA, and $V_p = -5$ V. Find the value of resistance $R_S$ for a drain current $I_{DS} = 6.4$ mA.

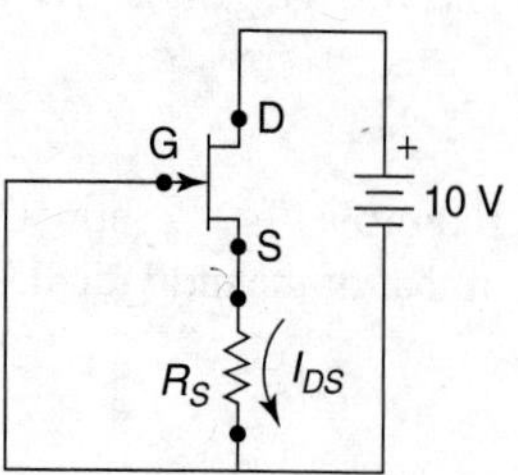


**Solution:**

Given,

$$I_{DSS} = 10 \text{ mA}, V_p = -5 \text{ V}$$

$$I_{DS} = 6.4 \text{ mA}, R_S = ?$$

$$6.4 = 10\left(1 - \frac{V_{GS}}{V_p}\right)^2$$

$$0.8 = \left(1 - \frac{V_{GS}}{V_p}\right)$$

$$V_{GS} = 0.2\ V_p = 0.2 \times (-5) = -1\ \text{V}$$

$\Rightarrow$
$$I_{DS}\, R_S = 1$$

Therefore,
$$R_s = \frac{1}{I_{DS}} = \frac{1}{6.4 \times 10^{-3}}\ 156\ \Omega$$

**Example 7-44** If $|I_{DSS}| = 4$ mA, $V_p = 4$ V, calculate the quiescent value of $I_D$, $V_{GS}$ and $V_{DS}$.

**Solution:**

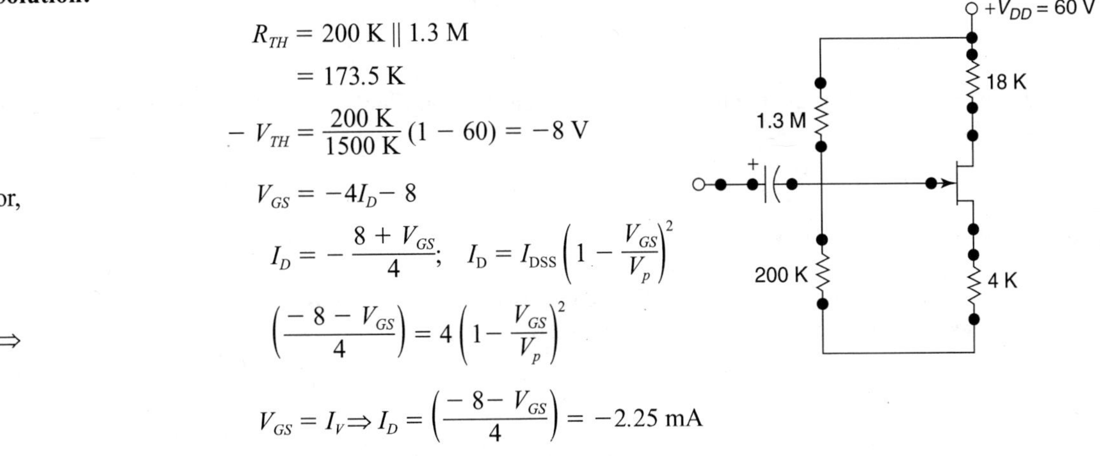


$$R_{TH} = 200\ \text{K} \parallel 1.3\ \text{M}$$
$$= 173.5\ \text{K}$$
$$-V_{TH} = \frac{200\ \text{K}}{1500\ \text{K}}(1 - 60) = -8\ \text{V}$$

or,
$$V_{GS} = -4I_D - 8$$
$$I_D = -\frac{8 + V_{GS}}{4};\quad I_D = I_{DSS}\left(1 - \frac{V_{GS}}{V_p}\right)^2$$

$\Rightarrow$
$$\left(\frac{-8 - V_{GS}}{4}\right) = 4\left(1 - \frac{V_{GS}}{V_p}\right)^2$$

$$V_{GS} = I_V \Rightarrow I_D = \left(\frac{-8 - V_{GS}}{4}\right) = -2.25\ \text{mA}$$

Solving, we get:
$$V_{GS} = I_V$$
$$I_D = -2.25\ \text{mA}$$

Hence,
$$V_{DS} = -60 - (18 + 4)(2.25)$$
$$= -60 + 49.5 = -10.5\ \text{V}$$

■

## 7-6 HIGH-FREQUENCY MOSFET MODEL

The high-frequency MOSFET model takes into account many capacitances of the device, such as inter-electrode capacitance, wiring capacitance, etc. It is advisable to peep into the details of these capacitances which arise due to many parameters related to the fabrication of the device. Among the many internal capacitances of the device, the gate-to-channel capacitance is the most common one.

There are basically two types of capacitances in the high-frequency MOSFET model.

(i) The gate capacitance effect: A parallel plate capacitor is formed due to the channel and the gate electrode where the silicon dioxide layer forms the dielectric substance.

(ii) The source–body and drain–body capacitive effect: These are capacitances which arise due to the reverse-biased *p*–*n* junctions formed by the $n^+$ source region and *p*-type substrate, and by the $n^+$ type drain region and substrate.

Effectively, there are five capacitances $C_{gs}$, $C_{gd}$, $C_{gb}$, $C_{sb}$ and $C_{db}$, where the subscripts locate the position of the terminals between which the capacitances are located.

### 7-6-1 Effective Capacitance of the Gate

With the gate capacitances, as listed in the previous section, it is advisable to revisit their expressions prior to investigating the operation of the device in the high-frequency region.

(i) When the MOSFET operates in the triode region at a relatively small voltage $v_{DS}$, the depth of the channel in the bulk is uniform. The equivalent gate-channel capacitance can be considered to be divided equally between the source and the gate ends. Thus, $C_{gs} = C_{gd} = WLC_{ox}/2$ where, the letters carry their usual meaning.

(ii) As the MOSFET operates in the saturation region, the channel is tapered into shape, and is gradually pinched-off at the drain end. Under such circumstances, the capacitances are given by:

$$C_{gs} = \frac{2WLC_{ox}}{3} \quad \text{and} \quad C_{gd} = 0$$

(iii) When the MOSFET is in the cut-off region, the channel disappears, and the capacitances take the values as given by:

$$C_{gs} = C_{gd} = 0 \quad \text{and} \quad C_{gb} = WLC_{ox}$$

### 7-6-2 The Junction Capacitance

The depletion-layer capacitance resulting from the two reverse-biased regions between the gate and the bulk, and the drain and the bulk respectively, is quantitatively given by:

$$C_{sb} = \frac{C_{sbo}}{\sqrt{1 + \frac{V_{sb}}{V_o}}} \tag{7-31}$$

where, $C_{sbo}$ is the value of the desired capacitance at zero body-source bias, $V_o$ is the junction built in voltage 0.7 V and $V_{sb}$ is the applied reverse-biased voltage.

### 7-6-3 The High-Frequency Models of the MOSFET

Figure 7-13 shows the high-frequency model of the MOSFET. This model is too complex to be analysed manually, but with the SPICE software, meant for circuit analysis, the procedure becomes exceedingly simple. This model is primarily used to predict the high-frequency response of the MOSFET.

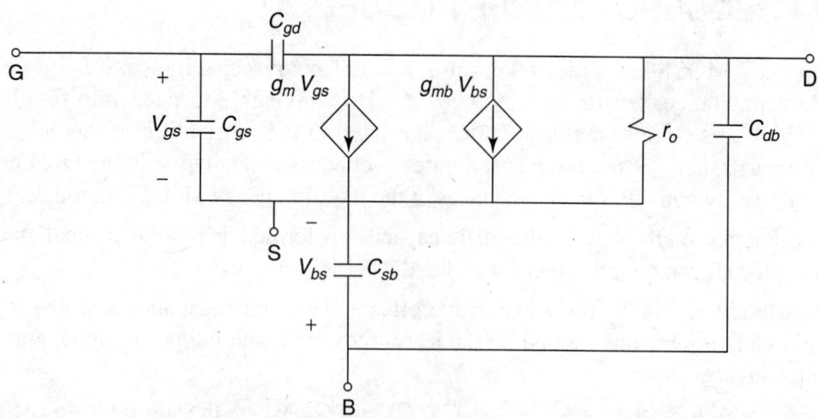


**Figure 7-13** High-frequency equivalent model of MOSFET

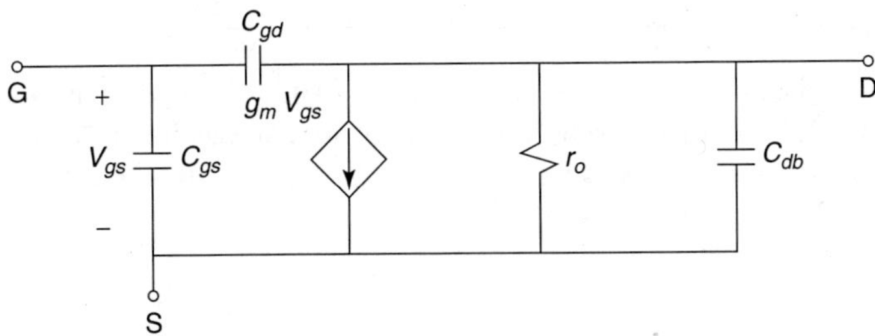


**Figure7-14** High-frequency MOSFET model when the source is connected to the base

When the condition so arises that the source is connected to the body, then the model, as shown in Fig. 7-13, gets simplified (see Fig 7-14). Again, in the model, as shown in Fig. 7-14, when $C_{db}$ is neglected, the resulting circuit is as given by Fig. 7-15.

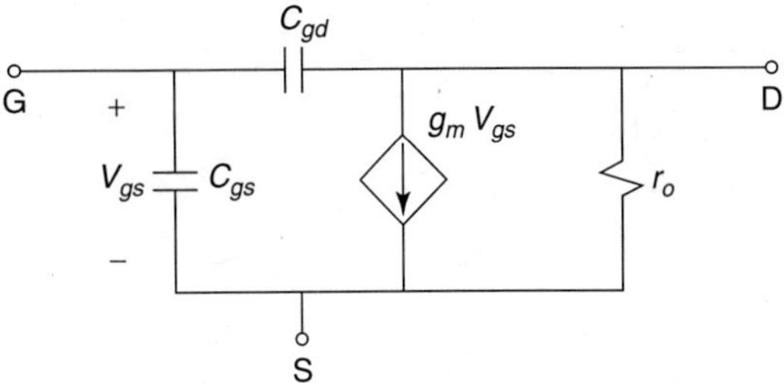


**Figure 7-15** Circuit obtained upon neglecting $C_{db}$

A parameter used to judge the operation of a high-frequency MOSFET as an amplifier, is the unity-gain bandwidth. The frequency at which short-circuit current gain of the common-source arrangement becomes unity, is known as the unity-gain frequency. This analysis is done using a hybrid $\pi$ model with a common-source configuration.

It can be noticed easily that the current in the short circuit is given by:

$$I_0 = g_m V_{gs} - sC_{gd}V_{gs} \tag{7-32}$$

where, $s$ is a complex variable.

Since $C_{gd}$ is very small, Eq. (7-32) can be written as $I_0 \approx g_m V_{gs}$.
And from Fig. 7-16, we get:

$$V_{gs} = \frac{I_i}{s(C_{gs} + C_{gd})} \tag{7-33}$$

Substituting, $I_0 = g_m V_{gs}$ we obtain:

$$\frac{I_0}{g_n} = \frac{I_i}{s(C_{gs} + C_{gd})}$$

$$\frac{I_0}{I_i} = \frac{g_m}{s(C_{gs} + C_{gd})} \tag{7-34}$$

Taking $s = j\omega$ (where, $\omega$ is the frequency of the applied voltage), and since the magnitude of current gain becomes unity at this frequency, we can write:

$$\omega_T = \frac{g_m}{(C_{gs} + C_{gd})} \tag{7-35}$$

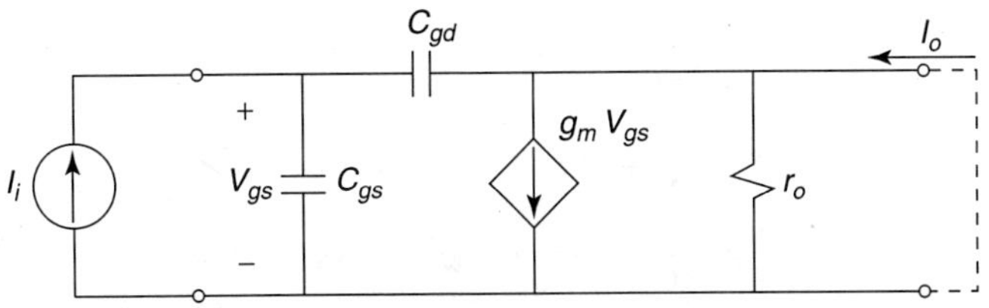


**Figure 7-16** Circuit representation for obtaining short-circuit current gain

## Solved Examples

**Example 7-45** Calculate the input admittance of an FET at $10^3$ and $10^6$ Hz when the total drain circuit impedance is (a) a resistance of 50 K and (b) a capacitive reactance of 50 K at each frequency. (Take the inter-electrode capacitances into consideration.) The FET parameters are $\mu = 20$, $r_d = 10$ K, $g_m = 2.0$ mA/V, $C_{gs} = 3.0$ pF, $C_{ds} = 1.0$ pF and $C_{gd} = 2.0$ pF. Express the results in terms of the input resistance and capacitance.

**Solution:**

We have two equations:

$$A_V = \frac{-g_m + Y_{gd}}{g_d + Y_d + Y_{ds} + Y_{gs}}$$

$$Y_1 = Y_{gs} + (1 - A_V)Y_{gd}$$

(a) $R_d = 50$ K

For, $f = 10^3$Hz.

$$Y_{gs} = j\omega C_{gs} = j2\pi \times 10^3 \times 3.0 \times 10^{-12} = j1.88 \times 10^{-6}$$

$$Y_{ds} = j\omega C_{ds} = j2\pi \times 10^3 \times 1.0 \times 10^{-12} = j0.628 \times 10^{-6}$$

$$Y_{gd} = j\omega C_{gd} = j2\pi \times 10^3 \times 2.0 \times 10^{-12} = j1.26 \times 10^{-6}$$

$$Y_d = 2 \times 10^{-6}\ g_d = 10^{-4}$$

Hence, $$A_V = \frac{-2 \times 10^{-3} + j1.26 \times 10^{-6}}{12 \times 10^{-6} + j1.88 \times 10^{-6}} \approx \frac{-2}{12} \times 10^2 = -16.7$$

and, $C_1 = 3.0 + 17.7 \times 2 = 38.4$ pF

For, $f = 10^6$Hz:

$$Y_{gs} = j1.88 \times 10^{-6}$$

$$Y_{ds} = j0.628 \times 10^{-6}$$

$$Y_{gd} = j1.26 \times 10^{-6}$$

Hence, $$A_V = \frac{-2 \times 10^{-3} + j1.26 \times 10^{-6}}{1.2 \times 1^{-5} + j1.88 \times 10^{-6}} = \frac{-200 + j1.26}{12 + j1.68} = -16.3 + j1.97$$

$$Y_1 = \frac{1}{R_1} + j\omega C_1; \quad C_1 = 37.6 \text{ pF}; \quad \frac{1}{R_1} = 1.26 \times 10^{-6} \times 1.97 = 2.48 \times 10^{-6}$$

$$R_1 = \frac{10^5}{2.48} = 40.4 \text{ K}$$

(b) $Z_L = j5 \times 10^4$; $Y_L = j2 \times 10^{-6}$

For, $f = 10^3$Hz:

$$A_V = \frac{-g_m}{g_d + Y_L} = \frac{-2 \times 10^{-3}}{10^{-4} + j2 \times 10^{-5}}$$

$$= \frac{-200}{10 + j2} = -19.2 + j3.84$$

$$Y_1 = \frac{1}{R_1} + j\omega C_1 = +j\omega C_{gs} + (20.2 - j3.84)j\omega C_{gd}$$

or, $C_1 = 30 + 20.2 \times 2.0 = 43.4$ pF

and, $$\frac{1}{R_1} = 4.84 \times 10^{-6}$$

or, $R_1 = +20.8$ M

For, $f = 10^6$ Hz:

$$A = \frac{-2 \times 10^{-3} + j1.26 \times 10^{-6}}{10^{-4} + j3.88 \times 10^{-6}}$$

$$= \frac{-200 + j1.26}{10 + j3.88}$$

$$= -17.4 + j6.85$$

$$C_1 = 3.0 + 18.4 \times 2.0 = 39.8 \text{ pF}$$

and, $$\frac{1}{R_1} = \frac{10^5}{8.64} = 11.6 \text{ K}$$

**Example 7-46** Starting with the low-frequency small-signal circuit model of the FET, show that, for the CG amplifier stage with $R_s = 0$ and $C_{ds} = 0$:

(a) $A_v = \dfrac{(g_m + g_d)R_d}{1 + R_d(g_d + j\omega C_{gd})}$

(b) $Y_i = g_m + g_d(1 - A_v) + j\omega C_{sg}$

(c) Repeat part (a), taking the source resistance $R_s$ into account.

(d) Repeat part (b), taking the source resistance $R_s$ into account.

**Solution:**

The small-signal equivalent of a CG ampliifer can be drawn as shown in the following diagram:

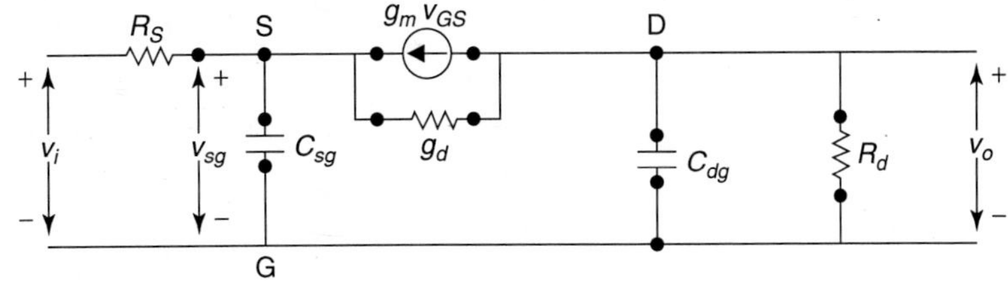


(a) $R_S = 0$, then $V_1 = V_{sg} = -V_{gs}\ A_V = \dfrac{v_o}{v_i} = \dfrac{v_o}{v_{sg}}$

where, $$v_o = I_{sc}Z \tag{1}$$

and, $$i_{sc} = -g_m v_{gs} + g_d v_{sg} = (g_m + g_d)v_{sg}$$

$$Z = \text{output impedance with } v_{sg} = 0$$

Now, $$Y = \frac{1}{Z} = \frac{1}{R_d} + g_d + j\omega C_{dg}$$

Substituting the values of $I_{sc}$ and $Z$ in Eq. (1):

$$v_o = \frac{g_m + g_d}{\frac{1}{R_d} + g_d + j\omega C_{dg}} \times v_i$$

Hence,

$$A_V = \frac{g_m + g_d}{\frac{1}{R_d} + g_d + j\omega C_{dg}} = \frac{(g_m + g_d)R_d}{1 + R_d(g_d + j\omega C_{dg})}$$

(b) KCL at node $S$ gives:

$$i_s = v_{sg}(j\omega C_{sg}) - g_m v_{gs} + g_d(v_{sg} - v_o)$$

$\therefore$

$$Y_1 = \frac{i_s}{v_{sg}} = j\omega C_{sg} + g_m + g_d(1 - A_V)$$

(c) With $R_S \neq 0$, $v_{sg} = v_i - i_s R_s$

where,

$$i_s = (g_m + g_d + j\omega C_{sg})v_{sg} + R_s g_d v_o$$

we get,

$$v_{sg} = v_i - R_s(g_m + g_d + j\omega C_{sg})v_{sg} + R_s g_d v_o$$

or,

$$v_{sg} = \frac{v_i + R_s g_d v_o}{1 + (g_m + g_d + j\omega C_{sg})R_s}$$

Now, KCL at node $D$ gives:

$$g_m v_{gs} + v_o\left(\frac{1}{R_d} + g_d + j\omega C_{dg}\right) = g_d v_{sg}$$

but, $v_{gs} = -v_{sg}$

so,

$$v_o\left(\frac{1}{R_d} + g_d + j\omega C_{dg}\right) = (g_m + g_d)v_{sg}$$

or,

$$v_o\left(\frac{1}{R_d} + g_d + j\omega C_{dg}\right) = \frac{(g_m + g_d)(v_i + R_s g_d v_o)}{1 + (g_m + g_d + j\omega C_{sg})R_s}$$

or,

$$v_o\left[\frac{1}{R_d} + g_d + j\omega C_{dg} - \frac{(g_m + g_d)R_s g_d}{1 + (g_m + g_d + j\omega C_{sg})R_s}\right] = \frac{(g_m + g_d)v_i}{1 + (g_m + g_d + j\omega C_{sg})R_s}$$

Hence,

$$A_V' = \frac{v_o}{v_i}$$

$$= \frac{(g_m + g_d)}{\left(\frac{1}{R_d} + g_d + j\omega C_{dg}\right)[1 + (g_m + g_d + j\omega C_{sg})R_s] - (g_m + g_d)R_s g_d}$$

(d) With $R_S \neq 0$:

$$Y_1' = \frac{1}{R_s + \frac{1}{Y_1}} = \frac{Y_1}{1 + R_s Y_1}$$

$$= \frac{j\omega C_{sg} + g_m + g_d(1 - A_V)}{1 + R_s[j\omega C_{sg} + g_m + g_d(1 - A_V)]}$$

**Example 7-47** (a) For the source follower with $g_m = 2$ mA/V, $R_s = 100$ K and $r_d = 50$ K, and with each inter-node capacitance 3pF, find the frequency at which the reactive component of the output admittance equals the resistive component. (b) At the frequency found in part (a), calculate the gain and compare it with the low-frequency value.

**Solution:**

(a) From the equation of input admittance, we have $Y_0 = g_m + g_d + j\omega C_T$

but, $$C_T = C_{gs} + C_{ds} + C_{sn} = 9\,pF$$

Hence, $$\omega C_T = g_m + g_d$$

or, $$\omega = \frac{2 \times 10^{-3} + 2 \times 10^{-5}}{9 \times 10^{-12}}$$

$$= \frac{2.02 \times 10^{-3}}{9 \times 10^{-12}} = 0.224 \times 10^{9}$$

or, $$f = 0.356 \times 10^{6} = 35.6 \text{ MHz}$$

(b) At low frequencies the gain is given by:

$$A_V = \frac{g_m R_s}{1 + (g_m + g_d)R_s}$$

$$= \frac{200}{1 + 202} = \frac{200}{203} = 0.985$$

For, $f = 35.6$ MHz, $\omega = 0.224 \times 10^{9}$.

$\therefore\ j\omega C_{gs} = j0.224 \times 10^{9} \times 3 \times 10^{-12} = j0.672 \times 10^{-3}$

$j\omega C_T = j0.224 \times 10^{9} \times 9 \times 10^{-12} = j2.02 \times 10^{-3}$

We have:

$$A_V = \frac{(1 + j0.672) \times 10^{-3} \times 10^{5}}{1 + 2.02(1 + j) \times 10^{-3} \times 10^{5}} = \frac{200 + j67.2}{202(1 + j)}$$

Hence, $$|A_V| = \frac{10^{2}\sqrt{4 + 0.45}}{202\sqrt{2}}$$

$= 0.738$ as compared to 0.985 at low frequencies.

**Example 7-48** The drain current (in milliamperes) of the enhancement-type MOSFET, as shown in the diagram, is given by $I_D = 0.2(V_{GS} - 3)^2$ in the region $V_{DS} \geq V_{GS} - V_P$. If $V_P = +3$ V, calculate the quiescent values $I_D$, $V_{GS}$ and $V_{DS}$.

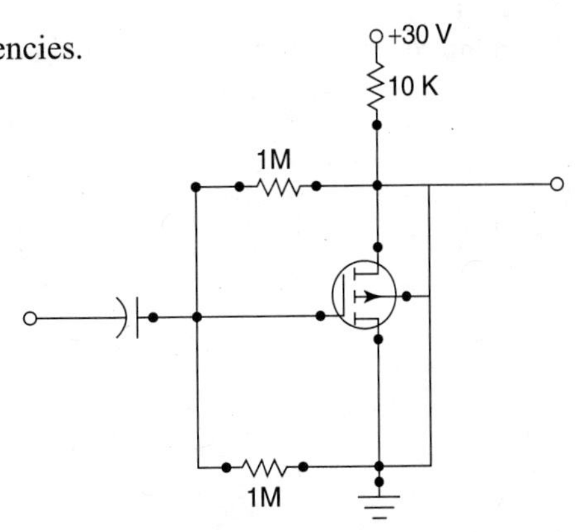


**Solution:**

Given, $I_D = 0.2(V_{GS} - 3)^2$

where, from the circuit we have:

$$V_{GS} = \frac{1}{2}(30 - 10I_D)$$

Substituting and solving we obtain:

$$I_D = 0.2(15 - 5I_D - 3)^2 = 0.2(144 - 120I_D + 25I_D^2)$$

or,

$$5I_D - 25I_D + 28.8 = 0$$

Then, $I_D = \frac{25 \pm \sqrt{625 - 576}}{10} = \frac{25 \pm \sqrt{49}}{10}$ and $I_D = 1.8$ mA or 3.2 mA.

The second value of $I_D$ is rejected since $V_{GS} = 3(30 - 32) = -1$V and the FET will be cut off. Therefore, $I_D = 1.8$ mA. Then:

$$V_{GS} = \frac{1}{2}(30 - 18) = +6 \text{ V}$$

and,

$$V_{DS} = 30 - 1.8 \times 10 = +12 \text{ V}$$

**Example 7-49** For the given FET circuit show that if load resistance $R_L << 1/h_{ob2}$, the voltage gain of the hybrid cascade amplifier stage is given to a very good approximation by $A_V = g_m h_{fb} R_L$ where, $g_m$ is the FET transconductance.

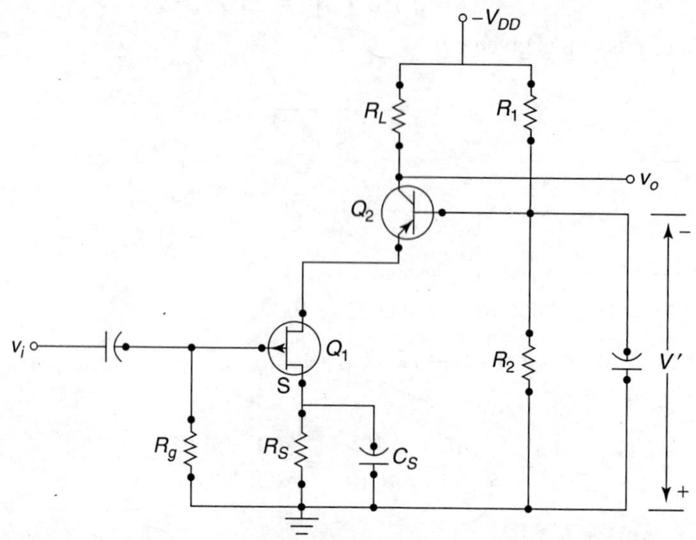


**Solution:**
From the equivalent circuit, as shown in the following diagram, we have:

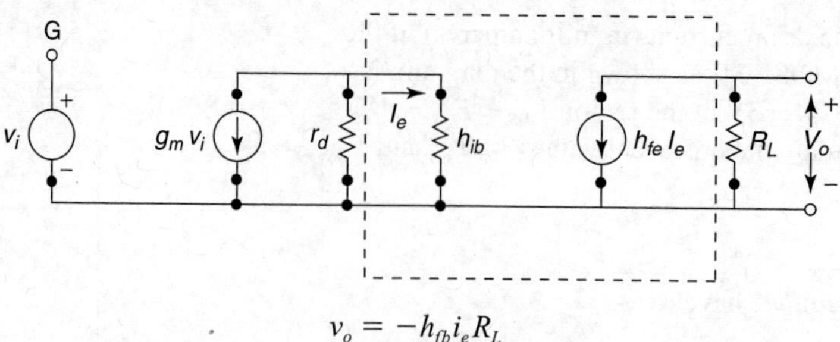


$$v_o = -h_{fb} i_e R_L$$

where, $$i_e = -g_m v_1 \frac{r_d}{r_d + h_{ib}} = -g_m v_1$$

since, $$r_d >> h_{ib}$$

Substituting, we obtain: $$A_V = \frac{v_o}{v_1} = -h_{fb} R_L(-g_m) = h_{fb} R_L g_m$$

**Example 7-50** For the circuit, as shown in the diagram, if $h_{ie} << R_d$, $h_{ie} << r_d$, $h_{fe} >> 1$, and $\mu >> 1$, show that (a) $A_{v1} = v_{o1}/v_i \approx g_m h_{fe} R_s / 1 + g_m h_{fe} R_s$ and (b) $A_{v2} = v_{o2}/v_i \approx g_m h_{fe}(R_s + R_c)/1 + g_m h_{fe} R_s$ where, $g_m$ is the FET transconductance.

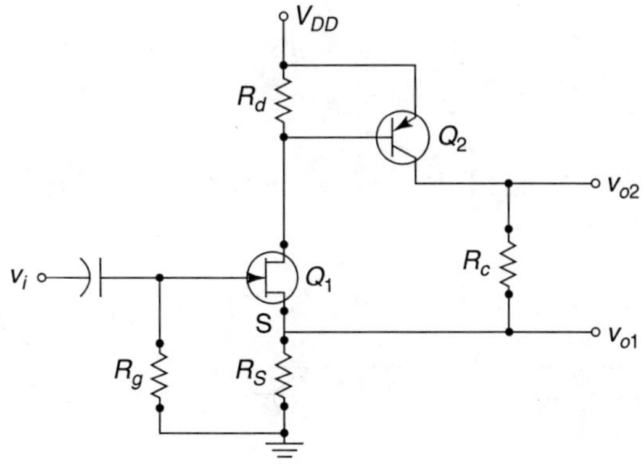


**Solution:**
The small-signal equivalent circuit is as shown in the following diagram.

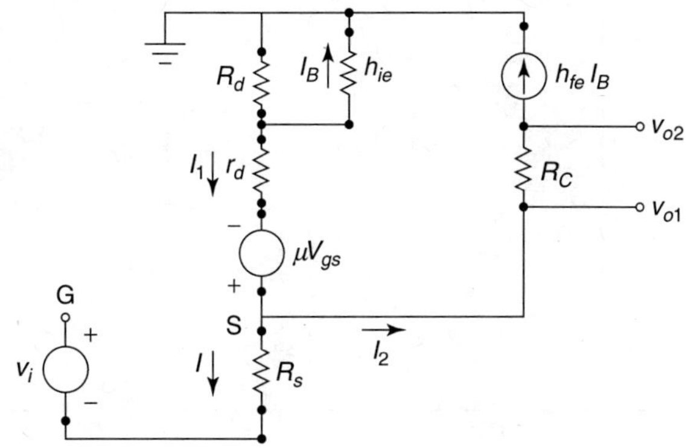


The approximate equivalent circuit is used for $Q_2$.
From the circuit, $v_{gs} = v_1 - IR_s$, where, $I = I_1 - I_2 = I_1 - h_{fe} I_B$.
Since $h_{ie} << R_D$, then:

$$I_B \approx -I_i,\ I = (1 + h_{fe})\, I_i$$

KVL in the FET loop gives:

$$[h_{ie} + i_d + R_s(1 + h_{fe})]\ I_1 = \mu v_{gs} = \mu v_1 = \mu(1 + h_{fe}) R_s I_1$$

or,
$$I_1 = \frac{\mu v_1}{h_{fe} + r_d + (\mu + 1)(1 + h_{fe})R_s} \approx \frac{g_m v_1}{1 + g_m h_{fe} R_s}$$

since
$$r_d >> h_{ie,} h_{fe} >> 1 \text{ and } \mu >> 1$$

then,
$$v_{o1} = IR_s = (1 + h_{fe})R_s I_1 \approx h_{fe} R_s I_1$$

$\therefore$
$$A_{V1} = \frac{v_{o1}}{v_1} \approx \frac{g_m h_{fe} R_s}{1 + g_m h_{fe} R_s}$$

and
$$v_{o2} = v_{o1} - h_{fe} I_B R_C \approx (h_{fe} R_s + h_{fe} R_C) I_1 = h_{fe}(R_s + R_C) I_1$$

hence,
$$A_{V2} = \frac{v_{o2}}{v_1} \approx \frac{g_m h_{fe}(R_s + R_C)}{1 + g_m h_{fe} R_s}$$

**Example 7-51** If $r_d >> R_1$, $R_2 >> h_{ib3}$, $1/h_{oe2} >> h_{ib3}$, $R' >> R_3$ and $1/h_{ob3} >> R_3$, show that the voltage gain at low frequencies is given by:

$$A_O = \frac{v_o}{v_i} = g_m(1 + h_{fe2}) h_{fb3} \frac{R_1 R_3}{R_1 + h_{ie2} + h_{ib3}(1 + h_{fe2})}$$

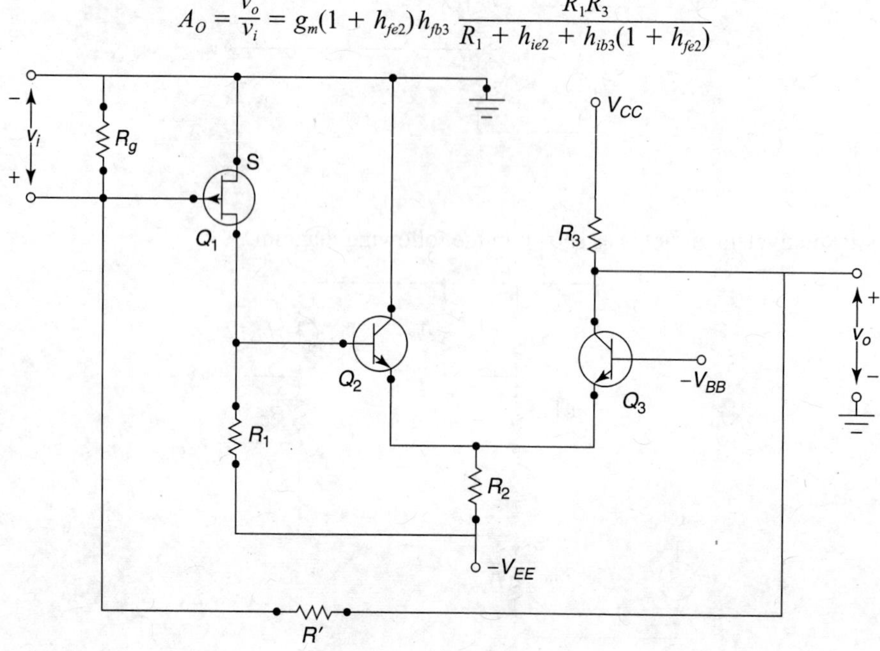


**Solution:**

A small-signal equivalent circuit of the amplifier is as shown in the following diagram.

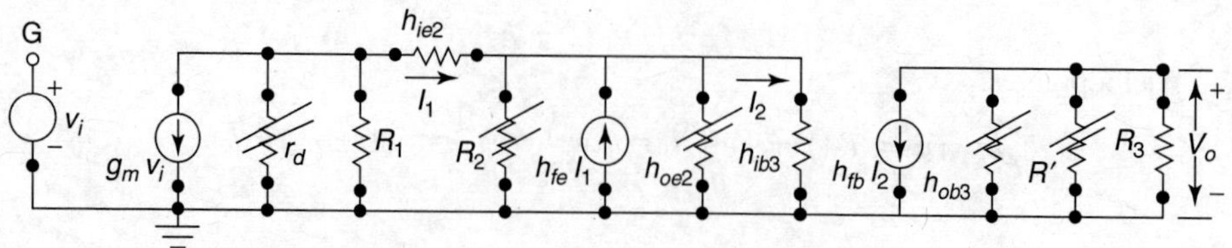

It is assumed that $h_r$ is negligible for $Q_2$ and $Q_3$.

Moreover, if $r_d >> R_1$, $R_2 >> h_{1bs}$, $\frac{1}{h_{0e2}} >> h_{1bs}$, $\frac{1}{h_{0bs}} >> R_s$ and $R' >> R_s$, we can cross out these elements as shown in the diagram.

Then, $$v_o = -h_{fbs} I_2 R_s \text{ where } I_s = (1 + h_{fe2}) I_1$$

and, $$I_1 = \frac{-g_m v_1 R_1 - (1 + h_{fe2}) h_{ibs} I_1}{h_{ie2} + R_1}$$

or, $$I_1 = \frac{-g_m v_1 R}{R_1 + h_{ie2} + (1 + h_{fe2}) h_{ibs}}$$

∴ $$v_o = \frac{g_m (1 + h_{fe2}) R_2 R_s h_{fb2} v_1}{R_1 + h_{ie2} + (1 + h_{fe2}) h_{ibs}}$$

Hence, $$A_o = \frac{v_o}{v_1} = g_m (1 + h_{fe2}) h_{fbs} \frac{R_1 R_s}{R_1 + h_{ie2} + h_{ibs}(1 + h_{fe2})}$$

Note that $A_o$ is negative, since $h_{fbs}$ is a negative number. Thus, the feedback provided by $R'$ is degenerative and stabilizing the amplifier at the same time.

**Example 7-52** (a) A MOSFET connected in the CS configuration works into a 100 K resistive load. Calculate the complex voltage gain and the input admittance of the system for frequencies of 100 and 100,000 Hz. Take the inter-electrode capacitances into consideration. The MOSFET parameters are $\mu = 100$, $r_d = 40$ K, $g_m = 2.5$ mA/V, $C_{gs} = 4.0$ pF, $C_{ds} = 0.6$ pF and $C_{gd} = 2.4$ pF. Compare these results with those obtained when the inter-electrode capacitances are neglected. (b) Calculate the input resistance and the capacitance.

**Solution:**

$Y_{gs} = j\omega C_{gs} = j\,2\pi \times 10^2 \times 4.0 \times 10^{-12} = j\,2.51 \times 10^{-9}$ mho

$Y_{ds} = j\omega C_{ds} = j\,2\pi \times 10^2 \times 0.6 \times 10^{-12} = j\,3.77 \times 10^{-10}$ mho

$Y_{gd} = j\omega C_{gd} = j\,2\pi \times 10^2 \times 2.4 \times 10^{-12} = j\,1.51 \times 10^{-9}$ mho

$$g_d = \frac{1}{r_d} = 2.5 \times 10^{-5} \text{ mho}$$

$$Y_d = \frac{1}{R_d} = 10^{-5} \text{ mho}$$

$$g_m = 2.5 \times 10^{-3} \text{ mho}$$

The gain of a one-stage amplifier is given by:

$$A_V = \frac{-g_m + Y_{gd}}{g_d + Y_d + Y_{ds} + Y_{gs}} = \frac{-2.5 \times 10^{-3} + j1.51 \times 10^{-9}}{2.5 \times 10^{-5} + 10^{-5} + j1.88 \times 10^{-9}}$$

It is seen that the $j$ terms are negligible in comparison with the real term. In that case:

$$A_V = -\frac{2.5}{3.5} \times 10^2 = -71.4$$

Since the gain is a real number, the input impedance consists of a capacitor whose value is given by:

$$\frac{Y_i}{j\omega} = C_i = C_{gs} + (1 + g_m R_d')C_{gd}$$

$$C_i = C_{gs} + (1 - A_V)C_{gd} = 4.0 + 72.4 \times 2.4 = 177.5 \text{ pF}$$

We repeat these calculations for $f = 10^6$ Hz.

Then,
$$Y_{gs} = j2.51 \times 10^{-6} \text{ mho}$$

$$Y_{ds} = j0.377 \times 10^6 \text{ mho}$$

$$Y_{gd} = j1.51 \times 10^{-6} \text{ mho}$$

Hence,
$$A_V = \frac{-2.5 \times 10^{-3} + j1.51 \times 10^{-9}}{3.5 \times 10^{-5} + j0.188 \times 10^{-8}} = \frac{-2.5 \times 10^{-3}}{3.5 + j0.188} \times 10^6$$

$$= \frac{-2.5 \times 3.5 \times 10^2}{12.30} + j\frac{2.5 \times 0.188 \times 10^2}{12.30}$$

$$= -71.2 + j3.82$$

From the equation, $Y_i = Y_{gs} + (1 - A_V)Y_{gd}$ we have:

$$G_1 + j\omega C_1 = j\omega C_{gs} + 72\, j\omega C_{gd} + 3.82\, \omega C_{gd}$$

or,
$$C_1 = C_{gs} + 72\, C_{gd} = 177 \text{ pF}$$

or,
$$G_1 = 2\pi \times 10^6 \times 2.4 \times 10^{-12} \times 3.82 = 0.580 \times 10^{-6} \text{ mho}$$

$$R_1 = 1.735 \times 10^6\ \Omega = 173.5 \text{ k}\Omega$$

**Example 7-53** An $n$-channel MOSFET having a $V_T$ of 2 V (threshold voltage) is used in the circuit, as shown in the following diagram. Initially T is OFF and in a steady state. At time $t = 0$, a step voltage of magnitude 4 V is applied to the input so that the MOSFET turns on instantaneously. Draw the equivalent circuit and calculate the time taken for the output $V_O$ to fall to 5 V. The device constant of the MOSFET is given by:

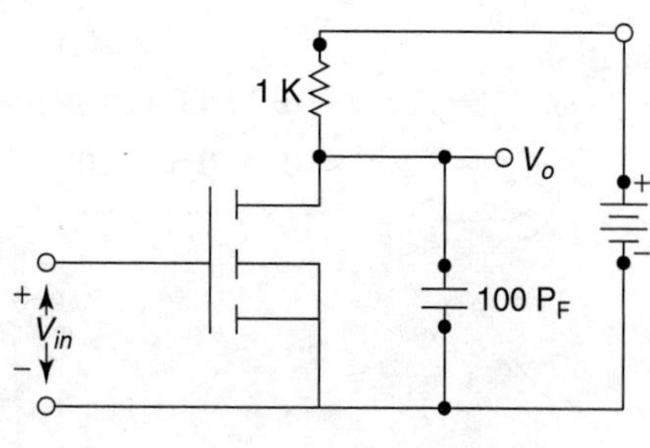


$$\frac{1}{2} k_n' \frac{W}{L} = K = 5 \text{ mA/V}^2,\ R_{DS} = \infty, C_{DS} = 0;\ C_{DG} = 0.$$

**Solution:**
Given $n$-channel MOSFET:

$$V_T = 2$$

T is in the OFF and steady state, therefore, MOS turns on:

$$K = 5 \text{ mA}/V^2,\ R_{DS} = \infty,\ C_{DS} = 0,\ C_{DG} = 0$$

Assuming that initally the capacitor is charged to 10 V, as the MOSFET is OFF:

$$V_{Gs} = 4 \text{ V}$$

$$I_D = K(V_{GS} - V_T)^2$$
$$= 5(4 - 2)^2 = 20 \text{ mA}$$

In the saturation region:

$$V_{GS} > V_T \quad \text{(satisfied)}$$
$$V_{GD} > V_T$$
$$\alpha_1 - V_o 2 \Rightarrow -V_o - 2$$

$\therefore$ $$V_o 2 \text{ V}$$

Since $V_o$ falls only to 5 V, the MOSFET is in the pinch-off region.

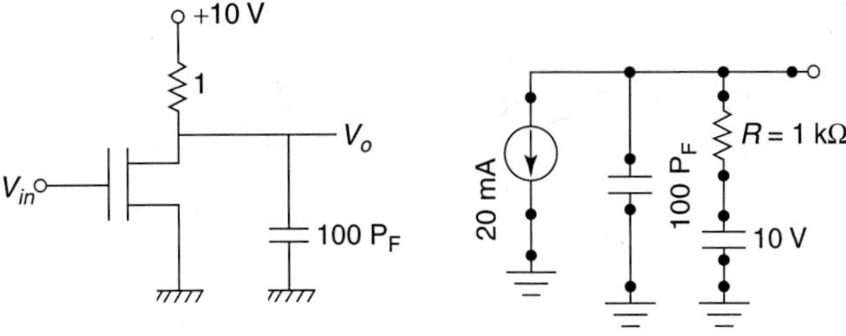


Application of KCL to the model gives:

$$2 \times 10^3 = -\frac{Cdv_o}{dt} + \frac{10 - V_o}{R}$$

Taking Laplace transform:

$$\frac{20 \times 10^{-3}}{5} = 10 \times 10^{12}[SV_o(s) - V_o(0)]$$
$$+ \frac{10 \times 10^{-3} V_o(S) \times 10^{-3}}{5} + \frac{10 \times 10^{-3}}{5}$$
$$= -V_o(S)[10 \times 10^{12} S - 10^3] + 100 \times 10^{12}$$
$$(S + 10^7)V_o(s) = -\frac{10^8}{5} - 10$$
$$V_o(S) = \frac{10^8}{S(S + 10^7)} = \frac{10}{S(S + 10^7)}$$
$$V_o(S) = \frac{10}{5} - \frac{10}{S + 10^7} - \frac{10}{S(S + 10^7)}$$

Inverse Laplace gives:

$$V_o(t) = 10^2 e^{-t}(10^7)$$
$$V_o = 5 \text{ V}$$

$\Rightarrow$ $$5 = 20(1^{-t} \times 10^7)$$

$\Rightarrow$ $$t = 10^{-7} \log_e^4 = 13.8 \times 10^{-8} = 138 \text{ ns}$$

**Example 7-54** An $n$-channel has $I_{DSS} = 1$ mA and $V_p = -5$ V. Find the maximum transconductance.

**Solution:**

Given, $I_{DSS} = 1 \text{ mA} \quad \text{and} \quad V_p = -5 \text{ V}$

It can be shown that the transconductance of the FET is:

$$g_m = \frac{2I_{DSS}}{V_P}\left(1 - \frac{V_{GS}}{V_P}\right) = \frac{2I_{DSS}}{|V_p|} = \frac{2 \times 1 \times 10^{-3}}{5} \text{ millimho} = 0.4 \text{ millimho}$$

**Example 7-55** An NMOS circuit is shown in the given diagram. The specifications of the circuit are as follows:

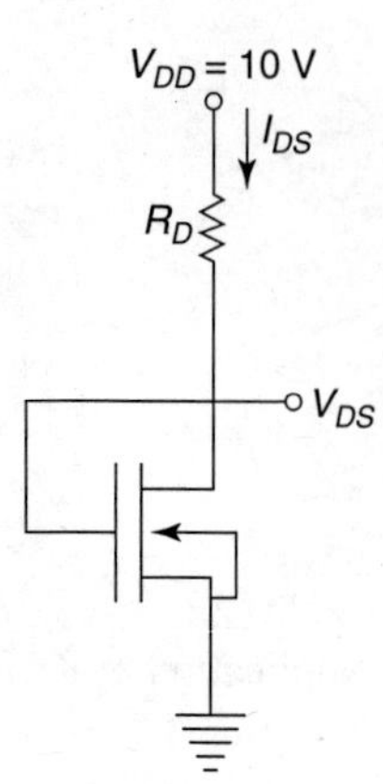


$$V_{DD} = 10 \text{ V}; \beta = k\left(= \frac{1}{2}k_n' \frac{W}{L}\right) = \mu n\, C_{ox}(W/L) = 10^{-4} \text{ A/V}_2$$

$V_T = 1$ V and $I_{DS} = 0.5$ mA. Evaluate $V_{DS}$ and $R_D$ for the circuit. Neglect body-effect for $V_T$.

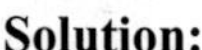

**Solution:**

Given an NMOS circuit, for which:

$$\beta = 10^{-4}\, A/V^2, \quad V_{DD} = 10\ V,\ V_T = 1\ V,\ I_{DS} = 0.5 \text{ mA}.$$

We know, in saturation:

$$I_D = K(V_{GS} - V_T)^2$$

$$5 \times 10^{-4} = 10^{-4} \times (V_{GS} - 1)^2$$

and, $$V_{DS} = V_{GS} = 3.24 \text{ V}$$

Hence, $$R_D = \frac{V_{DD} - V_{DS}}{I_{DS}} = \frac{10 - 3.24}{0.5} = 13.5 \text{ K}$$

**Example 7-56** A JFET amplifier with a voltage divider biasing circuit, as shown in the following diagram, has the following parameters: $V_p = -2$ V, $I_{DSS} = 4$ mA $R_D = 910\ \Omega$, $R_S = 3$ k$\Omega$, $R_1 = 12$ M$\Omega$, $R_2 = 8.57$ M$\Omega$ and $V_{DD} = 24$ V. Find the value of the drain current $I_D$ at the operating point. Verify whether the FET will operate in the pinch-off region.

$I_D$ $R_D$ $R_1$ $V_{DS}$ $V_{GG}$ $R_2$ $R_S$

**Solution:**

$$V_{GG} = V_D \frac{R_2}{R_1 + R_2} = 24 \times \frac{8.57 \times 10^6}{(12 + 8.57)10^6} = 10 \text{ V}$$

From the gate-source loop:

$$V_{GS} = V_{GG} - I_D R_S \qquad (1)$$

Using Eq. (1) we have:

$$I_D = I_{DSS}\left(1 - \frac{V_{GS}}{V_P}\right)^2 = 4\left(1 - \frac{10 - I_D \times 3}{2}\right)^3$$

$$9I_D^2 - 73I_D + 144 = 0$$

$$I_D = 3.39 \text{ mA} \quad \text{or} \quad 4.72 \text{ mA}$$

$$I_D = 4.72 \text{ mA} > 4 \text{ mA} = I_{DSS}$$

This value is inappropriate so, $I_{DQ} = 3.39$ mA is selected.

$$V_{GSQ} = V_{GG} - I_{DQ} R_S$$

$$= 10 - (3.39 \times 10^{-3} \times 10^3) = -0.17 \text{ V}$$

$$V_{DSQ} = V_{DD} - I_{DQ}(R_D + R_s)$$

$$= 24 - 3.39 \times 10^{-3}(0.91 + 3) \times 10^3 = 10.74$$

$$V_{DGQ} = V_{DSQ} - V_{GSQ} = 10745 + 0.17 = 10.915 \text{ V}$$

which is greater than $|V_D| = 2$ V
Hence the FET is in the pinch-off region.

**Example 7-57** (a) Calculate the voltage gain $A_V = V_o / V_i$ at 5 KHz for the circuit, as shown in the diagram. The FET parameters are $g_m = 2$ mA/V and $r_d = 10$ K. Neglect capacitances.(b) Repeat part (a) if the capacitance 0.025 $\mu$F is taken under consideration.

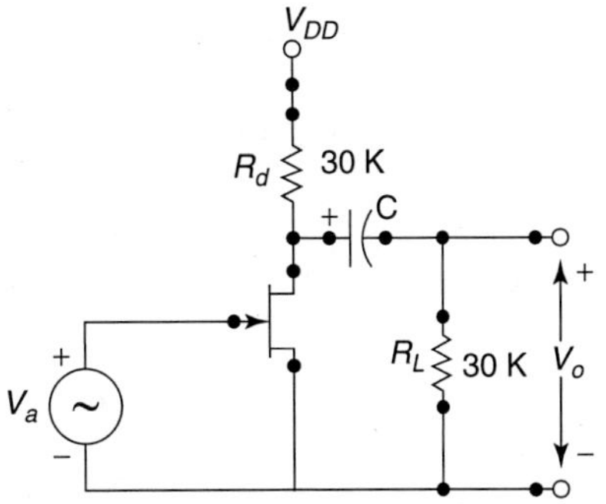

**Solution:**

$$g_m = \frac{2\text{mA}}{V}; \; r_d = 10 \text{ K}$$

(a)

$$R_L' = 30 \text{ K} \| 30 \text{ K} = \frac{30 \times 30}{30 + 30} = 15 \text{ k}\Omega$$

$$A_V = \frac{V_o}{V_i} = \frac{-\mu R_L'}{R_L' + r_d + R_s(1 + \mu)} = -\frac{g_m r_d R_L'}{R_L' + r_d}$$

$$= \frac{-(2 \times 10^{-3})(10 \times 10^3)(15 \times 10^3)}{(15 \times 10^{-3}) + (10 \times 10^3)}$$

$$A_V = -29.99 \approx -30$$

(b)

$$C = 0.025 \; \mu\text{F}$$

$$f_1 = \frac{1}{2\pi(r_d \| R_d + R_L)C_C}$$

$$f_1 = \frac{1}{2\pi(10 \text{ K} \| 30 \text{ K}) + 30 \text{ K})0.025 \times 10^{-6}} = 169.7 \text{ Hz}$$

$$\therefore \quad A_{VL} = \frac{A_v}{\sqrt{1 + \left(\frac{f_1}{f}\right)^2}} = \frac{-30}{1 + \left(\frac{169.7}{5k}\right)^2}$$

**Example 7-58** Consider the JFET circuit as given in the diagram.

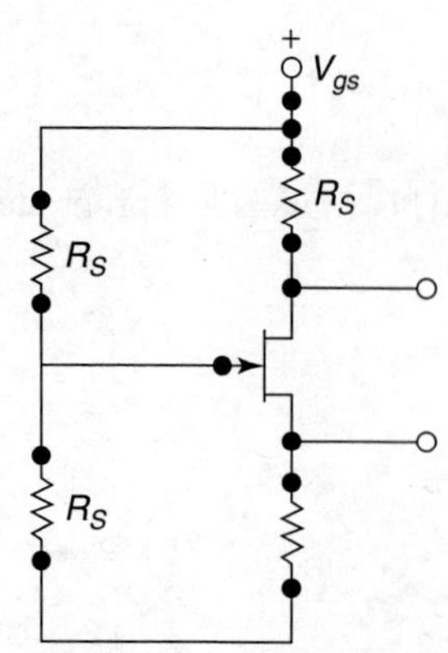


(a) Determine the resistance seen from terminal 2 and the ground at low frequencies.

(b) Valuate the resistance in part (a) for $R_D = 4$ kΩ, $R_S = 2.5$ kΩ, $R_1 = 20$ kΩ, $R_2 = 100$ kΩ, $g_m = 2.5$ ms and $r_d = 60$ kΩ.

(c) Repeat part (b) if $R_D = 0$.

**Solution:**

The small-signal model of the circuit is as shown in the given diagram. Use of Thevenin's equivalent converts the circuit, as shown in the following diagram.

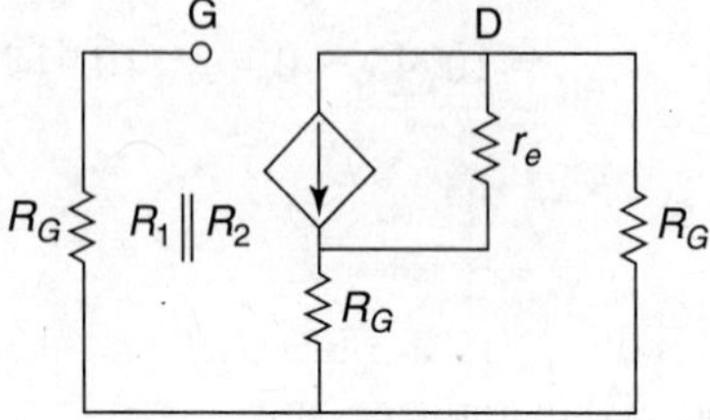


If an input signal of 1 V is applied:

$$V_{gs} = 1 - R_S\left(\frac{\mu V_{gs}}{r_d + R_S + R_D}\right)$$

$$V_{gs} = \left(\frac{r_d + R_S + R_D}{r_d(1+\mu)R_S + R_D}\right) \quad (\because \; V_i = 1 \text{ V})$$

Open circuit voltage $V_{2G}$ (O.C.) at terminal 2:

$$= \left(\frac{\mu V_{gs}}{r_d + R_S + R_D}\right) \times R_S = \left(\frac{\mu R_S}{r_d + (1+\mu)R_S + R_D}\right)$$

Short circuit current (terminal 2 shorted to G):

$$V_{gs} = V_1 = 1 \text{ V (as } R_S \text{ is shorted)}$$

$$I_{2G}(\text{S.C.}) = \frac{\mu V_{gs}}{r_d + R_D} = \frac{\mu V_{gs}}{r_d + R_D}$$

(a) Resistance as seen from terminal 2 and G:

$$R_o = \frac{V_{2G}(\text{O.C})}{I_{2G}(\text{S.C.})}$$

$$= \frac{\mu R_s}{r_d + (1 + \mu R_S) + R_D} \times \frac{(r_d + R_d)}{\mu} = \frac{1}{1 + \dfrac{(1+\mu)R_S}{r_d + R_D}}$$

(b) $$R_o = \frac{2.5}{1 + \frac{(1 + 2.5 + 60)2.5}{60 + 4}} = 362\ \Omega$$

(c) $$R_D = 0$$

$$R_o = \frac{2.5}{1 + \frac{(1 + 2.5 + 60)2.5}{60 + 4}} = 343\ \Omega$$

**Example 7-59** Design a source follower circuit at *Q*-point having $V_{DS} = 14$ V, $I_{DQ} = 3$ mA, $V_{DD} = 20$ V, $g_m = 2$ ms, $r_d = 50$ k, $V_{GS} = -1.5$ V.

**Solution:**
The circuit of the bootstrapped source follower may be considered for this.

Therefore, $$20\text{ V} = 14\text{ V} + 3\text{ mA}\,(R_1 + R_2)$$

$$R_1 + R_2 = 2\text{ K}$$

$$-1.5/-3\text{ mA}R_1 \Rightarrow R_1 = 0.5\text{ K}$$

$$R_2 = 2 - 0.5 = 1.5\text{ K}$$

Output resistance: $$R_o = \frac{1}{g_m} = \frac{1}{2\text{ ms}} = 0.5\text{ k}\Omega$$

$$V_s = A_v^1 V_g \frac{(R_1 + R_2)}{R_1 + R_2 - \frac{1}{g_m}} = A_v^1 V_g \frac{2\text{ k}}{2\text{ k} + 0.5\text{ k}}$$

$$A_V = \frac{V_s}{V_g} = 0.8\,A_v^1$$

The effective input resistance is expressed as:

$$R_1 = \frac{R_3}{1 + \frac{V_s}{V_g}\left(\frac{R_2}{R_1 + R_2}\right)}$$

$$= \frac{R_3}{1 - 0.8A_V \times \frac{0.5\text{ K}}{2\text{ K}}}$$

$$= \frac{R_3}{1 - 0.8 \times 0.25} = \frac{R_3}{0.8} = 1.25\,R_3$$

■

## 7-7 ADDITIONAL FET CIRCUITS

FET circuits are used extensively in the modern semiconductor industry. The Fin FET, for example, is very popular among microprocessor manufacturers. In the Fin FET, the source/drain region forms “fins” on the silicon surface; thus, the name. The FREDFET (fast-reverse epitaxial diode field-effect transistor) and the EOSFET (electrolyte-oxide-semiconductor field-effect transistor) are FET circuits that are used in motors and neurochips respectively. The OFET (organic field-effect transistor) uses an organic semiconductor compound and it can be used as a light-emitting device. Apart from these there are numerous modern FET circuits available in the market today. Some of these are discussed in the following sections.

### 7-7-1 MOS Differential Amplifiers

MOS differential amplifiers are one of the most important building blocks in MOS analog circuits. Figure 7-17 depicts the MOS differential pair.

Figure 7-17 shows two matched MOSFETs $Q_1$ and $Q_2$, biased with a constant current source. Here, the loads of the differential amplifiers are not shown. A relation needs to be found between the drain current and the input voltage. Also, it is assumed that the two MOSFETs operate in the saturation region.

The drain currents of the two transistors are given by:

$$i_{D1} = \frac{1}{2} k_n' \frac{W}{L} (V_{gs1} - V_T)^2 \qquad (7\text{-}36a)$$

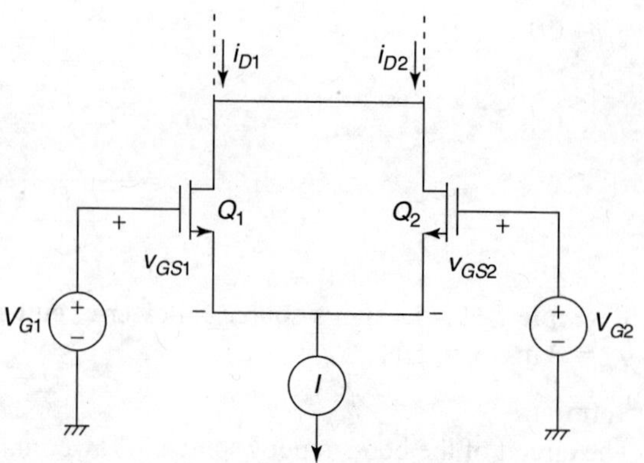


**Figure 7-17** Circuit representation of a differential amplifier

and,
$$i_{D2} = \frac{1}{2} k_n' \frac{W}{L} (V_{gs2} - V_T)^2 \qquad (7\text{-}36b)$$

Subtracting Eq. (7-36b) from Eq. (7-36a), and replacing $v_{gs1} - v_{gs2} = v_{id}$, we obtain:

$$\sqrt{i_{D1}} - \sqrt{i_{D2}} = \sqrt{\frac{1}{2} k_n' \frac{W}{L}}\, v_{id} \qquad (7\text{-}37)$$

The constraint imposed by the current source is given by:

$$i_{D1} + i_{D2} = I \qquad (7\text{-}38)$$

Solving Eq. (7-37) and Eq. (7-38) we obtain:

$$i_{D1} = \frac{I}{2} + \sqrt{k_n' \frac{W}{L} I}\, \frac{v_{id}}{2} \sqrt{1 - \frac{(v_{id}/2)^2}{I/k_n' \frac{W}{L}}}$$

$$i_{D2} = \frac{I}{2} - \sqrt{k_n' \frac{W}{L} I}\, \frac{v_{id}}{2} \sqrt{1 - \frac{(v_{id}/2)^2}{I/k_n' \frac{W}{L}}}$$

Also, when the circuit is at the quiescent point, we have:

$$i_{D1} = i_{D2} = \frac{I}{2} \quad \text{and} \quad v_{GS1} = v_{GS2} = V_{GS}$$

where, $\frac{I}{2} = \frac{1}{2} k_n' \frac{W}{L} (V_{GS} - V_T)^2$

We know that:

$$g_m = \frac{2I_D}{(V_{GS} - V_T)}$$

In this case, we have:

$$g_m = \frac{2I_D}{(V_{GS} - V_T)}, \text{ (since we already have } i_{D1} = i_{D2} = i/2)$$

### 7-7-2 Current Source Circuits

The concept of a power supply should be clear before we proceed to study the current source circuits. A practical current source is a voltage source where the resistance, in series with the voltage source, is zero. It is a current supply with a resistance in parallel. An ideal current source provides a constant current regardless of the load connected to it. Constant current sources are built using FET devices, as shown in Fig. 7-18.

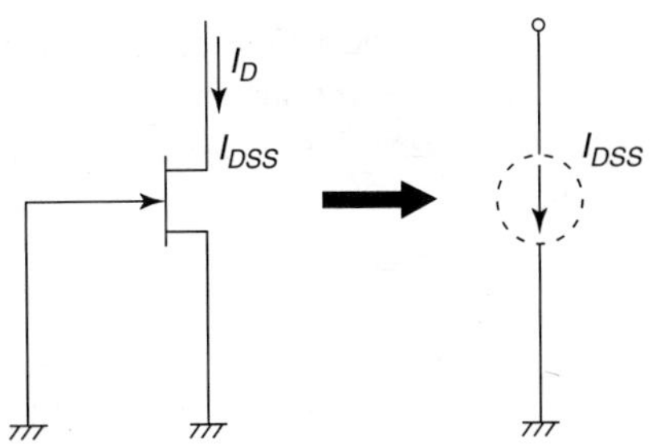


**Figure 7-18** MOSFET current source

If the voltage $V_{GS}$ is set to zero, the drain current is fixed at a constant current. Thus, we find that the essential condition for a MOSFET to operate as a current source is to operate the device in the saturation region. This can be easily found by a detailed look at the transfer characteristics.

## 7-8 COMPARISON BETWEEN THE FET AND THE BJT

Now that we have studied the field-effect transistor and the bipolar-junction transistor in detail, let us map their respective characteristics and draw a comparison between the two.

1. In the BJT, the carriers are transported by the process of diffusion but in the FET, it is the drift mechanism that helps the movement of carriers.
2. In the FET, carriers of only one type—either electron or hole (majority carrier)—are responsible for the conduction, but for the BJT both types of carriers—electrons and holes (majority and minority)—are involved in current conduction.
3. The FET is thermally more stable than the BJT, which is the primary reason behind the extensive usage of the FET.
4. The FET is a voltage-controlled device or voltage amplifier, whereas the BJT is a current-controlled device or current amplifier.
5. The input impedance offered by the FET is much higher than that offered by the BJT.
6. The FET is easy to fabricate, and hence all the ICs use the FET as their basic technology.
7. Fabrication of the FET requires less space compared to the BJT; therefore, the FET is usually preferred for the VLSI design.
8. The FET is less noisy compared to the BJT, which facilitates their extensive usage in communication devices.
9. The FET offers high power gain compared to the BJT.

## POINTS TO REMEMBER

1. The FET is a voltage-controlled unipolar device with high input impedance.
2. As long as the device is in the active region, the FET characteristic equations do not change with each network connection.
3. The network or the biasing assembly simply defines the level of current and voltage associated with the operating point through its own set of equations.
4. The self-bias arrangement eliminates the need for two voltage supplies.

5. To operate the MOSFET as an amplifier, it must be biased in the linear region where we obtain a linear variation between the input voltage and the output voltage.
6. Under proper operation of the device as an amplifier, the signal quantities are superimposed on dc quantities. Thus, the total drain current under any such arrangements of the MOSFET in a circuit, is the superimposition of the dc current and the small-signal current. Similar is the case for output voltage.
7. A simple circuit transformation causes the small-signal ac model of a MOSFET to lead to the formation of the T equivalent circuit.
8. The high-frequency model of the MOSFET takes into account the various capacitances of the device.
9. The parameter to judge the high-frequency operation of the MOSFET as an amplifier is given by the unity-gain frequency. It is defined as the frequency at which the short-circuit gain of the common-source arrangement becomes unity.

## IMPORTANT FORMULAE

1. For the depletion-type MOSFET and for the JFET, the basic relation guiding the input and output quantities is given by Shockley's equation:
$$I_D = I_{DSS}\left(1 - \frac{V_{GS}}{V_P}\right)^2$$
2. For the enhancement-type MOSFET, Shockley's equation gets modified and is given by:
$$I_D = k\,(V_{GS} - V_T)^2$$
3. The basic equations for the JFET/depletion-type MOSFETs.

   (a) Fixed bias configuration:
$$V_{GS} = -V_{GG} = V_G$$
   (b) Self bias configuration:
$$V_{GS} = -I_D R_S$$
   (c) Voltage divider biasing configuration:
$$V_G = \frac{R_2}{R_1 + R_2} V_{DD}$$
$$V_{GS} = V_G - I_D R_s$$
4. The dc bias point required to operate the MOSFET device as an amplifier is obtained by solving the basic equation as given by:
$$I_D = \frac{1}{2} k_n' \frac{W}{L} (V_{GS} - V_T)^2$$
(neglecting channel width modulation). And the drain voltage of the circuit is:
$$V_D = V_{DD} - I_D R_D$$
5. The instantaneous drain current is obtained after the bias point is introduced and is given by:
$$I_D = \frac{1}{2} k_n' \frac{W}{L} (V_{GS} + v_{gs} - V_t)^2$$
$$= \frac{1}{2} k_n' \frac{W}{L} (V_{GS} - V_t)^2 - k_n' \frac{W}{L} (V_{GS} - V_t) v_{gs}$$
$$+ \frac{1}{2} k_n' \frac{W}{L} + v_{gs}^2$$
6. For small-signal analysis, we have the instantaneous drain current as:
$$i_D = I_D + i_d$$
and,
$$i_d = k_n' \frac{W}{L} (V_{GS} - V_t) v_{gs}$$
This is obtained only on the condition that:
$$\frac{1}{2} k_n' \frac{W}{L} v_{gs}^2 \ll k_n' \frac{W}{L} (V_{GS} - V_t) v_{gs}$$
which results in $v_{gs} \ll 2(V_{GS} - V_t)$
7. Equivalent transconductance is given by:
$$g_m = \frac{i_d}{v_{gs}} = k_n' \frac{W}{L} (V_{GS} - V_T)$$
8. Small-signal component of drain voltage is:
$$v_d = -i_d R_D = -g_m R_D v_{gs} \Rightarrow \frac{v_d}{v_{gs}} = -g_m R_D$$
9. Voltage gain for a MOSFET amplifier is given by:
$$\frac{v_d}{v_{gs}} = -g_m (R_D \| r_o)$$
where, $r_o = \dfrac{|V_A|}{I_D}$
10. Current gain in the high-frequency model of the MOSFET is given by:
$$\frac{I_o}{I_i} = \frac{g_m}{s(C_{gs} + C_{gd})}$$

11. Unity-gain frequency (bandwidth) is given by:

$$\omega_T = \frac{g_m}{(C_{gs} + C_{gd})}$$

12. The currents at the drain terminals of the differential amplifier are given by:

$$i_{D1} = \frac{I}{2} + \sqrt{k_n' \frac{W}{L} I} \frac{v_{id}}{2} \sqrt{1 - \frac{\left(\frac{v_{id}}{2}\right)^2}{\frac{I}{k_n' \frac{W}{L}}}}$$

$$i_{D2} = \frac{I}{2} - \sqrt{k_n' \frac{W}{L} I} \frac{v_{id}}{2} \sqrt{1 - \frac{\left(\frac{v_{id}}{2}\right)^2}{\frac{I}{k_n' \frac{W}{L}}}}$$

And the transconductance in this case is given by:

$$g_m = \frac{I_D}{(V_{GS} - V_T)}$$

## OBJECTIVE QUESTIONS

1. For an $n$-channel JFET, with a constant drain-source voltage, if the gate-source voltage is increased (more negative) pinch-off would occur for:
   (a) High values of drain current
   (b) Saturation value of drain current
   (c) Zero drain current
   (d) Gate current equal to drain current

2. For a junction FET in the pinch-off region, as the drain voltage is increased the drain current:
   (a) Becomes zero
   (b) Abruptly decreases
   (c) Abruptly increases
   (d) Remains constant

3. In modern the MOSFET, the material used for the gate is:
   (a) High-purity silicon
   (b) High-purity silica
   (c) Heavily doped polycrystalline silicon
   (d) Epitaxially grown silicon

4. The threshold voltage of an $n$-channel MOSFET can be increased by:
   (a) Increasing the channel dopant concentration
   (b) Reducing the channel dopant concentration
   (c) Reducing the gate oxide thickness
   (d) Reducing the channel length

5. An $n$-channel JFET has a pinch-off voltage of $V_P = -5$ V, $V_{DS(\max)} = 20$ V and $g_m = 2$ mA/V. The minimum 'ON' resistance is achieved in $n$th JFET for:
   (a) $V_{GS} = -7$ V and $V_{DS} = 0$ V
   (b) $V_{GS} = 0$ V and $V_{DS} = 0$ V
   (c) $V_{GS} = 0$ V and $V_{DS} = 20$ V
   (d) $V_{FGS} = -7$ V and $V_{DS} = 20$ V

6. A JFET has $I_{DSS} = 10$ mA and $V_P = 5$ V. The value of the resistance $R_S$ for a drain current $I_D = 6.4$ mA is:
   (a) 150 Ω (b) 470 Ω
   (c) 560 Ω (d) 1 kΩ

7. FET is:
   (a) Bipolar (b) Unipolar
   (c) Tripolar (d) None of the above

8. In an FET, the transconductance $g_m$ is proportional to:
   (a) $I_{DS}$ (b) $I^2_{DS}$
   (c) $\sqrt{I_{DC}}$ (d) $1/I_{DC}$

9. In an FET, $g_{mo}$ is pel to:
   (a) $I_{DSS}$ (b) $\sqrt{I_{DSS}}$
   (c) $I^2_{DSS}$ (d) $1/I_{DSS}$

10. In a JFET, transconductance $g_m$ is of the order of:
    (a) 1 mS (b) 100 mS
    (c) 1 S (d) 100 S

11. In an FET, dynamic drain resistance $r_d$ is of the order of:
    (a) 1 kΩ (b) 10 kΩ
    (c) 100 Ω (d) 100 kΩ

12. In a MOSFET, dynamic drain resistance $r_d$ is of the order of:
(a) 10 kΩ (b) 1 kΩ
(c) 10 Ω (d) 100 kΩ

13. The input resistance $r_{gs}$ in the small-signal model of the MOSFET is of the order of:
(a) 100 kΩ (b) 1 MΩ
(c) 100 MΩ (d) $10^4$ MΩ

14. The feedback resistance $r_{gd}$ in the small-signal model of the MOSFET is of the order of:
(a) 1 MΩ (b) 100 MΩ
(c) $10^4$ MΩ (d) $10^6$ MΩ

15. The feedback capacitance $C_{gd}$ in the small-signal model of the JFET is of the order of:
(a) 5 pF (b) 15 pF
(c) 500 pF (d) 1 pF

16. The drain-to-source resistance $C_{ds}$ in the high-frequency model of the JFET is of the order of:
(a) 1 pF (b) 10 pF
(c) 100 pF (d) 1000 pF

17. The gate-to-source resistance $C_{gs}$ in the high-frequency model of the JFET is of the order of:
(a) 5 pF (b) 50 pF
(c) 500 pF (d) 5000 pF

18. The gain bandwidth of an FET amplifier w.r.t. a BJT amplifier is:
(a) Low (b) High
(c) Equal (d) Zero

19. Voltage gain of a common-gate amplifier with $\mu = 15$, $r_d = 20$ K, and $R_L = 2$ K, internal resistance of the voltage source is:
(a) 0.64 (b) 6.4
(c) 0.89 (d) 0.9

20. The input resistance of a BJT amplifier w.r.t. its FET counterpart is:
(a) More (b) Less
(c) Equal (d) None of the above

21. Total input capacitance across the input terminals of a CS amplifier at high frequency with $r_d = 20$ K, $R_D = 5$ K, $g_m = 5$ ms, $C_{gs} = 8$ pF, $C_{gd} = 4$ pF and $C_{ds} = 2$ pF is:
(a) 44 pF (b) 76 pF
(c) 26 pF (d) 22 pF

22. CMOS is equal to:
(a) NMOS + PMOS (b) PMOS + PMOS
(c) VMOS + NMOS (d) None of the above

23. CMOS is formed by the:
(a) Twin tub method (b) CZ method
(c) LPE method (d) None of the above

## REVIEW QUESTIONS

1. Define the following:
(a) Transconductance
(b) Drain resistance
(c) Amplification factor of an FET
2. Show the small-signal model of the FET at (a) low frequencies and (b) high frequencies. Account for the drastic changes in this model.
3. Point out and discuss the capacitances of the general device that have been included in the high-frequency model.
4. Draw the MOSFET NOR circuit and discuss its operation.
5. Draw a MOSFET NAND circuit.
6. Explain how a MOSFET acts as a load.
7. Draw the CMOS NAND circuit and discuss its operation.
8. Draw the three biasing circuits of the MOSFET and discuss their respective operations. Calculate the location of the $Q$-point under these biasing arrangements.
9. In which portion of the drain characteristics is the $Q$-point of an FET amplifier usually selected?
10. Draw the drain-to-gate bias circuit diagram of an enhancement-type MOSFET and show how the gate-source junction is forward-biased.

11. Deduce the relationship: $\mu$ = gain= $r_d g_m$

12. Why does the FET give better low-frequency amplification than the BJT?

13. Discuss the small-signal operation of a common-source $n$-channel JFET.

14. Derive an expression for the small-signal voltage gain of a common-source FET amplifier.

15. Define the importance of unity-gain frequency and also derive its expression.

16. Draw a comparison between the BJT and the FET with respect to their operations.

17. Compare the gate terminal and the bulk region as control electrodes.

18. How can the bulk region act as a gate?

19. Does it make sense to talk about the current-gain of a MOSFET? Why?

20. What is $f_T$ for a small-signal MOSFET amplifier?

21. What are the dominant parasitic elements in determining $f_T$?

## PRACTICE PROBLEMS

1. For the circuit, as shown in the diagram, find the values of $R_D$ and $R_S$ for establishing a drain current of 1mA and drain voltage. Given that:

$V_T = 2$ V, $\mu_n C_{OX} = 20\ \mu$A/V$^2$, $L = 10\ \mu$m and $W = 400\ \mu$m.

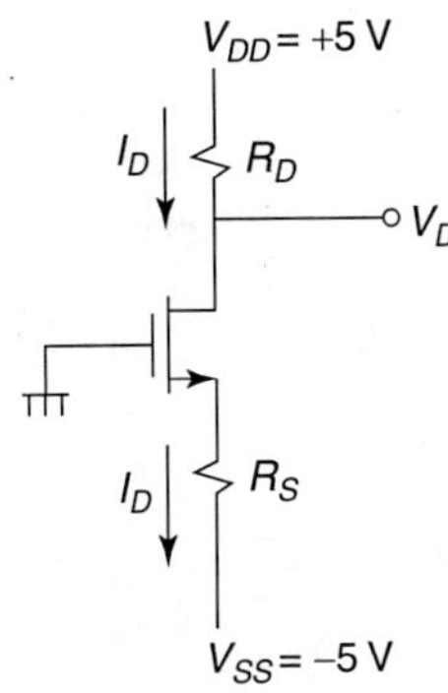


2. A small-signal CS FET amplifier has a load resistance $R_L$. For what value of load resistance is the ac power dissipation maximum? (HINTS: $r_d$)

3. Consider the circuit given in the diagram. Given that:

$V_T$=2 V, $\mu_n\ C_{ox} = 20\ \mu$A/V$^2$,

$L_1 = L_2 = 10\ \mu$m $W_1 = 100\ \mu$m and $\lambda = 0$.

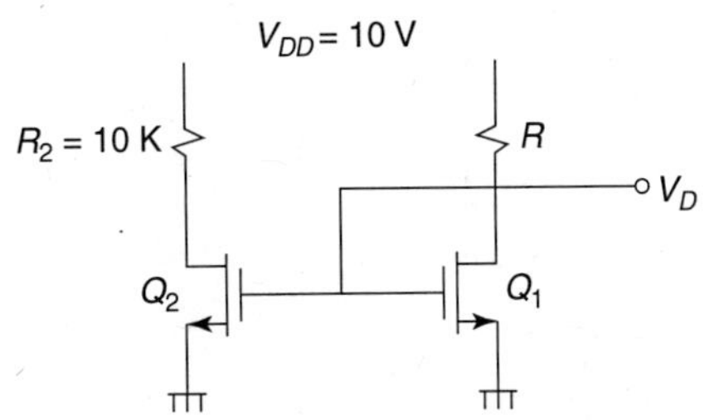


With this data find the value of $R$ required to establish a current of 0.1 mA in $Q_1$. Also find $W_2$ for $Q_2$ operating in the saturation region with a current of 0.5 mA.

4. The PMOS circuit, as shown in the diagram, has:

$V_t = -2$ V, $\mu_p C_{ox} = 8\,\mu$A/V$^2$, $L = 10\,\mu$m

and $\lambda = 0$.

Find the values required for $W$ and $R$ to establish a drain current of 0.1 mA and voltage $V_D$ of 7 V.

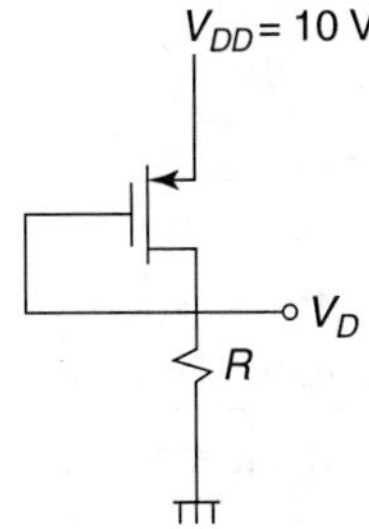


5. The NMOS circuit, as shown in the diagram, has the following specifications:

$V_t = 2$ V, $\mu_n C_{ox} = 20\ \mu$A/V$^2$, $L_1 = L_2 = 10\ \mu$m and $\lambda = 0$.

Find the values of the gate width for transistors, $Q_1$ and $Q_2$. Also find the value of $R$ required to obtain the voltage and current values as indicated.

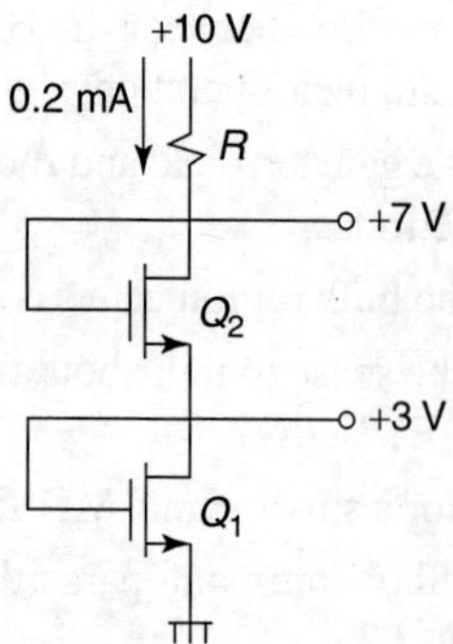


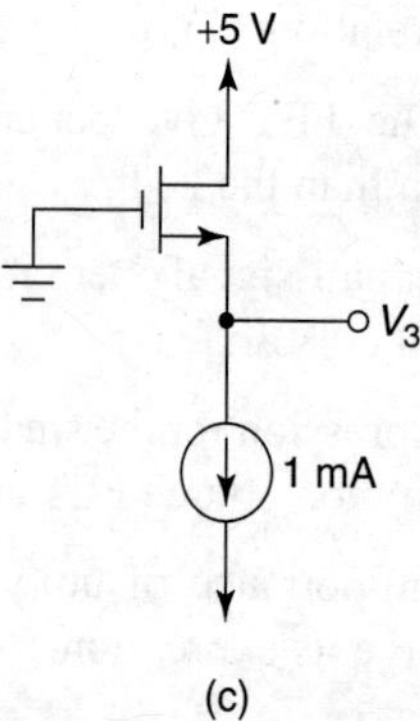


(c)

6. The NMOS transistors, as shown in the diagram, have the following specifications:
$V_t = 2$ V, $\mu_n C_{ox} = 20\ \mu A/V^2$ and
$L_1 = L_2 = L_3 = 10\ \mu m$.
Find the required values of gate width for the three transistors required to obtain the voltages and current as indicated.

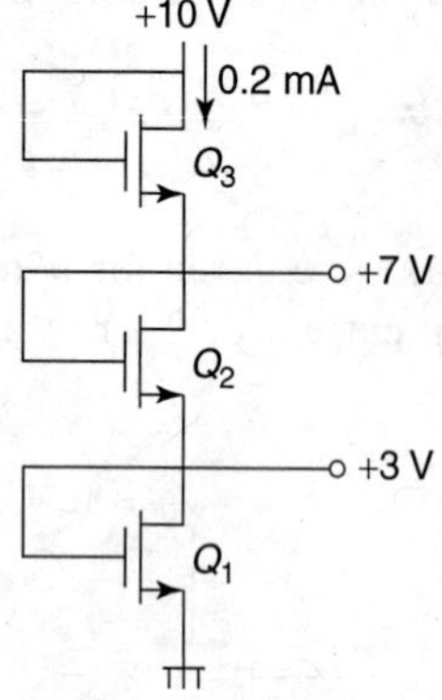


7. For the circuits, as labeled in the diagram, find the node voltages. For all the transistors:

$$k_n' \frac{W}{L} = 0.5\ mA/V^2, \text{ and } V_T = 2\ V.$$

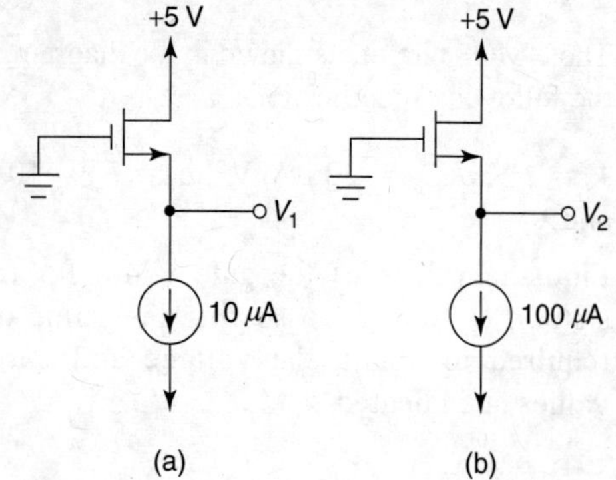


8. For the circuits shown, as in the diagram, find the drain voltages assuming that $k_n'(W/L) = 200\ \mu A/V^2$. $V_t = 2$ V and $V_A = 20$ V.

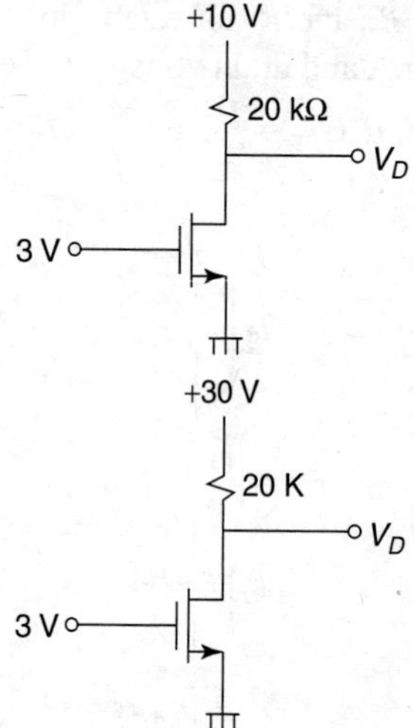


9. For the PMOS transistor, as shown in the circuit, $k_p' = 8\ \mu A/V^2$, $W/L = 25$ and $|V_{tp}| = 1$ V. With $I = 100\mu A$, find the voltages $V_{SD}$ and $V_{SG}$ for $R = 0, 10\,k\Omega, 30\,k\Omega$ and $100\,k\Omega$. And, for what value of $R$ is $V_{SD} = V_{SG}$?

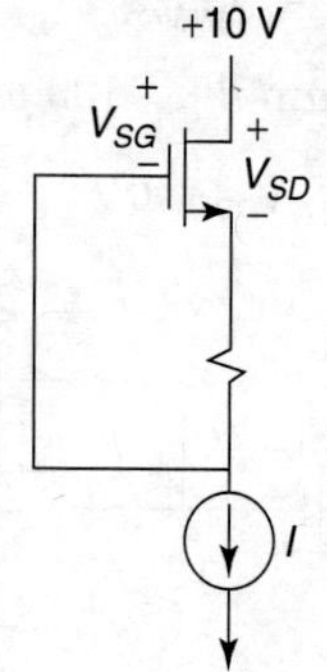

10. For the circuits, as shown in the diagram, $\mu_n C_{ox} = 2.5\,\mu_p C_{ox} = 20\,\mu\,\text{A/V}^2, |V_t| = 1\text{ V}, \lambda = 0,$ $\gamma = 0, L = 10\,\mu\text{m}$ and $W = 30\,\mu\text{m}$.

    Find the labeled values of currents and voltages.

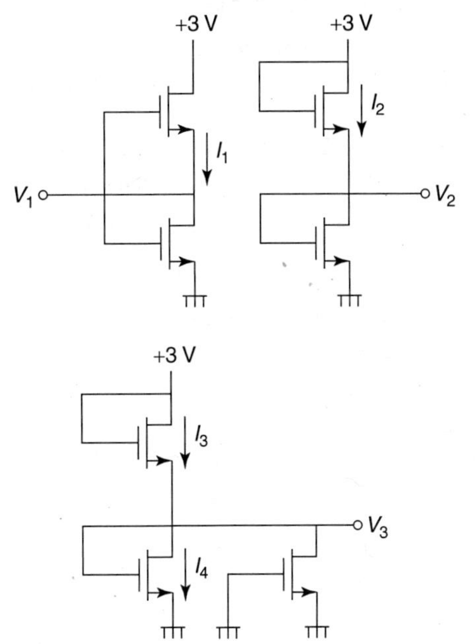


11. An NMOS amplifier is to be designed so as to provide a 0.50 V peak output signal across a 50 kΩ resistor that can be used as a drain resistor. For a gain of at least 5, what value of $g_m$ is required?

12. In Problem 11, using the transistor with a dc supply of 3 V and $V_t = 0.9\text{V}$, what values of $I_D$ and $V_{GS}$ would you choose? What value of $W/L$ is required for $\mu_n C_{ox} = 100\ \mu\text{A/V}^2$?

13. Consider an NMOS transistor having $V_t = 2$ V, $k_n'(W/L) = 1\text{ mA/V}^2$. Let the transistor be biased with voltage $V_{GS} = 4$ V. What is the dc bias current for operation in saturation?

14. If a +0.1 V signal is superimposed on $V_{GS}$, find the corresponding increment in collector current by evaluating the total collector current $i_D$ and subtracting the dc bias $I_D$.

15. Consider the given diagram under the following conditions:

    $V_t = 2\text{ V}, k_n'(W/L) = 1\text{ mA/V}^2, V_{GS} = 4\text{ V},$

    $V_{DD} = 10\text{ V}$ and $R_D = 3.6\text{ k}\Omega$.

    Find the dc quantities, $I_D$ and $V_D$.

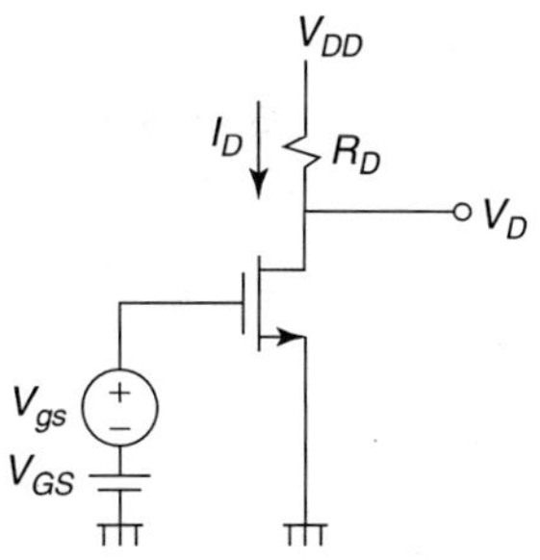


16. From Problem 15 find the transconductance at the dc bias point. If the MOSFET has $\lambda = 0.01\text{V}^{-1}$, find $r_o$ at the bias point and then calculate the gain.

17. Consider two identical transistors in current mirror circuits with the following specifications: $k_n' W/L = 40\ \mu\text{A/V}^2$, $V_t = 0.8$ V and $V_A = 20$ V. Let $I_{REF} = 10\ \mu\text{A}$. What is the output voltage at which $I_o$ is exactly equal to $I_{REF}$? What will be the change in output current corresponding to a +2 V increase in the output current?

18. The given diagram shows a circuit formed by cascading two common source stages. Assuming that the biasing current sources have very high output resistance, find the overall gain of $Q_1$ and $Q_2$ in terms of $g_m$ and $r_o$.

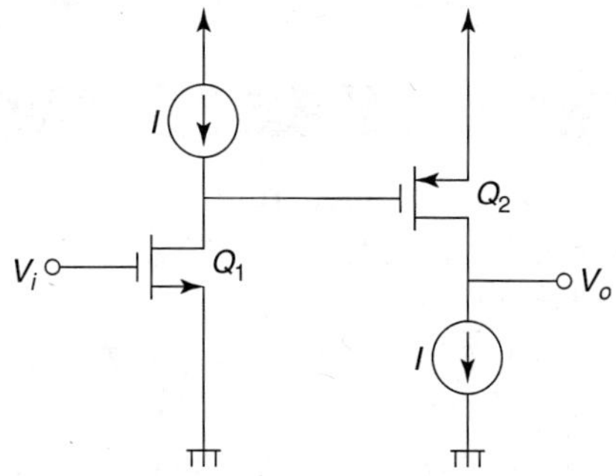


19. The MOSFETs, as shown in the diagram, are matched with the following data: $k_n'(W/L)_1 = k_p'(W/L)_2 = 50\ \mu\text{A/V}^2$ and $|V_t| = 2$ V. The resistance of $R_2$ is10 MΩ. For $G$ and $D$ open, what are the drain currents of the two transistors?

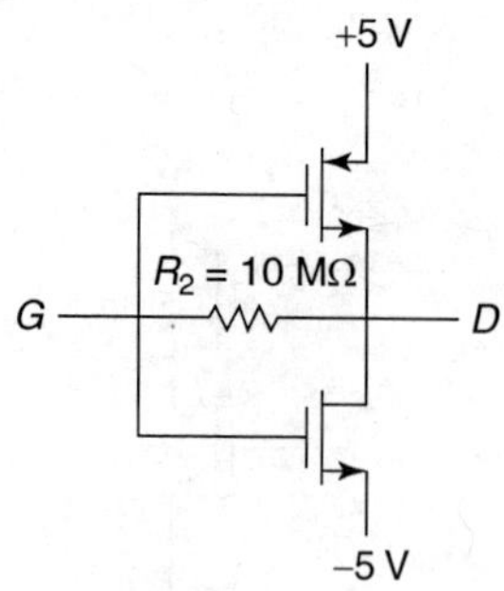


20. From Problem 19, for $r_o = \infty$, find the voltage gain of the amplifier from $G$ to $D$? For finite $r_o$ ($r_o = |V_A|/I_D$, $|V_A| = 180$), what is the voltage gain from $G$ to $D$ and the input resistance at $G$?

21. For the circuit in the given figure, let $|V_t| = 2$ V. For each of the following cases:

(a) $k_p'(W/L)_2 = k_n'(W/L)_1$

(b) $k_p'(W/L)_2 = 0.1k_n'(W/L)_1$

find $v_o$ for different value of $v_i$ 0 V, 3 V and 9 V.

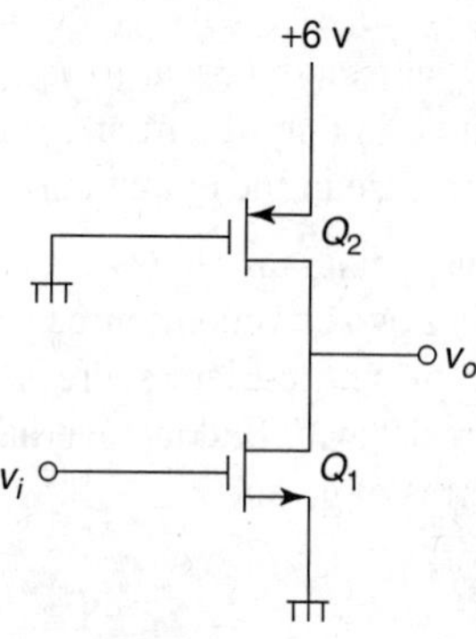


22. A JFET with $V_P = -1$ V and $I_{DSS} = 1$ mA shows an output resistance of 100 kΩ when operated in pinch-off with $v_{gs} = 0$ V. What is the value of output resistance when the device is operated in pinch-off with $v_{gs} = -0.5$ V?

23. Consider the source follower circuit, as shown in the diagram, when fabricated in the technology with $k_n' = 20\ \mu\text{A/V}^2$, $V_t = 1$ V and $V_A = 100$ V. Let $W/L = 10$ for all transistors, $I_{REF} = 300\ \mu$A and $\chi = 0.1$. Find $g_{m1}$, $r_{01}$, $r_{02}$ and $A_v$.

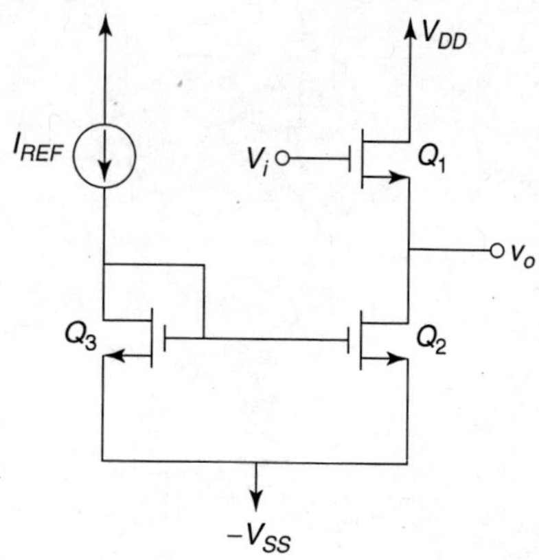


24. Calculate the ac transconductance and drain resistance of an FET from the following data.

| $V_{GS}$ | $V_{DS}$ | $I_D$ |
|---|---|---|
| 0 V | 5.9 V | 11 mA |
| 0 V | 13.9 V | 10 mA |
| 0.3 V | 14V | 9.9 mA |

## SUGGESTED READINGS


1. Mead, C.A. February 1966. "Schottky Barrier Gate Field Effect Transistor". *Proceedings of the IEEE* 2: 307–308.
2. Grove, A. S. 1967. *Physics and Technology of Semiconductor Devices*. New York: Wiley.
3. Kahng, D. July 1976. "A Historical Perspective on the Development of MOS Transistors and Related Devices". *IEEE Transactions on Electron Devices* 7: 655–7.
4. Hofstein, S. R. and F. P. Heiman. September 1963. "The Silicon Insulated-Gate Field-Effect Transistor". *Proceedings of the IEEE* 9: 1190–1202.
5. Sze, S. M. 1981. *Physics of Semiconductor Devices*. New York: Wiley.
6. Singh, J. 1994. *Semiconductor Devices: An Introduction*. New York, NY: McGraw-Hill.
7. Streetman, B. G. and S. Banerjee. 2000. *Solid State Electronic Devices*. New Delhi: Pearson Education.
8. Pierret, R. F. and G.W. Neudeck. 1989. *Modular Series on Solid State Devices*. Boston, MA: Addison Wesley.
9. Singh, B. P. and Rekha Singh. 2006. *Electronic Devices and Integrated Circuits*. New Delhi: Pearson Education.

# 8

# Special Semiconductor Devices

## Outline

8-1 Introduction
8-2 Silicon-Controlled Rectifier (SCR)
8-3 Triode AC Switch (TRIAC)
8-4 Diode AC Switch (DIAC)
8-5 Unijunction Transistor (UJT)
8-6 Insulated-Gate Bipolar Transistor (IGBT)
8-7 Real-Life Applications

## Objectives

In this chapter various semiconductor devices are dealt with in detail. Power electronic devices play a major role in modern electronic design. High-power semiconductor devices have better switching speed, they are smaller in size and cost of production is also low. The silicon-controlled rectifier (SCR), an important special semiconductor device that is used as a power-control device, is discussed with emphasis on the constructional features, physical operations and characteristics, and its simple applications are also provided. Subsequently, the TRIAC, DIAC, UJT and IGBT are discussed in a similar manner with respect to their real-life applications.

## 8-1 INTRODUCTION

Proceeding from the two- and three-terminal basic devices that we have studied in the previous chapters, we will now examine some special types of devices that have multiple layers and different switching behaviors. The SCR is the most important special semiconductor device that is used as a power-control device. This device is popular for its forward-conducting and reverse-blocking characteristics. Moreover the SCR can be used in high-power devices. For example, in the central processing unit of the computer, the SCR is used in switch mode power supply (SMPS). To understand computer SMPS, the study of SCR is essential. The DIAC, a combination of two Shockley dodes, and the TRIAC, a combination of two SCRs connected anti-parallelly are important power-control devices. The UJT has a different structure as compared to others but it is also used as an efficient switching device.

## 8-2 SILICON-CONTROLLED RECTIFIER (SCR)

The silicon-controlled rectifier or semiconductor-controlled rectifier is a two-state device used for efficient power control. The SCR is the parent member of the thyristor family and is used in high-power electronics. Its constructional features, physical operation and characteristics are explained in the following sections.

### 8-2-1 Constructional Features

The SCR is a four-layer structure, either *p–n–p–n* or *n–p–n–p*, that effectively blocks current through two terminals until it is turned ON by a small-signal at a third terminal. The SCR has two states: a high-current low-impedance ON state and a low-current high-impedance OFF state.

The basic transistor action in a four-layer *p–n–p–n* structure is analysed first with only two terminals, and then the third control input is introduced.

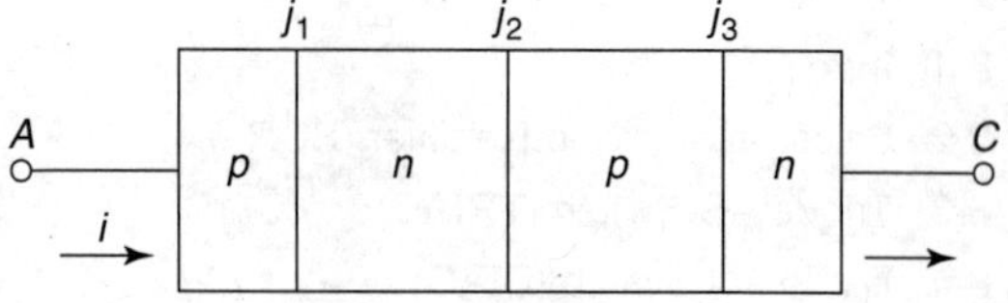


**Figure 8-1** Basic structure of two-terminal *p–n–p–n* device

A four-layer *p–n–p–n* diode with a cathode terminal *C* outside the *n*-region and with an anode terminal *A* at the outside *p*-region is shown in Fig. 8-1. The junction nearest to the anode is denoted as junction $J_1$, the center junction of *n–p* is denoted as $J_2$ and the junction nearest to the cathode is denoted as $J_3$. The device is in a forward-biased state when the anode *A* is biased positively with respect to the cathode terminal.

### 8-2-2 Physical Operation and Characteristics

The physical operation of the SCR can be explained clearly with reference to the current–voltage characteristics. The forward-bias condition and reverse-bias condition illustrate the conducting state and the reverse blocking state respectively. Based on these two states a typical I–V characteristic of the SCR is shown in Fig. 8-2.

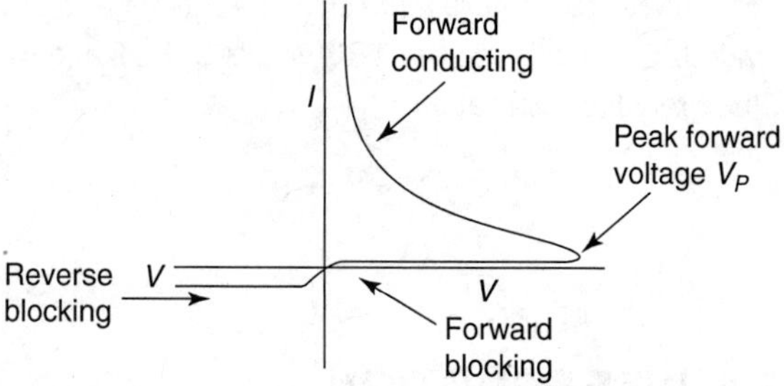


**Figure 8-2** I–V characteristics of a two terminal *p–n–p–n* device

**SCR in forward bias**

There are two different states in which we can examine the SCR in the forward-biased condition:

(i) The high-impedance or forward-blocking state

(ii) The low-impedance or forward-conducting state

At a critical peak forward voltage $V_p$, the SCR switches from the blocking state to the conducting state, as shown in Fig. 8-2.

A positive voltage places junction $j_1$ and $j_3$ under forward-bias, and the centre junction $j_2$ under reverse-bias. The forward voltage in the blocking state appears across the reverse-biased junction $j_2$ as the applied voltage *V* is increased. The voltage from the anode *A* to cathode *C*, as shown in Fig. 8-1, is very small after switching to the forward-conducting state, and all three junctions are forward-biased. The junction $j_2$ switches from reverse-bias to forward-bias.

### SCR in reverse bias

In the reverse-blocking state the junctions $j_1$ and $j_3$ are reverse-biased, and $j_2$ is forward-biased. The supply of electrons and holes to junction $j_2$ is restricted, and due to the thermal generation of electron–hole pairs near junctions $j_1$ and $j_2$ the device current is a small saturation current. In the reverse-blocking condition the current remains small until avalanche breakdown occurs at a large reverse-bias of several thousand volts.

The two-transistor analogy of the SCR is shown in Fig. 8-3. An SCR *p–n–p–n* structure is equivalent to one *p–n–p* transistor and one *n–p–n* transistor sharing some common terminals.

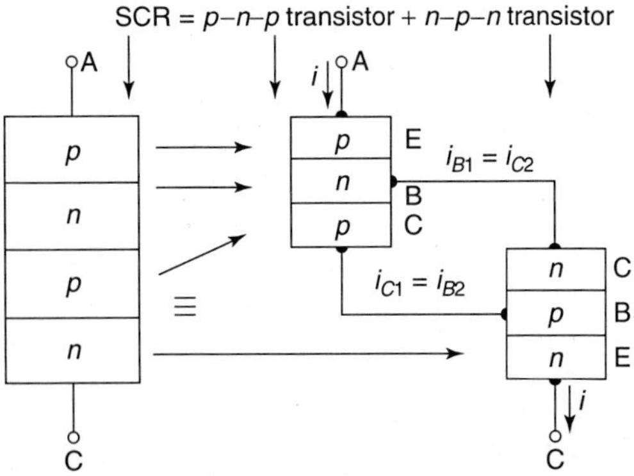

**Figure 8-3** An SCR *p–n–p–n*: a combination of one *p–n–p* transistor and one *n–p–n* transistor

The collector region of the *n–p–n* serves as the base of the *p–n–p*, and the base of the *n–p–n* serves as the collector region of the *p–n–p*. Thus, the *p–n–p* transistor's collector current $i_{C1}$ drives the base current $i_{B2}$ of the *n–p–n*, and the base current $i_{B1}$ of the *p–n–p* is dictated by the collector current $i_{C2}$ of the *n–p–n*.

With the collector current $i_{C1} = \alpha_1 i + I_{CO1}$ having a transfer ratio $\alpha_1$ for the *p–n–p*, and the collector current $i_{C2} = \alpha_2 i + I_{CO2}$ having a transfer ratio $\alpha_2$ for the *n–p–n*, where $I_{CO1}$ and $I_{CO2}$ stand for the respective collector-saturation currents, we have:

$$i_{C1} = \alpha_1 i + I_{CO1} = i_{B2} \tag{8-1}$$

$$i_{C2} = \alpha_2 i + I_{CO2} = i_{B1} \tag{8-2}$$

The total current through the SCR is the sum of $i_{C1}$ and $i_{C2}$:

$$i_{C1} + i_{C2} = i \tag{8-3}$$

Substituting the values of collector current from Eqs. (8-1) and (8-2) in Eq. (8-3) we get:

$$i(\alpha_1 + \alpha_2) + I_{CO1} + I_{CO2} = i$$

$$i = \frac{I_{CO1} + I_{CO2}}{1-(\alpha_1 + \alpha_2)} \tag{8-4}$$

Inference. Using Eq. (8-4) we can speculate upon the following cases:

**Case I:** When $\alpha_1 + \alpha_2 \rightarrow 1$, then the SCR current $i \rightarrow \infty$

As the sum of the values of alphas tends to unity, the SCR current *i* increases rapidly. The derivation is no longer valid as $\alpha_1 + \alpha_2$ equals unity. Since the junction $j_2$ becomes forward-biased in the forward-conducting state, both transistors become saturated just after switching. The two transistors remain in saturation state while the device is in the forward-conducting state held in the saturation region by the device current.

**Case II:** When $\alpha_1 + \alpha_2 \rightarrow 0$, i.e., when the summation value of alphas goes to zero, the SCR resultant current can be expressed as:

$$i = I_{CO1} + I_{CO2} \tag{8-5}$$

The current, *i*, passing through the SCR is very small. It is the combined collector-saturation currents of the two equivalent transistors as long as the sum $\alpha_1 + \alpha_2$ is very small or almost near zero.

The emitter-to-collector current transfer ratio ($\alpha$) of the transistor is the product of the base transport factor ($B$) and the emitter injection efficiency ($\gamma$), i.e., $\alpha = B\gamma$. The value of $\alpha$ increases with an increase in either of these factors or both. As the SCR current is increased, injection across the junction begins to dominate over recombination within the transition region and as a result $\gamma$ increases. The base transport factor $B$ increases with injection.

## 8-2-3 I–V Characteristics of the SCR

### Forward-blocking state

When the device is biased in the forward-blocking state, as shown in Fig. 8-4(a), the applied voltage appears primarily across the reverse-biased junction $j_2$. Although the junctions $j_1$ and $j_3$ are forward-biased, the current is small.

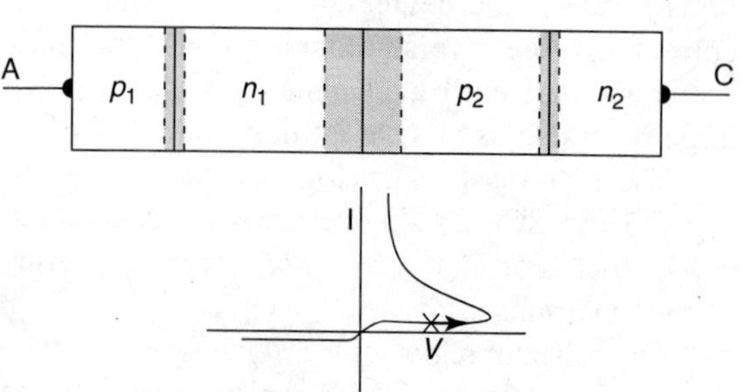


**Figure 8-4 (a)** The forward-blocking state of the SCR

We consider the supply of electrons available to $n_1$ and the supply of holes to $p_2$. At junction $j_1$ let us assume that a hole is injected from $p_2$ into $n_1$. If the hole recombines with an electron in $n_1$ (or in the $j_1$ transition region) then that electron must be re-supplied to the $n_1$ region in order to maintain space-charge neutrality. However, the supply of electrons in this case is severely restricted by the fact that $n_1$ is terminated in $j_2$, a reverse-biased junction. In a normal *p–n* diode the *n*-region is terminated in an ohmic contact.

The electron supply is restricted essentially to those electrons generated thermally within the diffusion length of the junction $j_2$. As a result the current passing through junction $j_1$ is almost the same as the reverse saturation current of the junction $j_2$. The same reason is valid for the current passing through the junction $j_3$. Holes required for injection into $n_2$ and to feed recombination in $p_2$ must originate in the saturation current of the center junction $j_2$.

The current crossing junction $j_2$ is strictly the thermally generated saturation current. This implies that electrons injected by the forward-biased junction $j_3$ do not diffuse across $p_2$ in any substantial numbers to be swept across the reverse-biased junction into $n_1$ by transistor action. The supply of holes to $p_2$ is primarily generated thermally since few holes injected at junction $j_1$ reach junction$j_2$ without recombination (i.e., $\alpha_1$ is small for the *p–n–p*). Thus, we can see that Eq. (8-4) implies a small current when $\alpha_1 + \alpha_2$ is small.

### Forward-conducting state of the SCR

As the value of $\alpha_1 + \alpha_2$ approaches unity through one of the mechanisms that we discussed, many holes injected at $j_1$ survive to be swept across $j_2$ into $p_2$. This process helps feed the recombination in $p_2$ and support the injection of holes into $n_2$. In a similar manner, the transistor action of electrons injected at $j_3$ and collected at $j_2$ supplies electrons for $n_1$. The current through the device can be much larger once this mechanism is put into effect. The process of transfer of injected carriers across $j_2$ is regenerative, and a greater supply of electrons to $n_1$ allows greater injection of holes at $j_1$ while maintaining the space-charge neutrality. This greater injection of holes further feeds $p_2$ by transistor action and the process continues to repeat itself.

If the value of $\alpha_1 + \alpha_2$ is large enough, many electrons are collected in $n_1$ and many holes are collected in $p_2$, and the depletion region at $j_2$ begins to decrease. The reverse-bias disappears across $j_2$ and is replaced by a forward-bias, as shown in Fig. 8-4(b). Two of these voltages nullify in the resultant voltage. The forward voltage drop of the device from anode to cathode in the conducting state is not much greater than that of a single *p–n* junction.

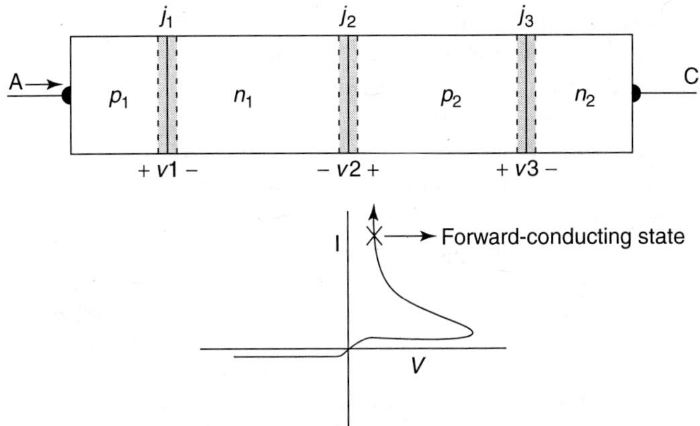


**Figure 8-4 (b)** Forward-conducting state of the SCR

As the SCR is biased in the forward-blocking state, a small current supplied to the gate can initiate switching to the conducting state; so the anode switching voltage $V_P$ decreases as the gate current $i_G$ applied to the gate is increased. This type of turn-on control makes the SCR useful in switching circuits.

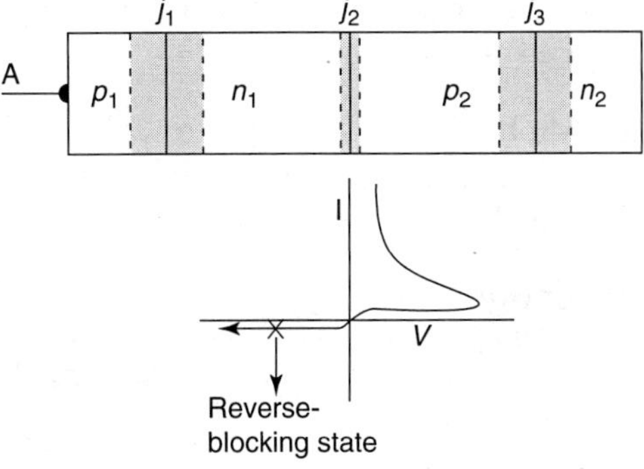


**Figure 8-4 (c)** Reverse-blocking state of the SCR

## Reverse-blocking state of the SCR

The SCR in reverse-biased condition allows almost negligible current to flow through it. This is shown in Fig. 8-4(c).

In the reverse-blocking state of the SCR, a small saturation current flows from anode to cathode. Holes will flow from the gate into $p_2$, the base of the $n$–$p$–$n$ transistor, due to positive gate current. The required gate current for turn-on is only a few milli amperes, therefore, the SCR can be turned on by a very small amount of power in the gate.

The change from the conducting state to the blocking state by reducing current $i$ below a critical value is called turning-off the SCR and it can be accomplished by keeping the $\alpha_1 + \alpha_2 = 1$ condition. As shown in Fig. 8-5, if the gate current is 0 mA, the critical voltage is higher, i.e., the SCR requires more voltage to switch to the conducting state. But as the value of gate current increases, the critical voltage becomes lower, and the SCR switches to the conducting state at a lower voltage. At the higher gate current $I_{G2}$, the SCR switches faster than at the lower gate current $I_{G1}$, because $I_{G2} > I_{G1}$.

Figure 8-5 clearly shows that at zero gate current the SCR requires more voltage compared to higher gate currents. As the gate current $I_{G2}$ is the highest, it requires less voltage to turn on compared to other values of gate current.

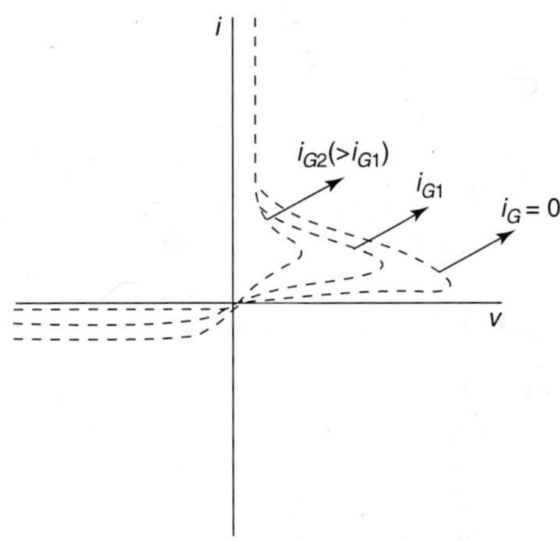


**Figure 8-5** I−V characteristics of SCR

In the SCR, gate turn-off is used , o reduce the sum of alpha from unity. If the gate voltage is reversed, holes are extracted from the $p_2$ base region. When the rate of hole extraction by the gate is sufficient to remove the *n–p–n* transistor from saturation, the device turns OFF.

Few SCRs have two gate leads, $G_2$ attached to $p_2$ and $G_1$ attached to $n_1$, as shown in Fig. 8-6. This configuration is called the semiconductor-controlled switch (SCS). The SCS, biased in the forward-blocking state, can be switched to the conducting state by a negative pulse at the anode gate $n_1$ or by a positive current pulse applied to the cathode gate at $p_2$.

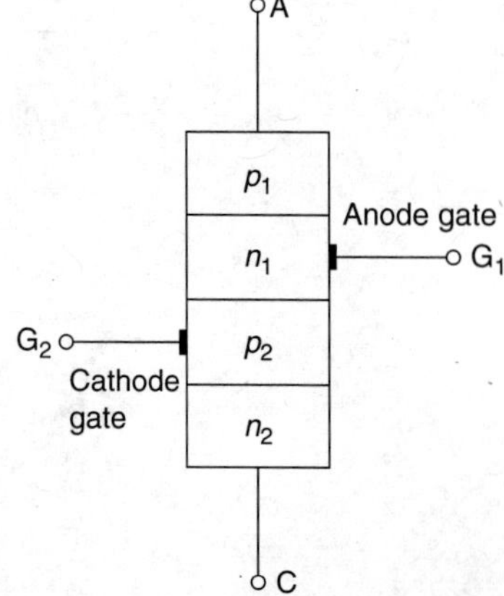


**Figure 8-6** Schematic diagram for a semiconductor-controlled switch

## 8-2-4 Simple Applications

The SCR is the most important member of the thyristor family. The SCR is a capable power device as it can handle thousands of amperes and volts. Generally the SCR is used in many applications such as in high-power electronics, switches, power-control and conversion mode. It is also used as surge protector.

### Static switch

The SCR is used as a switch for power-switching in various control circuits. This device can handle currents from a few milli amperes to hundreds of amperes.

### Power control

Since the SCR can be turned on externally, it can be used to regulate the amount of power delivered to a load. A very common example of this application is the light-dimmer switch used in many homes.

The SCR as a switch is turned on and off repetitively so that all or only part of each power cycle is delivered to the lights. As a result, the light intensity can be continuously varied from full intensity to minimum intensity. The same working principle can be applied to electronic motors, electronic heaters and many other electronic devices.

### Surge protection

In an SCR circuit, when the voltage rises beyond the threshold value, the SCR is turned on to dissipate the charge or voltage quickly. For example, in computer surge protector, SMPS and for the protection of telephone equipment exposed to lightning.

### Power conversion

The SCR is also used for high-power conversion and regulation. This includes conversion of power source from ac to ac, ac to dc and dc to ac.

## Solved Examples

**Example 8-1** For an SCR the gate-cathode characteristic has a straight line slope of 130. For a trigger source voltage of 14 V and an allowable gate dissipation of 0.5 watts, calculate the gate source resistance.

**Solution:**

From the given data:

$$V_g I_g = 0.5\ W$$

and, $$\frac{V_g}{I_g} = 130$$

$$130\, I_g^{\,2} = 0.5$$

$\therefore$ $$I_g = 63 \text{ mA}$$

Therefore, the gate voltage $V_g = 130 \times 0.063 = 8.06$ V.

For gate circuit:

$$E_S = I_g R_S + V_g$$

$\Rightarrow$ $$14 = 0.062\, R_S + 8.06$$

$\Rightarrow$ $$5.94 = 0.062\, R_S$$

$\Rightarrow$ $$R_S = \frac{5.94}{0.062}$$

$\Rightarrow$ $$R_S = 95.3\ \Omega$$

The gate to source resistance is:

$$R_S = 95.3\ \Omega.$$

**Example 8-2** In the SCR, as shown in the following diagram, the latching current is 50 mA. The duration of the firing pulse is 50 $\mu$s. Will the thyristor get fired?



**Solution:**

When the SCR is triggered the current will rise exponentially in the inductive circuit. Therefore:

$$i(t) = \frac{V}{R}(1 - e^{-t/\tau})$$

where, $$\tau = \frac{L}{R} = \frac{0.25}{10} = 0.025 \text{ sec}$$

or, $$i(t) = \frac{50}{10}(1 - e^{-t/0.025})$$

At, $t = 50\ \mu$ sec: $$i(50\mu s) = \frac{50}{10}(1 - e^{-50\times E-6/0.025}) = 9.99 \text{ mA}$$

Since the calculated circuit value is less than the given latching current value of SCR, it will not get fired.

**Example 8-3** If the latching current in the circuit is 14 mA, obtain the minimum width of the gating pulse required to properly turn on the SCR.

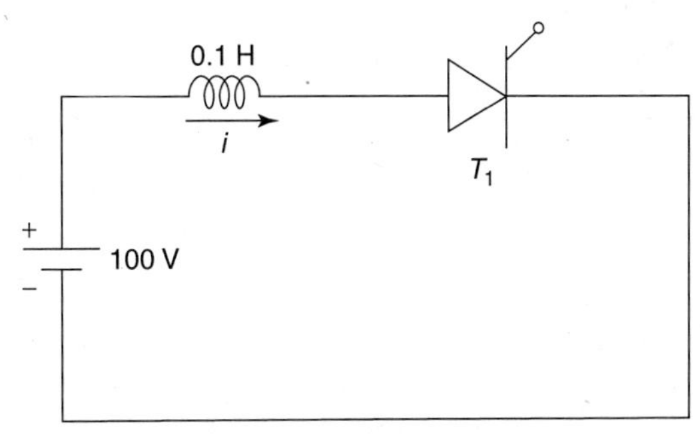


**Solution:**

The circuit equation is $V = L\dfrac{di}{dt}$ where, $i$ is the latching current and $t$ is the pulse width.

Therefore:

$$dt = \frac{L}{V}di$$

Integrating both sides:

$$t = \frac{L}{V}i$$

Therefore:

$$t_{min} = \frac{0.1}{100} \times 4 \times 10^{-3} = 4\ \mu s$$

**Example 8-4** If the $V_g - I_g$ characteristics of an SCR is assumed to be a straight line passing through the origin with a gradient of $3 \times 10^3$, calculate the required gate source resistance.

Given, $E_{gs} = 10$ V and allowable $P_g = 0.012\ W$.

**Solution:**

The allowable $P_g = 0.012\ W$.

$$\therefore \quad V_g I_g = 0.012$$

Also, gradient $\frac{V_g}{I_g} = 3 \times 10^3$.

$$\therefore \quad V_g = 10^3 \times I_g$$

$$\Rightarrow \quad I_g^2 = \frac{0.012}{3 \times 10^3}$$

so,

$$I_g = 2\ \text{mA}$$

and,

$$V_g = 6\ \text{V}$$

**Example 8-5** For an SCR, the gate cathode characteristics is given by a straight line with a gradient of 16 volts per amp passing through the origin; the maximum turn on time is 4 $\mu$s and the minimum gate current required to obtain this quick turn on is 500 mA. If the gate source voltage is 15 V calculate the resistance to be connected in series with the SCR gate.

**Solution:**

(a) Given $I_{g\,min} = 500\ \text{mA} = 0.5\ \text{A}$

$$\frac{V_g}{I_g} = 16\ \text{V/A}$$

$$\therefore \quad V_g = 16 \times 0.5 = 8\ \text{V}$$

$$\therefore \quad R_g = \frac{E_{gs} - V_g}{I_g} = \frac{15 - 8}{0.5} = 14\ \Omega$$

**Example 8-6** The limiting value of the charging current to turn on the thyristor is 32 mA. The capacitance of reverse-biased junction $J_2$ in a thyristor is $C_{j2} = 40$ pF and can be assumed to be independent of the off-state voltage. Determine the critical value of $dv/dt$.

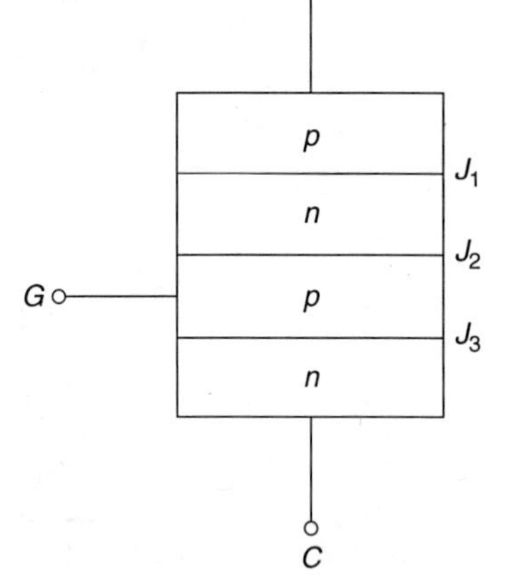


**Solution:**

$$C_{j2} = 40 \text{ pF}$$

$$i_{j2} = 32 \text{ mA}$$

Since,

$$\frac{dC_{j2}}{dt} = 0$$

The critical value of $\frac{dv}{dt} = \frac{i_{j2}}{C_{j2}} = \frac{32 \times 10^{-3}}{40 \times 10^{-12}} = 800 \text{ V}/\mu\text{s}$

**Example 8-7** An SCR has half-cycle surge current rating of 3000 A for a 50 Hz supply. Calculate its one-cycle and sub-cycle surge current rating and $I^2t$ rating.

**Solution:**
Let $I$ and $I_{sb}$ be the one-cycle and sub-cycle surge current ratings of the SCR respectively. Equating the energies involved, we get:

$$i^2t = i^2_{sb}t$$

or,

$$i^2 \times \frac{1}{100} = (3000)^2 \times \frac{1}{200}$$

or,

$$I = \frac{3000}{\sqrt{2}} = 2121.32 \text{ A}$$

$I^2t$ rating

$$= I^2_{\text{rms}} \times \frac{1}{2f} = \left(\frac{3000}{\sqrt{2}}\right)^2 \times \frac{1}{100}$$

$$= 45000 \text{ A}^2\text{s}$$

■

## 8-3 TRIODE AC SWITCH (TRIAC)

The term TRIAC is derived by combining the first three letters of the word "TRIODE" and the word "AC". An SCR is a unidirectional device as it can conduct from anode to cathode and not from cathode to anode. A TRIAC is capable of conducting in both the directions. The TRIAC, is thus, a bidirectional thyristor with three terminals. It is widely used for the control of power in ac circuits.

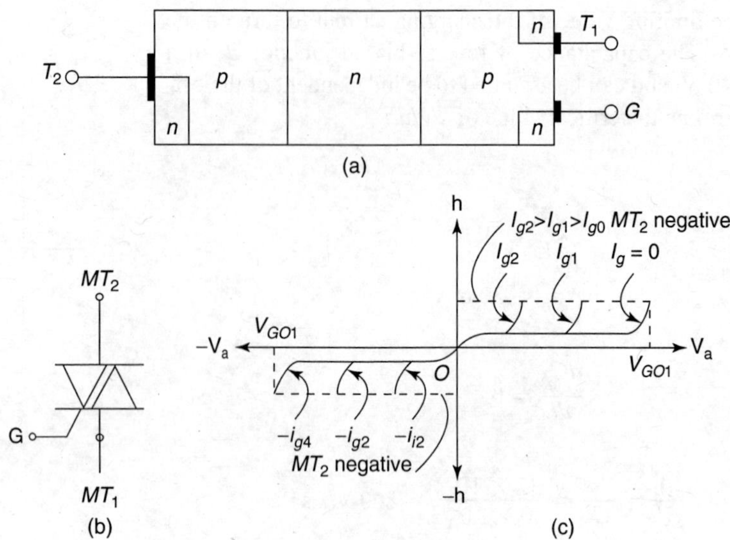


**Figure 8-7** (a) Structure of TRIAC (b) Circuit symbol (c) Static I−V characteristics

## 8-3-1 Constructional Features

The TRIAC is equivalent to two SCRs connected in an anti-parallel manner. The device structure, circuit symbol and its characteristics are shown in Fig. 8-7

Since the TRIAC is capable of conducting in both directions, the term anode and cathode are not applicable to it. Its three terminals are usually designated as $MT_1$ (main terminal one) and $MT_2$ (main terminal two) and the gate $G$. The gate is near terminal $MT_1$. The first quadrant is the region where $MT_2$ is positive with respect to $MT_1$ and vice versa for the third quadrant.

Depending upon the polarity of the gate pulse and the biasing conditions, the main four-layer structure that turns ON by a regenerative process could be one of $p_1 n_1$, $p_2 n_2$, $p_1 n_1 p_2 n_3$, or $p_2 n_1 p_1 n_4$, as shown in Fig. 8-8.

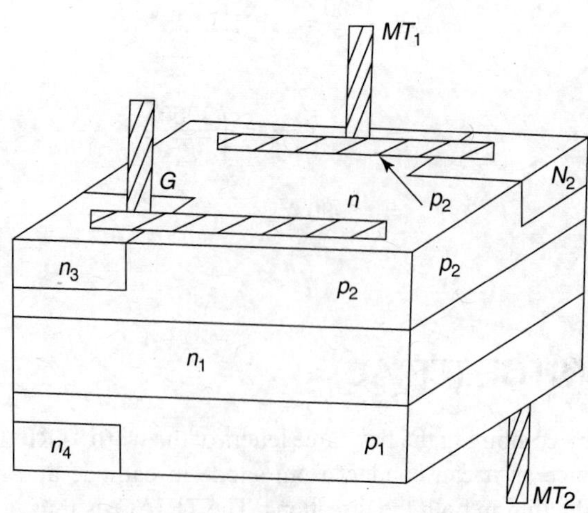


**Figure 8-8** Cross-sectional view of the TRIAC

## FOR ADVANCED READERS

### Physical Operation and Characteristics of the Triac

In this section we will discuss the different triggering modes of TRIAC, and how these modes explain the behavior of TRIAC under different physical conditions.

### Triggering Modes of the Triac

The TRIAC can be turned ON with positive or negative gate current keeping the $MT_2$ terminal at positive or negative potential. Let us examine the turn-on processes of the TRIAC.

**Mode I: $Mt_2$ is Positive and Gate Current is Positive**

When $MT_2$ is positive with respect to $MT_1$, junctions $p_1 n_1$ and $p_2 n_2$ are forward-biased but junction $p_2 n_1$ is reverse-biased. When the gate terminal is positive with respect to $MT_1$, the gate current flows mainly through $p_2 n_2$ junction like an ordinary SCR, as shown in Fig. 8-9.

In the TRIAC, the gate current requirement is higher for turn-on at a particular voltage. Because of ohmic contacts of the gate and $MT_1$ terminals on the $p_2$ layer, some more gate current flows from the gate lead $G$ to the main terminal $MT_1$ through the semi conductor $p_2$ layer without passing through the $p_2 n_2$ junction.

When the gate current has injected sufficient charge into the $p_2$ layer, the layer is flooded with electrons and when the gate current flows across the $p_2$–$n_2$ junction. These electrons diffuse into the edge of junction $j_2$, and are collected by the $n_1$ layer. Therefore, the electrons build a space charge in the $n_1$-region, and more holes from $p_1$ diffuse into $n_1$ to neutralize the negative space charge. These holes arrive at junction $j_2$. They produce a positive space charge in the $p_2$-region which results in more electrons being injected from $n_2$ into $p_2$. This results in positive regeneration and ultimately the $p_1n_1 p_2n_2$ structure conducts the external current. The device is more sensitive in this mode. It is a recommended method of triggering the conduction that is desired in the first quadrant.

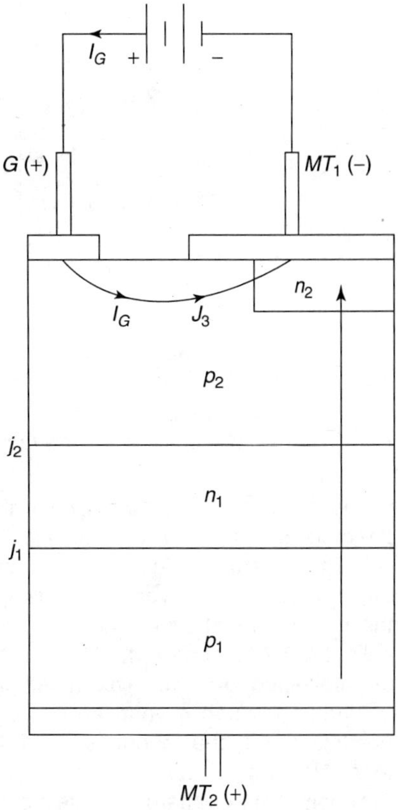

Figure 8-9 $MT_2$ positive and positive gate current

**Mode II: $Mt_2$ Is Positive and Gate Current is Negative.**

When the terminal $MT_2$ is positive and the gate terminal is negative with respect to terminal $MT_1$, the gate current flows through $p_2$–$n_3$ junction and this gate current $I_G$ forward-biases the gate junction $p_2$–$n_1$ of the auxiliary $p_1n_1p_2n_3$ structure. As a result, initially the TRIAC starts conducting through $p_1n_1p_2n_3$ layers. With the conduction of $p_1n_1p_2n_3$ the voltage drop across it falls, but potential of layer between $p_2$ $n_3$ rises towards the anode potential of $MT_2$. As the right hand portion of $p_2$ is clamped at the cathode potential of $MT_1$, a potential gradient exists across layer $p_2$, its left-hand region being at higher potential than its right-hand region, as shown in Fig. 8-10.

A current is established in layer $p_2$ from left to right and forward-biases the $p_2n_2$ junction. Finally the main structure $p_1n_1p_2n_2$ begins to conduct. The anode current of the pilot SCR serves as the gate current for the main SCR. Compared to the turn-on process discussed with respect to Mode 1, the device with $MT_2$ positive but gate current negative is less sensitive, and therefore, requires more gate current.

**Mode III: $Mt_2$ Is Negative and Gate Current is Positive**

When terminal $MT_2$ is negative and terminal $MT_1$ is positive, the device can be turned on by applying a positive voltage between the gate and the terminal $MT_1$. In this mode, the device operates in the third quadrant when triggered into conduction. The turn-on is initiated by a remote gate control. The main structure

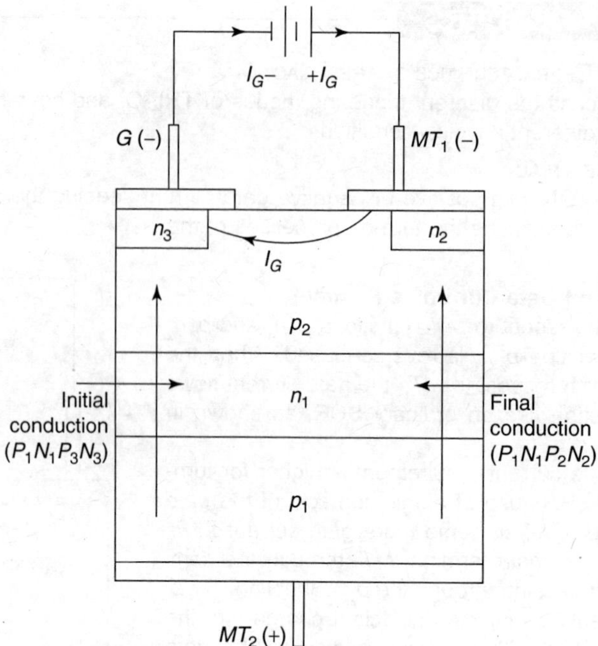


**Figure 8-10** $MT_2$ positive and negative gate current

that leads to the turn-on is $p_2n_1p_1n_4$ with $n_2$ acting as a remote gate, as shown in Fig. 8-11.

The external gate current $I_G$ forward-biases the $p_2n_2$ junction. Layer $n_2$ injects electrons into the $p_2$ layer, as shown by dotted arrows, and are collected by junction $p_2n_1$. The electrons from $n_2$ are collected by $p_2n_1$. The holes injected from $p_2$ diffuse through $n_1$ and arrive at $p_1$. Hence, a positive space charge builds up in the $p_1$ region. More electrons from $n_4$ diffuse into $p_1$ to neutralize the positive space charge. These electrons arrive at junction $j_2$. They produce a negative space charge in the $n_1$ region, which results in more holes being injected form $p_2$ into $n_1$. This regenerative process continues until the structure $p_2n_1p_1n_4$ completely turns on and conducts the current which is limited by the external load. As the TRIAC is turned on by remote gate $n_2$, the device is less sensitive in the third quadrant with respect to the positive gate current.

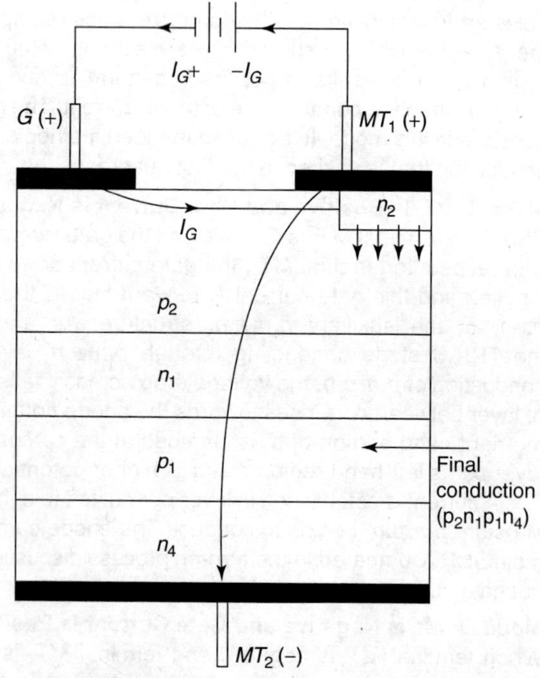


**Figure 8-11** $MT_2$ negative and positive gate current

### Mode IV: $Mt_2$ is Negative and Gate Current is Negative

A cross-sectional view of the structure is shown in Fig. 8-12.

In this mode of operation, $n_3$, acts as a remote gate. The external gate current $I_G$ forward-biases the $p_2n_3$ junction and the electrons are injected. These

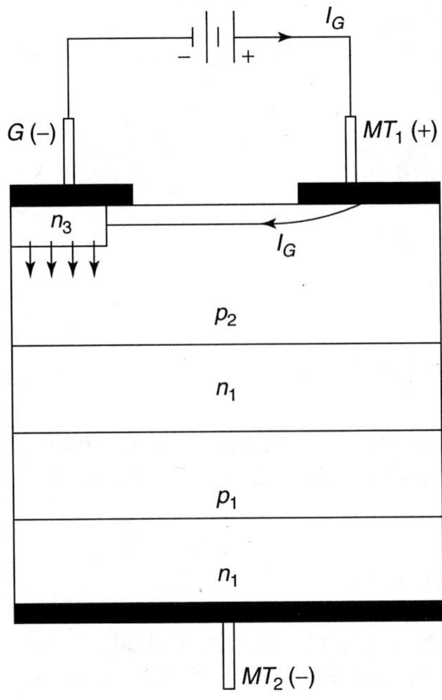


**Figure 8-12** $MT_2$ negative and negative gate current

electrons from $n_3$ collected by $p_2n_1$ cause an increase in the current across $p_1n_1$. The structure $p_2n_1p_1n_4$ turns on by the regeneration. The device will turn on due to the increased current in layer $n_1$.The following points can, therefore, be concluded from this analysis.

(i) Sensitivity of the TRIAC is greatest in the first quadrant when it is turned on by the positive gate current and in the third quadrant when it is turned on by the negative gate current.

(ii) Sensitivity of the TRIAC is low in the first quadrant when it is turned on by the negative gate current and in the third quadrant when it is turned on by the positive gate current.

## Advantages of the TRIAC

The TRIAC has the following advantages:

(i) They can be triggered with positive- or negative-polarity voltage.

(ii) They need a single heat sink of slightly larger size.

(iii) They need a single fuse for protection, which simplifies their construction.

(iv) In some dc applications, the SCR has to be connected with a parallel diode for protection against reverse voltage, whereas a TRIAC may work without a diode, as safe breakdown in either direction is possible.

## Disadvantages of the TRIAC

The disadvantages of the TRIAC are as follows:

(i) TRIACs have low *dv*/*dt* ratings compared to SCRs.

(ii) Since TRIACs can be triggered in either direction, the trigger circuits with TRIACs needs careful consideration.

(iii) Reliability of TRIACs is less than that of SCRs.

**Simple applications of the TRIAC**

The TRIAC as a bidirectional thyristor has various applications. Some of the popular applications of the TRIAC are as follows:

(i) In speed control of single-phase ac series or universal motors.
(ii) In food mixers and portable drills.
(iii) In lamp dimming and heating control.
(iv) In zero-voltage switched ac relay.

## 8-4 DIODE AC SWITCH (DIAC)

The DIAC is a combination of two diodes. Diodes being unidirectional devices, conduct current only in one direction. If bidirectional (ac) operation is desired, two Shockley diodes may be joined in parallel facing different directions to form the DIAC.

### 8-4-1 Constructional Features

The DIAC is a three-layer (*p–n–p* or *n–p–n*) bidirectional diode. It can be switched from OFF state to ON state by changing the polarity of the diode terminal. The DIAC is a uniformly doped *p–n–p* layer, as shown in Fig. 8-13. The uniform doping provides bidirectional symmetry in the I–V characteristics of this two-electrode DIAC.

At a glance, the construction of DIAC looks like a transistor but there are major differences. They are as follows:

(i) All the three layers, *p–n–p* or *n–p–n,* are equally doped in the DIAC, whereas in the BJT there is a gradation of doping. The emitter is highly doped, the collector is lightly doped, and the base is moderately doped.
(ii) The DIAC is a two-terminal diode as opposed to the BJT, which is a three-terminal device.

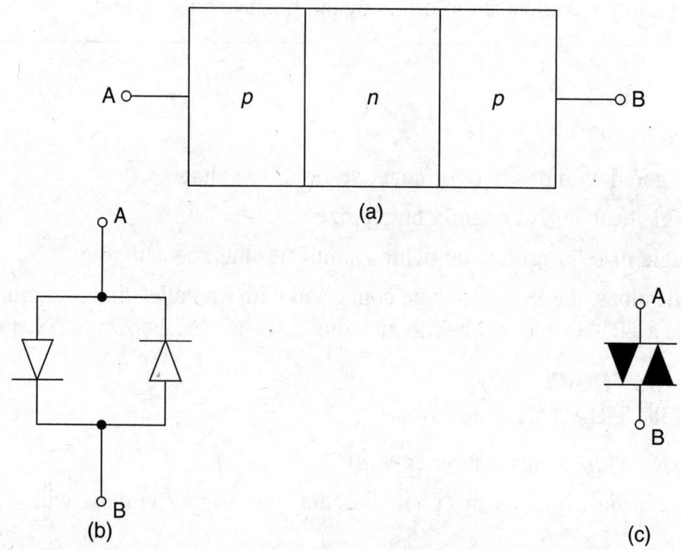


**Figure 8-13** (a) Basic structure of the DIAC (b) Equivalent circuit of the DIAC (c) Symbol of the DIAC

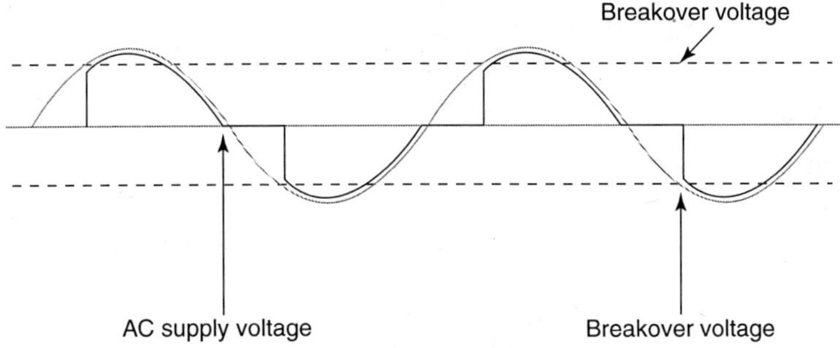


**Figure 8-14** Current waveform in the DIAC

## 8-4-2 Physical Operation and Characteristics

The DIAC, operated with a dc voltage across it, behaves in exactly the same manner as a Shockley diode. The behaviour is different in an ac operation because the alternating current repeatedly reverses directions; DIACs will not stay latched for more than one half-cycle. If a DIAC becomes latched, it will continue to conduct current only as long as there is voltage available to push enough current in that direction. If the ac polarity reverses twice per cycle, the DIAC will drop out due to insufficient current necessitating another break over before it conducts again. The result is a current waveform, as shown in Fig. 8-14

The DIAC is designed to trigger an SCR or a TRIAC. The voltage drop across the DIAC snaps back, typically several volts creating a break over current sufficient to trigger a TRIAC or an SCR. A typical I−V characteristic with its schematic symbol is shown in Fig. 8-15.

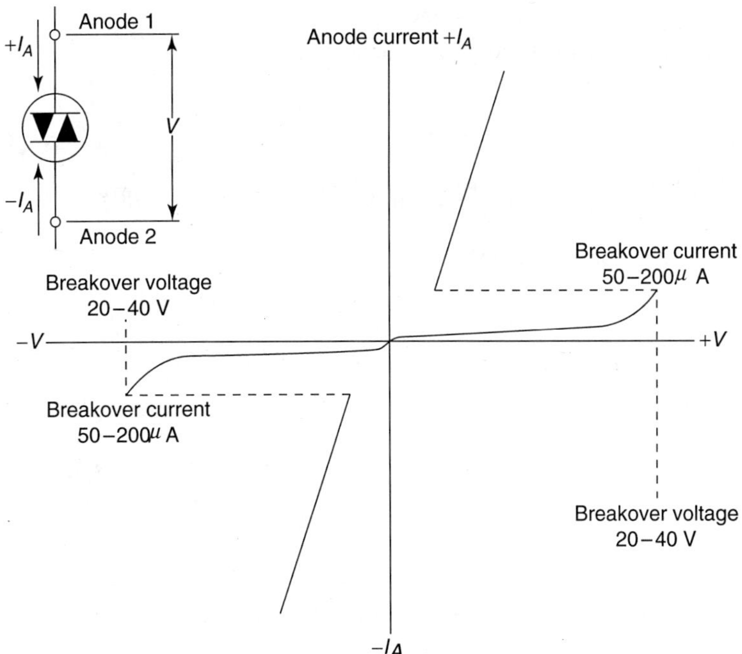


**Figure 8-15** I−V characteristics of the DIAC

The main characteristics are of the DIAC are as follows:

(i) Break over voltage
(ii) Voltage symmetry
(iii) Break-back voltage
(iv) Break over current
(v) Lower power dissipation

Although most DIACs have symmetric switching voltages, asymmetric DIACs are also available. Typical DIACs have a power dissipations ranging from 1/2 to 1 watt.

## 8-4-3 Applications

The DIAC is used in conjunction with other thyristor devices. It is also used to trigger devices in the phase control circuits of the TRIAC; mostly used in motor speed control, light-dimming and heat-controlling applications.

## Solved Examples

**Example 8-8** A DIAC with a breakdown voltage of 40 V is used in a circuit, with $R_1$ variable from 1−25 kΩ, $C$ = 470 nF and $E$ = 240 V rms at 50 Hz. What will be the maximum and minimum firing delays with this arrangement?

**Solution:**
Impedance of capacitor:

$$\frac{1}{\omega C} = 6773\ \Omega$$

The current through $R_1$ and when the DIAC is not conducting is:

$$I_d = 240\sqrt{2}\ \sin\frac{(\omega t + \phi)}{Z_d}$$

where, $Z_d = \sqrt{R_1^{2+1/\omega 2C2}}$ and $\phi = \tan^{-1}\left(\frac{1}{\omega RC}\right)$ with $R_1$ = 1000 Ω, $Z_d$ = 6846 Ω.

The voltage across the capacitor is given by:

$$V_c = I_d Z_c$$

$$V_c = 335.8\sin(\omega t - 8.4)$$

When the DIAC conducts: $V_c = 40$ V

$$\text{Minimum delay} = \sin^{-1}\left(\frac{40}{335.8}\right) + 8.4 = 15.24°$$

with, $R_1 = \tan^{-1}\left(\frac{1}{\omega RC}\right)$

$$R_1 = 25\ \text{k}\Omega,\ Z_d = 25901\ \Omega$$

The voltage across the capacitor is:

$$I_d - Z_C, \text{ as } V_C = 8.76\sin(\omega t - 74.84)$$

When the DIAC conducts: $V_c = 40$ V

Hence, maximum delay: $\sin^{-1}\left(\dfrac{40}{88.6}\right) + 74.84 = 101.6°$

■

# 8-5 UNIJUNCTION TRANSISTOR (UJT)

The unijunction transistor is a three-terminal single-junction device. The switching voltage of the UJT can be easily varied. The UJT is always operated as a switch in oscillators, timing circuits and in SCR/TRIAC trigger circuits.

## 8-5-1 Constructional Features

The UJT structure consists of a lightly doped *n*-type silicon bar provided with ohmic contacts on either side, as shown in Fig. 8-16. The two end connections are called base $B_1$ and base $B_2$. A small heavily doped *p*-region is alloyed into one side of the bar. This *p*-region is the UJT emitter ($E$) that forms a *p–n* junction with the bar.

Between base $B_1$ and base $B_2$, the resistance of the *n*-type bar called inter-base resistance ($R_B$) and is in the order of a few kilo ohm. This inter-base resistance can be broken up into two resistances—the resistance from $B_1$ to the emitter is $R_{B1}$ and the resistance from $B_2$ to the emitter is $R_{B2}$. Since the emitter is closer to $B_2$ the value of $R_{B1}$ is greater than $R_{B2}$.

Total resistance is given by:

$$R_B = R_{B1} + R_{B2} \tag{8-6}$$

The circuit symbol for the UJT and its equivalent circuit are shown in Fig. 8-16

The diode represents the *p–n* junction between the emitter and the base point. $R_B$ is variable, since during normal operation it typically ranges between a few kilo ohms to a few ohms.

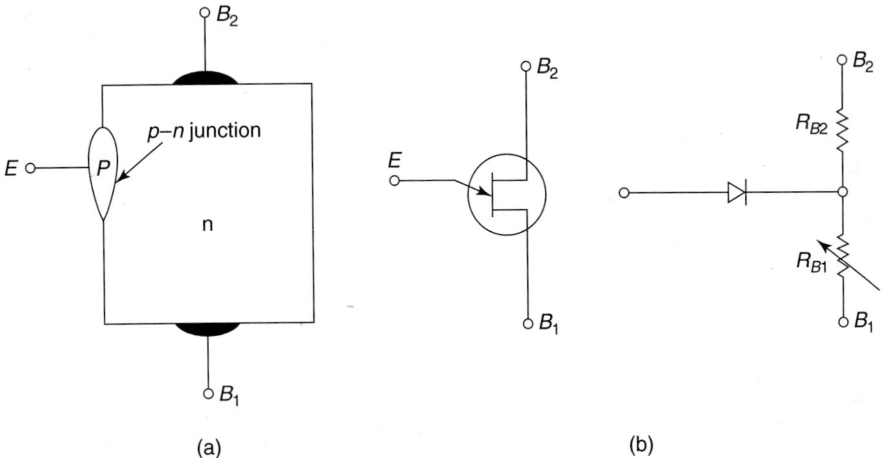


**Figure 8-16** (a) Basic UJT structure (b) UJT symbol and equivalent circuit

## 8-5-2 Physical Operation and Characteristics

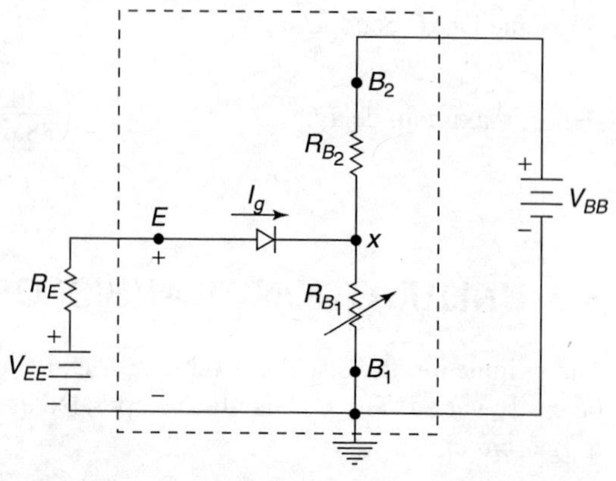


Figure 8-17 Equivalent circuit for UJT analysis

The UJT circuit operation has two states—ON state and OFF state. The change from the ON state to the OFF state can be performed very easily by simply controlling the emitter bias. When the emitter diode becomes forward-biased the resistance $R_{B1}$ drops to a very low value and the total resistance between $E$ and $B_1$ becomes very low allowing the emitter current to flow readily. This is the ON state. When the emitter diode is reverse-biased only a very small emitter current flows. Under this condition $R_{B1}$ is at its normal high value. This is the OFF state.

The $V_{BB}$ source is generally fixed and provides a constant voltage from $B_2$ to $B_1$. The UJT is normally operated with both $B_2$ and $E$ positive biased relative to $B_1$, as shown in Fig. 8-17. $B_1$ is always the UJT reference terminal and all voltages are measured relative to $B_1$ . $V_{EE}$ is a variable voltage source.

In the ON state, $I_E > 0$ and $V_E = V_p$; and in the OFF state, $I_E = 0$ and $V_E = V_{EE}$. The $V_E - I_E$ curve of the UJT is shown in Fig. 8-18.

### OFF state of the UJT circuit

When a voltage $V_{BB}$ is applied across the two base terminals $B_1$ and $B_2$, the potential of point $p$ with respect to $B_1$ is given by:

$$V_P = \frac{V_{BB}}{R_{B1} + R_{B2}} \times R_{B1} = \eta R_{B1} \tag{8-7}$$

where,

$$\eta = \frac{R_{B1}}{R_{B1} + R_{B2}} \tag{8-8}$$

$\eta$ is called the intrinsic stand off ratio with its typical value lying between 0.5 and 0.8.

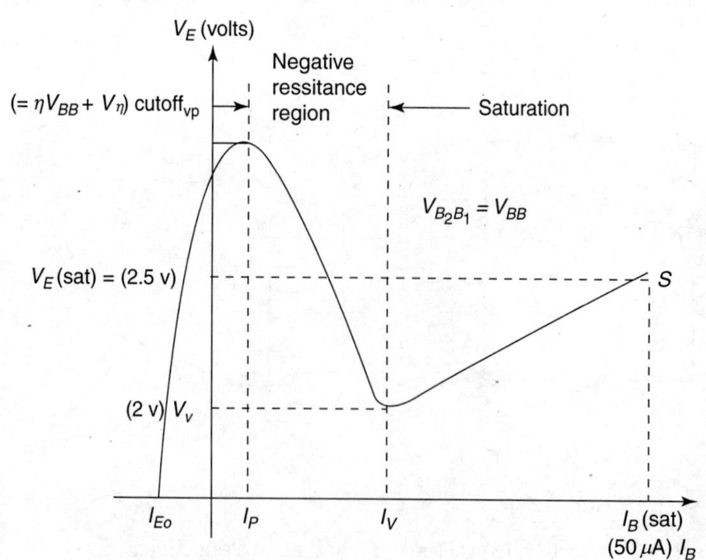


Figure 8-18 Typical UJT V–I characteristic curves

The $V_{EE}$ source is applied to the emitter which is the $p$-side. Thus, the emitter diode will be reverse-biased as long as $V_{EE}$ is less than $V_x$. This is OFF state and is shown on the $V_E - I_E$ curve as being a very low current region. In the OFF the UJT has a very high resistance between E and $B_1$, and $I_E$ is usually a negligible reverse leakage current. With no $I_E$, the drop across $R_E$ is zero and the emitter voltage equals the source voltage.

The UJT OFF state, as shown on the $V_E - I_E$ curve, extends to the point where the emitter voltage exceeds $V_x$ by the diode threshold voltage ($V_d$), which is needed to produce forward current through the diode. The emitter voltage and this point $p$ is called the peak point voltage $V_p$, and is given by $V_P = V_x + V_D = \eta V_{BB} + V_B$.

### ON state of the UJT circuit

As $V_{EE}$ increases, the UJT stays in the OFF state until $V_E$ approaches the peak point value $V_P$. As $V_E$ approaches $V_P$, the $p$–$n$ junction becomes forward-biased and begins to conduct in the opposite direction. As a result $I_E$ becomes positive near the peak point $P$ on the $V_E - I_E$ curve. When $V_E$ exactly equals $V_P$ the emitter current equals $I_P$ . At this point holes from the heavily doped emitter are injected into the $n$-type bar, especially into the $B_1$ region. The bar, which is lightly doped, offers very little chance for these holes to recombine. The lower half of the bar becomes replete with additional current carriers (holes) and its resistance $R_B$ is drastically reduced; the decrease in $B_{B1}$ causes $V_x$ to drop. This drop, in turn, causes the diode to become more forward-biased and $I_E$ increases even further. The larger $I_E$ injects more holes into $B_1$ and so on. When this regenerative process ends $R_B$ has dropped to a very small value and $I_E$ can become very large, limited mainly by external resistance $R_E$.

### UJT ratings

Maximum peak emitter current. This represents the maximum allowable value of a pulse of emitter current.

Maximum reverse emitter voltage. This is the maximum reverse-bias that the emitter base junction $B_2$ can tolerate before breakdown occurs.

Maximum inter base voltage. This limit is caused by the maximum power that the $n$-type base bar can safely dissipate.

Emitter leakage current. This is the emitter current which flows when $V_E$ is less than $V_p$ and the UJT is in the OFF state.

## 8-5-3 Applications

The UJT is very popular today mainly due to its high switching speed. A few select applications of the UJT are as follows:

(i) It is used to trigger SCRs and TRIACs
(ii) It is used in non-sinusoidal oscillators
(iii) It is used in phase control and timing circuits
(iv) It is used in saw tooth generators
(v) It is used in oscillator circuit design

## Solved Examples

**Example 8-9** Derive the expression for periodic time $T = RC \log_e\left(\frac{1}{1-\eta}\right)$ of the UJT relaxation circuit.

**Solution:**
Voltage across the capacitor is given by:

$$V_P = V_{BB}(1 - e^{-t/RC})$$

When $V_P = \eta V_{BB} + V_o$, the capacitor will discharge through $R_1$:

$$\eta V_{BB} + V_o = V_{BB}(1 - e^{-t/RC})$$

$$\Rightarrow \quad e^{-t/RC} = 1 - \eta$$

$$\Rightarrow \quad e^{t/RC} = \frac{1}{1-\eta}$$

$$\Rightarrow \quad \frac{t}{RC} = \log_e\left(\frac{1}{1-\eta}\right)$$

$$\therefore \quad t = RC \log_e\left(\frac{1}{1-\eta}\right)$$

Here, $t = \mathrm{T}$

$$\therefore \quad T = RC \log_e\left(\frac{1}{1-\eta}\right)$$

**Example 8-10** Design the triggering circuit for a UJT where $V_s = 30$ V, $\eta = 0.51$, $I_P = 10\ \mu$A, $V_u = 3.5$ V and $I_V = 10$ mA. The frequency is $f = 60$ Hz and the width of the triggering pulse is $t_g = 50\ \mu$s.

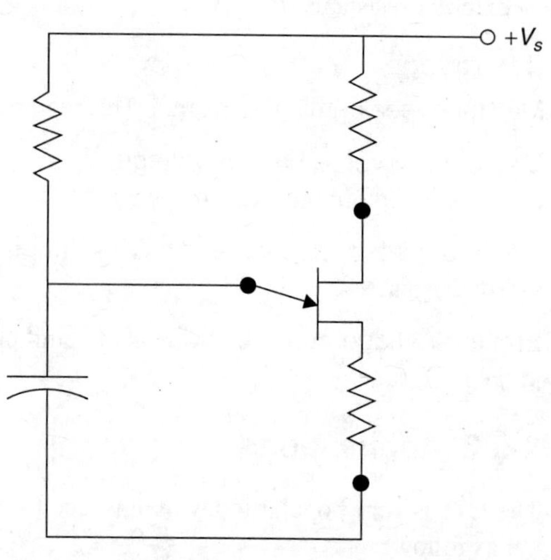


**Solution:**

$$T = \frac{1}{f} = \frac{1}{60} = 16.67 \text{ ms}$$

We know that:

$$V_P = \eta V_S + V_D = 0.51 \times 30 + 0.5 = 15.8 \text{ V}$$

Let, $C = 0.5\ \mu F$, $R < \dfrac{V_S - V_P}{I_P}$

$$= \frac{30 - 15.8}{10\ \mu\text{A}} = 1.42\ \text{M}\Omega$$

and, $R > \dfrac{V_S - V_v}{I_v} = \dfrac{30 - 3.5}{10 \text{ mA}} = 2.65\ \text{M}\Omega$

Also, $\tau = RC \ln\left(\dfrac{1}{1-\eta}\right)$

or, $16.67 \text{ ms} = R \times 0.5 \text{ m} \times \ln\left(\dfrac{1}{1-0.51}\right)$

or, $R = 46.7\ \text{k}\Omega$

The peak gate voltage:

$$V_{B1} = V_P = 15.8 \text{ V}$$

$$T_g = R_{B1}C \Rightarrow R_{B1} = \frac{50\ \mu s}{0.5\ \mu F} = 100\ \Omega$$

$$R_{B2} = \frac{10^4}{\eta V_S} = \frac{10^4}{0.15 \times 30} = 654\ \Omega$$

**Example 8-11** If $R_E = 1\ k\Omega$ and $I_V = 5$ mA, determine the value of $V_{EE}$ which will cause the UJT to turn off.

**Solution:**
At the valley point $V_E = V_V = 2$ V, $I_E = I_V = 5$ mA.

$V_{EE} = I_E R_E + V_E = 7$ V is the value of $V_{EE}$ below which the UJT will switch back to the OFF state. ■

## 8-6 INSULATED-GATE BIPOLAR TRANSISTOR (IGBT)

The insulated-gate bipolar transistor is a recent model of a power-switching device that combines the advantages of a power BJT and a power MOSFET. Both power MOSFET and IGBT are the continuously controllable voltage-controlled switch.

### 8-6-1 Constructional Features

The structure of an IGBT cell is shown in Fig. 8-19. The $p^+$ region acts as a substrate which forms the anode region, i.e., the collector region of the IGBT. Then there is a buffer layer of $n^+$ region and a bipolar-base

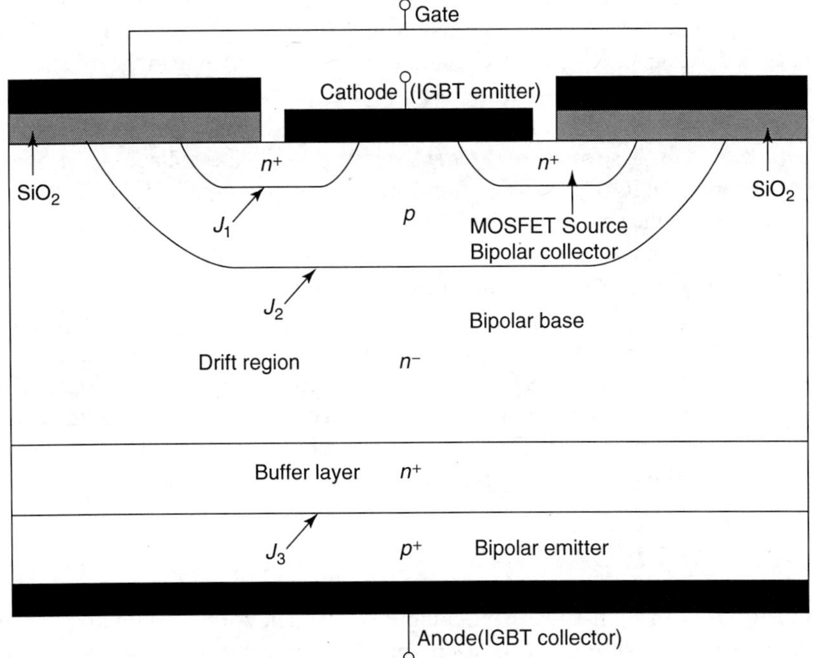


**Figure 8-19** Schematic diagram of insulated-gate bipolar transistor

drift region. The $p$-region contains two $n^+$ regions and acts as a MOSFET source. An inversion layer can be formed by applying proper gate voltage. The cathode, i.e., the IGBT emitter is formed on the $n^+$ source region.

To explain the operation the two layers of $n^+$ and $n^-$ can be considered as a single $n$-layer. The location of the emitter corresponds to that of the source in the power MOSFET. The collector is a power terminal of the IGBT switch and similar to the drain of the power MOSFET. The other power terminal is the IGBT emitter $E$. The control terminal is gate $G$. The switching control voltage is applied between emitter $E$ and gate $G$.

The circuit symbol for the IGBT is shown in Fig. 8-20. It is similar to the symbol for an $n$–$p$–$n$ bipolar-junction power transistor with the insulated-gate terminal replacing the base.

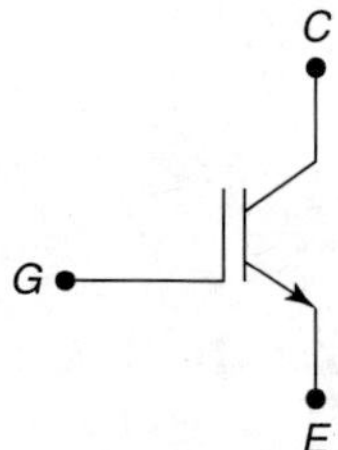


**Figure 8-20** Circuit symbol of insulated-gate bipolar transistor

## 8-6-2 Physical Operation and Characteristics

The principle behind the operation of an IGBT is similar to that of a power MOSFET.
The IGBT operates in two modes: (i) the blocking or non-conducting mode and (ii) the ON or conducting mode.

The resistance of the $n$-region on the top is quite small. When the device carries current in the ON state, the injection of holes from the top $p^+$ region into the $n$-region accounts for this low resistance. This effect is referred to as the conductivity modulation of the $n$-region, which substantially reduces the ON state voltage drop. So the current rating in this case is high.

The collector is at a positive voltage with respect to the emitter. If the gate voltage is zero or below the threshold value, the junction between the middle $n$-zone and the $p$-zone is reverse-biased. The only current that flows is the reverse leakage current of this junction, which is negligibly small. This is the OFF state of the switch.

When a positive gate-to-emitter control voltage (that is above the threshold value) is applied, the device goes into the ON state, where it carries a large current with a small voltage drop across it. The above-threshold gate voltage induces an $n$-channel. This channel connects the $n^+$ emitter zone to the middle $n$-region. A $p$–$n$–$p$ transistor is formed by the top $p^+$ region, the middle $n$-region and the lower $p$-island. The top $p^+$ region, i.e., the collector of the IGBT serves as the emitter of this $p$–$n$–$p$ transistor with normal voltage polarities.

The middle $n$-region is the base of the $p$–$n$–$p$ transistor. When the gate-voltage produces a channel, current flows from the collector of the IGBT across the forward-biased top $p$–$n$ junction, across the middle $n$-region and through the channel into the emitter terminal. The current passing through the channel is the base current of the $p$–$n$–$p$ transistor. The emitter current in this transistor is the result of a large-scale injection of holes across the top $p$–$n$ junction. These injected holes cause the conductivity modulation of the middle $n$-region. Two current paths now lead to the emitter terminal. One is through the middle $n$-region and the channel, the other is across the collector junction of the $p$–$n$–$p$ transistor in question and through the lower $p$-region.

In the ON state, the voltage drop is less than that of a power required by a conventional MOSFET. The voltage requirement is small in a junction power transistor. The IGBT cannot block large reverse voltages; the maximum safe reverse voltage is less than 10 V. The voltage and current ratings of the IGBT is quite high. The current rating is as high as few hundred amperes (approximately) and voltage rating can be in the order of kilovolts.

# 8-7 REAL-LIFE APPLICATIONS

The IGBT is mostly used in high-speed switching devices. They have switching speeds greater than those of bipolar power transistors. The turn-on time is nearly the same as in the case of a power MOSFET, but the turn-off time is longer. Thus, the maximum converter switching frequency of the IGBT is intermediate between that of a bipolar power transistor and a power MOSFET.

## POINTS TO REMEMBER

1. A thyristor is a multilayer *p–n* terminal electronic device used for bi-stable switching.
2. The SCR has two states:
   (a) High-current low-impedance ON state
   (b) Low-current OFF state
3. Latching current is defined as a minimum value of anode current which is a must in order to attain the turn-on process required to maintain conduction when the gate signal is removed.
4. Holding current is defined as a minimum value of anode current below which it must fall for turning off the thyristor.
5. The TRIAC is a bidirectional thyristor with three terminals. It is used extensively for the control of power in ac circuits.
6. The DIAC is an *n–p–n* or *p–n–p* structure with a uniformly doped layer.
7. Applications of the UJT:
   (a) As trigger mechanism in the SCR and the TRIAC
   (b) As non-sinusoidal oscillators
   (c) In saw-tooth generators
   (d) In phase control and timing circuits
8. The UJT operation can be stated as follows:
   (a) When the emitter diode is reverse-biased, only a very small emitter current flows. Under this condition $R_{B1}$ is at its normal high-value. This is the OFF state of the UJT.
   (b) When the emitter diode becomes forward-biased $R_{B1}$ drops to a very low value so that the total resistance between $E$ and $B_1$ becomes very low, allowing emitter current to flow readily. This is the ON state.
9. The IGBT is mostly used in high-speed switching devices.

## IMPORTANT FORMULAE

1. In the SCR the equivalent current is given by:

$$i = \frac{I_{CO1} + I_{CO2}}{1-(\alpha_1 + \alpha_2)}$$

2. The intrinsic standoff ratio $\eta$ is expressed as:

$$\eta = \frac{R_{B1}}{R_{B1} + R_{B2}}$$

## OBJECTIVE QUESTIONS

1. Fabricated layers of SCR are:
   (a) *n–p–n–p* (b) *p–n–p–n*
   (c) *n–p–n* (d) Either (a) or (b)
2. Fabricated layers of the TRIAC are:
   (a) *n–p–n–p* (b) *p–n–p–n*
   (c) *n–p–n* (d) *p–n*
3. Fabricated layers of SCR are:
   (a) *n–p–n–p* (b) *n–p–n*
   (c) *n–p* (d) None of the above
4. Holding current of a thyristor is:
   (a) Less than latching current
   (b) More than latching current
   (c) Equal to latching current
   (d) Zero
5. Anode current in a thyristor is made up of:
   (a) Electrons only
   (b) Electrons or holes
   (c) Electrons and holes
   (d) Holes only

6. For normal SCRs switching turn on depends on:
   (a) Gate current (b) Cathode current
   (c) Anode current (d) None of the above
7. A thyristor, when triggered, will change from forward-blocking state to conduction state if its anode to cathode voltage is equal to:
   (a) Peak repetitive off state forward voltage
   (b) Peak working off state reverse voltage
   (c) Peak working off state forward voltage
   (d) Peak non-repetitive off state forward voltage
8. A DIAC can be termed as:
   (a) Diode AC switch (b) Triode AC switch
   (c) Both (a) and (b) (d) None of the above
9. Turn-on time of an SCR can be reduced by using a:
   (a) Rectangular pulse of high amplitude and narrow width
   (b) Rectangular pulse of low amplitude and wide width
   (c) Triangular pulse
   (d) Trapezoidal pulse
10. For an SCR, *dv/dt* protection is achieved through the use of:
    (a) *RL* in series with SCR
    (b) *RC* across SCR
    (c) *L* in series with SCR
    (d) *RC* in series with SCR
11. A forward voltage can be applied to an SCR after its:
    (a) Anode current reduces to zero
    (b) Gate recovery time
    (c) Reverse recovery time
    (d) Anode voltage reduces to zero
12. In a UJT, with $V_{nn}$ as the voltage across two base terminals, the emitter potential at peak point is given by:
    (a) $V_{BB}$ (b) $V_o$
    (c) $V_{nn} + V_D$ (d) None of the above
13. A UJT exhibits negative resistance region:
    (a) Before the peak point
    (b) Between peak and valley points
    (c) After the valley point
    (d) Both (a) and (c)
14. In a UJT maximum value of charging resistance is associated with:
    (a) Peak point
    (b) Valley point
    (c) Any point between peak and valley point
    (d) After the valley point
15. A UJT is used for triggering an SCR then the wave shape of the voltage obtained from the UJT circuit is:
    (a) Saw-tooth wave (b) Sine wave
    (c) Trapezoidal wave (d) Square wave
16. In synchronized UJT triggering of an SCR voltage $V_c$ across capacitor reaches UJT threshold voltage thrice in each half-cycle so that there are three firing pulses during each half-cycle. The firing angle of the SCR can be controlled:
    (a) Thrice in each half-cycle
    (b) Once in each half-cycle
    (c) Twice in each half-cycle
    (d) None of the above
17. A resistor connected across the gate and cathode of an SCR increases its:
    (a) Turn off time (b) Holding current
    (c) Noise immunity (d) *dv/dt* rating
18. A TRIAC can be used only in:
    (a) Chopper
    (b) Inverter
    (c) Multi quadrant chopper
    (d) None of the above
19. SCR is a:
    (a) Unidirectional switch
    (b) Bidirectional switch
    (c) Four-directional
    (d) None of the above
20. TRIAC is a:
    (a) Unidirectional Switch
    (b) Bidirectional switch
    (c) Four-directional
    (d) None of the above

21. Inverter converts:
    (a) DC to AC
    (b) AC to DC
    (c) Both (a) and (b)
    (d) None of the above

22. SCR is a:
    (a) Two-terminal device
    (b) Three-terminal device
    (c) Four-terminal device
    (d) None of the above

## REVIEW QUESTIONS

1. Explain the constructional features, physical operation, characteristics and simple applications of the SCR.
2. Explain the constructional features, physical operation, characteristics and simple applications of the DIAC.
3. Explain the constructional features, physical operation, characteristics and simple applications of the TRIAC.
4. Explain the constructional features, physical operation, characteristics and simple applications of the UJT.
5. Explain the constructional features, physical operation, characteristics and simple applications of the IGBT.
6. Explain the importance of thyristor as a switching device.
7. What are the differences between a DIAC and a TRIAC?
8. What are the differences between a DIAC and a SCR?
9. What are the basic similarities and differences between a transistor and a DIAC?
10. Give the two-transistor analogy of the SCR, and explain the I−V characteristics.

## PRACTICE PROBLEMS

1. For an SCR, the gate-cathode characteristic has a straight line slope of 110. For trigger source voltage of 17 V and allowable gate dissipation of 0.7 watts, calculate the gate-to-source resistance.
2. The source voltage of a UJT is $V_{BB}$ = 32 Volt and firing voltage is in the order of 12 Volt. The value connected series $RC$ is $C$ = 1000 pF and $R$ = 100 kΩ. Calculate the time period of the saw tooth voltage across the capacitor.
3. An SCR has a half-cycle surge current rating of 2500 A for 50 Hz supply. Calculate its one-cycle and sub-cycle surge current rating and $I^2t$ rating.
4. Latching current of a thyristor circuit is 50 mA. The duration of the firing pulse is 50 $\mu$s. Will the thyristor get fired?
5. The gate cathode characteristic For an SCR has a straight line slope of 110. For trigger source voltage of 12 V and allowable gate dissipation of 0.75 watts, calculate the gate source resistance.
6. If the $V_g - I_g$ characteristic of an SCR is assumed to be a straight line passing through the origin with a gradient of $2 \times 10^3$, calculate the required gate source resistance. Given, $E_{gs}$ = 10 V and allowable $P_g$ = 0.02 W.
7. An SCR has a half-cycle surge current rating of 2000 A for 50 Hz supply. Calculate its one-cycle and sub-cycle surge current rating and $I^2t$ rating.
8. For an SCR, the gate cathode characteristics is given by a straight line with a gradient of 14 volts per amp passing through the origin, the maximum turn on time is 5 $\mu$s and the minimum gate current required to obtain this quick turn on is 600 mA. If the gate source voltage is 12 V:
    (a) Calculate the resistance to be connected in series with the SCR gate.
    (b) Compute the gate power dissipation, given that the pulse width is equal to the turn on time and the average gate power dissipation

is 0.35 W. Also, compute the maximum triggering frequency that will be possible when pulse firing is used.

9. The capacitance of reverse-biased junction $J_2$ in a thyristor is $C_{j2} = 40$ pF and can be assumed to be independent of the OFF state voltage. The limiting value of the charging current to turn on the thyristor is 16 mA. Determine the critical value of $dv/dt$.

10. Design the triggering circuit for an UJT with the following specifications:
$V_s = 30$ V, $\eta = 0.51$, $I_P = 10\ \mu$A, $V_u = 3.5$ V and $I_V = 10$ mA. The frequency is $f = 60$ Hz and the width of the triggering pulse is $t_g = 50\ \mu$s.

11. In a thyristor circuit, as shown in the given figure, the latching current is 20 mA. The duration of the firing pulse is 100 $\mu$s. Will the thyristor get fired?

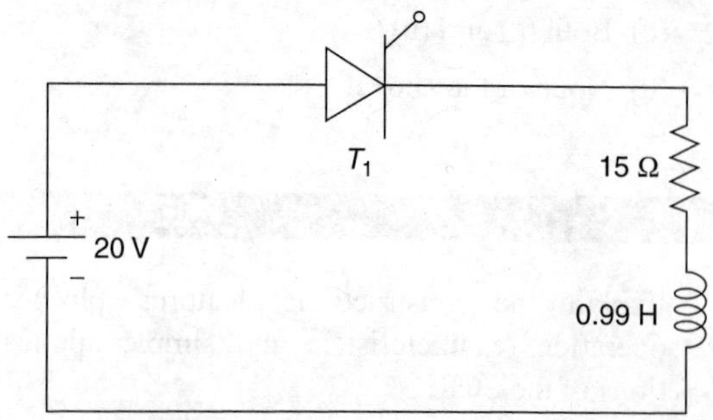


## SUGGESTED READINGS

1. Singh, J. 1994. *Semiconductor Devices: An Introduction*. New York, NY: McGraw-Hill.
2. Streetman, B.G. and S. Banerjee. 2000. *Solid State Electronic Devices*. 5th Ed. New Delhi: Pearson Education.
3. Millman, Jacob and Christos C. Halkias. 1986. *Integrated Electronics: Analog and Digital Circuits and Systems*. New Delhi: McGraw Hill Book Company.
4. Pierret, R. F and G. W. Neudeck. 1989. *Modular Series on Solid State Devices*. Boston, M.A.: Addison Wesley.
5. Singh, B. P. and Rekha Singh. 2006. *Electronic Devices and Integrated Circuits*. New Delhi: Pearson Education.
6. Boylsted, R. L and Louis Nashelsky. 2007. *Electronic Devices and Circuit Theory*. 9th Ed. New Delhi: Pearson Education.

# 9

# Feedback Amplifier

## Outline



## Objectives

In this chapter we will address the concept of feedback. Feedback is the fundamental concept in the design of a stable amplifier and an unstable oscillator circuit. Beginning with the conceptual development of feedback through block diagrams, this chapter explains both negative and positive feedback, and their effects on different circuit parameters. Calculations of open-loop gain and closed-loop gain have been done in detail, followed by a discussion on the effects of feedback on gain, input and output impedances. An overview of the practical implementation of feedback topologies, and the sensitivity and bandwidth stability of the feedback amplifier has also been provided. The chapter ends with an examination of the effects of positive feedback with emphasis on the Nyquist and Barkhausen criteria.

## 9-1 INTRODUCTION

Feedback is one of the fundamental processes in electronics. It is defined as the process whereby a portion of the output signal is fed to the input signal in order to form a part of the system-output control. This action tends to make the system self-regulating. Feedback is used to make the operating point of a transistor insensitive to both manufacturing variations in $\beta$ as well as temperature.

The feedback system has many advantages especially in the control of impedance levels, bandwidth improvement, and in rendering the circuit performance relatively insensitive to manufacturing as well as to environmental changes. This is of principal importance in modern electronics because a controlled and precise circuit performance can be obtained without resorting to costly precision components.These are the advantages of negative or degenerative feedback in which the signal feedback from output to input is 180° out of phase with the applied excitation. It increases bandwidth and input impedance, and lowers the output impedance.

There is another type of feedback called positive or regenerative feedback in which the overall gain of the amplifier is increased. Positive feedback is useful in oscillators and while establishing the two stable states of flip-flop.

## 9-2 CONCEPTUAL DEVELOPMENT THROUGH BLOCK DIAGRAMS

The block diagram of a basic feedback amplifier is shown in Fig. 9-1. It consists of five basic elements. These are input and output signals, the measure of the output, comparison between input and the sampled feedback, which gives rise to four possible types of feedback circuits, and the processing of the compared signal by the basic amplifier.

### 9-2-1 Input Signal

The block in Fig. 9-1 represents the signal to be amplified. The signal source is modeled either by a voltage source $V_s$ in series with a resistance $R_s$, or by a current source $I_s$ in parallel with a resistance $R_s$.

### 9-2-2 Output Signal

The output can either be the voltage across the load resistance or the current through it. It is the output signal that is desired to be independent of the load and insensitive to parameter variations in the basic amplifier.

### 9-2-3 Sampling Network

The function of the sampling network is to provide a measure of the output signal, i.e., a signal that is proportional to the output. Two sampling networks are shown in Fig. 9-2. In Fig. 9-2(a) the output voltage is sampled by connecting the output port to the feedback network in parallel with the load. This configuration is called shunt connection. In Fig. 9-2(b) the output current is sampled and the output port of the feedback network is connected in series with the load. This is a series connection.

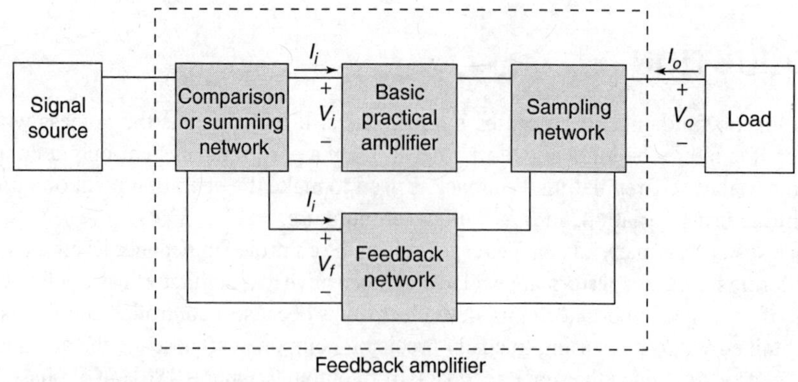


**Figure 9-1** Block diagram of a basic feedback amplifier

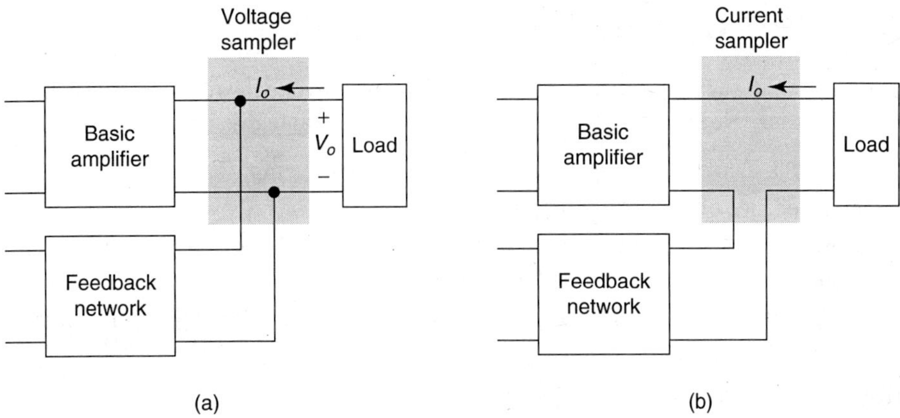


**Figure 9-2** Feedback connections at the output of a basic amplifier (a) the measure of the output voltage (b) the measure of the output current

## 9-2-4 Comparison or Summing Network

The two very common networks used for the summing of input and feedback signals are displayed in Fig. 9-3.

The circuit shown in Fig. 9-3(a) is a series connection and it is used to compare the signal voltage $V_s$ and feedback signal $V_f$. The amplifier input signal $V_i$ is proportional to the voltage difference $V_s - V_f$ that results from the comparison. A differential amplifier is used for comparison as its output voltage is proportional to the difference between the signals at the two inputs. A shunt connection is shown in Fig. 9-3(b) in which the source current $I_s$ and feedback current $I_f$ are compared. The amplifier input current $I_i$ is proportional to the difference $I_s - I_f$.

## 9-2-5 Basic Amplifier

The basic amplifier is one of the important parts of the feedback amplifier. The circuit amplifies the difference signal that results from comparison and this process is responsible for de-sensitivity and control of the output in a feedback system.

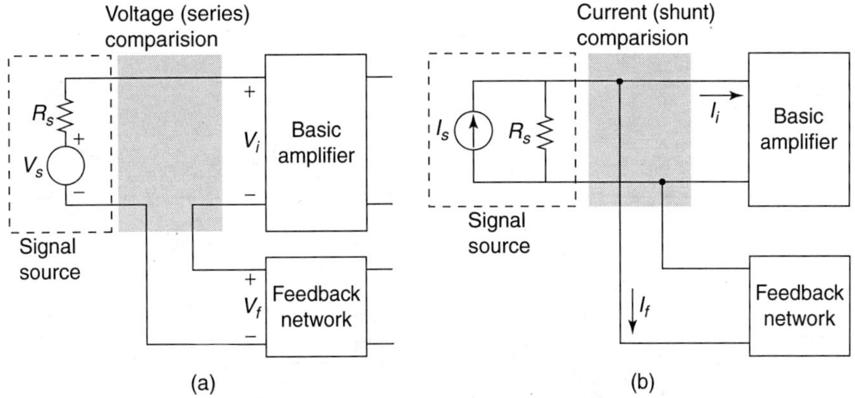


**Figure 9-3** Feedback connections at the input of a basic amplifier (a) voltage summing (series comparison) (b) current summing (shunt comparison)

## 9-3 PROPERTIES OF NEGATIVE FEEDBACK

A comparative study of the advantages and disadvantages of negative feedback illustrates the basic properties of negative feedback. Negative feedback has the following advantages:

(i) Negative feedback increases the input impedance of the voltage amplifier.

(ii) The output impedance of the voltage amplifier can be further lowered by negative feedback.

(iii) The transfer gain $A_f$ of the amplifier with a feedback can be stabilized against the variations of $h$ or hybrid parameters of the transistors, or the parameters of the other active devices used in the amplifier.

(iv) Negative feedback increases the frequency response and the bandwidth of the amplifier.

(v) Negative feedback increases the linear range of operation of the amplifier.

(vi) Negative feedback causes reduction in noise.

(vii) Phase distortion is reduced.

However, the gain $A_f$ of the amplifier with negative feedback is lower compared to an amplifier without a feedback. This is the only disadvantage of a negative feedback system.

Negative feedback has a lot of advantages which nullify its disadvantages. Negative feedback is used in amplifiers with the objective of obtaining stability of operation. It is important to design amplifiers with negative feedback correctly because inaccuracy in design can cause undesired oscillation.

## 9-4 CALCULATIONS OF OPEN-LOOP GAIN, CLOSED-LOOP GAIN AND FEEDBACK FACTORS

The general block diagram of an ideal feedback amplifier indicating basic amplifier, feedback network, external load and corresponding signals is shown in Fig. 9-4.

The ideal feedback amplifier can have any of the four configurations as listed in Table 9-1.

The input signal $X_s$, the output signal $X_o$, the feedback signal $X_f$ and the difference signal $X_i$ each represent either a voltage or a current. These signals and the transfer ratios $A$ and $\beta$ are summarized in Table 9-1 for different feedback topologies. The symbol indicated by the circle with the summation sign $\Sigma$ enclosed within (see Fig. 9-4), represents the summing network whose output is the algebraic sum of inputs.

Thus, for a positive feedback, we get:

$$X_i = X_s + X_f \qquad (9\text{-}1)$$

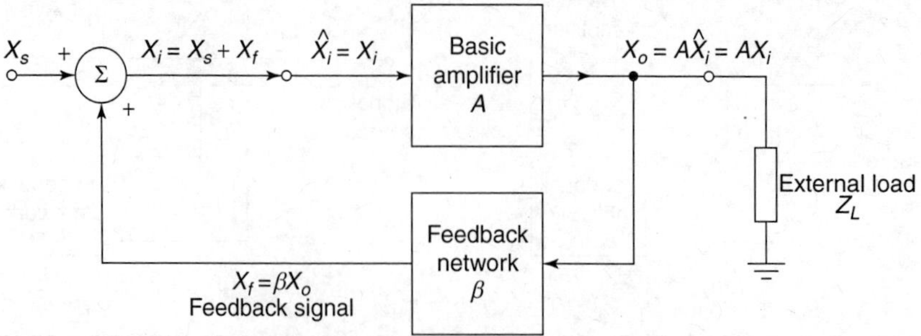


**Figure 9-4** Block diagram of ideal feedback amplifier

**Table 9-1** Signals and transfer ratios in feedback amplifiers

| | *Feedback Topology* | | | |
|---|---|---|---|---|
| *Signals* | *Series-Shunt (Voltage-Series)* | *Series-Series (Current-Series)* | *Shunt-Series (Current-Shunt)* | *Shunt-Shunt (Voltage-Shunt)* |
| $X_o$ | Voltage | Current | Current | Voltage |
| $X_s$, $X_i$, $X_f$ | Voltage | Voltage | Current | Current |
| *Ratio or Gain* | | | | |
| $A$ | $V_o/V_i$ | $I_o/V_i$ | $I_o/I_i$ | $V_o/I_i$ |
| $\beta$ | $V_f/V_o$ | $V_f/I_o$ | $I_f/I_o$ | $I_f/V_o$ |
| $A_f$ | $V_o/V_s$ | $I_o/V_s$ | $I_o/I_s$ | $V_o/I_s$ |

The signal $X_i$, representing the output of the summing network is the amplifier input $X_i$. If the feedback signal $X_f$ is 180° out of phase with the input $X_s$—as is true in negative feedback systems—then $X_i$ is a difference signal. Therefore, $X_i$ decreases as $|X_f|$ increases. The reverse transmission of the feedback network $\beta$ is defined by:

$$\beta = \frac{X_f}{X_o} \tag{9-2}$$

The transfer function $\beta$ is a real number, but in general it is a function of frequency. The gain of the basic amplifier $A$ is defined as:

$$A = \frac{X_o}{X_i} \tag{9-3}$$

Now, from Eq. (9-1), we get: $X_i = X_s + X_f$

Substituting the value of $X_f$ from Eq. (9-2) as $X_f = \beta X_o$ in Eq. (9-1), we get:

$$X_i = X_s + X_f = X_s + \beta X_o \tag{9-3a}$$

From Eq. (9-3) we get:

$$X_o = AX_i \tag{9-3b}$$

Substituting the value of $X_i$ from Eq. (9-3a), we get:

$$X_o = AX_i = A(X_s + \beta X_o) = AX_s + A\beta X_o$$

or,

$$X_o(1 - A\beta) = AX_s$$

or,

$$\frac{X_o}{X_s} = \frac{A}{1 - A\beta} \tag{9-3c}$$

The feedback gain $A_f$ is obtained from Eq. (9-3c) as:

$$A_f = \frac{X_o}{X_s} = \frac{A}{1 - A\beta} \tag{9-4}$$

The gain in Eq. (9-3) represents the transfer function without feedback. If $\beta=0$; eliminating the feedback signal, no feedback exists and Eq. (9-4) reduces to Eq. (9-3). Frequently $A$ is referred to as the open-loop gain and is designated by $A_{OL}$. When $\beta \neq 0$, a feedback loop exists and $A_F$ is often called the closed-loop gain.

If $|A_f| < |A|$, the feedback is negative. If $|A_f| > |A|$, the feedback is positive. In case of a negative feedback, $|1 - A\beta| > 1$. *The reason behind this being the Barkhausen criterion.*

Similarly for negative feedback:

$$X_i = X_s - X_f$$

Calculating in the same manner as done for positive feedback, we can represent the feedback gain as:

$$A_f = \frac{A}{1 + A\beta} \tag{9-4a}$$

For negative feedback, $|1 + A\beta| > 1$. So, $A_f < A$, i.e., $A_f$ decreases. Therefore, the general equation of feedback can be written as:

$$A_f = \frac{A}{1 \pm A\beta}$$

### 9-4-1 Loop Gain or Return Ratio

The signal $\hat{X}_i$ in Fig. 9-4 is multiplied by gain $A$ when passing through the amplifier and by $\beta$ in transmission through the feedback network. Such a path takes us from the amplifier input around the loop consisting of the amplifier and the feedback network. The product, $A\beta$, is called the loop gain or return ratio $T$. Equation (9-4) can be written in terms of $A_{OL}$ and $T$ as:

$$A_F = \frac{A}{1 - A\beta} = \frac{A_{OL}}{1 + T} \tag{9-5a}$$

For negative feedback, $-A\beta = T > 0$.

We can give a physical interpretation for the return ratio by considering the input signal $X_s = 0$, and keeping the path between $X_i$ and $\hat{X}_i$ open. If a signal $\hat{X}_i$ is now applied to the amplifier input, then $X_i = X_f = A\beta$.

$$T = -A\beta = -\left.\frac{X_i}{\hat{X}_i}\right|_{X_S = 0} \tag{9-5b}$$

The return ratio is then the negative of the ratio of the feedback signal to the amplifier input. Often the quantity $F = 1-A\beta=1+T$ is referred to as the return difference. If negative feedback is considered then both $F$ and $T$ are greater than zero.

## 9-5 TOPOLOGIES OF THE FEEDBACK AMPLIFIER

There are four basic amplifier types. Each of these is being approximated by the characteristics of an ideal controlled source. The four feedback topologies are as follows:

1. Series-shunt feedback
2. Series-series feedback
3. Shunt-series feedback
4. Shunt-shunt feedback

These designations correspond to the output- and input-port connections of the feedback network with the basic amplifier. For example, in the shunt-series amplifier, the input port of the feedback network is connected in series with the output port of the amplifier. An alternative nomenclature is based on the quantity sampled

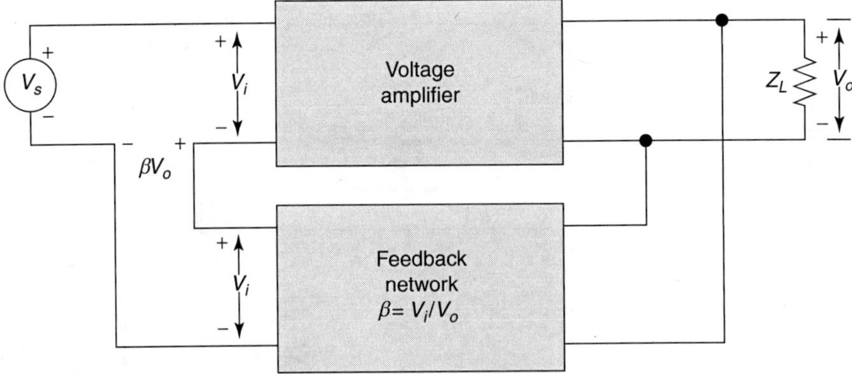


**Figure 9-5** Voltage amplifiers with voltage-series feedback

and input connection used. Thus, a current-shunt topology corresponds to the shunt-series connection. The alternative nomenclature used is as follows:

1. Voltage-series or series-shunt feedback
2. Current-series or series-series feedback
3. Current-shunt or shunt-series feedback
4. Voltage-shunt or shunt-shunt feedback

## 9-5-1 Voltage-Series or Series-Shunt Feedback

The configuration of voltage-series or series-shunt feedback circuit is illustrated in Fig. 9-5.

The input voltage $V_i$ of the basic amplifier is the algebraic sum of input signal $V_s$ and the feedback signal $\beta V_o$, where $V_o$ is the output voltage.

## 9-5-2 Current-Series or Series-Series Feedback

The current-series or series-series feedback topology is illustrated in Fig. 9-6.

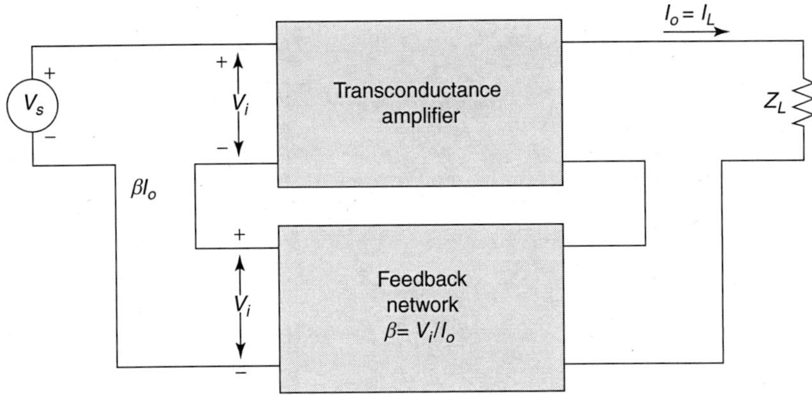


**Figure 9-6** Transconductance amplifier with current-series feedback

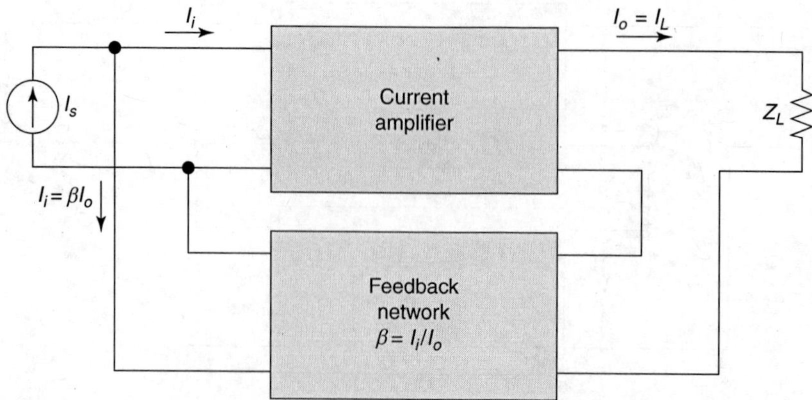


**Figure 9-7** Current amplifiers with current-shunt feedback

As mentioned in Table 9-1, the transconductance feedback amplifier provides an output current $I_o$ which is proportional to the input voltage $V_s$. The feedback signal is the voltage $V_f$, which is added to $V_s$ at the input of the basic amplifier.

## 9-5-3 Current-Shunt or Shunt-Series Feedback

The current-shunt or shunt-series feedback amplifier, as shown in Fig. 9-7, supplies an output current $I_o$ which is proportional to the input current $I_i$. This makes it a current amplifier.

The feedback signal is the current $I_f$. The input current of the basic amplifier is $I_i = I_s + I_f$ and the output current is $I_o = I_L$.

## 9-5-4 Voltage-Shunt or Shunt-Shunt Feedback

The voltage-shunt or shunt-shunt feedback amplifier is illustrated in Fig. 9-8.

This provides an output voltage $V_o$ in proportion to the input current $I_s$. The input current $I_i$ of the basic amplifier is the algebraic sum of $I_s$ and the feedback current $I_f$.

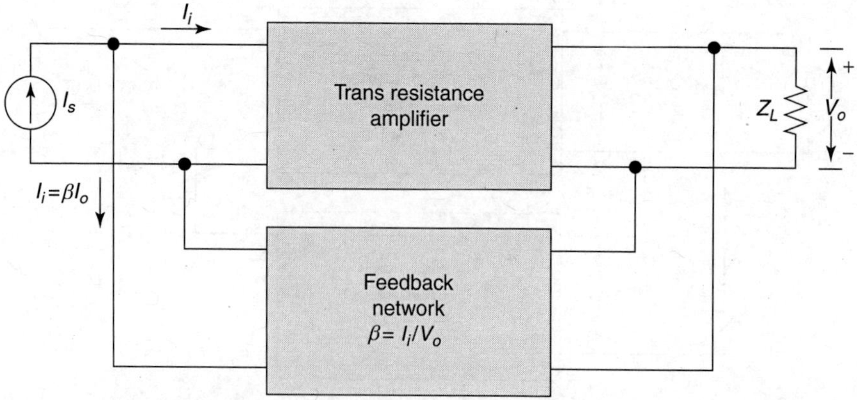


**Figure 9-8** Trans resistance amplifier with voltage-shunt feedback

# 9-6 EFFECT OF FEEDBACK ON GAIN, INPUT AND OUTPUT IMPEDANCES

Feedback is applied with the objective of improving the performance of an amplifier. The operation of an amplifier is regulated by controlling the gain and impedance. The effect of feedback on gain and impedance for the different topologies—voltage-series, current-series, current-shunt, voltage-shunt—are discussed in the following sections.

## 9-6-1 Effect of Feedback on Input Impedance

### Voltage-series feedback

Fig. 9-9 shows the equivalent Thevenin's model of the voltage-series amplifier of Fig. 9-5. In this circuit $A_v$ represents open-circuit voltage gain taking $Z_S$ into account. From Fig. 9-9 the input impedance with the feedback is:

$$Z_{if} = \frac{V_S}{I_i} \tag{9-6}$$

and,

$$V_S = I_i Z_i + V_f = I_i Z_i + \beta V_o \tag{9-7}$$

Using voltage divider rule, we get:

$$V_o = \frac{A_V V_i Z_L}{Z_o + Z_L} = A_V I_i Z_L \tag{9-8}$$

where,

$$I_i = \frac{V_i}{Z_o + Z_L}$$

Now,

$$V_o = A_V I_i Z_L = A_V V_i$$

or,

$$A_V = \frac{V_o}{I_i}$$

From Fig. 9-9, the input impedance without feedback is:

$$Z_i = \frac{V_i}{I_i} \tag{9-9}$$

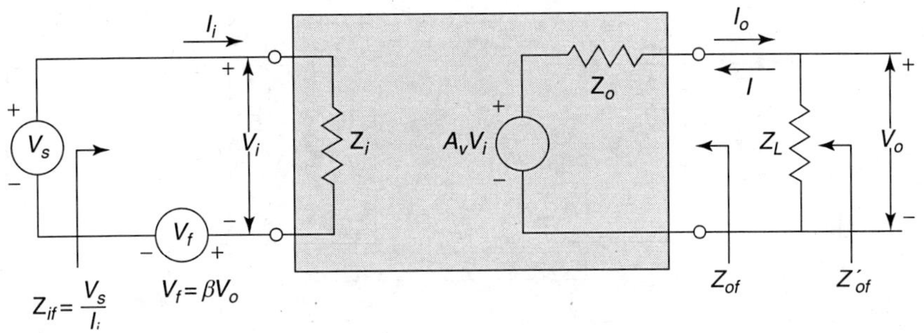


**Figure 9-9** Voltage-series feedback circuit used to calculate input and output resistance

Now, $$Z_{if} = \frac{V_S}{I_i} = \frac{V_i(1 + A_V\beta)}{I_i}$$

From Eqs. (9-6) and (9-7) we have:

$$Z_{if} = \frac{V_S}{I_i} = Z_i(1 + \beta A_V) \tag{9-10}$$

Thus, the input impedance is increased.

Although $A_v$ represents the open-circuit voltage gain without feedback, Eq. (9-6) indicates that $A_V$ is the voltage gain without feedback taking the load $Z_L$ into account.

### Current-series feedback

In a similar manner as for voltage series, for current series feedback as shown in Fig. 9-6, we obtain:

$$Z_{if} = Z_i(1 + \beta Y_M) \tag{9-11a}$$

where, $Y_M$ is the short-circuit transadmittance without feedback considering the load impedance, and is given by:

$$Y_M = \frac{I_o}{V_i} = \frac{Y_m Z_o}{Z_o + Z_L} \tag{9-11b}$$

where, $Y_m$ is the short-circuit transadmittance without feedback.

From Eq. (9-11a) it is clear that for series mixing $Z_{if} > Z_i$.

### Current-shunt feedback

Figure 9-10 shows the current-shunt feedback in which the amplifier is replaced by its Norton equivalent circuit. If $A_i$ is the short-circuit current gain then from Fig. 9-10:

$$I_s = I_i + I_f = I_i + \beta I_0 \tag{9-12}$$

and, $$I_o = \frac{A_i I_i}{Z_o + Z_L} = A_I I_i \tag{9-13}$$

where, $$A_i = \frac{I_o}{I_i} = \frac{A_i Z_o}{Z_o + Z_L} \tag{9-14}$$

From Eqs. (9-12) and (9-13) we have:

$$I_S = I_i(1 + \beta A_I) \tag{9-15}$$

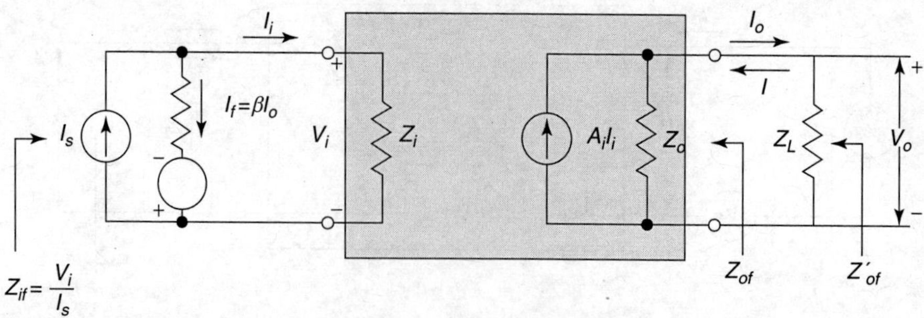


**Figure 9-10** Current-shunt feedback circuit used to calculate input and output resistance

From Fig. 9-10: $Z_{if} = \frac{V_i}{I_S}$ and $Z_i = \frac{V_i}{I_i}$

Using Eq. (9-15) we obtain:

$$Z_{if} = \frac{V_i}{I_i(1 + \beta A_I)} = \frac{Z_i}{1 + \beta A_I} \tag{9-16}$$

where, $A_i$ represents the short-circuit current gain.

### Voltage-shunt feedback

For voltage-shunt feedback, proceeding in a similar way as we have done in the previous sections, we obtain:

$$Z_{if} = \frac{Z_i}{1 + \beta Z_M} \tag{9-17a}$$

where, $Z_M$ is the transimpedance without feedback considering the load, and is given by:

$$Z_M = \frac{V_o}{I_i} = \frac{Z_m Z_L}{Z_o + Z_L} \tag{9-17b}$$

where, $Z_m$ is the open-circuit transimpedance without feedback.

From Eq. (9-17 b) it is clear that for shunt comparison $Z_{if} < Z_i$.

## 9-6-2 Effect of Feedback on Output Impedance

### Voltage-series feedback

To find the output resistance with feedback $Z_{of}$—looking into output terminals with $Z_L$ disconnected—external signals must be removed ($V_s = 0$ or $I_s = 0$). Let $Z_L = \infty$ impress a voltage $V$ across the output terminals which delivers current $I$.

Therefore:

$$Z_{of} = \frac{V}{L} \tag{9-18}$$

Replacing $V_0$ by $V$ in Fig. 9-10 we get:

$$I = \frac{V - A_V V_i}{Z_o} = \frac{V + \beta A_V V}{Z_o} \tag{9-19}$$

with, $V_s = 0,\ V_i = -V_f = -\beta V$

Hence:

$$Z_{of} = \frac{V}{I} = \frac{Z_o}{1 + \beta A_V} \tag{9-20}$$

The output resistance with feedback $Z'_{of}$, which includes $Z_L$, is given by $Z_{of}$ in parallel with $Z_L$. So:

$$Z'_{of} = \frac{Z_{of} Z_L}{Z_{of} + Z_L} = \frac{Z_o Z_L}{1 + \beta A_V} \frac{1}{Z_o|(1 + \beta A_V) + Z_L}$$

$$= \frac{Z_o Z_L}{Z_o + Z_L + \beta A_V Z_L} = \frac{Z_o Z_L / Z_o + Z_L}{1 + \beta A_V Z_L / (Z_o + Z_L)} \tag{9-21}$$

It should be noted that $Z'_o = Z_0 \| Z_L$ is the output resistance without feedback.

Using Eq. (9-9) in Eq. (9-21) we obtain:

$$Z'_{of} = \frac{Z'_o}{1 + \beta A_V} \tag{9-22}$$

**Voltage-shunt feedback**

Proceeding in the similar manner, we have:

$$Z_{of} = \frac{Z_o}{1 + \beta Z_m} \tag{9-23}$$

and,

$$Z'_f = \frac{Z'_0}{1 + \beta Z_M} \tag{9-24}$$

For voltage sampling it is clear that $Z_{of} < Z_o$.

**Current-shunt feedback**

In Fig. 9-10, replacing $V_0$ by $V$, we have:

$$I = \frac{V}{Z_o} - A_i I_i \tag{9-25}$$

with,

$$I_s = 0, I_i = -I_f = -\beta I_0 = \beta I.$$

hence,

$$I = \frac{V}{Z_o} - \beta A_i I_i$$

or,

$$I(1 + \beta A_i) = \frac{V}{Z_o} \tag{9-26}$$

$$Z_{of} = \frac{V}{I} = Z_o(1 + \beta A_i) \tag{9-27}$$

Eq. (9-27) is the expression for the output impedance with feedback, and without load resistance $R_L$. To find $R'_{of}$:

$$Z'_{of} = \frac{Z_{of} Z_L}{Z_{of} + Z_L} = \frac{Z_o(1 + \beta A_i)Z_L}{Z_o(1 + \beta A_i) + Z_L}$$

$$= \frac{Z_o Z_L}{Z_o + Z_L} \frac{1 + \beta A_i}{1 + \beta A_i Z_o/(Z_o + Z_L)} \tag{9-28}$$

Using Eq. (9-27), and with $Z'_o = Z_o \| Z_L$, we have:

$$Z'_{of} = Z'_o \frac{1 + \beta A_i}{1 + \beta A_i} \tag{9-29}$$

**Current-series feedback**

Proceeding in the similar manner we have:

$$Z_{of} = Z_o(1 + \beta Y_m) \quad \text{(9-30a)}$$

and,

$$Z'_{of} = Z'_o \frac{1 + \beta Y_m}{1 + \beta Y_m} \quad \text{(9-30b)}$$

From Eqs. (9-30a) and (9-30b) we see that for current sampling $Z_{of} > Z_0$.

## 9-7 PRACTICAL IMPLEMENTATIONS OF THE FEEDBACK TOPOLOGIES

The feedback topologies discussed so far have a variety of practical implementations. We will now proceed to discuss the applications of these different topologies with examples of transistors with different configurations.

### 9-7-1 Voltage-Series Feedback Using Transistor

The emitter-follower circuit, as shown in Fig. 9-11, is an example of voltage-series feedback.

The feedback signal $V_f$ and the output signal $V_o$ are both voltage quantities. According to Table 9-1, this is a voltage-series feedback circuit. To determine the gain of the basic amplifier without feedback we should consider $V_o = 0$ for the input loop and $I_b = 0$ for the output loop so that we obtain the approximate hybrid equivalent circuit, as given in Fig. 9-12.

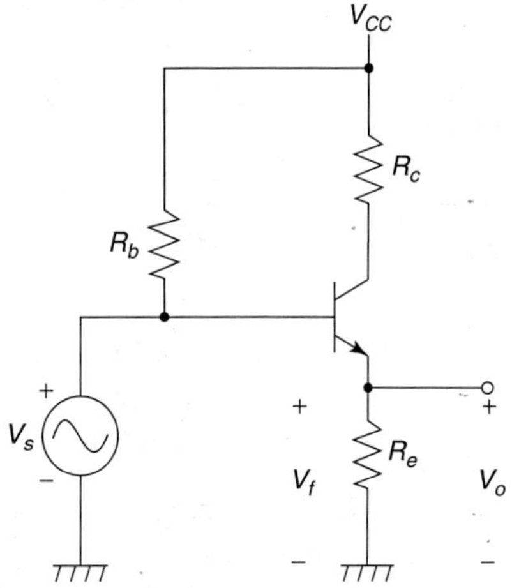


**Figure 9-11** Voltage-series feedback circuit

$$A_v = \frac{V_o}{V_s} = \frac{h_{fe} I_b Z_e}{V_s} = \frac{h_{fe} Z_e}{h_{ie}} \quad \text{(9-31a)}$$

and,

$$\beta = \frac{V_f}{V_o} = 1 \quad \text{(9-31b)}$$

Using Eqs. (9-9), (9-10), and (9-22) we can calculate $A_V$, $Z_{if}$, and $Z'_f$.

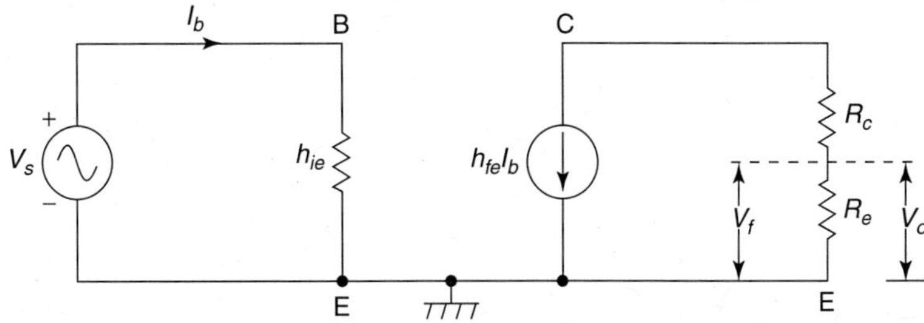


**Figure 9-12** Approximate hybrid equivalent circuit of practical voltage-series feedback amplifier

## 9-7-2 Current-Series Feedback Using Transistor

An example of current-series feedback is given in Fig. 9-13.

In the circuit shown in Fig. 9-13, the output signal is the load current $I_o$ and the feedback signal is the voltage $V_f$ across $R_e$. Referring to Table 9-1 it can be concluded that this is a current-series feedback circuit. To obtain the transadmittance of the basic amplifier we consider $V_f = 0$. The approximate $h$-parameter equivalent is illustrated in Fig. 9-14.

According to Fig. 9-14, the transfer gain of the basic amplifier is:

$$Y_m = \frac{I_o}{V_s} = \frac{-I_b h_{fe}}{I_b(h_{ie} + R_e)} = \frac{-h_{fe}}{h_{ie} + R_e} \qquad (9\text{-}31c)$$

and,

$$\beta = \frac{V_f}{I_o} = \frac{-I_o R_e}{I_o} \qquad (9\text{-}32d)$$

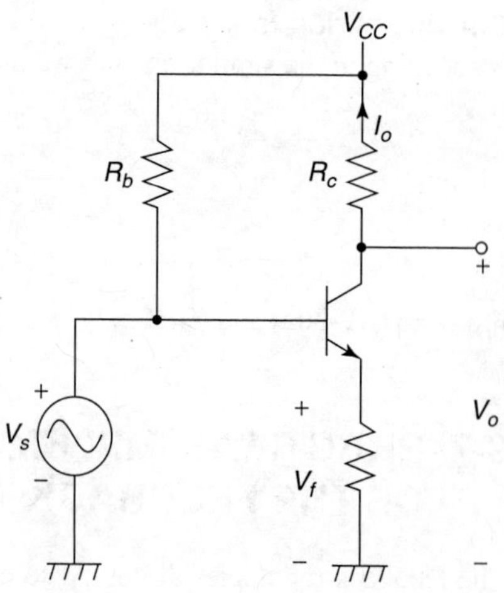


**Figure 9-13** Current-series feedback circuit using transistor

Hence, from Eqs. (9-11a), (9-11b) and (9-31) the expressions of $Y_M$, $Z_{if}$ and $Z'_{of}$ can be obtained.

## 9-7-3 Voltage-Shunt Feedback Using Transistor

The circuit diagram of a voltage-shunt feedback topology is given in Fig. 9-15.

In the circuit given in Fig. 9-15, the input current is proportional to the output voltage $V_o$. To determine the gain of the basic amplifier we consider that $R_f$ is open-circuited and we can draw the approximate $h$-parameter equivalent circuit as shown in Fig. 9-16.

From Fig. 9-16 we can write:

$$I_b = I_s \times \frac{R_s}{R_s + h_{ie}}$$

or,

$$\frac{I_b}{I_s} = \frac{R_s}{R_s + h_{ie}}$$

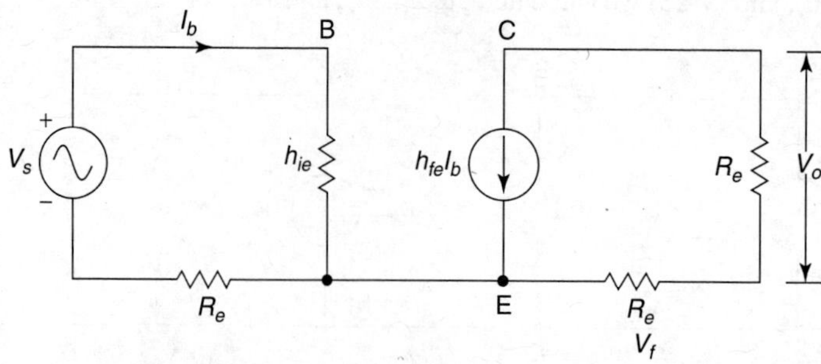


**Figure 9-14** Simplified h-parameter circuit of the current-series feedback amplifier

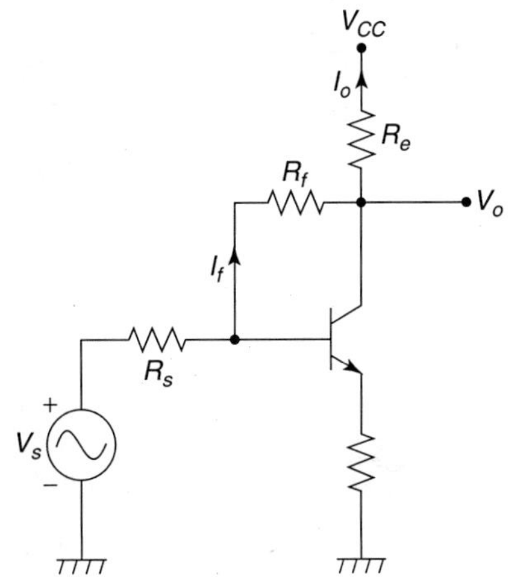


**Figure 9-15** Implementation of voltage-shunt feedback

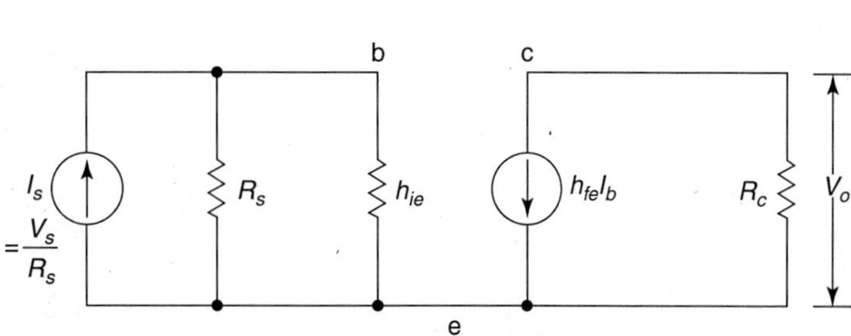


**Figure 9-16** Approximate *h*-parameter equivalent circuit for voltage-shunt feedback circuit

The gain of the amplifier is given by:

$$Z_m = \frac{V_o}{I_s} = \frac{-h_{fe} I_b R_c}{I_s} = \frac{-h_{fe} R_c R_s}{R_s + h_{ie}} \tag{9-31e}$$

Considering the resistance $R_f$ in Fig. 9-16 we obtain:

$$I_f = \frac{h_{fe} I_b R_c}{h_{ie} + R_c + R_f} \tag{9-31f}$$

From Eq. (9-31f)

$$\beta = \frac{I_f}{V_o}$$

$$\beta = \frac{h_{fe} I_b R_c}{h_{ie} + R_c + R_f} \times \frac{-1}{h_{fe} I_b R_c} = \frac{-1}{h_{ie} + R_c + R_f} \tag{9-31g}$$

From Eqs. (9-31e) and (9-31g), and using Eqs. 9-17(a), 9-17(b) and 9-24, we can obtain $Z_M$, $Z_{if}$, $Z'_{of}$.

## 9-7-4 Current-Shunt Feedback Using Transistor

A simple current-shunt feedback amplifier is shown in Fig. 9-17.

Without feedback $I_s = I_e$, we can draw the approximate *h*-parameter circuit as shown in Fig. 9-18.

From Fig. 9-18 we have:

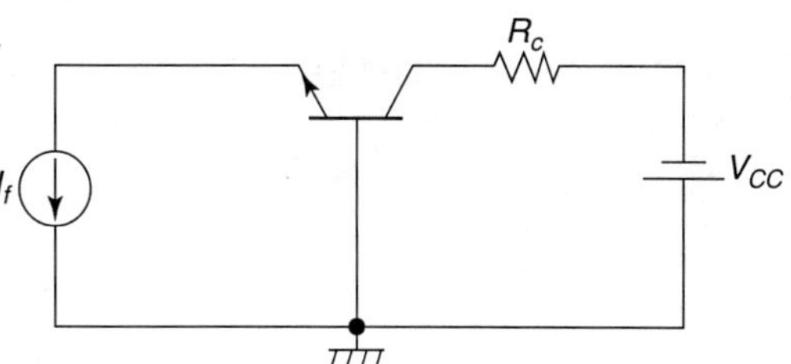


**Figure 9-17** Current-shunt feedback circuit

$$I_s = -(h_{fe} + 1)I_b \tag{9-31h}$$

and

$$I_o = -h_{fe} i_b \tag{9-31i}$$

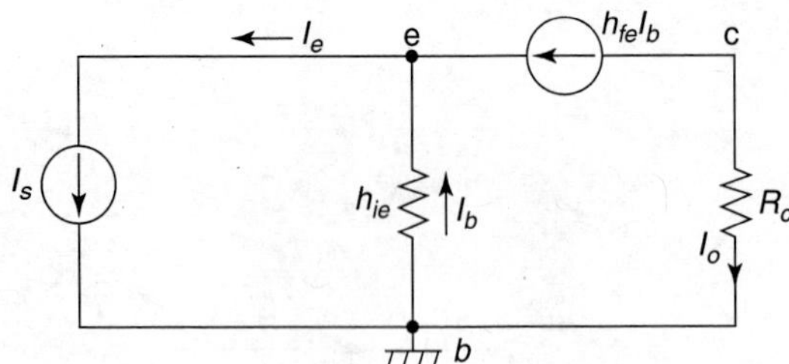


**Figure 9-18** Approximate *h*-parameter circuit of current-shunt feedback circuit

Substituting the values of $I_s$ and $I_o$ from Eqs. (9-31h) and (9-31i) in Eq. 9-14, we get:

$$A_I = \frac{I_o}{I_s} = \frac{h_{fe}}{1 + h_{fe}} = \frac{A_I Z_0}{Z_0 + Z_L} \tag{9-31j}$$

Hence, we can write $A_i = h_{fe}$ and $\beta = 1$. The values of $Z_{if}$ and $Z'_{if}$ can be determined from Eqs. (9-16) and (9-29) using the value of $A_i$ and $\beta$.

## 9-8 SENSITIVITY

The sensitivity of transfer gain of the feedback amplifier $A_F$ with respect to the variations in the internal amplifier gain $A$ is defined as the ratio of the fractional change in gain with the feedback to the fractional change in gain without the feedback. The gain sensitivity $S$ of the feedback amplifier is given by:

$$S = \frac{\dfrac{dA_f}{A_f}}{\dfrac{dA}{A}} \tag{9-32}$$

where, $dA_f/A_f$ = fractional change in gain with the feedback; $dA/A$ = fractional change in gain without the feedback.

From Eq. (9-4a) we have: $$A_f = \frac{A}{1 + A\beta}$$

Differentiating with respect to $A$: $$\frac{dA_f}{A_f} = \frac{1}{1 + A\beta}\frac{dA}{A}$$

$\therefore$ $$S = \frac{\dfrac{dA_f}{A_f}}{\dfrac{dA}{A}} = \frac{1}{1 + A\beta} \tag{9-33}$$

From Eq. (9-33), we can say that $1/1 + A\beta$ is the sensitivity. The inverse or reciprocal of sensitivity is called de-sensitivity.

$$S = \frac{1}{1 + A\beta} = \frac{1}{D} \tag{9-34}$$

Again from Eq. (9-4) we have:

$$A_f = \frac{A}{1 + A\beta} = \frac{A}{D} \tag{9-35}$$

$$D = \frac{A}{A_f} \tag{9-36}$$

Equation (9-36) shows that de-sensitivity ($D$) indicates the fraction by which the voltage gain has been reduced due to feedback.

## 9-9 BANDWIDTH STABILITY

The transfer gain of an amplifier having the feedback is given by:

$$A_f = \frac{A}{1 + \beta A}$$

If $|\beta A| >> 1$, then:

$$A_f \approx \frac{A}{\beta A} = \frac{1}{\beta} \tag{9-37}$$

From Eq. (9-37) we can directly conclude that the transfer gain can be made dependent entirely on the feedback network $\beta$. However, it is important to consider that while $\beta$ is a constant term, the gain $A$ is not constant and depends on the frequency. This means that at certain high or low frequencies $|\beta A|$ will be much larger than unity. To study the dependence we will consider the single-pole transfer function. The gain $A$ of single-pole transfer function is given by:

$$A = \frac{A_0}{1 + j(f/f_H)} \tag{9-38}$$

$A_0$ is the mid-band gain without the feedback and $f_H$ is the high frequency (where $A_0$ is decreased by 3 dB). The gain $A$ of the single pole amplifier with the feedback is obtained from Eqs. (9-4) and (9-38) as:

$$A_f = \frac{A_0/[1 + j(f/f_H)]}{1 + \beta A_0/[1 + j(f/f_H)]} = \frac{A_0}{1 + \beta A_0 + j(f/f_H)} \tag{9-39}$$

By dividing the numerator and denominator by $1 + \beta A_0$, Eq. (9-39) can be written as:

$$A_f = \frac{A_{0f}}{1 + j(f/f_{Hf})} \tag{9-40}$$

where,

$$A_{0f} \equiv \frac{A_0}{1 + \beta A_0} \tag{9-41}$$

and,

$$f_{Hf} \equiv f_H(1 + \beta A_0) \tag{9-42}$$

It should be noted that $A_{0f}$ is the mid-band gain with the feedback, and $f_{Hf}$ is the high 3 dB frequency with the feedback.

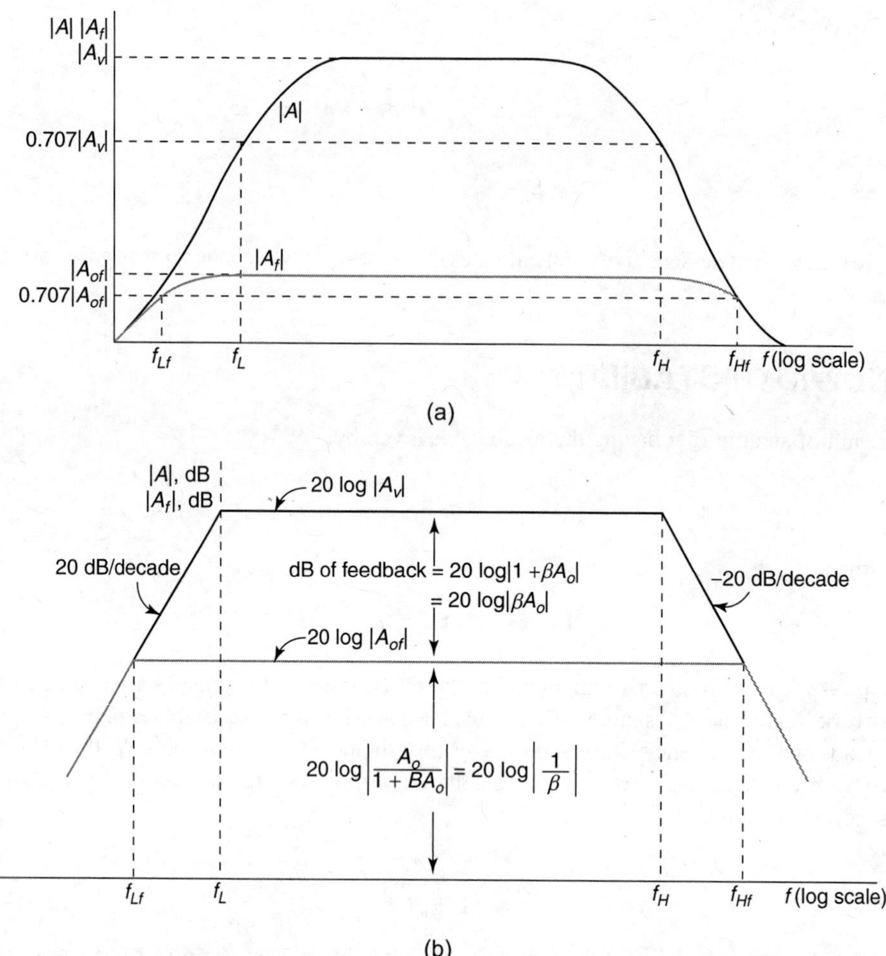


**Figure 9-19** (a) Transfer gain is decreased and bandwidth is increased for an amplifier using negative feedback. (b) Idealized bode plot

From Eq. (9-41), we have:

$$A_{0f} f_{Hf} = A_0 f_H \tag{9-43}$$

Similarly, it can be shown that the low 3 dB frequency with the feedback is given by:

$$f_{Lf} = \frac{f_L}{1 + \beta A_0} \tag{9-44}$$

where, $f_L$ is the low 3 dB frequency without the feedback.

For an video amplifier $f_H >> f_L$; therefore, the bandwidth is $f_H - f_L \approx f_H$. Under these circumstances, Eq. (9-43) may be interpreted to mean that the gain bandwidth product is constant with or without the feedback.

Also, from Eqs. (9-41) and (9-44) it is evident that the bandwidth has improved a lot. This can be shown by plotting the frequency versus gain in the logarithmic scale, as shown in Fig. 9-19.

## 9-10 EFFECT OF POSITIVE FEEDBACK

If $|1 + \beta A| > 1$, it is considered to be negative feedback, and if $|1 + \beta A| < 1$, it is considered to be positive feedback. In second case, the resultant transfer gain $A_f$ will be greater than $A$—the nominal gain without feedback—since $|A_f| = |A|/|1 + \beta A| > |A|$. Positive feedback increases the amplification but at the cost of reduced stability.

To illustrate the instability in an amplifier with positive feedback, we will consider the following situation. No signal is applied, but because of some transient disturbance a signal $X_0$ appears at the output terminals. A portion of this signal $-\beta X_0$ will be the feedback to the input circuit, and will appear in the output as an increased signal $-A\beta X_0$. If this term just equals $X_0$, then the spurious output has regenerated itself. In other words, if $-A\beta X_0 = X_0 \Rightarrow -A\beta = 1$, the amplifier will oscillate. If an attempt is made to obtain a large gain by making $|\beta A|$ almost equal to unity, there is a possibility that the amplifier may break into spontaneous oscillation. This situation may be created by the processes like variation in supply voltages, ageing of transistors, etc.

### 9-10-1 Instability and Oscillation

If an amplifier is designed to have negative feedback in a particular frequency range but breaks into oscillation at some high or low frequency, it is useless as an amplifier. While designing the amplifier, it must be ensured that the circuit is stable at all frequencies and not merely over the frequency range of interest.

The stability of a circuit lies in the pole of the transfer function of the circuit, which also determines the transient response of the circuit. A pole existing with a positive real part will result in a signal disturbance increasing with time. So the condition to be satisfied, if a system is to be stable, is that the poles of the transfer function must all lie in the left-hand half of the complex-frequency plane.

### 9-10-2 Nyquist Criterion

Harry Nyquist obtained a condition for stability, which may be expressed in terms of the steady-state or frequency-response characteristics. The loop gain factor $A\beta$ is a complex number; it may be represented as a point in the complex plane, the real component being plotted along the $x$-axis and the $j$ component along $y$-axis. Also $A\beta$ is the function of the frequency. Consequently, points in the complex plane are obtained for the values of $A\beta$ corresponding to all values of $f$ from $-\infty$ to $+\infty$. The locus of all such points forms a closed curve. The Nyquist criterion states that the amplifier is unstable if this curve encloses $-1 + j0$, and the amplifier is stable if the curve does not enclose this point.

The criterion for positive or negative feedback is represented in the complex plane. Figure 9-20 shows that $|1 + A\beta| = 1$ represents a circle with radius equal to unity and centre at the point $-1 + j0$. For any frequency, $A\beta$ extends outside this circle and the feedback is negative; therefore $|1 + A\beta| > 1$. If the feedback is positive and the $A\beta$ value falls within the circle, then $|1 + A\beta| < 1$. Also, when the feedback is positive the system will not oscillate unless the Nyquist criterion is satisfied.

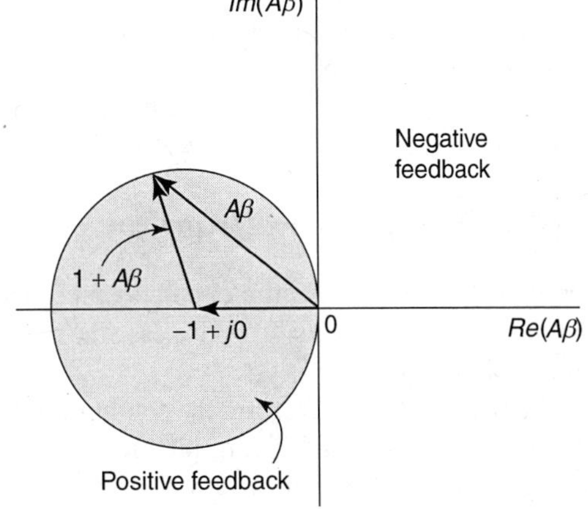


**Figure 9-20** The locus of $|1 + A\beta| = 1$ is a circle of unit radius with centre at $-1 + j0$ (when the vector $A\beta$ ends in the shaded region, the feedback is positive)

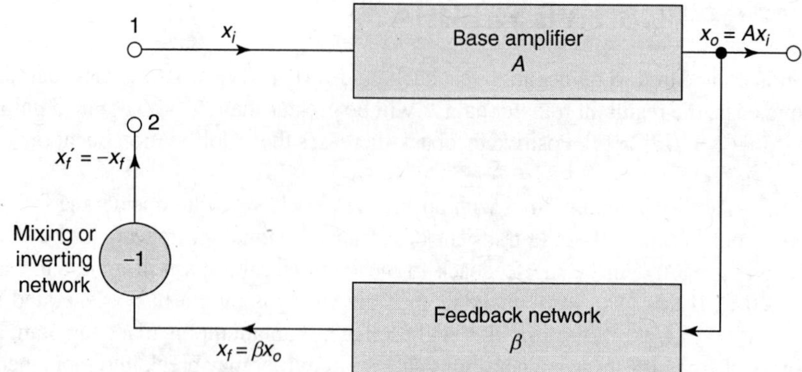


**Figure 9-21** An amplifier with feedback gain *A* and feedback network *β* not yet connected to the network

## 9-10-3 Condition of Oscillation

To investigate the oscillation of the circuit consider Fig. 9-21. It shows an amplifier, a feedback network and an input mixing circuit *not connected to form a closed loop*. The amplifier provides an output signal $x_0$ as a consequence of the signal $x_i$ applied directly to the amplifier input terminal. The output of feedback network is $x_f = \beta x_0 = A\beta x_i$ and the output of the mixing circuit is:

$$x'_f = -x_f = -A\beta x_i$$

From Fig. 9-21, loop gain can be written as:

$$\frac{x'_f}{x_i} = \frac{-x_f}{x_i} = -A\beta \tag{9-45}$$

Suppose that the signal $x'_f$ is adjusted such that it equals the externally applied signal $x_i$. Since the amplifier has no means of distinguishing the source of the input signal applied to it, it would appear that if the external source were removed and if terminal 2 were connected to terminal 1 the amplifier would continue to provide the same output signal $x_0$ as before.

The statement, $x'_f = -x_i$, means that the instantaneous values of $x'_f$ and $x_i$ are exactly equal at all times. The condition $x'_f = x_i$ is equivalent to $-A\beta = 1$ or, the loop gain must be unity. This is the condition of oscillation.

## 9-10-4 Barkhausen Criterion

Let us consider that the entire circuit operates linearly and that the feedback or the amplifier network or both contain reactive elements. In such a case the only periodic waveform which will maintain its form is the sinusoidal waveform.

For a sinusoidal waveform the condition $x_i = x'_f$ is equivalent to the condition that the amplitude, phase and frequency of $x_i$ and $x'_f$ are identical. As the phase shift is introduced, any signal that is transmitted through a reactive network is always a function of the frequency. Hence we have the following important principle:

*The frequency at which a sinusoidal oscillator will operate is the frequency for which the total shift introduced, as a signal proceeds from the input terminals, through the amplifier and feedback network, and back again to the input, is precisely zero or an integral multiple of 2π. Stated more simply, the frequency of a sinusoidal oscillator is determined by the condition that the loop gain phase shift is zero.*

It might be noted parenthetically that it is not inconceivable that this condition might be satisfied for more than a single frequency. In such a contingency there is a possibility of simultaneous oscillations at several frequencies or an oscillation at one of the allowed frequencies.

The condition given here determines the frequency, provided that the circuit will oscillate at all. Another condition, which must clearly be met, is that the magnitude of $x_i$ and $x'_f$ must be identical. This condition is then embodied in the following principle:

*Oscillations will not be sustained if, at the oscillator frequency, the magnitude of the product of the transfer gain of the amplifier and the magnitude of the feedback factor of the feedback network (the magnitude of the loop gain) are less than unity.*

*The condition of unity loop gain $A\beta = 1$ is called the Barkhausen criterion.* This condition implies, of course, that $|A\beta| = 1$, and that the phase of $-A\beta$ is zero. The two principles stated previously are consistent with the feedback formula $A_f = A/(1 + \beta A)$. If $-\beta A = 1$ then $A_f \rightarrow \infty$, which may be interpreted to mean that there exists an output voltage even in the absence of an externally applied signal voltage. Therefore, the conditions for Barkhausen criteria for oscillation are as follows:

1. Positive feedback
2. Loop gain is unity, $A\beta = 1$; therefore, the feedback gain is infinite, $A_f = \infty$
3. Phase variation is zero or integral multiple of 360°

## Solved Examples

**Example 9-1** An amplifier has an open-loop gain of 500 and a feedback of 0.1. If open-loop gain changes by 20% due to the temperature find the percentage change in the closed-loop gain.

**Solution:**

Given: $A = 500, \beta = 0.1, \dfrac{dA}{A} = 20.$

Change in closed-loop gain:

$$\frac{dA_f}{A_f} = \frac{dA}{A} \times \frac{1}{1 + \beta A}$$

$$= 20 \times \frac{1}{1 + 500 \times 0.1}$$

$$= 0.3921 = 39.21\%$$

**Example 9-2** An amplifier has a voltage gain of 200. This gain is reduced to 50 when negative feedback is applied. Determine the reverse transmission factor and express the amount of feedback in dB.

**Solution:**

Given: $A = 200, A_f = 50.$

We know that:

$$A_f = \frac{A}{1 + \beta A}$$

$\Rightarrow$

$$50 = \frac{200}{1 + \beta \times 200}$$

$\therefore$

$$\beta = 0.015$$

Feedback in dB:

$$N = 20\log_{10}\left|\frac{A_f}{A}\right| = 20\log_{10}\left(\frac{50}{200}\right) = 20\log_{10}\left(\frac{1}{4}\right)$$

$$= -12.042 \text{ dB}$$

**Example 9-3** An amplifier has a voltage gain of 100. The feedback ratio is 0.05. Find:

(a) The voltage gain with feedback in dB
(b) Feedback factor
(c) The output voltage, if input voltage is 1 volt
(d) The feedback voltage

**Solution:**

(a) Voltage gain is $A_f = \dfrac{A}{1+\beta A} = \dfrac{100}{1 + 100 \times 0.05} = \dfrac{100}{6} = 16.67$ dB

$$F = 20\log_{10}\left|\frac{A_f}{A}\right| = 20\log_{10}\left|\frac{1}{6}\right| = -15.56 \text{ dB}$$

(b) $A\beta = 100 \times 0.05 = 5$

(c) $V_o = A_f V_i = 100 \times 0.05 = 5$

(d) $V_f = \beta V_0 = 0.05 \times 5 = 0.25$ Volt

**Example 9-4** For the circuit shown in the diagram, $R_c = 4$ kΩ, $R_L = 4$ kΩ, $R_B = 20$ kΩ, $R_s = 1$ kΩ and the transistor parameters are:

$$h_{ie} = 1 \text{ k}\Omega,\ h_{fe} = 50,\quad h_{re} = 2.5 \times 10^{-4} \quad\text{and}\quad h_{oe} = 24\ \mu\text{S}$$

Find:

(a) The current gain
(b) The voltage gain
(c) The transconductance
(d) The transresistance
(e) The input resistance seen by the source
(f) The output resistance seen by the load. Neglect all capacitive effects.

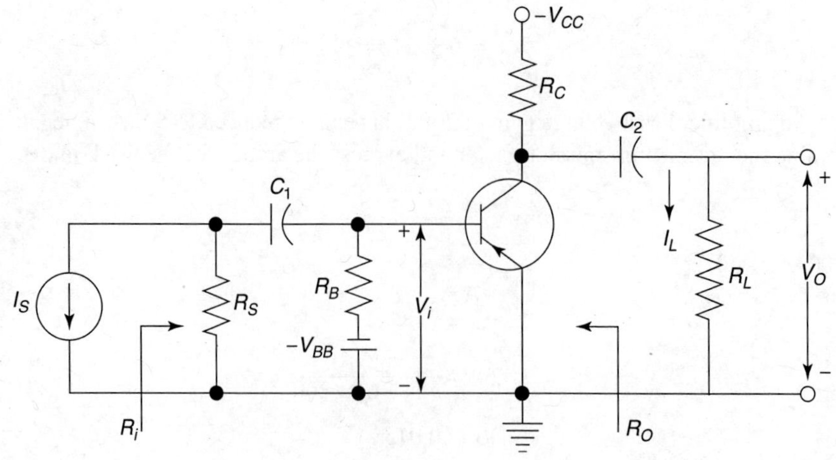

**Solution:**

Given: $R_c = 4\text{ k}\Omega, R_L = 4\text{ k}\Omega, R_B = 20\text{ k}\Omega, R_s = 1\text{ k}\Omega$

The ac equivalent of the circuit is shown in the following figure:

(a) Current gain: $A_I = \dfrac{I_L}{I_s} = \dfrac{I_i}{I_s}\dfrac{I_b}{I_i}\dfrac{I_L}{I_b}$

$$\frac{I_i}{I_s} = \frac{R_s}{R_s + R_i}$$

Input resistance:

$$R_i = R_b || h_{ie} = 20\text{ k}||1.1\text{ k}$$

$$= \frac{20 \times 1.1}{20 + 1.1} = 1.04\text{ k}\Omega$$

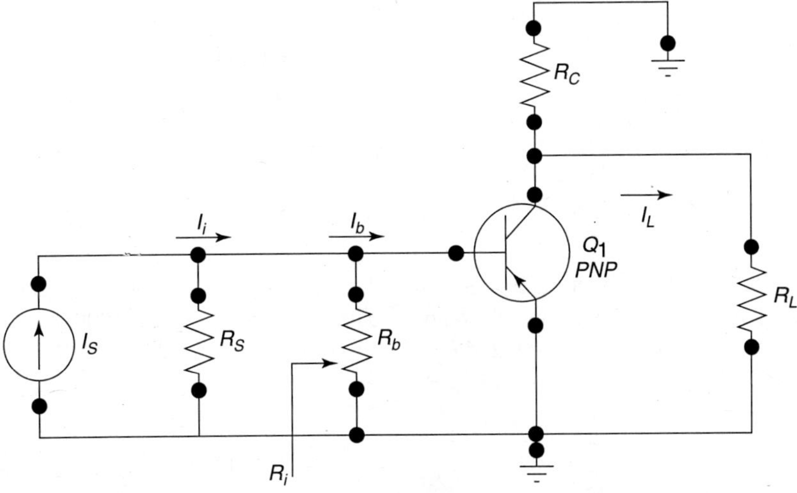


Then:

$$\frac{I_i}{I_s} = \frac{1\text{ k}}{1\text{ k} + 1.04\text{ k}} = \frac{1}{2.04}$$

$$\frac{I_b}{I_i} = \frac{R_B}{R_B + h_{ie}}$$

$$= \frac{20}{20 + 1.1} = 0.95$$

$$\frac{I_L}{I_b} = -h_{fe}\frac{R_c}{R_c + R_L}$$

$$= -50 \times \frac{4}{4 + 4} = -25$$

$$A_I = \frac{I_L}{I_s} = \frac{1}{2.04} \times 0.95 \times (-25)$$

$$= -11.65$$

(b) Voltage gain:

$$A_V = \frac{V_0}{V_s} = \frac{I_L R_L}{I_s R_s}$$

$$= (-11.65) \times \frac{4\text{ k}}{1\text{ k}} = -46.6$$

(c) Transconductance:

$$G_m = \frac{I_L}{V_s} = \frac{V_0}{R_L}\frac{1}{V_s} = \frac{V_0}{V_s}\frac{1}{R_L}$$

$$= \frac{-46.6}{4k} = -11.65 \text{ mA/V}$$

(d) Transresistance:

$$R_m = \frac{V_0}{I_s} = \frac{R_s}{V_s}V_0 = \frac{V_0}{V_s}R_s$$

$$= 1\text{ k} \times (-46.6) = -46.6 \text{ k}\Omega$$

(e) Input resistance: $R_i = 1.04$ kΩ

(f) Output resistance:

$$R_0 = R_c \left|\right| \frac{i}{h_{oe}} = 4\text{ k}||40\text{ k} = 3.64 \text{ k}\Omega$$

**Example 9-5** For the circuit, as given in the diagram, show that (a) the ac voltage $V_i$ is a function of $V_s$ and $V_f$. Assume that the inverting amplifier input resistance is infinite, that $A_v = A_V = -1000$, $\beta = V_f/V_0 = 1/100$, $R_s = R_c = 1$ kΩ $R_E$, $h_{ie} = 1$ kΩ, $h_{re} = h_{oe} = 0$ and $h_{fe} = 100$. (b) Find $A_{Vf} = V_0/V_i = AV_i/V_s$.

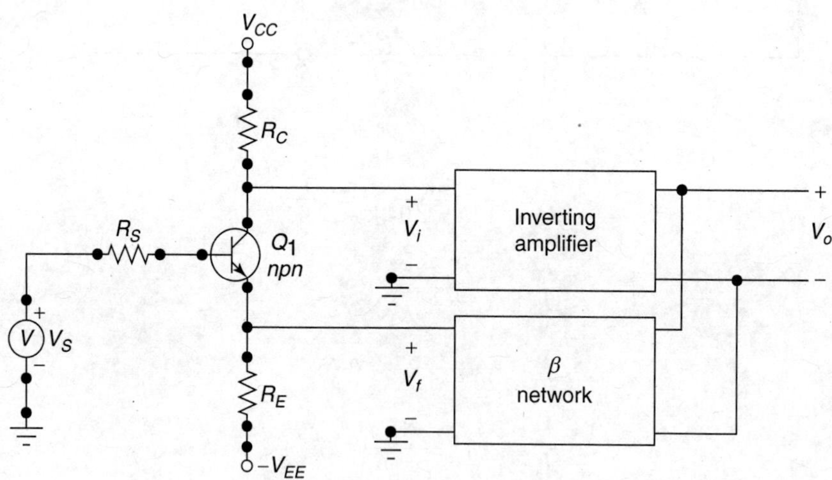


**Solution:**

Assume that the $\beta$ network can be represented by an ideal controlled voltage source with $V_f = \beta V_0$. From the equivalent circuit we have:

$$V_i = -h_{fe} I_b R_c \quad \text{where,} \quad I_b = \frac{V_s - V_f}{R_s + h_{ie}}$$

$$V_i = -h_{fe} R_c \frac{V_s - V_f}{R_s + h_{ie}}$$

$$= -100k \times 1\text{ k} \frac{V_s - V_f}{(1 + 1)\text{k}}$$

$$= -50(V_s - V_f)$$

(b) With the output of the inverting amplifier connected to the input of the $\beta$ network, we have:

$$V_f = \beta V_0 \text{ and } V_0 = AV_i = A_V V_i$$

$$V_0 = A_V \times (-50)(V_s - V_f) = -A_V \times 50(V_s - \beta V_0)$$

$$\Rightarrow \quad V_0 = 5 \times 10^4(V_s - 0.001 V_0)$$

$$A_{vf} = \frac{V_0}{V_s} = 100$$

**Example 9-6** An amplifier with open-loop voltage gain $AV = 1000 \pm 100$ is available. It is necessary to have an amplifier whose voltage gain varies by no more than $\pm$ 0.1%. Find (a) the feedback ratio and (b) the gain with feedback.

**Solution:**

(a) We know that:

$$\frac{dA_f}{A_f} = \frac{dA}{A} \times \frac{1}{1 + \beta A}$$

$$\Rightarrow \quad \frac{0.1}{100} = \frac{1}{1 + \beta A} \times \frac{100}{1000} = \frac{1}{1 + \beta A} \times \frac{1}{10}$$

$$\therefore \quad \beta = 0.099$$

(b) Voltage gain with feedback is given by:

$$A_f = \frac{A}{1 + \beta A} = \frac{1000}{1 + 0.099 \times 1000} = 10$$

**Example 9-7** An amplifier without feedback gives a fundamental output of 36 V with 7% second harmonic distortion when the input is 0.028 V.

(a) If 1.2% of the output is feedback into the input in a negative voltage-series feedback circuit, what is the output voltage?

(b) If the fundamental output is maintained at 36 V but the second harmonic distortion is reduced to 1%, what is the input voltage?

**Solution:**

$$D/D' = 7, V_f = 1.2\% \, V_0$$

(a) Voltage gain:

$$|A| = \frac{V_0}{V_i} = \frac{36}{0.028} = 1285$$

Feedback ratio: $$\beta = \frac{V_f}{V_0} = \frac{1.2}{100} = 0.012$$

We know that: $$A_f = \frac{A}{1 + \beta A} = \frac{1285}{1 + 0.012 \times 1285} = 78.2$$

Therefore, output voltage:

$$V_0' = A_f V_s = 78.2 \times 0.028 = 2.19 \text{ V}$$

(b) If the output is maintained constant at 36 V then the distortion generated by the device is unchanged. The reduction of the total distortion is caused by the feedback.
We know that:

$$D' = \frac{D}{1 + \beta A}$$

$$\Rightarrow \quad 1 + \beta A = \frac{D}{D'} = 7$$

$$\Rightarrow \quad \beta A = 6$$

and $$A_f = \frac{A}{1 + \beta A} = \frac{1285}{7}$$

and $$V_s = \frac{V_0}{V_f} = \frac{36}{1285/7} = 0.196 \text{ V}$$

**Example 9-8** The output resistance of voltage-series feedback amplifier is 10 Ω. If the gain of the basic amplifier is 100 and the feedback fraction is 0.01, what is the output resistance without feedback?

**Solution:**

$$R_{of} = 10\ \Omega,\ A = 100,\ \beta = 0.01$$

We know that:

$$R_{of} = \frac{R_0}{1 + \beta A}$$

$$R_0 = R_{of}(1 + \beta A)$$

$$= 10(1 + 0.01 \times 100)$$

$$= 20\ \Omega$$

**Example 9-9** The signal and output voltages of an amplifier are 1 mV and 1 V respectively. If the gain with negative feedback is 100 and the input resistance without the feedback (voltage-series) is 2 kΩ, find the feedback fraction and input resistance with the feedback.

**Solution:**

$$V_s = 1 \text{ mV},\quad V_0 = 1 \text{ V},\quad A_f = 100,\quad R_i = 20 \text{ k}\Omega$$

We know that,

$$A = \frac{V_0}{V_s} = \frac{1\text{ V}}{1\text{ mV}} = 1000$$

$$A_f = \frac{A}{1 + \beta A}$$

$$\Rightarrow \qquad 1 + \beta A = \frac{A}{A_f} = \frac{1000}{100} = 10$$

$$\therefore \qquad \beta = 0.009$$

Input resistance with the feedback:

$$R_{if} = R_i(1 + \beta A)$$

$$= 2\text{ k}\Omega \times 10 = 20\text{ k}\Omega$$

**Example 9-10** If an amplifier has a bandwidth of 200 kHz and voltage gain of 80, what will be the new bandwidth and gain if 5% of negative feedback is introduced?

**Solution:**

$$BW = 200\text{ kHz}, \quad A = 80, \quad \beta = 5\% = 0.05$$

We know that:

$$A_f = \frac{A}{1 + \beta A}$$

$$= \frac{80}{1 + 0.05 \times 80} = 16$$

With the feedback:

$$A \times BW = A_f \times BW_f$$

$$\Rightarrow \qquad BW_f = \frac{A \times BW}{A_f}$$

$$= \frac{80 \times 200}{16} = 1\text{ Mz}$$

**Example 9-11** The open-loop gain of an amplifier is –100 and distortion voltage is of 0.2 Volt. Tolerance voltage is 0.04 Volt. Find out the transmission factor.

**Solution:**
Output distortion voltage with feedback is:

$$V_{Df} = \frac{V_D}{1 + A\beta}$$

$$1 + A\beta = \frac{V_D}{V_{Df}} = \frac{0.1}{0.04} = \frac{10}{4} = 2.5$$

or
$$A\beta = 2.5 - 1 = 1.5$$

$$\beta = \frac{1.5}{-100} = -0.015$$

**Example 9-12** An amplifier has an open-loop gain of 500 and a feedback ratio of 0.1. If open-loop gain changes by 20% due to the temperature, find the percentage change in the closed-loop gain.

**Solution:**

$$A = 500, \beta = 0.1, \frac{dA}{A} = 20$$

Change in closed-loop gain:

$$\frac{dA_f}{A_f} = \frac{dA}{A} \times \frac{1}{1 + \beta A}$$

$$= 20 \times \frac{1}{1 + 500 \times 0.1}$$

$$= 0.3921 = 39.21\%$$

**Example 9-13** Discuss how gain of an amplifier is stabilized with negative feedback. The open loop gain of an amplifier changes by 5% if 10 dB negative feedback is applied. Calculate the percentage change of the closed loop gain.

**Solution:**

$$\frac{dA}{A} = 5\%$$

We know that feedback is negative.

Thus,

$$20\log_{10}\frac{A_f}{A} = -10$$

or,
$$2\log_{10}\frac{A_f}{A} = -1$$

or,
$$2\log_{10}\frac{1}{1 + \beta A} = -1$$

or,
$$2\log_{10}(1 + \beta A) = -1$$

or,
$$\log_{10}(1 + \beta A)^2 = 1$$

or,
$$(1 + \beta A)^2 = 10$$

or,
$$(1 + \beta A) = \sqrt{10}$$

We know:

$$\frac{dA_f}{A_f} = \frac{1}{1 + \beta A}\frac{dA}{A}$$

$$\frac{dA_f}{A_f} = \frac{1}{\sqrt{10}}5 = 1.6\%$$

**Example 9-14** An amplifier consists of three identical stages connected in cascade. The output voltage is sampled and returned to the input in series opposing. If it is specified that the relative change $dA_f/A_f$ in the closed-loop voltage gain $A_f$ must not exceed $y/f$, show that the minimum value of the open-loop gain $A$ of the amplifier is given by:

$$A = 3A_f\left|\frac{\psi_1}{\psi_f}\right|$$

where, $\psi \equiv dA_1/A_1$ is the relative change in the voltage gain of each stage of the amplifier.

**Solution:**

From:
$$\left|\frac{dA_f}{A_f}\right| = \frac{1}{|1+\beta A|}\left|\frac{dA}{A}\right|$$

we obtain
$$\left|\frac{dA_f}{A_f}\right| \equiv |\psi_f| = \frac{1}{|1+\beta A|} = \left|\frac{dA}{A}\right|$$

where, $A = A_1^3$

or,
$$dA = 3A_1^2\, dA_1$$

Then,
$$\left|\frac{dA}{A}\right| = 3\frac{dA_1}{A_1} = 3\psi_1$$

∵
$$A_f = \frac{A}{1+\beta A}$$

$$\frac{A_f}{A} = \frac{1}{1+\beta A}$$

$$|\psi_f| = \frac{A_f}{A}3|\psi_1|$$

or,
$$A = 3A_f\frac{|\psi_1|}{|\psi_f|}$$

**Example 9-15** Discuss how gain of an amplifier is stabilized with negative feedback. The open loop gain of an amplifier changes by 5% if 10 dB negative feedback is applied. Calculate the percentage change of the closed loop gain.

**Solution:**

$$\frac{dA}{A} = 5\%$$

We know that feedback is negative.

Thus,
$$20\log_{10}\frac{A_f}{A} = -10$$

or,
$$2\log_{10}\frac{A_f}{A} = -1$$

or, $$2\log_{10}\frac{1}{1+\beta A} = -1$$

or, $$2\log_{10}(1+\beta A) = -1$$

or, $$\log_{10}(1+\beta A)^2 = 1$$

or, $$(1+\beta A)^2 = 10$$

or, $$(1+\beta A) = \sqrt{10}$$

We know: $$\frac{dA_f}{A_f} = \frac{1}{1+\beta A}\frac{dA}{A}$$

$$\frac{dA_f}{A_f} = \frac{1}{\sqrt{10}}5 = 1.6\%$$

**Example 9-16** If an amplifier has a bandwidth of 200 kHz and voltage gain of 80, what will be the new bandwidth and gain if 5% of negative feedback is introduced?

**Solution:**

$$BW = 200\text{ kHz},\quad A = 80,\quad \beta = 5\% = 0.05$$

We know that:

$$A_f = \frac{\text{A}}{1+\beta A}$$

$$= \frac{80}{1+0.05\times 80} = 16$$

With the feedback:

$$A\times BW = A_f\times BW_f$$

⇒ $$BW_f = \frac{A\times BW}{A_f}$$

$$= \frac{80\times 200}{16} = 1\text{ Mz}$$

**Example 9-17** The output resistance of voltage-series feedback amplifier is 10 Ω. If the gain of the basic amplifier is 100 and the feedback fraction is 0.01, what is the output resistance without the feedback?

**Solution:**

$$R_{of} = 10\ \Omega,\quad A = 100,\quad \beta = 0.01$$

We know that:

$$R_{of} = \frac{R_0}{1+\beta A}$$

$$R_0 = R_{of}(1+\beta A)$$

$$= 10(1+0.01\times 100)$$

$$= 20\ \Omega$$

**Example 9-18** An amplifier with an open-loop voltage gain $AV = 1000 \pm 100$ is available. It is necessary to have an amplifier whose voltage gain varies by no more than $\pm$ 0.1%.

(a) Find the reverse transmission factor $\beta$ of the feedback network used.

(b) Find the gain with the feedback.

**Solution:**

(a) From: $\left|\frac{dA_f}{A_f}\right| = \frac{1}{|1+\beta A|}\left|\frac{dA}{A}\right|$

$$\frac{0.1}{100} = \frac{A}{1+\beta A}\,\frac{100}{1000}$$

$$1 + A\beta = 100$$

or,

$$\beta A = 99$$

∴

$$\beta = \frac{99}{1000} = 0.099$$

(b)

$$A_f = \frac{A}{1+\beta A} = \frac{1000}{1+99} = 10$$

**Example 9-19** The transistors in the feedback amplifier shown in the diagram are identical. Make reasonable approximations whenever appropriate, and neglect the reactance of the capacitors. Calculate:

$$R_{if} = \frac{V_s}{I_1},\quad A_{if} = -\frac{I}{I_1},\quad A'_{vf} = \frac{V_0}{V_1},\quad A_{vf} = \frac{V_0}{V_s}\quad \text{and}\quad R_{of}'$$

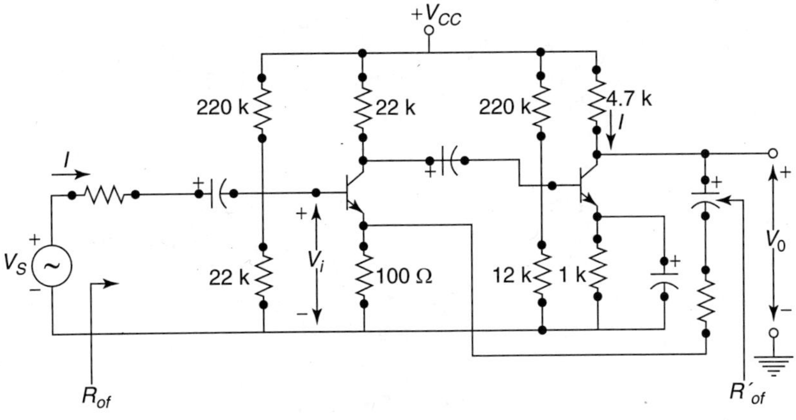

+VCC
220 k
22 k
220 k
4.7 k
I
I
+
+
Vi
−
Vs
22 k
100 Ω
12 k
1 k
+
+
+
Vo
−
Rof
R′of

**Solution:**

**Step 1:** Consider the type of feedback. We sample the output voltage and $V_f = V_0\, 0.1/10.1 = bV_0$; hence, it is a voltage series feedback with $b = 0.0099 - 0.01$

**Step 2:** We obtain the basic amplifier without the feedback to find the input circuit we short circuit the output and to find the output circuit we open the input loop. Thus, we have the amplifier, as shown in the following figure.

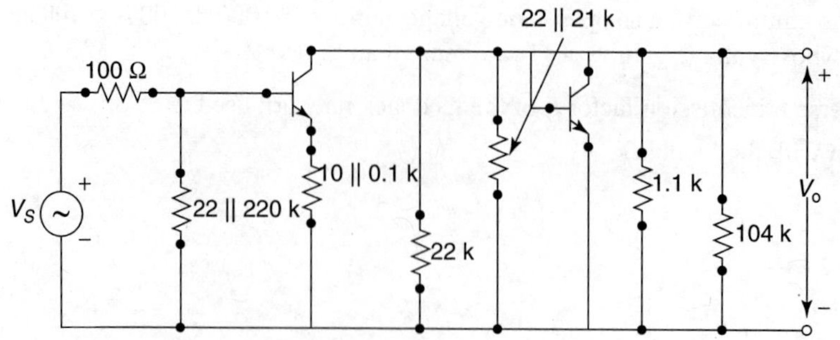


**Step 3:** Replacing the transistors with the appropriate hybrid equivalent small signal model we obtain the next stage of the amplifier, as shown in the following figure.

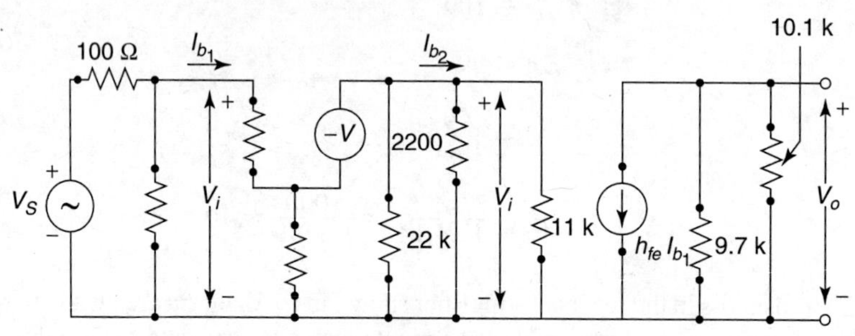


Let:
$$AV_1 = \frac{V_0}{V_f} = \frac{V_0}{V_1} \times \frac{V_1}{V_f}$$

$$\frac{V_0}{V_f} = -h_{fe}\frac{4.7\|10.1}{h_{ie}} = -50 \times \frac{3.2}{1.1} = -145.5$$

$$\frac{V_1}{V_f} = -h_{fe}\frac{R_{L1}}{R_{f1}}$$

But,
$$R_{f1} = h_{ie} + (1 + h_{fo})R_e = 1.1 + 51 \times 0.1 = 6.2 \text{ K}$$

and
$$R_{L1} = 22\|20.2\|h_{ie} = 990\ \Omega$$

Hence,
$$\frac{V_1}{V_f} = -50 \times \frac{0.99}{6.2} = -8.0$$

Now that
$$A_{VI}\, n = (-8) \times (-145.5) = 1100$$

∵
$$D = 1 + \beta A_{VI} = 1 + 11.60 - 12.6$$

$$A_{Vf}' = \frac{A_{VI}}{D} = \frac{1160}{12.6} = 92.1$$

where,
$$\beta = 0.01, \quad R_{if}' = R_{f1}D.$$

$$R_{11} = 6.2 \text{ k} \quad \text{hence,} \quad R_{if} = 0.2 \times 12.0 = 78 \text{ k}$$

Then, $$\frac{V_i}{V_s} = \frac{78\|20.2}{0.1 + 78 \| 20.2} = \frac{16.1}{16.2} = 0.993$$

∴ $$A_{vf} = A'_{vf} = 0.993 - 91.5$$

We have that, $$R_{if} \times L_1 = V_s$$

∴ $$R_{if} = 78\|20.2 - 02 = 16.2 \text{ k}$$

$$A_{vf} = -\frac{V_0}{V_s} = -\frac{IR_L}{I_1 R_{if}} = A_{if} = 91.5 \times \frac{10.2}{4.7} = 314$$

Finally, $$R'_0 = 4.7 \| 10.1 - 3.2 \text{ k}$$

$$R'_{of} = \frac{R'_Q}{D} = \frac{3.2}{12.6} = 253\ \Omega$$

**Example 9-20** An amplifier has a voltage gain of 100. The feedback fraction is 0.04. Find the voltage gain with the feedback.

**Solution:**

Using the formula: $$A_v = \frac{A_v}{1 + \beta A_v} = \frac{100}{1 + 0.04 \times 100} = \frac{100}{5} = 20$$

Hence, the voltage gain of the feedback is 20

**Example 9-21** The signal and output voltages of an amplifier are 1 mV and 1 V respectively. If the gain with negative feedback is 100 and the input resistance without the feedback (voltage-series) is 2 kΩ, find the feedback fraction and input resistance with the feedback.

**Solution:**

$$V_s = 1 \text{ mV},\quad V_0 = 1 \text{ V},\quad A_f = 100,\quad R_i = 20 \text{ k}\Omega$$

We know that: $$A = \frac{V_0}{V_s} = \frac{1 \text{ V}}{1 \text{ mV}} = 1000$$

$$A_f = \frac{A}{1 + \beta A}$$

⇒ $$1 + \beta A = \frac{A}{A_f} = \frac{1000}{100} = 10$$

∴ $$\beta = 0.009$$

Input resistance with the feedback:

$$R_{if} = R_i(1 + \beta A)$$
$$= 2 \text{ k}\Omega \times 10 = 20 \text{ k}\Omega$$

**Example 9-22** An amplifier has a voltage gain of 200. This gain is reduced to 50 when negative feedback is applied. Determine the reverse transmission factor and express the amount of feedback in dB.

**Solution:**

$$A = 200, A_f = 50$$

We know that: $$A_f = \frac{A}{1 + \beta A}$$

$\Rightarrow$ $$50 = \frac{200}{1 + \beta \times 200}$$

$\therefore$ $$\beta = 0.015$$

Feedback in dB: $$N = 20\log_{10}\left|\frac{A_f}{A}\right| = 20\log_{10}\left(\frac{1}{1+\beta A}\right)$$

$$= 20\log_{10}\left(\frac{1}{1 + 200 \times 0.015}\right) = 20\log_{10}\left(\frac{1}{4}\right)$$

$$= -12.042 \text{ dB}$$

**Example 9-23** An amplifier without the feedback gives a fundamental output of 36 V with 7% second harmonic distortion when the input is 0.028 V.

(a) If 1.2% of the output is feedback into the input in a negative voltage-series feedback circuit, what is the output voltage?

(b) If the fundamental output is maintained at 36 V but the second harmonic distortion is reduced to 1%, what is the input voltage?

**Solution:**

$$D/D' = 7, V_f = 1.2\% V_0$$

(a) Voltage gain: $$|A| = \frac{V_0}{V_i} = \frac{36}{0.028} = 1285$$

Feedback ratio: $$\beta = \frac{V_f}{V_0} = \frac{1.2}{100} = 0.012$$

We know that: $$A_f = \frac{A}{1+\beta A} = \frac{1285}{1 + 0.012 \times 1285} = 78.2$$

Output voltage: $$V'_0 = A_f V_s = 78.2 \times 0.028 = 2.19 \text{ V}$$

(b) If the output is maintained constant at 36 V then the distortion generated by the device is unchanged. The reduction of the total distortion is caused by the feedback.

We know that: $$D' = \frac{D}{1+\beta A}$$

$\Rightarrow$ $$1 + \beta A = \frac{D}{D'} = 7$$

$\Rightarrow$ $$\beta A = 6$$

$$A_f = \frac{A}{1+\beta A} = \frac{1285}{7}$$

$$V_s = \frac{V_0}{V_f} = \frac{36}{1285/7} = 0.196 \text{ V}$$

**Example 9-24** When a negative feedback is applied to an amplifier of gain 200, the overall gain becomes 50.

(a) Calculate the value of the feedback factor.

(b) If feedback factor remains same, calculate the value of amplifier gain, so that the overall gain becomes 30.

**Solution:**

Amplifier gain = 200

Overall gain (gain with the feedback) = 50

Feedback factor can be calculated from the formula:

$$A_f = \frac{A}{1 + \beta A}$$

or,

$$A_f(1 + \beta A) = A$$

$$(1 + \beta A) = \frac{A}{\beta A}$$

or,

$$\beta A = \frac{A}{A_f} - 1$$

$$A = \frac{1}{A}\left(\frac{A}{A_f} - 1\right)$$

Putting the values:

$$\frac{1}{200}\left(\frac{200}{50} - 1\right) = \frac{3}{200}$$

(b) If $\beta$ remains same:

$$\beta = \frac{3}{200} \quad \text{and} \quad A_f = 50$$

Then:

$$A_f = \frac{A}{(1 + \beta A)}$$

$$A_f(1 + \beta A) = A$$

$$\Rightarrow \quad A_f = A(1 - \beta A_f)$$

$$\Rightarrow \quad A = \frac{A_f}{1 - \beta A_f} = \frac{30}{1 - \frac{3 \times 50}{200}} = 120$$

**Example 9-25** An amplifier without the feedback gives a fundamentals output of 36 V with 7% second-harmonic distortion when the input is 0.028 V.

(a) If 1.5% of the output is feedback into the input in a negative voltage-series feedback circuit, what is the output voltage?

(b) For an output of 36 V with 1% second-harmonic distortion, what is the input voltage?

**Solution:**

(a) $A = \frac{36}{0.028} = 1285$

$\beta = 0.012$

From: $$A_f = \frac{A}{1+\beta A}$$

$$= \frac{1285}{1+(0.015)(1285)} = 63.38$$

$$\therefore \quad V_0 = A_f V_s = 63.38 \times 0.028 = 1.77 \text{ V}$$

(b) If the output remains constant at 36 V, then the distortion produced within the active devices of the amplifier is unchanged. However since the distortion at the output is less than in part a by a factor of 7, follows that the feedback now increased by 9 and hence, the voltage gain decreases by 9. Thus the input signal required to produce the same output—as in part (a) without FB—must be $V_s = 9(0.028 \text{ V}) = 0.252$ V.

**Example 9-26** An amplifier with an open-loop voltage gain of 2000 delivers 20 W of output power at 10% second-harmonic distortion within the input signal is 10 mV. A 40 dB negative voltage-series feedback is applied and the output power is to remain at 10 W. Determine: (a) the required input signal, (b) the percent harmonic distortion.

**Solution:**

(a) $$-40 \text{ dB} = 20 \log_{10} \frac{1}{|1+\beta A|}$$

$$= -20 \log_{10} |i + \beta A|$$

$$\therefore \quad |1+\beta A| = 100$$

But, $$|A_f| = \frac{|A|}{|1+\beta A|} = \frac{2000}{100} = 20$$

When the amplifier delivers 20 W, its output voltage is:

$$V_0 = A_{Vs} = 2000\,(10 \times 10 - 3) = 20 \text{ V}$$

If the output power is to remain at 10 W, then the output voltage also must remain at 10 V. Hence, the input signal required when $I_B$ is applied will be $V_s = V_o/A_f = 20v/20$.

(b) The distortion of the amplifier with FB will be reduced by FB factor:

$$|1+\beta A| = 100$$

Hence, $$D_4 = \frac{10\%}{100} = 0.1\%$$

**Example 9-27** For the circuit, as shown in the figure below, find:

(a) $A_{If} = I_0/I_s$, (b) $R_{if}$, (c) $A_{Vf} = V_0/V_s$, (where $I_s = V_s/R_s$), (d) $A_{Vf}' = V_0/V_s$, and (e) $R_{of}'$.

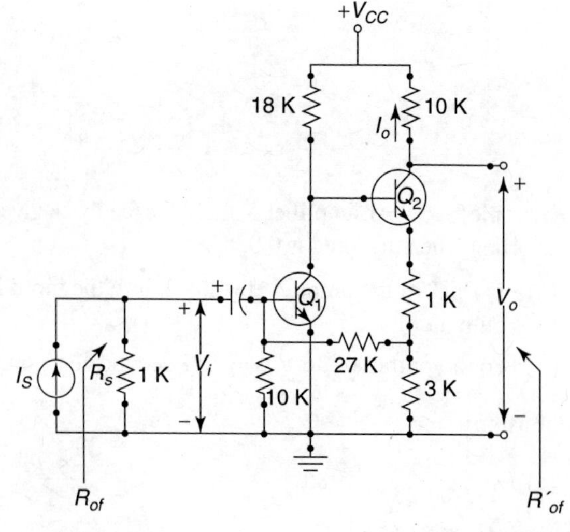

**Solution:**

(a) **Step 1:** Type of feedback: the mixing is of shunt type and $I_f = V_1 - V/27$ K (where, $V$ is the voltage drop across the 3 K resistor). Since $V \gg V_1$:

$$I_1 = -\frac{V}{27\text{ K}} = \frac{3}{27}(I_0 - I_1) \quad \text{or} \quad I_f = \frac{3}{30}I_0 - 0.1I_0 = 8I_0$$

Hence, this is again a current-shunt feedback.

**Step 2:** To obtain the basic amplifier without feedback we open the output loop $I_0 = 0$, and we thus have at the input a resistor 27 + 3 = 30 K in parallel with $R_s$. Let $V_1 = 0$; and we have at the output loop a resistor 27||3 K = 2.7 K. The resulting equivalent circuit is as shown in following figure.

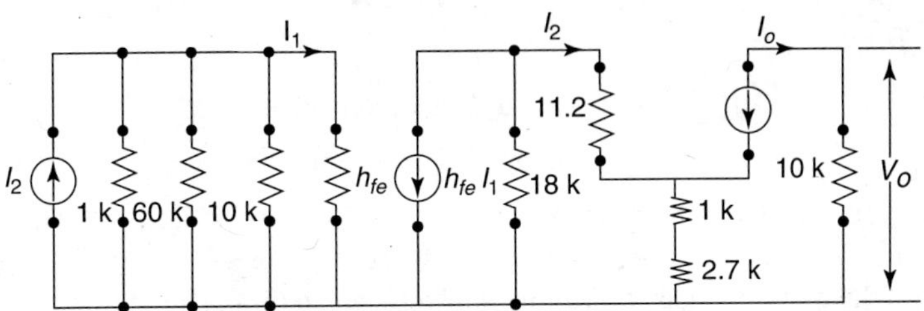


**Step 3:** Since the circuit is a current-shunt feedback amplifier the current gain will be stabilized. We now find $A_1$ for the basic amplifier without the feedback.

$$A_1 = \frac{I_0}{I_2}\frac{I_2}{I_1}\frac{I_1}{I_s}, \quad \frac{I_0}{I_s} = -h_{fe} = -50, \quad \frac{I_2}{I_1} = -h_{fe}\frac{18}{18 + R_{if}}$$

where, $\quad R_{if} = h_{fe} + (1 + h_{fe})\,3.7 \quad \text{or} \quad R_{if} = 2 + 51 \times 3.7 = 190\text{ K}$

or,
$$\frac{I_1}{I_s} = -50 \times \frac{18}{208} = -4.32$$

$$\frac{I_1}{I_s} = \frac{1||10||30}{2 + (1||10||30)}$$

But,
$$1||10||30 = \frac{1}{1 + 0.1 + 0.39} = \frac{1}{1.133} = 0.0884\text{ K}$$

or,
$$\frac{I_1}{I_s} = \frac{0.884}{2.884} = 0.308$$

Hence,
$$A_1 = \frac{A_1}{D}\ bit = D = 1 + \beta A_1 = 7.66 \quad \text{or} \quad A_{if} = \frac{66.6}{7.66} = 8.69$$

(b) $R_1 = 1||10||30||h_{fe} = 0.884||2 = \frac{1.768}{2.884} = 0.615$

$$R_{if} = \frac{6.15}{7.66} = 80.2\text{ W}$$

(c) $A_{vf} = \frac{V_0}{V_1} = \frac{I_0 R_C}{I_s R_s} = A_{if} = \frac{10}{1} = 86.9$

(d) $A_{vf} = \frac{V_0}{V_1} = \frac{V_0}{V_s}\frac{V_s}{V_1}$

Now with feedback,

$$I_s R_{if} = V_1$$

$$I_s = \frac{V_s}{R_s}$$

$$\frac{V_s}{V_1} = \frac{R_s}{R_{11}} = \frac{1}{0.080} = 12.5$$

Therefore,

$$A_{vf} = 86.0 \times 12.5 = 1070$$

(e) $R_{of}' = R_{of}\left(\frac{1 + \beta A_1}{1 + \beta A_1}\right)$

Since,

$$A_1 - \lim/R_L \rightarrow 0 \text{ and } A_1 \text{ is independent of } R_L:$$

$$A_1 - A_1 \quad \text{and} \quad R_{of}' - R_0' = 10\ \text{K}$$

**Example 9-28** The circuit shown in the diagram has $R_c = 4$ kΩ, $R_L = 4$ kΩ, $R_B = 20$ kΩ, $R_s = 1$ kΩ, and the transistor parameters are: $h_{ie} = 1$ kΩ, $h_{fe} = 50$, $h_{re} = 2.5 \times 10^{-4}$ and $h_{oe} = 24\ \mu$S. Find: (a) the current gain, (b) the voltage gain, (c) the transconductance, (d) the transresistance, (e) the input resistances by the source and (f) the output resistance seen by the load. Neglect all capacitive effects.

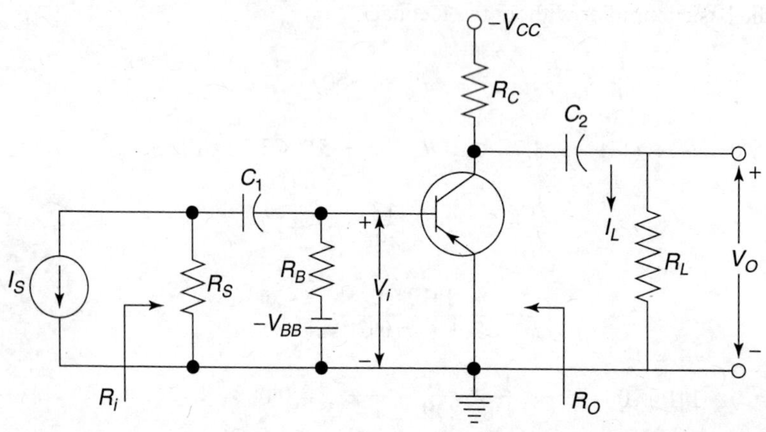


**Solution:**

$$R_c = 4\ \text{k}\Omega,\ R_L = 4\ \text{k}\Omega$$

$$R_B = 20\ \text{k}\Omega,\ R_s = 1\ \text{k}\Omega$$

The ac equivalent of the circuit is shown in the following figure.

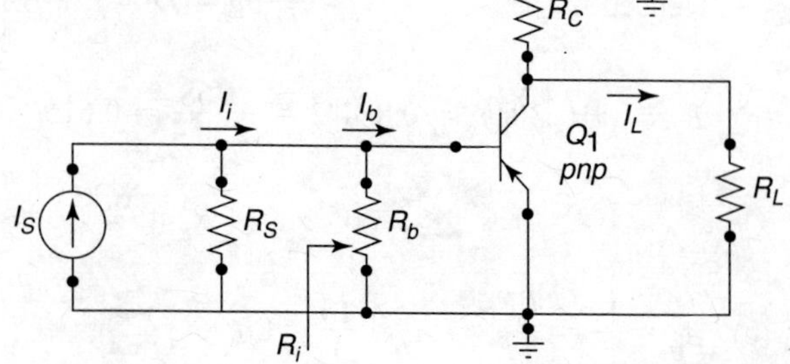


(a) Current gain, $A_I = \frac{I_L}{I_s} = \frac{I_i}{I_s}\frac{I_b}{I_i}\frac{I_L}{I_b}$

$$\frac{I_i}{I_s} = \frac{R_s}{R_s + R_i}$$

Input resistance: $R_i = R_b \| h_{ie} = 20\text{ k}\|1.1\text{ k}$

$$= \frac{20 \times 1.1}{20 + 1.1} = 1.04\text{ k}\Omega$$

Then: $\frac{I_i}{I_s} = \frac{\text{T1 k}}{1\text{ k} + 1.04\text{ k}} = \frac{1}{2.04}$ and $\frac{I_b}{I_i} = \frac{R_B}{R_B + h_{ie}}$

$$= \frac{20}{20 + 1.1} = 0.95$$

$$\frac{I_L}{I_b} = -h_{fe} \frac{R_c}{R_c + R_L}$$

$$= -50 \times \frac{4}{4 + 4} = -25$$

$$A_I = \frac{I_L}{I_s} = \frac{1}{2.04} \times 0.95 \times (-25)$$

$$= -11.65$$

(b) Voltage gain, $A_V = \frac{V_0}{V_s} = \frac{I_L R_L}{I_s R_s}$

$$= (-11.65) \times \frac{4\text{ k}}{1\text{ k}} = -46.6$$

(c) Transconductance, $G_m = \frac{I_L}{V_s} = \frac{V_0}{R_L}\frac{1}{V_s} = \frac{V_0}{V_s}\frac{1}{R_L}$

$$= \frac{-46.6}{4\text{ k}} = -11.65\text{ mA/V}$$

(d) Transresistance, $R_m = \frac{V_0}{I_s} = \frac{R_s}{V_s} V_0 = \frac{V_0}{V_s} R_s$

$$= 1\text{ k} \times (-46.6) = -46.6\text{ k}\Omega$$

(e) Input resistance, $R_i = 1.04\text{ k}\Omega$

(f) Output resistance, $R_0 = R_c \left\| \frac{i}{h_{oe}} \right. = 4\text{ k}\|40\text{ k} = 3.64\text{ k}\Omega$

■

## POINTS TO REMEMBER

1. Feedback is defined as the process by which a portion of the output is returned to the input to form part of the system excitation.
2. The four feedback topologies are:
   (a) Shunt-shunt
   (b) Shunt-series
   (c) Series-shunt
   (d) Series-series
3. The main advantage of negative feedback is stability. Its main application is in the design of a stable amplifier.
4. Positive feedback produces instability in the system. Its main application is in the design of an oscillator.
5. Conditions for Barkhausen criteria:
   (a) Positive feedback
   (b) Loop gain is unity, i.e., $A\beta=1$; therefore, feedback gain is infinite ($A_f = \infty$).
   (c) Phase variation is zero or integral multiple of 360.
6. The frequency of a sinusoidal oscillator is determined by the condition that the loop gain phase shift is zero.

## IMPORTANT FORMULAE

1. Gain with feedback:

$$A_f = \frac{A}{1 + A\beta}$$

2. dB of feedback:

$$N = 20\log_{10}\left|\frac{A_f}{A}\right| = 20\log_{10}\left(\frac{1}{1 + \beta A}\right)$$

3. Sensitivity:

$$S = \frac{\frac{dA_f}{A_f}}{\frac{dA}{A}} = \frac{1}{1 + A\beta}$$

4. Voltage-series feedback:

$$Z_{if} = \frac{V_s}{I_i} = Z_i(1 + \beta A_V)$$

$$Z_{of} = \frac{V}{I} = \frac{Z_o}{1 + \beta A_v}$$

5. Current-series feedback:

$$Z_{if} = Z_i(1 + \beta Y_M)$$

$$Z_{of} = Z_o(1 + \beta Y_m)$$

6. Current-shunt feedback:

$$Z_{if} = \frac{Z_i}{1 + \beta A_I}$$

$$Z_{of} = \frac{V}{I} = Z_o(1 + \beta A)$$

7. Voltage-shunt feedback:

$$Z_{if} = \frac{Z_i}{1 + \beta Z_M}$$

$$Z_{of} = \frac{Z_o}{1 + \beta Z_m}$$

8. Distortion with feedback:

$$D' = \frac{D}{1 + A_v\beta}$$

## OBJECTIVE QUESTIONS

1. Open-loop gain of an amplifier is given by:
   (a) $A$ (b) $A\beta$
   (c) $\beta$ (d) None of the above
2. Loop gain is given by:
   (a) $A$ (b) $A\beta$
   (c) $\beta$ (d) None of the above
3. In a feedback amplifier, sensitivity $D$ is equal to:
   (a) $A\beta$ (b) $1-A\beta$
   (c) $1+A\beta$ (d) $1/(A\beta+1)$
4. In a negative feedback amplifier, voltage sampling:
   (a) Tends to decrease the output resistance
   (b) Tends to increase to output resistance

(c) Does not alter the output resistance

(d) Produces the same effect on output resistance as current sampling

5. In a negative feedback amplifier, current sampling:

(a) Tends to increase the output resistance

(b) Tends to decrease the output resistance

(c) Does not alter the output resistance

(d) Produces the same effect on input resistance as voltage sampling

6. In a negative feedback amplifier, series mixing:

(a) Tends to increase the input resistance

(b) Tends to decrease the input resistance

(c) Does not alter the input resistance

(d) Produces the same effect on input resistance as shunt mixing

7. In a negative feedback amplifier, shunt mixing:

(a) Tends to increase the input resistance

(b) Tends to decrease the input resistance

(c) Does not alter the input resistance

(d) Produces the same effect on input resistance as the series mixing

8. Negative feed back in an amplifier improves:

(a) The signal to noise ratio at the output

(b) Reduces distortion

(c) Both (a) and (b)

(d) None of the above

9. For a shunt-shunt negative feedback amplifier

(a) Input impedance decreases but output impedance increases

(b) Both input impedance and output impedance increases

(c) Both input impedance and output impedance decreases

(d) None of the above

10. An amplifier with the negative feedback:

(a) Controls the gain

(b) Reduces the noise

(c) Reduces phase distortion

(d) All of the above

11. An amplifier with resistive negative feedback has two left half plane poles in its open-loop transfer junction. The amplifier will be:

(a) Stable for all frequencies

(b) Unstable for all frequencies

(c) Stable for a particular frequencies

(d) Unstable for a particular frequencies

12. Barkhusen criteria is:

(a) Positive feedback, $A\beta = 1$, $\theta = 0$ or multiple 360

(b) Negative feedback, $A\beta = 1$, $\theta = 0$ or multiple 360

(c) Positive feedback, $A\beta = 0$, $\theta = 0$ or multiple 360

(d) Negative feedback, $A\beta = 1$, $\theta = 180$

## REVIEW QUESTIONS

1. Draw and explain the concept of feedback in detail with the help of a block diagram. Explain the operation of each block.
2. What do you mean by positive and negative feedback?
3. Calculate the expression for feedback gain for an amplifier?
4. Explain the following terms

   (a) Feedback ratio (b) Feedback factor (c) Open-loop gain (d) Closed-loop gain
5. Classify the different topologies of a feedback network. Explain each topology with the help of block diagrams and proper circuit diagrams.
6. Explain the advantages and disadvantages of negative feedback.
7. Explain the advantages and disadvantages of positive feedback.

8. "Gain bandwidth product is constant." Comment on this statement with respect to negative feedback.
9. What is Barkhausen criterion? Explain the use of it.
10. Negative feedback reduces the gain of an amplifier, but why is it used in an amplifier design?
11. What is Nyquist Criterion? Why this is so important in oscillator design?

## PRACTICE PROBLEMS

1. The open-loop gain of an amplifier changes by 20% due to changes in the parameters of the active amplifying devices. If gain changes by 5%, what type of feedback has to be applied? Find the minimum value of the feedback ratio and open-loop gain for the feedback gain of –30.
2. The open-loop gain of an amplifier is –1000 and gives an output distortion voltage of 0.15 Volt. For negative feedback the tolerable output distortion voltage is 0.05 Volt. Find out the reverse transmission factor.
3. An amplifier has a voltage gain of 100. The feedback ratio is 0.15. Find:
   - (a) The voltage gain with feedback band feedback in dB.
   - (b) The feedback factor.
   - (c) The output voltage, if input voltage is 1.15 Volt.
   - (d) The feedback voltage.
4. An amplifier has a bandwidth of 300 kHz and voltage gain of 100, what will be the new bandwidth and gain if 10% of negative feedback is introduced?
5. The signal and output voltages of an amplifier are 5 mV and 1 V, respectively. If the gain with negative feedback is 200 and the input resistance without feedback (voltage-series) is 2 kΩ, find the feedback fraction and input resistance with the feedback.
6. An amplifier has a bandwidth of 30 kHz and voltage gain of 100, what will be the new bandwidth and gain if 9% of negative feedback is introduced?
7. The signal and output voltages of an amplifier are 12 mV and 2 V, respectively. If the gain with negative feedback is 60 and the input resistance without feedback is 8 kΩ, calculate the feedback fraction and input resistance with feedback.
8. An amplifier uses negative feedback having an open-loop gain of –30 and voltage amplification of –100. Calculate the overall gain and reverse transfer ratio.
9. If open-loop gain of an amplifier changes by 20% calculate the percentage change of the closed-loop gain if a 10 dB negative feedback is applied.
10. The open-loop gain of an amplifier is 99 and distortion voltage is in the order of 0.35 volt. Tolerance voltage is 0.06 volt. Calculate the transmission factor.

## SUGGESTED READINGS


1. Singh, J. 1994. *Semiconductor Devices: An Introduction*. New York, NY: McGraw-Hill.
2. Streetman B.G., Banerjee S. 2000. *Solid State Electronic Devices*. 5th Ed. New Delhi: Pearson Education.
3. Millman, Jacob, Halkias Christos C. 1986. *Integrated Electronics: Analog and Digital Circuits and Systems*. New Delhi: McGraw Hill Book Company.
4. Pierret, R. F and G.W. Neudeck. 1989. *Modular Series on Solid State Devices*. Boston, M.A.: Addison Wesley.
5. Singh, B. P. and Rekha Singh. 2006. *Electronic Devices and Integrated Circuits*. New Delhi: Pearson Education.

# 10

# Fundamentals of Integrated Circuit Fabrication

## Outline



## Objectives

This chapter deals with the fundamentals of integrated circuit fabrication. In the field of nanotechnology, this basic idea of fabrication is very important from the point of view of semiconductor device engineering. Semiconductor device fabrication is the process used to create chips for devices that are a part of our everyday use. It is a multiple-step sequence of photographic and chemical processing during which electronic circuits are gradually created on a wafer substrate made of pure semi-conducting material. In this chapter, readers will learn how integrated circuits are designed and fabricated. The topics include electronic fundamentals, crystal growth and wafer types, transistor types, basic design concepts, mask making, wafer processing, materials used for fabricating the wafer, equipments used for processing the wafer, clean room standards, testing and reliability control, assembly and packaging, and different IC types.

## 10-1 INTRODUCTION

In the world of semiconductor devices and integrated circuits, silicon is the commonly used semiconductor material, and it is mostly used as a substrate material. For example, solar cells made of silicon wafers are cheaper and can be used for practical applications by common people. Of late, it has been observed that GaAs fabrication has advanced compared to silicon due to its greater thermal stability, direct band gap and higher electron mobility. However, GaAs fabrication is a difficult process and the cost incurred is also very high.

Today compound semiconductors are the main point of attraction. IV–IV, III–V and II–VI are the common compound semiconductors used in research and industry. Among these the compound semiconductor III–V is widely used as it has multiple applications in LED, LASER, etc. Excimer lasers are ideal sources for large-area patterning, lithographic processes, TFT display annealing and LED laser lift-off. Micromachining is a broad application sector in which both excimer and DPSS lasers are used. Silicon, sapphire, polymers, CVD diamond, III–V semiconductors—gallium arsenide (GaAs), indium phosphide (InP), gallium phosphide (GaP) and III-nitrides gallium nitride (GaN) and aluminum nitride (AlN)—are materials routinely machined by both these lasers in applications such as structuring, drilling, dicing and cutting. Optoelectronic devices using III–V compound semiconductor materials are increasingly becoming central to optical communications. One of the III–V materials systems on which they are based—indium phosphide (InP)—is of particular importance because it enables devices with the crucial cost and performance characteristics that system vendors require and will be required in the future. InP provides monolithic integration of multiple functions (e.g., lasers, modulators, and optical attenuators) on a chip, and offers high-performance low-cost solutions. InP is a very compact, low-cost technology for 2.5 Gbit/sec and medium-reach systems operating at 10 Gbit/sec for metro transmission.

Band-gap energies of II–VI semiconductors span a wide range of electromagnetic spectrum, with the band-gap energy of HgTe (0 eV) at the lower end of the spectrum to BeSe (6 eV) at the other extreme of the spectrum. The large band-gap energy II–VI compounds are sensitive to wavelengths in the visible and ultraviolet region of the electromagnetic spectrum. Its main applications lie in electroluminescent phosphorous, solar cells, IR sensors based on HgCdTe, and blue laser diode using ZnSe. In recent years the grand success in the field of GaN-based III–V material has reduced the interest in the development of the ZnSe-based blue laser diode. Moreover doping is a big problem in the II–VI material. It has been observed that the II–VI-based diodes are not stable for long-duration applications, and their usage in solar cells results in lower efficiency. Nowadays II–VI semiconductors are popular in the investigation of spintrinics and dilute magnetic semiconductors (DMS). The most common DMS are CdTe, CdSe, ZnSe and CdS used with transition metal ions such as Mn, Fe and Co forming the alloys like CdXTe, ZnXSe where, X = Mn, Fe and Co. These structures are suitable for studying carrier transport. But the main problem of II–VI is that it uses III–V as the only suitable substrate for the growth of the nanostructures.

## 10-2 FUNDAMENTALS OF INTEGRATED CIRCUITS

The fabrication of integrated circuits is an art. From the age of LSI (large-scale integration) to VLSI (very-large-scale integration), the device dimension has reduced by a large amount, but at the same time its switching speed has increased. ULSI (ultra-large-scale integration) is used for low-dimensional nanostructure devices. Nowadays, the use of MBE (molecular beam epitaxy), MOCVD (metal-organic chemical vapour deposition) and CBD (chemical bath deposition), make it possible to fabricate the nanostructure semiconductor devices in the form of Quantum Well, Quantum Wire and Quantum Dot. Nano devices are typically 1–100 nm (in each direction) in dimension. The smaller the devices, greater is the challenge of fabrication due to the new electronic properties of these devices.

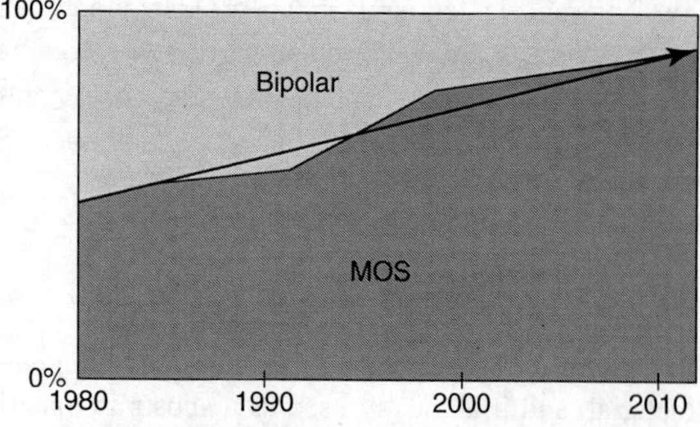


**Figure 10-1** Comparative diagram of development of the BJT/MOS vs. year

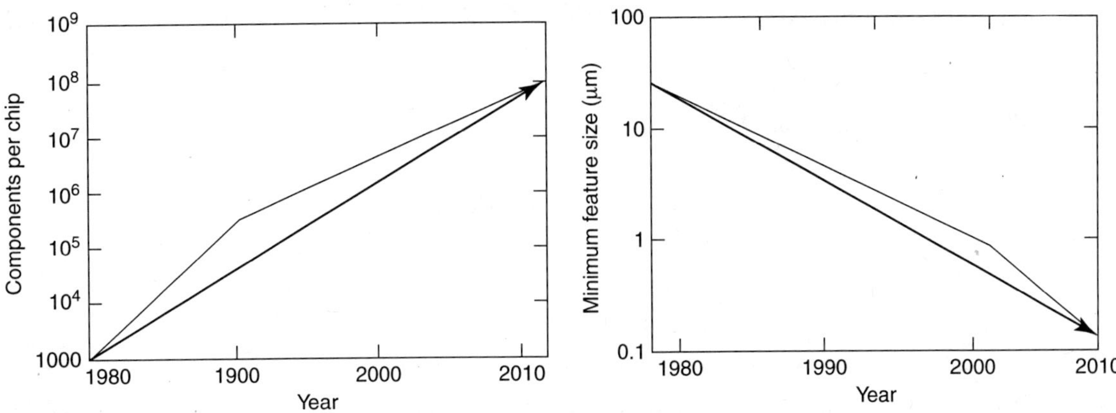


**Figure 10-2** Components/chip vs. year

**Figure 10-3** Minimum size vs. year

As shown in Fig. 10-1, there has been a shift from the bipolar technology to the MOS technology. A projection up to the year 2010 shows that an approximate 95 percent of the market will be composed of MOS devices, and the remaining 5 percent will be bipolar. The bold arrowhead line shows the average trend line in the figure.

With devices getting smaller and smaller, the number of their components/chips has been increasingly getting bigger and bigger with time. A plot in Fig. 10-2 projected up to year 2010 shows the number of components/chip vs. time (year). The bold arrowhead line shows the average trend line in the figure.

Figure 10-3 shows the drop in feature size projected up to the year 2010. The bold arrowhead line shows the average trend line in the figure.

## 10-3 TYPES OF INTEGRATED CIRCUITS

### 10-3-1 Monolithic IC

In the monolithic IC the entire circuit is built into a single piece of semiconductor chip, which consists of active and passive components. The most commonly used integrated circuits, microprocessors, memories, etc., are all monolithic.

### 10-3-2 Hybrid IC

In hybrid IC the electronic circuit is generally integrated in the ceramic substrate using various components and then enclosed in the single package. Therefore, the hybrid IC consists of several monolithic ICs connected by metallic interconnects mounted on a common substrate. Hybrid integrated circuits simultaneously form an electrical, mechanical and thermal bond.

Hybrid IC technology bonds various substrates either at the die level or at the wafer level. This technology streamlines the connections between different semiconductor chips by replacing wire bonding. This process achieves an accelerated, streamlined and less costly process. Hybrid IC allows increase in communication bandwidth and facilitates higher system yield.

## 10-4 ADVANTAGES AND DISADVANTAGES OF INTEGRATED CIRCUITS

The advantages of integrated circuits are as follows:

1. Small in size due to the reduced device dimension
2. Low weight due to very small size
3. Low power requirement due to lower dimension and lower threshold power requirement
4. Low cost due to large-scale production
5. High reliability due to the absence of a solder joint
6. Facilitates integration of large number of devices
7. Improves the device performance even at high-frequency region

The research and development in the field of IC technology is going on in different parts of the world but still there are several limitations. The disadvantages of integrated circuits are as follows:

1. IC resistors have a limited range
2. Generally inductors (L) cannot be formed using IC
3. Transformers cannot be formed using IC

## 10-5 SCALE OF INTEGRATION

Historically, the first semiconductor IC chips held one diode/transistor. Advancement of technology enabled us to add more and more transistors. The first to arrive was small-scale integration (SSI), then improvements in technique led to devices with hundreds of logic gates—large-scale integration (LSI). Present day micro-processors have millions of logic gates and transistors. Intel co-founder, Gordon E. Moore, in 1965 published a paper on the future projection of IC technology. Moore's Law is responsible for "smaller, cheaper and more efficient IC". Gordon Moore's empirical relationship is cited in a number of forms, but its essential thesis is that the number of transistors that can be manufactured on a single die will double every 18 months. The Moore's low representation of number of components per chip versus year is shown is Fig. 10-2.

### 10-5-1 Types of IC Chips

#### Analog

Analog chips are small transistor count precision circuits. Amplifiers, filters, sensors, etc. fall into this category.

#### ASIC or application specific integrated circuits

The development in the fabrication of IC's has enabled us to create powerful circuits in nano devices. ASIC brings a lot more of functionality into the same area. These are IC's that are created for specific purposes—each device is created to do a particular job. The common application area for this is Digital Signal Processing—signal filters, image compression, etc.

#### SoC or systems on a chip

These are highly complex mixed signal circuits (digital and analog all on the same chip). A network processor chip or a wireless radio chip is an example of a SoC. The planar technology for IC fabrication consists of the following processes:

1. Crystal growth of the wafer
2. Epitaxial growth
3. Oxidation
4. Photolithography

5. Etching
6. Diffusion
7. Ion implantation
8. Metallization
9. Testing and packaging

## 10-6 CRYSTAL GROWTH AND WAFER PREPARATION

Crystal growth and wafer preparation is the most fundamental step in device fabrication. Various crystal growth techniques are used to form the bulk crystal. One of the methods is the Czochralski process to form silicon wafers. A small seed crystal of silicon is attached to the top of a rod and lowered into a crucible of molten silicon to which acceptor impurities have been added. As the rod is very slowly pulled out of the 'melt' under carefully controlled conditions, we find that a single *p*-type or *n*-type crystal ingot of the order of several inches has grown, as shown in Fig. 10-4. This technique is referred to as the Czochralski or *CZ* process. The ingot is subsequently sliced into round wafers to form the substrate on which all integrated components are fabricated. One side of each wafer is lapped and polished to eliminate surface imperfections before proceeding with the next process.

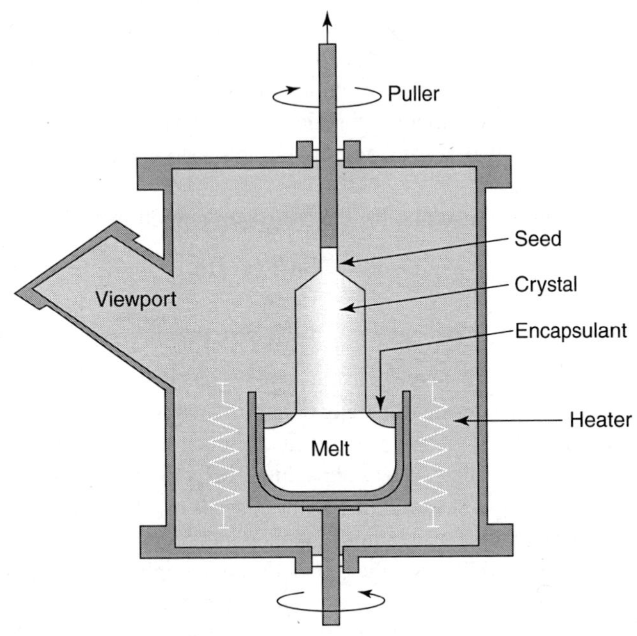


**Figure 10-4** Growth of bulk silicon crystal by Czochralski method

The crystal ingot in Fig.10-5(a) and the cylindrical Si crystal in Fig. 10-5(b) show that the crystal is properly shaped in a cylindrical form using a diamond cutter, and ultimately the cylindrical shaped crystal is cut into smaller pieces to form the wafer, as shown in Fig. 10-5(c).

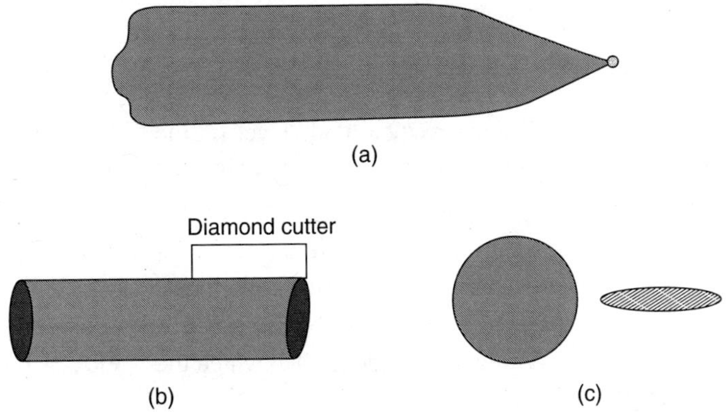


**Figure 10-5** Preparation of wafers from ingot (a) Ingot of Si (b) Cylindrical Si crystal (c) Wafer preparation

## 10-7 EPITAXIAL GROWTH

Epitaxy, derived from Greek *epi* 'upon' and *taxis* 'arrangement', refers to the growth of crystals on a crystalline substrate that determines their orientation. Epitaxy involves the extension of the substrate lattice by the overgrowth of a layer of identical material. This is known as homoepitaxy or autoepitaxy. For example, Si on Si or Ga As on Ga As. On the other hand, if the epitaxial layer and the substrate are chemically and often crystallo graphically different, then this is called heteroepitaxy. For example, Si (diamond lattice) on sapphire (hexagonal) or Ga As on Si. The epitaxial process is used to form a layer of single-crystal silicon on an existing crystal wafer of the same or different material. Epitaxial growth is performed in a special furnace called the reactor. Silicon wafers are inserted in the reactor and heated to a temperature of 900–1000°C. The temperature varies from substrate to substrate. Production technology uses the hydrogen reduction of gases like silane ($SiH_4$) or silicon tetrachloride ($SiCl_4$) as the source for the silicon to be grown.

Silane has two advantages. It requires a lower temperature and has a faster growth rate than silicon tetrachloride.

The chemical reaction for the hydrogen reduction of $SiCl_4$ is:

$$SiCl_4 + 2H_2 \xrightarrow{1200°C} Si + 4HCl$$

And that for $SiH_4$ is:

$$SiH_4 \xrightarrow{H_2 \text{ atmosphere, } 1000°C} Si + 2H_2$$

For example, an *n*-type epitaxial layer, typically from few nanometre to few micrometre thick, is grown into a *p*-type substrate having resistivity of approximately few Ω-cm, corresponding to $N_A = 1.4 \times 10^{15}$ atoms/cm$^3$.

Since epitaxial growth requires the production of epitaxial films of impurity concentrations, it is essential to introduce impurities, such as phosphine ($PH_3$) for *n*-type doping or diborane ($B_2H_6$) for *p*-type doping into the $SiCl_4$-hydrogen gas stream. The reactor consists of a long cylindrical quartz tube encircled by a radio-frequency induction coil and allows precise impurity control for the production of an epitaxial layer. The silicon wafers are loaded on a rectangular graphite rod called boat. The boat is inserted into the reactor and the graphite is heated to a temperature above 1500°C. The temperature is raised keeping in mind that the melting point of Si is 1420°C. A control console permits the introduction and removal of various gases required for the growth of suitable epitaxial layers. As a result, it is possible to form an almost abrupt step *p–n* junction as shown in Fig. 10-6.

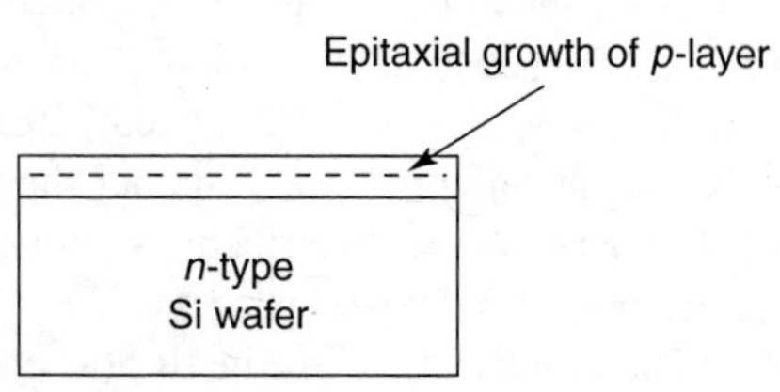


**Figure 10-6** Epitaxial growth on Si wafer

## 10-8 OXIDATION FOR ISOLATION

Silicon technology has the natural ability to grow an oxide layer on the silicon surface, as shown in Fig. 10-7. $SiO_2$ as a passivating layer has the following characteristics:

1. The impurities that are used to dope the silicon do not penetrate the silicon dioxide. Thus, when used with the masking techniques, selective doping of specific regions of the chip is accomplished.
2. It is capable of being etched by hydrogen fluoride (HF) to which the underlying silicon is impervious.

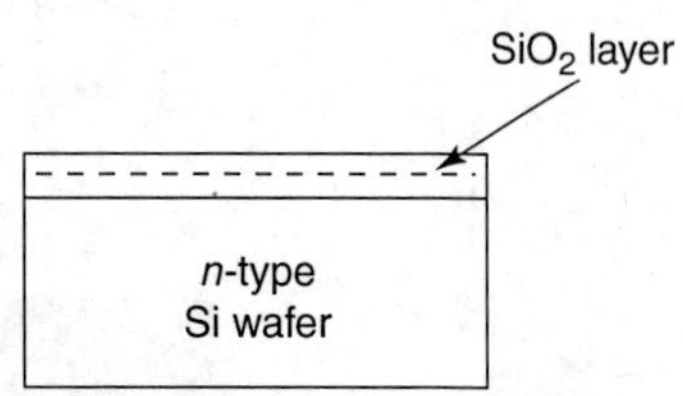


**Figure 10-7** Growth of oxide layer

Thermal oxidation of silicon is achieved in the presence of water vapour; this process is called *wet oxidation.* The chemical reaction for this process is:

$$Si + 2H_2O \rightarrow SiO_2 + 2H_2$$

Dry oxidation of oxygen: $Si + O_2 \rightarrow SiO_2$

The thickness of the oxide layers is generally in the order of 0.02 $\mu$m to 2 $\mu$m; the specific value selected depends on the barrier required to prevent dopant penetration. Process temperature, impurity concentration, and processing time are several of the factors that influence the thickness of the $SiO_2$ layer. The outer silicon dioxide layer is obtained by chemical vapour deposition (CVD).

## 10-9 PHOTOLITHOGRAPHY FOR PATTERN TRANSFER

Photolithography is the optical process of transferring geometric shapes on a mask to the surface of a silicon wafer. The steps for the photolithographic process are: photo resist coating, soft-baking, mask alignment, UV exposure and development, and hard-baking. The process for the formation of a thin uniform layer of viscous liquid photo resist on the wafer surface is shown in Fig. 10-8(a). In the starting of the photolithographic process, the wafer is coated with a uniform film of a photosensitive emulsion. The photo resist is slightly hardened by soft-baking and then selectively removed by projection of UV light through mask. The monolithic fabrication technique requires the selective removal of the insulator layer of $SiO_2$ to form openings through which impurities can be diffused.

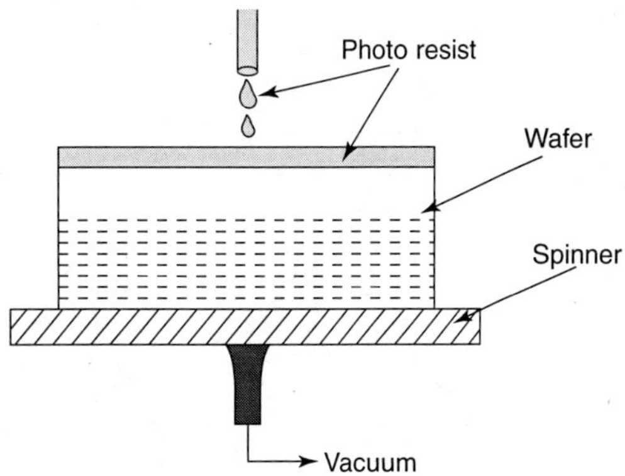


**Figure 10-8(a)** Wafer on a spinner and photo resist

There are two types of photo resists used—positive and negative photo resist [see Fig. 10-8(b)]. For positive resists, the resist is exposed with UV light. Exposure to the UV light changes the chemical structure of the resist and it becomes more soluble in the developer. The developer solution washes the exposed resist part, forming windows on the substrate. The mask contains an exact copy of the pattern which is to be transferred on the wafer surface.

The behaviour of the negative resists is just the opposite of the positive photo resist. When the negative resist is exposed to the UV light it becomes polymerized, which is not possible to dissolve. Negative photo resist masks contain the inverse of the pattern to be transferred on the surface of substrate. The developer solution removes only the unexposed portions and the negative resist remains on the surface wherever it is exposed.

The photo resist coatings become photosensitive, or image-able, only after soft-baking. Soft-baking is performed during which almost all of the solvents are removed from the photo resist coating. Under-soft-baking will prevent light from reaching the sensitizer. This under-soft-baked positive resist is then readily attacked by the developer in both exposed and unexposed areas, causing less etching resistance. Over-soft-baking will degrade the photosensitivity of resists by either reducing the developer solubility or actually destroying a portion of the sensitizer.

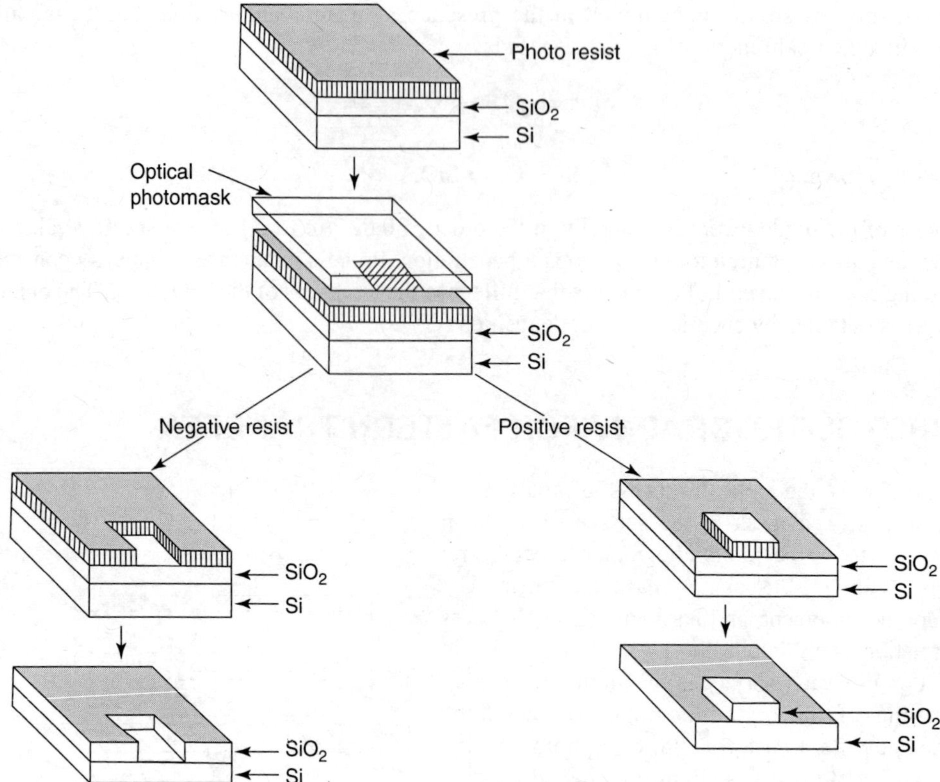


**Figure 10-8(b)** Photolithography processes using negative and positive photo resist

## 10-9-1 Mask Alignment and UV Exposure

The most important and complicated step in the photolithography process is mask alignment. In general, photo-mask is a plastic or glass plate with a pattern of emulsion. The mask is aligned on the top surface of the wafer, so that the pattern can be transferred on it. Once the mask has been accurately aligned with the pattern on the wafer's surface, the photo resist is exposed through the pattern on the mask with a high intensity ultraviolet light. There are three primary exposure methods.

### Contact printing

In this technique the resist-coated silicon wafer is brought into physical contact with the photo-mask, as shown in Fig. 10-8(c). The wafer is held on a vacuum system. The system is exposed with UV light from the top of the mask as the wafer is in contact position with the mask. As the photo resist and the mask are in direct contact, a high resolution is possible in contact printing. But the disadvantage with contact printing is that the dust trapped between the resist and the mask can damage the mask and cause defects in the pattern. As the photo-mask is in direct contact with photo resist it cannot be reused.

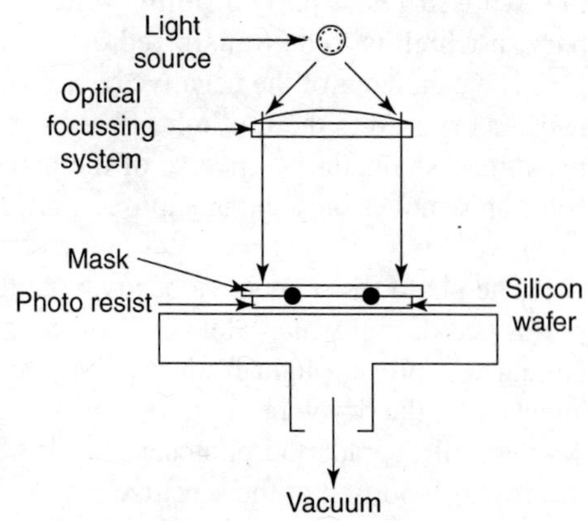


**Figure 10-8(c)** Contact printing

### Proximity printing

This technique is almost similar to contact printing technique except that a small (10–15 $\mu$m) gap is maintained between the mask and the wafer [see Fig. 10-8(d)]. It is very important to note that the gap minimizes, but may not eliminate the damage to the mask.

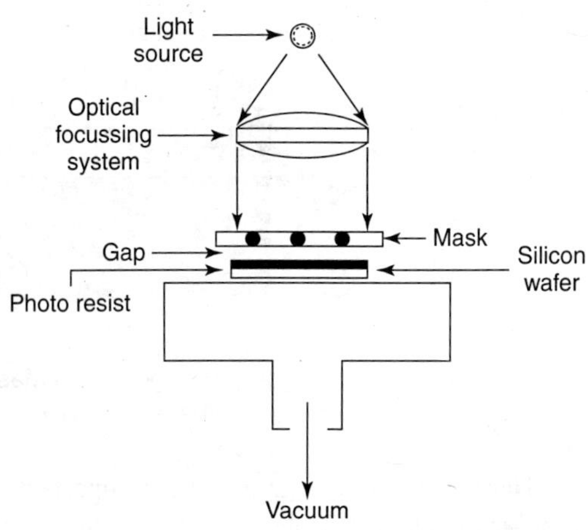


**Figure 10-8(d)** Proximity printing

### Projection printing

Projection printing is the most advanced technique, it avoids mask damage. An image of the patterns on the mask is projected onto the resist-coat which is a certain height away as shown in Fig. 10-9. To achieve high resolution only a small portion of the mask is imaged. This small region image field is scanned over the surface of the wafer. Projection printers that step the mask image over the wafer surface are called step-and-repeat systems having several nano-metre resolution.

Figure 10-10 (a) shows response curves for negative and positive resist after the exposure and development.

The negative resist is completely soluble in the developer solution at low-exposure energies. As the exposure is increased above a threshold energy more of the resist film remains after development.

For positive resists, the resist solubility in its developer is finite even at zero-exposure energy. The solubility gradually increases until, at some threshold, it becomes completely soluble. These curves are affected by all the resist processing variables: initial resist thickness, pre-bake conditions, developer chemistry, developing time, and others.

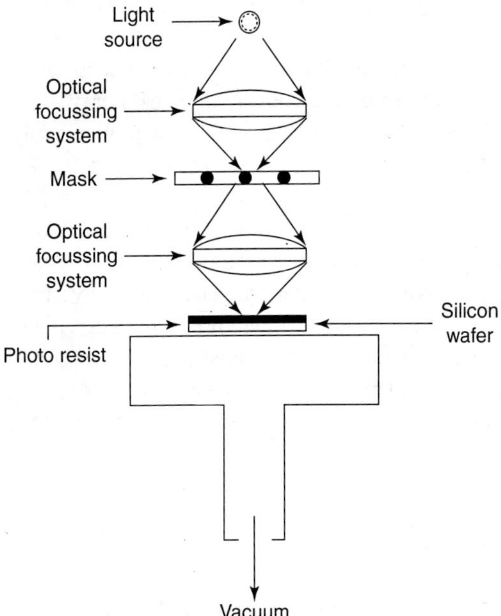


**Figure 10-9** Projection printing

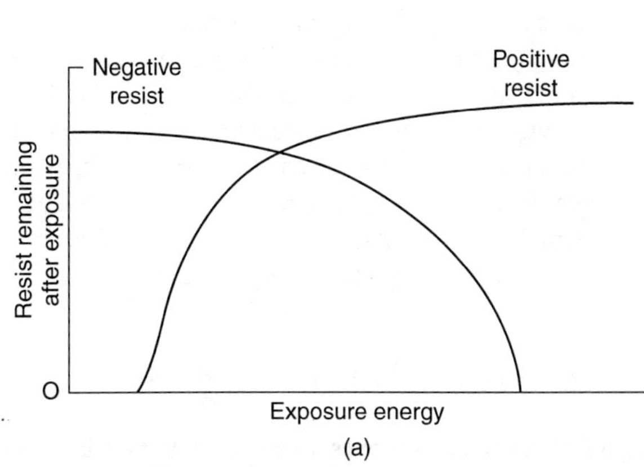


**Figure 10-10(a)** Resist exposure characteristics of the negative and positive photo resist

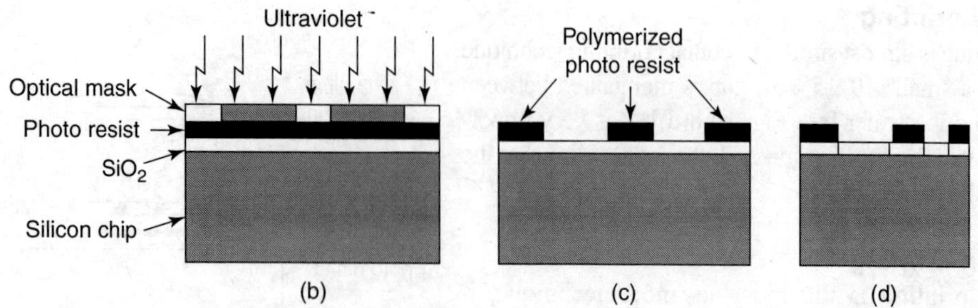


**Figure 10-10** Photolithography process (b) Masking and exposure to UV radiation (c) Development (d) Etching for desired pattern

Hard-baking is the final and the important step in the photolithographic process. This step is essential in order to harden the photo resist and improve adhesion of the photo resist to the wafer surface. A large black-and-white layout of the desired pattern openings is prepared, and then reduced photographically. This negative of the required dimensions is placed as a mask over the photo resist, as shown in Fig. 10-10(b). By exposing the emulsion to ultraviolet light through the mask, the photo resist becomes polymerized under the transparent regions of the mask. The mask is now removed, and the wafer is developed using a chemical (such as trichloroethylene), which dissolves the unexposed, i.e., unpolymerized portions of the photo resist film, and leaves a surface pattern [see Fig. 10-10(c)]. The emulsion, which was not removed in development, is now fixed, so that it becomes resistant to the corrosive etches used next [see Fig. 10-10(d)].

Those portions of $SiO_2$ which are protected by the photo resist are unaffected by the acid as shown in Fig. 10-10(c). After diffusion of impurities, the resist mask is removed (stripped) with a chemical solvent (such as hot $H_2SO_4$) coupled with a mechanical abrasion process. A negative photo resist is used in the process described above. Positive photo resist is also employed in which the exposed portion of the polymer is washed away and thus, the unexposed material is retained. The remaining processing steps are identical, and independent of the type of photo resist used.

Mask design for photolithographic process is an art. The designing of a photographic mask involves complicated and expensive processes. After the circuit layout has been determined, a large-scale drawing is made showing the locations of the openings to be etched in the $SiO_2$ for a particular process step. Invariably, chip layout is obtained by computer-aided design. The drawing is made to a magnified scale of about 1000:1, and results in dimensions more easily managed by a drafter.

The composite drawing of the circuit is partitioned into several levels called masking levels, used in fabricating the chip. For example the gate patterns for the MOS/CMOS devices are on one level, the source and drain contact windows on another level. Today's computer-driven optical-pattern generators convert the patterns into digital information and transfer the geometric layout of the chip onto a photosensitive glass plate.

The smallest features that can be formed by the photolithographic process are limited by a wavelength of UV light. Electron beams have much smaller wavelengths than optical radiation, and are capable of defining much smaller areas. A narrow electron beam scans a mask covered with an electron-sensitive resist and the pattern is written on the mask with the scanning controlled by a computer.

## 10-10 ETCHING FOR DESIGN

Etching selectively removes layers of $SiO_2$, metals, and poly-silicon, according to the desired patterns delineated by the resist. The two major methods of etching are wet chemical etching or dry etching. The windows are created by the photolithographic process and subsequent etching process in the desired pattern implemented by the masking process is shown in Fig. 10-11.

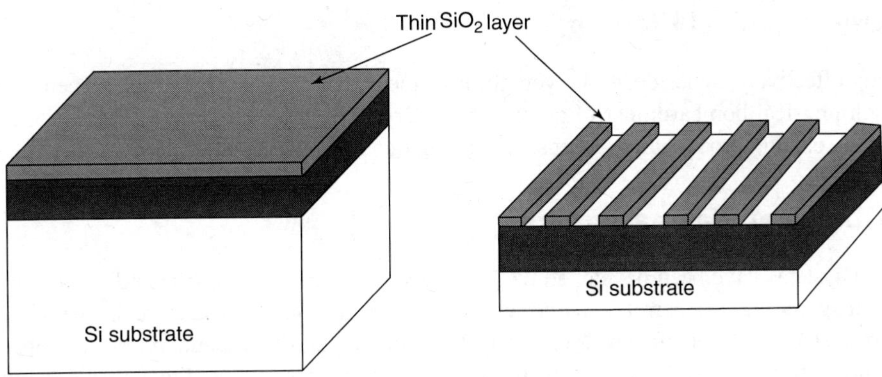


**Figure 10-11** Etching of selected section

## 10-10-1 Wet Chemical Etching

Wet etching is accomplished by partial submersion of the wafer in an acid bath having some definite concentration. Etching solutions are housed in a temperature-controlled baths. The baths are usually equipped with ventilation or a slotted exhaust at the back of the etch station. Vertical laminar-flow hoods are generally used to supply uniformly-filtered air to the top surface of the etch baths. The difficulty of the wet etching is that the chemicals enter just below the photo resists and etch more regions below the feature size and sometimes damage the devices with unnecessary extra etching. Common etching solutions are given in Table 10-1.

**Table 10-1** Common etching solutions

| *Materials to Etch* | *Etchants* |
|---|---|
| **Silicon** | |
| Polycrystalline Silicon (Si) | Hydrofluoric, nitric, acetic acids and iodine<br>Potassium hydroxide<br>Ethylene diamine |
| Silicon Dioxide ($SiO_2$) | Buffered oxide etch (BOE) – Hydrofluoric acid and ammonium fluoride<br>BOE, ethylene glycol, monomethyl ether<br>Hydrofluoric acid and nitric (P-etch) acid |
| Silicon Nitride ($Si_3N_4$) | Phosphoric and hydrofluoric acid |
| CVD Oxide or Pad Etch | Ammonium fluoride, acetic and hydrofluoric acids |
| **Metals** | |
| Aluminium (Al) | Phosphoric, nitric, acetic and hydrochloric acids |
| Chromium/Nickel (Cr/Ni) | Ceric ammonium nitrate and nitric acid<br>Hydrochloric and nitric acids (aqua regia) |
| Gold (Au) | Hydrochloric and nitric acids (aqua regia)<br>Potassium iodide (KI)<br>Potassium cyanide (KCN) and hydrogen peroxide ($H_2O_2$)<br>Ferric chloride ($FeCl_3$) and hydrochloric acid |
| Silver (Ag) | Ferric nitrate ($FeNO_3$) and ethylene glycol<br>Nitric acid |

## 10-10-2 Dry Chemical Etching

Dry processing effectively etches desired layers through the use of gases, using either, a chemically reactive gas, or through physical bombardment of argon atoms. Dry etching is commonly used due to its ability to better control the etching process and decrease contamination levels.

## 10-10-3 Chemical Plasma Etching

Plasma etching systems have been developed that can effectively etch silicon, silicon dioxide, silicon nitride, aluminium, gold, glass, etc. The barrel or cylindrical, and the parallel plate or planar are the two kinds of plasma etching reactor systems are generally used. The typical reactor consists of a vacuum reactor chamber made usually of aluminium, glass, or quartz. A radiofrequency (RF) energy source is used to activate fluorine-based or chlorine-based gases, which act as etchants. Wafers are loaded into the chamber, a pump evacuates the chamber, and the reagent gas is introduced in the chamber. The RF energy ionizes the gas and forms the etching plasma, which reacts with the wafers.

### Physical bombardment

Plasma or sputter etching process is the physical technique involving ion impact and energy transfer. In a glow-discharge environment the wafer to be etched is attached to a negative electrode. The dislocation of the surface atom occurs due to the positive argon ions bombarded on the wafer surface. An RF energy source provides the power in the system as shown in Fig.10-12.

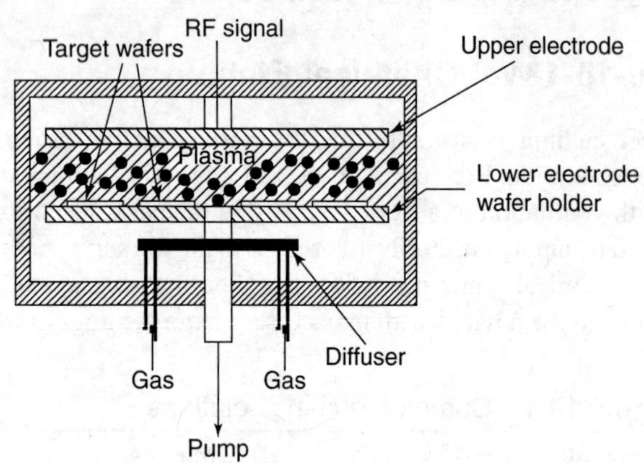


**Figure 10-12** Plasma etching system

## 10-10-4 Ion Beam Etching

Ion beam etching is similar physical etching processes which use a beam of low-energy ions to remove material. By an electrical discharge the ion beam is extracted from an ionized gas such as argon or argon/oxygen or plasma. An electrical discharge creates an ion plasma with the energy of a few hundred electron volts.

## 10-10-5 Reactive Ion Etching

Reactive ion etching (RIE) is a combination of chemical and physical etching. During RIE, a wafer is placed in a chamber with an atmosphere of chemically reactive gas either $CF_4$ or $CCl_4$ at a low pressure. The ions strike the wafer surface vertically.

# 10-11 DIFFUSION FOR DOPING

Diffusion is a process of introduction of impurities into the substrate layer in the planar process. The introduction of controlled impurity concentrations is performed in a diffusion furnace at a temperature of about 1000°C over several hours for silicon substrate. The diffusion oven accommodates few wafers in a quartz carrier inside a quartz tube. The temperature must be controlled carefully so that it becomes uniform

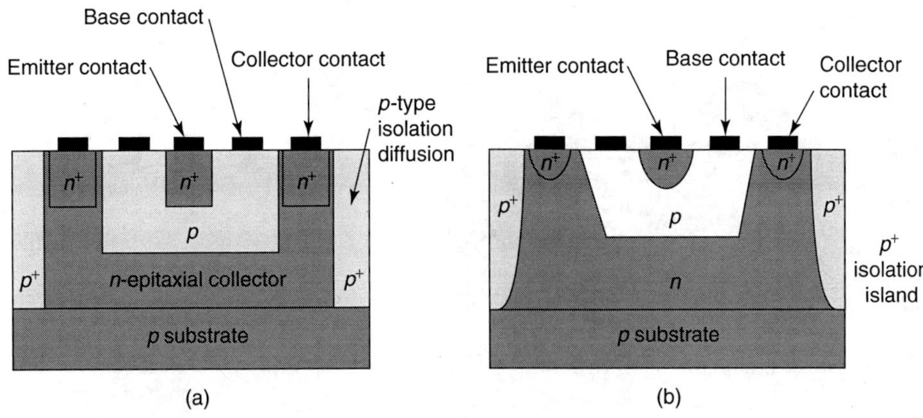


**Figure 10-13** Cross section of an IC transistor (a) Idealized (b) Actual

over the entire hot zone of the furnace. Impurity sources may be gases, liquids, or solids. Gaseous impurities used are generally the hydrides of boron, arsenic, and phosphorus. An inert gas (nitrogen) transports the impurity atoms to the surface of the wafers where they diffuse into the wafer substrate.

The cross-sectional diagrams with diffusion are shown in Fig. 10-13(a). When a window is opened in the $SiO_2$ impurities are introduced, they will laterally diffuse the same distance vertically. Hence, the impurity will spread out under the passivating oxide surface layer. The junction profiles should be drawn more realistically in this case, as shown in Fig.10-13(b).

## 10-12 ION IMPLANTATION FOR DOPING

Ion implantation is the most widely used technique to introduce dopant impurities into semiconductors. The ionized particles are accelerated through an electrical field and targeted at the semiconductor wafer. In a vacuum, a beam of appropriate ions, like phosphorus for *n*-type or boron for *p*-type, are accelerated by energies between several hundreds of keV as shown in Fig. 10-14. The depth of penetration of these ions in the substrate is determined by the accelerating energy, the beam current, and the concentration of dopant ions.

This process is used wherever thin layers of doped silicon are required—the channel in a MOSFET, the emitter region of a BJT, and the gate region of a JFET. Ion implantation process is specially designed for such narrow regions; it permits the controlled doping concentrations than diffusion. The $SiO_2$ passivation layer forms an effective barrier against implanted ions, so that only the photo-lithographically defined regions are doped.

A second and the most important advantage of ion implantation is that it is performed at low temperatures. As a result,

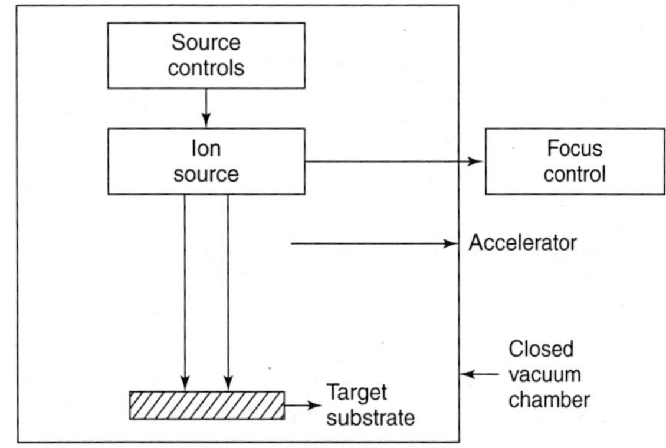


**Figure 10-14** Ion Implantation process in a vacuum chamber

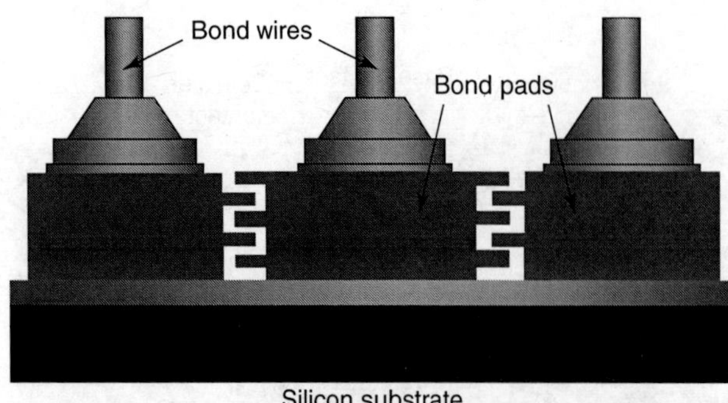


**Figure 10-15(a)** Metallization for different contact points

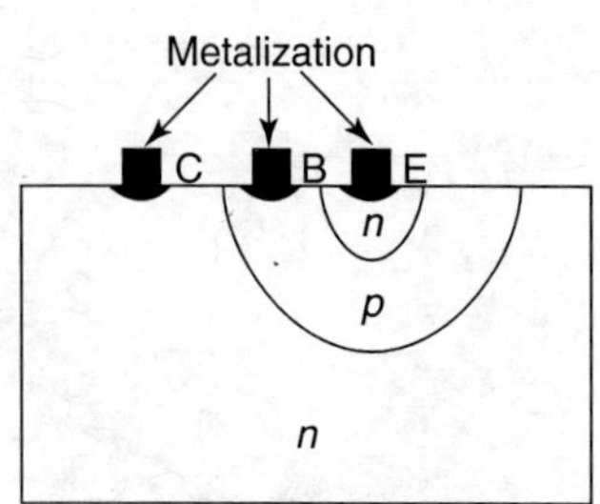


**Figure 10-15(b)** Metallization for contact points for an *n–p–n* transistor

previously implanted regions have a lesser tendency for lateral spreading. The third advantage of the ion-implantation process is that both the beam current and the accelerating potential are electrically controlled outside of the apparatus in which the implantation occurs.

## 10-13 METALLIZATION FOR INTERCONNECTION

The metallization process is used to form the contacts and interconnections of the components on the devices/chips. These are formed by the deposition of a thin layer of good conducting pure metal like aluminium or gold over the selected surface of the chip using a metallic mask. The metallization is performed in a high vacuum (~$10^{-6}$ milibar) enclosure in order to avoid oxidation. Metal deposition is achieved by high-vacuum evaporation inside a bell jar. The metal (aluminium or gold) is heated until it vaporizes, and the gaseous molecules formed uniformly radiate in all directions and cover the wafer surface. A metal mask with a definite pattern is used to define the connection pattern between the components, and the unwanted metal is etched and removed. The metal connection with wire bond is shown in Fig. 10-15(a).

Metal connection in semiconductor layer of a transistor emitter, base and collector region is shown in Fig. 10-15(b).

## 10-14 TESTING FOR RELIABILITY

After all the fabrication processes are done successfully, the semiconductor device is tested thoroughly through the different characterization processes. IC's are fabricated by the batch process, so some of the IC may not perform according to standard. An electronic tester presses tiny probes against the chip to check its functional property; for example, checking of truth table for digital logic gates. ICs are often designed with "self testability features" for example, built-in self-test (BIST) for speed testing. These processes reduce test costs and time. The standard checking can be done by the reliability testing process by an IC tester.

## 10-15 PACKAGING PROTECTION

The last step of this process is packaging. Packaging is done according to the dimension of the product type, and the requirements of the manufacturer. The wafer is certified then the wafer is broken into small individual dies. Small wires are bonded to connect pads to the external connection pins. The most common packaging is dual input packaging or DIP, mostly used in digital IC. Pin grid array (PGA) and leadless chip carrier (LCC) packages are used in modern VLSI design. After testing and packaging, the product is sent to the market for sale.

## 10-16 IC SYMBOLS

The standard symbols of different types of IC like passive components, active components and logic gates are shown in Fig. 10-16.

Pasives

| | | | | |
|---|---|---|---|---|
| Resistor | Potentiometer | Resistor, Photo | Thermistor | VDR |
| or | Capacitor, Polarized | Coil, Air core | Coil, Iron core | Coil powdered iron or ferite core |
| Fuse | Varistor | Crystal | Resistor, Trimer | Lamp |
| NC push button switch | NC push button switch | Switch DPDT | Switch DPST | Switch SPDT |
| Switch SPST | Switch, Thermal NC | Switch, Thermal NO | Switch NO | Switch, NC |

Transistors

| | | | | |
|---|---|---|---|---|
| Biopolar *n-p-n* | Bipolar *p-n-p* | *n-p-n* Darlington | *p-n-p* Darlington | Phototransistor |
| JFET *n*-ch | JFET *p*-ch | MOSFET *n*-ch | MOSFET *p*-ch | MOSFET Nch Enhancment |
| MOSFET Pch Enhancment | Dual Gate MOSFET | | | |

**Figure 10-16** Standard IC symbols *(Contd.)*

**Figure 10-16** (*Contd.*)

Thyristors

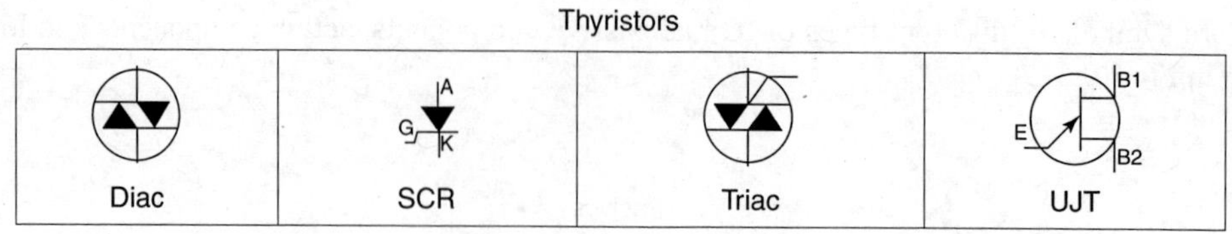


Logic, Standard

| | | | | |
|---|---|---|---|---|
| AND | OR | NOT or INVERTER | NAND | NOR |
| XOR | XNOR | Buffer | S Q R $\bar{Q}$ RS Flip Flop | J Q C K $\bar{Q}$ JK Flip Flop |
| S Q C R $\bar{Q}$ RS Flip Flop w/Clock | D Q C $\bar{Q}$ D Flip Flop | | | |

Diodes

| | | | | |
|---|---|---|---|---|
| K A Diode | K A Diode, Photo Conductive of Photo Voltaic | Tunel | Varactor | Zener |

# 10-17 FABRICATION STEPS FOR DIFFERENT CIRCUITS

## 10-17-1 Fabrication of Resistors in Integrated Circuits

Resistors in IC can be made by the base diffusion method. The resistor is made up of a $p$-layer within one $n$-type island or the reverse, as shown in Fig. 10-17.

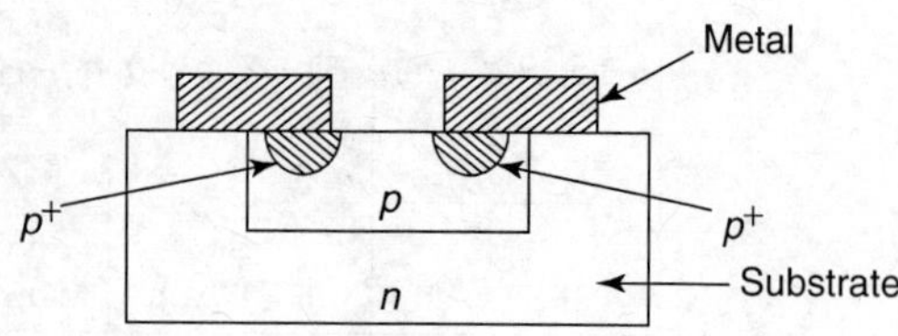


**Figure 10-17** Resistor in IC

Here $p^+$ diffusions are useful for low value resistors to form the ohmic contact. The value of resistance can be changed by changing the length and width, i.e., the dimension of the device, and the resistivity of the diffused material. Resistance values can vary from a few ohms to kilo ohms by the diffusion technique; tolerance of the resistor value is very poor, about 30% approximately. Thin film resistors which have reduced space are used in order to get high value resistors.

$$\text{Resistance, } R = \rho \frac{l}{A}$$

where, $\rho$ is the resistivity, $l$ = length and $A$ = area.

Resistance value depends on $l$, $A$ and $\rho$. The drawback for these resistors is substantial parasitic junction capacitance, which causes a lot of problems for high frequency operations. In MOS devices the poly-silicon layer on the oxide is used as an improved useful resistor, since the thin layer provides a better surface area matching, and hence more accurate resistor ratio.

### Difference between ohomic contact and Schottky contact

Both ohomic contact and Schottky contact are types of metal-semiconductor contact. Schottky diode or contact is a metal-semiconductor diode or contact having characteristics similar to a $p$–$n$ junction. When the metal is deposited directly on the semiconductor surface, the Schottky contact is produced. The fabrication structure is shown in Fig. 10-18.

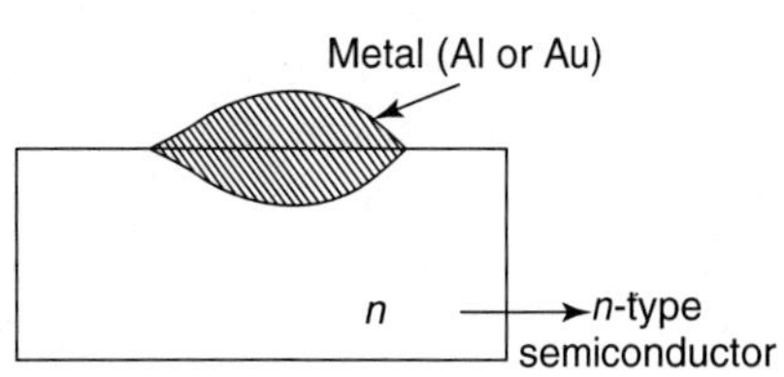


**Figure 10-18** Schottky contact

The direct metal-semiconductor contact produces a contact potential barrier known as Schottky barrier. This barrier produces the rectifying behaviour. Thus, the Schottky barrier diode is designed. In case of ohomic contact the metal is not directly deposited on the semiconductor surface. Instead of that a high doped layer (either $n^+$ or $p^+$) is placed between the semiconductor and the metal to make the gradual change in barrier potential. So, there is no rectifying behaviour.

The junction of the metal semiconductor, the same metal-doped layer, produces a linear behaviour, i.e., ohomic behaviour. The junction that follows ohm's law, i.e., linear nature in I–V characteristics forms an ohmic contact (see Fig. 10-19).

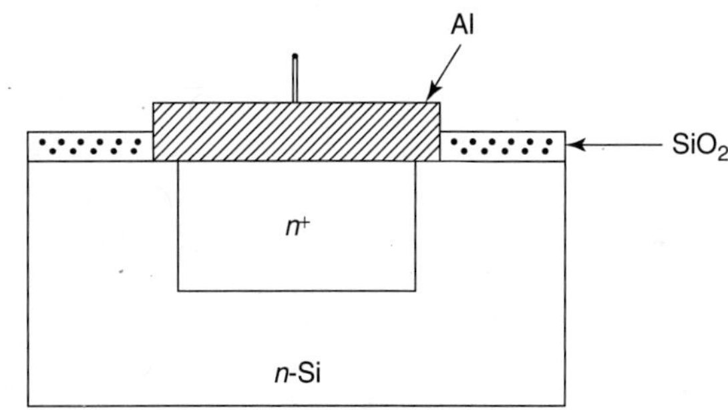


**Figure 10-19** Ohomic contact

The $n^+$ region is formed with highly doped Al; the same metal is used for metallic contact.

### The applications of Schottky contact

Electron availability is very high in the metal as the Schottky diode is made with the directly deposited metal on semiconductor surface. Therefore, the Schottky diode is a majority carrier device. As the time of injection of the electrons from conduction band to metal decreases, the switching becomes faster. This diode has low reverse and forward impedances. So, it is used in high-speed switching devices.

## 10-17-2 Steps of Fabrication of Capacitors

In ICs, capacitors are fabricated using the depletion region capacitance of a reverse-bias $p$–$n$ junction, or by MOS structure. In case of junction capacitance, the magnitude depends on the dimension and type of the junction, and the applied bias. A depletion capacitance can be formed using MOS structure. A parallel plate capacitor is obtained by MOS technique.

A pair of $n^+$ layers is deposited to form source (S) and drain (D), as shown in Fig. 10-20. A thin silicon dioxide ($SiO_2$) layer acts as the dielectric layer. The heavily doped $n$-region and the Al layers act as the two plates of the capacitor. The magnitude of the capacitance is generally very small, in the order of picofarad.

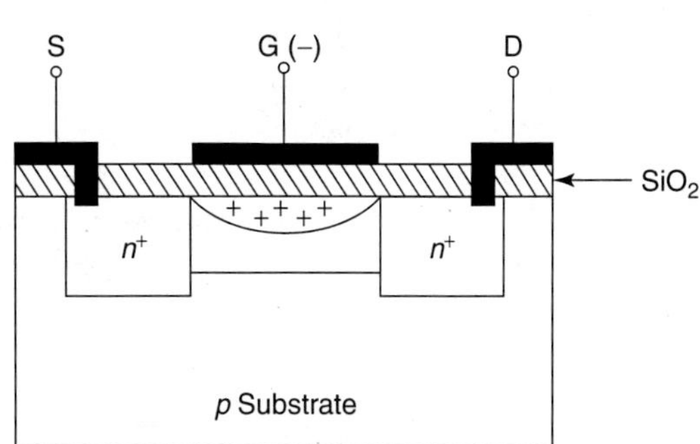


**Figure 10-20** Capacitor on IC

## 10-17-3 Steps of Fabrication of the Transistor Circuit

A step-by-step process will now be followed for the fabrication of a transistor circuit, as shown in Figs. 10-21(a) and 10-21(b).

**Step I:** On the *p*-type substrate a layer of *n*-type semiconductor is formed. However, in the oxidation process a thin layer of $SiO_2$ is formed for isolation.

**Step II:** Photolithographic process using UV ray source is shown in Fig. 10-21(c). The optical mask of a desired pattern is placed in between UV source and the Kodak photo resist (KPR). The mask consists of a dark pattern. The UV ray, which is allowed to fall on the selected region, becomes solid polymerized; other unexposed regions remain liquid, which are then removed.

**Step III:** The $p^+$ diffusion at the window region is done by ion implantation technique or by chemical deposition, as shown in Fig. 10-21(d).

**Step IV:** Again, to form a *p*-region within an *n*-region, another set of masks are used, and the second step of photolithography is performed, as shown in Fig. 10-21(e).

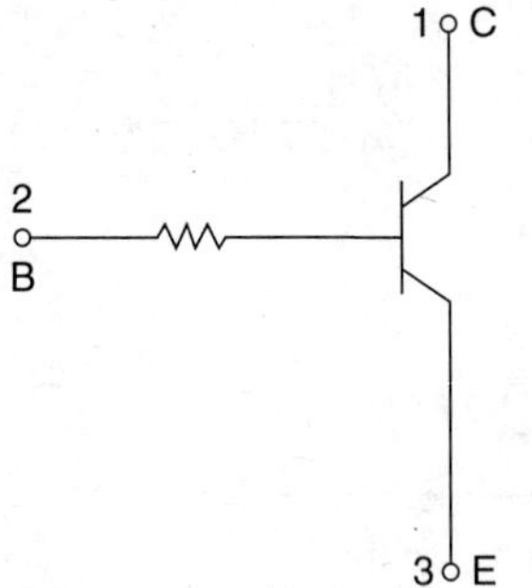


**Figure 10-21(a)** Transistor circuit

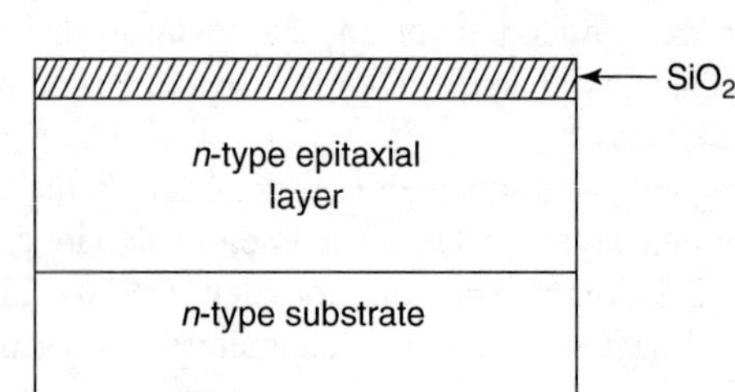


**Figure 10-21(b)** Epitaxial layer growth and oxidation

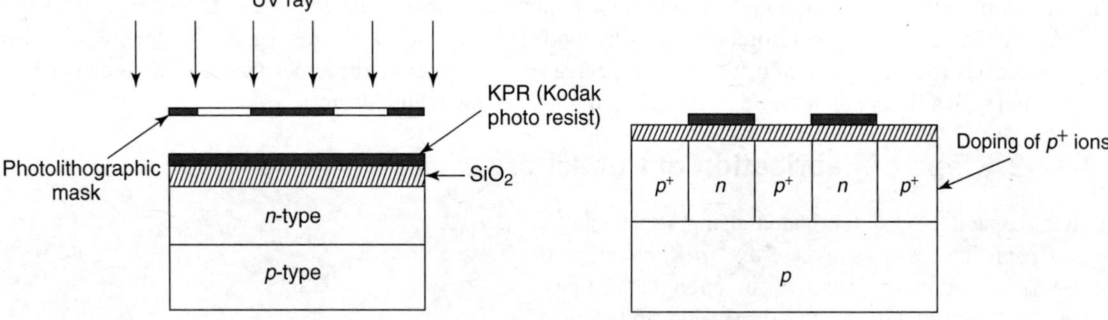


**Figure 10-21(c)** Photolithography process

**Figure 10-21(d)** Etching process

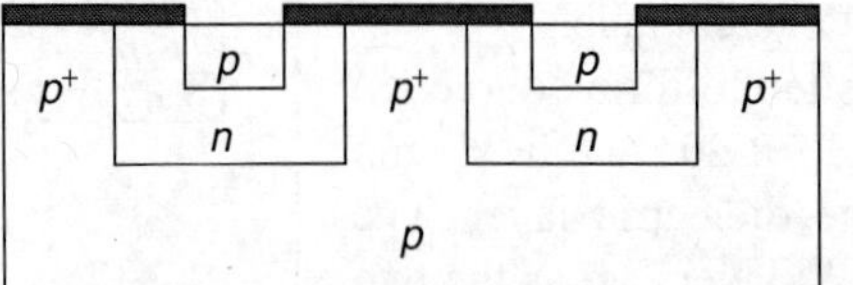


**Figure 10-21(e)** Second level of photolithography and doping

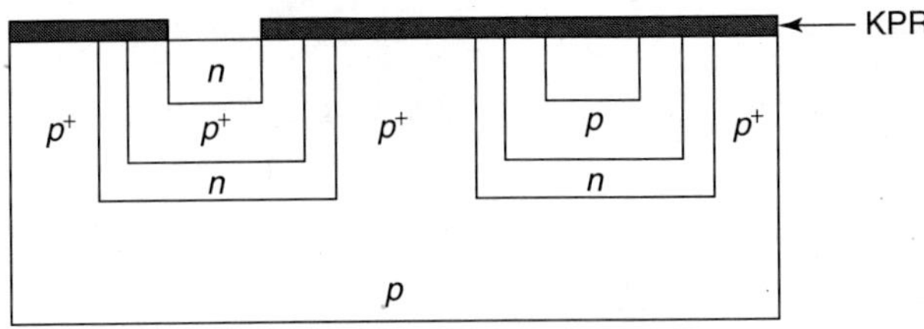


Figure 10-21(f) Third level of doping

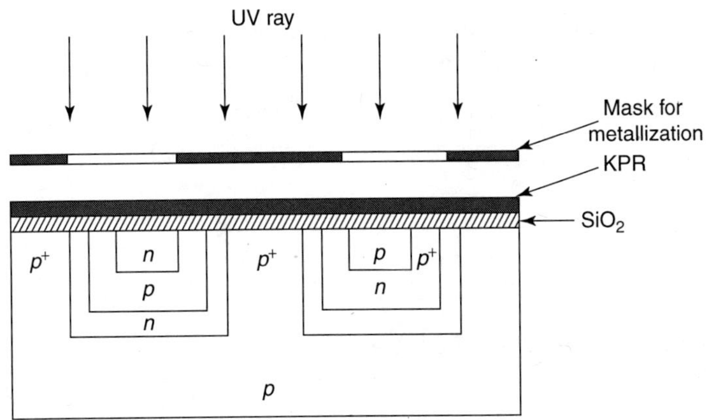


Figure 10-21(g) Multilevel pholithography

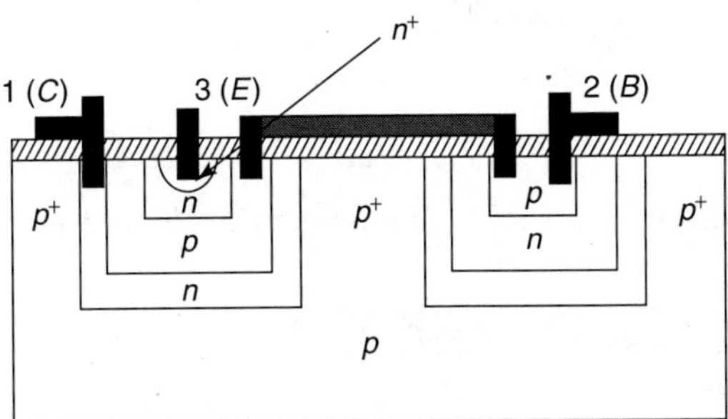


Figure 10-21(h) Metal interconnections for different contact points

**Step V:** The third-level mask is used to get the pattern, as shown in Fig. 10-21(f). Multiple types of masks and subsequent photolithography is done to achieve the desired complicated circuit.

**Step VI:** Before metallization, the metal mask of the desired pattern is used. In the mask, the blank space is kept where we want to go for metallization [see Fig. 10-21(g)].

**Step VII:** Metal interconnections for different contact points are used, as shown in Fig. 10-21(h). The metallization is performed at a very low pressure, at about $10^{-6}$ forr. Special pumps are used to achieve low pressure and $O_2$ free environment in order to avoid oxidation. Total metallization is preformed in a closed chamber and the room should be clean. High pure gold or Al is used for metallization. Purity is 99.9999% and the cost of the metal is very high.

## Solved Examples

**Example 10-1** Fabricate the circuit, as shown in the following diagram.

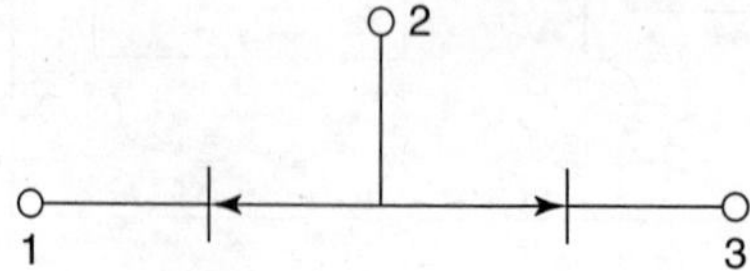


**Solution:**

In the circuit shown in the diagram, two diodes are connected back to back with a common *p*-type. Oxidation is done on the *p–n* layer, where *p* is the substrate.

**Step I:** Oxidation is done on the *p–n* layer, where *p* is the substrate.

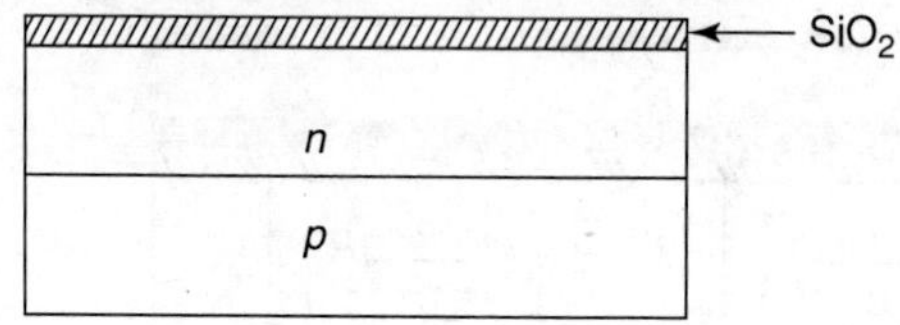


**Step II:** Photolithography is performed using UV source.

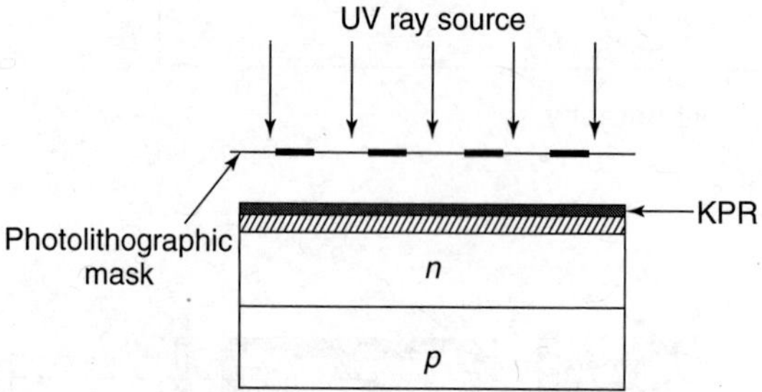


**Step III:** After photolithography the desired window region is achieved, and the remaining $SiO_2$ is removed by chemical etching.

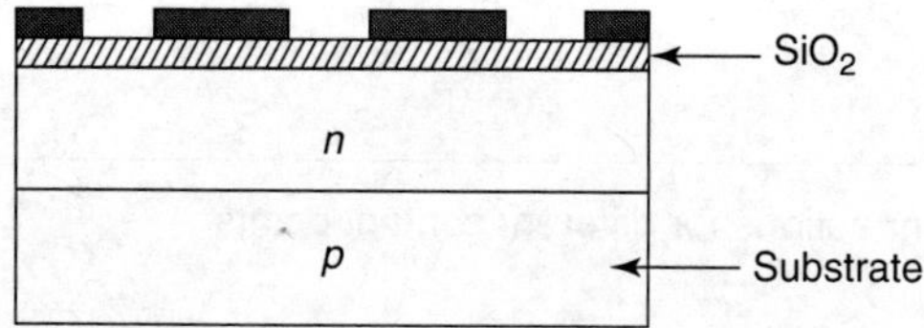


**Step IV:** In the window region, $p^+$ ions are doped for the following desired pattern. By doping, a $p^+$ region isolated of the *n*-region can be achieved.

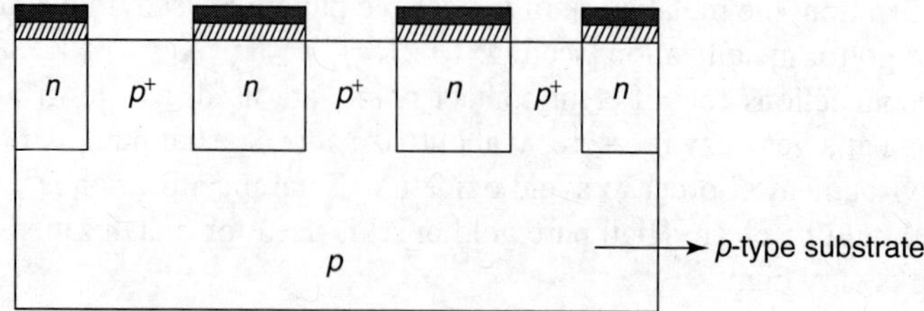

**Step V:** Using the metal masking, subsequent metallization is performed at a very low pressure in order to avoid oxidation.

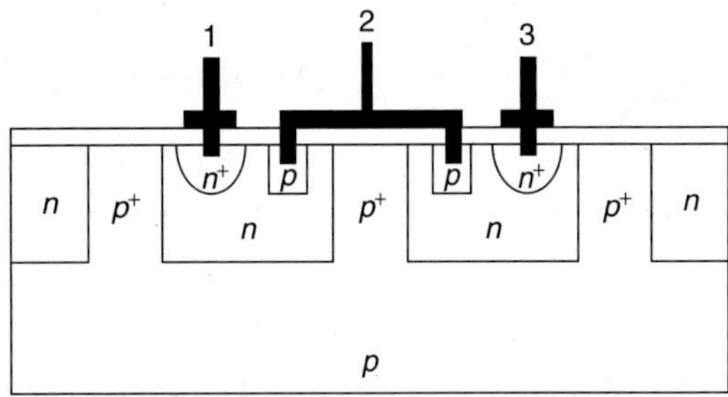


Thus, the desired pattern is achieved in these steps.

**Example 10-2** Fabricate the circuit, as shown in the following diagram.

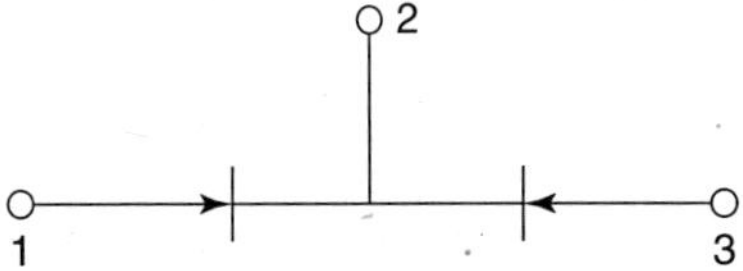


**Solution:**

Two front-to-front connected diodes are given in the circuit, as shown in the diagram.

**Step I:** The *n*-type layer is grown on the *p*-type substrate.

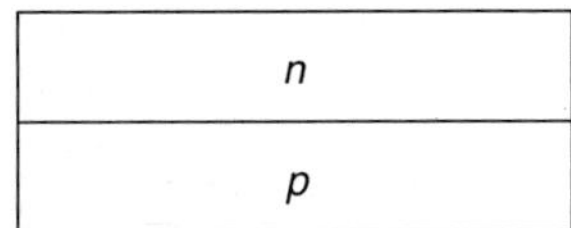


**Step II:** By wet or dry oxidation, the $SiO_2$ layer is grown on top of the *n*-type layer.

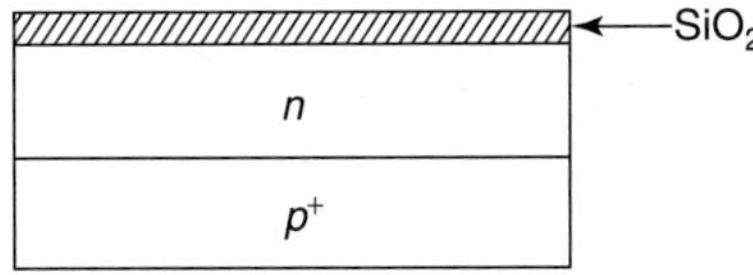


**Step III:** Photolithography is performed using mask.

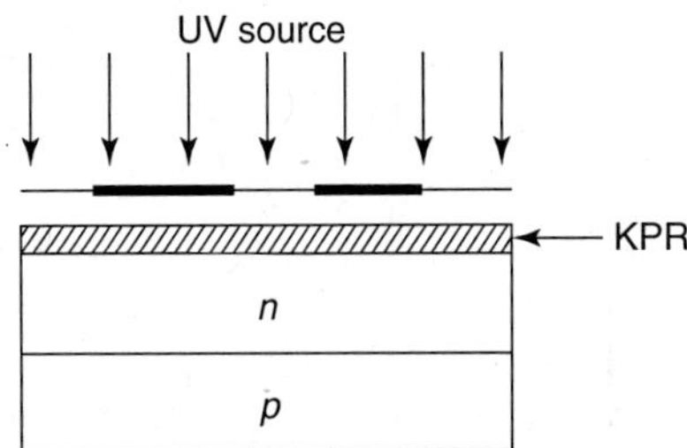

**Step IV:** The $SiO_2$ layer is removed by chemical or wet etching.

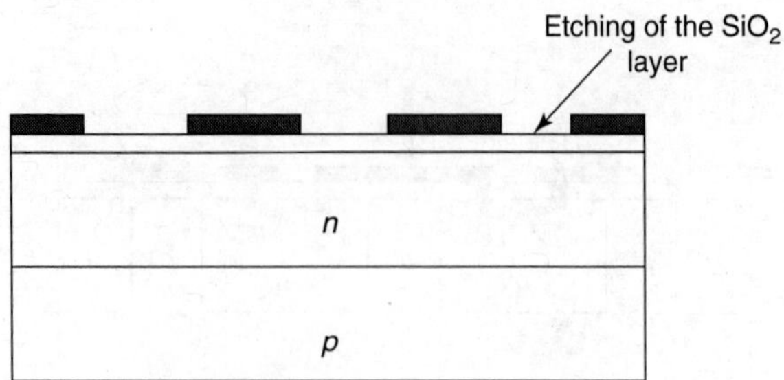


**Step V:** After removing the oxide layer, the *p*-type impurity within the window region is doped.

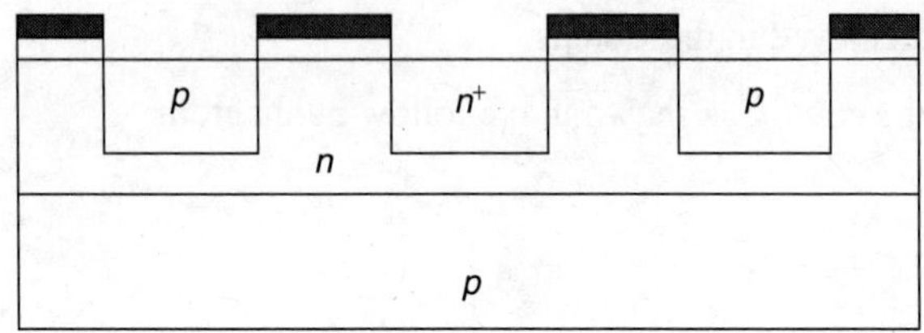


**Step VI:** Metallization is performed to get the contact points 1, 2 and 3. Terminals 1 to 2 form one diode, and 2 to 3 form another diode placed front-to-front.

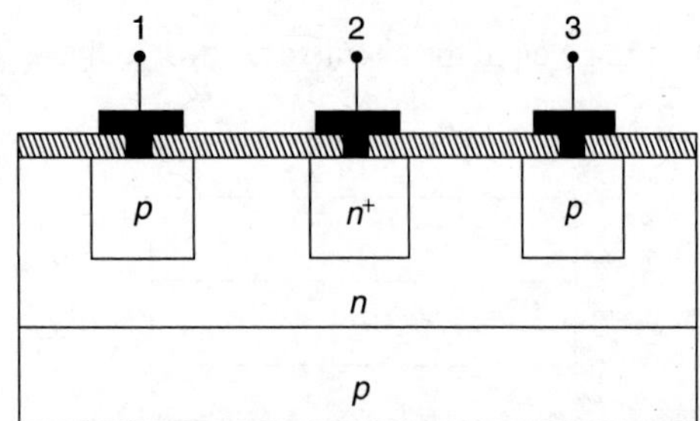


**Example 10-3** Draw the cross-sectional diagram of the circuit, as shown in the following diagram.

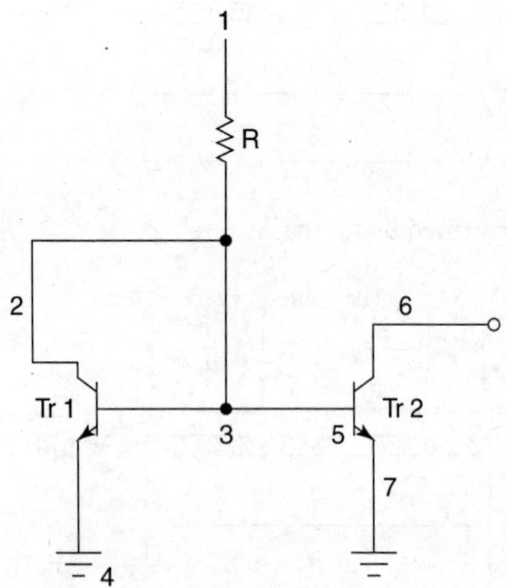

**Solution:**
The given diagram shows a transistor circuit. The following diagram shows the cross sectional view of the transistor.

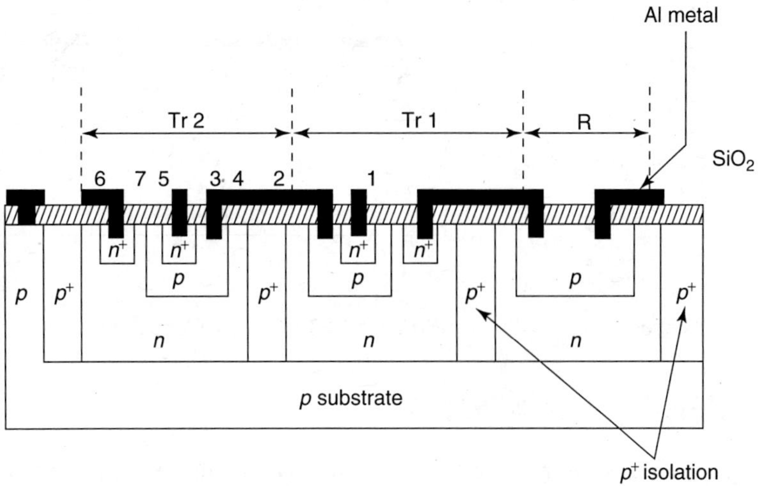


■

## 10-17-4 The Schottky Diode

In Fig. 10-22, terminal 1 forms the ohmic contact, because Al is deposited on the $n^+$ region and not on the $n^-$ region, but terminal 2 forms the Schottky contact. The Schottky contact together with the two terminals forms the Schottky diode

## 10-17-5 Schematic Diagram of a CMOS Circuit

CMOS or complimentary MOS consists of *n*MOS and *p*MOS. Figure 10-23(a) shows the CMOS circuit.

Figure 10-23(b) shows that one *n*MOS and one *p*MOS are placed on a common substrate. It looks like two tubs are placed on a same base. The source, gate and drain are indicated by *S*, *G* and *D* respectively.

The CMOS circuit contains both *n*MOS and *p*MOS; it is either *n*MOS or *p*MOS for a complete cycle of input signal and it is operated for either a +ve or a −ve cycle. Therefore, the CMOS is ON for both the half-cycles. With the help of twin-tub method, a complete CMOS can be realized.

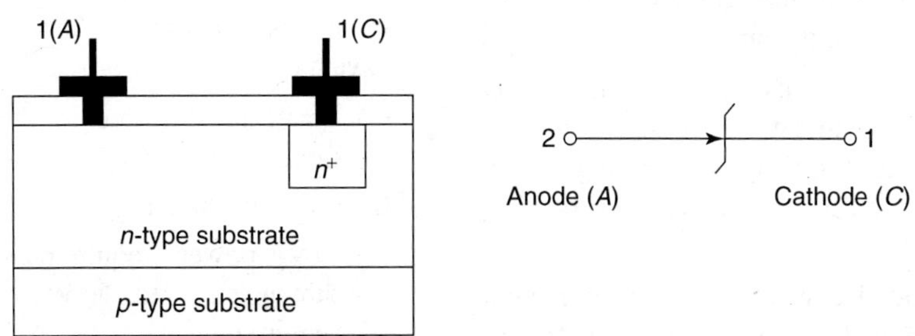


**Figure 10-22** Cross-sectional view of a Schottky diode

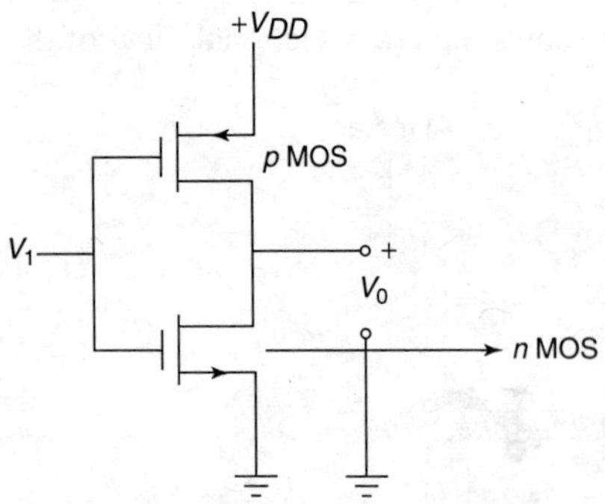


Figure 10-23(a) CMOS circuit

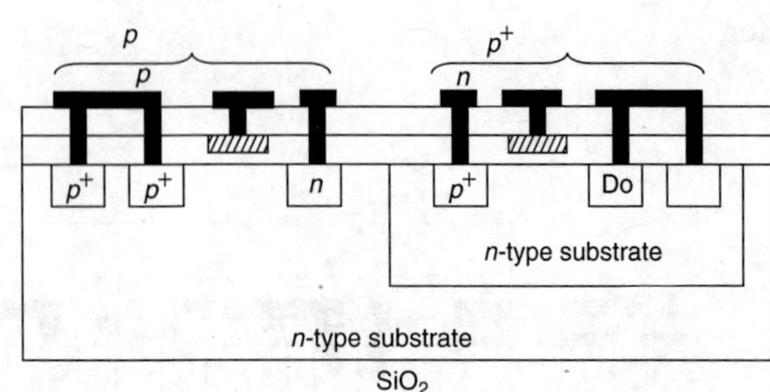


Figure 10-23(b) Cross sectional view of twin tub CMOS

## 10-18 REAL-LIFE APPLICATIONS

In our daily life integrated circuits are applied everywhere. In addition to their wide use in computers and mobile phones, ICs are mostly used in optoelectronic devices—LED, LASER, modulator, demodulator, set-top box, etc. Even a calling bell uses an IC with music encoded in it. Solar cells are very popular as an alternative source of energy. They are also very popular in the photovoltaic industry today. These also use IC chips.

Since the early 1990s, many new electronic materials have been incorporated into silicon CMOS transistors for enhancing their device performance and efficiency. This trend will continue, and it is expected that more non-silicon-based materials will be incorporated into CMOS transistors and integrated onto the silicon substrate in future technology nodes. With the advent of new fabrication techniques, 90 nm, 65 nm and 45 nm device fabrication is possible. This lower dimension technology starts a new era of small and powerful devices and also makes technology integration possible. For example, we were not happy with mobile phones equipped only with voice service, but nowadays cell phones come with numerous features—camera, music, radio, bluetooth, etc. This technology integration is possible due to the advancement in fabrication and chip-processing technology. We are looking forward for many more advanced features as time progresses.

## POINTS TO REMEMBER

1. In monolithic circuits, the entire circuit is built into a single piece of semiconductor chip consisting of passive and active components; physical properties of the semiconductor determine performance of the circuit.
2. Hybrid integrated circuits are devices that apply standard semiconductor processing technology to individual ICs, and fuse them together to simultaneously form an electrical, mechanical, and thermal bond.
3. Hybrid integrated circuit offers a new a paradigm in integrated circuit and system designs and architectures by permitting:
   (a) Increase in communication bandwidth
   (b) Modular chip design process
   (c) Higher system yields with more reliability
4. Integrated circuits have the following advantages:
   (a) Small in size due to the reduced device dimension
   (b) Low weight due to very small size
   (c) Low power requirement due to lower dimension and lower threshold power requirement
   (d) Low cost due to large-scale production

5. The Czochralski process is used to grow ingot and subsequently to design the wafer.
6. The epitaxial process is used to form a layer of single-crystal silicon on an existing crystal wafer of the same or different material.
7. Thermal oxidation of silicon is achieved in the presence of water vapour; this process is called wet oxidation.
8. The process for pattern definition by applying thin uniform layer of viscous liquid photo resist on the wafer surface is the photolithography process.
9. The composite drawing of the circuit is partitioned into several levels called masking levels, used in fabricating the chip.
10. Etching is the process of removing the unwanted portion of the layer or region from the surface during the fabrication process. Etching can be of two types:
    (a) Dry etching
    (b) Chemical etching
11. Diffusion of impurities into substrate layer is the basic step in the planar process.
12. The ionized particles are accelerated through an electrical field and targeted at the semiconductor wafer.
13. The metallization process is used to form the interconnections of the components on the chip.
14. IC testing is performed before packaging.

## OBJECTIVE QUESTIONS

1. Monolithic IC consists of:
    (a) Active components
    (b) Passive components
    (c) Both active and passive components
    (d) None of the above
2. In monolithic IC:
    (a) Performance depends on the substrate
    (b) Performance does not depend on the substrate
    (c) Performance depends on interconnects
    (d) Performance depends on packaging
3. Oxidation is used for:
    (a) Isolation (b) Interconnection
    (c) Doping (d) None of the above
4. Doping means:
    (a) Addition of impurity material in semiconductor band structure
    (b) Removing of impurity material in semiconductor band structure
    (c) Cleaning the surface
    (d) None of the above
5. Metallization is used for:
    (a) Interconnection (b) Protection
    (c) Packaging (d) None of the above
6. Packaging is used for:
    (a) Protection (b) Safety
    (c) Both (a) and (b) (d) None of the above
7. Testing is used for:
    (a) Checking reliability
    (b) Quality control
    (c) Both (a) and (b)
    (d) None of the above
8. Optical masking is used for:
    (a) Pattern transfer
    (b) Protection
    (c) Cleaning
    (d) None of the above
9. Etching is used for:
    (a) Selective removal of the unwanted surface
    (b) Cleaning
    (c) Interconnection
    (d) None of the above
10. System on chip means:
    (a) It consists of both analog and digital IC
    (b) It consists of only analog IC
    (c) It consists of only digital IC
    (d) None of the above

11. Advantages of IC are:
    (a) Small size
    (b) Low cost
    (c) High packing density
    (d) None of the above

12. Inductor design in an IC:
    (a) Is not possible
    (b) Is possible
    (c) Is possible with discrete components
    (d) None of the above

## REVIEW QUESTIONS

1. What is an IC? Explain its application in modern life.
2. Classify the different types of IC. Explain their uses.
3. What is the difference between Ohomic contact and Schottky barrier contact?
4. Explain the process of crystal growth.
5. Explain Moore's law.
6. Classify different types of oxidation process. Explain each process in detail.
7. What is photolithography?
8. Classify different types of photolithographic processes and explain their working principle in detail.
9. What are the differences between positive and negative mask?
10. What are the differences between dry etching and wet etching?
11. Explain the process of ion implantation.
12. Explain the process of metallization.
13. Explain the impact of the use of IC in modern life.

## PRACTICE PROBLEMS

1. Draw the schematic diagram of a MOS transistor.
2. Draw the schematic diagram of a bipolar junction transistor.
3. Draw the schematic diagram of fabrication of a diode with a resistance in series.
4. Draw the schematic diagrams of CMOS circuit.

## SUGGESTED READINGS


1. Singh, J. 1994. *Semiconductor Devices: An Introduction.* New York, NY: McGraw-Hill.
2. Streetman, B.G. and S. Banerjee. 2000. *Solid State Electronic Devices.* New Delhi: Pearson Education.
3. Millman, Jacob and Christos C. Halkias. 1986. *Integrated Electronics: Analog and Digital Circuits and Systems.* New Delhi: McGraw Hill Book Company.
4. Pierret, R. F and G.W. Neudeck. 1989. *Modular Series on Solid State Devices.* Boston, M.A.: Addison Wesley.
5. Gandhi, S. K. 2002. *VLSI Fabrication Principles.* New Delhi: Wiley & Sons.
6. Singh, B. P. and Rekha Singh. 2006. *Electronic Devices and Integrated Circuits.* New Delhi: Pearson Education.

# 11

# Operational Amplifier

## Outline



## Objectives

In this chapter we examine the operational amplifier, which is the basic component of analog computers. The most commonly used op-amp IC is the 8 pin IC 741C. The internal structure of the op-amp, and the basic fabrication steps of integrated circuits are discussed in detail. All the basic terminal properties of the op-amp are also explained in this chapter and some practical applications of the op-amp are illustrated with real-life examples.

## 11-1 INTRODUCTION

The operational amplifier (popularly known as op-amp) is an active device used to design circuits that perform useful operations, such as generating sine waves or square waves; amplifying, combining, integrating, differentiating and removing noise; and transforming alternating current into direct current and vice-versa. It can also change the shape of a waveform, produce a change in the output when an input signal reaches a certain level, provide constant voltage or current, and perform various other important circuit operations. Op-amp circuits are very important as we develop a valuable perception about how electronic circuits work in general.

An op-amp is a very high-gain differential amplifier with high input impedance and low output impedance. Figure 11-1 shows a basic op-amp with two inputs and one output. The negative terminal is known as the inverting input terminal (Input 1), and the positive terminal is known as the non-inverting input terminal (Input 2). Each input results in an output, which further depends upon the input that is being applied to positive (+) or negative (−) input terminals. The op-amp is known as the differential amplifier because it amplifies the voltage difference of the inverting and non-inverting terminals.

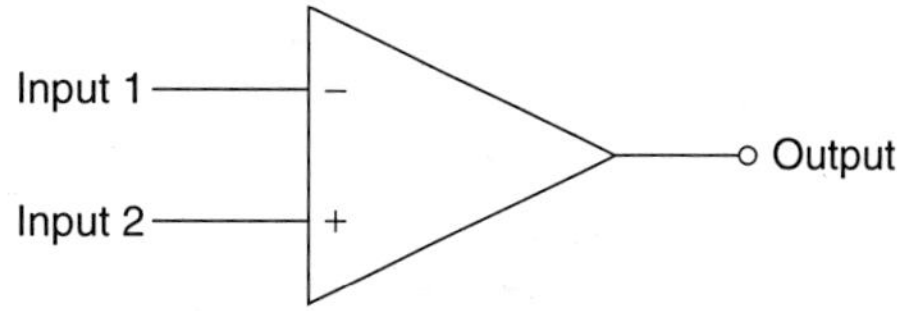


**Figure 11-1** Basic op-amp

**Table 11-1** Comparison between the parameters of an ideal op-amp and a practical op-amp

| *Op-amp Parameters* | *Ideal Op-amp Parameters* | *Practical Op-amp Parameters* |
|---|---|---|
| Gain | Infinite | $10^3$ to $10^6$ order |
| Output voltage | Zero (0 volt) | Few volts (in $\mu$V or nV), due to offset |
| Input resistance | Infinite ($\alpha$) | $10^3\ \Omega$ to $10^6\ \Omega$ order |
| Output resistance | Zero (0 $\Omega$) | Few ohm ($\Omega$) order |
| CMRR | Infinite ($\alpha$) | 100 dB order |
| Bandwidth | Infinite ($\alpha$) | Mega Hz order |
| Slew rate | Infinite ($\alpha$) | 0.5 V/$\mu$s order |

## 11-2 PROPERTIES OF THE IDEAL OPERATIONAL AMPLIFIER

An ideal op-amp should have the following properties:

1. Gain must be infinite
2. Output voltage must be zero when input voltages are same or when both are zero
3. The input resistance must be infinite
4. The output resistance must be zero
5. The common mode rejection ratio (CMMR) must be infinite
6. Infinite bandwidth, i.e., it must allow all frequencies to pass
7. Op-amp characteristics should not drift with temperature

All these parameters for an ideal op-amp are different from those of a practical op-amp, as illustrated by the comparison given in Table 11-1.

## 11-3 SPECIFICATIONS OF IC 741C

The op-amp popularly used in the laboratory is IC 741C. It is an eight (8) pin DIP (dual input package) IC, as shown in the Fig. 11-2.

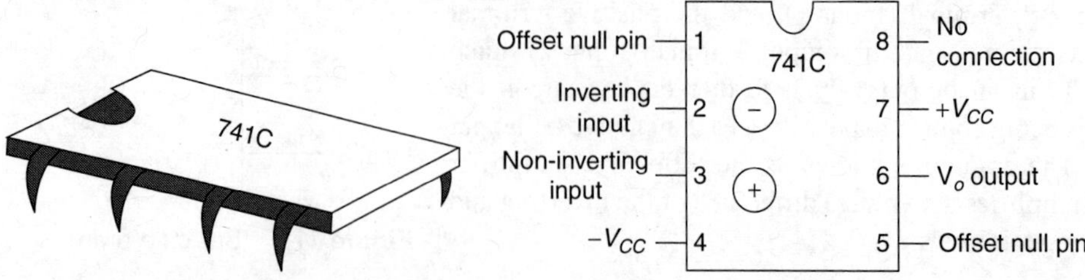


**Figure 11-2** Block diagram of 8 pin IC 741C

### 11-3-1 Description of Op-Amp 741 IC Pins

Pins 1 and 5: These two pins are used for offset null process.

Pin 2: Inverting input terminal, i.e., when a sinusoidal signal is applied to the input pin 2, inverted output is obtained at the output terminal 6.

Pin 3: Non-inverting input terminal, i.e., when a sinusoidal signal is applied to the input pin 3, waveform of same phase output is obtained.

Pin 4: $-V_{cc}$, i.e., negative terminal of supply voltage is connected to this pin.

Pin 6: Output terminal.

Pin 7: $+V_{cc}$, i.e., positive terminal of supply voltage is connected to this pin.

Pin 8: No electrical connection is there in this pin; this pin is just for balance and the symmetric dual-input package look.

## 11-4 OPERATIONAL AMPLIFIER AND ITS TERMINAL PROPERTIES

An ideal operational amplifier is perfectly balanced with the output voltage $V_o = 0$ for $V_1 = V_2$ i.e., the op-amp output is zero when the same or zero (ground) voltage is applied to both the inverting and non-inverting terminals, as shown in Fig. 11-3(a). A complete representation of the op-amp is shown in Fig. 11-3(b).

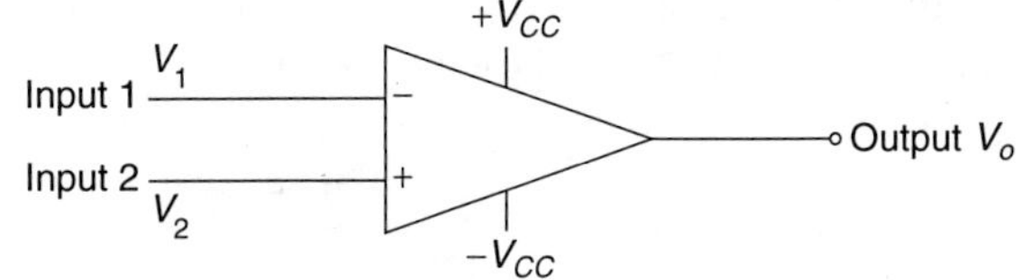

**Figure 11-3(a)** Op-amp input and output terminals

A practical op-amp exhibits an unbalance caused by a mismatch of the input transistors ($Q_1$ and $Q_2$). This mismatch results in the flow of unequal bias currents through the input terminals. This in turn, causes a change in the various parameters of an op-amp. The circuit diagram of a differential amplifier is shown in Fig. 11-4.

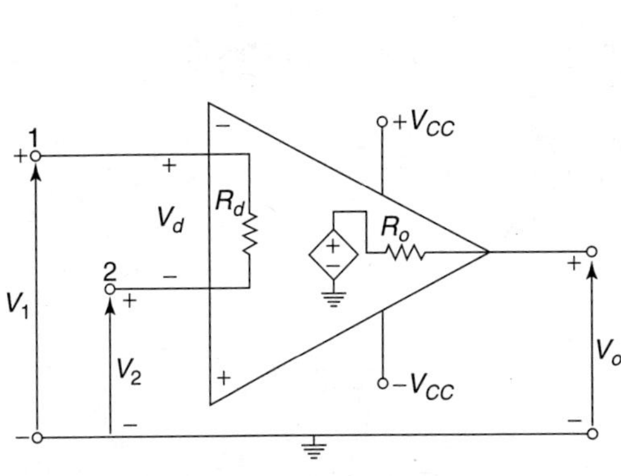

**Figure 11-3(b)** Complete representation of op-amps

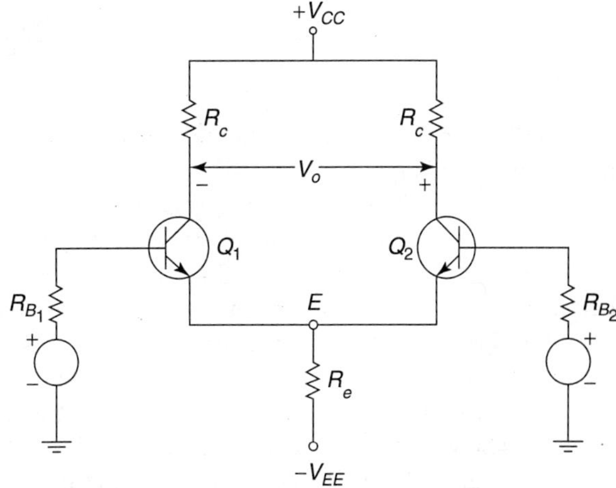

**Figure 11-4** Circuit diagram of a differential amplifier

## 11-4-1 Input Offset Voltage and Output Offset Voltage

The input offset voltage, $V_{io}$, is the differential input voltage that exists between two input terminals of an op-amp without any external inputs applied. In other words, it is the amount of input voltage that should be applied between the two input terminals in order to force the output voltage to zero, as shown in Fig. 11-5.

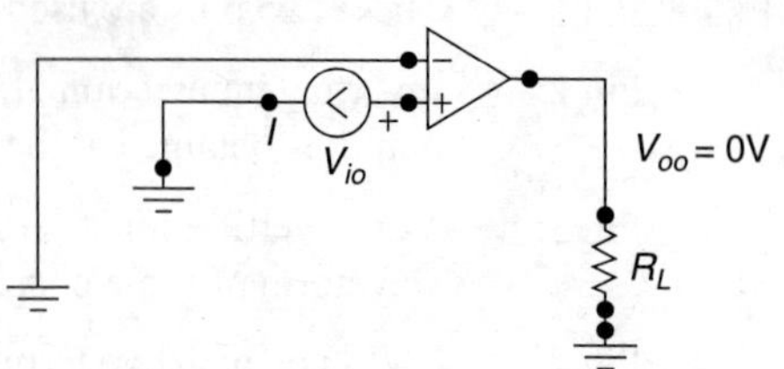


**Figure 11-5** Input offset voltage in an op-amp

The output offset voltage, $V_{oo}$, is a dc voltage; it may be positive or negative in polarity depending upon whether the potential difference between two input terminals is positive or negative. It is impossible to predict the polarity of the output offset voltage as it is dependent on the mismatching between the two input terminals.

To reduce $V_{oo}$ to zero we need to have a circuit at the input terminals of the op-amp that will give us the flexibility of obtaining $V_{io}$ with proper amplitude and polarity. Such a circuit is called *input offset voltage-compensating network*, as shown in Fig. 11-5. Before we apply external input to the op-amp, with the help of an offset voltage-compensating network, we reduce the output offset voltage $V_{oo}$ to zero; the op-amp is said to be nulled or balanced. The total output offset voltage can be expressed as:

Output offset voltage $V_{oo}$ = offset due to input voltage $V_{io}$ + offset due to input offset current.

The output offset voltage diagram is given in Fig. 11-6.

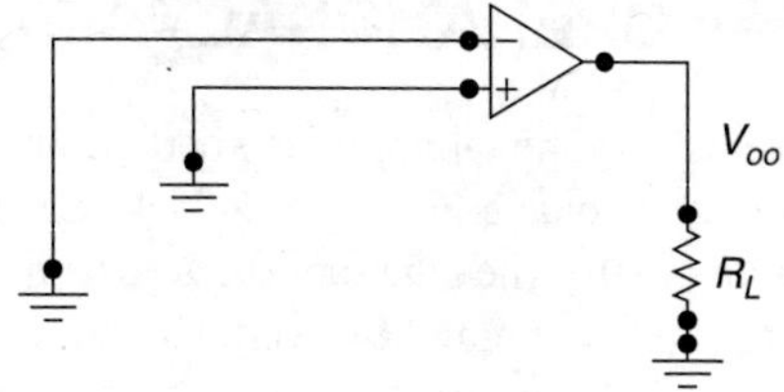


**Figure 11-6** Output offset voltage in an op-amp

## 11-4-2 Input-Bias Current

The input-bias current $I_B$ is defined as the average of the two input-bias currents, $I_{B1}$ and $I_{B2}$ as shown in Fig. 11-7, and is given by:

$$I_B = \frac{I_{B1} + I_{B2}}{2} \tag{11-1}$$

where, $I_{B1}$ is the dc bias current flowing into the non-inverting input, and $I_{B2}$ is the dc bias current flowing into the inverting input.

In the Fig. 11-7, both the input terminals are grounded so that no input voltage is applied to the op-amp. However, the plus-minus sign supply voltages are necessary to bias the op-amp properly.

The input-bias currents $I_{B1}$ and $I_{B2}$ are the base bias currents of the two transistors in the input differential amplifier stage of the op-amp. Even though both of the input transistors are identical, it is not possible to have $I_{B1}$ and $I_{B2}$ exactly equal to each other because of the external imbalance between the two inputs.

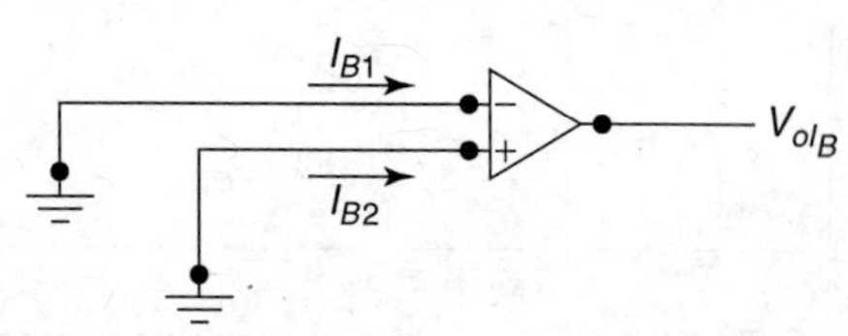


**Figure 11-7** Input-bias current in an op-amp

Although very small, the input-bias current $I_B$ can cause a significant output offset voltage in circuits using relatively large feedback resistors. This output offset voltage may not be as large as that caused by the input offset voltage, but certain precautions must be taken to minimize it.

Here we obtain the expression for the output offset voltage caused by the input-bias current $I_B$ in the inverting and non-inverting amplifiers, and then devise some scheme to eliminate or minimize it. The non-inverting or inverting amplifier with $V_{in} = 0$ is shown in Fig. 11-18. Let us assume that the input offset voltage $V_{io}$ is zero, and there is no offset voltage due to $V_{io}$. Let $V_{oI_B}$ be the output voltage due to the input-bias current $I_B$.

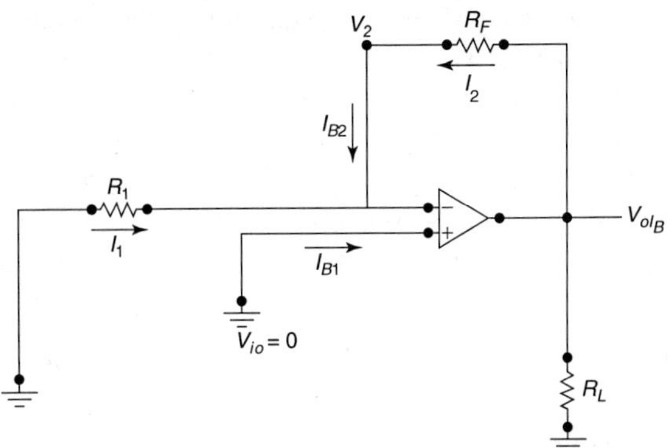


**Figure 11-8** Output offset voltage due to input-bias current in an inverting or non-inverting amplifier

In Fig. 11-8 the input-bias currents $I_{B1}$ and $I_{B2}$ are flowing into the non-inverting and inverting input leads, respectively. The non-inverting terminal is connected to the ground; therefore, the voltage $V_1 = 0$ V. The controlled voltage source $AV_{io} = 0$, since $V_{io} = 0$ is assumed. With output resistance $R_0$ negligibly small, the right end of $R_F$ is essentially at ground potential; that is resistors $R_1$ and $R_F$ are in parallel, and the bias current $I_{B2}$ flows through them. Therefore, the voltage at the inverting terminal is:

$$V_2 = (R_1 \parallel R_F)\, I_{B2} \tag{11-2}$$

$$V_2 = \frac{R_1 R_F}{R_1 + R_F} I_{B2} \tag{11-3}$$

Writing the node voltage equation for node $V_2$, we get:

$$I_1 + I_2 = I_{B2}$$

$$\frac{0 - V_2}{R_1} + \frac{V_{oI_B} - V_2}{R_F} = \frac{V_2}{R_i} \tag{11-4}$$

where, $V_{oI_B}$ is the output offset voltage due to input-bias current, and $R_i$ is the input resistance of the op-amp. Re-arranging Eq. (11-4), we get:

$$\frac{V_{oI_B}}{R_F} = V_2\left(\frac{1}{R_1} + \frac{1}{R_F} + \frac{1}{R_i}\right)$$

Since $R_i$ is extremely high (ideally ∞) $\frac{1}{R_i} \cong 0$. Therefore:

$$\frac{V_{oI_B}}{R_F} = V_2\left(\frac{R_1 + R_F}{R_1 R_F}\right) \tag{11-5}$$

Substituting the value of $V_2$ from Eq. (11-3) in Eq. (11-5), we get:

$$V_{oI_B} = \frac{R_1 R_F I_{B2}}{R_1 + R_F}\left(\frac{R_1 + R_F}{R_1}\right)$$

$$V_{oI_B} = R_F I_{B2} \tag{11-6}$$

Since, $I_{B1} = I_{B2} = I_B$, we can write Eq. (11-6) as:

$$V_{oI_B} = R_F I_B \tag{11-7}$$

According to Eq. (11-7), the amount of output offset voltage $V_{oI_B}$ is a function of feedback resistor $R_F$ for a specified value of input-bias current $I_B$. The amount of $V_{oI_B}$ is increased by the use of relatively large feedback resistors. Therefore, the use of small feedback resistors is recommended.

To eliminate or to reduce the output offset voltage $V_{oI_B}$ due to input-bias current $I_B$, we have to introduce some scheme at the input by which voltage $V_1$ can be made equal to voltage $V_2$. In other words, if voltages $V_1$ and $V_2$—caused by the currents $I_{B1}$ and $I_{B2}$—can be made equal, there will be no output voltage $V_{oI_B}$.

From Eq. (11-3) we have:

$$V_2 = R_P I_{B2} \tag{11-8}$$

where,

$$R_P = \frac{R_1 R_F}{R_1 + R_F}$$

Equation (11-8) implies that we must express voltage $V_1$ at the non-inverting input terminal as a function of $I_{B1}$ and some specific resistor $R_{OM}$. This can be accomplished as follows. The input-bias current $I_{B1}$ does not produce any voltage at the non-inverting input terminal, because this terminal is directly connected to the ground. If we could connect the proper value of resistor $R_{OM}$ to the non-inverting terminal, the voltage $V_1$ would be:

$$V_1 = R_{OM} I_{B1} \tag{11-9}$$

To have voltage $V_1$ equal to $V_2$, the right hand side of Eqs. (11-8) and (11-9) must be equal, therefore:

$$R_P I_{B2} = R_{OM} I_{B1} \tag{11-10}$$

or,

$$\frac{R_1 R_F}{R_1 + R_F} = R_{OM} \tag{11-11}$$

Thus, the proper value required of a $R_{OM}$ resistor connected to the non-inverting terminal is the parallel combination of resistors $R_1$ and $R_F$. However, the use of $R_{OM}$ may eliminate the output offset voltage $V_{oI_B}$ because the currents $I_{B1}$ and $I_{B2}$ are not exactly equal. Nevertheless, the use of $R_{OM}$ will minimize the amount of output offset voltage $V_{oI_B}$; therefore, the $R_{OM}$ resistor is referred to as the *offset minimizing resistor.*

## 11-4-3 Input Offset Current and Output Offset Current

We have seen that the use of $R_{OM}$ in series with the non-inverting terminal reduces the output offset voltage $V_{oI_B}$ caused by $I_B$. However, the value of $R_{OM}$ was derived on the assumption that the input-bias currents $I_{B1}$ and $I_{B2}$ are equal. In practice, these currents are not equal because of the internal imbalances in the op-amp circuitry. The algebraic difference between the individual currents entering into the inverting and non-inverting terminals of a balanced amplifier is referred to as input offset current $I_{io}$. The input offset current $I_{io}$ is used as an indicator of the degree of mismatch between these two currents. Therefore, the value of $I_{io}$ is the difference between two input-bias currents. The input offset current for the 741C is 200 nA maximum.

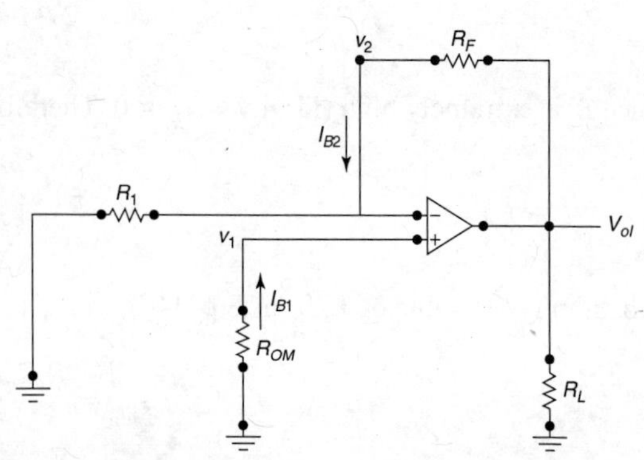


**Figure 11-9** Resistor reducing the output offset voltage caused by the input-bias current

From Fig. 11-9, we can write the expression for input offset current as:

$$I_{io} = |I_{B1} - I_{B2}| \quad (11\text{-}12)$$

In the circuit shown in Fig. 11-10, there will be an output offset voltage due to the input-bias currents $I_{B1}$ and $I_{B2}$. In other words, the output offset voltage can be expressed as a function of input offset current $I_{io}$. Let $V_{OI_{io}}$ be the output offset voltage caused by the input offset current $I_{io}$. To separate the effect of the input offset current from that of the input offset voltage, assume that $V_{io} = 0$ V.

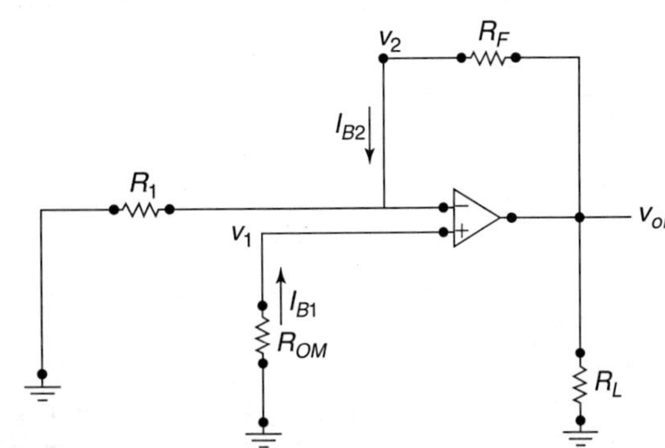


**Figure 11-10** Output offset voltage caused by the input offset current in an inverting or non-inverting amplifier

Referring to Fig. 11-10, and expressing the voltages $V_1$ and $V_2$ in terms of a function of $I_{B1}$ and $I_{B2}$ for a given value of $R_1$ and $R_F$, we get:

$$V_1 = R_{OM} I_{B1} \quad (11\text{-}13)$$

$$V_2 = R_P I_{B2} \quad (11\text{-}14)$$

where, $$R_{OM} = \frac{R_1 R_F}{R_1 + R_F}$$

Applying the superposition theorem, we can now find the output offset voltage due to $V_1$ and $V_2$ in terms of $I_{B1}$, $I_{B2}$ and $R_F$. From Eq. (11-6) we have:

$$V_{oI_{B2}} = -R_F I_{B2}$$

Here the negative sign is used because voltage $V_2$ is the voltage at the inverting input terminal. This output offset voltage $V_{oI_{B2}}$ is due to voltage $V_2$, only in terms of $I_{B2}$, and $R_F$. Similarly, the output offset voltage $V_{oI_{B1}}$ due to $V_1$, only in terms of $I_{B1}$ and $R_F$, can be obtained as follows:

$$V_{oI_{B1}} = V_1 \left(1 + \frac{R_F}{R_1}\right) \quad (11\text{-}15)$$

where, $V_1$ is the voltage at the non-inverting terminal, and $\left(1 + \frac{R_F}{R_1}\right)$ is the gain of the non-inverting amplifier.

Substituting the value of $V_1$ from Eq. (11-13) in Eq. (11-15), we get:

$$V_{oI_{B1}} = R_{OM} I_{B1} \left(1 + \frac{R_F}{R_1}\right)$$

$$= \frac{R_1 R_F}{R_1 + R_F} I_{B1} \frac{R_1 + R_F}{R_1}$$

$$V_{oI_{B1}} = R_F I_{B1} \quad (11\text{-}16)$$

Therefore, the maximum magnitude of the output offset voltage due to $I_{B1}$ and $I_{B2}$ is:

$$V_{oI_{B1}} + V_{oI_{B2}} = R_F I_{B1} - R_F I_{B2}$$

$$= R_F (I_{B1} - I_{B2})$$

$$V_{oI_{io}} = R_F I_{io} \qquad (11\text{-}17)$$

where, $V_{oI_{B1}} + V_{oI_{B2}} = V_{oI_{io}}$ is the output offset voltage due to $I_{io} = |I_{B1} - I_{B2}|$, the input offset current. Thus, for a given value of the input offset current $I_{io}$, the amount of output offset voltage $V_{oI_{io}}$ depends on the value of feedback resistor $R_F$.

The output offset voltage $V_{oo}$ caused by $V_{io}$ could be either positive or negative with respect to the ground. Similarly, the output offset voltage $V_{oI_B}$ caused by $I_B$ could also be positive or negative with respect to the ground. If these output offset voltages are of different polarities, the resultant output offset will be very little. On the other hand, if both of these output offset voltages were of the same polarity, the maximum amplitude of the total output offset would be:

$$V_{ooT} = V_{oo} + V_{oI_B}$$

$$= \left(1 + \frac{R_F}{R_1}\right) V_{io} + R_F I_B \qquad (11\text{-}18)$$

## 11-4-4 Input Offset Null Voltage

Input offset null voltage is the voltage that must be applied between the two input terminals of an op-amp to null the output or to balance the amplifier. In Fig. 11-11, $V_{dc1}$ and $V_{dc2}$ are dc voltages, and $R_s$ represents the source resistance. Input offset voltage is represented by $V_{io}$. This voltage $V_{io}$ could be positive or negative (its absolute value is listed in the datasheet). For IC 741C, the maximum value of $V_{io}$ is 6 mV. The smaller the value of $V_{io}$, the better the input terminals are matched.

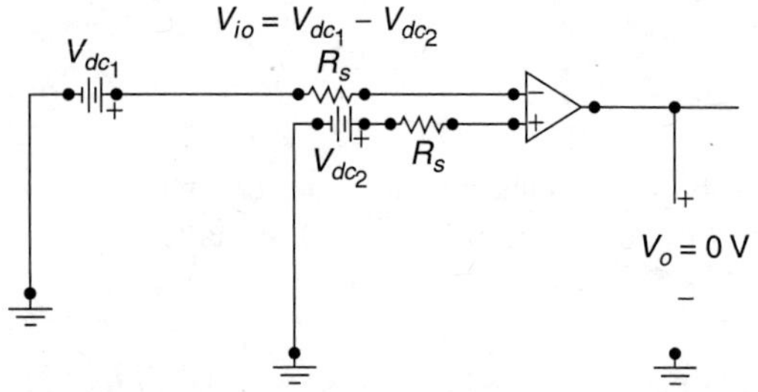


**Figure 11-11** Input offset voltage $V_{io}$

## 11-4-5 Differential Input Resistance

Differential input resistance or input resistance $R_i$ is the equivalent resistance that can be measured either at the inverting or at the non-inverting input terminal with the other terminal connected to the ground. For 741C, the input resistance is a relatively high, 2 MΩ.

## 11-4-6 Input Capacitance

Input capacitance $C_i$ is the equivalent capacitance that can be measured either at the inverting or at the non-inverting terminal with other terminal connected to the ground. A typical value of $C_i$ is 1.4 pF for 741C.

## 11-4-7 Offset Voltage Adjustment Range

One of the features of the 741 family of op-amp is an offset voltage null capability. The 741 op-amp has pins 1 and 5 marked as offset null for this purpose. A potentiometer can connect offset null pins 1 and 5, and the wiper of the potentiometer can be connected to the negative supply $-V_{EE}$. By varying the potentiometer, the output offset voltage can be reduced to zero. Thus, offset voltage adjustment range is the range through which the input offset voltage can be adjusted by varying the potentiometer. For 741C, the offset voltage adjustment range is ±15 mV.

### 11-4-8 Input Voltage Range

When the same voltage is applied to both input terminals, the voltage is called a common-mode voltage $V_{cm}$ and the op-amp is said to be operating in common-mode configuration. For the 741C, the range of the input common-mode voltage is ±13 V at the maximum. This means that the common-mode voltage applied to both input terminals can be as high as +13 V and as low as −13 V, without disturbing the proper functioning of the op-amp. In other words, the input voltage range is the range of the common-mode voltage over which the offset specifications apply. The common-mode configuration is used only for test purposes to determine the degree of matching between the inverting and non-inverting input terminals.

### 11-4-9 Common-Mode Rejection Ratio (CMRR)

If $V_1$ and $V_2$ are the signal voltages applied to the non-inverting and inverting terminals of an op-amp respectively, and $V_o$ is the voltage appearing at the output terminal, then we can define the two terms related to the voltage.

1. Difference signal voltage:

$$V_d = V_1 - V_2$$

2. Common-mode signal voltage:

$$V_c = \frac{V_1 + V_2}{2} \quad (11\text{-}19)$$

The unwanted common-mode signal $V_c$ is the cause of distortion in the output voltage. The differential amplifier has the property of rejecting the noise. The output voltage $V_o$ at the output terminal of the differential op-amp will be:

$$V_o = A_1 V_1 + A_2 V_2 \quad (11\text{-}20)$$

where, $A_1$ is the voltage gain when the inverting terminal is grounded, and $A_2$ is the voltage gain when the non-inverting terminal is grounded.

From Eqs. (11-20) and (11-21) we get:

$$V_1 = V_c + \frac{V_d}{2} \quad (11\text{-}21)$$

$$V_2 = V_c - V_d \quad (11\text{-}22)$$

Using Eqs. (11-22) and (11-23) in Eq. (11-21) we can write:

$$V_o = \frac{(A_1 - A_2)}{2} V_d + (A_1 + A_2)\, V_c$$

$$V_o = A_d V_d + A_c V_c$$

where, $A_d = (A_1 - A_2)/2$ is the difference signal voltage gain, and $A_C = A_1 + A_2$ is the common-mode voltage gain.

Considering the ideal case we find at $A_c = 0$ that $A_1 = -A_2$. Therefore:

$$A_d = \frac{[A_1 - (-A_1)]}{2} = A_1$$

The value of $A_d$ can be infinitely large. In practice it is not true. So, the CMRR is defined as the ratio of the differential voltage gain $A_d$ to the common-mode voltage gain $A_{cm}$. Therefore:

$$\text{CMRR} = \frac{\text{Differential voltage gain}}{\text{Common-mode voltage gain}} = \frac{A_d}{A_{cm}} \quad (11\text{-}23a)$$

The value of CMRR can be expressed in logarithmic form as:

$$\text{CMRR} = 20 \log_{10} \left| \frac{A_d}{A_{cm}} \right| \quad (11\text{-}23b)$$

The differential voltage gain $A_d$ is the same as the large signal voltage gain $A$, however, the common-mode voltage gain is given by:

$$A_{cm} = \frac{V_{ocm}}{V_{cm}} \quad (11\text{-}24)$$

where, $V_{ocm}$ is the output common-mode voltage, and $V_{cm}$ is the input common-mode voltage.

Generally, $A_{cm}$ is very small and $A_d = A$ is very large; therefore, CMRR is very large. Being very large, CMRR is expressed in decibels (dB). For the 741C, CMRR is 90 dB. The higher the value of CMRR, the better is the matching between the two input terminals and the smaller is the output common-mode voltage.

## 11-4-10 Supply Voltage Rejection Ratio (SVRR)

The change in an op-amp's input offset voltage $V_{io}$ caused by a variation in one supply voltage when the other supply voltages remain constant in the circuit is called the supply voltage rejection ratio (SVRR). It is also referred to as power supply rejection ratio (PSRR) or power supply sensitivity (PSS). These are expressed in terms of microvolts per volt or in decibels. If we denote the change in supply voltages by $\Delta V$, and the corresponding change in the input offset voltage by $\Delta V_{io}$, then SVRR can be defined as:

$$\text{SVRR} = \frac{\Delta V_{io}}{\Delta V} \quad (11\text{-}25)$$

For the 741C, SVRR=150 $\mu$V/V. On the other hand, for the 741C:

$$\text{SVRR} = 20 \log_{10} \left( \frac{\Delta V}{\Delta V_{io}} \right) = 104 \text{ dB} \quad (11\text{-}26)$$

This means, that the lower the value of SVRR in microvolts/volt, the better it is for the op-amp performance.

### 11-4-11 Large Signal Voltage Gain

Since the op-amp amplifies the difference voltage between the two input terminals, the voltage gain of amplifier is defined as:

$$\text{Voltage gain} = \frac{\text{Output voltage}}{\text{Differential input voltage}}$$

$$A = \frac{V_o}{V_{id}} \qquad (11\text{-}27)$$

Under the test conditions $R_L \geqslant 2\ \text{k}\Omega$ and $V_o = \pm 10$ V; the large signal voltage gain of 741C is 200,000 typically.

### 11-4-12 Output Voltage Swing

The output voltage swing $V_{o_{\max}}$ of the 741C is guaranteed between $-12$ V and $+12$ V for $R_L \geqslant 2\ \text{k}\Omega$, i.e., giving a 26 V peak-to-peak undistorted sine wave for ac input signals. In fact, the output voltage swing indicates the values of positive and negative saturation voltages of the op-amp. The output voltage never exceeds these limits for the given supply voltages, $+V_{cc}$ and $-V_{cc}$.

### 11-4-13 Output Resistance

The output resistance $R_o$ is the equivalent resistance that can be measured between the output terminal of the op-amp and the ground. For an ideal op-amp, the output resistance is zero. But it is typically 50 Ω or less for the 741C op-amp.

### 11-4-14 Supply Voltage

The op-amp has a dual dc input supply voltage. Supply voltage $V_{cc}$ used in op-amp is generally of a few volts. Typically it varies from $-12$ V to $+12$ V. In IC 741C, pin 4 is connected with the negative supply voltage, say $-12$ V Volt ($-V_{cc}$), and pin 7 is connected with the positive supply, say $+12$ V ($+V_{cc}$). Voltage transfer characteristics of an op-amp depends upon the input supply voltage. The transfer characteristics of an op-amp are shown in Fig. 11-12.

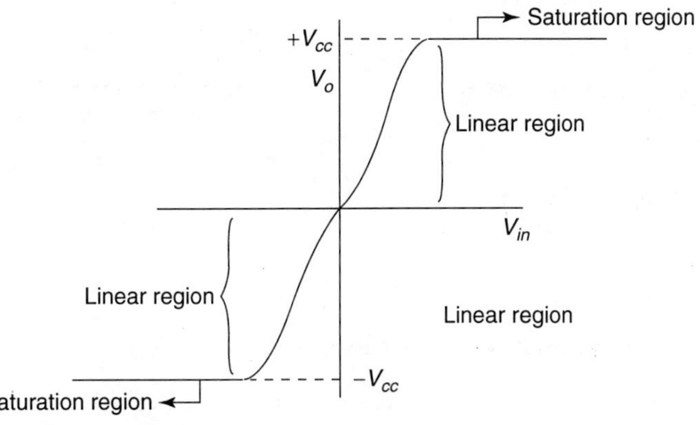


**Figure 11-12** Transfer characteristics of an op-amp

### 11-4-15 Supply Current

Supply current $I_s$ is the current drawn by the op-amp from the power supply. For the 741C the supply current is typically $I_s = 2.8$ mA.

### 11-4-16 Power Consumption

Power consumption ($P_c$) is the amount of quiescent power ($v_{in} = 0$ V) that must be consumed by the op-amp in order to operate properly. The amount of power consumed by the 741C is 85 mW.

### 11-4-17 Slew Rate

Slew rate (SR) is defined as the maximum rate of change in output voltage per unit of time, and is expressed in volts per $\mu$-seconds. The slew rate can be expressed as:

$$\text{SR} = \left.\frac{dV_o}{dt}\right|_{\text{maximum}} \text{V}/\mu\text{s} \quad (11\text{-}28)$$

Slew rate indicates how rapidly the output of an op-amp changes in response to changes in the input frequency. The slew rate changes with change in voltage gain and is normally specified at unity (+1) gain. The slew rate of an op-amp is fixed. Therefore, if the slope requirements of the output signal are greater than the slew rate, then distortion occurs. Thus, slew rate is one of the important factors in selecting the op-amp for ac applications; particularly at relatively high frequencies. One of the drawbacks of the 741C is its low slew rate (0.5V/$\mu$s), which limits its use in relatively high frequency applications, especially in comparators, oscillators, and filters.

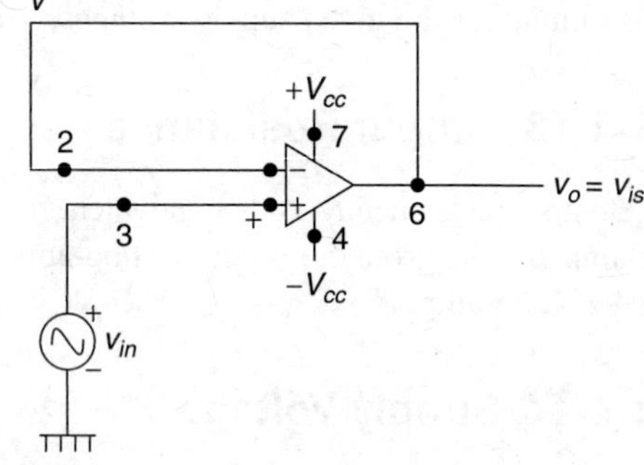


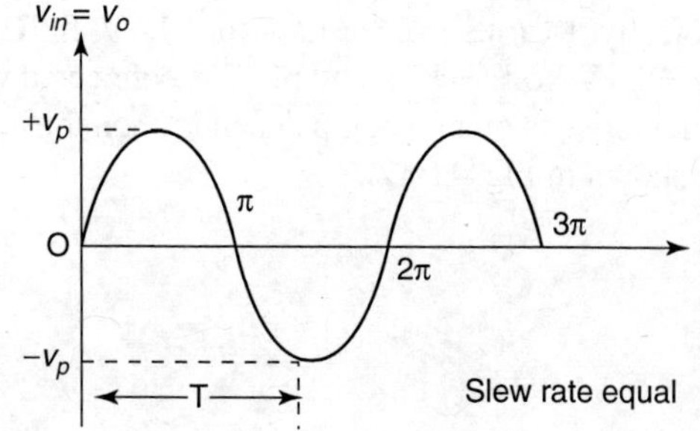


**Figure 11-13** Schematic diagrams for slew rate

For the purpose of calculating the slew rate equation, consider the voltage follower circuit, as shown in Fig. 11-13. From this we can assume that the amplitude of the input voltage is large and of a high-frequency sine wave.

So, we can express the input voltage $V_{in}$ as follows:

$$V_{in} = V_p \sin \omega t \quad (11\text{-}29a)$$

And the output voltage $V_o$ can be written as:

$$V_o = V_p \sin \omega t \quad (11\text{-}29b)$$

Therefore, the rate of change of the output voltage is:

$$\frac{dV_o}{dt} = V_p \omega \cos \omega t \quad (11\text{-}29c)$$

So, the maximum rate of change of the output occurs when:

$$\cos \omega t = 1$$

$$\therefore \qquad \left.\frac{dV_o}{dt}\right|_{max} = V_p\omega$$

or,
$$SR = 2\pi f V_p \text{ volts/sec} \tag{11-29d}$$

or,
$$SR = \frac{2\pi f V_p}{10^6} \text{ volts/}\mu\text{sec} \tag{11-29e}$$

where, $SR$ is the slew rate in volts/sec, $f$ is the input frequency, and $V_p$ is the peak value of the output sine wave in volts.

It can be concluded from Eq. (11-29e), that the small value of input frequency $f$ determines the maximum undistorted output voltage swing.

### Application of slew rate

Slew rate has an important application in both open-loop and closed-loop op-amp circuit. The open-loop configuration is shown in Fig. 11-14. The output voltage rises from −14 V to +14 V because the open-loop voltage gain is very high. Calculating the slew rate of op-amp 741C we can calculate the time taken by the output to reach from −14 V to +14 V. The slew rate can be determined from the slope of the voltage follower from the large signal pulse response curve of Fig. 11-14.

For a high-frequency operation the slew rate has to be increased. From Eq. (11-29d) we can write:

$$SR = 2\pi f_{max} V_p$$

$$f_{max} = \frac{SR}{2\pi V_p} \quad \text{where, } V_p \text{ is fixed} \tag{11-29f}$$

$$SR = \left.\frac{dV_c}{dt}\right|_{max} = \frac{I_{max}}{C} \tag{11-29g}$$

So, from Eq. (11-29g) we can conclude that:

$$SR = I_{max}$$
$$= g_m V_t$$

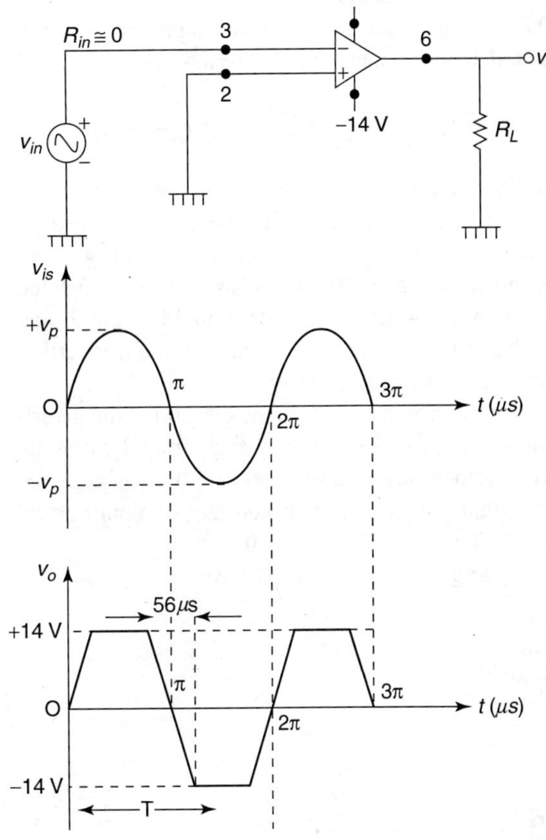

**Figure 11-14** Open-loop configuration using IC 741C and large signal pulse response curve

Therefore, for better slew rate, the first stage differential amplifiers are composed of FETs, and are independent of $g_m$, unlike transistors. Hence, operational amplifiers are hybrid BiFETs. It is important to note that the slew rate is a small signal phenomenon. At the output there is no distortion, but only attenuation of magnitude.

### 11-4-18 Gain Bandwidth Product

The gain bandwidth product (GB) is the bandwidth of the operational amplifier when the voltage gain is 1. For the 741C, it is found to be 1 MHz.

### 11-4-19 Virtual Ground

Virtual ground is defined as a node that has zero voltage, i.e., it is maintained at a steady reference potential, without being physically grounded. Sometimes the reference potential is considered to be the surface of the earth, from which "ground" is derived. Virtual ground is a concept that is used in operational amplifiers and depends on very large voltage gain used to calculate the overall voltage gain of the amplifier.

## 11-5 APPLICATIONS OF THE OPERATIONAL AMPLIFIER

The op-amp is used either in inverting mode or non-inverting mode. In many practical applications the op-amp is used as adder, subtractor, intergrator, integrator, differentiator, voltage follower, phase changer, etc.

### 11-5-1 Inverting Mode of Operation

In the inverting mode of operation, the input supply $V_{in}$ is connected to the inverting terminal, and the non-inverting terminal is directly connected to the ground, as shown in Fig. 11-15. The resistor $R_1$ is connected to the inverting terminal of op-amp 741C pin 2, and feedback resistor $R_F$ is connected between inverting pin 2 and output pin 6, to the circuit.

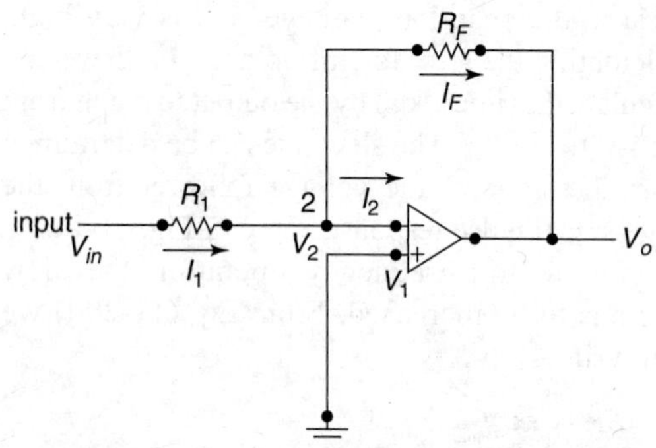


**Figure 11-15** Operational amplifier in the inverting mode of operation

Our aim here is to calculate the gain of the inverting amplifier. As the non-inverting terminal is directly connected to the ground, we get $V_1 = 0$, a consequence of virtual ground; $V_2 = 0$; and the terminal current $I_2 = 0$. Therefore, $V_1 = V_2 = 0$.

Using KCL at node 2, we have:

$$I_1 = I_F + I_2$$

or,
$$\frac{V_{in} - V_2}{R_1} = \frac{V_2 - V_o}{R_F} + 0$$

or,
$$\frac{V_{in} - 0}{R_1} = \frac{0 - V_o}{R_F} \qquad \text{(since } V_2 = V_1 = 0\text{)}$$

or,
$$\frac{V_{in}}{R_1} = -\frac{V_o}{R_F}$$

or,
$$V_o = -\left(\frac{R_F}{R_1}\right) V_{in} \tag{11-30}$$

From Eq. (11-30), it is clear that the negative sign indicates that the output voltage is in the opposite phase with that of the input. Therefore, this mode is said to be the inverting mode of operation.

Feedback gain of the inverting amplifier is represented as:

$$A_F = \frac{V_o}{V_{in}} = -\frac{R_F}{R_1} \tag{11-31}$$

## 11-5-2 Non-Inverting Mode of Operation

In the non-inverting mode of operation the power supply $V_{in}$ is connected to the non-inverting terminal of op-amp 741C pin 3 and the inverting terminal is directly connected to the ground, as shown in Fig. 11-16. The resistor $R_1$ is connected to the inverting terminal of op-amp 741C pin 2, and feedback resistor $R_F$ is connected between inverting the pin 2 and the output pin 6, to the circuit.

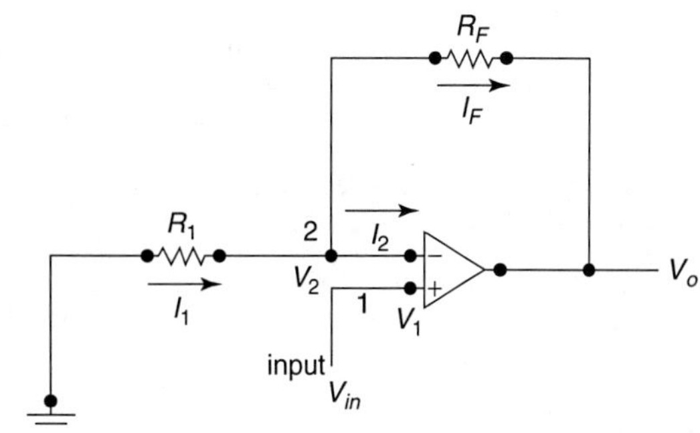


**Figure 11-16** Op-amp in non-inverting configuration

Our aim here is to calculate gain of the non-inverting amplifier shown in Fig. 11-16. As the inverting terminal is directly connected to the ground and the terminal current $I_2 = 0$. Input voltage is applied to the non-inverting terminal of op-amp 741C pin 3.

$$V_1 = V_2 = V_{in}$$

Using KCL at node 2 in Fig. 11-26 we have:

$$I_1 = I_F + I_2$$

$$\frac{V_2 - V_{in}}{R_1} = \frac{V_{in} - V_o}{R_F} + 0$$

or,

$$\frac{0 - V_{in}}{R_1} = \frac{V_{in} - V_o}{R_F} + 0$$

or,

$$-\frac{V_{in}}{R_1} = \frac{V_{in} - V_o}{R_F}$$

or,

$$\frac{V_o}{R_F} = \frac{V_{in}}{R_F} + \frac{V_{in}}{R_1}$$

or,

$$V_o = V_{in}\left(1 + \frac{R_F}{R_1}\right) \tag{11-32}$$

Equation (11-32) shows that the output has the same phase as that of the input due to the same sign. Therefore, this mode of operation is said to be the non-inverting mode of operation. The feedback gain is given by:

$$A_F = \frac{V_o}{V_{in}} = 1 + \frac{R_F}{R_1} \tag{11-33}$$

Equation (11-33) indicates that the voltage gain is greater than or equal to unity. The feedback gain is unity if $R = 0$ or $R = \alpha$. In this case, the non-inverting mode of the op-amp acts as a voltage follower.

## 11-5-3 Voltage Summing, Difference, and Constant Gain Multiplier

### Inverting configuration

Figure 11-17 shows the inverting configuration with three inputs $V_a$, $V_b$ and $V_c$. Depending on the relationship between $R_a$, $R_b$ and $R_c$, the circuit can be used as a summing amplifier/scaling amplifier or as an averaging amplifier.

Let $V_o$ be the output voltage. Using KCL at node $V_2$, we have:

$$I_a + I_b + I_c = I_B + I_F \tag{11-34}$$

Since the op-amp is ideal, we have:

$$\frac{V_a - V_2}{R_a} + \frac{V_b - V_2}{R_b} + \frac{V_c - V_2}{R_c} = \frac{V_2 - V_o}{R_F}$$

$$I_B = 0 \text{ and } V_1 = V_2 \cong 0 \text{ V}$$

$\therefore$

$$V_o = -\left(\frac{R_F}{R_a}V_a + \frac{R_F}{R_b}V_b + \frac{R_F}{R_c}V_c\right)$$

or,

$$\frac{V_a}{R_a} + \frac{V_b}{R_b} + \frac{V_c}{R_c} = -\frac{V_o}{R_F} \tag{11-35}$$

Now if $R_a = R_b = R_c = R_F = R$, we have from Eq. (11-35)

$$V_o = -(V_a + V_b + V_c) \tag{11-36}$$

Hence, the amplifier shown in Fig. 11-17 acts as the summing amplifier.

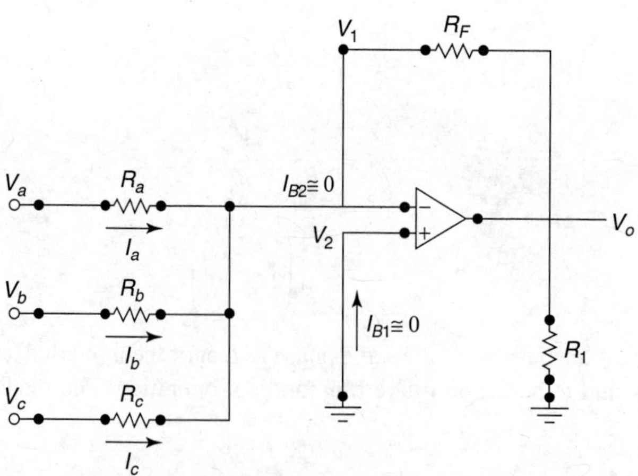


**Figure 11-17** Inverting configurations with three inputs

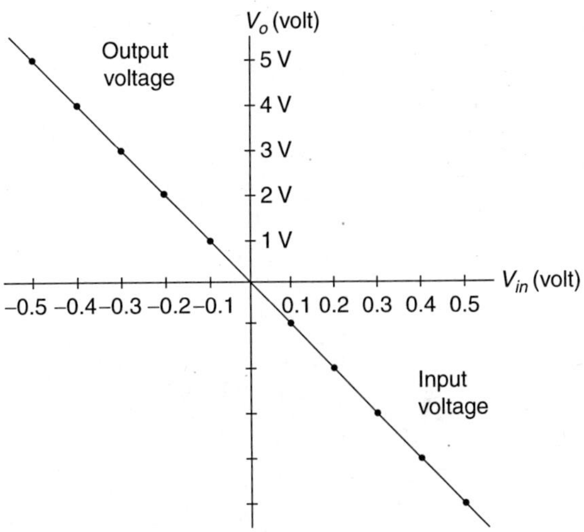


**Figure 11-18** Output characteristics of an inverting amplifier

The output characteristic of an inverting amplifier, as shown in Fig. 11-18 is that the output voltage is of opposite polarity to the input voltage. Hence, it is called the inverting amplifier.

### Non-inverting configuration

From Fig. 11-19, using superposition theorem, the voltage $V_1$ at the non-inverting terminal is derived as:

$$V_1 = \frac{\frac{R_a}{2}}{R_a + \frac{R_a}{2}} V_a + \frac{\frac{R_b}{2}}{R_b + \frac{R_b}{2}} V_b + \frac{\frac{R_c}{2}}{R_c + \frac{R_c}{2}} V_c$$

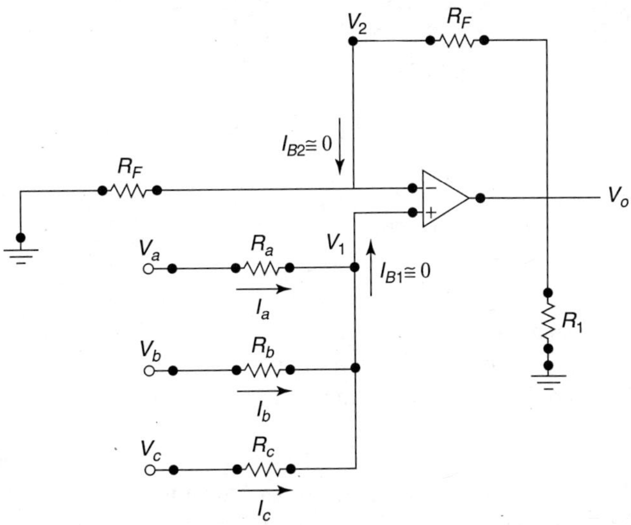


**Figure 11-19** Non-inverting configurations with three inputs

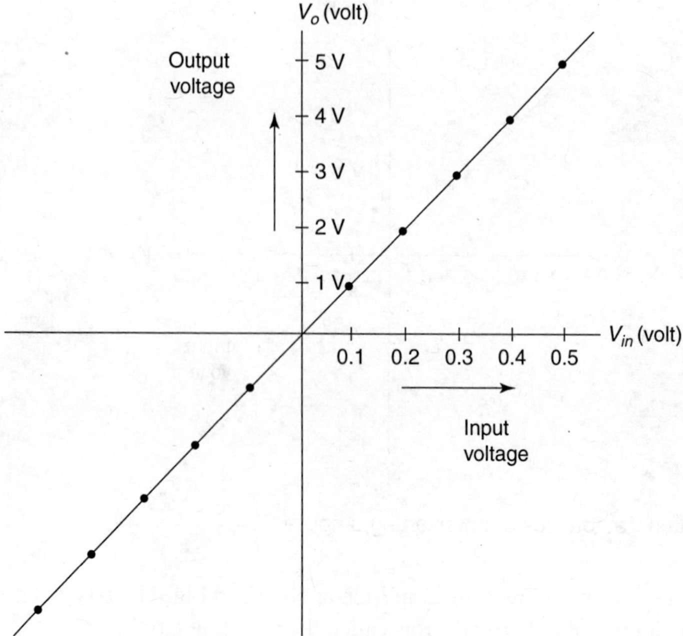


**Figure 11-20** Output characteristics of a non-inverting amplifier

or,

$$V_1 = \frac{V_a}{3} + \frac{V_b}{3} + \frac{V_c}{3} = \frac{V_a + V_b + V_c}{3} \tag{11-37}$$

Hence, the output voltage $V_o$ is given by:

$$V_o = V_1\left(1 + \frac{R_F}{R_1}\right)$$

$$= \left(1 + \frac{R_F}{R_1}\right)\frac{V_a + V_b + V_c}{3} \tag{11-38}$$

If, $1 + \frac{R_F}{R_1} = 3$, then the output voltage is given by:

$$V_o = V_a + V_b + V_c$$

Hence, the amplifier acts as a summing amplifier.

The output characteristics of a non-inverting amplifier as shown in the Fig. 11-20 is that output voltage is of the same polarity as the input voltage. Hence, the name of the amplifier is non-inverting amplifier.

## Subtractor

A basic differential amplifier can be used as a subtractor as shown in Fig. 11-21. The input signals can be scaled to the desired values by selecting appropriate values for the external resistors.

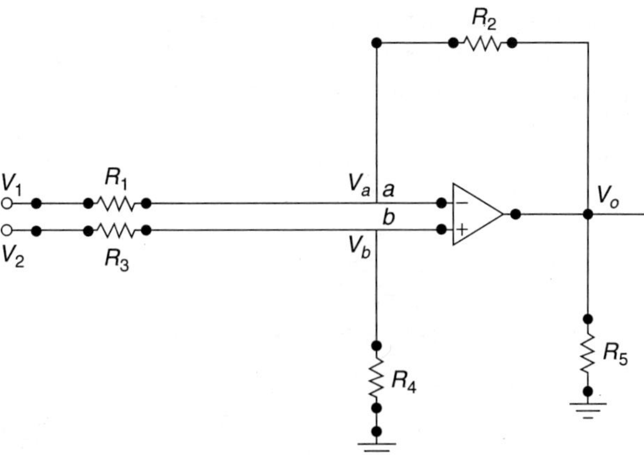


**Figure 11-21** Basic differential amplifier used as subtractor

In Fig. 11-21 all the resistors are of the equal value so that the gain of the amplifier is equal to 1. Using the superposition theorem, the output due to voltage $V_a$ is given by:

$$V_{oa} = -\frac{R}{R}V_a$$
$$= -V_a$$

Here,
$$V_1 = V_b \frac{R}{R+R} = \frac{V_b}{2}$$

The output voltage due to voltage $V_b$ is:

$$V_{ob} = \left(1 + \frac{R}{R}\right)V_1$$
$$= 2V_1 = V_b$$

Therefore, output voltage $V_o$ is given by:

$$V_o = V_{oa} + V_{ob}$$
$$= V_b - V_a \tag{11-39}$$

Thus, the output voltage $V_o$ is equal to the voltage $V_b$ applied to the non-inverting terminal minus the voltage $V_a$ applied to the inverting terminal. Hence, the circuit is called the subtractor.

## 11-5-4 Voltage Follower or Unity Gain Amplifier

The lowest gain that can be obtained from the non-inverting amplifier with feedback is 1. When the non-inverting amplifier is configured for unity gain, it is called a voltage follower because the output voltage is equal to, and also in phase with the input. In other words, in the voltage follower the output follows the input voltage.

It is similar to the discrete emitter follower, but the voltage follower is preferred because it has much higher input resistance, and the output amplitude is exactly equal to the input. To obtain the voltage follower from the non-inverting amplifier $R_1$ is simply made open, and $R_F$ is shorted. Voltage follower is also called non-inverting buffer because, when placed between two networks, it removes the loading on the first network.

In the non-inverting mode of operation the power supply $V_{in}$ is connected to the non-inverting terminal, and the inverting terminal is connected to the ground, as shown in Fig. 11-22. The resistor $R_1$ and feedback resistor $R_F$ are replaced by a short circuit. We rewrite the Eq. (11-32) as follows:

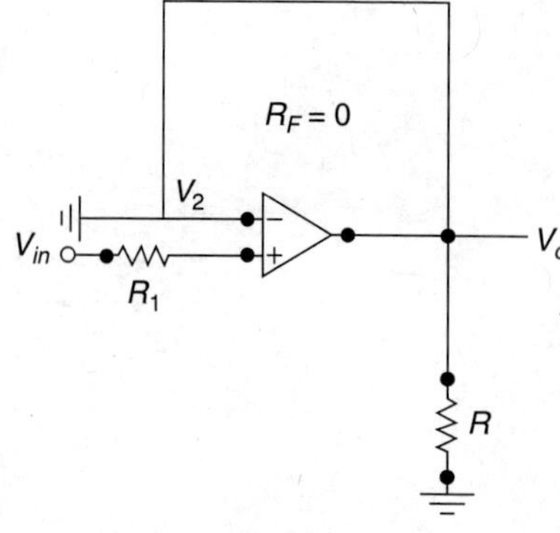


**Figure 11-22** Voltage follower circuit

$$V_o = V_{in}\left(1 + \frac{R_F}{R_1}\right)$$

In this case, voltage follower $R_F = 0$, so the output voltage will become equal to input voltage; hence the name voltage follower becomes more appropriate. And the voltage gain becomes unity.

$$A_F = \frac{V_o}{V_{in}} = 1 + \frac{R_F}{R_1} = 1 + \frac{0}{R_1} = 1$$

## 11-5-5 Comparator

The comparator in the non-inverting mode is shown in Fig. 11-23.

A fixed reference voltage ($V_{ref}$) is applied to the $(-)$ve input, and the other time-varying signal voltage $v_{in}$ is applied to the $(+)$ve input. When $v_{in}$ is less than $V_{ref}$, the output voltage $v_o$ is at $-V_{sat}$ $(\cong -V_{EE})$, because the voltage at $(-)$ve input is higher than that at the $(+)$ve input. On the other hand, when $v_{in}$ is greater than $V_{ref}$, the $(+)$ve input becomes positive with respect to the $(-)$ve input, and $v_o$ goes to $+V_{sat}$ $(\cong +V_{CC})$. Thus, $v_o$ changes from one saturation level to another whenever $v_{in} \cong V_{ref}$, as shown in Fig.11-23. In short, the comparator is a type of analog-to-digital converter. At any given time, the $v_o$ waveform shows whether $v_{in}$ is greater or less than $V_{ref}$. The comparator is sometimes also called a voltage level detector, because for a desired value of $V_{ref}$, the voltage level of the input $v_{in}$ can be detected.

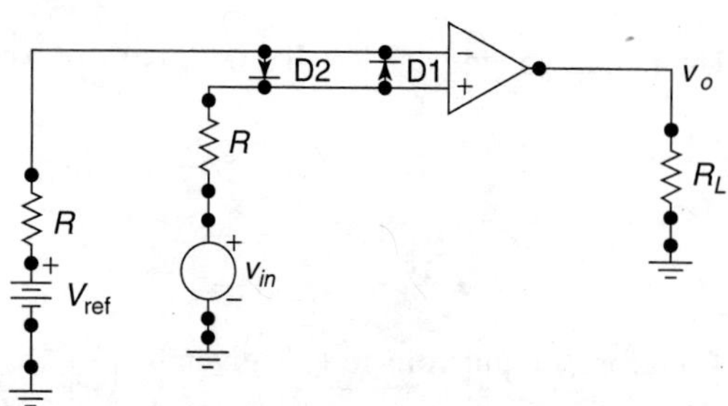


**Figure 11-23** Non-inverting comparator

In Fig. 11-23, the diodes $D_1$ and $D_2$ protect the op-amp from damage caused by the excessive input voltage $v_{in}$. Because of these diodes, the difference input voltage $v_{id}$ of the op-amp is clamped to either 0.7 V or − 0.7 V; hence, the diodes are called clamp diodes.

If the reference voltage $V_{ref}$ is negative with respect to the ground, with sinusoidal signal applied to the $(+)$ve input, the output waveform will be as shown in Fig. 11-24. When $v_{in} > V_{ref}$, $v_o$ is at $+V_{sat}$; on the other hand, when $v_{in} < V_{ref}$, $v_o$ is at $-V_{sat}$. The amplitude of $v_{in}$ must be large enough to pass through $V_{ref}$ if the switching action is to take place.

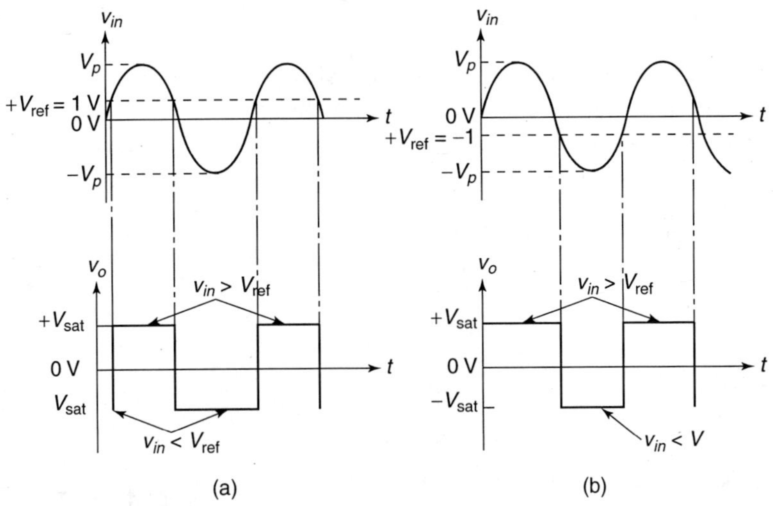


**Figure 11-24** (a) Output if $V_{ref}$ is positive (b) Output if $V_{ref}$ is negative

Figure 11-25 shows an inverting comparator in which the reference voltage $V_{ref}$ is applied to the (+)ve input and $v_{in}$ is applied to the (−)ve terminal. In this circuit, $V_{ref}$ is obtained by using a potentiometer that forms a voltage divider with the dc supply voltages $+V_{CC}$ and $-V_{EE}$, and the wiper is connected to the (+)ve input. As the wiper is moved towards $-V_{EE}$, $V_{ref}$ becomes more negative. When it is moved towards $+V_{CC}$ $V_{ref}$ becomes more positive (see Fig. 11-26). Thus, the reference voltage $V_{ref}$, of desired amplitude and polarity can be obtained by simply adjusting the potentiometer.

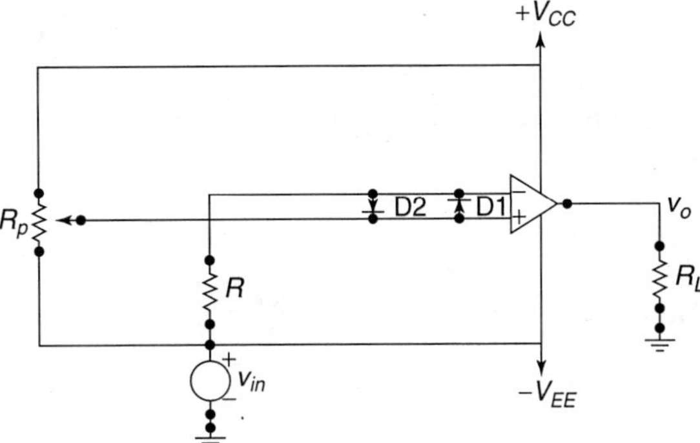


**Figure 11-25** Inverting comparator

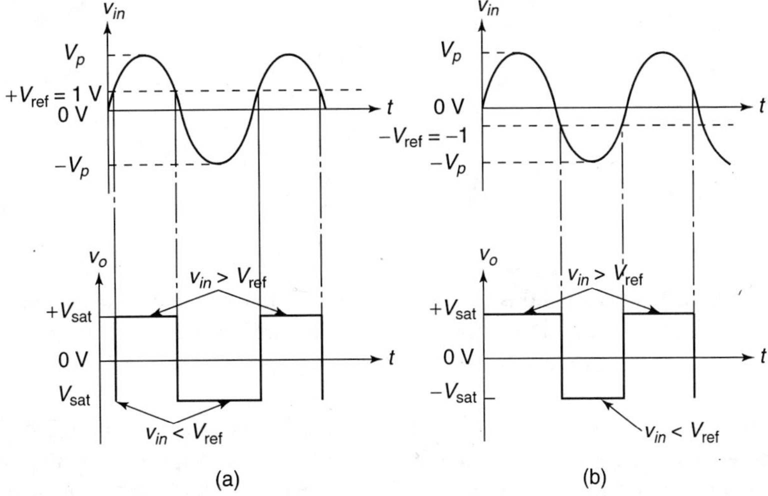


**Figure 11-26** (a) Output if $V_{ref}$ is positive (b) Output if $V_{ref}$ is negative

### 11-5-6 Integrator

A circuit in which the output voltage waveform is the integral of the input voltage waveform is the integrator or the integration amplifier. Such a circuit is obtained by using a basic inverting amplifier configuration if the feedback resistor $R_F$ is replaced by a capacitor $C_F$, as shown in Fig. 11-27.

The expression for the output voltage $v_o$ can be obtained by using KCL at node $v_2$:

$$i_1 = I_B + i_F$$

Since, $I_B$ is negligibly small:

$$i_1 \cong i_F$$

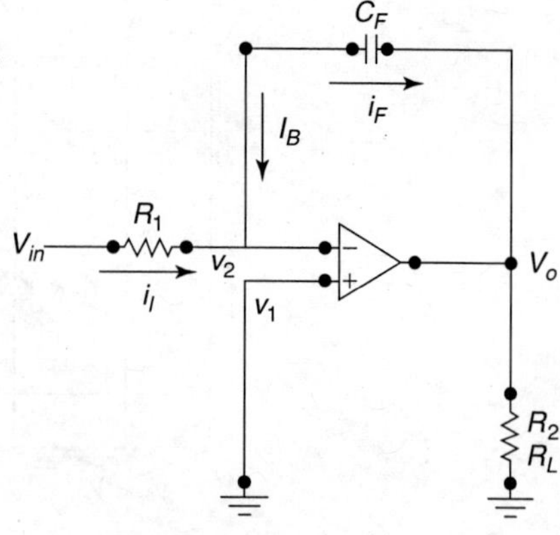


**Figure 11-27** The integrator circuit

Relation between the current through and voltage across the capacitor is given by:

$$i_C = C\,\frac{dv_C}{dt}$$

From Fig. 11-27, applying KCL, we get:

$$\frac{v_{in} - v_2}{R_1} = C_F\,\frac{d}{dt}\,(v_2 - v_0)$$

However, $v_1 = v_2 \cong 0$, because $A$ is very large. Therefore:

$$\frac{v_{in}}{R_1} = C_F\left(\frac{d}{dt}\right)(-v_0)$$

The output voltage can be obtained by integrating both sides with respect to time:

$$\int_0^t \frac{v_{in}}{R_1}dt = \int_0^t C_F\left(\frac{d}{dt}\right)(-v_0)dt$$

$$= C_F(-v_0) + v_0|_{t=0}$$

or,

$$v_0 = -\frac{1}{R_1C_F}\int_0^t v_{in}\,dt + c \qquad (11\text{-}40)$$

where, $C$ is the integration constant and is proportional to the value of the output voltage $v_0$ at time $t = 0$.

Equation (11-40) indicates that the output voltage is directly proportional to the negative integral of the input voltage and inversely proportional to the time constant $R_1C_F$. The corresponding input and output waveforms are shown in Fig. 11-28.

When $v_{in} = 0$, the integrator shown in Fig. 11-29 works as an open-loop amplifier. This is because the capacitor $C_F$ acts as an open circuit ($X_{CF} = \infty$) to the input offset voltage $V_{io}$, i.e., the input offset voltage $V_{io}$ and the part of the input current charging capacitor $C_F$ produce the error voltage at the output of the integrator.

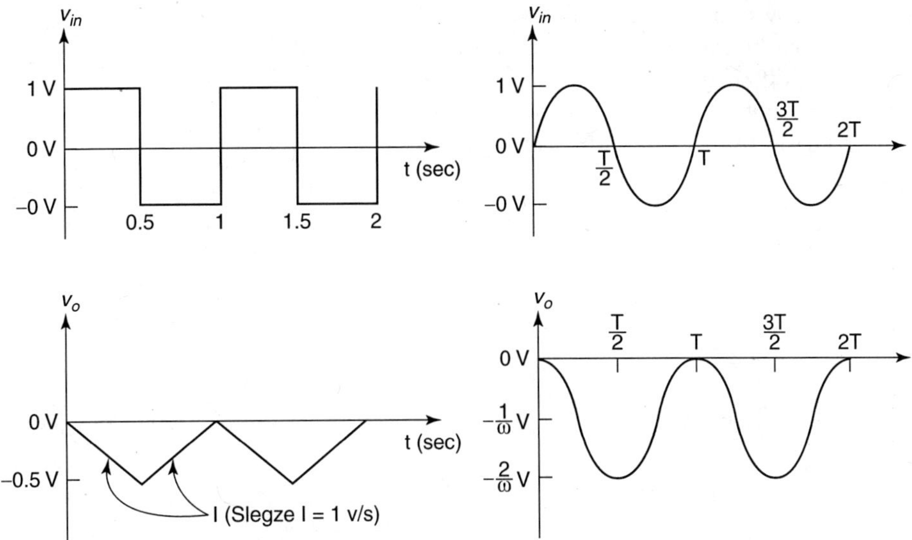


**Figure 11-28** Input and output waveforms

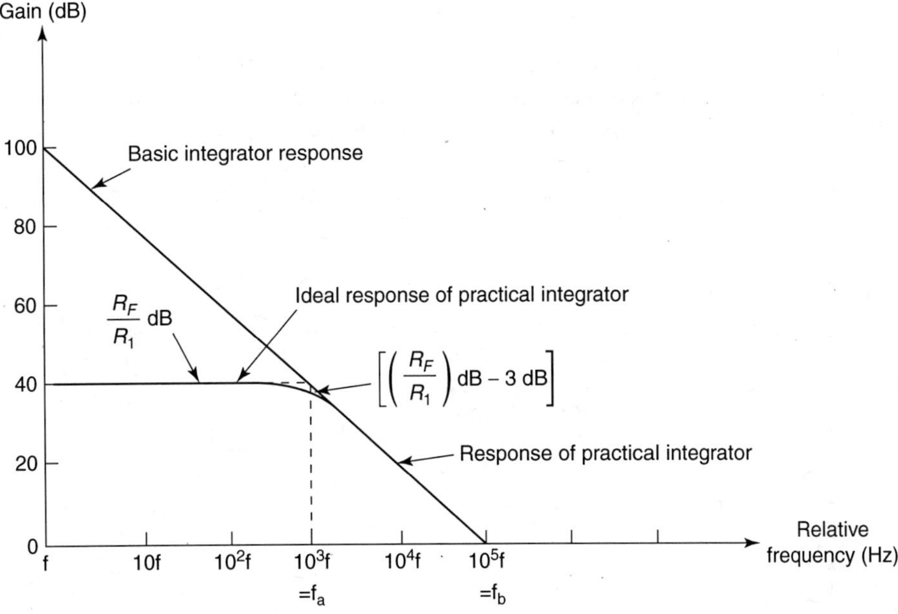


**Figure 11-29** Frequency response of integrator

## 11-5-7 Differentiator

The op-amp circuit whose output voltage is proportional to the differential form of the input voltage is known as differentiator.

Figure 11-30 shows the differentiator or differentiating amplifier. As the name suggests, the circuit performs the mathematical operation of differentiation; so, the output waveform is the derivative of the input waveform. The differentiator may be constructed from a basic inverting amplifier if an input resistor $R_1$ is replaced by capacitor $C_1$.

The expression for the output voltage can be obtained from KCL at node $v_2$ as follows,

$$i_C = I_B + i_F$$

Since, $i_B \cong 0$:

$$i_C = i_F$$

$$C_1 \frac{d}{dt}(v_{in} - v_2) = \frac{v_2 - v_o}{R_F}$$

Figure 11-30 The differentiator

However, $v_1 = v_2 \cong 0$ V, because $A$ is very large. Therefore:

$$C_1 \frac{dv_{in}}{dt} = -\frac{v_o}{R_F}$$

or,

$$v_o = -R_F C_1 \frac{dv_{in}}{dt} \tag{11-41}$$

Thus, the output $v_o$ is equal to $R_F C_1$ times the negative instantaneous rate of change of the input voltage $v_{in}$ with time. Since the differentiator performs the reverse of the integrator's function, a cosine wave input will produce a sine wave output, or a triangular input will produce a square wave output as shown in Fig. 11-31.

Table 11-2 shows the comparison between the output waveform of an integrator and a differentiator.

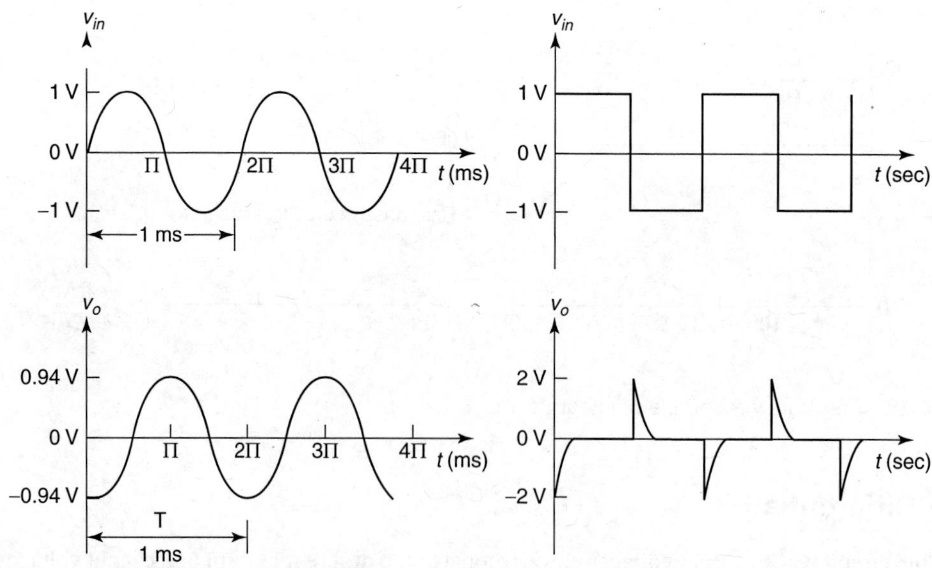


Figure 11-31 Input and output waveform

**Table 11-2** Comparison between the output waveform of an integrator and a differentiator

| *Input Signal Waveform* | *Integrator Output Waveform* | *Differentiator Output Waveform* |
|---|---|---|
| Sine | Minus cosine | Cosine |
| Square wave | Triangular wave | Pulse train |
| Triangular wave | Modified parabola | Square wave |
| Saw tooth wave | Parabolic ramp | Step function |
| Step function | Ramp function | Single pulse |
| Ramp function | Parabola | Step function |

## 11-5-8 Logarithmic Amplifier

If the feedback resistor $R_f$ in the op-amp circuit of Fig 11-32 is replaced by a diode, we obtain a logarithmic amplifier giving an output voltage $V_o$ that changes as the logarithm of the input voltage $V_1$.

The voltage-ampere characteristic of the diode is given by:

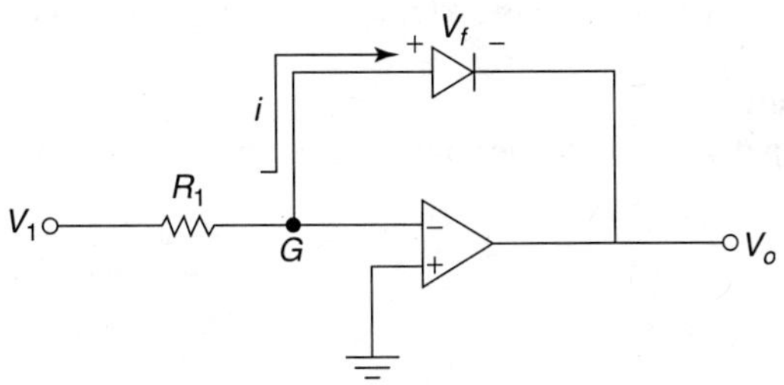

$$i = I_S\left[\exp\left(\frac{eV_f}{\eta k_B T}\right) - 1\right]$$

**Figure 11-32** Logarithmic amplifier for positive input

where, $e = 1.6 \times 10^{-19}\text{C}$

$k_B = 1.38 \times 10^{-23}\ \text{J/K}$

$T = 300\text{K}$

Here, $i$ is the diode current for the forward voltage $V_f$. If $eV_f/(\eta\, k_B t) >> 1$, then $i >> I_S$.

We have:

$$i = I_S\left[\exp\left(\frac{eV_f}{\eta k_B T}\right)\right]$$

or,

$$\ln\left(\frac{i}{I_S}\right) = \frac{eV_f}{\eta k_B T}$$

Therefore,

$$V_f = \frac{\eta k_B T}{e}\ln\left(\frac{i}{I_S}\right)$$

Since $G$ is the virtual ground in Fig. 11-32, we have $i = V_1/R_1$, and the output voltage is:

$$V_o = -V_f = -\frac{\eta k_B T}{e}\ln\left(\frac{V_1}{R_1 I_S}\right)$$

Hence, $V_o$ responds to the logarithm of $V_1$.

## Solved Examples

**Example 11-1** For an op-amp having a slew rate of SR = 2 V/$\mu$s, what is the maximum closed-loop voltage gain that can be used when the input signal varies by 0.5 V/$\mu$s?

**Solution:**

Since $V_o = A_{CL} V_i$, we have:

$$SR = \frac{dV_o}{dt} = A_{CL}\frac{dV_i}{dt}$$

From this:

$$A_{CL} = \frac{\frac{dV_o}{dt}}{\frac{dV_i}{dt}} = \frac{SR}{\frac{dV_i}{dt}} = \frac{2\text{ V}/\mu\text{s}}{0.5\text{ V}/10\ \mu\text{s}} = 40$$

**Example 11-2** A non-inverting amplifier has an input voltage of 1 V. The input resistance $R_1$ = 1 kΩ and feedback resistance is 5 kΩ. Find the output voltage and voltage gain of the amplifier.

**Solution:**
Output voltage of a non-inverting amplifier is given by:

$$v_o = \left(1 + \frac{R_F}{R_1}\right) v_i$$

$$v_o = \left(1 + \frac{5}{1}\right) 1\text{ V}$$

$$v_o = 6\text{ volt}$$

$$A_v = \frac{v_o}{v_{in}} = \frac{6}{1} = 6\text{ V}$$

**Example 11-3** An inverting amplifier with input voltage is 1 V. Input resistance is $R_1$ = 1 kΩ and feedback resistance is 5 kΩ. Find the output voltage and voltage gain of the amplifier.

**Solution:**
Output voltage of a non-inverting amplifier is given by:

$$v_o = -\frac{R_F}{R_1} v_i$$

$$v_o = -\frac{5}{1} \times 1\text{ V}$$

$$v_o = -5\text{ volt}$$

$$A_v = \frac{v_o}{v_{in}} = -\frac{5}{1} = -5\text{ V}$$

**Example 11-4** For an adder circuit, feedback resistance is 10 kΩ. Input arm resistances are all 1 kΩ. If the input voltages are 0.2 V, 0.3 V and 0.4 V, find the output voltage of the adder circuit.

**Solution:**
Output of an adder is given by:

$$V_o = -\left(\frac{R_F}{R_a}V_a + \frac{R_F}{R_b}V_b + \frac{R_F}{R_c}V_c\right)$$

Since, $R_a = R_b = R_c = R$

$$V_o = -\frac{R_F}{R}(V_a + V_b + V_c)$$

$$V_o = -\frac{10}{1}(0.2 + 0.3 + 0.4)$$

$$V_o = -10\,(0.9) = -9\text{ V}$$

**Example 11-5** For an adder circuit feedback resistance is 10 kΩ. Input arm resistances are 1 kΩ, 2kΩ and 3.3 kΩ. If the input voltage is 0.2 V, 0.06 V and 0.33 V, find the output voltage of the adder.

**Solution:**
Output of an adder is given by:

$$V_o = -\left(\frac{R_F}{R_a}V_a + \frac{R_F}{R_b}V_b + \frac{R_F}{R_c}V_c\right)$$

$$V_o = -\left[\frac{10}{1}(0.2) + \frac{10}{2}(0.06) + \frac{10}{3.3}(0.33)\right]$$

$$V_o = -(2 + 0.3 + 1)\text{ V}$$

$$V_o = -\,3.3\text{ V}$$

**Example 11-6** Determine the ideal voltage gain for the circuit shown in the following diagram.

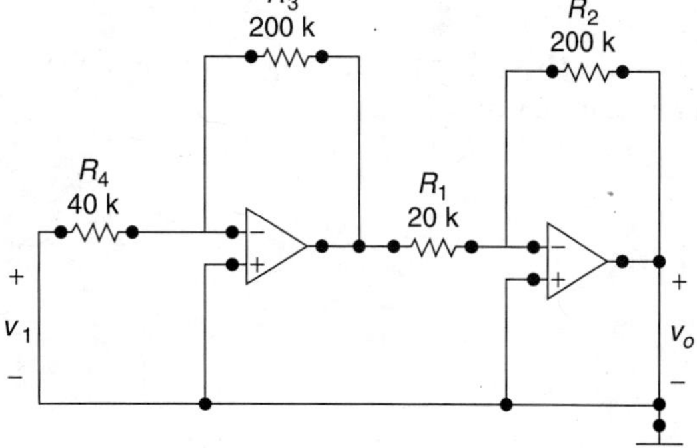


**Solution:**
Both the op-amps, as given in the diagram, are in an inverting configuration. With $v_m$ as the output of the first op-amp:

$$v_m = -\frac{R_{F1}}{R_1}v_1 = -\frac{200\text{ k}\Omega}{40\text{ k}\Omega}v_1 = -5v_1$$

Then, with $v_m$ taken as the input of the second op-amp:

$$v_o = -\frac{R_{F1}}{R_1}v_m = -\frac{200\text{ k}\Omega}{20\text{ k}\Omega}v_m = -10v_m$$

Since $v_m = -5v_1$: $\quad \frac{v_o}{v_1} = -10(-5) = 50$

**Example 11-7** For the circuit, as shown in the following diagram, find the ratio $v_o/v_i$.

**Solution:**

Applying KCL at node 1:

$$\frac{v_1 - v_i}{R_1} + \frac{v_1 - v_o}{R_2} + \frac{v_1 - v_a}{R_3} = 0$$

Since $v_1 = v_+ = v_- = 0$ for the first op-amp:

$$-\frac{v_i}{R_1} = \frac{v_o}{R_2} + \frac{v_a}{R_3}$$

Using a voltage divider:

$$v_b = \left(\frac{R_4}{R_4 + R_5}\right)v_o$$

Since $v_a = v_+ = v_- = v_b$ for the second op-amp:

$$-\frac{v_i}{R_1} = \frac{v_o}{R_2} + \frac{1}{R_3}\left(\frac{R_4}{R_4 + R_5}\right)v_o$$

Then:

$$\frac{v_o}{v_i} = -\frac{1}{R_1\left[\frac{1}{R_2} + \frac{R_4}{R_3(R_4 + R_5)}\right]}$$

or,

$$\frac{v_o}{v_i} = -\frac{R_2R_3(R_4 + R_5)}{R_1(R_3R_4 + R_3R_5 + R_2R_4)}$$

**Example 11-8** For the circuit, as shown in the following diagram, determine the output voltage $v_o$ if the input voltage $v_i = 1.2$ V.

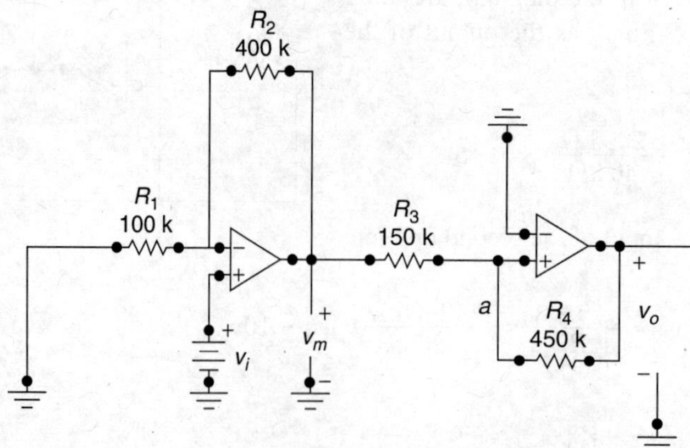

**Solution:**

These op-amps are in inverting configuration with $v_m$ designated as the output of the first op-amp.

$$v_m = \left(1 + \frac{R_{F,1}}{R_{1,1}}\right) v_i = \left(1 + \frac{400\ \text{k}\Omega}{100\ \text{k}\Omega}\right)(1.2\ \text{V}) = 6\ \text{V}$$

Then, with $v_0$ designated as the input to the second op-amp and applying the superposition theorem, voltage at node $a$ due to $v_m$, we get:

$$v_{a,m} = v_m \left(\frac{450\ \text{k}\Omega}{450\ \text{k}\Omega + 150\ \text{k}\Omega}\right) = 6 \times 0.75 = 4.5\ \text{V}$$

Again voltage at node $a$ due to $v_0$:

$$v_{a,o} = v_o \frac{150\ \text{k}\Omega}{150\ \text{k}\Omega + 450\ \text{k}\Omega} = 0.25 v_o$$

But, $$v_{a,o} = v_{a,m}$$

or, $$0.25 v_o = 4.5\ \text{V}$$

$\therefore$ $$v_o = \frac{4.5}{0.25} = 18\ \text{V}$$

**Example 11-9** The circuit, as given in the diagram, shows an ideal op-amp connected as a differential amplifier. Both the inverting and non-inverting terminals are used. Determine the expression for the output voltage $v_o$ as a function of $v_{i,1}$ and $v_{i,2}$ $[v_o = f(v_{i,1}, v_{i,2})]$. Then determine $v_o = f(v_{i,1}, v_{i,2})$ if $R_1 = R_2$.

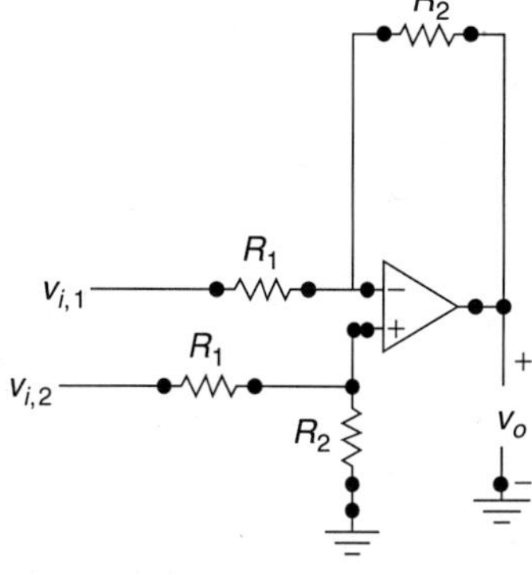


**Solution:**

For the non-inverting terminal at (+)ve node, KCL gives:

$$\frac{V_+}{R_2} + \frac{v_+ - v_{i,2}}{R_1} = 0$$

or, $$\left(\frac{1}{R_1} + \frac{1}{R_2}\right) v_+ = \frac{V_{i,2}}{R_1}$$

So that: $$v_+ = \left(\frac{R_2}{R_1 + R_2}\right) v_{i,2}$$

For the inverting terminal, at (–)ve node, KCL provides:

$$\frac{v_- - v_{i,1}}{R_1} + \frac{v_- - v_o}{R_2} = 0$$

or, $$\left(\frac{1}{R_1} + \frac{1}{R_2}\right) v_- = \frac{v_{i,1}}{R_1} + \frac{v_o}{R_2}$$

$v_o$ is given by:
$$v_o = \left(\frac{R_1 + R_2}{R_1}\right) v_- - \frac{R_2}{R_1} v_{i,1}$$

But,
$$v_+ = v_-$$

$\therefore$
$$v_o = \left(1 + \frac{R_2}{R_1}\right)\left(\frac{R_2}{R_1 + R_2}\right) v_{i,2} - \frac{R_2}{R_1} v_{i,1}$$
$$= \frac{R_2}{R_1} v_{i,2} - \frac{R_2}{R_1} v_{i,1}$$

So if, $R_1 = R_2$:
$$v_o = v_{i,2} - v_{i,1}$$

**Example 11-10** The following diagram shows an ideal op-amp that can be used as a strain gage, which is based on the fact that the value of resistance $\Delta R$ will change slightly when the resistor is bent or twisted. Determine the value of $\Delta R$ as a function of the input voltage $v_i$ and the two resistances $R_a$ and $R_b$.

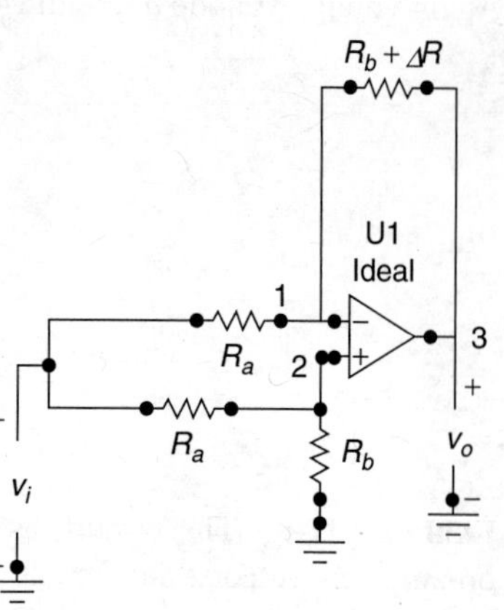


**Solution:**
Here, the node equation at 1 and 2 is given by:
$$\frac{v_- - v_i}{R_a} + \frac{v_2 - v_o}{R_b + \Delta R} = 0$$

and,
$$\frac{v_+}{R_b} + \frac{v_+ - v_i}{R_a} = 0$$

Then:
$$v_- = \left(\frac{R_b + \Delta R}{R_a + R_b + \Delta R}\right) v_i + \left(\frac{R_a}{R_a + R_b + \Delta R}\right) v_o$$

or,
$$v_- = -\left(\frac{R_a}{R_a + R_b + \Delta R} - 1\right) v_i + \left(\frac{R_a}{R_a + R_b + \Delta R}\right) v_o$$

and,
$$v^+ = \left(\frac{R_b}{R_a + R_b}\right) v_i$$

But,
$$v_- = v_+$$

or,
$$-\left(\frac{R_a}{R_a + R_b + \Delta R} - 1\right) v_i + \left(\frac{R_a}{R_a + R_b + \Delta R}\right) v_o = \left(\frac{R_b}{R_a + R_b}\right) v_i$$

or,
$$\left(\frac{R_a}{R_a + R_b + \Delta R}\right) v_o = \left(\frac{R_b}{R_a + R_b} + \frac{R_a}{R_a + R_b + \Delta R} - 1\right) v_i$$

After simplifying the previous expression we have:
$$-R_a \Delta R v_i = R_a (R_a + R_b) v_o$$

or,
$$\Delta R = -(R_a + R_b) \frac{v_o}{v_i}$$

**Example 11-11** The circuit, as shown in the diagram, shows how an ideal op-amp can be put together using resistors that have relatively small resistance values. If $R_1 = 2000\ \Omega$, determine the value of a single feedback resistor to produce a gain of $-1200$, and then with $R_1 = 2000\ \Omega$ and $R_b = 50\ \Omega$, determine the value of $R_a$ to provide a gain of $-1200$.

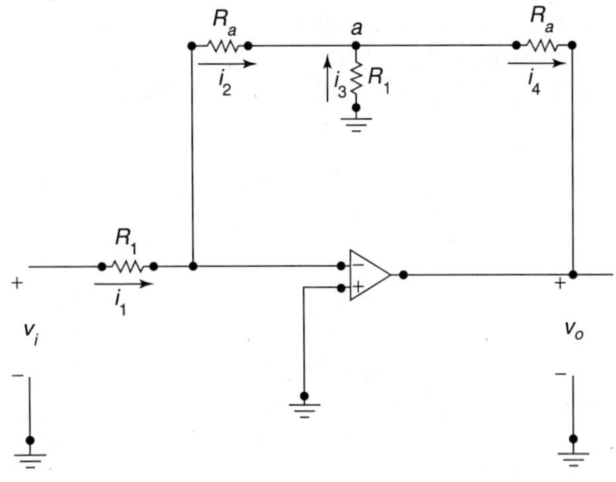


**Solution:**

This ideal op-amp is in the inverting configuration.

If $R_1 = 2000\ \Omega$ and $v_o/v_i = -1200$, then:

$$R_F = -R_1 \frac{v_o}{v_i} = -2000 \times (-1200) = 2.4\ \text{M}\Omega$$

With $v_+ = v_- = 0$:

$$i_1 = \frac{v_i - v_-}{R_1} = \frac{v_1}{R_1}$$

and because the ideal op-amp does not draw current:

$$i_2 = i_1 = \frac{v_1}{R_1}$$

The voltage at node *a* will be:

$$v_a = v_1 - R_a i_2 = 0 - R_a i_2 = -\frac{R_a}{R_1} v_i$$

Noting that $i_3$ flows upward:

$$i_3 = \frac{0 - v_a}{R_b} = \left(\frac{R_a}{R_1 R_b}\right) v_i$$

An application of KCL at node *a* gives:

$$i_4 = i_2 + i_3 = \frac{1}{R_1} v_i + \left(\frac{R_a}{R_1 R_b}\right) v_i = \left(\frac{R_a + R_b}{R_1 R_b}\right) v_i$$

so that:

$$v_o = v_a - R_a i_4 = -\frac{R_a}{R_1} v_i - \left(\frac{R_a(R_a + R_b)}{R_1 R_b}\right) v_i$$

or,

$$v_o = -\frac{R_a}{R_1}\left(1 + \frac{R_a + R_b}{R_b}\right) v_i$$

Now if $v_o/v_i = -1200$, $R_b = 50\ \Omega$ and $R_1 = 2000\ \Omega$:

$$-1200 = -\frac{R_a}{200\ \Omega}\left(1 + \frac{R_a + 50\ \Omega}{50\ \Omega}\right)$$

Hence, we have:

$$R_a = 10{,}900\ \Omega$$

Notice that four resistors with valves less than 12,000 Ω can be used to take the place of a single feedback resistor of 2.4 MΩ.

**Example 11-12** The following diagram illustrates the use of the op-amp as a negative impedance converter. Determine the input resistance.

**Solution:**

In this case: $R_{in} = \dfrac{v_{in}}{i_d}$

Since the ideal op-amp does not draw current at either input terminal and because $v_+ = v_-$:

$$i_b = \frac{v_1}{R_1}$$

KCL at node 1 gives: $i_e = i_b = \dfrac{v_1}{R_1}$

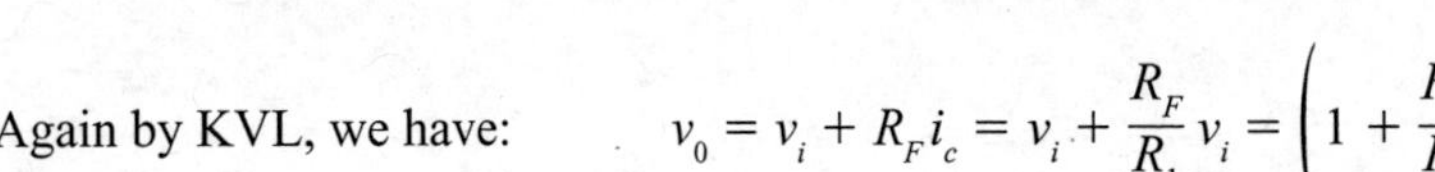

Again by KVL, we have: $v_0 = v_i + R_F i_c = v_i + \dfrac{R_F}{R_1} v_i = \left(1 + \dfrac{R_F}{R_1}\right) v_i$

By Ohm's law: $i_d = \dfrac{v_i - v_0}{R} = \dfrac{v_i}{R} - \dfrac{1}{R}\left(1 + \dfrac{R_F}{R_1}\right) v_i = -\dfrac{R_F}{RR_1} v_i$

Hence, the input resistance is: $R_{in} = \dfrac{v_i}{i_d} = -\dfrac{R_1}{R_F} R$

**Example 11-13** An ideal op-amp in the inverting configuration is to have a gain of −125 and input resistance as high as possible. If no resistance in the op-amp circuit is to have a value higher than 5 MΩ, how is this design achieved using just two resistors?

**Solution:**

Here $R_{in}$ is as high as possible, thus:

$$R = R_1 = \text{high}$$

With gain $G = -\dfrac{R_F}{R_1}$, $R_F$ should be set at 5 MΩ.
Thus $R_f = 5$ MΩ.

$$\therefore \quad R_1 = -\frac{R_F}{G} = -\frac{5\ \text{M}\Omega}{-125} = 40000\ \Omega$$

**Example 11-14** In the cascade of ideal op-amps, as shown in the following diagram, if $v_i = 2$ V and $v_o = 30$ V, determine the value of $R$.

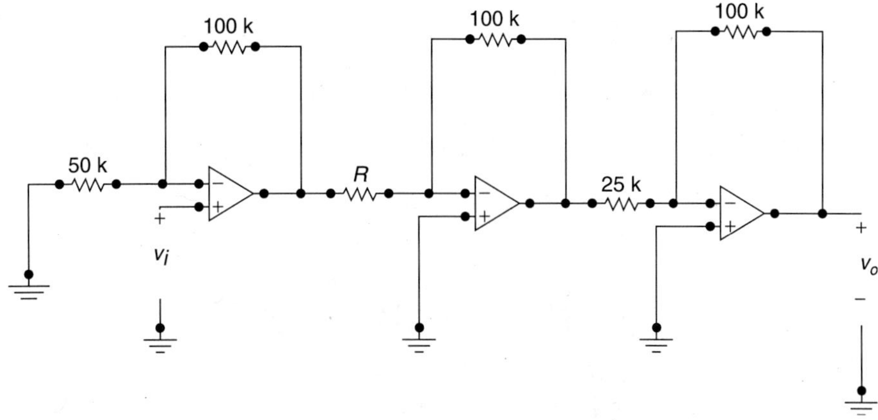


**Solution:**

Observe that the first op-amp is in a non-inverting configuration, and the rest are in an inverting configuration. With $v_{o1}$ as the output of first op-amp and $v_{o2}$ as the output of second op-amp, we have:

$$\frac{v_o}{v_i} = \left(\frac{v_{o1}}{v_i}\right)\left(\frac{v_{o2}}{v_{o1}}\right)\left(\frac{v_o}{v_{o2}}\right)$$

$$= \left(1 + \frac{R_{f,1}}{R_{1,1}}\right)\left(-\frac{R_{f,2}}{R_{1,2}}\right)\left(-\frac{R_{f,3}}{R_{1,3}}\right)$$

Thus,

$$\frac{30}{2} = \left(1 + \frac{100\,\text{k}\Omega}{50\,\text{k}\Omega}\right)\left(-\frac{100\,\text{k}\Omega}{R}\right)\left(-\frac{100\,\text{k}\Omega}{25\,\text{k}\Omega}\right)$$

Upon solving we get:

$$R = 80\,\text{k}\Omega$$

**Example 11-15** Find the relationship between $v_o$, $v_{s1}$ and $v_{s2}$ in the circuit, as shown in the following diagram.

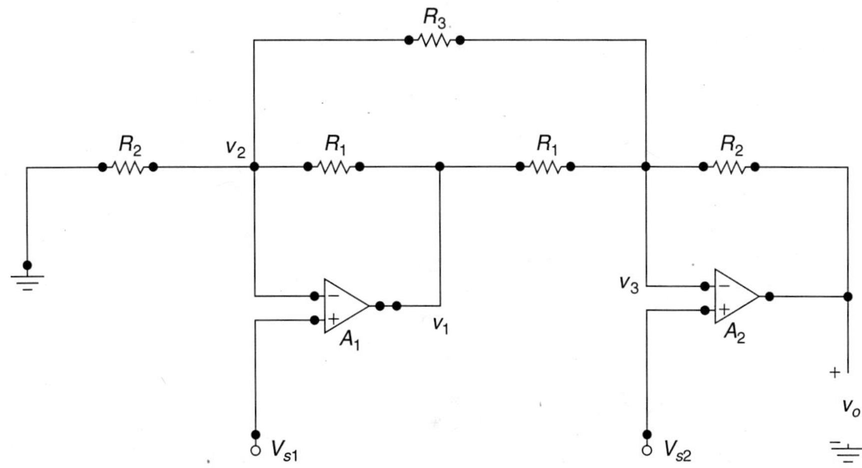

**Solution:**

Using the superposition theorem and with $v_{s1} = 0$:

$$v_2 = 0 \text{ (voltage constraint applied to } A_2)$$

$$v_1 = -\frac{R_1}{R_3} v_{s2} \text{ (inverting amplifier)}$$

Using superposition theorem again, and with $v_2 = v_1$ and $v_3$ as ground referenced voltages:

$$v_o' = -\frac{R_2}{R_1} v_1 + \left(1 + \frac{R_2}{R_3 \| R_1}\right) v_{s2}$$

$$= -\left(\frac{R_2}{R_1}\right)\left(\frac{R_1}{R_3}\right) v_{s2} + \left(1 + \frac{R_2(R_1 + R_3)}{R_1 R_3}\right) v_{s2}$$

$$= \left(1 + \frac{R_2}{R_1} + \frac{2R_2}{R_3}\right) v_{s2}$$

For $v_{s2} = 0$ and $v_3 = 0$:

$$v_2 = v_{s1}$$

$$v_1 = \left(1 + \frac{R_1}{R_3 \| R_2}\right) v_{s2}$$

$$v_o'' = -\frac{R_2}{R_1} v_{s1} - \frac{R_2}{R_1} v_1$$

$$= -v_{s1}\left(1 + \frac{R_2}{R_1} + \frac{2R_2}{R_3}\right)$$

And with,

$$v_o = v_o' + v_o'':$$

$$= \left(1 + \frac{R_2}{R_1} + \frac{2R_2}{R_3}\right)(v_{s2} - v_{s1})$$

**Example 11-16** Find the $v_1$ and $v_o$ in terms of $v_{s1}$ and $v_{s2}$.

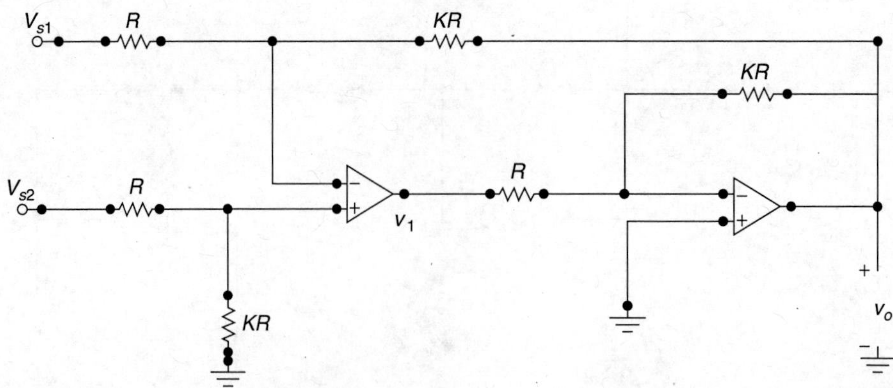

**Solution:**

Using the superposition theorem and noticing that the cascade combination of U1 and the inverting circuit of U2 acts like an ideal op-amp with infinite open-loop gain:

$$v_o' = -Kv_{s1}\ (v_{s2} = 0,\ \text{inverting amplifier})$$

$$v_o'' = v_{s2}\frac{KR}{R + KR}\left(1 + \frac{KR}{R}\right) \quad \text{(voltage division and non-inverting amplifier)}$$

$$= Kv_{s2}$$

$$v_o = v_o' + v_o''$$

$$= K(v_{s2} - v_{s1})$$

Since $v_o = -Kv_1$: $\quad v_1 = \dfrac{v_o}{-K} = v_{s1} - v_{s2}$

**Example 11-17** Find $I_s$ in terms of $v_c$.

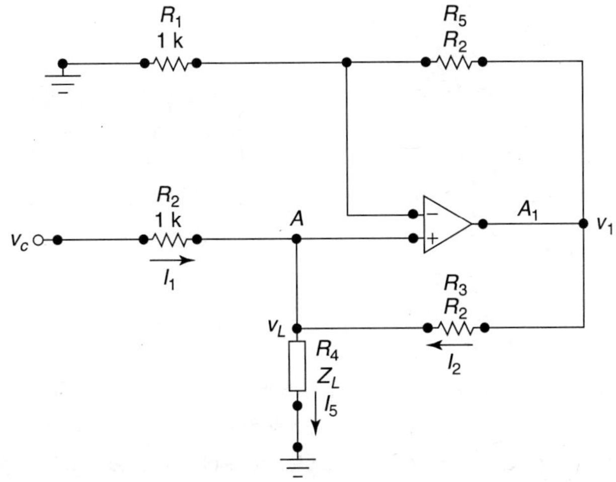


**Solution:**

Using voltage division and the voltage and current constraints for $A_1$ we get:

$$v_1 = 2v_L\ \text{(referenced to ground)}$$

Using Ohm's law: $$I_1 = \frac{v_c - v_L}{R}$$

and, $$I_2 = \frac{v_c - v_L}{R} = \frac{v_L}{R}$$

Applying KCL at node $A$: $$I_s = I_1 + I_2$$

$$= \frac{v_c}{R}$$

The circuit is a voltage-controlled current source. It is independent of $Z_L$ and is a function of $v_c$ and R.

**Example 11-18** Using the op-amps, as shown in the diagram, find:

(a) $I_s$ in terms of $v_c$

(b) The necessary circuit changes to reverse the actual direction of $I_s$

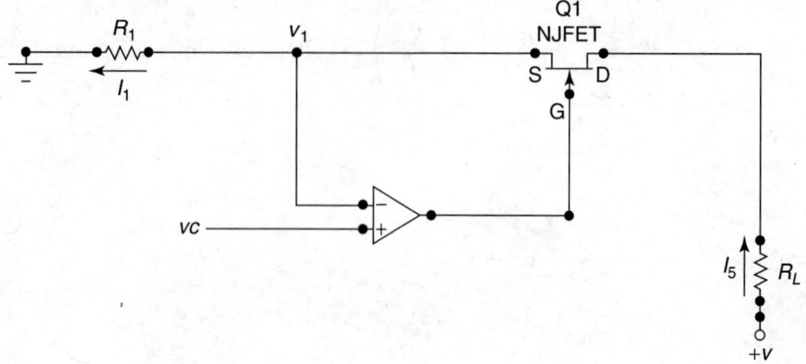


**Solution:**

(a) Because of the input voltage constraint $v_1 = v_c$ of the op-amps, and applying Ohm's law to $R_1$:

$$I_1 = \frac{v_c}{R_1}$$

Due to the input current constraint of the op-amps, and because the source current $I_s$ and the drain current $I_D$ of a JFET transistor are equal:

$$I_1 = I_s = I_D$$

$$\therefore \quad I_s = \frac{v_c}{R_1}$$

The JFET is a depletion mode device; $V_{GS}$ will be negative for the direction of $I_s$ shown. The output of the amplifier (or the gate voltage) will be at a value less positive than the source voltage (or $v_c$).

(b) To reverse the actual direction of $I_s$ replace Q1, an *n*-channel device by a *p*-channel device, return $R_L$ to $V^-$, and reverse the polarity of $v_c$.

**Example 11-19** Find $v_o/v_s$ for the circuit, as shown in the diagram. What mathematical function does the circuit perform for $\omega << 1/R_1C$?

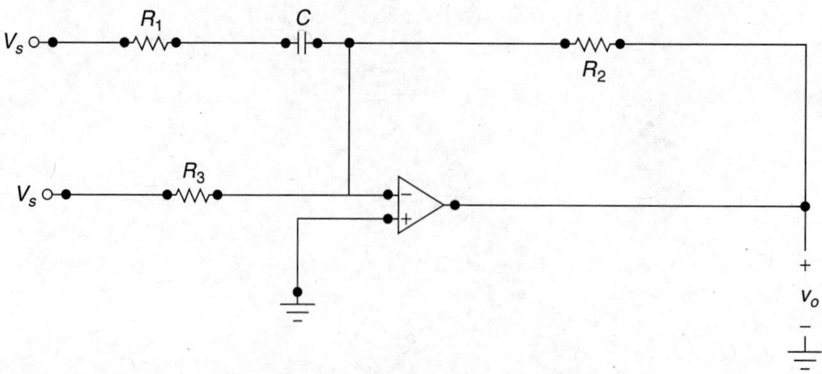

**Solution:**
Using the superposition theorem and treating $v_s$ as two equal-valued sources:

$$v_o = -\frac{R_2}{R_3}v_s - \left(\frac{R_2}{R_1 + \frac{1}{sC}}\right)v_s$$

$$v_o = -\frac{R_2}{R_3}v_s - \left[\frac{\left(\frac{R_2}{R_1}\right)s}{s + \frac{1}{R_1C}}\right]v_s$$

For $\omega << \frac{1}{R_1C}$:

$$v_o = -\frac{R_2}{R_3}v_s - sCR_2v_s$$

The output voltage is a function of the input voltage and its derivative.

**Example 11-20** Find the relationship between $v_0$ and the inputs $I_{s1}$ and $I_{s2}$.

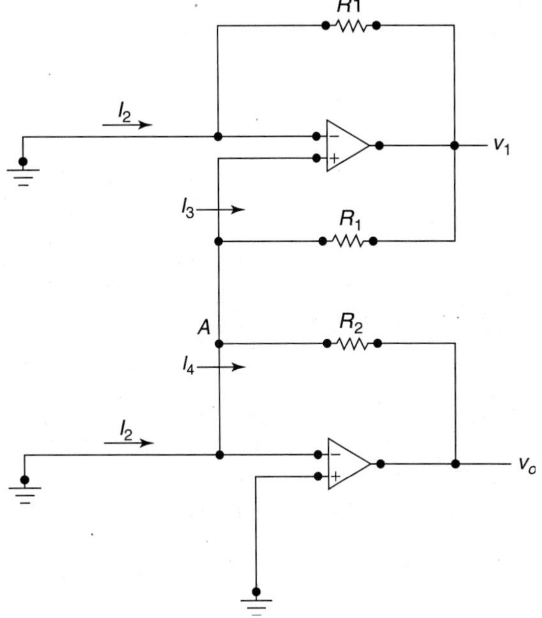


**Solution:**
Because the non-inverting input of U1 is at the ground and because of the voltage constraints of U1 and U2, the inputs of U1 and U2 are at zero volts.

Applying the current constraint to the inverting input of U2 and the Ohm's law:

$$v_2 = -I_{s2}R_1$$

tApplying the current constraint and Ohm's law:

$$I_3 = -\frac{v_2}{R_1} = I_{s2}$$

Applying KCL to node $A$:

$$I_4 = I_{s1} - I_3 = I_{s1} - I_{s2}$$

Using Ohm's law:

$$v_o = -(I_{s1} - I_{s2})\,R_2 = R_2\,(I_{s2} - I_{s1})$$

The output voltage of the circuit is proportional to the difference between the input currents.

**Example 11-21** Using the following diagram, determine:

(a) $\omega_0$ (b) K (c) The minimum value of K that will sustain oscillation

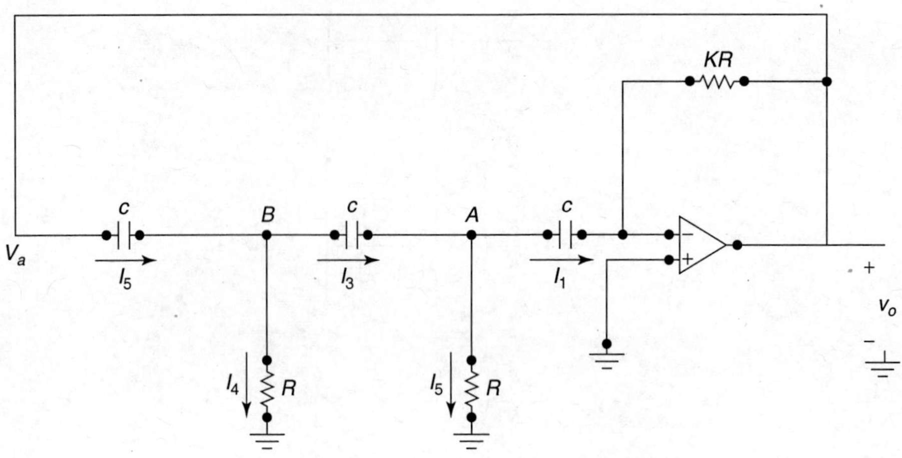


**Solution:**

(a) The frequency of oscillation is obtained from loop 1. The loop gain is obained from loop 2. The frequency where the loop gain is one.

Loop gain:

$$L_G(s) = \frac{v_o}{v_\alpha}$$

Using the op-amps input constraints and Ohm's law:

$$I_1 = -\frac{v_o}{KR}$$

$$V_A = I_1\frac{1}{sC} = -\frac{v_o}{sKRC}$$

$$I_A = \frac{V_A}{R}$$

Using KVL, KCL and Ohm's law:

$$I_3 = I_1 + I_2 = I_1 + \frac{V_A}{R} = -\frac{v_o}{KR} - \frac{v_o}{sKR^2C}$$

$$V_B = V_A + I_3\frac{1}{sC}$$

$$I_5 = I_3 + I_4 = I_3 + \frac{V_B}{R}$$

$$V_\alpha = V_B + I_5\frac{1}{sC}$$

Using the expression for $I_5$ in terms of $v_o$, solving for $L_G(s)$, and replacing $s$ by $j\omega$:

$$\frac{v_o}{v_\alpha}(j\omega) = \frac{K\omega^2R^2C^2}{4 + j\left[3\omega RC - \frac{1}{(\omega RC)}\right]}$$

For $L_G(j\omega_o) = 1$:

$$\omega_o = \frac{1}{\sqrt{3}RC}$$

(b) For loop gain, $v_o/v\alpha = 1$: $K = 12$

(c) The minimum value of K, which is necessary to sustain the oscillation, is 12.

**Example 11-22** The circuit, as shown in the following diagram is a simulated inductor. Find $L_{eq}$.

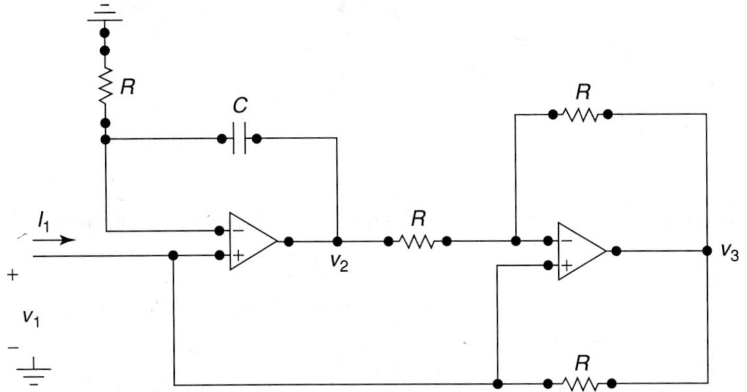


**Solution:**

The U1 circuit is non-inverting amplifier where:

$$V_2 = \frac{sRC + 1}{sRC}V_1$$

Using superposition theorem for the U2 circuit:

$$V_3 = -V_2 + 2V_1$$

$$= \frac{sRC - 1}{sRC}V_1$$

Using Ohm's law and the input current constraint for U2:

$$I_1 = \frac{V_1 - V_3}{R} = \frac{V_1}{sR^2C}$$

The network's input impedance is:

$$\frac{V_1}{I_1} = s(R^2C) = sL_{eq}$$

where, $L_{eq} = R^2C$

**Example 11-23** Find $T(s) = V_o/V_s$ for the above circuit. Determine the relationships between $R_1$, $R_2$, $R_3$ and $R_4$ for $T(s)$ to have an LP response, an HP response and an AP response.

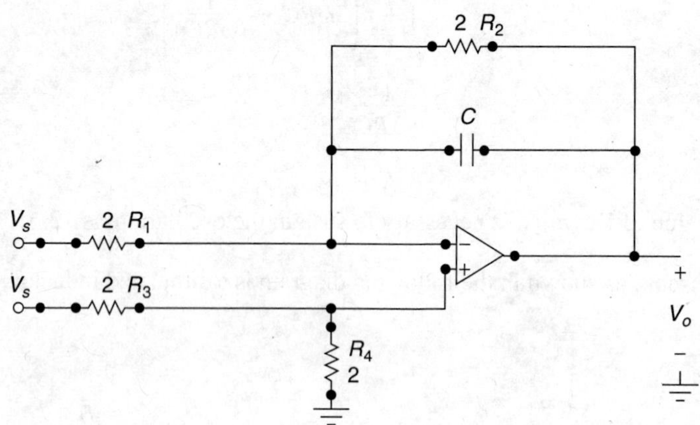


**Solution:**

$T(s)$ is found by splitting $V_s$ into two equal valued sources and using superposition theorem:

$$V_o' = -\frac{V_s(R_2/R_1)}{1 + sR_2C}$$

$$V_o'' = -V_s\frac{R_3}{R_3 + R_4}\,\frac{sR_1R_2C + R_1 + R_2}{sR_1R_2C + R_1}$$

$$V_o = V_o' + V_o''$$

$$= V_s\frac{-R_2\,(sR_2C + 1)\,K\,(sR_2C + 1)(sR_1R_2C + R_1 + R_2)}{(sR_2C + 1)(sR_1R_2C + R_1)}$$

where, $K = \dfrac{R_4}{R_3 + R_4}$

Solving for $T(s)$ produces:

$$T(s) = \frac{R_3}{R_3 + R_4}\,\frac{s + \left\{\left[\dfrac{1}{(K_1C)}\right]\left(\dfrac{R_1}{R_2} - \dfrac{R_3}{R_4}\right)\right\}}{s + \dfrac{1}{(R_2C)}}$$

For an LP response, $R_3 = 0$. For an HP response, $R_1R_4 = R_2R_3$ and for an LP response, $R_2R_3 = 2R_1R_4$.

■

## 11-6 REAL-LIFE APPLICATIONS

The op-amp is the basic component of analog computers. The typical uses of op-amp include providing amplitude changes in oscillators, active filter circuits and amplifier circuits of electronic instruments. In analog computers it is used to solve mathematical equations, simulate physical systems and to control physical process. The programming of analog computer is to arrange op-amps in different modes to solve mathematical, logical problems of certain equations. Mathematical functions for the solution of equation include integration, differentiation, summation, subtraction, etc. Op-amp is also used in many electronic devices as an amplifier, voltage multiplier, etc. Calculators, phase changer circuits also use the op-amp. It should be kept in mind that the op-amp is only used for low-power devices. It cannot be used in high-power applications.

## POINTS TO REMEMBER

1. The operational amplifier is a differential amplifier.
2. Important properties of ideal op-amp:
   (a) Gain of an ideal op-amp is infinite.
   (b) Output voltage is zero when input voltages are same or when both are zero.
   (c) Input resistance infinite.
   (d) Output resistance is zero.
   (e) CMRR must be infinity.
   (f) Infinite bandwidth.
3. Offset voltage occurs due to the mismatch of op-amp internal transistor.
4. Common-mode rejection ratio (CMRR) is defined as the ratio of the differential voltage gain $A_d$ to the common-mode voltage gain $A_{cm}$.
5. Slew rate (SR) is defined as the maximum rate of change of output voltage per unit of time and is expressed in volts per micro seconds.

## IMPORTANT FORMULAE

1. Output voltage of a non-inverting amplifier is given by:

$$v_o = \left(1 + \frac{R_{F,1}}{R_{1,1}}\right) v_i$$

2. Output of an inverting amplifier is given by:

$$V_o = -\left(\frac{R_F}{R_1}\right) V_{in}$$

3. Output of an adder is given by:

$$V_o = -\left(\frac{R_F}{R_a} V_a + \frac{R_F}{R_b} V_b + \frac{R_F}{R_c} V_c\right)$$

4. The common-mode rejection ratio is given by:

$$\text{CMRR} = \frac{\text{Differential voltage gain}}{\text{Common mode voltage gain}} = \frac{A_d}{A_{cm}}$$

5. Slew rate is given by:

$$SR = \left.\frac{dV_o}{dt}\right|_{\text{maximum}} \text{V}/\mu\text{s}$$

6. Differentiator output is given by:

$$v_o = -R_F C_1 \frac{dv_{in}}{dt}$$

7. Integrator output is given by:

$$v_o = -\frac{1}{R_1 C_F} \int_0^t v_{in}\, dt + c$$

## OBJECTIVE QUESTIONS

1. Op-amp is a/an:
   (a) Differential amplifier (b) Oscillator
   (c) Rectifier (d) None of the above
2. Op-amp operates at a:
   (a) High voltage (~100 KV)
   (b) Medium voltage (~220 Volt)
   (c) Low voltage (~12 Volt)
   (d) Very high voltage (~10 mili Volt)
3. Voltage gain of an ideal op-amp is:
   (a) Infinite (b) Very high
   (c) Low (d) Very low
4. Bandwidth of an ideal op-amp is:
   (a) Infinite (b) Very high
   (c) Low (d) Very low
5. Output Impedance of an ideal op-amp is:
   (a) Infinite (b) Very high
   (c) Low (d) Zero
6. Input impedance of an ideal op-amp is:
   (a) Infinite (b) Very high
   (c) Low (d) Zero
7. CMRR of an ideal op-amp is:
   (a) Infinite (b) Very high
   (c) Low (d) Very low
8. Slew rate of an ideal op-amp is:
   (a) Infinite (b) Very high
   (c) Low (d) Very low
9. Op-amp is a:
   (a) Voltage-controlled voltage source (VCVS)
   (b) Voltage-controlled current source (VCCS)
   (c) Current-controlled voltage source (CCVS)
   (d) Current-controlled current source (CCCS)
10. Op-amp uses:
    (a) Only +ve voltage
    (b) Only −ve voltage
    (c) Dual supply, i.e., $\pm V_{CC}$
    (d) None of the above
11. Virtual ground of an op-amp is:
    (a) The terminal is grounded directly
    (b) The terminal is not physically grounded but terminal voltage is zero as the other terminal is connected to the ground due to op-amp properties
    (c) Both (a) and (b)
    (d) None of the above
12. Op-amp uses:
    (a) Negative feedback (b) Positive feedback
    (c) No feedback (d) None of the above
13. Common-mode gain signifies:
    (a) That the ability to reject the common mode signals like noise, interference
    (b) Increase in the noise
    (c) Increase in the distortion
    (d) All of the above
14. Slew rate is defined as the:
    (a) Maximum rate of change of output voltage with time
    (b) Minimum rate of change of output voltage with time
    (c) Moderate rate of change of output voltage with time
    (d) None of the above
15. Op-amp integrator uses:
    (a) Capacitor as feedback element
    (b) Resistor as feedback element
    (c) Inductor as feedback element
    (d) A simple wire as feedback element
16. Match the following table columns w.r.t an integrator:

| **Input signal waveform** | **Integrator output waveform** |
|---|---|
| (a) Square wave | (i) Modified parabola |

(b) Triangular wave (ii) Triangular wave

(c) Saw-tooth wave (iii) Parabolic ramp

(d) Step function (iv) Ramp function

17. Match the following table columns w.r.t a differentiator:

| **Input signal waveform** | **Differentiator output waveform** |
|---|---|
| (a) Sine | (i) Square wave |
| (b) Square wave | (ii) Pulse train |
| (c) Triangular wave | (iii) Cosine |
| (d) Saw-tooth wave | (iv) Step function |

18. The high input impedance of an IC op-amp is achieved by using:

(a) FET (b) CE transistor stages
(c) MOSFET. (d) Darlington connection

19. The closed-loop gain of an inverting op-amp is:

(a) Equal to unity (b) Greater than unity
(c) Less than unity (d) Zero

20. The open-loop gain of an op-amp is:

(a) Large (b) Zero
(c) Small (d) Anything

## REVIEW QUESTIONS

1. What is an operational amplifier? Provide the circuit symbol of an op-amp.
2. (a) List the characteristics of an ideal op-amp.
   (b) Explain clearly how an op-amp can be used as: (i) a summing amplifier (ii) an integrator.
3. Explain with the help of a diagram, how two supply voltages $+V$ and $-V$ are obtained from a single dual power supply.
4. Why is the gain of an amplifier circuit independent of the gain of the op-amp employed?
5. In circuit analysis of an ideal op-amp with feedback, what values are assumed for $V_i$ and $I_i$? Why?
6. What is differential amplifier? Explain with a diagram, the action of a differential amplifier.
7. What is meant by CMRR? Explain the significance of a relatively large value of CMRR.
8. Explain why open-loop op-amp configurations are not used in linear applications.
9. Explain clearly, how an op-amp can be used for the following applications:
   (a) Inverting amplifier
   (b) Differentiator
   (c) Integrator
   (d) Summing amplifier
   Provide appropriate diagrams.
10. List the differences between the integrator and the differentiator.
11. Explain how an op-amp is checked by using it as a buffer.
12. Explain virtual ground.
13. The gain of a buffer amplifier is unity. Explain this statement.
14. Write the ideal values for each of the following parameters. There are three voltage sources $V_1$, $V_2$ and $V_3$. It is required to obtain the sum of these signals without the change in the magnitude and sign. Write a suitable circuit and explain its operation.
15. What is the difference between open-loop and closed-loop gains of an op-amp?
16. Show the internal block schematic of an op-amp while mentioning the role of each stage.
17. Define the following terms for an op-amp:
   (a) CMRR
   (b) Slew rate
   (c) Virtual ground
   (d) Schmitt trigger
   (e) Voltage-to-current converter
   (f) Current-to-voltage converter

18. Explain how a triangular wave can be generated by using op-amp.
19. Describe the use of an op-amp as a stable multi-vibrator.
20. What is a regenerative comparator?
21. Draw a schematic diagram of a scale changer and explain its working principle.
22. Describe the significance of virtual ground.

## PRACTICE PROBLEMS

1. Calculate the voltage gain of an inverting amplifier, given that $R_I = 8$ kΩ and $R_f = 56$ kΩ.
2. For an inverting amplifier, $R_1 = 2$ kΩ and $R_f =$ 1 MΩ. Determine the following circuit values:
   (a) $A_v$ (b) $R_i$ (c) $R_o$
3. For inverting amplifier, let $R_f = 250$ kΩ, $R_1 =$ 10 kΩ and $V_i = 0.5$ V. Calculate:
   (a) $I$ (b) The voltage across $R_f$
   (c) $V_o$
4. Design an inverting amplifier with a gain of −5 and an input resistance of 10 kΩ.
5. Calculate the output voltage of an inverting amplifier, if $R_I = 50$ kΩ, $R_f = 500$ kΩ and $V_i =$ 20.4 V.
6. Calculate the output voltage of non-inverting amplifier, if $R_I = 50$ kΩ, $R_f = 500$ kΩ and $V_i = 0.4$ V.
7. Calculate the value of the feedback resistor given that $A_v = -100$ and $R_I = 1$ kΩ, in the case of an inverting amplifier.
8. Design an amplifier with a gain of +10.
9. Calculate the voltage gain of non-inverting amplifier 0 input is 0 mV.
10. Calculate the output voltage of a three input adder for the following values: $R_1 = 20$ kΩ, $R_2 = 40$ kΩ, $R_3 = 60$ kΩ and $R_f = 100$ kΩ, $V_I = 20$ mv, $V_2 =$ 40 mV and $V_3 = 60$ mV.
11. Calculate the output voltage of an adder for the following values: $R_I = 250$ kΩ, $R_2 = 500$ kΩ, $R_3 = 1$ MΩ, $V_I = -3$ V, $V_2 = 3$ V and $V_3 = 2$ V.
12. In the subtractor circuit $R_I = 5$ kΩ, $R_f = 10$ kΩ, $V_I = 4$ V and $V_2 = 5$ V. Find the value of output voltage.
13. Calculate the output voltage of subtractor circuit if $R_I = 2$ kΩ, $R_f = 10$ kΩ, $V_I = 3$ V and $V_2 = 6$ V.
14. The input of the differentiator is a sinusoidal voltage of peak value 5 mV and frequency 1 kHz. Find the output if $R = 100$ kΩ and $C = 1$ F.
15. A 10 mV, 5 kHz sinusoidal signal applied to the input of an op-amp integrator for which $R = 100$ Ω and $C =$ 1F. Find the output voltage.
16. Design an adder using an op-amp to get the output expression as:

$$V_o = -(0.1\ V_1 + V_2 + 20\ V_3)$$

where, $V_1$, $V_2$ and $V_3$ are the inputs.

## SUGGESTED READINGS

1. Singh, J. 1994. *Semiconductor Devices: An Introduction*. New York, NY: McGraw-Hill.
2. Streetman, B.G. and S. Banerjee. 2000. *Solid State Electronic Devices*. New Delhi: Pearson Education.
3. Millman, Jacob and Christos C. Halkias. 1986. *Integrated electronics: Analog and Digital Circuits and Systems*. New Delhi: McGraw Hill Book Company.
4. Pierret, R. F and G.W. Neudeck. 1989. *Modular Series on Solid State Devices*. Boston, M.A.: Addison Wesley.
5. Singh, B.P. and Rekha Singh. 2006. *Electronic Devices and Integrated Circuits*. New Delhi: Pearson Education.

# 12

# Oscillators

## Outline



## Objectives

In this chapter we will explore the working principle of the oscillator. Generally speaking, the oscillator produces sinusoidal and other waveforms. Beginning with a detailed circuit analysis of the oscillator, we will proceed to discuss the conditions and frequency of oscillation. Following this, the different types of oscillators—Tuned oscillator, Hartley oscillator, Colpitts oscillator, Clapp oscillator, Phase-shift oscillator, Crystal oscillator and Wien-bridge oscillator—will be examined with detailed mathematical analysis and illustrations. The chapter ends with an overview of the applications of the oscillator.

## 12-1 INTRODUCTION

An oscillator is an electronic system. It comprises active and passive circuit elements and sinusoidal produces repetitive waveforms at the output without the application of a *direct external input signal* to the circuit. It converts the dc power from the source to ac power in the load. A rectifier circuit converts ac to dc power, but an oscillator converts dc noise signal/power to its ac equivalent. The general form of a harmonic oscillator is an electronic amplifier with the output attached to a narrow-band electronic filter, and the output of the filter attached to the input of the amplifier. When the power supply to the amplifier is just switched on, the amplifier's output consists only of noise. The noise travels around the loop, being filtered and re-amplified till it increasingly resembles the desired signal level. In this chapter, the oscillator analysis is done in two methods—first by a general analysis, considering all other circuits are the special form of a common generalized circuit and second, using the individual circuit KVL analysis. The difference between an amplifier and an oscillator is shown in Fig. 12-1. Figure 12-1(a) shows how an amplifier produces output based on the given input. Figure 12-1(b) shows how an oscillator produces output without any direct input.

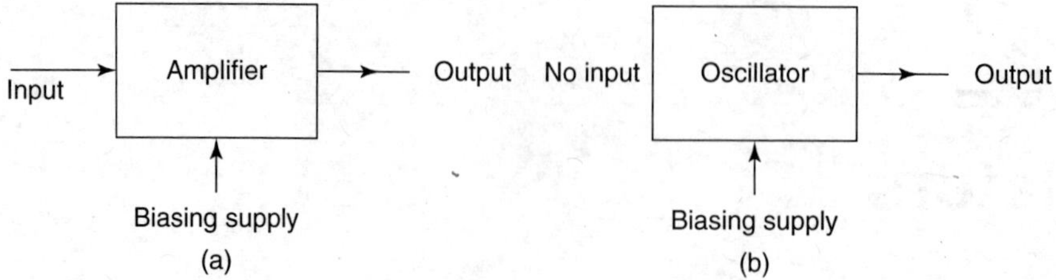


**Figure 12-1** Schematic block diagrams showing the difference between an amplifier and an oscillator

**Table 12-1** Different types of oscillators and their frequency ranges

| *Type of Oscillator* | *Frequency Range Used* |
|---|---|
| Audio-frequency oscillator | 20 Hz–20 kHz |
| Radio-frequency oscillator | 20 kHz–30 MHz |
| Very-high-frequency oscillator | 30 MHz–300 MHz |
| Ultra-high-frequency oscillator | 300 MHz–3 GHz |
| Microwave oscillator | 3 GHz–30 GHz |
| Millimeter wave oscillator | 30 GHz–300 GHz |

## 12-2 CLASSIFICATIONS OF OSCILLATORS

Oscillators are classified based on the type of the output waveform. If the generated waveform is sinusoidal or close to sinusoidal (with a certain frequency) then the oscillator is said to be a sinusoidal oscillator. If the output waveform is non-sinusoidal, which refers to square/saw-tooth waveforms, the oscillator is said to be a relaxation oscillator. In this chapter the oscillator circuits are designed by transistors, but oscillators also can be designed using the FET and op-amp.

An oscillator has a positive feedback with the loop gain infinite. Feedback-type sinusoidal oscillators can be classified as LC (inductor-capacitor) and RC (resistor-capacitor) oscillators. The classification of various oscillators is shown in Table 12-1.

Generally no direct external signal is applied to an oscillator circuit. During the period when the power supply to the system is switched on, a noise voltage is generated which triggers the oscillations. A negative resistance must be provided in an oscillator by an external positive feedback to make the gain infinite.

## 12-3 CIRCUIT ANALYSIS OF A GENERAL OSCILLATOR

This section discusses the general oscillator circuit with a simple generalized analysis using the transistor, as shown in Fig. 12-2. An impedance $z_1$ is connected between the base $B$ and the emitter $E$, an impedance $z_2$ is connected between the collector $C$ and emitter $E$. To apply a positive feedback $z_3$ is connected between the collector and the base terminal. All the other different oscillators can be analysed as a special case of the generalized analysis of oscillator.

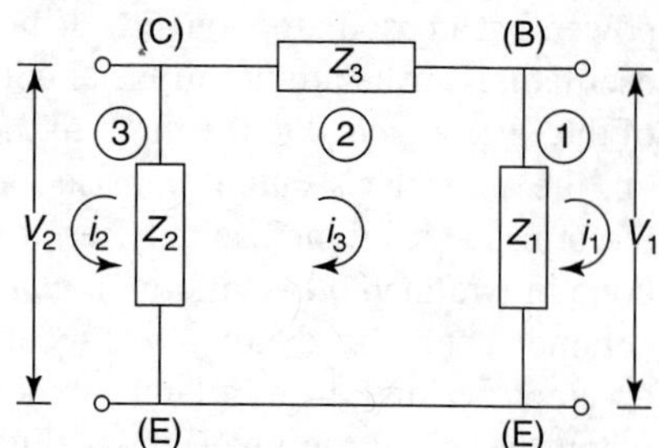


**Figure 12-2** A generalized oscillator circuit analysis

The above generalized circuit of an oscillator is considered using a simple transistor-equivalent circuit model. The equivalent model of transistor, as calculated in Chapter 5, where the current voltage expressions are expressed as follows:

$$v_1 = h_i i_1 + h_r v_2 \simeq h_i i_1 \tag{12-1}$$

As the numerical value of $h_r v_2$ is negligible:

$$v_1 = h_i i_1 \tag{12-2}$$

$$i_2 = h_f i_1 + h_0 v_2 \simeq h_f i_1 \tag{12-3}$$

As the numerical value of $h_0 v_2$ negligible the Eq. (12-3) can be written as:

$$i_2 = h_f i_1 \tag{12-4}$$

Applying KVL at loop (1) of Fig. 12-2 by considering that current through the impedance $z_1$ is $(i_1 - i_3)$, we get:

$$v_1 + z_1 (i_1 - i_3) = 0$$

or,
$$v_1 = -z_1(i_1 - i_3) = z_1 (i_3 - i_1) \tag{12-5}$$

Substituting the value of voltage $v_1$ from Eq. (12-2) in Eq. (12-5) we get:

$$h_i i_1 + z_1 i_1 - z_1 i_3 = 0$$

or,
$$i_1 (h_1 + z_1) - z_i i_3 = 0 \tag{12-6}$$

Applying KVL at loop (3) by considering voltage across the impedance $z_2$:

$$v_2 + z_2(i_3 + i_2) = 0 \tag{12-7}$$

Substituting the value of current $i_2$ we get:

$$v_2 = -z_2(h_f i_1 + i_3)$$

or,
$$z_2 h_f i_1 + z_2 i_3 + v_2 = 0 \tag{12-8}$$

Applying the KVL at loop (2) by considering voltage across $z_3$ we get,

$$i_3 z_3 + (i_2 + i_3) z_2 + (i_3 - i_1) z_1 = 0 \tag{12-9}$$

or,
$$i_3 z_3 - v_2 + v_1 = 0 \tag{12-10}$$

or,
$$i_3 z_3 = v_2 - v_1 \tag{12-11}$$

Substituting the value of $v_1$ in Eq. (12-11) we get:

$$i_3 z_3 = v_2 - h_i i_1 \tag{12-12}$$

or,
$$i_3 z_3 - v_2 + h_i i_1 = 0 \tag{12-13}$$

or,
$$-(v_2 - i_3 z_3 - h_i i_1) = 0 \tag{12-14}$$

or,
$$v_2 - h_i i_1 - z_3 i_3 = 0 \tag{12-15}$$

Eq. (12-6), Eq. (12-8) and Eq. (12-15) can be rewritten as:

$$i_1(h_i + z_1) + 0.\,v_2 + (-z_1)i_3 = 0$$
$$-i_1 z_2 h_f + 1.\,v_2 + z_2 i_3 = 0$$
$$-i_1 h_i + 1.\,v_2 + (-z_3)i_3 = 0$$

Eliminating three variables $i_1$, $v_2$, $i_3$ using Camers rule, and from Eqs. (12-6), (12-8) and (12-15), we get the following matrix:

$$\begin{vmatrix} (h_i + z_1) & 0 & -z_1 \\ z_2 h_f & 1 & z_2 \\ -h_i & 1 & -z_3 \end{vmatrix} = 0\,\cdot$$

or, $$(h_i + z_1)[-z_3 - z_2] + 0 + (-z_1)[z_2 h_f + h_i] = 0 \qquad (12\text{-}16)$$

or, $$-z_3 h_i - z_2 h_i - z_1 z_3 - z_1 z_2 - z_1 z_2 h_f - z_1 h_i = 0$$

or, $$-h_i[z_3 + z_2 z_1] - z_1 z_2\,[1 + h_f] - z_1 z_3 = 0$$

or, $$h_i[R + jx] + z_1 z_2\,[1 + h_f] + z_1 z_3 = 0$$

Let, $$z_1 = R_1 + jx_1 \simeq jx_1 \qquad \Theta \;\; R_1 \simeq x_1$$
$$z_2 = R_2 + jx_2 \simeq jx_2 \qquad R_2 \simeq x_2$$
$$z_3 = R_3 + jx_3 \simeq jx_3 \qquad \& \;\; R_3 \simeq x_3$$

By adding, we get:

$$(z_1 + z_2 + z_3) = (R + jx)$$

where, $R = (R_1 + R_2 + R_3)$ is not negligible in comparison with $x = x_1 + x_2 + x_3$ as we shall see $x = 0$, at frequency of oscillation.

∴ $$z_1 z_2\,(h_f + 1) + (z_1 + z_2 + z_3)h_i + z_1 z_3 = 0 \qquad (12\text{-}17)$$
$$-x_1 x_2(h_f + 1) + (R + jx)\,h_i - x_1 x_3 = 0$$
$$-x_1[x_2(h_f + 1) + x_3] + (R + jx)h_i = 0 \qquad (12\text{-}18)$$

Equating imaginary parts $[\Theta\, jx_1\, jx_2 = -x_1 x_2]$:

$$jxh_i = 0 \qquad (+\text{ve}) \text{ inductive impedance}$$

∴ $$x = 0 \qquad (-\text{ve}) \text{ capacitive impedance}$$

$$x_1 + x_2 + x_3 = 0 \qquad (12\text{-}19)$$

Equating real parts we get:

$$-x_1 x_2[h_f + 1] + R h_i - x_1 x_3 = 0 \qquad (12\text{-}20)$$

$$x_1 x_2 h_f + x_1(x_2 + x_3) - R \cdot h_i = 0 \qquad (12\text{-}21)$$

From the Eq. (12-19) we get: $$x_2 + x_3 = -x_1 \quad (12\text{-}22)$$

Substituting the value of Eq. (12-22) in Eq. (12-19) we get:

$$x_1 x_2 h_f - x_1^2 - R \cdot h_i = 0$$

$$h_f = \frac{x_1}{x_2} + \frac{R \cdot h_i}{x_1 x_2} \quad (12\text{-}23)$$

This is the general condition for oscillation for an oscillator.

Different types of oscillator circuits with different configurations can be analysed through this general method. This makes the analysis simpler.

## 12-3-1 Hartley Oscillator

Hartley oscillator contains two inductors and one capacitor, as shown in Fig. 12-3 where, $x_1$ and $x_2$ are inductances, and $x_3$ is a capacitance, i.e., $x_1 = \omega L_1, x_2 = \omega L_2, x_3 = -1/\omega C$.

Substituting the values in Eq. (12-23) we get the condition for oscillation, considering $R$ is small.

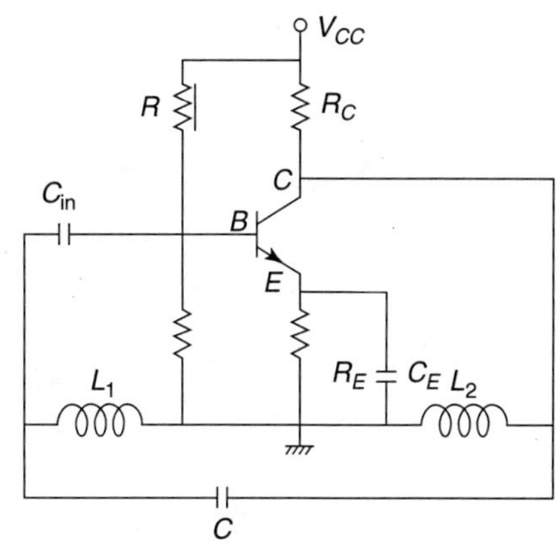


**Figure 12-3** Hartley Oscillator

$$h_f = \frac{\omega L_1}{\omega L_2} + \frac{R \cdot h_i}{\omega^2 L_1 L_2}$$

$$\therefore \quad h_f = \frac{L_1}{L_2}$$

$$h_f = \frac{L_1}{L_2} + \frac{RCh_i}{L_{11}} \quad (12\text{-}24)$$

Where,

$$L_{11} = \frac{L_1 L_2}{L_1} + L_2$$

$$L_{11} = L_1 L_2 / L_1 + L_2$$

$$L_{11}(L_1 + L_2) = L_1 L_2$$

$$L_{11} \times \frac{1}{\omega 2C} = L_1 L_2$$

$$\Theta\, x_1 + x_2 = -x_3$$

$$\omega L_1 + \omega L_2 = -\left(\frac{-1}{\omega C}\right)$$

$$L_1 + L_2 = -\frac{1}{\omega^2 C}$$

$$\therefore \quad \frac{L_1}{L_2} + \frac{Rh_i}{\omega^2 L_1 L_2} = \frac{L_1}{L_2} + \frac{Rh_i}{\omega^2 L_1 L_2} = \frac{L_1}{L_2} + \frac{Rh_i C}{L_{11}}$$

Frequency of oscillation can be calculated as follows:

$$\omega L_1 + \omega L_2 = -\left(-\frac{1}{\omega C}\right)$$

$$\omega^2 (L_1 + L_2) = \frac{1}{c}$$

$$\omega = \frac{1}{\sqrt{C(L_1 + L_1)}}$$

$$2\pi f = \frac{1}{\sqrt{C(L_1 + L_1)}}$$

$$f = \frac{1}{2\pi}\sqrt{C(L_1 + L_1)} \quad (12\text{-}25)$$

This is the required frequency of oscillation for Hartley oscillator.

## 12-3-2 Colpitts Oscillator

Colpitts oscillator contains two capacitors and one inductor, as shown in Fig. 12-4. $X_1$ and $X_2$ are capacitances, $X_3$ is inductance, $Z_1$ and $Z_2$ are capacitors, $C_1$ and $C_2$ are capacitances, and $Z_3$ is an inductor of inductance $L$.

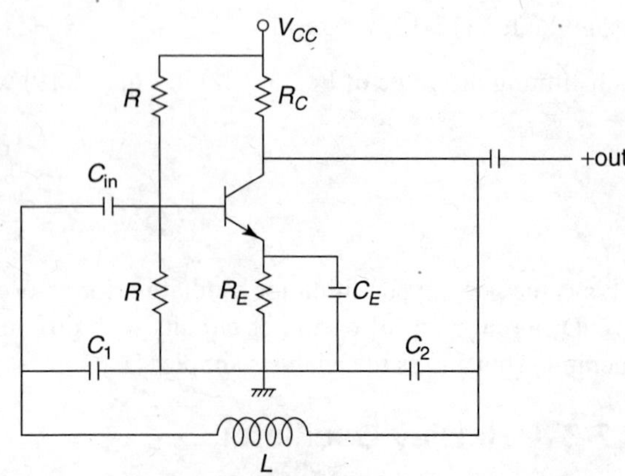


**Figure 12-4** Colpitts oscillator

$$X_1 = -\frac{1}{\omega C_1}$$

$$X_2 = -\frac{1}{\omega C_2}$$

$$X_3 = \omega L$$

$$X_1 + X_2 + X_3 = 0$$

$$-\frac{1}{\omega C_1} - \frac{1}{\omega C_2} + \omega L = 0$$

$$\frac{1}{\omega}\left(\frac{1}{C_1} + \frac{1}{C_2}\right) = \omega L$$

$$\frac{1}{L}\left(\frac{1}{C_1} + \frac{1}{C_2}\right) = \omega^2, \qquad \omega = \sqrt{\frac{1}{L}\left(\frac{1}{C_1} + \frac{1}{C_2}\right)}$$

Frequency of oscillation:

$$2\pi f = \frac{1}{\sqrt{LC}} \Rightarrow f = \frac{1}{2\pi\sqrt{LC}} \qquad (12\text{-}26)$$

where, $\frac{1}{C'} = \frac{1}{C_1} + \frac{1}{C_2}$

$$h_f = \frac{X_1}{X_2} + \frac{Rh_i}{X_1 X_2} \qquad (12\text{-}27)$$

Therefore, condition for oscillation:

$$h_f = \frac{C_2}{C_1} + Rh_i\, \omega^2 C_1 C_2 \qquad (12\text{-}28)$$

$$= \frac{C_2}{C_1} + R_{hi}\, \omega^2 C_1 C_2$$

$R$ = Resistance of the coil 2

$$R = \frac{C_2}{C_1} + Rh_i \frac{1}{L}\, \frac{C_1 + C_2}{C_1 C_2}\, C_1\, C_2 \left[\because \omega^2 = \frac{1}{L}\left(\frac{1}{C_1} + \frac{1}{C_2}\right)\right]$$

$$R = \frac{C_2}{C_1} + Rh_i \frac{1}{L} (C_1 + C_2)\, Rh_i \qquad (12\text{-}29)$$

$$R = \frac{C_2}{C_1}\left[\text{neglecting } \frac{Rh_i}{L}(C_1 + C_2)\right]$$

The circuit diagram of Colpitts oscillator is shown in Fig. 12-4.

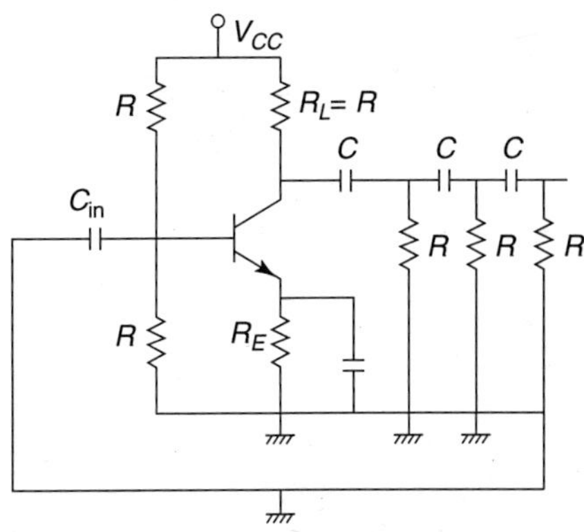


**Figure 12-5(a)** Phase-shift oscillator: equivalent circuit using the approximate equivalent circuit of the transistor

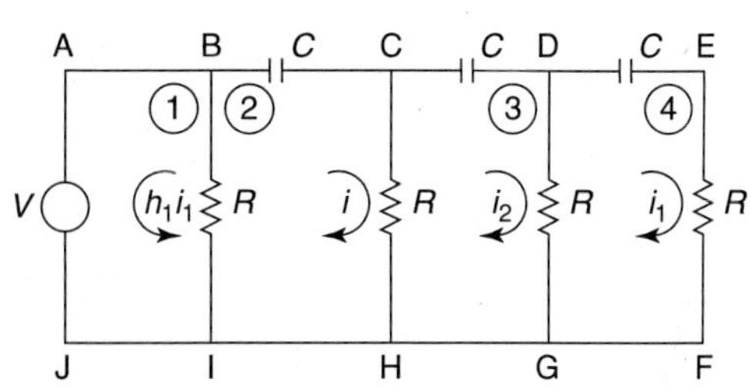


**Figure 12-5(b)** Equivalent circuit representation of a phase-shift oscillator

## 12-3-3 Phase-Shift Oscillator

The circuit diagram of a phase-shift oscillator with three pairs of RC combination is shown in Fig. 12-5(a).

The equivalent circuit representation of phase-shift oscillator is shown in Fig. 12-5(b). By applying KVL in the circuit in Fig. 12-5(b) we have the mesh ABCHIJ at loop (2).

$$(i+ h_f i_1)\, R + (i - i^1)\, R + \frac{i}{jwc} = 0$$

$$\left(2R + \frac{1}{jwc}\right) i + Rh_f i_1 - R\, i^1 = 0$$

$$(2R + jx_c)\, i + Rh_f i_1 - Ri^1 = 0 \qquad (12\text{-}30)$$

At mesh CDGH [at loop (3)]:

$$(i^1 - i)\, R + \frac{1}{j\omega c}\, i^1 + (i^1 - i_1)\, R = 0$$

$$(2R + jx_c) i^1 - Ri - Ri_1 = 0 \qquad (12\text{-}31)$$

At mesh CDEFGH [at loop (4)]:

$$(i_1 - i^1)R + jx_c i_1 + Ri_1 = 0$$

$$(2R + jx_c)\, i_1 - Ri^1 = 0 \qquad (12\text{-}32)$$

Eliminating $i_1$, $i$, $i^1$, order wise Eqs. (12-30), (12-31) and (12-32) will be:

$$\begin{vmatrix} i_1 & i & i^1 \\ (2R + jx_c) & 0 & -R \\ Rh_f & (2R + jx_c) & -R \\ -R & -R & (2R + jx_c) \end{vmatrix}$$

Dividing each element of the determinant by $R$:

$$\therefore \quad \frac{1}{R}\begin{vmatrix} R(2 + jx_c/R) & 0 & -R \\ R_{hf} & R(2 + jx_c/R) & -R \\ -R & -R & R(2 + jx_c/R) \end{vmatrix} = 0$$

Let $$\frac{X_C}{R} = a$$

$$\therefore \quad \begin{vmatrix} (2 + ja) & 0 & -1 \\ h_f & (2 + ja) & -1 \\ -1 & -1 & (2 + ja) \end{vmatrix} = 0$$

$$(2 + ja)\,[(2 + ja)^2 - 1] + 0 + (-1)\,[-h_f + 2 + ja] = 0$$

$$(2 + ja)\,[4 + 4ja - a^2 - 1] + h_f - 2 - ja = 0$$

$$8 + 8ja - 2a^2 - 2 + 4ja - 4a^2 - ja^3 - ja + h_f - 2 - ja = 0$$

$$- ja^3 + 8 + 12ja - 6a^2 - 4 - 2ja + h_f = 0$$

$\therefore$ Equating the imaginary parts:

$$j\,(-a^3 - 2a + 12a) = 0$$

$$a\,(10 - a^2) = 0$$

$$a^2 - 10 = 0$$

$$\therefore \quad a = \sqrt{10}$$

$$\frac{XC}{R} = \sqrt{10}$$

$$\frac{X_c^2}{R^2} = 10$$

$$\frac{1}{\omega^2 C^2 R^2} = 10$$

or, $$\omega^2 = \frac{1}{10C^2R^2}$$

or, $$\omega = \frac{1}{\sqrt{10}\,CR}$$

$\therefore$ Frequency of oscillation is: $f = \dfrac{1}{2\pi\sqrt{10}\,CR}$ (12-33)

Equating the real parts we get: $8 - 6a^2 - 4 + h_{fe} = 0$

$$h_{fe} = 4 + 6a^2 - 8 = 4 + 6.10 - 8$$

$$= 4 + 60 - 8$$

$$= 56$$

For sustained oscillations, $h_{fe}$ of 56 for $R = R_L$

The equivalent diagram of a phase-shift oscillator is shown in Fig. 12-6.

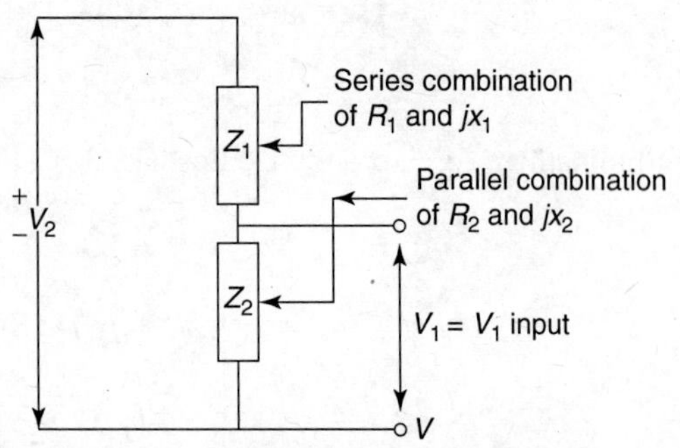


**Figure 12-6** Equivalent diagram of a phase-shift oscillator

## 12-3-4 Wien-Bridge Oscillator

Wien-bridge oscillator is the series and parallel combination of a resistance $R$ and a capacitor $C$. According to Barkhausen criteria, $A_v\beta = 1$.

Since, $A_v\beta = 1$

$$\beta = \frac{1}{A_v} = \frac{v_{ir}}{v_o} = \frac{v_{zi}}{(z_1 + z_2)i}$$

$$A_v = \frac{1}{\beta} = \frac{z_1 + z_2}{z_2} = 1 + \frac{z_1}{z_2} \qquad (12\text{-}34)$$

$$z_1 = R + jx_1 \text{ (series combination)}$$

$$\frac{1}{z_2} = \frac{1}{R_2} + \frac{1}{jx_2} \text{ (parallel combination)}$$

$$A = 1 + (R_1 + jx_1)\left(\frac{1}{R_2} + \frac{1}{jx_2}\right)$$

$$= 1 + \left(\frac{R_1}{R_2} + \frac{x_1}{x_2}\right) + j\left(\frac{x_1}{R_2} - \frac{R_1}{x_2}\right) \qquad (12\text{-}35)$$

The two-stage $RC$ coupled amplifier can be used by equating real and imaginary parts. Considering only the real parts, we get:

$$A = 1 + \frac{R_1}{R_2} + \frac{x_1}{x_2} \qquad (12\text{-}36)$$

Considering only the imaginary parts, we get:

$$\frac{x_1}{R_2} - \frac{R_1}{X_2} = 0 \qquad (12\text{-}37)$$

$X_1X_2 = R_1R_2$ (frequency of oscillation)

$$R_1R_2 = \frac{1}{w^2c_1c^2}$$

$$w^2 = \frac{1}{C_1C_2R_1R_2} \qquad (12\text{-}38)$$

If $R_1 = R_2 = R$ & $C_1 = C_2 = C$

$$A = 1 + 1 + 1 = 3$$

and, $$w^2 = \frac{1}{C^2R^2} \Rightarrow w = \frac{1}{CR} \qquad (12\text{-}39)$$

$$f = \frac{1}{2\pi CR} \qquad (12\text{-}40)$$

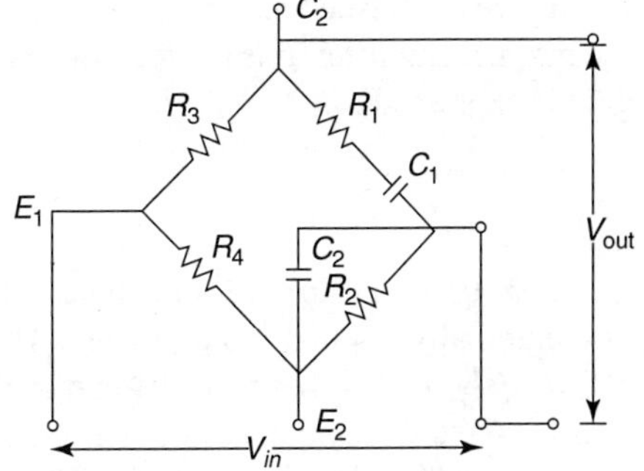

**Figure 12-7** Wien-bridge oscillator

At balance condition: $$\frac{R_3}{R_4} = \frac{Z_1}{Z_2} \text{ (for oscillation)}$$

From the circuit diagram of the Wien-bridge oscillator, as given in Fig. 12-7, we get:

$$\frac{R_3}{R_4} = \left(R_1 + \frac{1}{jwc_1}\right)\left(\frac{1}{R_2} + jwc_2\right) \qquad (12\text{-}41)$$

$$= \left(\frac{R_1}{R_2} + \frac{C_2}{C_1}\right) + j\left(\omega C_2R_1 - \frac{1}{\omega C_1R_2}\right)$$

Equating imaginary parts we get:

$$\omega c_2 R_1 = \frac{1}{\omega C_1 R_2}$$

$$\omega^2 = \frac{1}{C^2 R^2} \quad (12\text{-}42)$$

$\because R_1 = R_2 = R$ and $C_1 = C_2 = C$

$$\therefore \quad \frac{R_3}{R_4} = \frac{R_1}{R_2} + \frac{C_2}{C_1}$$

$$\frac{R_3}{R_4} = \frac{R}{R} + \frac{C}{C} = 1 + 1 = 2 \quad (12\text{-}43)$$

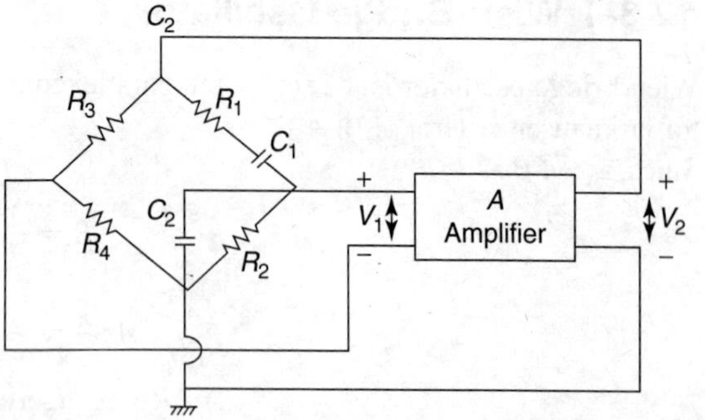

**Figure 12-8** Wien-bridge oscillator with an amplifier

The Wien-bridge oscillator with an amplifier is shown in Fig. 12-8.

### Advantages of Wien-Bridge Oscillator

The Wien-bridge oscillator has the following advantages:

1. The frequency of oscillation can be easily varied just by changing *RC* network
2. High gain due to two-stage amplifier
3. Stability is high

### Disadvantages of Wien-Bridge Oscillator

The main disadvantage of the Wien-bridge oscillator is that a high frequency of oscillation cannot be generated.

## 12-4 CONDITIONS FOR OSCILLATION: BARKHAUSEN CRITERION

Earlier we established that the overall gain of a feedback amplifier is $A_f = A/(1 + A\beta)$, where $A$ is the gain of the internal amplifier, $\beta$ is the feedback ratio, and $-A\beta$ is the loop gain. For positive feedback, the feedback gain is expressed as:

$$A_f = \frac{A}{1 - A\beta} \quad (12\text{-}44)$$

For $A\beta = 1$, Eq. (12-44) yields $A_f = \infty$. The amplifier then produces an output voltage without any externally applied input voltage. Thus, the amplifier becomes an oscillator. When the signal equals $V_o' \Rightarrow A\beta V_o = V_o$ or $A\beta = 1$, the output voltage regenerates itself and the amplifier oscillates. This condition is called the Barkhausen criterion. This condition means that $|A\beta| = 1$ and the phase angle of $A\beta$ is zero or an integral multiple of 360. The basic conditions for oscillation in a feedback amplifier are: (1) the feedback must be regenerative, (2) the loop-gain must be unity, and (3) the phase difference must be zero or an integral multiple of 360.

### 12-4-1 Nyquist Criterion for Oscillation

The loop gain factor $A\beta$ is a complex quantity and it is function of frequency. In case of positive feedback, the amplifier breaks into oscillations. The system is unstable due to circuit non-linearity. The condition of instability of an amplifier is expressed by the Nyquist criterion.

In Fig. 12-9, $X$ is a point on the locus of $A\beta$. If $X$ is inside the circle of unit radius $(1 + j0)$, we have $|1 - A\beta| < 1$ and positive feedback occurs. If $X$ is outside the circle, we have $|1 - A\beta| > 1$ and the feedback is negative. Nyquist criterion states that if this closed curve passes through or encloses the point $(1 + j0)$, the amplifier becomes unstable and oscillates. It is important to note that a positive feedback amplifier will not oscillate unless the Nyquist criterion is satisfied.

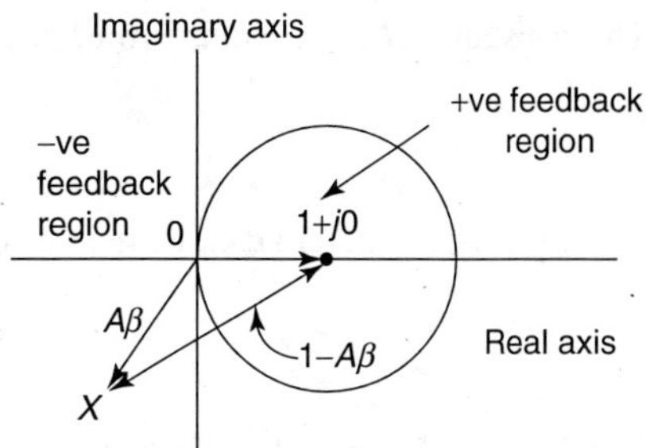


**Figure 12-9** Regions of positive and negative feedback in the complex plane

In the steady state condition the loop gain becomes unity and the oscillations are sustained, the frequency of oscillations is controlled by the frequency-determining network of the oscillator. The RC and LC combination circuits are used in oscillators to serve as the frequency-determining network.

Let us summarize the key necessities of a feedback oscillator.

1. Amplifier with positive feedback produces a negative resistance in the system.
2. A frequency-determining network creates oscillations at certain required frequencies.
3. System non-linearity introduced by the devices contain the amplitude of oscillation.

## 12-5 TUNED OSCILLATOR

The circuit diagram of a tuned oscillator is shown in Fig. 12-10(a). The emitter by pass capacitor $C_E$ shunts the ac so that $R_E$ is omitted from the ac equivalent circuit of Fig. 12-10(b). The dc operating point of the transistor is determined by the resistances $R_1$, $R_2$ and $R_E$, and supply voltage. The transistor gives a phase-shift of 180°.

### 12-5-1 Circuit Analysis

The frequency of oscillation is approximately given by the natural resonant frequency of the $LC$ tank circuit. Thus:

$$f = \frac{1}{2\pi\sqrt{LC}} \tag{12-45}$$

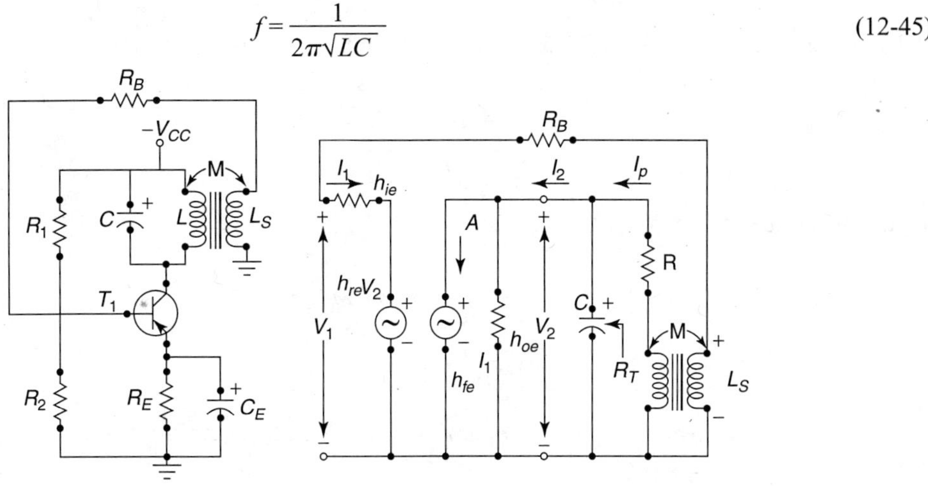


**Figure 12-10** (a) Circuit diagrams of a tuned oscillator (b) ac equivalent circuit of tuned circuit

The resistance $R$ of the transformer primary is small at resonance; the resistance is:

$$R_T = \frac{L}{CR} \tag{12-46}$$

From the $h$-parameter ac equivalent circuit of Fig. 12-10(b), we obtain:

$$V_2 = -I_2 R_T = -\frac{I_2 L}{CR} \tag{12-47}$$

Upon application of KCL at the point we have:

$$I_2 = h_{oe} V_2 + h_{fe} I_1 \tag{12-48}$$

Substituting for $V_2$ from Eq. (12-47) into Eq. (12-48), we have:

$$I_2 = -\frac{I_2 L}{CR} h_{oe} + h_{fe} I_1$$

or,

$$I_2\left(1 + h_{oe}\frac{L}{CR}\right) = h_{fe} I_1 \tag{12-49}$$

If M is the mutual inductance, then the voltage in the transformer secondary is $j\omega MI_p$. $I_P$ is the primary current. The secondary circuit impedance is $R_B + h_{ie}$. Applying Kirchhoff's voltage law in the secondary circuit, we obtain:

or,

$$I_1 = \frac{j\omega M I_p - h_{re} V_2}{R_B + h_{ie}} \tag{12-50}$$

From Fig. 12-10(b), we also obtain:

$$I_p = \frac{1/(j\omega C)}{R + j\omega L + \dfrac{1}{j\omega C}} I_2 \tag{12-51}$$

Substituting $I_p$ and $V_2$ from Eqs. (12-51) and (12-47), in Eq. (12-50) yields:

$$I_1 = \frac{\left(\dfrac{\omega^2 ML}{R} + \dfrac{h_{re} L}{CR}\right) I_2}{R_B + h_{ie}} \tag{12-52}$$

Putting this value of $I_1$ in Eq. (12-52) gives:

$$1 + h_{oe}\frac{L}{CR} = \frac{h_{fe}}{R_B + h_{ie}}\left(\omega^2 M + \frac{h_{re}}{C}\right)\frac{L}{R} \tag{12-53}$$

Substituting $\omega = 1/\sqrt{LC}$ in Eq. (12-53) we get:

$$M = \frac{R_B}{h_{fe}}(CR + h_{oe} L) + CR\frac{h_{ie}}{h_{fe}} + L\frac{\left(h_{ie} h_{oe} - h_{fe} h_{re}\right)}{h_{fe}} \tag{12-54}$$

The value of M is used to calculate the condition for sustained oscillation.

## 12-6 CRYSTAL OSCILLATOR

Crystal oscillator is most commonly used oscillator with high-frequency stability. Crystal oscillators are used for laboratory experiments, communication circuits and biomedical instruments. Crystal oscillators are usually, fixed frequency oscillators where stability and accuracy are the primary considerations. In order to design a stable and accurate LC oscillator for the upper HF and higher frequencies it is absolutely necessary to have a crystal control; hence, the reason for crystal oscillators. Crystal oscillators are oscillators where the primary frequency determining element is a quartz crystal. Because of the inherent characteristics of the quartz crystal the crystal oscillator may be held to extreme accuracy of frequency stability. Temperature compensation may be applied to crystal oscillators to improve thermal stability of the crystal oscillator. Crystal oscillators are usually, fixed frequency oscillators where stability and accuracy are the primary considerations. A typical crystal oscillator is shown in Fig. 12-11(a). A series LCR circuit shunted by a capacitor $C'$, as shown in Fig. 12-11(a), represents the equivalent circuit of a piezoelectric crystal, shown in Fig. 12-11(a). The crystal size and cut determine the values of $L$, $C$, $R$ and $C'$. The resistance $R$ is the friction of the vibrating crystal, capacitance $C$ is the compliance, and inductance $L$ is the equivalent mass. The capacitance $C'$ is the electrostatic capacitance between the mounted pair of electrodes with the crystal as the dielectric.

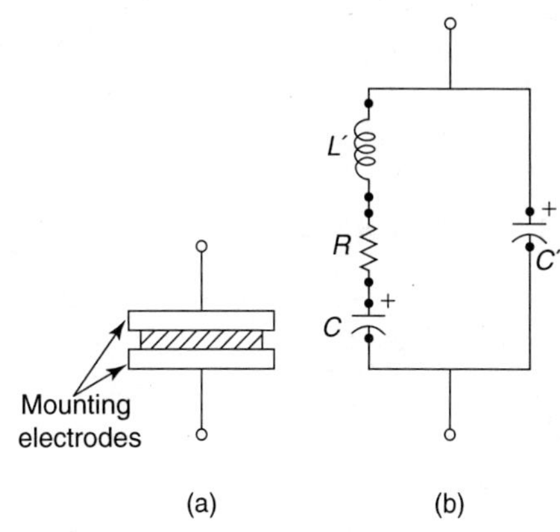


**Figure 12-11** (a) Symbol of a vibrating piezoelectric crystal (b) Its equivalent electrical circuit

The circuit of Fig. 12-11(b) has two resonant frequencies. At the series resonant frequency $f_s$ the reactance of the series $LC$ arm is zero, that is:

$$\omega_s L - \frac{1}{\omega_s C} = 0$$

or,
$$\omega_s = \frac{1}{\sqrt{LC}} \tag{12-55}$$

$\omega_p$ is the parallel resonant frequency of the circuit greater than $\omega_s$, where:

$$\left(\omega_p L - \frac{1}{\omega_p C}\right) = \frac{1}{\omega_p C'}$$

or,
$$\omega_p^2 = \frac{1}{2}\left(\frac{1}{C} + \frac{1}{C'}\right)$$

or,
$$\omega_p = \sqrt{\frac{1}{2}\left(\frac{1}{C} + \frac{1}{C'}\right)} \tag{12-56}$$

Therefore, $\omega_p$ and $\omega_s$ are as shown in Fig. 12-12. At the parallel, resonant frequency, the impedance offered by the crystal to the internal circuit is very high.

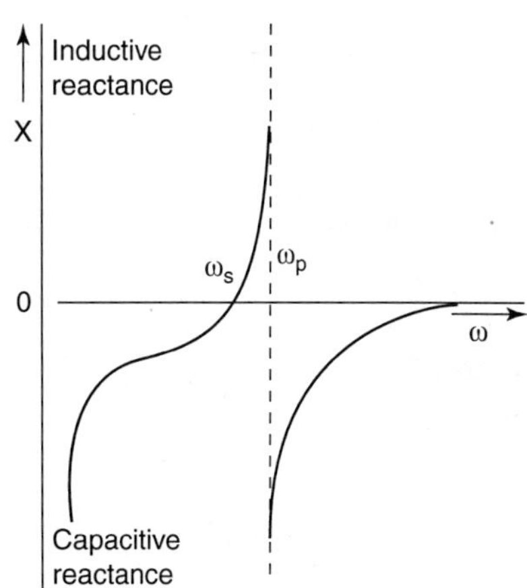


**Figure 12-12** Reactance vs. frequency graph

The resonant frequencies of a crystal vary inversely as the thickness of the cut.

$$f = \frac{1}{t}$$

Crystal oscillators are generally used in the frequency range from about 10 kHz to 10 MHz. For lower frequencies, the size of the quartz crystal is inconveniently large. At higher frequencies, the thickness of the crystal is so small that it becomes very much fragile. The actual circuit diagram of the crystal oscillator is shown in Fig. 12-13.

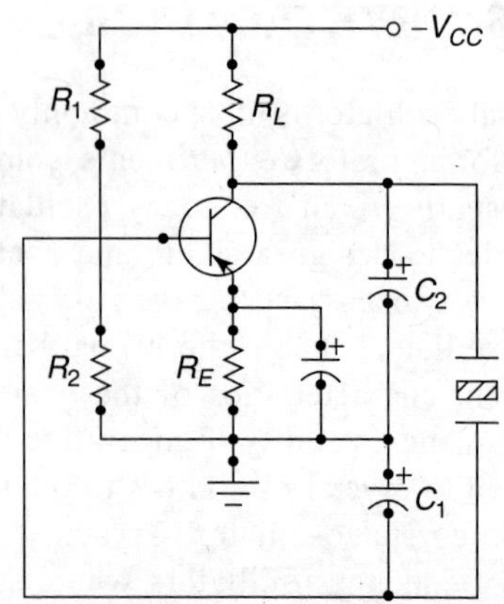


**Figure 12-13** Circuit of a crystal oscillator

## Solved Examples

**Example 12-1** Find the operating frequency of the transistor of a Colpitts oscillator if $C_1 = 0.001\ \mu$F, $C_2 = 0.01\ \mu$F and $L = 15\ \mu$H.

**Solution:**
For Colpitts oscillator:

$$f = \frac{1}{2\pi\sqrt{C_T L}}$$

$$\Theta\ C_T = \frac{C_1 \times C_2}{C_1 + C_2}$$

$$= \frac{0.01 \times 0.001 \times 10^{-6} \times 10^{-6}}{(0.01 + 0.001) \times 10^{-6}}$$

$$C_T = 909.09 \times 10^{-12}\ \text{F}$$

$$\therefore \quad f = \frac{1}{2\pi\sqrt{909.09 \times 10^{-12} \times 15 \times 10^{-6}}} = \frac{10^9}{2\pi\sqrt{909.09}}$$

$$\therefore \quad f = 1.36\ \text{MHz}$$

**Example 12-2** A crystal has a thicknes reduced by 1%. What happens to the oscillations?

**Solution:**

Frequency, $f = \frac{K}{t}$

$$\therefore \quad f \propto \frac{1}{t}$$

If the of the crystal is reduced by 1%, the frequency of oscillations will increase by 1%.

**Example 12-3** The ac equivalent circuit of a crystal has these values: $L = 1$ H, $C = 0.01$ pF, $R = 1000\ \Omega$ and $C_m = 20$ pF. Calculate $f_S$ and $f_P$ of the crystal.

**Solution:**
Given:

$$L = 1\text{H}, C = 0.01 \text{ pF}, R = 1000\ \Omega, C_m = 20 \text{ pF}$$

For series resonance:

$$f_S = \frac{1}{2\pi\sqrt{LC}} = \frac{1}{2\pi \text{ R } 1 \times 0.01 \times 10^{-12}}$$

$\therefore$ $$f_S = 1.59 \text{ mHz}$$

For parallel resonance:

$$f_P = \frac{1}{2\pi\sqrt{LC_T}}$$

$$C_T = \frac{C \times C_m}{C + C_m} = \frac{0.01 \times 20 \times 10^{-24}}{(0.01 + 20) \times 10^{-12}}$$

$$C_T = 0.009995 \times 10^{-12} \text{ F}$$

$$f_p = \frac{1}{2\pi\sqrt{0.009995 \times 10^{-12} \times 1}}$$

$\therefore$ $$f_p = 1.592 \text{ MHz}$$

If this crystal is used in an oscillator, the frequency of oscillations will be between 1.59 and 1.592 MHz.

**Example 12-4** Calculate the value of $\beta$ in order for oscillation to occur if $A = 30$ and the amplification with feedback.

**Solution:**
According to Barkhausen criteria:

$$A\beta = 1$$

$\because$ $$A = 30$$

$\therefore$ $$\beta = \frac{1}{A} = \frac{1}{30}$$

$\therefore$ $$\beta = 0.033$$

Amplification with feedback:

$$A' = \frac{A}{1 - A\beta} = \frac{30}{1 - 30 \times 0.0333}$$

$\therefore$ $$A' = \infty$$

The amplification with feedback becomes infinite and it should be controlled by non-linearity.

**Example 12-5** A Clapp oscillator has the following circuit components $C_1$ =10000 pF, $C_2$ = 1000 pF, $L_3$ = 50 $\mu$H and $C_3$ is 5–150 pF, variable capacity. Find the tuning frequency range and minimum gain for oscillation.

**Solution:**

Frequency of oscillation is by:

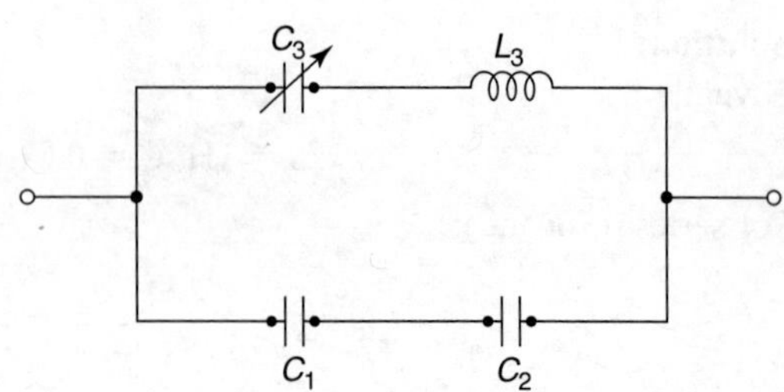


$$f_o = \frac{1}{2\,\pi\sqrt{L_3 C_{eq}}}$$

$$C_{eq} = \frac{C_1 \times C_2 \times C_3}{C_1 C_2 + C_2 C_3 + C_1 C_3}$$

For $C_3 = 5$ pF

$$C_{eq1} = \frac{10000 \times 1000 \times 5}{10000 \times 1000 + 1000 \times 5 + 5 \times 10000}$$

$\therefore$ $$C_{eq1} = 4.97 \text{ pF}$$

$$f_{o1} = \frac{1}{2\pi\sqrt{50 \times 10^{-6} \times 4.97 \times 10^{-12}}}$$

$\therefore$ $$f_{o1} = 10.096 \text{ MHz}$$

For $C_3 = 150$ pF

$$C_{eq2} = \frac{10000 \times 1000 \times 150}{10000 \times 1000 + 150 \times 1000 + 10000 \times 150}$$

$\therefore$ $$C_{eq2} = 128.75 \text{ pF}$$

$$f_{o2} = \frac{1}{2\pi\sqrt{50 \times 10^{-6} \times 128.75 \times 10^{-12}}}$$

$\therefore$ $$f_{o2} = 1.98 \text{ MHz}$$

$\therefore$ The tuning range of capacitor is from 1.98 to 10.096 MHz.
For minimum range for oscillation:

$$A\beta = 1$$

$$A_{\text{loop.}} \geq 1$$

$\therefore$ $$A_{\text{loop}} = A_{vo} \times \frac{C_2}{C_1} = \frac{10000}{1000}$$

$\therefore$ $$A_{vo} = 10$$

**Example 12-6** A Hartley oscillator uses a FEET with $g_m$ of 3 ms and $r_d = 20$ kΩ. The total coil inductance $C_e$ is 20 $\mu$H with a turns ratio of input side to output side of 1:10. It is turned with a 20 pF capacitor. Find the frequency of oscillation and the amplifier gain margin in dB.

**Solution:**

For Hartley oscillator:

$$f_o = \frac{1}{2p\sqrt{L_T . L_3}} = \frac{1}{2\pi\sqrt{20 \times 10^{-6} \times 20 \times 10^{-12}}}$$

∴ Frequency of oscillation: $f_o = 7.95$ MHz

$$\mu = g_m r_d = 3 \times 10^{-3} \times 20 \times 10^3 = 60$$

For oscillation: $$A_{loop} = \mu \times \frac{N_1}{N_2} = 60 \times \frac{1}{10} = 6$$

∴ Excess gain $$20 \log 6 = 15.56 \text{ dB}$$

**Example 12-7** Prove that in a crystal the ratio of frequencies in series and parallel resonance is given by:

$$1 + \frac{1}{2} \times \frac{C}{C'}$$

**Solution:**

The equivalent circuit is as shown in the given diagram.

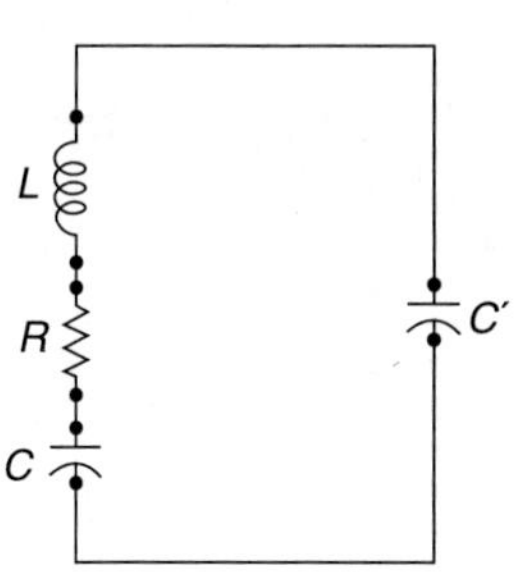


For series resonance:

$$w_1^2 = \frac{1}{LC}$$

For parallel resonance:

$$w_2^2 = \frac{1}{C_{eq} \times L}$$

$$C_{eq} = \frac{C \times C'}{C + C'}$$

$$\therefore \quad \frac{\omega_2^2}{\omega_1^2} = \frac{1}{C_{eq}L} \times \frac{LC}{1}$$

$$\frac{\omega_2^2}{\omega_1^2} = \frac{C + C'}{L \times C \times C'} \times \frac{LC}{1} = 1 + \frac{C}{C'}$$

$$\therefore \quad \frac{\omega_2}{\omega_1} = \sqrt{1 + \frac{C}{C'}} = \left(1 + \frac{C}{C'}\right)^{1/2}$$

On expanding and neglecting higher powers:

$$\therefore \quad \frac{\omega_2}{\omega_1} = 1 + \frac{1}{2} \times \frac{C}{C'} \text{ (Proved)}$$

**Example 12-8** Find the value of $R'$ in the circuit of the figure for generally sinusoidal oscillations. Find the frequency of oscillations.

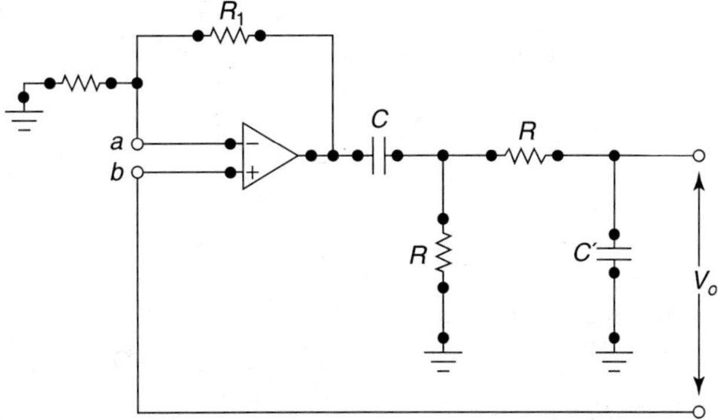

**Solution:**

At node $a$:

$$\frac{V_a}{R} + \frac{V_a - V_c}{R'} = 0$$

$$V_c = V_a\left(1 + \frac{R'}{R}\right) \quad (1)$$

At node $b$:

$$V_a = V_b$$

$$V_c = V_b\left(1 + \frac{R'}{R}\right) \quad (2)$$

$$0 - (V_b)DC + \frac{V_b - V_d}{R} = 0$$

$$V_b(1 + SCR) = V_d \quad (3)$$

At node $d$:

$$(V_d - V_c)SC + \frac{V_d}{R} + \frac{(V_d - V_b)}{R} = 0$$

$$V_d(2 + SCR) = V_c(\mathrm{SCR}) + V_b \quad (4)$$

From Eqs. (2), (3) and (4):

$$V_b(1 - SCR)(2 + SCR) = V_b\left(1 + \frac{R'}{R}\right)(SCR) - V_b$$

$$2 - 3\,SCR + S^2C^2R^2 - \left(1 + \frac{R'}{R}\right)(SCR) - 1 = 0$$

$$1 + S^2C^2R^2 + \left(2 - \frac{R'}{R}\right)(SCR) = 0$$

For oscillation: $$R' = 2R,\ \omega = \frac{1}{CR}$$

**Example 12-9** In a Wien-bridge oscillator, the capacitors are of the variable type. The maximum and minimum values of capacitance are 900 pF and 90 pF respectively. The value of $R = 100\ \mathrm{k\Omega}$.

(a) Determine the range of the operating frequency of the oscillator.

(b) Determine the value of $R_3$ if $R_4 = 10\ \mathrm{k\Omega}$, so that oscillation can be maintained.

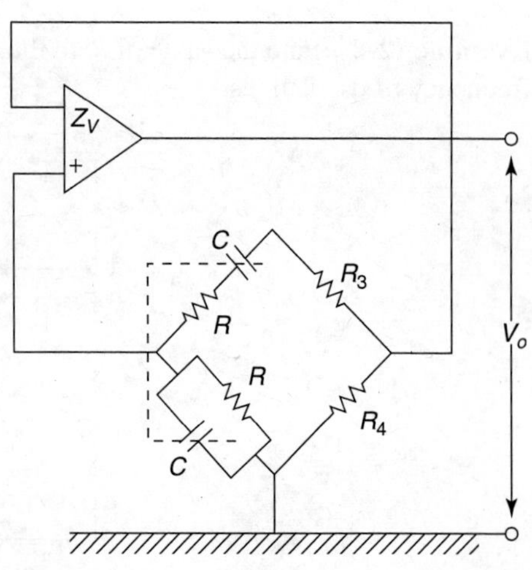


**Solution:**

(a) The frequency of oscillation is given by:

$$f_o = \frac{1}{2\pi RC}$$

The capacitor $C$ is variable, $C_{maximum} = 900$ pF and $C_{minimum} = 90$ pF. When the $C$ is the minimum, we get the maximum frequency, and:

$$f_{o\ maximum} = \frac{1}{2\pi RC_{minimum}}$$

$$= \frac{1}{2\pi\ 100\ k \times 90\ \text{pF}} = 17.68\ \text{KHz}$$

When the $C$ is the maximum, we get the minimum frequency as:

$$f_{o\ minimum} = \frac{1}{2\pi RC_{maximum}}$$

$$= \frac{1}{2\pi 100 k \times 900\ \text{pF}} = 1.768\ \text{KHz}$$

Therefore, the range of oscillation frequency is from 1.768 to 17.68 KHz.

(b) Using,

$$\frac{R_3}{R_3 + R_4} \leq \frac{1}{3}$$

$$R_3 = 2R_4$$

$$R_3 = 2 \times 10\ \text{k}\Omega = 20\ \text{k}\Omega$$

**Example 12-10** Find the minimum voltage gain and the frequency of oscillation for a Colpitts oscillator with $C_1 = 0.004\ \mu$F, $C_2 = 0.003\ \mu$F and $L = 4.0$ mH.

**Solution:**

$$A_v \geq \frac{C_2}{C_1}$$

$$A_v \geq \frac{0.03 \times 10^{-6}}{0.04 \times 10^{-6}} \geq 7.5$$

so,

$$2\pi f_o = \sqrt{\frac{C_1 + C_2}{LC_1C_2}}$$

$$f_o = \frac{1}{2\pi}\sqrt{\frac{0.004 \times 10^{-6} + 0.003 \times 10^{-6}}{(4 \times 10^{-3})(0.004 \times 10^{-6})(0.003 \times 10^{-6})}}$$

$$= \frac{1}{2\pi}\sqrt{7.08 \times 10^{10}} = 42.4 \times 10^3\ \text{Hz} = 42.4\ \text{KHz}$$

**Example 12-11** A transistor LC oscillator circuit is as shown in the following diagram. Assume that the transistor has very high impedance. Derive an equation governing the circuit operation and find the frequency of oscillation. Also state the gain condition required for oscillations to start.

**Solution:**

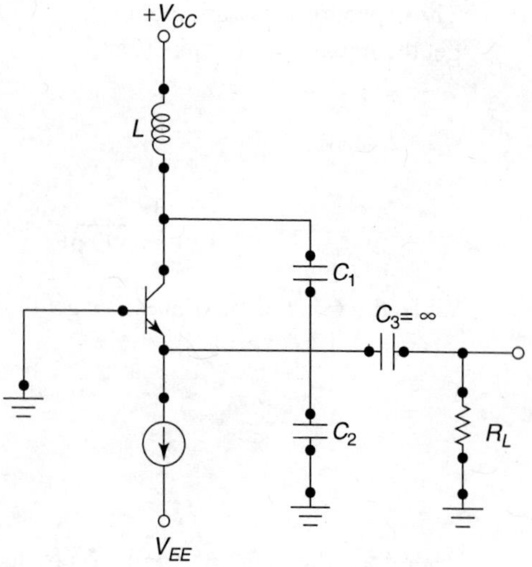

$$Z_1 = \frac{1}{j\omega C_1} = \frac{-j}{\omega C_1}$$

$$Z_2 = \frac{1}{j\omega C_2} = \frac{-j}{\omega C_2}$$

$$Z_3 = j\omega L_1$$

$$\frac{-1 + h_{fe}}{\omega^2 C_1 C_2} + j\omega\left[L_1 - \frac{1}{\omega^2}\left(\frac{1}{C_1} + \frac{1}{C_2}\right)\right]h_{fe} + \frac{L_1}{C_2} = 0$$

$$\omega^2 L_1 = \frac{1}{C_1} = \frac{1}{C_2} = \frac{C_1 + C_2}{C_1 C_2}$$

$$X_C^2 = \frac{C_1 + C_2}{L_1 C_1 C_2}$$

Equating real part to zero:

$$1 + h_{fe} - \frac{1}{\omega_0^2 C_1 C_2 + \dfrac{L_1}{C_2}} = 0$$

$$(1 + h_{fe}) = \omega^2 C_1 L_1 = \frac{C_1 + C}{C_1 C_2} C_1$$

$$h_{fe} = \frac{C_1}{C_2} \text{ (condition of oscillation)}$$

**Example 12-12** A Hartley oscillator, as shown in the following diagram, has the circuit parameters: $L_1 = 500\ \mu$H, $L_2 = 5000\ \mu$H, M = 300 $\mu$H and $C = 150$ pF.

(a) Determine the frequency of oscillation.

(b) If the transistor has the following parameters, determine whether the circuit will oscillate or not. Given: $R_L = 10$ kΩ, $g_m = 8$ mA/V, $r_o = 50$ kΩ.

**Solution:**

(a) Frequency of oscillation:

$$f_o = \frac{1}{2\pi\sqrt{L_T C}}$$

$$L_T = L_1 + L_2 + 2M$$

$$= 500\ \mu\text{H} + 5000\ \mu\text{H} + 2 \times 300\ \mu\text{H} = 6.1\ \text{mH}$$

and, $C = 150$ pF

$$f_o = \frac{1}{2\pi\sqrt{6.1\ \text{mH} \times 150\ \text{pF}}} = 166.4\ \text{KHz}$$

The voltage gain using the approximate formula is:

$$A_v = -g_m R_L'$$

(b) The effective $R'_L = R_L \| r_o = 10\ \text{k}\Omega \| 50\ \text{k}\Omega = 8.33\ \text{k}\Omega$

$$A_v = -8\ \text{mA/V} \times 8.33\ \text{k}\Omega = -66.64$$

$$|A_v| = 66.64$$

From the equation:

$$\frac{N_2}{N_1} = \frac{L_2 + M}{L_1 + M} = \frac{500\ \mu\text{H} + 300\ \mu\text{H}}{500\ \mu\text{H} + 300\ \mu\text{H}} = 6.625$$

Therefore, the condition for maintaining sustained oscillation is given by:

$$A_v \frac{N_2}{N_1} \geq 1$$

$|A_v| \dfrac{N_2}{N_1} = 66.64 \times 6.625 = 441.5$, which is much larger than 1.

Hence, the circuit will oscillate.

**Example 12-13** A piezo-oscillator has the following circuit and the device parameters: $C_{gs} = 5$ pF $C_{ds} = 1$ pF, gm = 10 mA/V, $r_d = 50$ k and REGISTER = 10 M. The crystal parameters are $L = 0.5$ H, $C_2 = 0.05$ pF, $R_{se} = 1$ K and $C_1 = 1$ pF. Find:

(a) The equivalent series-resonant capacitance $C_r$

(b) The frequency of oscillations $f$

(c) The series and the parallel resonant frequencies $f_S$ and $f_P$, and $Q$ of the crystal

(d) The loop gain AB at zero bias

(e) The bias time constant

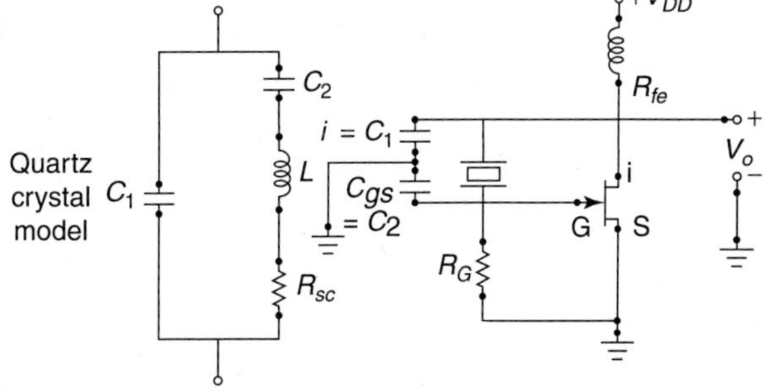


**Solution:**

(a) From the figure, we obtain the total capacitance $C_T$ in series with $L$ as:

$$C_T = C_2\,[C_1 \| (C_{ds} \text{ in series with } C_{gs})]$$

$$= \frac{1}{\dfrac{1}{0.05} + \dfrac{1}{1 + \left(\dfrac{1 \times 5}{1 + 5}\right)}} = 0.049\ \text{pF}$$

(b) The series resonant frequency of the Clapp oscillator like the piezo-oscillator is:

$$f_o = \frac{\sqrt{2}}{2\pi\sqrt{LC_T}} = \frac{\sqrt{2}}{2\pi\sqrt{0.5 \times 0.049 \times 10^{-12}}} = 1.017\ \text{MHz}$$

(c) Parallel resonant frequency:

$$\omega_p = \frac{1}{\left(\dfrac{LC_1C_2}{C_1 + C_2}\right)^{1/2}}$$

$$f_p = \frac{1}{2\pi\left(\frac{0.5 \times 10^{-12} \times 0.05 \times 10^{-12}}{1 \times 10^{-12} + 0.05 \times 10^{-12}}\right)^{1/2}} = 1.3 \text{ MHz}$$

Series resonant frequency:

$$\omega_s = \frac{1}{(LC_2)^{1/2}}$$

$$f_s = \frac{1}{2\pi(0.5 \times 0.05 \times 10^{-12})^{1/2}} = 1.0065 \text{ MHz}$$

Quality factor $Q$ of the crystal:

$$Q = \frac{(L/C_2)^{1/2}}{R_{se}}$$

$$Q = \frac{1}{10^3}\left(\frac{0.5}{0.05 \times 10^{-12}}\right)^{1/2} = 316$$

(d) The loop-gain:

$$A\beta = g_m r_d \frac{C_{ds}}{C_{gs}} = 10 \times 10^{-3} \times 50 \times 10^3 \times \frac{1 \text{ pF}}{5 \text{ pF}} = 100$$

(e) $T_{\text{bias}} = R_G C_G = R_G(C_{gs} \parallel C_{ds}) = 10 \times 10^6 (5 - 1) \times 10^{12} = 60\ \mu\text{s}$

$$\frac{1}{\omega_O} = \frac{1}{2\pi \times 1.017 \times 10^6} = 0.45\ \mu\text{s}$$

Thus, $T_{\text{bias}} >> \frac{1}{\omega_O}$ (as required for proper operation)

**Example 12-14** A Colpitts oscillator, as shown in the following diagram, has a coil with inductance of $C_1 = 100$ pF and $C_2 = 1000$ pF. Find:

(a) Frequency of oscillation

(b) The minimum gain required of the amplifier to maintain oscillation

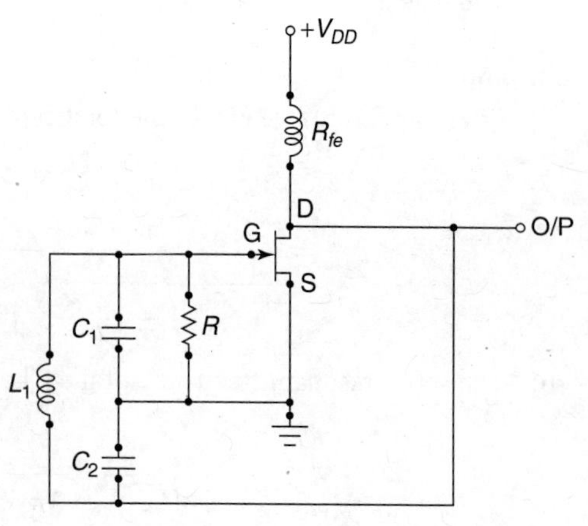


**Solution:**

The frequency of oscillation is:

$$f_o = \frac{1}{2\pi\sqrt{L_1 C_{eq}}}$$

$$L_1 = 100\ \mu\text{H} = 100 \times 10^{-6}$$

$$C_{eq} = \frac{C_1 \times C_2}{C_1 + C_2} = \frac{200 \times 10^{-12} \times 1000 \times 10^{-12}}{200 \times 10^{-12} + 1000 \times 10^{-12}}$$

$$= 166.67 \times 10^{-12} \text{ F}$$

$$f_o = \frac{1}{2\pi\sqrt{100 \times 10^{-6} \times 166.67 \times 10^{-12}}}$$

$$f_o = 123.3 \text{ KHz}$$

The minimum gain required is:

$$|A_v| \geq \frac{C_2}{C_1} = \frac{1000 \text{ pF}}{200 \text{ pF}} = 5$$

**Example 12-15** In the Hartley oscillator, $L_2 = 0.4$ mH and $C = 0.04$ mF. If the frequency of the oscillation is 120 kHz, find the value of $L_1$. Neglect the mutual inductance.

**Solution:**

Frequency of Hartley oscillator is:

$$f = \frac{1}{2\pi\sqrt{(L_1 + L_2)C}}$$

$$L_1 = \frac{1}{4\pi^2 f^2 C} - L_2$$

$$L_1 = \frac{1}{4\pi^2(120 \times 10^3)^2 \times 0.004 \times 10^{-6}} - 0.4 \times 10^{-3}$$

$$L_1 = 0.04 \text{ mH}$$

**Example 12-16** A crystal has $L = 0.33$ H, $C = 0.065$ pF and $C_1 = 1.0$ pF and $R = 5.5$ K.

(a) Calculate the series resonant frequency.
(b) Calculate the series resonant frequency.
(c) By what percent does the parallel resonant frequency exceed the resonant frequency?
(d) Find $Q$ of the crystal.

**Solution:**

(a) $f_s = \frac{1}{2\pi}\left(\frac{1}{LC}\right)^{\frac{1}{2}} = \frac{1}{2\pi}\left(\frac{10^{12}}{0.33 \times 0.065}\right)^{\frac{1}{2}} = 1.09$ MHz

(b) $\frac{f_p}{f_s} = \left(1 + \frac{C}{C_1}\right)^{\frac{1}{2}} = \left(1 + \frac{0.065}{1.0}\right)^{\frac{1}{2}} \approx 1.033$ MHz

(c) $f_p$ exceeds $f_s$ by 3.3%

(d) $Q = \frac{\omega_s L}{R} = \frac{1}{R}\sqrt{\frac{L}{C}} = \frac{1}{5.5 \times 10^3}\left(\frac{0.33}{0.065 \times 10^{-12}}\right)^{\frac{1}{2}} = 410$

■

## 12-7 REAL-LIFE APPLICATIONS

Oscillators are a common element of almost all electronic circuits. They are used in various applications, and their use makes it possible for circuits and subsystems to perform numerons useful functions. In oscillator circuits, oscillation usually builds up from zero when power is first applied under linear circuit operation. The oscillator's amplitude is kept from building up by limiting the amplifier saturation and various non-linear effects. Oscillator design and simulation is a complicated process. It is also extremely important and crucial to design a good and stable oscillator. Oscillators are commonly used in communication circuits. All the communication circuits for different modulation techniques—AM, FM, PM—the use of an oscillator is must. Add-all digital modulators use oscillators.

Oscillators frequently consist of one or two transistors, an inductor (*L*), and a capacitor (*C*) in an LC tank circuit, followed by a buffering amplifier. An oscillator circuit may be implemented with a tuned amplifier having positive feedback from the amplifier's output terminal to its input terminal, in which design takes advantage of the instability possible in circuits having such a feedback loop. Oscillators are used as stable frequency sources in a variety of electronic applications. Oscillator circuits are used in computer peripherals, counters, timers, calculators, phase-locked loops, digital multi-metres, oscilloscopes, and numerous other applications. An oscillator circuit may act as an active device, such as a transistor, to produce power gain—routing a sufficient amount of the active device's output signal to an input of the active device, to sustain oscillations. An oscillator circuit may be used to provide a clock signal, or to produce an accurate waveform. In communication systems, oscillators are employed to provide a stable frequency reference signal for translating information signals to a desired frequency band.

## 12-7-1 Voltage-Controlled Oscillator

A common oscillator implementation is the voltage-controlled oscillator (VCO) circuit, where an input tuning voltage is applied to an oscillator circuit and the tuning voltage adjusted to set the frequency at which the circuit oscillates. The VCO is the most widely used oscillator circuit and it produces an oscillatory output voltage. It provides a periodic signal, where the frequency of the periodic signal is related to the level of an input voltage control signal supplied to the VCO. A VCO is simply an oscillator having a frequency output that is proportional to an applied voltage.

VCOs are the basic building blocks of many electronic systems especially phase-locked loops and may be found in computer disk drives, wireless electronic equipment such as cellular telephones, and other systems in which oscillation frequency is controlled by an applied tuning voltage. The centre frequency of a VCO is the frequency of the periodic output signal formed by the VCO when the input control voltage is set to a nominal level. The VCO has a characteristic gain, which often is expressed as a ratio of the VCO output frequency to the VCO input voltage. VCO typically utilizes a variable control voltage input to produce a frequency output. A VCO is capable of varying an oscillating frequency in response to a change in control voltages. The VCO in an integrated circuit comprises of a monolithic amplifier section with a resonant circuit external to the amplifier or a fully integrated solid state device, such as a ring oscillator, that does not include a resonator with reactive components. VCOs often have a tuning stub that is used to fine-tune the frequency of operation of the VCO. A stub is employed to establish the frequency of operation of the VCO.

In some VCOs, a varactor diode is employed since the space-charge capacitance of the varactor changes as a function of control voltage, thus changing the capacitance of the tank circuit. Ring-type oscillator is one type of VCO.

VCOs are used in many applications to produce an oscillating signal having a frequency defined by an input voltage. A VCO is a critical component in almost every digital communications systems. VCOs are often used to generate local oscillator (*LO*) signals, which are used by transmitter and receiver subsystems for frequency up-conversion and down-conversion, respectively. In cellular telephone applications, VCOs are used to establish a channel frequency within one or two bands according to the GSM digital telephone standard. In order to provide local oscillator signals, as well as transmit carriers, tunable VCOs are implemented in a frequency synthesizer application. The VCO is an important building block in phase-locked loops, clock recovery circuits, and frequency synthesizers. High-frequency and radio-frequency (RF) VCOs can be implemented monolithically as LC oscillators, as relaxation oscillators and ring oscillators. Some applications require the VCO to rapidly change the carrier frequency. These types of oscillators are referred to as agile VCOs. VCOs are utilized within many synthesizer and tuner circuits, such as those found in TVs and in wireless communication devices.

## 12-7-2 Cascode Crystal Oscillator

The cascode crystal oscillator is composed of a Colpitts crystal oscillator and a base-common buffer amplifier in mobile circuits. In the cascode crystal oscillator, a temparature-independent voltage source biases the buffer amplifier and the bypass capaciter gets eliminated. GSM phones, set-top boxes and digital audio broadcasting equipments use oscillators. and digital audio broadcasting equipment use oscillators. VCOs are used in wireless modems, coaxial cables, voltage regulators, wireless speakers, RF and microwave attenuators, cordless phones, integrated electronic circuits, frequency converters, computer motherboards, thin-film capacitors, wireless routers, cell phone batteries, printed circuit boards (PCB), RF power amplifiers, two-way radios, membrane switches, wireless access points, microprocessors, radio-frequency modulators, wireless keyboards, electronic doors, network analyzers, microcontroller, bandpass filters, crystal oscillators, GPS car navigation systems, cable modems, spectrum analyzers, flexible printed circuits (FPC), TV antennae, field-effect transistors, sound cards, power dividers/combiners, variable resistors, GPS antennae, car remote control, network processors, RFID reader, DIP switches, satellite phones, BNC connectors, graphics processors, video cards, MEMs, PIN diodes, SAW filters, digital potentiometers, SMA connectors, PCMCIA PC cards, pulse generators, vacuum circuit breakers, digital-to-analog converters, memory cards, compact flash, memory stick, SD cards, PCB connectors, Real time clocks (RTC), wireless telemetry systems, field programmable gate array (FPGA), low-noise amplifiers (LNA), Magnetic resonance imaging (MRI) systems, embedded computer systems, electrical fuses, phase-locked loop (PLL), remote keyless entry systems, Schottky diode, tire pressure monitoring systems (TPMS), programmable logic controller (PLC), system on chip (SOC), RF directional couplers, wireless headphones, earphones, variable capacitors (varactor), overload relay, phase shifter, application specific integrated circuits (ASIC), EMI shielding solutions, air circuit breakers, electronic circuit design, RF transmitter modules, dielectric resonators, precision resistors, PCB assembly, fabrication, microstrip patch antennae, miniature circuit breakers, digital delay generators, ground fault circuit interrupters (GFCI), etc.

## POINTS TO REMEMBER

1. Oscillator converts dc to ac.
2. Oscillator has no input signal.
3. Oscillator behaviour is opposite to that of a rectifier.
4. The conditions and frequencies of oscillation are classified as:

| *Types of Oscillation* | *Condition of Oscillator* | *Frequency of Oscillation* |
|---|---|---|
| Hartley Oscillator | $h_f = \frac{\omega L_1}{\omega L_2} + \frac{Rh_i}{\omega^2 L_1 L_2}$ | $f = \frac{\omega}{2\pi} = \frac{1}{2\pi} \frac{1}{[(h_{oe} L_1 L_2 / h_{ie}) + C(L_1 + L_2)]^{1/2}}$<br>Simply,<br>$f = \frac{1}{2\pi\sqrt{C(L_1 + L_2)}} = \frac{1}{2\pi\sqrt{LC'}}$ |
| Colpitts Oscillator | $h_f = \frac{C_2}{C_1} + Rh_i \cdot \omega^2 C_1 C_2$ | $f = \frac{\omega}{2\pi} = \frac{1}{2\pi}\left(\frac{h_{oe}}{h_{ie} C_1 C_2} + \frac{1}{LC_1} + \frac{1}{LC_2}\right)^{1/2}$ |
| Phase-Shift Oscillator | The transistor should have an $h_{fe}$ of 56 when $RL = R$. | $f = \frac{1}{2\pi\sqrt{10}CR}$ |
| Wein-Bridge Oscillator | $\frac{R_1}{R_2} = 2$ | $f = \frac{\omega}{2\pi} = \frac{1}{2\pi RC}$ |
| Crystal Oscillator | — | $f_p = \frac{\omega_P}{2\pi} = \frac{1}{2\pi}\sqrt{\frac{C + C'}{LCC'}}$ |

## IMPORTANT FORMULAE

1. General condition for oscillation for an oscillator:

$$h_f = \frac{x_1}{x_2} + \frac{R_{hi}}{x_1 x_2}$$

2. Frequency of oscillation for a Hartley oscillator:

$$f = \frac{1}{2\pi}\sqrt{C(L_1 + L_2)}$$

3. Condition for oscillation for a Colpitts oscillator:

$$h_f = \frac{C_2}{C_1} + Rh_i\,\omega^2 C_1 C_2$$

4. Frequency of oscillation for a phase-shift oscillator:

$$f = \frac{1}{2\pi\sqrt{10}\,CR}$$

5. Frequency of oscillation for a Wien-bridge oscillator:

$$f = \frac{1}{2\pi CR}$$

6. If the feedback signal aids the externally applied input signal, the overall gain is given by:

$$Af = \frac{A}{1 - A\beta}$$

7. Value of $M$ required for sustained oscillations is given by:

$$M = \frac{R_B}{h_{fe}}(CR + h_{oe}L) + CR\frac{h_{ie}}{h_{fe}} + L\frac{\Delta_{he}}{h_{fe}}$$

8. Oscillation frequency of a Clapp oscillator is given by:

$$f_o = \frac{1}{2\pi}\sqrt{\frac{1}{L}\left(\frac{1}{C_o} + \frac{1}{C_1} + \frac{1}{C_2}\right)}$$

9. Condition for sustained oscillation for a phase-shift oscillator is given by:

$$h_{fe} = 23 + 29\frac{R}{R_L} + 4\frac{R_L}{R}$$

## OBJECTIVE QUESTIONS

1. A distorted sinusoid has the amplitudes $A_1$, $A_2$, $A_3$—of the fundamental, second harmonic and third harmonic—respectively. The total harmonic distortion is:
   - (a) $(A_2 + A_3 + \cdots)/A_1$
   - (b) $\sqrt{A_2^2 + A_3^2 + \cdots /A_1}$
   - (c) $\sqrt{(A_2^2 + A_3^2 + \cdots)/\left(\sqrt{A_1^2 + A_2^2 + A_2^3 + \cdots}\right)}$
   - (d) $1 - \left(\sqrt{A_2^2 + A_3^2 + \cdots /A_1}\right)$

2. An amplifier is assumed to have a single-pole high-frequency transfer function. The rise time of its output response to a step function input is 35 nsec. The upper 3 dB frequency (in MHz) for the amplifier to a sinusoidal input is approximately at:
   - (a) 4.55 (b) 10
   - (c) 20 (d) 28.6

3. Frequency of oscillation of a Wein-bridge oscillator is given by:
   - (a) $1/6\pi RC$ (b) $1/2\pi RC$
   - (c) $2\pi RC$ (d) $1/\pi RC$

4. Which of the following is not true?
   - (a) An oscillator is a circuit that converts dc to ac.
   - (b) An oscillator is an amplifier that supplies it own input signal.
   - (c) All oscillators generate sine wave.
   - (d) In-phase feedback is called positive feedback.

5. What do phase-shift oscillators, twin-T oscillator and Wein-bridge oscillators have in common?
   - (a) They use $RC$ frequency control
   - (b) They have a sinusoidal output
   - (c) They use amplifier gain to overcome feedback loss
   - (d) All of the above

6. The stability of frequency of oscillation is high if:
   - (a) $\frac{d\phi}{dt} = 0$ (b) $\frac{d\phi}{dt} = \infty$
   - (c) $\frac{d\phi}{d} = 0$ (d) $\frac{d\phi}{dw} = \infty$

7. Match List I with List II and select the correct answer using the codes as provided.

| List I (Oscillator) | List II (Characteristics/Features) |
|---|---|
| (a) Wein-bridge | 1. RF oscillator; two inductances and one capacitance in the reactance network |
| (b) Colpitts | 2. LC oscillator for RF; three capacitances and one inductance in the reactance network. |
| (c) Hartley | 3. RC oscillator for audio-frequency applications. |
| (d) Clapp | 4. RF oscillator; two capacitances and one inductance in the reactance network. |

**Code**

(a) a b c d
2 1 4 3

(b) a b c d
2 4 1 3

(c) a b c d
3 4 1 2

(d) a b c d
3 1 4 2

8. The primary advantage of a crystal oscillator is that:
   (a) It can oscillate at any frequency
   (b) It gives a high output voltage
   (c) Its frequency of oscillation remains almost constant
   (d) It operates on a very low dc supply voltage

9. A Hartley oscillator circuit uses:
   (a) A tapped inductor
   (b) A tapped capacitor
   (c) Both the above
   (d) A tapped inductor for inductive feedback

10. A Colpitts oscillator uses:
   (a) A tapped inductor
   (b) A tapped capacitor
   (c) Both (a) and (b)
   (d) None of the above

11. An oscillator circuit using a quartz crystal has:
   (a) High-frequency stability
   (b) Low-frequency stability
   (c) Medium-frequency stability
   (d) None of these

12. An oscillator is basically an amplifier with:
   (a) Zero gain (b) Very large gain
   (c) Infinite gain (d) Very low gain

13. In a feedback oscillator, constant amplitude oscillation are obtained when loop loop gain $-A\beta$ equals:
   (a) 0 (b) –1
   (c) 1 (d) ∞

14. Crystal oscillator uses:
   (a) Silicon crystal
   (b) Germanium crystal
   (c) Crystal diode
   (d) Piezo-electric quartz crystal

15. Effective $Q$ of the equivalent electrical circuit of a quartz crystal is of theorder of:
   (a) 200 (b) 2000
   (c) 20, 000 (d) $10^5$

16. Barkhausen criterion for sustained oscillation gives:
   (a) $-A\beta = 1$ (b) $A\beta = 0$
   (c) $A = \beta$ (d) $A = 1/\beta$

17. Quartz crystal oscillators are popularly used because of:
    (a) High $Q$ and high-frequency stability
    (b) Low $Q$ and high-frequency stability
    (c) Low $Q$ and large output power
    (d) High $Q$ and large output power
18. The crystal oscillator frequency is highly stable due to:
    (a) Crystal structure
    (b) High $Q$ of the crystal
    (c) Vibration of the crystal
    (d) Rigidity of the crystal
19. The most likely application of a crystal oscillator is:
    (a) As an RF test oscillator
    (b) As an electronic organ for home use
    (c) As a hi-fi test audio-frequency sweep generator
    (d) In commercial radio transmitter
20. In RC phase phase-shift oscillator using FET and 3-section RC phase phase-shift network, the condition for sustained oscillation is:
    (a) $\beta > 6$
    (b) $\beta > 29$
    (c) $\beta > 4n + 23 + 29/n$, $n = R_d/R$
    (d) $\beta > 23 + 29/n$
21. FET RC phase-shift oscillator has angular frequency of oscillation $\omega$ equal to:
    (a) 1/3 $RC$ (b) 1/6 $RC$
    (c) $1/\sqrt{6}\,RC$ (d) $1/\sqrt{3}RC$
    where, $R$ and $C$ refer to each section of three section $RC$ phase phase-shift network.
22. In an FET Hartley oscillator, the frequency of oscillation $\omega$:
    (a) $\approx \omega_0$ (b) $<< \omega_0$
    (c) $>> \omega_0$ (d) $\omega_0 (L_1/L_2)$
    where, $\omega_0$ is the frequency of resonance.
23. In tuned collector oscillator, frequency $f\,[= \omega/(2\pi)]$ of oscillation is:
    (a) $= \omega_0$ (b) $< \omega_0$
    (c) $>\omega_0$ (d) $\omega/h_{fe}$
    where, $\omega_0$ is frequency of resonance.
24. The feedback factor $\beta$ at frequency of oscillation of Wein-bridge oscillator is:
    (a) 1/3 (b) 3
    (c) 1/29 (d) –1/29
25. The minimum number of $RC$ sections required in phase phase-shift oscillator is:
    (a) Two (b) Three
    (c) Four (d) None
26. The current amplification factor in radian square of Colpitts oscillator is:
    (a) $C_1/C_2$ (b) $C_1C_2$
    (c) $C_1 + C_2$ (d) $C_1 - C_2$
27. Electronic oscillator is better than a mechanical one because:
    (a) It has better frequency stability
    (b) It has higher efficiency
    (c) It can produce 20–200 Hz
    (d) None
28. Wein-bridge oscillator is most often used whenever:
    (a) Wide range of high purity sine waves is to be generated
    (b) High feedback ratio is needed
    (c) Square output waves are required
    (d) Extremely high resonant frequencies are required
29. If Barkhausen criterion is not fulfilled by an oscillator circuit, it will:
    (a) Stop oscillating
    (b) Produce damped waves continuously
    (c) Become an amplifier
    (d) Produce high-frequency whistles
30. To generate a 1 MHz signal, the most suitable circuit is:
    (a) Phase-shift oscillator
    (b) Wein-Bridge oscillator
    (c) Colpitts oscillator
    (d) None of the above

31. In oscillator circuit, the energy feedback to its input terminal from the output is:
   (a) 180° out of phase with input signal
   (b) In phase with the input signal
   (c) 90° out- of- phase with the input signal
   (d) None

32. For sustaining oscillations in an oscillator:
   (a) Feedback factor should be unity
   (b) Phase shift should be 0°
   (c) Feedback should be negative
   (d) Both (a) and (b)

## REVIEW QUESTIONS

1. What is an oscillator?
2. Classify the different types of oscillators.
3. Define Barkhausen criteria?
4. What are the basic differences between a rectifier and an oscillator?
5. What are the applications of oscillators?
6. What is the source of input of an oscillator circuit?
7. Explain the working principle of a Hartley oscillator and find its condition and frequency of oscillation.
8. Explain the working principle of a Colpitts oscillator and find its condition and frequency of oscillation.
9. Explain the working principle of phase-shift oscillator and find its condition and frequency of oscillation.
10. Explain the working principle of tuned oscillator, and find its condition and frequency of oscillation.
11. Explain the working principle of crystal oscillator, and find its condition and, frequency of oscillation.

## PRACTICE PROBLEMS

1. Find the operating frequency of a transistor in Colpitts oscillator if $C_1 = 0.002\ \mu$F, $C_2 = 0.011\ \mu$F and $L = 10.5\ \mu$H.
2. Find the operating frequency of a transistor Colpitts oscillator if $C_1 = 0.022$ F, $C_2 = 0.051$ F and $L = 20.5$ H.
3. A crystal's thickness is reduced by 3%. What happens to its oscillations?
4. The ac equivalent circuit of a crystal has these values: $L = 1.5$ H, $C = 0.01$ pF, $R = 2000\ \Omega$ and $C_m = 200$ pF. Calculate $f_S$ and $f_P$ of the crystal.
5. The ac equivalent circuit of a crystal has these values: $L = 1$ H, $C = 0.01$ pF, $R = 1000\ \Omega$ and $C_m = 20$ pF. Calculate $f_S$ and $f_P$ of the crystal.
6. Calculate the value of $\beta$ in order for oscillation to occur if $A = 10$ and the amplification is with feedback.
7. A Clapp oscillator has the following circuit components $C_1 = 10000$ pF, $C_2 = 1000$ pF, $L_3 = 50\ \mu$H and $C_3$ is 5–150 pF variable capacity. Find the tuning frequency range and minimum gain for oscillation.
8. A Hartley oscillator uses a FET with $g_m$ of 3 ms and $r_d = 21.00$ kΩ. The total coil inductance $C_e$ is 200 $\mu$H with a turns ratio of 1:10 (input side to output side). It is turned with a 20 pF capacitor. Find the frequency of oscillation and the amplifier gain margin in dB.
9. Prove that in a crystal the ratio of frequencies in series and parallel resonance is given by:

$$1 + \frac{1}{2} \times \frac{C}{C'}$$

10. In a Wien-bridge oscillator, the capacitors are of a variable type. The maximum and minimum values of capacitance are 1000 pF and 90 pF respectively. The value of $R = 100$ kΩ.
   (a) Determine the range of the operating frequency of the oscillator.
   (b) Determine the value of $R_3$ if $R_4 = 10$ kΩ, so that oscillation can be maintained.

11. Find the minimum voltage gain and the frequency of oscillation for a Colpitts oscillator with, $C_1$ = 0.004 $\mu$F, $C_2$ = 0.03 $\mu F$ and $L$ = 4.0 mH.

12. A Hartley oscillator has the following circuit parameters:

    $L_1$ = 500 $\mu$H, $L_2$ = 5000 $\mu$M, $M$ = 300 $\mu$H and $C$ = 150 pF

    (a) Determine the frequency of oscillation.

    (b) If the transistor has the following parameters: $R_L$ = 10 k$\Omega$, $g_m$ = 8 mA/V and $r_o$ = 50 k$\Omega$, determine whether the circuit will oscillate or not

13. A Colpitts oscillator has a coil with inductance of $C_1$ = 100 pF and $C_2$ = 1000 pF. Find (a) frequency of oscillation; (b) the minimum gain required of the amplifier to maintain oscillation.

14. A crystal has $L$ = 0.33 H, $C$ = 0.065 pF and $C_1$ = 1.0 pF and $R$ = 5.5 K

    (a) Find the series resonant frequency.

    (b) By what percent does the parallel resonant frequency exceed the resonant frequency?

    (c) Find $Q$ of the crystal.

15. In the Hartley oscillator, $L_2$ = 0.4 mH and $C$ = 0.04 mF. If the frequency of the oscillation is 120 KHz. Find the value of $L_1$. Neglect the mutual inductance.

## SUGGESTED READINGS


1. Millman, Jacob and Christos C. Halkias. 1986. *Integrated Electronics: Analog and Digital Circuits and Systems*. New Delhi: McGraw Hill Book Company.

2. Pierret, R. F and G.W. Neudeck. 1989. *Modular Series on Solid State Devices*. Boston and M.A.: Addison Wesley.

3. Singh, B.P. and Rekha Singh. 2006. *Electronic Devices and Integrated Circuits*. New Delhi: Pearson Education.

4. Chattopadhyay, D. and P.C. Rakshit. 2008. *Electronics Fundamentals and Applications*. New Delhi: New Age International Publishers.

# 13

# Digital Electronic Principles

## Outline



## Objectives

This chapter covers the fundamentals of digital electronics with a comprehensive overview of number systems, binary codes, logic gates and the applications of digital circuits. Beginning with an introduction to the history of digital systems, the chapter proceeds to discuss the number system and the conversion from binary to decimal. This is followed by a detailed analysis of Boolean algebra and its laws, supplemented with relevant examples. De Morgan's theorem and logic gate circuits are also examined with emphasis on practical applications. The chapter ends with a brief analysis of the real-life applications of digital circuits.

## 13-1 INTRODUCTION

The journey of digital electronics started in the year 1946 with the digital computer using vacuum tube technology. The term "digit" is derived from the counting operation of the computer. The twenty-first century is the digital world. The devices common people use daily are mostly "digital" in their functions. Modern computer systems consist of digital components, starting from microprocessors to all other peripheral components. A binary digital circuit operates in two modes/states: ON state [1 (one)] and OFF state [0 (zero)]. These states are called *binary states*. These binary digits are called *bits*. In the binary digital world the logic lies between these two states; there is no intermediate state. This makes the digital states discrete in nature, whereas the analog states are continuous.

The evolution of digital electronics is a process of simultaneous development and simplification. The English scientist, George Boole, is known to be the inventor of Boolean algebra. Boolean algebra is the foundation of all contemporary computer arithmetic. In hindsight, Boole is regarded as one of the founders of computer science. Boolean algebra, developed by Boole in 1830, is a logical calculus of truth values.

The mathematician, Lewis Carroll, had formulated truth tables as early as 1894. Augustus De Morgan, the Indian-born British scientist, formulated the famous De Morgan's theorem, which was a simplification of the Boolean expressions. In 1952, the Karnaugh map was invented by Edward W. Veitch. The Karnaugh map is used to reduce the need for calculations. It was further developed in 1953 by Maurice Karnaugh, a telecommunications engineer at Bell Labs, to help simplify digital electronic circuits.

Digital logic can be of two different types—positive logic system and negative logic system. Generally, the digital circuits are connected with a dc supply battery (typically 5 V) and a ground terminal of 0 V. If the 5 V refers to logic 1 or ON state and the ground 0 V refers to logic 0 or OFF state, the logic is referred to as *positive logic*. On the contrary, if 5 V referrers to logic 0 and ground 0 V referrers to logic 1, then this logic is called *negative logic*.

## 13-2 NUMBER SYSTEM

The number system digits occupy certain relative positions having their relative positional significance. The left-most digit is called the most significant digit (MSD) and the right-most digit is called the lowest significant digit (LSD). Number systems are classified on the basis of radix/base, as shown in Table 13-1.

## 13-3 CONVERSION OF NUMBER SYSTEM

Conversion from one number system to another is an important aspect in digital electronics especially with respect to conversion from binary to decimal, decimal to binary, decimal to octal, decimal to hexadecimal, etc. Representation of a number in a system with base (radix) $N$ may only consist of digits that are less than $N$.

The base or radix of these number systems depends on the number of digit present in each number system. The binary system contains 0 and 1, so its base is 2. The decimal base is 10 and the hexadecimal base is 16. The binary number system, having only two states, is the simplest way of calculating.

More accurately, if:

$$M = a_k N^k + a_{k-1} N^{k-1} + \cdots + a_1 N^1 + a_0$$

With $0 \leq a_i < N$ we have a representation of $M$ in base $N$ system, thus:

$$M = (a_k a_{k-1} \ldots a_0)_N$$

**Table 13-1** Classification of number systems

| | |
|---|---|
| Binary | 0, 1 |
| Ternary | 0, 1, 2 |
| Quaternary | 0, 1, 2, 3 |
| Octal | 0, 1, 2, 3, 4, 5, 6, 7 |
| Decimal | 0, 1, 2, 3, 4, 5, 6, 7, 8, 9 |
| Hexadecimal | 0, 1, 2, 3, 4, 5, 6, 7, 8, 9, A, B, C, D, E, F |

### 13-3-1 Binary to Decimal

Conversion from binary to decimal using decimal arithmetic is accomplished by simply summing the powers of 2 corresponding to the powers of 1 in the binary number.

#### Conversion

$$11\ 11 = 1 \times 10^3 + 1 \times 10^2 + 1 \times 10^1 + 1 \times 10^0$$

$$= 1 \times 2^3 + 1 \times 2^2 + 1 \times 2^1 + 1 \times 2^0$$

$$= 8 + 4 + 2 + 1 = 15$$

Similarly,

$$11\ 10 = 1 \times 10^3 + 1 \times 10^2 + 1 \times 10^1 + 0 \times 10^0$$

$$= 1 \times 2^3 + 1 \times 2^2 + 1 \times 2^1 + 0 \times 2^0$$

$$= 8 + 4 + 2 + 0 = 13$$

For example, the binary number:

$$10111 = b_4\ b_3\ b_2\ b_1\ b_0$$

This is converted to decimal by adding:

$$b_4 \times 2^4 + b_3 \times 2^3 + b_2 \times 2^2 + b_1 \times 2^1 + b_0 \times 2^0$$

$$= b_4 \times 16 + b_3 \times 8 + b_2 \times 4 + b_1 \times 2 + b_0 \times 1$$

$$= 1 \times 16 + 0 \times 8 + 1 \times 4 + 1 \times 2 + 1 \times 1$$

$$= 16 + 4 + 2 + 1$$

$$= (23) \text{ decimal}$$

$$= (23)_{10}$$

### 13-3-2 Decimal to Binary

Conversion from decimal to binary using decimal arithmetic is accomplished by repeated division of the decimal number by two. After each division the remainder is the next bit of the binary number starting from the least significant.

For example, the decimal number (26):

$$26/2 = 13 \quad R = 0 = b_0$$

$$13/2 = 6 \quad R = 1 = b_1$$

$$6/2 = 3 \quad R = 0 = b_2$$

$$3/2 = 1 \quad R = 1 = b_3$$

$$1/2 = 0 \quad R = 1 = b_4$$

$$0/2 = 0$$

So (26) decimal = (11010) binary.

$$(26)_{10} = (11010)_2$$

The decimal number (15):

$$15/2 = 7\ R = 1 = b_0$$
$$7/2 = 3\ R = 1 = b_1$$
$$3/2 = 1\ R = 1 = b_2$$
$$1/2^{1/2} = 1\ R = 1 = b_3$$
$$0/2 = 0$$

so, $$(15)_{10} = (1111)_2$$

## 13-3-3 Numer System Conversions

Conversions between binary and other number systems requires simple arithmetic computation. For example, the decimal number (19):

$$19 = 1 \times 10 + 9 \times 1$$

This is written as a binary number with:

$$10011 = 1 \times 2^4 + 0 \times 2^3 + 0 \times 2^2 + 1 \times 2^1 + 1 \times 2^0$$
$$= 1 \times 16 + 0 \times 8 + 0 \times 4 + 1 \times 2 + 1 \times 1$$

If this number is interpreted in groups of three bits, then each group of three can be interpreted as an octal digit, i.e., multiplying a power of 8.

$$(10011)_{binary} = 1 \times 16 + 0 \times 8 + 0 \times 4 + 1 \times 2 + 1 \times 1$$
$$= (1 \times 16 + 0 \times 8) + (0 \times 4 + 1 \times 2 + 1 \times 1)$$
$$= 2 \times 8 + 3 \times 1$$
$$= (23)_{octal}$$

Similarly, when groups of 4 bits are used, each group can be interpreted as a hexadecimal digit, i.e., multiplying a power of 16.

$$(10011)_{binary} = 1 \times 16 + 0 \times 8 + 0 \times 4 + 1 \times 2 + 1 \times 1$$
$$= (1 \times 16) + (0 \times 8 + 0 \times 4 + 1 \times 2 + 1 \times 1)$$
$$= 1 \times 16 + 3 \times 1$$
$$= (13)_{hexadecimal}$$

The octal numeral system is the base-8 number system, and uses the digits 0 to 7. Numerals can be made from binary numerals by grouping consecutive digits into groups of three. For example, the binary representation for decimal 74 is 1001010, which groups into 001 001 010 so the octal representation is 112. In decimal systems each decimal place is a base of 10.

$\therefore \quad 74 = 7 \times 10^1 + 4 \times 10^0$

In octal numerals each place is a power with base 8.

$\therefore \quad 112 = 1 \times 8^2 + 1 \times 8^1 + 2 \times 8^0$

By performing the calculation decimal system we see how 112 in octal is equivalent to $64 + 8 + 2 = 74$ in decimal.

The conversion from decimal to hexadecimal to octal to binary is shown in Table 13-2.

**Table 13-2** Conversion of decimal to hexadecimal to octal to binary

| *DEC* | *HEX* | *OCT* | *BIN* |
|---|---|---|---|
| 0 | 0 | 000 | 00000000 |
| 1 | 1 | 001 | 00000001 |
| 2 | 2 | 002 | 00000010 |
| 3 | 3 | 003 | 00000011 |
| 4 | 4 | 004 | 00000100 |
| 5 | 5 | 005 | 00000101 |
| 6 | 6 | 006 | 00000110 |
| 7 | 7 | 007 | 00000111 |
| 8 | 8 | 010 | 00001000 |
| 9 | 9 | 011 | 00001001 |
| 10 | A | 012 | 00001010 |
| 11 | B | 013 | 00001011 |
| 12 | C | 014 | 00001100 |
| 13 | D | 015 | 00001101 |
| 14 | E | 016 | 00001110 |
| 15 | F | 017 | 00001111 |

## 13-4 BOOLEAN ALGEBRA

The working principle of the digital circuit is guided by Boolean algebra. Boolean algebra functions through addition, subtraction and its five basic laws.

### 13-4-1 Addition

Boolean addition uses the addition process with binary numbers—0's and 1's. Various combinations of binary addition are shown in Table 13-3. The general form of addition of two binary numbers, say $A$ and $B$ produce a sum ($S$) and a carry ($C_y$) as shown here.

$$\begin{array}{r} A \\ +\ B \\ \hline C_y\ \ S \end{array}$$

(Carry) (Sum)

**Table 13-3** Binary addition

| 0 | 0 | 1 | 1 |
|---|---|---|---|
| +0 | +1 | +0 | +1 |
| 0 0 | 0 1 | 0 1 | 1 0 |

Examples of binary addition:

(i) $A(10001) + B(11101) = C(101110)$

| | | | | | | | |
|---|---|---|---|---|---|---|---|
| | 1 | | | | 1 | | → Carry |
| | | 1 | 0 | 0 | 0 | 1 | $A$ |
| + | | 1 | 1 | 1 | 0 | 1 | $B$ |
| | 1 | 0 | 1 | 1 | 1 | 0 | $C$ |

(ii) $A(101101) + B(11001) = C(1000110)$

| | | | | | | | | |
|---|---|---|---|---|---|---|---|---|
| | 1 | 1 | 1 | | | 1 | | → Carry |
| | | 1 | 0 | 1 | 1 | 0 | 1 | $A$ |
| + | | | 1 | 1 | 0 | 0 | 1 | $B$ |
| | 1 | 0 | 0 | 0 | 1 | 1 | 0 | $C$ |

(iii) $A(1011001) + B(111010) = C(10010011)$

| | | | | | | | | | |
|---|---|---|---|---|---|---|---|---|---|
| | 1 | 1 | 1 | 1 | | | | | → Carry |
| | | 1 | 0 | 1 | 1 | 0 | 0 | 1 | $A$ |
| + | | | 1 | 1 | 1 | 0 | 1 | 0 | $B$ |
| | 1 | 0 | 0 | 1 | 0 | 0 | 1 | 1 | $C$ |

(iv) $A(1110) + B(1111) = C(11101)$

| | | | | | | |
|---|---|---|---|---|---|---|
| | 1 | 1 | 1 | | | → Carry |
| | | 1 | 1 | 1 | 0 | $A$ |
| + | | 1 | 1 | 1 | 1 | $B$ |
| | 1 | 1 | 1 | 0 | 1 | $C$ |

(v) $A(10111) + B(110101) = C(1001100)$

| | | | | | | | | |
|---|---|---|---|---|---|---|---|---|
| | 1 | 1 | | 1 | 1 | 1 | | → Carry |
| | | | 1 | 0 | 1 | 1 | 1 | $A$ |
| + | | 1 | 1 | 0 | 1 | 0 | 1 | $B$ |
| | 1 | 0 | 0 | 1 | 1 | 0 | 0 | $C$ |

(vi) $A(11011) + B(1001010) = C(1100101)$

| | | | | | | | | |
|---|---|---|---|---|---|---|---|---|
| | | 1 | 1 | | 1 | | | → Carry |
| | | | 1 | 1 | 0 | 1 | 1 | $A$ |
| + | 1 | 0 | 0 | 1 | 0 | 1 | 0 | $B$ |
| | 1 | 1 | 0 | 0 | 1 | 0 | 1 | $C$ |

**Table 13-4** Binary subtraction

| 0 | 1 | 0 | 1 |
|---|---|---|---|
| −0 | −0 | −1 | −1 |
| 0 0 | 0 1 | 1 0 | 0 0 |

## 13-4-2 Subtraction

Boolean subtraction binary numbers—0's and 1's, is shown in Table 13-4. The general form of subtraction of two binary numbers $A$ and $B$ produce a difference ($D$) and a borrow, as shown here.

$$\begin{array}{r} A \\ -\ B \\ \hline B_r\ \ D \end{array}$$

(Borrow) (Difference)

Examples of binary subtraction:

(i) $A(1011011) - B(10010) = C(1001001)$

| | | | | | | | | |
|---|---|---|---|---|---|---|---|---|
| | 1 | 0 | 1 | 1 | 0 | 1 | 1 | *A* |
| − | | | 1 | 0 | 0 | 1 | 0 | *B* |
| | 1 | 0 | 0 | 1 | 0 | 0 | 1 | *C* |

(ii) $A(1010110) - B(101010) = C(101100)$

| | | | | | | | | |
|---|---|---|---|---|---|---|---|---|
| | 0 | | 0 | | | | → | Borrow |
| | ×1 | 10 | ×1 | 10 | 1 | 1 | 0 | *A* |
| − | | 0 | 0 | 1 | 0 | 1 | 0 | *B* |
| | | 1 | 0 | 1 | 1 | 0 | 0 | *C* |

(iii) $A(1000101) - B(101100) = C(11001)$

| | | | | | | | | |
|---|---|---|---|---|---|---|---|---|
| | 0 | 1 | 1 | | | | → | Borrow |
| | ×1 | ×10 | ×10 | 10 | 1 | 0 | 1 | *A* |
| − | | 1 | 0 | 1 | 1 | 0 | 0 | *B* |
| | | | 1 | 1 | 0 | 0 | 1 | *C* |

(iv) $A(100010110) - B(1111010) = C(10011100)$

| | | | | | | | | | | |
|---|---|---|---|---|---|---|---|---|---|---|
| | 0 | 1 | 1 | 1 | 10 | | | | → | Borrow |
| | ×1 | ×10 | ×10 | ×10 | ×1 | 10 | 1 | 1 | 0 | *A* |
| − | | | 1 | 1 | 1 | 1 | 0 | 1 | 0 | *B* |
| | | 1 | 0 | 0 | 1 | 1 | 1 | 0 | 0 | *C* |

(v) $A(101101) - B(100111) = C(110)$

| | | | | | | | |
|---|---|---|---|---|---|---|---|
| | | | 0 | 10 | | → | Borrow |
| | 1 | 0 | ×1 | ×1 | (1)0 | 1 | *A* |
| − | 1 | 0 | 0 | 1 | 1 | 1 | *B* |
| | | | | 1 | 1 | 0 | *C* |

(vi) $A(1110110) - B(1010111) = C(11111)$

| | | | | | | | | |
|---|---|---|---|---|---|---|---|---|
| | | 0 | 10 | 1 | 10 | 10 | | → Borrow |
| | 1 | ×1. | ×1 | ×10 | ×1 | ×1 | 10 | *A* |
| − | 1 | 0 | 1 | 0 | 1 | 1 | 1 | *B* |
| | | | 1 | 1 | 1 | 1 | 1 | *C* |

### 13-4-3 Basic Boolean Laws

Every law has two expressions—(a) and (b). This is known as duality. These are obtained by changing every AND (.) to OR (+); every OR (+) to AND (.); and all 1's to 0's and vice-versa. It has become conventional to drop (.)—the AND symbol, i.e., *A.B* is written as *AB*. Some of the very important laws are as follows:

- Commutative Law

  (a) $A + B = B + A$

  (b) $A\,B = B\,A$

- Associate Law

  (a) $(A + B) + C = A + (B + C)$

  (b) $(A\,B)\,C = A\,(B\,C)$

- Distributive Law

  (a) $A\,(B + C) = A\,B + A\,C$

  (b) $A + (B\,C) = (A + B)\,(A + C)$

- Identity Law

  (a) $A + A = A$

  (b) $A\,A = A$

- Redundance Law

  (a) $A + A\,B = A$

  (b) $A\,(A + B) = A$

- Inverse Law

  (a) $A + \overline{A} = 1$

  (b) $A\,.\,\overline{A} = 0$

## 13-5 LOGIC GATES

The basic elements of digital circuits are logic gates. Logic circuits are generally designed with BJT, FET and CMOS circuits having many inputs and a single output. The output will be either logic high (1) or logic low (0) depending on the combination of input logic high and low. Logic gates have the ability of making logic decisions by producing a particular output under certain input. The functional behaviour of a logic gate is realized by the three processes, as explained in the following sections.

1. **Truth Table:** Truth table is a prescribed specification table that explains the input−output relation for all possible combination of inputs.

**Table 13-5** Truth table of AND logic

| *Input* | | *Output* |
|---|---|---|
| *A* | *B* | *Y* |
| 0 | 0 | 0 |
| 0 | 1 | 0 |
| 1 | 0 | 0 |
| 1 | 1 | 1 |

The output *Y* of the AND gate is high or 1 when both the inputs *A* and *B* are high. If any of the two inputs is low, the output becomes low.

Logic equation for AND gate is $Y = AB$

2. **Logic Equation:** In logic equations, the output is expressed in terms of input according to the truth table. Logic equations are generally unique as truth tables vary from logic to logic.
3. **Timing Diagram:** Timing diagram of a logic gate indicates the variation of the output waveform with respect to the input waveform. It is a pictorial representation of the time-varying input and output of the logic gate.

## 13-5-1 AND Gate

The AND gate has two or more inputs. Its output is logic high (1) only when both the inputs are at logic high (1). The truth table of an AND gate is shown in Table 13-5 and logic symbol is shown in Fig. 13-1

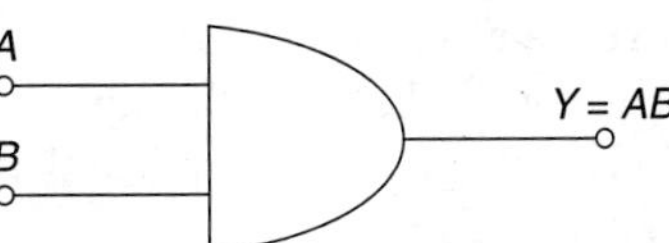


**Figure 13-1** Logic symbol for AND gate

## 13-5-2 OR Gate

The OR gate is a two or more input logic gate. Its output is logic high (1) if any of the inputs are high (1). The truth table of an OR gate is shown in Table 13-6 and logic symbol is shown in Fig. 13-2.

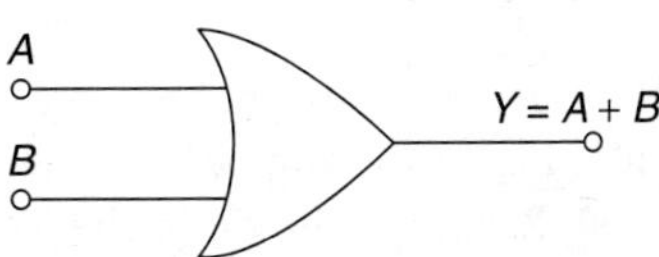


**Figure 13-2** Logic symbol of OR Gate

**Table 13-6** Truth table of OR logic

| *Input* | | *Output* |
|---|---|---|
| *A* | *B* | *Y* |
| 0 | 0 | 0 |
| 0 | 1 | 1 |
| 1 | 0 | 1 |
| 1 | 1 | 1 |

The output of the OR gate *Y* is high when any of the inputs *A* and *B* is high or when both the inputs are high.

Logic equation for OR gate is $Y = A + B$

**Table 13-7** Truth table of NOT logic

| *Input* | *Output* |
|---|---|
| 0 | 1 |
| 1 | 0 |

The output $Y$ of the NOT gate is the complement of the input $A$

Logic equation for NOT gate is $Y = \overline{A}$

## 13-5-3 NOT Gate

The NOT gate is a single input and single output gate, which performs a basic logical inversion/complementation operation. The truth table of a NOT gate is shown in Table 13-7 and the logic symbol is shown in Fig. 13-3.

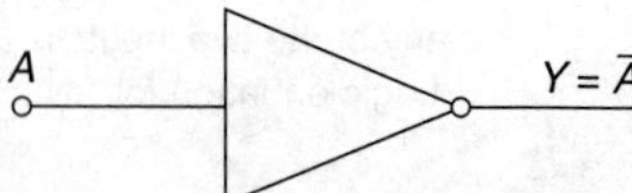


**Figure 13-3** Logic symbol of NOT gate

## 13-5-4 NAND Gate

The NAND gate is a two or more input logic gate. Its output is logic low (0) only when both the inputs are at logic (1). The truth table of a NAND gate is shown in Table 13-8 and logic symbol is shown in Fig. 13-4.

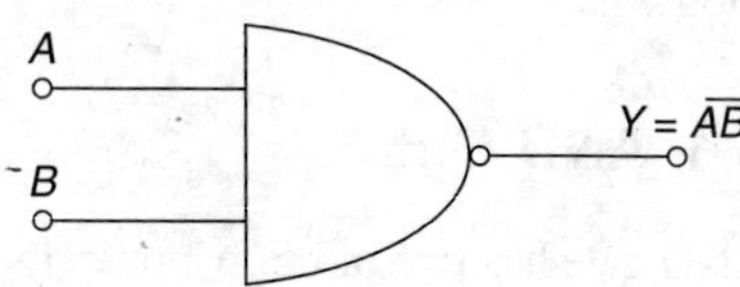


**Figure 13-4** Logic symbol of NAND gate

## 13-5-5 NOR Gate

The NOR gate is a two or more input logic gate. Its output is logic high (1) both the inputs are at logic low (0). The truth table of a NOR gate is shown in the Table 13-9 and the logic symbol is shown in Fig. 13-5.

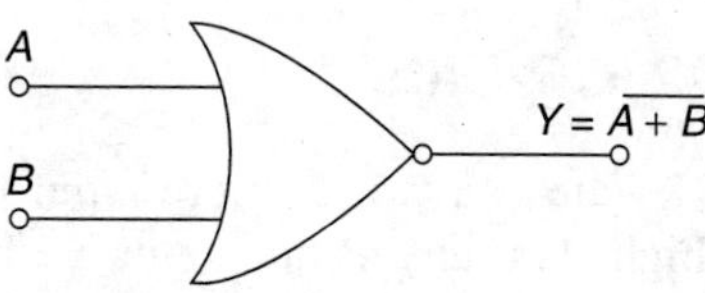


**Figure 13-5** Logic symbol of NOR gate

**Table 13-8** Truth table of NAND logic

| *Input* | | *Output* |
|---|---|---|
| *A* | *B* | *Y* |
| 0 | 0 | 1 |
| 0 | 1 | 1 |
| 1 | 0 | 1 |
| 1 | 1 | 0 |

NAND, which is a contraction of NOT and AND (AND gate followed by an inverter) gates, has a low output, $Y$, when the two inputs $A$ and $B$ are both high. For all other inputs the output is high.

Logic equation for NAND gate is $Y = \overline{AB}$

**Table 13-9** Truth table of NOR logic

| *Input* | | *Output* |
|---|---|---|
| *A* | *B* | *Y* |
| 0 | 0 | 1 |
| 0 | 1 | 0 |
| 1 | 0 | 0 |
| 1 | 1 | 0 |

NOR is the contraction between NOT and OR gates (OR gate followed by an inverter). The output Y of the NOR gate is high only when the two inputs *A* and *B* are low.

Logic equation for NOR gate is $Y = \overline{A + B}$

## 13-5-6 XOR Gate

XOR gate is a two or more input logic gate. Its output is logic high (1) if only one of the input is at logic high. The truth table of a XOR gate is shown in Table 13-10 and the logic symbol is shown in Fig. 13-6.

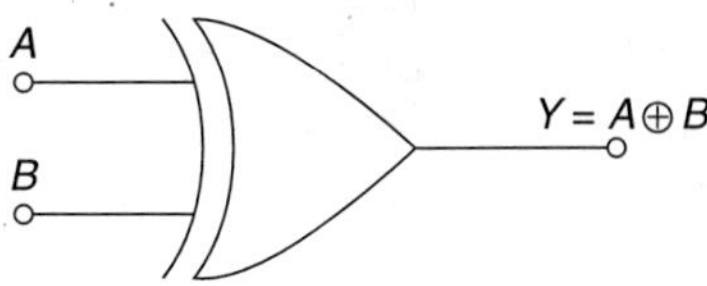


**Figure 13-6** Logic symbol of XOR Gate

## 13-5-7 XNOR Gate

XNOR gate is a two or more input logic gate. Its output is logic high (1) if both the inputs are either logic high or logic low. The truth table of XNOR gate is shown in Table 13-11 and the logic symbol is shown in Fig. 13-7.

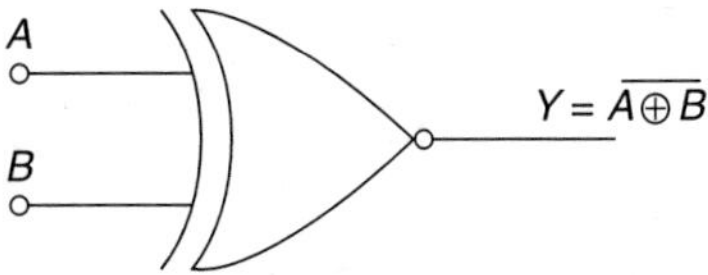


**Figure 13-7** Logic symbol of XNOR gate

## 13-5-8 Universal Gate

NAND and NOR gates are called universal, as all the basic logic AND, OR and NOT gates can be designed/realized through a combination of NAND/NOR. The design of AND, OR and NOT gates using NAND and NOR gates is shown in Figs. 13-8 and 13-9 respectively.

**Table 13-10** Truth table of XOR logic

| *Input* | | *Output* |
|---|---|---|
| *A* | *B* | *Y* |
| 0 | 0 | 0 |
| 0 | 1 | 1 |
| 1 | 0 | 1 |
| 1 | 1 | 0 |

The exclusive-OR gate or the XOR gate has a high output state only when one and only one of the two inputs *A* and *B* is high.

Logic equation for XOR gate is $Y = \overline{A}B + A\overline{B}$

**Table 13-11** Truth table of XNOR logic

| *Input* | | *Output* |
|---|---|---|
| *A* | *B* | *Y* |
| 0 | 0 | 1 |
| 0 | 1 | 0 |
| 1 | 0 | 0 |
| 1 | 1 | 1 |

The Exclusive-NOR gate or XNOR is an XOR gate followed by an inverter. The output *Y* is high if both the inputs *A* and *B* have the same logic state.

Logic equation for XNOR gate is $Y = AB + \overline{BA}$

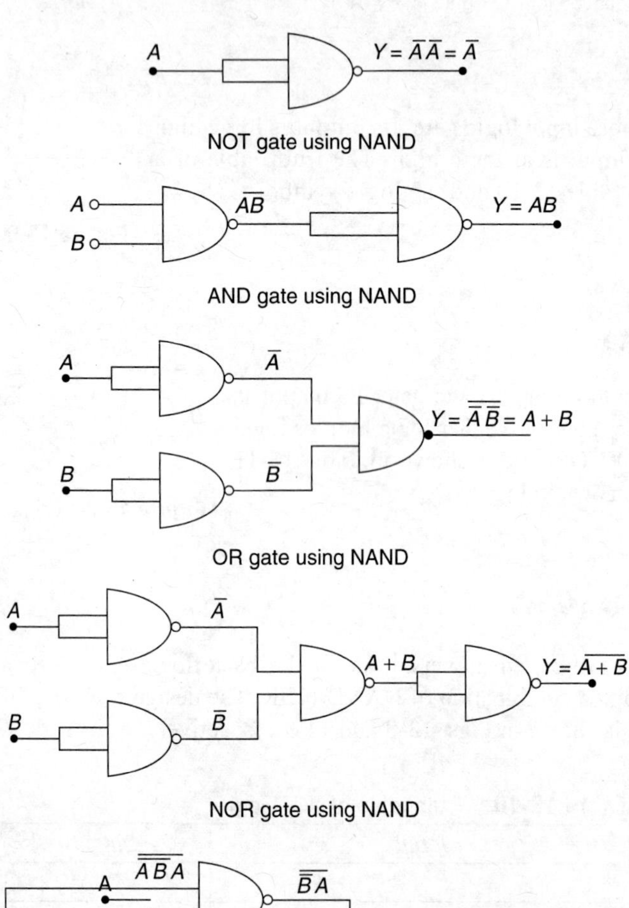


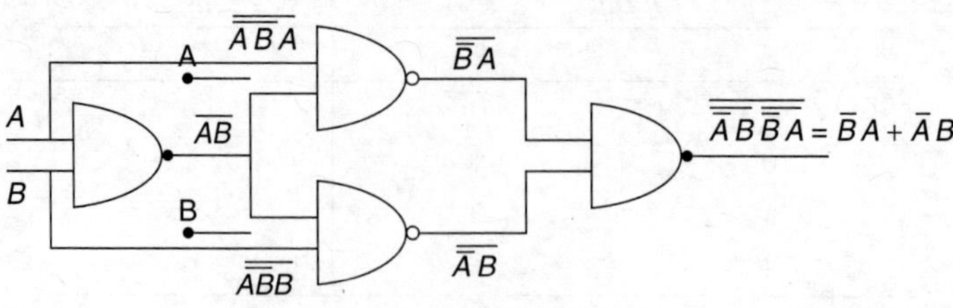


**Figure 13-8** NAND as universal gate

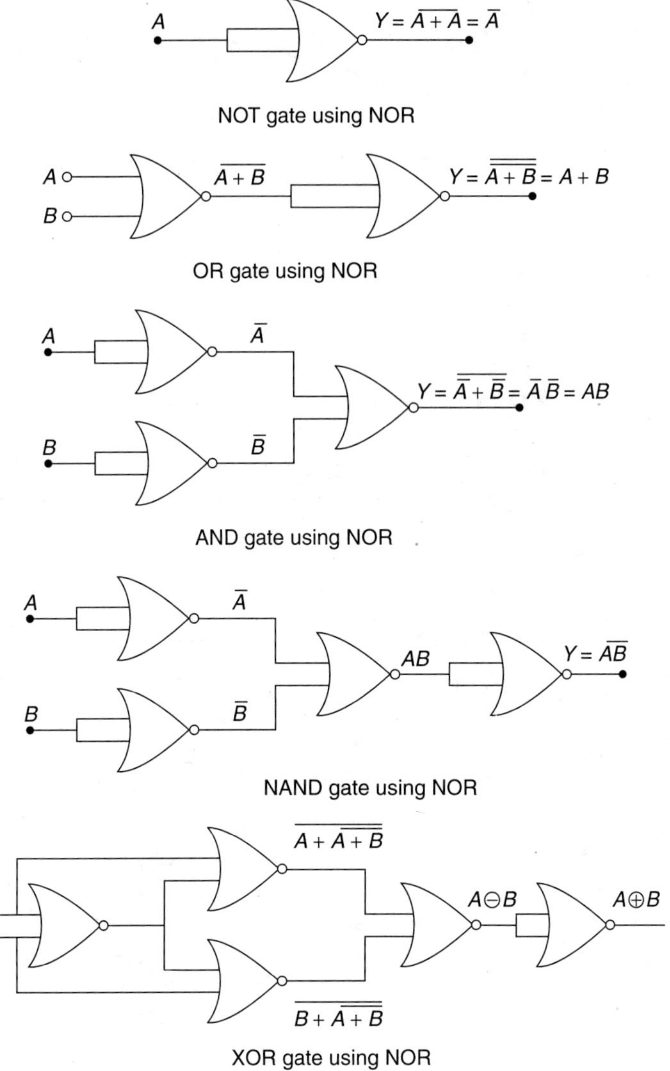


**Figure 13-9** NOR as universal gate

## 13-5-9 Characteristics of Logic Gates

The characteristic features of logic gates are as follows:

### Fan-out

The measure of the maximum number of logic gates that can be driven by a single logic gate without affecting the specified operational characteristics of the driving gate is called fan-out. A standard fan-out for TTL is ten (10).

### Fan-in

The measure of the maximum possible number of inputs that can be connected to a logic gate without affecting the specified operational characteristics of the driven logic gate is called *fan-in*. A standard fan-in for TTL is ten (10).

### Propagation delay

The time taken by an input signal to pass through a logic gate and emerge from the output is known as *propagation delay*. A typical delay time for a TTL logic gate is 30 nsec.

### Power dissipation

The power consumed by a logic gate is known as *power dissipation*.

### Noise

The unwanted input signal in a logic gate is known as *noise*.

### Noise margin

The maximum level of noise voltage allowed at the input without affecting the output is the *noise margin*.

## 13-6 DE MORGAN'S THEOREM

De Morgan developed a pair of important rules concerning group complementation in Boolean algebra. Through group complementation, the complement of the logical sum of a number of binary variables is equal to the logical product of the complements of all the individual variables. For all elements $A$ and $B$ of the set $S$:

(i) $\overline{AB} = \overline{A} + \overline{B}$

(ii) $\overline{A + B} = \overline{A}\,\overline{B}$

### 13-6-1 Using Basic Logic Gates

Figure 13-10 illustrates the proof of De Morgan's theorem.

$$\overline{AB} = \overline{A} + \overline{B}$$

A B AB $\overline{AB}$ ≡ A $\overline{A}$ B $\overline{B}$ $\overline{A}+\overline{B}$

In other way:

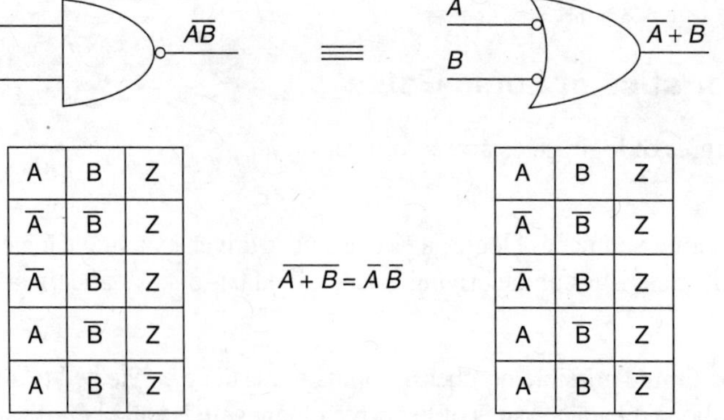


| A | B | Z |
|---|---|---|
| $\overline{A}$ | $\overline{B}$ | Z |
| $\overline{A}$ | B | Z |
| A | $\overline{B}$ | Z |
| A | B | $\overline{Z}$ |

| A | B | Z |
|---|---|---|
| $\overline{A}$ | $\overline{B}$ | Z |
| $\overline{A}$ | B | Z |
| A | $\overline{B}$ | Z |
| A | B | $\overline{Z}$ |

**Figure 13-10** Proof of De Morgan's theorem

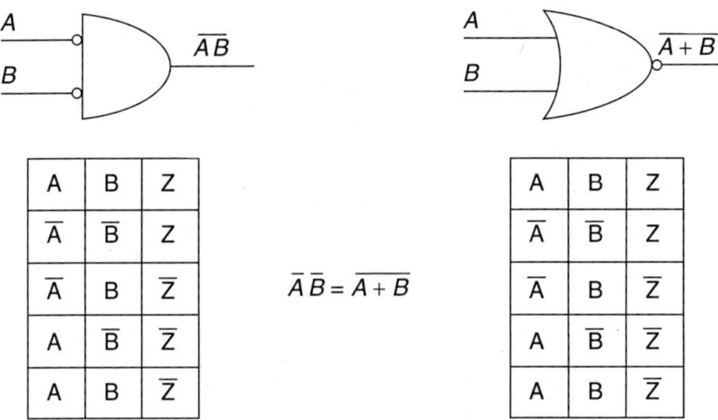


**Figure 13-11** Proof of De Morgan's theorem

The complement of the logical product of a number of binary variables is equal to the logical sum of the complements, as shown in Fig. 13-11.

$$\overline{AB} = \overline{A} + \overline{B}$$

### 13-6-2 Application of De Morgan's Theorem

This theorem is used for the simplification of Boolean expressions in a simplest and smallest form of equivalent circuit. De Morgan's theorem describes the equality between gates with inverted inputs and gates with inverted outputs. For example, a NAND gate is equivalent to a negative OR gate, and a NOR gate is equivalent to a negative AND gate. When "breaking" a complementation bar in a Boolean expression, the operation directly underneath the break reverses, and the broken bar pieces remain over the respective terms. Complementation bars function as grouping symbols. So, when a bar is broken, the terms underneath it must remain grouped.

## 13-7 SIMPLIFICATION OF BOOLEAN EXPRESSION

Simplification of Boolean expressions can be done by the algebraic method or by the Karnough Map method. Karnough Map is one of the simplest methods of solving the four variable algebraic equations. In this chapter we will discuss simplification using the algebraic method only.

(i)

$$\begin{aligned} & A + AC \\ &= A(1 + C) \\ &= A1 \\ &= A \end{aligned}$$

(ii)

$$\begin{aligned} & A + \overline{A}B \\ &= (A + \overline{A})(A + B) \\ &= 1(A + B) \\ &= (A + B) \end{aligned}$$

(iii)

$$\begin{aligned}
& A + \overline{A}B + AB\overline{C} \\
&= A(1 + B\overline{C}) + \overline{A}B \\
&= A1 + \overline{A}B \\
&= A + \overline{A}B \\
&= (A + \overline{A})(A + B) \\
&= 1(A + B) \\
&= (A + B)
\end{aligned}$$

(iv)

$$\begin{aligned}
& A + \overline{A}B + ABC + \overline{A}C \\
&= (A + \overline{A})(A + B) + C(AB + \overline{A}) \\
&= 1(A + B) + C(A + \overline{A}) \\
&= (A + B) + C \\
&= A + B + C
\end{aligned}$$

(v)

$$\begin{aligned}
& \overline{A}\overline{B}C + AC + ABC + AB \\
&= \overline{A}\overline{B}C + AC + AB(C + 1) \\
&= \overline{A}\overline{B}C + AC + AB \\
&= C(\overline{A}\overline{B} + A) + AB \\
&= C(\overline{A} + A)(\overline{B} + A) + AB \\
&= C1(\overline{B} + A) + AB \\
&= C(\overline{B} + A) + AB \\
&= C\overline{B} + CA + AB
\end{aligned}$$

(vi)

$$\begin{aligned}
& AC + \overline{A}\overline{C} \\
&= \overline{A}C + \overline{A} + \overline{C} \\
&= \overline{A}(C + 1) + \overline{C} \\
&= \overline{A}1 + \overline{C} \\
&= \overline{A} + \overline{C}
\end{aligned}$$

(vii)

$$\begin{aligned}
& AB + BC + B\overline{C} + AC \\
&= AB + AC + B(C + \overline{C}) \\
&= AB + AC + B \\
&= B(A + 1) + AC \\
&= B1 + AC \\
&= B + AC
\end{aligned}$$

(viii)

$$\begin{aligned}
& \overline{A\overline{B} + \overline{A}B} \\
&= \overline{A\overline{B}}\;\overline{\overline{A}B} \\
&= (\overline{\overline{A}} + \overline{B})(\overline{A} + \overline{\overline{B}}) \\
&= (A + \overline{B})(\overline{A} + B) \\
&= A\overline{A} + \overline{A}\overline{B} + AB + B\overline{B} \\
&= \overline{A}\overline{B} + AB
\end{aligned}$$

(ix) $\overline{AB + \overline{AB} + A}$

$= \overline{\overline{AB}}\ \overline{\overline{A}}\,\overline{AB}$

$= ABA\ (\overline{A} + \overline{B})$

$= AB\,(\overline{A} + \overline{B})$

$= A\overline{A}B + AB\overline{B}$

$= 0 + 0 = 0$

(x) $AB + \overline{A} + \overline{AB}$

$= AB + \overline{A} + \overline{A} + \overline{B}$

$= AB + \overline{A} + \overline{B}$

$= AB + \overline{AB}$

$= 1$

(xi) $\overline{(\overline{A} + C)(B + \overline{D})}$

$= \overline{(\overline{A} + C)}\,\overline{(B + \overline{D})}$

$= \overline{\overline{A}}\,\overline{C} + \overline{B}\,\overline{\overline{D}}$

$= A\overline{C} + \overline{B}D$

(xii) $(B + \overline{C})\,(\overline{B} + C) + \overline{\overline{A} + B + \overline{C}}$

$= B\overline{B} + \overline{B}\,\overline{C} + C\,\overline{C} + BC + \overline{\overline{A}}\,\overline{B}\,\overline{\overline{C}}$

$= \overline{B}\,\overline{C} + BC + A\overline{B}C$

$= \overline{B}\,\overline{C} + (B + A\overline{B})C$

$= \overline{B}\,\overline{C} + (B + \overline{B})\,(B + A)\,C$

$= \overline{B}\,\overline{C} + (B + A)C$

$= \overline{B}\,\overline{C} + BC + AC$

(xiii) $\overline{\overline{A + B} + \overline{C}}$

$= A\overline{(\overline{A + B})}\ \overline{\overline{C}}$

$= (A + B)\,C = AC + BC$

## 13-8 LOGIC GATE CIRCUITS

Logic gate circuits are divided into two categories based on whether they are with feedback sequential logic circuit or without feedback combinational logic circuit. Thus, digital electronics is classified into combinational logic and sequential logic. Combinational logic output depends on the inputs levels, whereas sequential logic output depends on stored levels of past data and also the present input levels.

### 13-8-1 Combinational Logic

Combinational logic circuit is used to realize different logic functions using different basic logic gates.

#### Adder

Adder is a combinational logic circuit is used to add two or more bits.

## Half adder

Half adder is a combinational logic circuit, which is used to add two bits and generate output as sum ($S$) and carry ($C_y$). The truth table of half adder is shown in Table 13-12 and the circuit diagram is shown in Fig. 13-12.

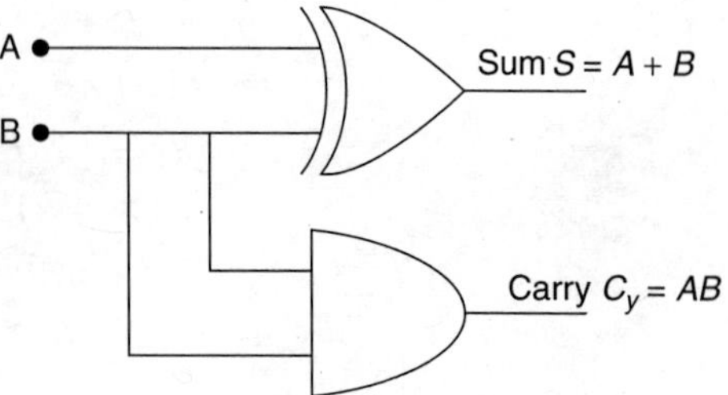


**Figure 13-12** Block diagram of half adder

**Table 13-12** Truth table of half adder

| Input | | Output | |
|---|---|---|---|
| *A* | *B* | $C_y$ | *S* |
| 0 | 0 | 0 | 0 |
| 0 | 1 | 0 | 1 |
| 1 | 0 | 0 | 1 |
| 1 | 1 | 1 | 0 |

## Full adder

Full adder is a combinational logic circuit used to add three or more bits. The reason for the name full adder is that it can add the carry bit as third bit ($C_{in}$) along with other two inputs $A$ and $B$. The truth table of full adder is shown in the Table 13-13 and circuit diagram is shown in Fig. 13-13.

From the truth table as given in Table 13-12, we can write sum:

$$S = A \oplus B$$

And carry: $$C_y = AB$$

A half adder has two inputs and the two bits to be added. The output of the XOR gate is the sum of the two bits and the output of the AND gate is the carry. The full adder produces a sum and carrier values, which are both binary digits. The expression of the sum and carry, is given by:

$$S = (A \oplus B) \oplus C_{in}$$

The full adder takes three inputs. The two inputs A and B are the two bits to be added while the third input is the carry from the previous lower significant position. The output of the XOR gate is the *sum* while the output of the OR gate is the *carry*.

A full adder can be constructed from two half adders by connecting $A$ and $B$ to the input of one half adder, connecting the sum from that to an input to the second adder, connecting $C_{in}$ to the other input and OR the two carry outputs.

**Table 13-13** Truth table of full adder

| Input | | | Output | |
|---|---|---|---|---|
| *A* | *B* | $C_{in}$ | $C_y$ | *S* |
| 0 | 0 | 0 | 0 | 0 |
| 0 | 0 | 1 | 0 | 1 |
| 0 | 1 | 0 | 0 | 1 |
| 0 | 1 | 1 | 1 | 0 |
| 1 | 0 | 0 | 0 | 1 |
| 1 | 0 | 1 | 1 | 0 |
| 1 | 1 | 0 | 1 | 0 |
| 1 | 1 | 1 | 1 | 1 |

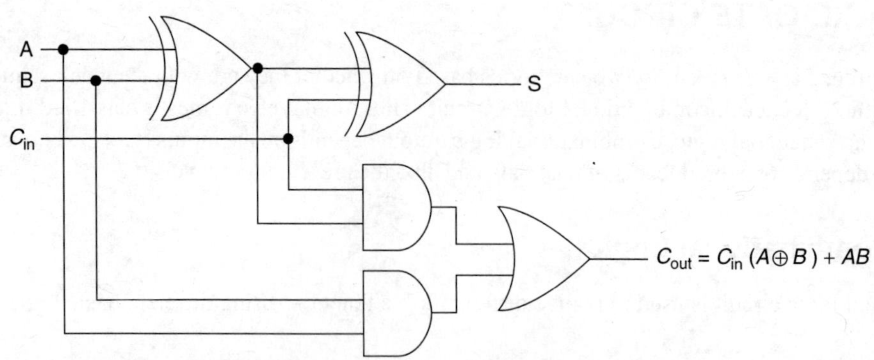


**Figure 13-13** Circuit diagram of full adder

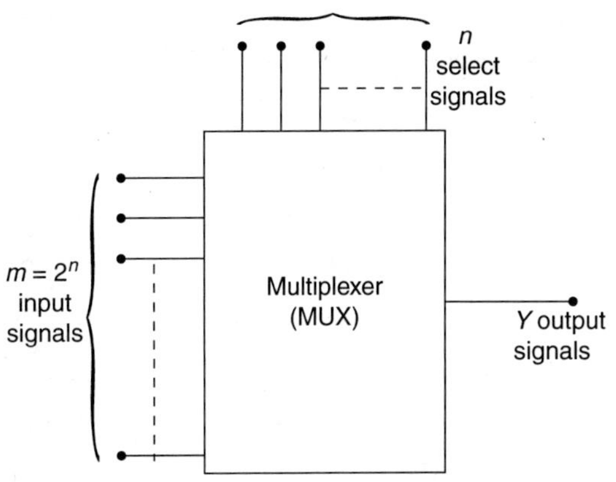


**Figure 13-14** Block diagram of a multiplexer

**Figure 13-15** Circuit diagram of a multiplexer

Equivalently, sum $S$ is expressed by a 3-bit XOR of $A$, $B$, and $C_{in}$; and $C_{out}$ could be made the 3-bit majority function of $A$, $B$, and $C_{in}$.

$$\begin{aligned} C_y &= \overline{A}BC_{in} + A\overline{B}C_{in} + AB\overline{C}_{in} + ABC_{in} \\ &= C_{in}(\overline{A}B + A\overline{B}) + AB(\overline{C}_{in} + C_{in}) \\ &= C_{in}(A \oplus B) + AB \end{aligned}$$

## Multiplexer

Multiplexer means "path selector"; it has many inputs, a single output and select input signals. A multiplexer has $n$ number of select inputs, $2^n$, inputs, and only one output. The truth table of a multiplexer is shown in the Table 13-14, and the block diagram is given in Fig. 13-14, The circuit diagram of a multiplexer is shown in Fig. 13-15.

If we introduce enable input $E_n$ then the equation becomes:

$$Y = (\overline{S}_1\overline{S}_0I_0 + \overline{S}_1S_0I_0 + S_1\overline{S}_0I_2 + S_1S_0I_3)\,E_n$$

A 4-to-1 multiplexer can be implemented by using two inverters, four AND gates and a 4-input OR gate. To each of the AND gates any of the data input lines and the corresponding select lines are applied, and the output of all the AND gates are applied, to the OR gate to get the output.

**Table 13-14** Truth table of multiplexer

| *Select Input* | | *Output* |
|---|---|---|
| $S_1$ | $S_0$ | Y |
| 0 | 0 | $I_0$ |
| 0 | 1 | $I_1$ |
| 1 | 0 | $I_2$ |
| 1 | 1 | $I_3$ |

## De-multiplexer

De-multiplexer operation is the opposite to that of the multiplexer. It has $n$ number of select inputs, $2^n$ outputs and only one select input. The truth table of de-multiplexer is shown in Table 13-15 and the block diagram is shown in Fig. 13-16. The circuit diagram is shown in Fig. 13-17.

**Table 13-15** Truth table of de-multiplexer

| *Select Input* | | *Output* | | | |
|---|---|---|---|---|---|
| $S_1$ | $S_1$ | $Y_0$ | $Y_1$ | $Y_2$ | $Y_3$ |
| 0 | 0 | *D* | 0 | 0 | 0 |
| 0 | 0 | 0 | *D* | 0 | 0 |
| 1 | 1 | 0 | 0 | *D* | 0 |
| 1 | 1 | 0 | 0 | 0 | *D* |

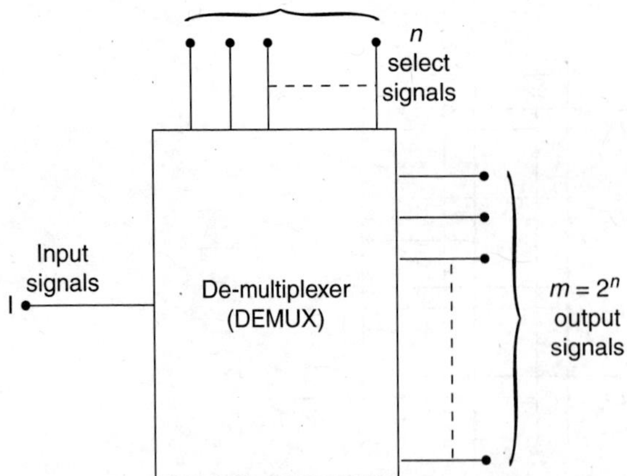


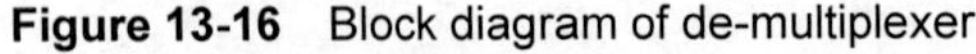

Figure 13-16 Block diagram of de-multiplexer

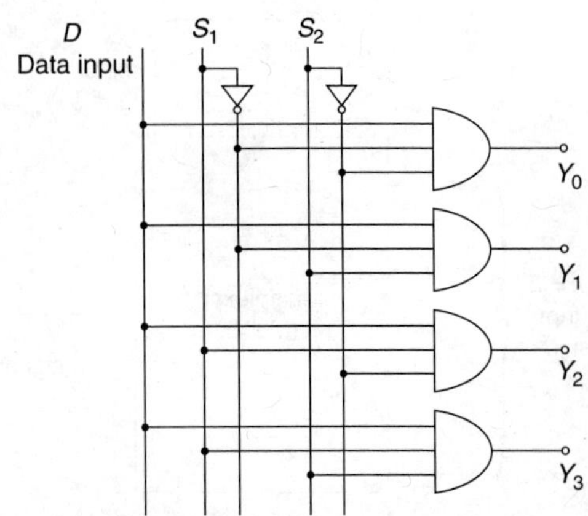


Figure 13-17 Circuit diagram of de-multiplexer

Considering a common data input $D$ for the de-multiplexer, we can write the following logic expressions for all four outputs from the truth table:

$$Y_0 = \overline{S}_1 \overline{S}_0 D$$

$$Y_1 = \overline{S}_1 S_0 D$$

$$Y_2 = S_1 \overline{S}_0 D$$

$$Y_3 = S_1 S_0 D$$

A 1-to-4 demultiplexer can be implemented by two inverters and four 3-input AND gates. The single input $D$ is applied to all the AND gates. The two select lines $S_1$, $S_0$ enable any one AND gate at a time and the data appears at the output of the selected AND gate as shown in Fig. 13-17.

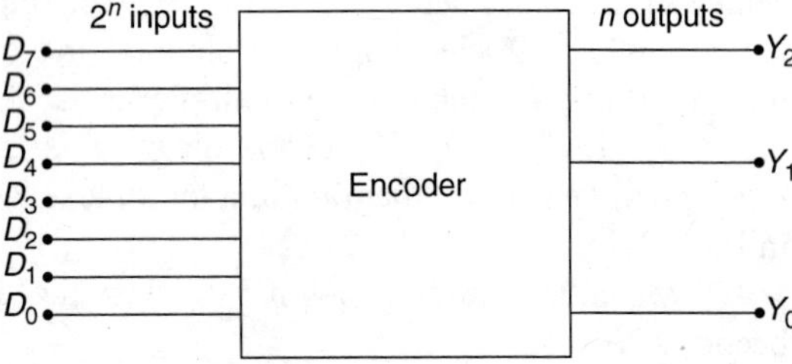


Figure 13-18 Block diagram of encoder

### Encoder

An encoder is a combinational logic circuit which converts non-digital data to digital data. An encoder has $2^n$ input lines and $n$ output lines.The output lines generate a binary code corresponding to the input value. For example a single bit 4-to-2 encoder takes in 4 bits and outputs 2 bits. An encoder combinational circuit that performs the inverse operation of a decoder. The truth table of an encoder is shown in Table 13-16 and the block diagram is shown in Fig. 13-18. The circuit diagram is shown in Fig. 13-19.

For an 8-to-3 binary encoder with inputs $D_0 - D_7$ the logic expressions of the outputs $Y_0 - Y_2$ are:

$$Y_0 = D_1 + D_3 + D_5 + D_7$$

$$Y_1 = D_2 + D_3 + D_6 + D_7$$

$$Y_2 = D_4 + D_5 + D_6 + D_7$$

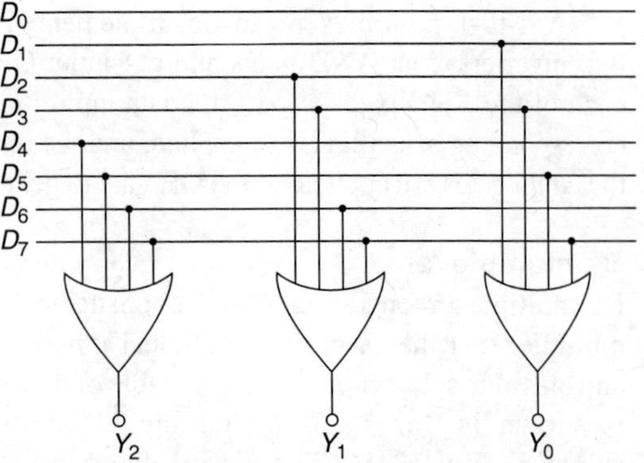


Figure 13-19 Circuit diagram of an octal-to-binary encoder

**Table 13-16** Truth table of encoder

| Input | | | | | | | | Output | | |
|---|---|---|---|---|---|---|---|---|---|---|
| $D_0$ | $D_1$ | $D_2$ | $D_3$ | $D_4$ | $D_5$ | $D_6$ | $D_7$ | $Y_2$ | $Y_1$ | $Y_0$ |
| 1 | 0 | 0 | 0 | 0 | 0 | 0 | 0 | 0 | 0 | 0 |
| 0 | 1 | 0 | 0 | 0 | 0 | 0 | 0 | 0 | 0 | 1 |
| 0 | 0 | 1 | 0 | 0 | 0 | 0 | 0 | 0 | 1 | 0 |
| 0 | 0 | 0 | 1 | 0 | 0 | 0 | 0 | 0 | 1 | 1 |
| 0 | 0 | 0 | 0 | 1 | 0 | 0 | 0 | 1 | 0 | 0 |
| 0 | 0 | 0 | 0 | 0 | 1 | 0 | 0 | 1 | 0 | 1 |
| 0 | 0 | 0 | 0 | 0 | 0 | 1 | 0 | 1 | 1 | 0 |
| 0 | 0 | 0 | 0 | 0 | 0 | 0 | 1 | 1 | 1 | 1 |

Octal-to-binary takes 8 inputs and provides 3 outputs. At any one time, only one input line has a value of 1.

An octal to binary encoder can be implemented using three 4-input OR gates. The encoder accepts a 3-bit input code and activates one of the eight output lines ($D_0-D_7$) corresponding to the input code.

### Decoder

Decoder is combinational logic circuit multiple-input, multiple-output logic circuit which converts digital data to non-digital data. The truth table of decoder is shown in Table 13-17 and the block diagram is shown in Fig. 13-20. The circuit diagram is shown in Fig. 13-21.

A 3-to-8 decoder is implemented using three inverters and eight 3-input AND gates, as shown in Fig. 13-21. The three inputs *A*, *B*, *C* -are decoded into eight outputs. Each one of the AND gates produce one minterms of the input variables.

### Applications of thė decoder

1. Any *n*-variable logic function, in canonical sum-of-minterms form can be implemented using a single $n$-to-$2^n$ decoder to generate the minterms, and an OR gate to form the sum.
2. The output lines of the decoder corresponding to the minterms of the function are used as inputs to the OR gate.
3. Any combinational circuit with *n* inputs and *m* outputs can be implemented with an $n$-to-$2^n$ decoder with *m* OR gates.

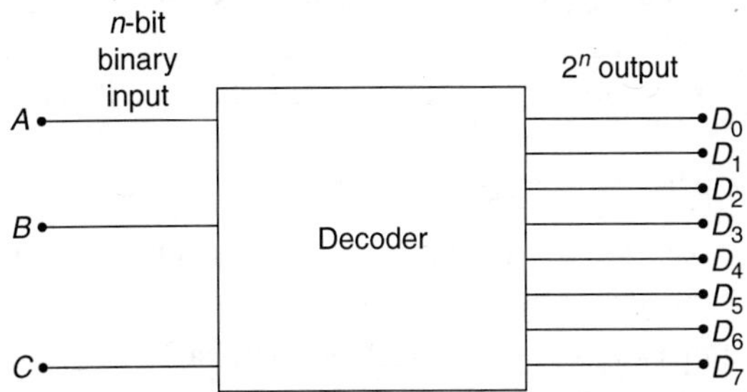


**Figure 13-20** Block diagram of a decoder

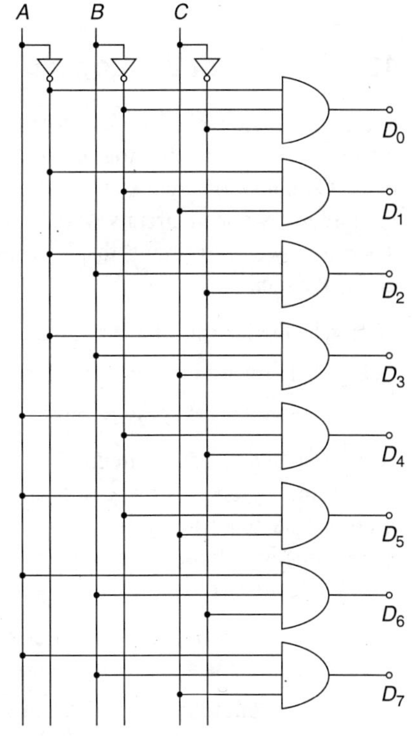


**Figure 13-21** Circuit diagram of a 3-to-8 decoder

**Table 13-17** Truth table of decoder

| *Input* | | | *Output* | | | | | | | |
|---|---|---|---|---|---|---|---|---|---|---|
| *A* | *B* | *C* | $D_0$ | $D_1$ | $D_2$ | $D_3$ | $D_4$ | $D_5$ | $D_6$ | $D_7$ |
| 0 | 0 | 0 | 1 | 0 | 0 | 0 | 0 | 0 | 0 | 0 |
| 0 | 0 | 1 | 0 | 1 | 0 | 0 | 0 | 0 | 0 | 0 |
| 0 | 1 | 0 | 0 | 0 | 1 | 0 | 0 | 0 | 0 | 0 |
| 0 | 1 | 1 | 0 | 0 | 0 | 1 | 0 | 0 | 0 | 0 |
| 1 | 0 | 0 | 0 | 0 | 0 | 0 | 1 | 0 | 0 | 0 |
| 1 | 0 | 1 | 0 | 0 | 0 | 0 | 0 | 1 | 0 | 0 |
| 1 | 1 | 0 | 0 | 0 | 0 | 0 | 0 | 0 | 1 | 0 |
| 1 | 1 | 1 | 0 | 0 | 0 | 0 | 0 | 0 | 0 | 1 |

A 3-to-8 decoder consists of three inputs and eight outputs.

4. A decoder is suitable when a circuit has many outputs, and each output function is expressed with few minterms. For example, we can implement a full adder with sum:

$$S(x, y, z) = (1, 2, 4, 7)$$

And carry:

$$C_y(x, y, z) = (3, 5, 6, 7)$$

## 13-8-2 Sequential Logic Circuit

Sequential logic circuit is a clock-driven feedback-based circuit, where the present output is a combination of the present input and previous output. Sequential circuits have loops that enable these circuits to receive feedback. Combinational circuits have no memory. In order to build sophisticated digital logic circuits, including computers, we need circuits whose output depends both upon the input of the circuit and its previous state. Therefore, we need circuits that have memory. For a device to serve as a memory, it must have these following characteristics:

1. The device must have two stable states
2. There must be a way to read the state of the device
3. There must be a way to set the state

It is possible to produce circuits with memory using the digital logic gates that use feedback (see Fig. 13-22). The memory elements are devices capable of storing binary information. The binary information stored in the memory elements at any given time defines the state of the sequential circuit. The present input and the state of the memory element determine the output.

There are two types of sequential circuits. Their classification depends on the timing of their signals.

1. Asynchronous sequential circuits: The outputs at any stage of this kind of circuit depend on the previous output stage, and all the outputs do not change their state at the same time.
2. Synchronous sequential circuits: All the stage flip-flops are connected together at the same time; and all stage outputs change at the simultaneously.

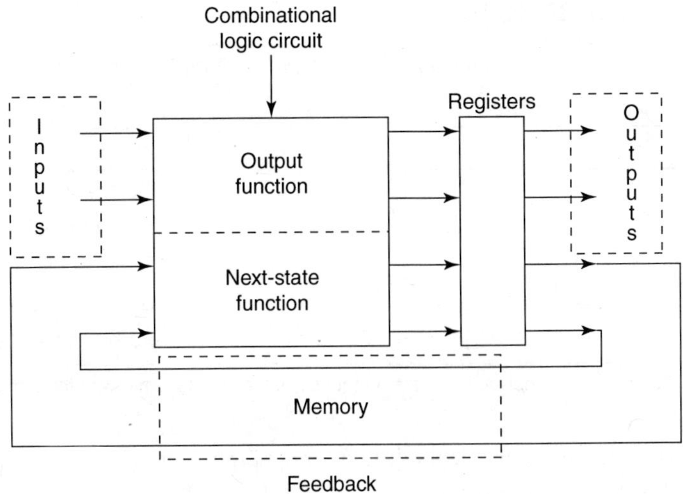


**Figure 13-22** Block diagram of a sequential circuit

## Latch

Latch is a level sensitive sequential circuit with bistable states. The simplest latch is an S-R latch having two inputs—$S$ and $R$, and output $Q$. The complement form of the S-R latch is shown in Fig. 13-23 and Fig. 13-24 shows the circuit diagram. The S-R latch circuit is a cross-coupled latch with direct feedback from its output $Q$ and $Q'$.

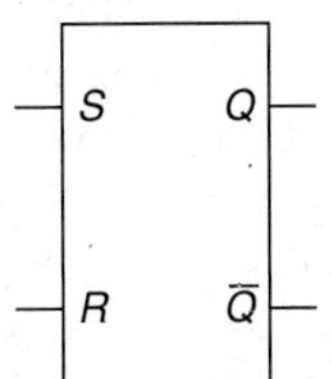


**Figure 13-23** Block diagram of latch

## Flip-flop

Flip-flop is an edge sensitive sequential logic circuit. All the flip-flops (FF) are driven by a clock—a feature that is not available in latch. Change of state in FF happens at the rising or falling edge of the clock pulse, but in case of latch it is level sensitive, i.e., in latch, change of state occurs depending upon 1 or 0 state only, but latch is not sensitive to rising or falling edge of the clock pulse. A general block diagram of FF is shown in Fig.13-25.

A sequential circuit which has only two states, 1 or 0, is a flip-flop. The two outputs are complimentary to each other.

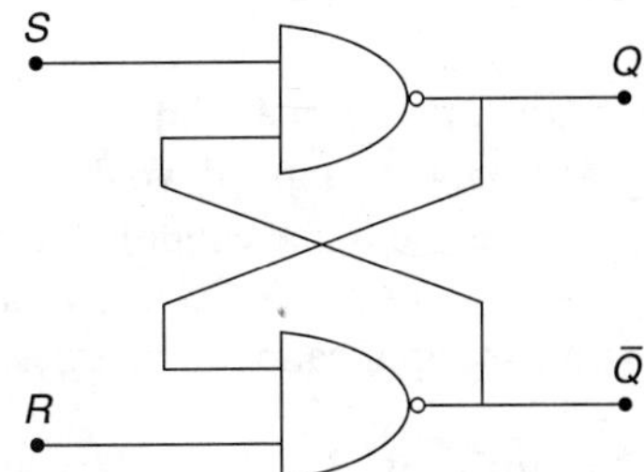


**Figure 13-24** Circuit diagram of latch

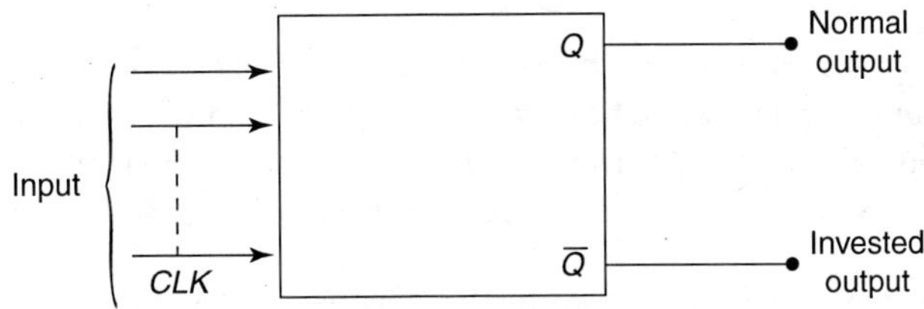


**Figure 13-25** Block diagram of flip-flop

**Table 13-18** State table of SR flip-flop

| Input | | Present Output State | Next Output Input State | Comments |
|---|---|---|---|---|
| S | R | Q | Q+ | |
| 0 | 0 | 0 | 0 | No change, i.e., Hold state/mode |
| 0 | 0 | 1 | 1 | No change, i.e., Hold state/mode |
| 0 | 1 | X | 0 | Reset state |
| 1 | 0 | X | 1 | Set state |
| 1 | 1 | X | 0 | Forbidden state |

The operation of S-R flip-flop has to be analysed with the 4 input combinations together with the 2 possible previous states.

There are different types of flips flops depending on their inputs and clock pulses that produce transition between two states. These are as follows:

1. **Set-Reset Flip-Flop or S-R FF:** The S-R FF has two inputs—*S* and *R*. *S* is called set and *R* is called reset. The *S* input is used to produce HIGH on *Q*, i.e., store binary 1 in flip-flop. The *R* input is used to produce LOW on *Q*, i.e., store binary 0 in flip-flop. $Q'$ is *Q* complementary output. The output of the S-R latch depends on current as well as previous stored output. The circuit and the truth table of the S-R latch is shown in Table 13-18 and circuit diagram is shown in Fig. 13-26.

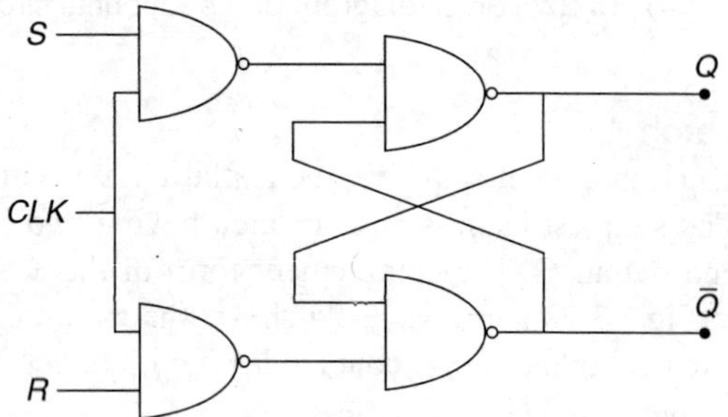


**Figure 13-26** Circuit diagram of flip-flop

**Condition I. S = 0 and R = 0:** If we assume $Q = 1$ and $Q' = 0$ as initial condition, then output *Q* after input is applied would be $Q = (R + Q')' = 1$ and $Q' = (S + Q)' = 0$. Assuming $Q = 0$ and $Q' = 1$ as the initial condition, then output *Q* after the input applied would be $Q = (R + Q')' = 0$ and $Q' = (S + Q)' = 1$. So it is clear that when both *S* and *R* inputs are LOW, the output is retained as before the application of inputs, i.e., there is no state change.

**Condition II. S = 1 and R = 0:** If we assume $Q = 1$ and $Q' = 0$ as, initial condition, then output *Q* after input is applied would be $Q = (R + Q')' = 1$ and $Q' = (S + Q)' = 0$. Assuming $Q = 0$ and $Q' = 1$ as the initial condition, then output *Q* after the input applied would be $Q = (R + Q')' = 1$ and $Q' = (S + Q)' = 0$. So in simple words when *S* is HIGH and *R* is LOW, output *Q* is HIGH.

**Condition III. S = 0 and R = 1:** If we assume $Q = 1$ and $Q' = 0$ as the initial condition, then output *Q* after input is applied would be $Q = (R + Q')' = 0$ and $Q' = (S + Q)' = 1$. Assuming $Q = 0$ and $Q' = 1$ as the initial condition, then output *Q* after the input applied would be $Q = (R + Q')' = 0$ and $Q' = (S + Q)' = 1$. So in simple words, when *S* is LOW and *R* is HIGH, output *Q* is LOW.

**Condition IV. S = 1 and R = 1:** No matter what state *Q* and $Q'$ are in, application of 1 at the input of NOR gate always results in 0 at the output which results in both *Q* and $Q'$ being set to LOW ($Q = Q'$). LOW in both the outputs basically is wrong, so this case is invalid.

**Table 13-19** State table of direct/delay flip-flop

| *Input* | *Present Output State* | *Next Output State* | *Comments* |
|---|---|---|---|
| *D* | Q | Q+ | |
| 1 | X | 1 | Set state |
| 0 | X | 0 | Reset state |

2. **Direct/Delay Flip-Flop or D-FF:** The S-R FF contains a forbidden/ambiguous state. To eliminate this condition we connect *S* and *R* together with an inverter, as shown in Fig. 13-27, and ensure that *S* and *R* are never equal. D-FF is almost the same as the S-R FF with only one input, as shown in Table 13-19. This input is called *D* or data input. The *D* latch is called *D* transparent latch. The output directly follows input with some delay so it is also known as *delay FF*.

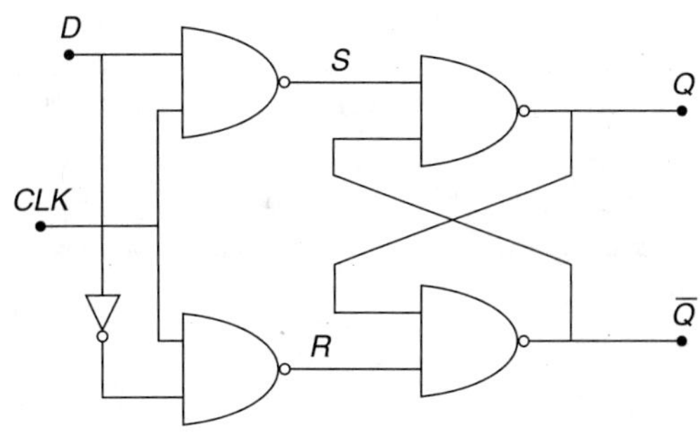


**Figure 13-27** Circuit diagram of flip-flop

3. **J-K Flip-Flops:** The forbidden/ambiguous state output in the S-R FF was eliminated in the D-FF by joining the inputs with an inverter. But the D-FF has a single input. The J-K latch is similar to S-R latch in that it has 2 inputs *J* and *K*, as shown in Fig. 13-28. When both inputs are high, output toggles and the ambiguous state is eliminated, as shown in the Table 13-20.

4. **Toggle Flip-Flop or T-FF:** When the two inputs of J-K latch are shorted as shown in Fig. 13-29, a T-FF is formed. When input is held HIGH, output toggles, as shown in the state Table 13-21.

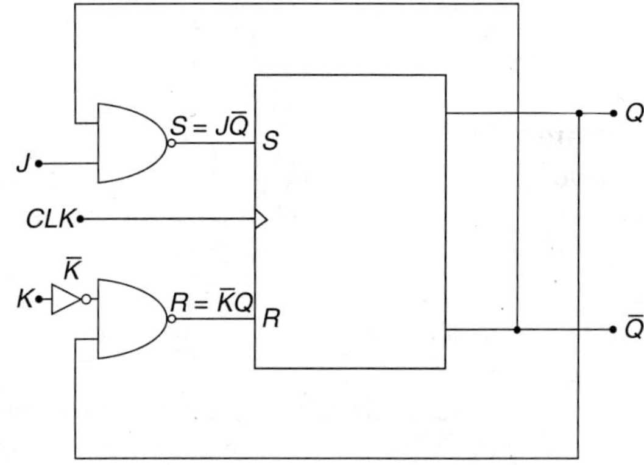


**Figure 13-28** Circuit diagram of flip-flop

**Table 13-20** State table of J-K flip-flop

| *Input* | | *Present Output State* | *Next Output Input State* | *Comments* |
|---|---|---|---|---|
| *J* | *K* | Q | Q+ | |
| 0 | 0 | 0 | 0 | No change, i.e., Hold state/mode |
| 0 | 0 | 1 | 1 | No change, i.e., Hold state/mode |
| 0 | 1 | X | 0 | Reset state |
| 1 | 0 | X | 1 | Set state |
| 1 | 1 | 1 | 0 | Toggle mode/state |
| 1 | 1 | 0 | 1 | Toggle mode/state |

**Table 13-21** State table of T flip-flop

| *Input* | *Present Output State* | *Next Output State* | *Comments* |
|---|---|---|---|
| *T* | *Q* | *Q+* | |
| 1 | 0 | 1 | Toggle mode |
| 1 | 1 | 0 | Toggle mode |
| 0 | 1 | 1 | Hold mode |
| 0 | 0 | 0 | Hold mode |

The main difference between latch and flip-flop is that latch is level sensitive but flip-flop is edge sensitive. Thus, in a flip-flop the output change occurs either at the rising edge or falling edge of the clock pulse depending upon the circuit configurations.

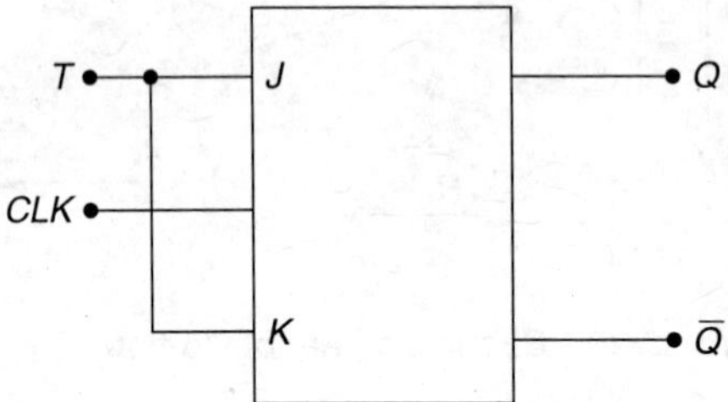


**Figure 13-29** Circuit diagram of flip-flop

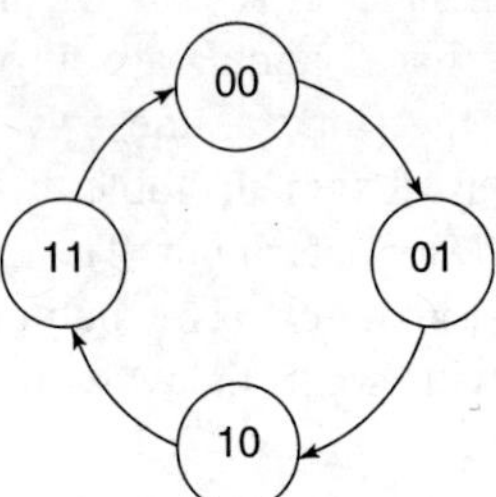


**Figure 13-30** State diagram of 2-bit up counter

## Counter circuit

Counter is a kind of sequential circuit used for counting the number of clock pulses. Counters are used to perform three main functions—timing, sequencing and counting. Counters can be divided into two categories: (1) Up counter and (2) Down counter. A 2-bit up counter counts the sequence 00 → 01 → 10 → 11, as illustrated by Table 13-22 and the state diagram given in Fig. 13-30.

State diagrams are used to describe the behaviour of a system. State diagrams can describe the possible states of events. Each diagram usually tracks the different states of its objects through the system. A 2-bit down counter counts the sequence 11 → 10 → 01 → 00, as shown in Table 13-23 and state diagram given in Fig. 13-31.

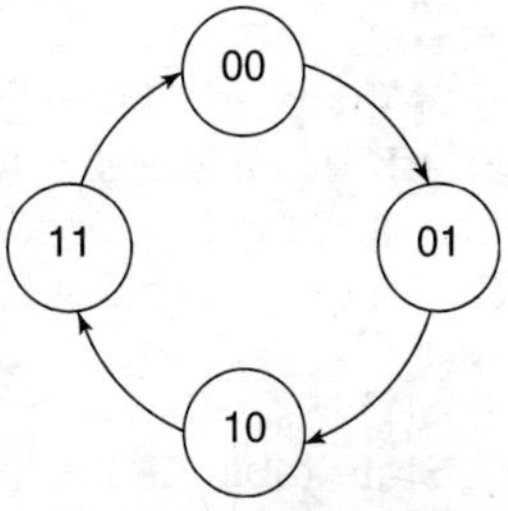


**Figure 13-31** State diagram of 2-bit down counter

**Table 13-22** State table of 2-bit up counter

| *A* | *B* | *Comments* |
|---|---|---|
| 0 | 0 | Decimal equivalent 0 |
| 0 | 1 | Decimal equivalent 1 |
| 1 | 0 | Decimal equivalent 2 |
| 1 | 1 | Decimal equivalent 3 |

**Table 13-23** State table of 2-bit down counter

| *A* | *B* | *Comments* |
|---|---|---|
| 1 | 1 | Decimal equivalent 3 |
| 1 | 0 | Decimal equivalent 2 |
| 0 | 1 | Decimal equivalent 1 |
| 0 | 0 | Decimal equivalent 0 |

**Table 13-24** State table of 4-bit up counter

| *Current or Present State* | | | | *Next State* | | | |
|---|---|---|---|---|---|---|---|
| *A*(*t*) | *B*(*t*) | *C*(*t*) | *D*(*t*) | *A*(*t*+1) | *B*(*t*+1) | *C*(*t*+1) | *D*(*t*+1) |
| 0 | 0 | 0 | 0 | 0 | 0 | 0 | 1 |
| 0 | 0 | 0 | 1 | 0 | 0 | 1 | 0 |
| 0 | 0 | 1 | 0 | 0 | 0 | 1 | 1 |
| 0 | 0 | 1 | 1 | 0 | 1 | 0 | 0 |
| 0 | 1 | 0 | 0 | 0 | 1 | 0 | 1 |
| 0 | 1 | 0 | 1 | 0 | 1 | 1 | 0 |
| 0 | 1 | 1 | 0 | 0 | 1 | 1 | 1 |
| 0 | 1 | 1 | 1 | 1 | 0 | 0 | 0 |
| 1 | 0 | 0 | 0 | 1 | 0 | 0 | 1 |
| 1 | 0 | 0 | 1 | 1 | 0 | 1 | 0 |
| 1 | 0 | 1 | 0 | 1 | 0 | 1 | 1 |
| 1 | 0 | 1 | 1 | 1 | 1 | 0 | 0 |
| 1 | 1 | 0 | 0 | 1 | 1 | 0 | 1 |
| 1 | 1 | 0 | 1 | 1 | 1 | 1 | 0 |

The state diagram of a 3-bit down counter is shown in Fig. 13-32. The state diagram of a 3-bit up-down counter is shown in Fig.13-33. The present/current state—next state table and state diagram of a 4-bit counter is shown in the Table 13-24 and Fig. 13-34, respectively.

Asynchronous counter. Asynchronous counter performs the counting operation where all the basic flip-flop blocks within the counter do not change the state at the same time. A 2-bit asynchronous

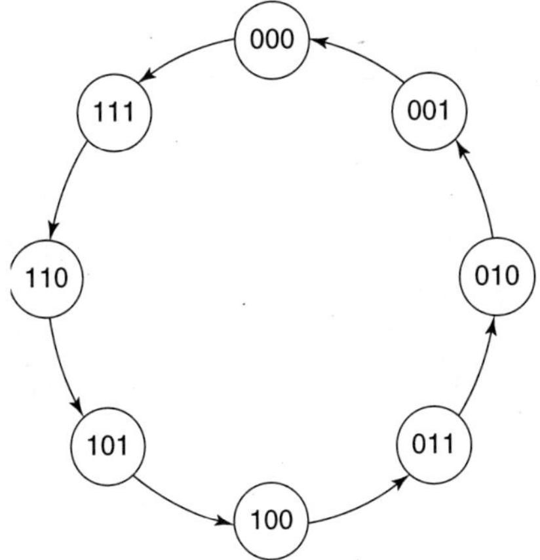


**Figure 13-32** State diagram of 3-bit down counter

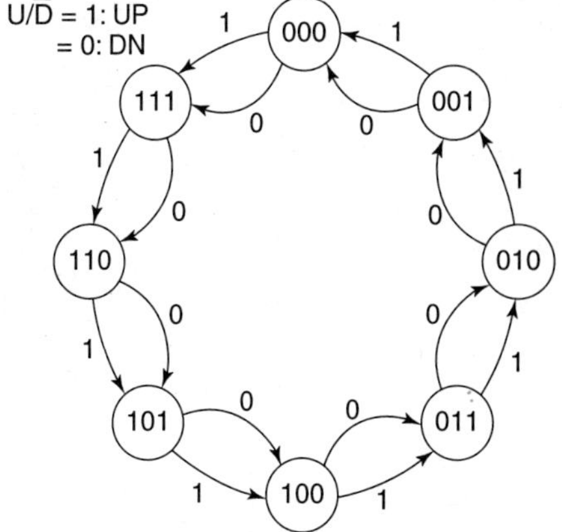


**Figure 13-33** Implementation of up down sequence by special control input U/D̄

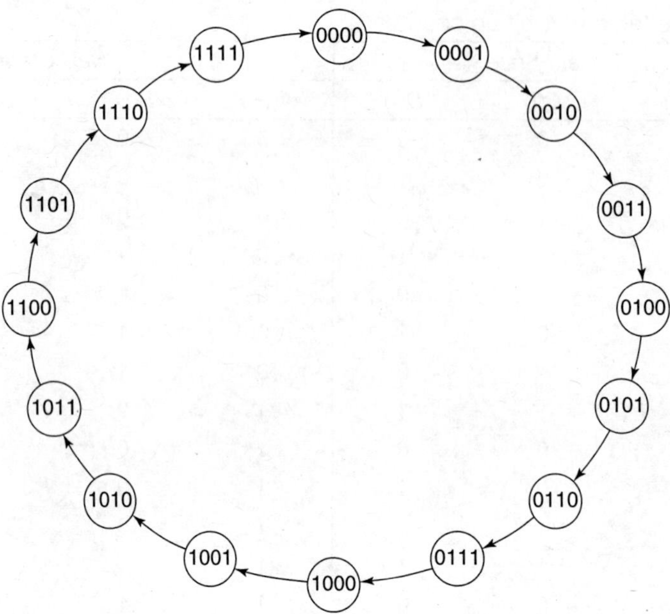


**Figure 13-34** State diagram of 4-bit up counter

counter is shown in Fig.13-35. The external clock is connected to the clock input of the first flip-flop only. The first flip-flop changes state at the falling edge of each clock pulse, but the second one changes only when triggered by the falling edge of the $Q$ output of the first one. Because of the inherent propagation delay through a flip-flop, the transition of the input clock pulse and a transition of the $Q$ output of the first flip-flop can never occur at exactly the same time. Therefore, the flip-flops cannot be triggered simultaneously, producing an asynchronous operation. Figure 13-36 shows a 3-bit asynchronous binary counter with its timing diagram using T-FF and J-K FF.

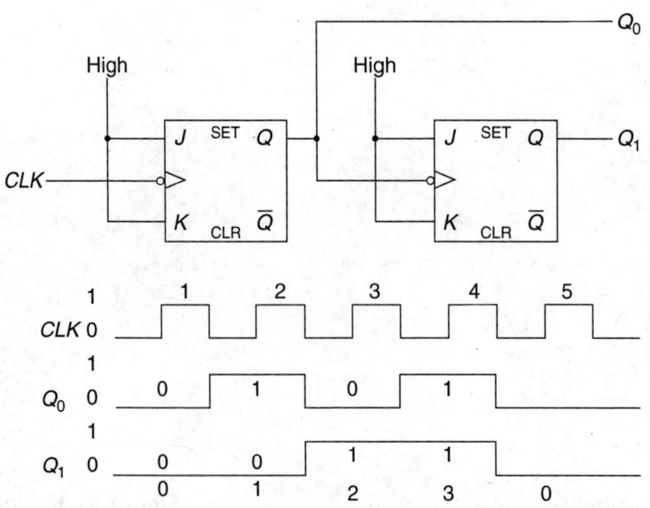


**Figure 13-35** A 2-bit asynchronous counter

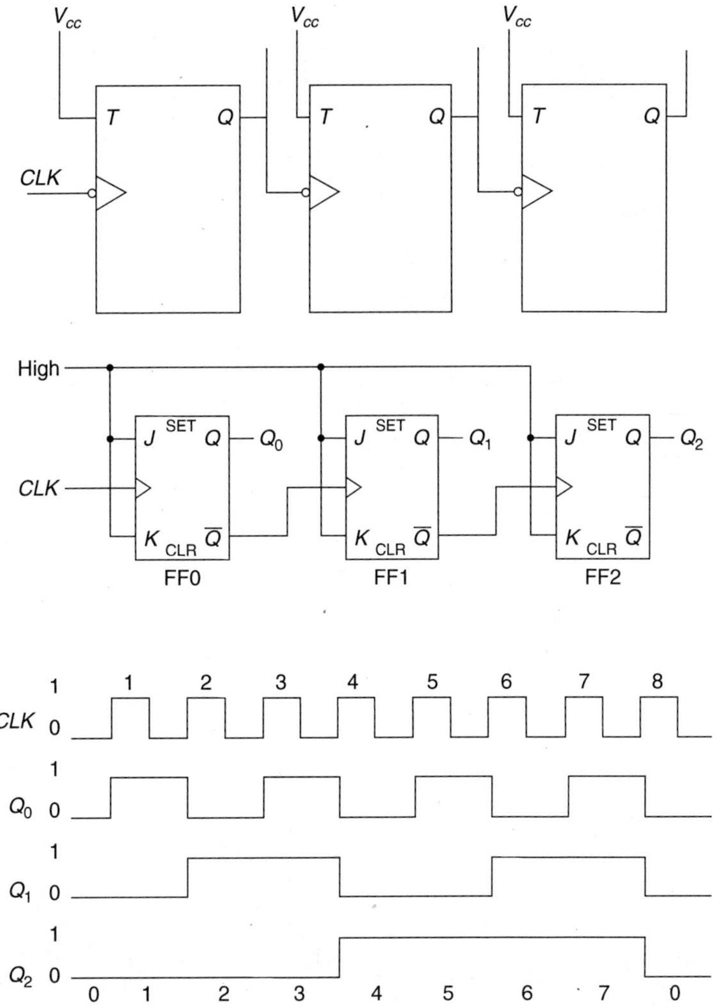


**Figure 13-36** Block diagram of 3-bit asynchronous/ripple counter

A 3-bit ripple counter is implemented using T flip-flop. The clock pulse is applied to the first flip-flop and the successive flip-flop is triggered by the output of the previous flip-flop. All the T inputs are connected to $V_{cc}$. The flip-flops toggle on the negative edge of the clock input.

**Four-bit asynchronous/ripple counters.** A 4-bit asynchronous binary counter with its timing diagram is shown in Fig. 13-37.

**Synchronous counter.** Synchronous counter performs the counting operation where all the basic flip-flop blocks within the counter clocks inputs are connected together, as shown in Fig.13-38, and change their states exactly at the same time.

### Shift register

Shift register is a type of sequential circuit used for storing and shifting of digital data. Shift registers consist of a number of single-bit flip-flops connected together in a chain arrangement so that the output from one

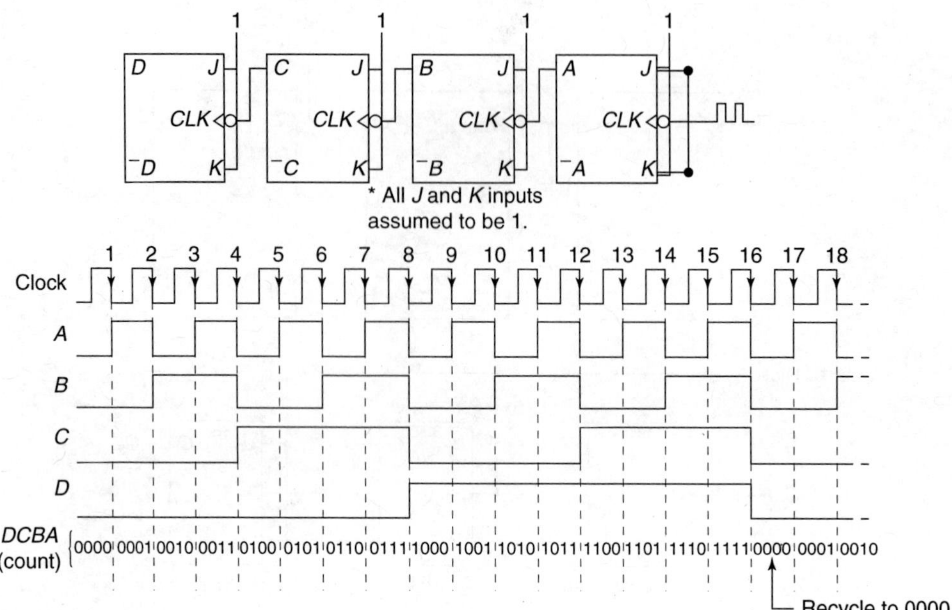


**Figure 13-37** Block diagram of 4-bit asynchronous/ripple counter

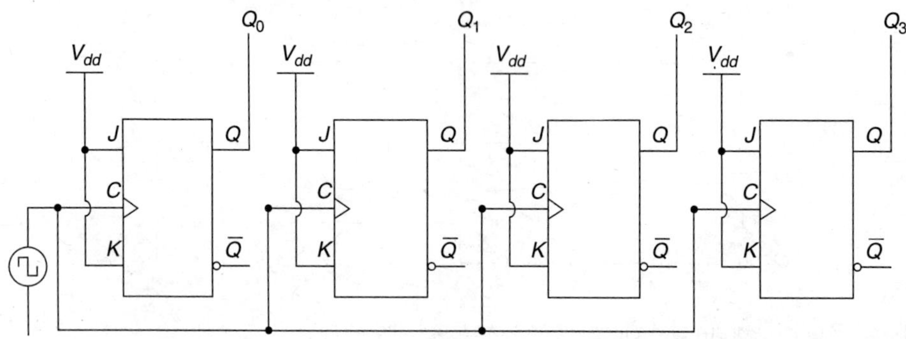


**Figure 13-38** Block diagram of synchronous counter

flip-flop becomes the input of the next one, thereby moving the stored data serially from either the left or the right direction. The numbers of individual data latches used to make up shift registers are determined by the number of bits to be stored. Shift registers are mainly used to store data and to convert data from either a serial to parallel or parallel to serial format with the entire flip-flop being driven by a common clock signal making them a synchronous circuit. They are generally provided with a clear or reset connection so that they can be "SET" or "RESET" as required. There are generally four types of shift registers.

**Serial input serial output (SISO) shift register.** SISO registers accept serial data from one input, and after storage and movement, output the data in a serial mode via one output. SISO registers shift data to the right. The block diagram of 4-bit SISO shift resistor is shown in Fig. 13-39.

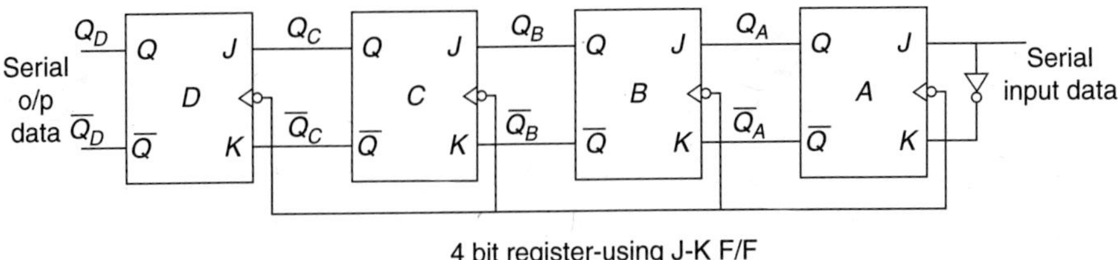


**Figure 13-39** Block diagram of SISO register

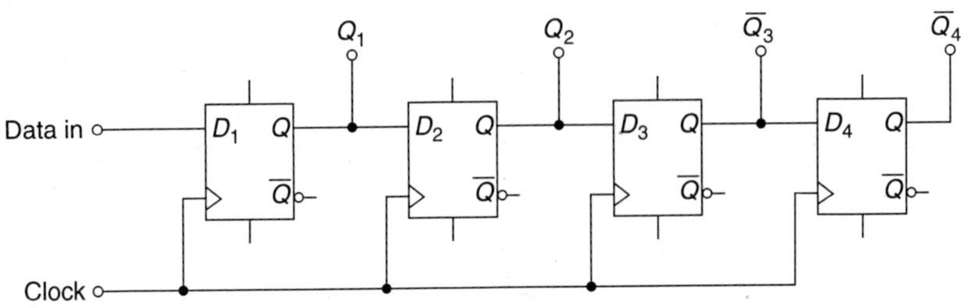


**Figure 13-40** Block diagram of SIPO register

This register accepts data serially and produces the stored information on its output line also in a serial form. It can be built using D-FF where the input of the rightmost flip-flop is used as a serial input line. The clock pulse is applied to all the flip-flops simultaneously. New data is entered into stage A while the data present in stage D are shifted out. Table 13-25 shows the state table of SISO entering data 1111 in the circuit and leaving out after nine cycles.

**Serial input parallel output (SIPO) shift register.** SIPO registers accept serial data from one input, and after storage and movement, output the data in a parallel mode via several outputs. The block diagram of 4-bit SIPO Shift resistor is shown in Fig.13-40.

This register consists of one serial input and outputs are taken from all the flip-flops parallel. The clock is applied simultaneously to all the flip-flops. Once the data is stored, each bit appears on its respective output line so that all the bits are available simultaneously.

**Table 13-25** State table of SISO shift register

| $Q_A$ | $Q_B$ | $Q_C$ | $Q_D$ |
|---|---|---|---|
| 0 | 0 | 0 | 0 |
| 1 | 0 | 0 | 0 |
| 1 | 1 | 0 | 0 |
| 1 | 1 | 1 | 0 |
| 1 | 1 | 1 | 1 |
| 0 | 1 | 1 | 1 |
| 0 | 0 | 1 | 1 |
| 0 | 0 | 0 | 1 |
| 0 | 0 | 0 | 0 |

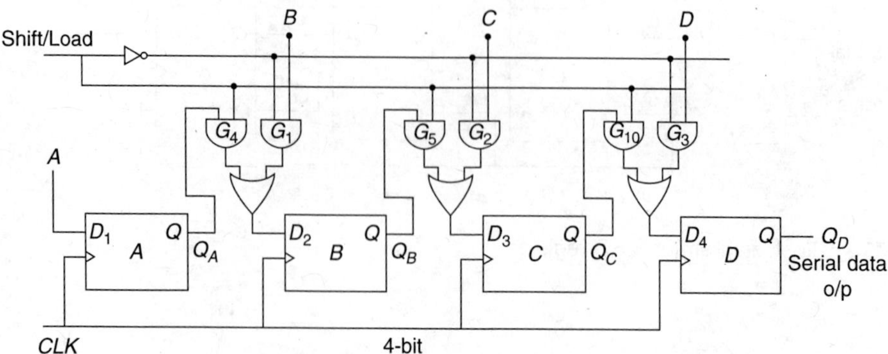


**Figure 13-41** Block diagram of PISO register

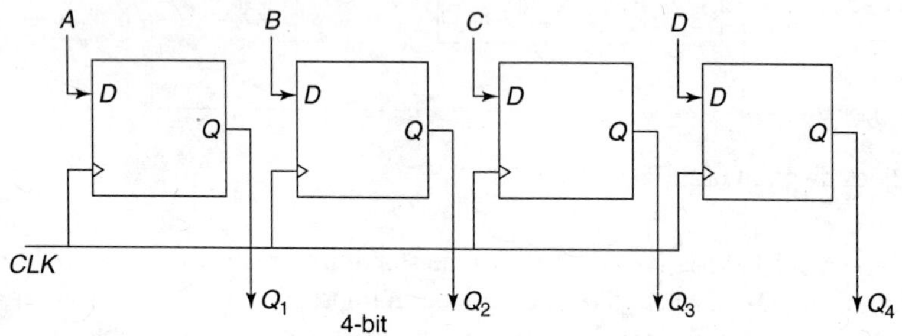


**Figure 13-42** Block diagram of PIPO register

**Parallel input serial output (PISO) shift register.** PISO registers accept parallel data from several inputs, and after storage and movement, output the data in a serial mode via one output. This PISO configuration has the data input on lines $D_1$, $D_2$, $D_3$ and $D_4$ connected in parallel format. To write the data to the register, the Write/Shift control line must be held LOW. To shift the data, the Write/Shift control line is brought HIGH and the registers are clocked. The arrangement now acts as a SISO shift register, with $D_1$ as the data input. The block diagram of 4-bit PISO shift resistor is shown in Fig. 13-41.

The four bits *A*, *B*, *C*, *D* are entered simultaneously into their respective flip-flops. Shift/Load is a control input that allows the four bits of data to enter into the register in parallel or shift data in serial.

**Parallel input parallel output (PIPO) shift register.** PIPO registers accept parallel data from several inputs, and after storage and movement, output the data in a parallel mode via several outputs. PIPO is the fastest shift register; it needs only one clock pulse to store the data through its parallel inputs and in the second clock pulse the data out through its parallel out lines. The loading and shifting of a data set, 1011, is shown in the Table 13-26. The block diagram of 4-bit PIPO shift register is shown in Fig.13-42.

**Table 13-26** State table of PIPO shift register

| *A* | *B* | *C* | *D* | *Comments* |
|---|---|---|---|---|
| 0 | 0 | 0 | 0 | Empty shift resistor |
| 1 | 0 | 1 | 1 | Storing of data 1011 in shiftresis-tor after the first clock pulse |
| 0 | 0 | 0 | 0 | Shifting of data 1011 after second clock pulse the. |

In this register, data can be shifted either in or out of the register in parallel. A 4-bit parallel-in-parallel-out register can be implemented using D-FFs. The parallel inputs are to be applied at *A*, *B*, *C* and *D* inputs, which are directly connected to delay inputs of respective flip flops.

## 13-9 REAL-LIFE APPLICATIONS

Nowadays digital circuits are used in almost all daily applications. These include computers, mobile phones, washing machines, and DVD players among other things. Digital devices and instruments give better performance over the analog ones. For example, a digitally equipped DTH provides studio-quality picture at home. New digital compression technologies generate very high quality photographs and movies.

## POINTS TO REMEMBER

1. Binary numbers consist of 1 and 0.
2. NAND and NOR gates are used as universal logic gates.
3. Multiplexer is called path selector.
4. Decoder with an enable input acts as a de-multiplexer.
5. Encoder converts non-digital data to digital data.
6. Latch is level sensitive whereas flip-flop is edge sensitive.
7. Counter is a kind of sequential circuit used for counting the number of clock pulses.
8. A 2-bit up counter counts the sequence 00 → 01 → 10 → 11.
9. Shift register is a kind of sequential circuit used for storing and shifting of digital data.

## IMPORTANT FORMULAE

1. $X+0=X$
2. $X+1=1$
3. $X+X=X$
4. $X.\overline{X}=1$
5. $X+\overline{X}=1$
6. $\overline{A+B}=\overline{A}.\overline{B}$
7. $\overline{A.B}=\overline{A}+\overline{B}$

## OBJECTIVE QUESTIONS

1. $\overline{\overline{A}}=$
   (a) A (b) 0
   (c) 1 (d) None of the above
2. $A+\overline{A}=$
   (a) 1 (b) 0
   (c) $A$ (d) $\overline{A}$
3. $\overline{A}+\overline{B}=$
   (a) $\overline{A.B}$ (b) $\overline{A}$
   (c) $\overline{B}$ (d) $A.B$
4. Universel logic gates are:
   (a) NAND and NOR (b) OR and AND
   (c) NOT and OR (d) OR and XOR
5. NAND gate is a combination of :
   (a) AND and NOT gates
   (b) AND and OR gates
   (c) AND and XOR gates
   (d) OR and NOR gates

6. Exclusive OR logic equation is:
   (a) $\bar{A}B + \overline{BA}$ (b) $\bar{A}B + \overline{B + A}$
   (c) $\bar{A}B + \bar{B}$ (d) $\bar{A}B + A\bar{B}$
7. Binary equivalent of decimal 13 is:
   (a) 1110 (b) 1111
   (c) 0111 (d) 1101
8. Decimal value of 1010 is:
   (a) 9 (b) 7
   (c) 19 (d) 10
9. How many minimum NAND gates are used in a half adder?
   (a) 5 (b) 7
   (c) 9 (d) 2
10. The minimum number of NAND gates used in a full adder is:
   (a) 5 (b) 7
   (c) 9 (d) 2
11. Decoder with an enable input can be used as a:
   (a) De-multiplexer (b) Encoder
   (c) XOR (d) Multiplexer
12. Latch is:
   (a) Logic level sensitive
   (b) Edge sensitive
   (c) Both (a) and (b)
   (d) None of the above
13. Flip-flop is:
   (a) Logic level sensitive
   (b) Edge sensitive
   (c) Both (a) and (b)
   (d) None of the above
14. Counters are used to count the:
   (a) Number of clock pulses
   (b) Number of glitches
   (c) Number of flip-flop
   (d) None of the above
15. Purpose of using a shift register is:
   (a) Shifting and storing
   (b) Only storing
   (c) Only shifting
   (d) None of the above
16. Fastest operating shift register is:
   (a) PIPO (b) SIPO
   (c) PISO (d) SISO
17. Slowest operating shift register is:
   (a) PIPO (b) SIPO
   (c) PISO (d) SISO

## REVIEW QUESTIONS

1. What is a logic circuit?
2. What are the differences between positive and negative logic?
3. Draw and explain the operation of AND, OR, NOT, NOR, XOR, XNOR, NAND using the truth table.
4. How does the XOR gate differ from the OR gate?
5. What is the universal gate? Give reasons.
6. State and explain De Morgan's theorem.
7. What is adder?
8. What are the differences between a half adder and a full adder?
9. Draw the block diagrams of half adder and full adder and explain their respective operations?
10. Design a full adder using a half adder.
11. What are the differences between an encoder and a decoder?
12. What is a multiplexer? Explain the operation of multiplexer using the truth table.

13. Explain the significance of the select input in a multiplexer.
14. What is a de-multiplexer? Explain its operation using a block diagram.
15. Why is a decoder with an enable input called a de-multiplexer?
16. What is latch?
17. What is flip-flop?
18. What are the basic differences between a latch and a flip-flop?
19. What is a counter?
20. Explain the working principle of counter.
21. What are the basic differences between asynchronous and synchronous counter?
22. What are the two important properties of shift register?
23. Explain the working principle of SISO?
24. Explain the working principle of SIPO?
25. Explain the working principle of PISO?
26. Explain the working principle of PIPO?

## PRACTICE PROBLEMS

1. Convert the following numbers into the other number system:
   (a) $(1111)_2 = (?)_8$
   (b) $(238)_8 = (?)_{10}$
   (c) $(111101)_2 = (?)_{16}$
   (d) $(67)_{10} = (?)_8$
   (e) $(1DFF)16 = (?)_{10}$
   (f) $(FF)\ 16 = (?)_2$
   (g) $(111.11)_2 = (?)_{10} = (?)_{16}$
   (h) $(455.67)_{10} = (?)_2$
   (i) $(445.67)_{10} = (?)_2$
   (j) $(111.11)_{10} = (?)_2$
   (k) $(1110.11)_2 = (?)_{10}$
   (l) $(45.675)_{10} = (?)_{Hex}$
   (m) $(451.167)_8 = (?)_2$
   (n) $(45.67)_{Hex} = (?)_2$
   (o) $(745.167)_8 = (?)_2$
2. Simplify the given equations and implement the results with logic gates.
   (a) $\overline{A} + \overline{A}\,B + AB\,\overline{C}$
   (b) $\overline{A} + \overline{A}\,\overline{B} + AB\,\overline{C}$
   (c) $AB\,\overline{C} + \overline{A}\,\overline{B} + AB\,\overline{C}$
   (d) $AB\,\overline{C} + A\,\overline{B}\,\overline{C}$
   (e) $A + \overline{A}\,B + A\,B\,\overline{C} + \overline{A\,B}$
   (f) $\overline{A}\,BC + \overline{A}\,\overline{B} + AB\,\overline{C}$
   (g) $\overline{A}\,BC + \overline{A}\,\overline{B} + A\,\overline{B}\,\overline{C} + \overline{A\,B}\,C$
3. Find the output logic expression from the given logic gates and then form a truth table.

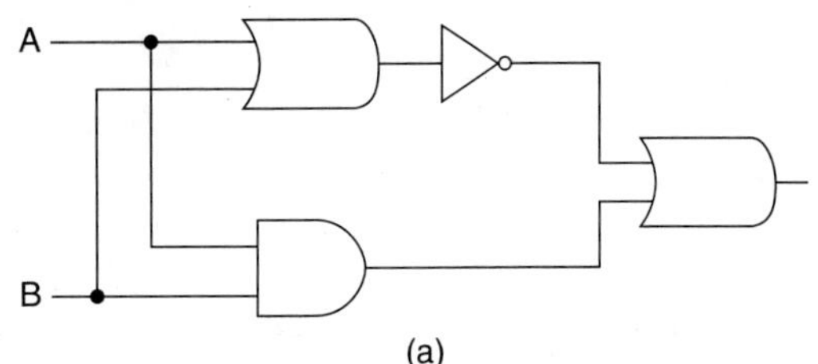

(a)

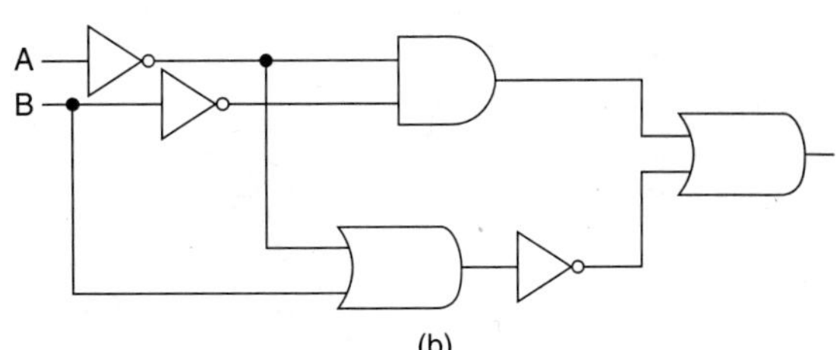

(b)

4. Draw the circuit diagram using any logic gate for the given equations:
   (a) $Q = A\,\overline{B} + \overline{A}\,B$
   (b) $Q = AA + AB + AC + BC$

5. Find the logic expression from the given truth table and implement the result with logic gates.

| *Input* | | | *Output* |
|---|---|---|---|
| *A* | *B* | *C* | × |
| 0 | 0 | 0 | 0 |
| 0 | 0 | 1 | 1 |
| 0 | 1 | 0 | 0 |
| 0 | 1 | 1 | 0 |
| 1 | 0 | 0 | 0 |
| 1 | 0 | 1 | 0 |
| 1 | 1 | 0 | 0 |
| 1 | 1 | 1 | 0 |

## SUGGESTED READINGS

1. Chattopadhyay, D. and P. C. Rakshit. 2006. *Electronics Fundamentals and Applications*. New Delhi: New Age International Publishers.
2. Ghosh, K. K. 2008. *Basic Electronics*. Kolkata: Platinum Publishers.
3. Chaudhury, Diptiman Ray. 2001. *Digital Circuits, Digital Electronics*. Kolkata: Eureka Publisher.
4. Floyd, Thomas L. 2007. *Digital Fundamentals*. New Delhi: Pearson Education.

# 14

# Electronic Instruments

## Outline



## Objectives

This final chapter discusses the key instruments of electronic measurement with special emphasis on the most versatile instrument of electronic measurement—the cathode-ray oscilloscope (CRO). The objective of this book will remain unrealized without a discussion on the CRO. The chapter begins with the details of construction of the CRO, and proceeds to examine the active and passive mode input–output waveforms for filter circuits and lead-lag network delay. This will be followed by a detailed study of the dual beam CRO and its uses in op-amp circuit integrator, differentiator, inverting and non-inverting circuits, comparative waveform study, and accurate measurement with impeccable visual display. In addition to the CRO, the chapter also examines the sweep frequency generator, the function generator, the sine wave generator, the square wave generator and the AF signal generator.

## 14-1 INTRODUCTION

The cathode-ray oscilloscope (CRO) is a multipurpose display instrument used for the observation, measurement, and analysis of waveforms by plotting amplitude along *y*-axis and time along *x*-axis. CRO is generally an *x-y* plotter; on a single screen it can display different signals applied to different channels. It can measure amplitude, frequencies and phase shift of various signals. Many physical quantities like temperature, pressure and strain can be converted into electrical signals by the use of transducers, and the signals can be displayed on the CRO. A moving luminous spot over the screen displays the signal. CROs are used to study waveforms, and other time-varying phenomena over different frequency ranges. The central unit of the oscilloscope is the cathode-ray tube (CRT), and the remaining part of the CRO consists of the circuitry required to operate the cathode-ray tube.

## 14-2 COMPONENTS OF THE CATHODE-RAY OSCILLOSCOPE

The heart of the oscilloscope, the cathode-ray tube (CRT), which generates the electron beam using the electron gun, accelerates the beam to a high velocity, deflects the beam according to the input signal and the electron beam ultimately becomes visible on the phosphor screen. The block diagram of the oscilloscope is shown in Fig. 14-1. The power supply block provides the necessary voltages for the CRT to generate and accelerate the electron beam and also to supply the required operating voltages for the other circuits of the oscillo scope. Comparatively high voltages, in the order of a few thousand volts, are required by CRTs for acceleration. However, a low voltage is also required for the heater of the electron gun.

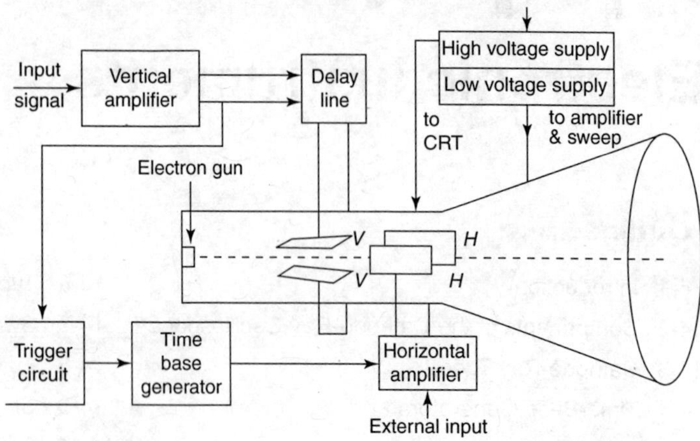


**Figure 14-1** Block diagram of a cathode-ray oscilloscope

The oscilloscope has a time-base generator, which generates the accurate voltage to be supplied to the cathode-ray tube for deflecting the spot at a constant time-dependent rate. The signal to be observed is fed into a vertical amplifier, which increases the potential of the input signal to a certain level in order to provide a usable deflection of the electron beam. To summarize, the CRO consists of the following:

(i) CRT
(ii) Vertical amplifier
(iii) Delay line
(iv) Horizontal amplifier
(v) Time-base generator
(vi) Triggering circuit
(vii) Power supply

## 14-3 CATHODE-RAY TUBE

The electron gun or electron emitter, the deflecting system and the fluorescent screen are the three major components of a general purpose CRT. A detailed diagram of the cathode-ray oscilloscope is given in Fig. 14-2.

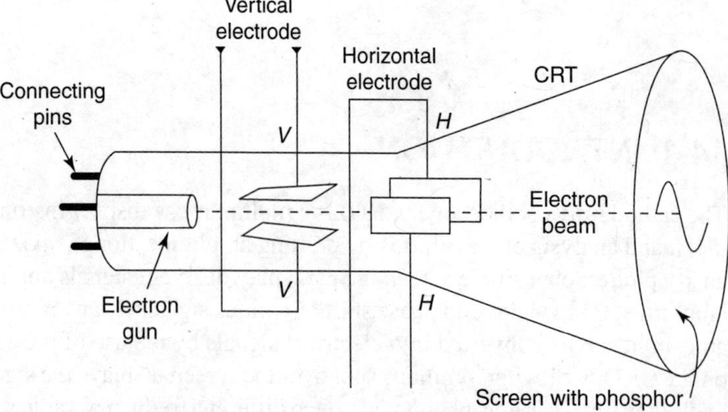


**Figure 14-2** Components of a cathode-ray oscilloscope

### 14-3-1 Electron Gun

In the electron gun of the CRT, electrons are emitted, converted into a sharp beam and focused upon the fluorescent screen. The electron beam consists of an indirectly heated cathode, a control grid, an accelerating electrode and a focusing anode. The electrodes are connected to the base pins. The cathode emitting the electrons is surrounded by a control grid with a fine hole at its centre. The accelerated electron beam passes through the fine hole. The negative voltage at the control grid controls the flow of electrons in the electron beam, and consequently, the brightness of the spot on the CRO screen is controlled.

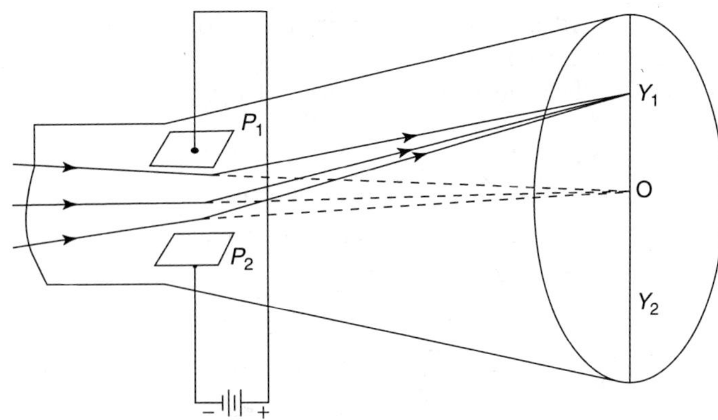


Figure 14-3(a) Deflecting system using parallel vertical plates

## 14-3-2 Deflection Systems

Electrostatic deflection of an electron beam is used in a general purpose oscilloscope. The deflecting system consists of a pair of horizontal and vertical deflecting plates. Let us consider two parallel vertical deflecting plates $P_1$ and $P_2$. The beam is focused at point $O$ on the screen in the absence of a deflecting plate voltage. If a positive voltage is applied to plate $P_1$ with respect to plate $P_2$, the negatively charged electrons are attracted towards the positive plate $P_1$, and these electrons will come to focus at point $Y_1$ on the fluorescent screen. The deflection is proportional to the deflecting voltage between the plates. If the polarity of the deflecting voltage is reversed, the spot appears at the point $Y_2$, as shown in Fig. 14-3(a). To deflect the beam horizontally, an alternating voltage is applied to the horizontal deflecting plates and the spot on the screen horizontally, as shown in Fig. 14-3(b). The electrons will focus at point $X_2$. By changing the polarity of voltage, the beam will focus at point $X_1$. Thus, the horizontal movement is controlled along $X_1OX_2$ line.

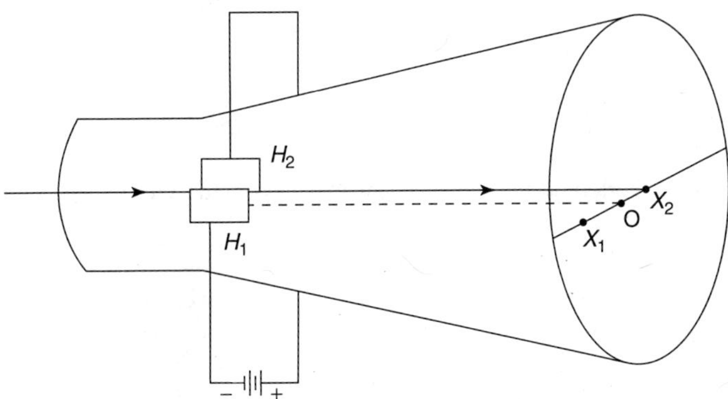


Figure 14-3(b) Deflecting system using parallel horizontal plate

### Spot beam deflection sensitivity

The deflection sensitivity of a CRT is defined as the distance of the spot-beam deflection on the screen per unit voltage. If $l_{total}$ is the total amount of deflection of the spot beam on the screen for the deflecting voltage $V_d$, as shown in Fig.14-4, the sensitivity can be expressed as:

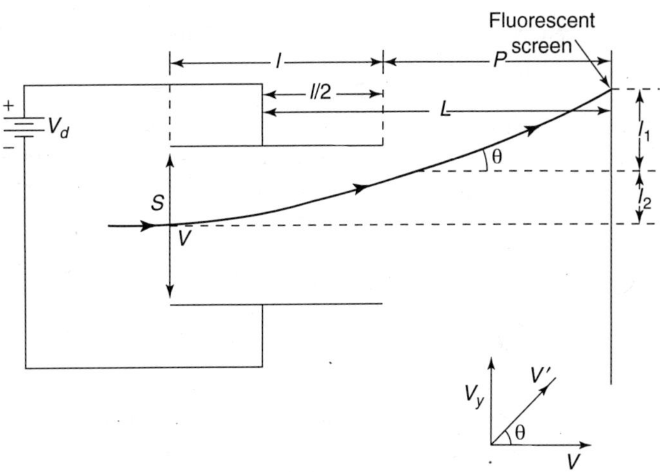


Figure 14-4 Schematic diagram of electrostatic deflection systems

$$S = \frac{l_{total}}{V_d} \qquad (14\text{-}1)$$

The unit of deflection sensitivity is milimeter/Voltage.

If $V_d = 1$, then $S = l_{total}$, i.e., the magnitude of the defection sensitivity is equal to the spot deflection distance for the applied unit deflection voltage.

**Electrostatic deflection**

$s$ = separation between deflecting plates

$P$ = distance between the plate and screen $S$

$l$ = length of each deflecting plate

$V_d$ = deflecting voltage applied across the plates

$m$ = mass of the electron

$e$ = charge of the electron

$v$ = velocity of the entering electron

$V_a$ = accelerating anode voltage

Thus:

$$\frac{1}{2}mv^2 = eV_a \tag{14-2}$$

$$v^2 = \frac{2eV_a}{m} \tag{14-3}$$

Force exerted on the electron towards the positive deflecting plate is:

$$Fs = eV_d$$

$$F = \frac{eV_d}{s} \tag{14-4}$$

$$mf = \frac{eV_d}{s}$$

Hence, acceleration is:

$$1f = \frac{eV_d}{ms} \tag{14-5}$$

Time taken by the electron to move through the deflecting plates is:

$$t = \frac{l}{v}$$

Therefore, the upward velocity acquired by the emerging electron is:

$$v_y = ft$$

$$v_y = \frac{fl}{v}$$

$$v_y = \frac{fl}{v} = \frac{eV_d}{sm}\frac{l}{v} \tag{14-6}$$

The vertical displacement of the electron in this case is derived directly from the well-known formula of mechanics:

$$D = ut + \frac{1}{2}ft^2$$

where, $D$ is the distance traversed by an electron, $u$ is the initial velocity, $f$ is the acceleration of an electron, and $t$ is the time taken.

As the electron is starting from rest, the initial velocity is zero, i.e., $u = 0$ and the distance travelled by the electron $D = l_2$.

Substituting this value of $D$ in the expression for $D$, from the formula of mechanics, we get:

$$l_2 = \frac{1}{2} ft^2 \tag{14-7}$$

Substituting the value of $t$ in Eq. (14-7) we get:

$$l_2 = \frac{1}{2} f(\frac{l}{v})^2 = \frac{eV_d}{2\,sm} (\frac{l}{v})^2 \tag{14-8}$$

$$\tan\theta = \frac{v_y}{v} = \frac{l_1}{P} \tag{14-9}$$

$$l_{\text{total}} = l_1 + l_2 = \frac{eV_d l}{smv^2} (\frac{l}{2} + P) \tag{14-10}$$

Here:

$$L = \left(\frac{l}{2} + P\right) \tag{14-11}$$

Substituting $v^2$ from Eq. (14-3) and $L$ from Eq. (14-11) in Eq. (14-10) we have:

$$l_{\text{total}} = \frac{lLV_d}{2sV_a} \tag{14-12}$$

The deflection sensitivity of the CRT is, by definition:

$$S = \frac{l_{\text{total}}}{V_d} = \frac{lL}{2sV_a} \text{ m/V} \tag{14-13}$$

The deflection factor of the CRT is:

$$G = \frac{1}{S} = \frac{2sV_a}{lL} \text{ V/m} \tag{14-14}$$

The deflection sensitivity can be increased by reducing the anode voltage; as a result the brightness of the spot gets reduced. This problem is solved by post-acceleration. The beam is accelerated after it has been deflected by an enhancer anode.

## 14-3-3 Fluorescent Screen

Phosphor is used as screen material on the inner surface of a CRT. Phosphor absorbs the energy of the incident electrons. The spot of light is produced on the screen where the electron beam hits. The bombarding electrons striking the screen, release secondary emission electrons. These electrons are collected or trapped by an aqueous solution of graphite called "Aquadag" which is connected to the second anode. Collection of the secondary electrons is necessary to keep the screen in a state of electrical equilibrium.

The type of phosphor used, determines the colour of the light spot. The brightest available phosphor isotope, P31, produces yellow–green light with relative luminance of 99.99%. The persistence of glow of the phosphor after the electrons have ceased bombarding plays an important role in the selection of the phosphor. The continuous excited spots on the screen form an unbroken line. High persistence results in an undesirable overlapping of a new trace with an old or previous trace.

Usually the CRT is kept at relatively lower brightness. A bright spot on the screen for a long time will cause a burnout. After hitting the screen, the electrons should return to the anode for closing the circuit.

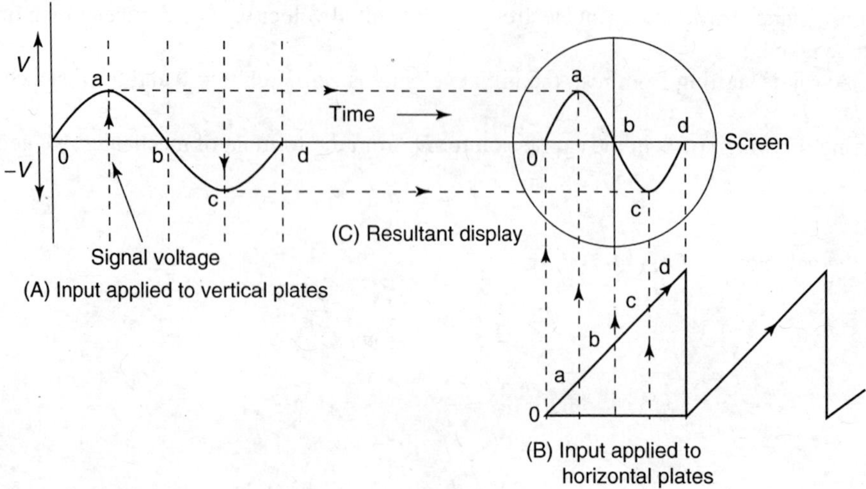


**Figure 14-5(a)** A typical display waveform on the screen

The negative charge accumulated near the screen will repel the electron beam, and there will be no display on the screen. Figure 14-5(a) illustrates a typical CRT display waveform on a fluorescent screen.

Figure 14-5(a) shows a sine wave applied to vertical deflecting plates and a repetitive ramp or saw-tooth applied to the horizontal plates. The ramp waveform at the horizontal plates causes the electron beam to be deflected horizontally across the screen. If the waveforms are perfectly synchronized then the exact sine wave applied to the vertical display appears on the CRO display screen. Similarly the display of the triangular waveform is as shown in Fig. 14-5(b).

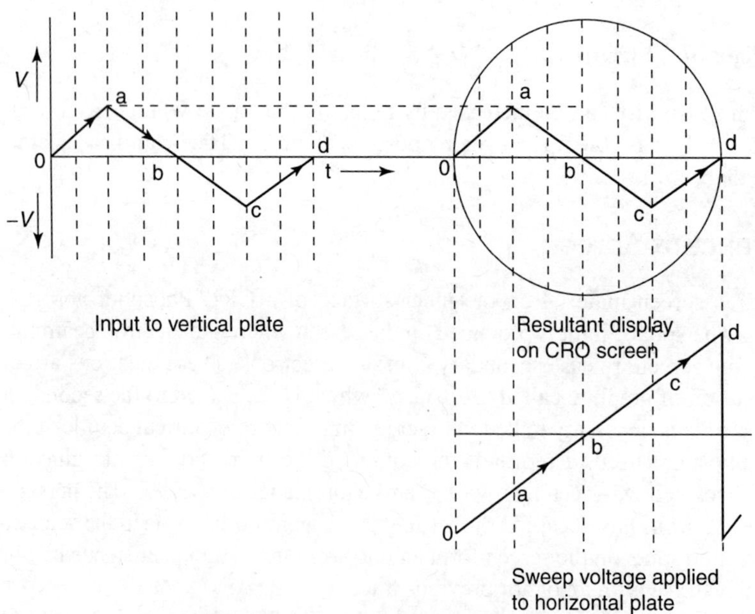


**Figure 14-5(b)** Triangular waveform input applied to the vertical deflecting plates of CRO

## Solved Examples

**Example 14-1** The electron beam in a CRT enters a magnetic deflection system after being accelerated through a potential difference of 1 kV. The deflection system employs a magnetic field of 160 gauss acting over an axial length of 1cm. Find the deflection of the spot on the fluorescent screen placed at a distance of 20 cm from the centre of the deflection system. Calculate the deflection sensitivity.

**Solution:**

Given: $B = 160$ gauss $= 1.6 \times 10^{-2}$ tesla, $V = 103$ volt, $l = 1$ cm $= 0.01$ m

$$L + \frac{1}{2} = 20 \text{ cm} = 0.2 \text{ m}$$

$\therefore$ The deflection of the spot is:

$$d = B\sqrt{\frac{e}{2Vm}}\, l(L + \frac{l}{2})$$

$$= 0.299 \text{ m} = 29.9 \text{ cm}$$

The magnetic deflection sensitivity is:

$$S = \frac{d}{B} = \frac{299}{150} = 1.99 \text{ mm/gauss}$$

**Example 14-2** The Lissajous figure is produced upon application of sinusoidal voltages to the two deflecting plates. The pattern gives four tangencies with the vertical and three tangencies with the horizontal. If the frequency of the horizontal signal is 1 kHz, what will be the frequency of the vertical signal?

**Solution:**

Let $f_v$ and $f_h$ be the frequencies of the vertical and horizontal signals respectively.

Then,
$$\frac{f_v}{f_h} = \frac{3}{4}$$

Given:
$$f_h = 4 \text{ kHz} = 4000 \text{ Hz}$$

$\therefore$
$$f_v = 3000 \text{ Hz} = 3 \text{ kHz}$$

**Example 14.3** The transit time of an electron through the deflection plates of a CRT should be kept below 0.2 per cycle. If the electrons enter the deflection system with a KE of 1 keV and the axial length of each deflection plate is 1.5 cm, calculate the highest frequency of the deflecting voltage.

**Solution:**

Axial velocity of the electron:

$$v = \sqrt{\frac{2eV_a}{m}} = 1.87 \times 10^7 \text{ m/s}$$

Transit time of the electron through the deflecting plates:

$$\tau = \frac{l}{v} = \frac{1.5 \times 10^{-2}}{1.87 \times 10^7} = 0.802 \times 10^{-9} \text{ s}$$

$T$ is the time period of the sinusoidal deflecting voltage:

$$\tau_{max} = (0.2/360) \times T$$

Hence,
$$\frac{0.2}{360} T = 0.802 \times 10^{-9}$$

$$T = \frac{2.887}{2} \times 10^{-6} \text{ s} = 1.4435 \times 10^{-6} \text{ s}$$

Therefore, the highest frequency of the deflecting voltage is:

$$f = \frac{1}{T} = 6.92 \times 10^5 \text{ Hz} = 692 \text{ kHz}$$

**Example 14-4** A trigger pulse is applied to the sweep generator every 10 ms. Compute the amplitude of the voltage $V_o$ across the capacitor when the trigger pulse is applied. The value of the capacitor is 0.2$\mu$F and resistance is 500 k$\Omega$. Supply voltage is 50 volt.

**Solution:**

Using standard formula: $$V_o = V_{cc}\left[1 - \exp\left(-\frac{t}{RC}\right)\right]$$

$$V_{cc} = 50 \text{ V}$$

$$T = 10 \text{ msec} = 10 \times 10^{-3} \text{ sec}$$

$$R = 500 \times 10^3 \ \Omega$$

$$C = 0.2 \times 10^{-6} \text{ Farad}$$

Putting the values we get: $V_o = 4.76$ Volt

Since $V_o$ is less than 10% of $V_{cc}$, the charge curve should still be quite at this point when $V_o$ is 4.76 V.

**Example 14-5** The time-base of a CRO is set at 0.1 ms/cm. A sinusoidal signal applied to the vertical deflection plate gives 5 cycles over a sweep width of 20 cm. Calculate the frequency of the signal.

**Solution:**

One cycle of signal occupies: $\frac{20}{5}$ cm $= 4$ cm

The time interval corresponding to 4 cm is $0.1 \times 4 = 0.4$ ms.

Hence, the frequency is: $f = \dfrac{1}{0.4 \text{ ms}} = \dfrac{10 \times 10^3}{4} = 2500 \text{ Hz} = 2.5 \text{ kHz}$

**Example 14-6** Two sinusoidal voltages of the same amplitude and frequency are applied simultaneously to the vertical and horizontal deflection systems of a CRO. An ellipse is traced on the fluorescent screen. The slope of the major axis is negative. The trace has a maximum vertical value of 3 divisions and it crosses the vertical axis at 1.1 divisions above the origin. What is the phase difference between the voltages?

**Solution:**
If $\theta$ is the phase difference between the two voltages, we have:

$$\sin\theta = \frac{v_{y0}}{A} = \frac{1.1}{3} = 0.366$$

$$\theta = 159°$$

As the major axis of the ellipse has negative slope, $\theta$ must lie between 90° and 180°.

**Example 14-7** The electron beam in a CRT is accelerated through a potential difference of 4000 V. The beam then travels through a pair of deflecting plates of axial length 2 cm, the separation between the plates being 5 mm. The potential difference applied between the plates is 20 V. The distance of the CRT screen from the centre of the deflecting plates is 25 cm. Determine:

(a) The transit time of the beam through the deflecting plates
(b) The transverse acceleration imparted to the electrons
(c) The deflection of the spot on the CRT screen. Also calculate deflection sensitivity

**Solution:**
The axial velocity of the electron is:

$$v = \sqrt{\frac{2eV_a}{m}} = 3.74$$

(a) Transit time: $t = \frac{l}{v} = 0.5035 \times 10^{-9}\,\text{s}$

(b) Transverse acceleration: $\frac{eV_d}{sm} = 0.702 \times 10^{15}\,\text{m/s}^2$

(c) Spot deflection: $d = \frac{lL}{2s}\frac{V_d}{V_a} = .25\,\text{cm}$

(d) Deflection sensitivity: $S = \frac{d}{V_d} = 0.5\,\text{mm/V}$

■

## 14-4 TIME-BASE GENERATORS

The CRO is used to display a waveform that varies as a function of time. If the wave form is to be accurately reproduced, the beam should have a constant horizontal velocity. As the beam velocity is a function of the deflecting voltage, the deflecting voltage must increase linearly with time. A voltage with such characteristics is called a ramp voltage. If the voltage decreases rapidly to zero—with the waveform repeatedly produced, as shown in Fig. 14-6—we observe a pattern which is generally called a saw-tooth waveform. The time taken to return to its initial value is known as flyback or return time.

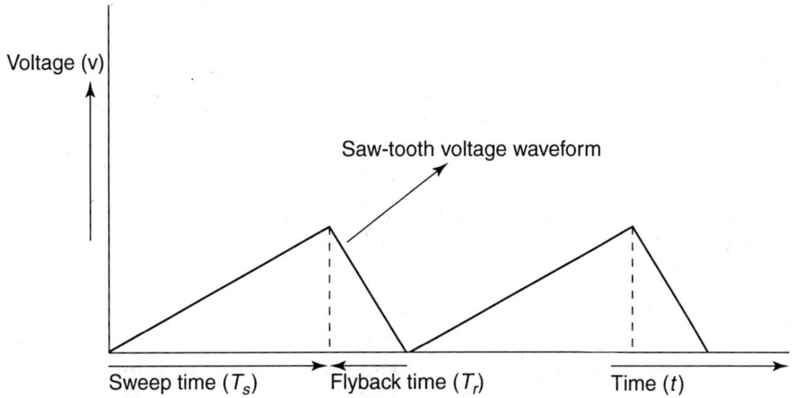


**Figure 14-6** Typical saw-tooth waveform applied to the horizontal deflection plates

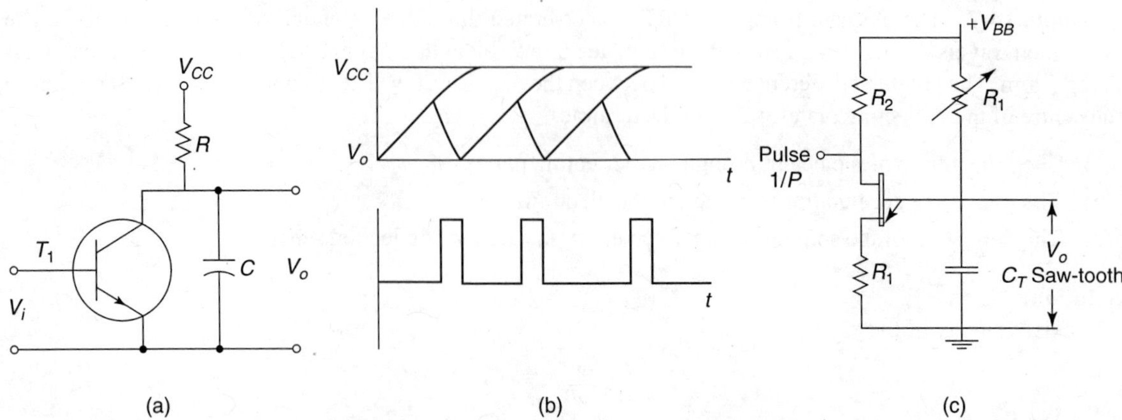


**Figure 14-7:** (a) Simple saw-tooth generator (b) Associated waveforms (c) Time-base generator using

During the sweep time $T_s$, the beam moves from left to right across the CRT screen. Hence the beam is deflected to the right by the increasing amplitude of the ramp voltage. As the positive voltage attracts the negative electrons, during the retrace time or fly back time $T_r$, the beam returns quickly to the left side of the screen. The control grid is generally "gated-off", which blanks out the beam during retrace time and prevents an undesirable retrace pattern from appearing on the screen.

Since signals of different frequencies can be observed using the oscilloscope, the sweep rate must be regulating. We can change the sweep rate in steps by switching different capacitors in the circuit. The front panel control for this adjustment is marked, time/div or sec/div. The sweep rate can be adjusted by the variable resistor $R$, as shown in Fig.14-7(a).

The circuit shown in Fig. 14-7(a) is a simple sweep circuit, in which the capacitor $C$ charges through the resistor $R$. The capacitor discharges periodically through the transistor $T_1$, which causes the waveform shown in Fig. 14-7(b) to appear across the capacitor. The signal voltage, $V_i$ which must be applied to the base of the transistor to turn it ON for short time intervals is also shown in Fig. 14-7(b). When the transistor is completely turned ON, it presents a low-resistance discharge path, through which the capacitor discharges quickly.

If the transistor is not turned ON, the capacitor will charge exponentially to the supply voltage $V_{cc}$ according to the equation:

$$V_o = V_{cc}\left(1 - \exp^{-\frac{t}{RC}}\right) \tag{14-15}$$

where, $V_o$ is the instantaneous voltage across the capacitor at time $t$, $V_{cc}$ is the supply voltage, $t$ is the time of interest, $R$ is the value of series resistor, and $C$ is the value of capacitor.

The continuous sweep CRO uses the UJT as a time-base generator. The UJT is used to produce the sweep. When power is first applied to the UJT, it is in the OFF state and $C_T$ changes exponentially through $R_T$. The UJT emitter voltage $V_E$ rises towards $V_{BB}$ and $V_E$ reaches the plate voltage $V_P$. The emitter-to-base diode becomes forward biased and the UJT triggers ON. This provides a low resistance discharge path and the capacitor discharges rapidly. When the emitter voltage $V_E$ reaches the minimum value rapidly, the UJT goes OFF. The capacitor recharges and the cycles repeat.

To improve the sweep linearity, two separate voltage supplies are used; a low voltage supply for the UJT and a high voltage supply for the $R_TC_T$ circuit. This circuit is as shown in Fig. 14-7(c). $R_T$ is used for continuous control of frequency within a range and $C_T$ is varied or changed in steps. They are sometimes known as timing resistor and timing capacitor.

## 14-4-1 Oscilloscope Amplifiers

The purpose of an oscilloscope is to produce a faithful representation of the signals applied to its input terminals. Considerable attention has to be paid to the design of these amplifiers for this purpose. The oscillographic amplifiers can be classified into two major categories.

(i) AC-coupled amplifiers
(ii) DC-coupled amplifiers

The low-cost oscilloscopes generally use ac-coupled amplifiers. The ac amplifiers, used in oscilloscopes, are required for laboratory purposes. The dc-coupled amplifiers are quite expensive. They offer the advantage of responding to dc voltages, so it is possible to measure dc voltages as pure signals and ac signals superimposed upon the dc signals. DC-coupled amplifiers have another advantage. They eliminate the problems of low-frequency phase shift and waveform distortion while observing low-frequency pulse train.

The amplifiers can be classified according to bandwidth use also:

(i) Narrow-bandwidth amplifiers
(ii) Broad-bandwidth amplifiers

If the frequency band does not extend up to the television colour sub-carrier frequency of 4.33 MHz, the amplifier can be classified as a narrow-bandwidth amplifier. If the frequency response curve of the amplifier is flat, beyond 4.33 MHz, the amplifier is classified as a broad-bandwidth amplifier. Laboratory oscilloscopes respond to frequencies in excess of 5 MHz.

## 14-4-2 Vertical Amplifiers

Vertical amplifiers determines the sensitivity and bandwidth of an oscilloscope. Sensitivity, which is expressed in terms of V/cm of vertical deflection at the mid-band frequency.

Generally, amplifiers have a certain gain bandwidth product, which is the product of the voltage gain of the amplifier and its bandwidth. Voltage gain may be sacrificed in favour of greater bandwidth or vice versa.

The gain of the vertical amplifier determines the smallest signal that the oscilloscope can satisfactorily measure by reproducing it on the CRT screen. The sensitivity of an oscilloscope is directly proportional to the gain of the vertical amplifier. So, as the gain increases the sensitivity also increases. This facilitates the study the smaller amplitude signals.

The vertical sensitivity measures how much the electron beam will be deflected for a specified input signal. The CRT screen is covered with a plastic grid pattern called a graticule. The spacing between the grids lines is typically 10 mm. Vertical sensitivity is generally expressed in volts per division. On the front panel of the oscilloscope a rotary switch is used for gain control. The gain could be designated V/Div. The rotary attenuator switch is electrically connected to the input attenuator network. The setting of the rotary switch indicates the amplitude of the signal required to deflect the beam vertically by one division. The vertical sensitivity of an oscilloscope measures the smallest deflection factor that can be selected with the rotary switch.

The bandwidth of an oscilloscope detects the range of frequencies that can be accurately reproduced on the CRT screen. The greater the bandwidth, the wider is the range of observed frequencies. The broadband gain should be constant near the higher range of frequencies which can be observed using an oscilloscope.

The bandwidth of an oscilloscope is the range of frequencies over which the gain of the vertical amplifier stays within 3 db of the mid-band frequency gain, as shown in Fig. 14-8. Rise time is defined as the time required for the edge to rise from 10–90% of its maximum amplitude. When an oscilloscope is used to observe a pulse or a square wave, the rise time of the instrument must be faster than the rise time of the pulse or square wave. An approximate relation is given as follows:

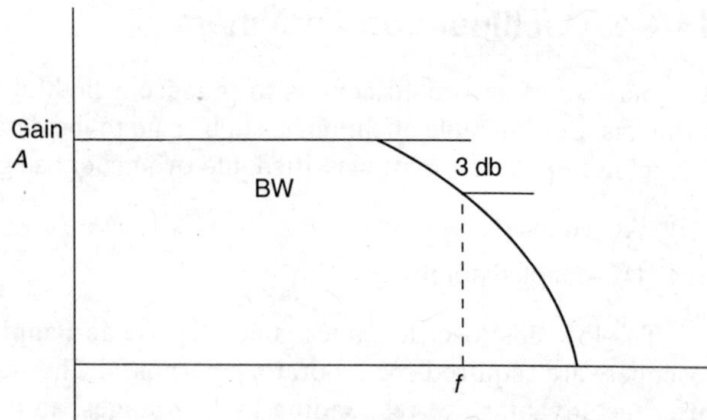


**Figure 14-8** Frequency response graphs

$$t_r \times BW = 0.35$$

where, $t_r$ is the rise time in seconds and $BW$ is the band width in Hertz.

## 14-5 MEASUREMENTS USING THE CATHODE-RAY OSCILLOSCOPE

The CRO is used to measure frequency, time, phase and different parameters of signals. In the following sections we discuss the methods of measuring frequency using Lissajous figures and time-base; measurement of phase using the double-beam CRO and Lissajous figures.

### 14-5-1 Measurement of Frequency

**Time-base measurement**

Time-base measurement helps to determine the frequency of a time-varying signal displayed on the CRT screen. If a time interval $t$ has $x$ complete cycles, then the time period of the signal is:

$$T = \frac{t}{x}$$

or,

$$f = \frac{1}{T} = \frac{x}{t}$$

Hence, the frequency is determined.

**Measurement using Lissajous figures**

The application of sinusoidal waves at the same time to the deflection plates produces various patterns. These patterns are generated on the basis of the relative amplitudes, frequencies and phases of the different waveforms and are known as Lissajous figures.

Figure 14-9 shows the Lissajous figure as a form of ellipse. Frequency can be determined from:

$$\frac{f_v}{f_h} = \frac{\text{Number of horizontal tangencies}}{\text{Number of vertical tangencies}}$$

where, $f_v$ and $f_h$ are the frequencies of the vertical and the horizontal signals, respectively. Example 14-8 and 14-11 show the practical ways of calculating frequency using Lissajous patterns.

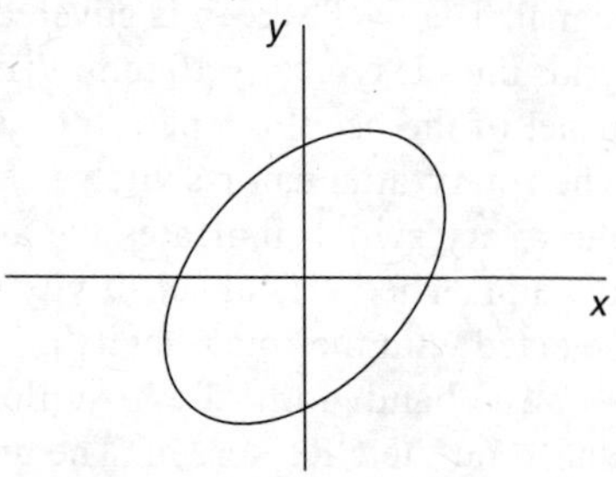


**Figure 14-9** Lissajous figure as a form of ellipse

## 14-5-2 Measurement of Phase

The phase difference of two different waveforms displayed on the CRT screen can be found from the time axis. Two sinusoidal signals of time period $T$ are in the same phase at time $t_1$ and $t_2$ respectively, and the phase difference between them is expressed as:

$$\varphi = \frac{2\pi}{T}(t_1 - t_2) \qquad (14\text{-}16)$$

Figure 14-10 shows the phase difference of two different waveforms.

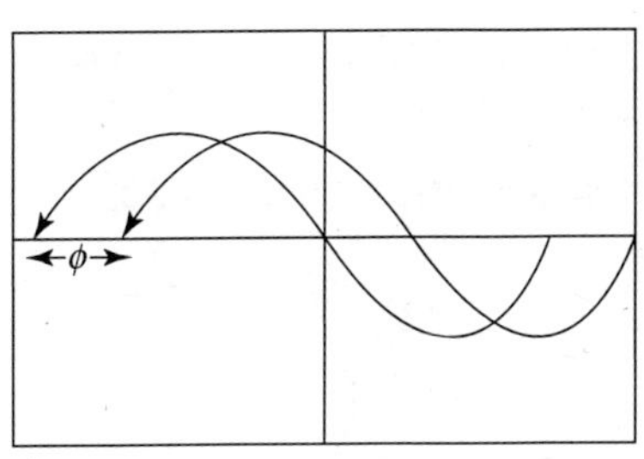


**Figure 14-10** Measurement of phase difference

## 14-5-3 Measurement of Phase Using Lissajous Figures

Lissajous figures are used to measure the phase difference between two sinusoidal voltages of the same amplitude and frequency. The signals are applied simultaneously to the horizontal and vertical deflection plates. The values of the deflection voltages are given by:

$$v_y = A \sin(\omega t + \varphi) \qquad (14\text{-}17)$$

and

$$v_x = A \sin \omega t \qquad (14\text{-}18)$$

Here A is the amplitude, $\omega$ is the angular frequency and $\varphi$ is the phase angle by which $v_y$ leads $v_x$. Eq. (14-17) can be expanded as:

$$v_y = A \sin \omega t \cos \varphi + A \cos \omega t \sin \varphi \qquad (14\text{-}19)$$

Equation (14-18) yields:

$$A \cos \omega t = \sqrt{A^2 - v_x^2} \qquad (14\text{-}20)$$

Substituting the sine and cosine terms from Eqs. (14-17) and (14-18) in Eq. (14-19), we get:

$$v_y = A \sin \omega t \cos \varphi + \sqrt{A^2 - v_x^2} \sin \varphi$$

$$v_y = v_x \cos \varphi + \sqrt{A^2 - v_x^2} \sin \varphi$$

$$v_y - v_x \cos \varphi = \sqrt{A^2 - v_x^2} \sin \varphi$$

$$(v_y - v_x \cos \varphi)^2 = (A^2 - v_x^2)\sin^2 \varphi$$

$$v_y^2 - 2v_x \cos \varphi\, v_y + v_x^2 \cos^2 \varphi = A^2 \sin^2 \varphi - v_x^2 \sin^2 \varphi$$

$$v_y^2 - 2v_x \cos \varphi\, v_y + v_x^2 \cos^2 \varphi\, v_x^2 \sin^2 \varphi = A^2 \sin^2 \varphi$$

$$v_y^2 - 2v_x \cos \varphi\, v_y + v_x^2 (\cos^2 \varphi + \sin^2 \varphi) = A^2 \sin^2 \varphi$$

$$v_y^2 - 2v_x \cos \varphi\, v_y + v_x^2 = A^2 \sin^2 \varphi$$

$$v_x^2 + v_y^2 - 2v_x v_y \cos\varphi = A^2 \sin^2 \varphi. \qquad (14.21)$$

The Lissajous figure is thus an ellipse represented by Eq. (14-21). The ellipse is depicted in Fig. 14-9.

**Case I: When $\varphi = 0°$, $\cos \varphi = 1$, $\sin\varphi = 0$**

Then, Eq. (14-21) reduces to:

$$v_x^2 + v_y^2 - 2v_x v_y = 0$$

$$(v_x - v_y)^2 = 0$$

$$v_x = v_y \qquad (14\text{-}22)$$

$\varphi = 0°$

y, Slope $m = 1$, 0, 45°, x

**Figure 14-11(a)** Lissajous figure at $\varphi = 0°$ is a straight line with slope $m = 1$

Equation (14-22) represents a straight line with slope 45°, i.e., $m = 1$. The straight line diagram is shown in Fig. 14-11(a).

**Case II: When $0 < \varphi < 90$, $\varphi = 45°$,**

$$\cos\varphi = \frac{1}{\sqrt{2}}, \quad \sin\varphi = \frac{1}{\sqrt{2}}$$

Then Eq. (14-21) reduces to:

$$v_x^2 + v_y^2 - \sqrt{2}v_x v_y = \frac{A^2}{2} \qquad (14\text{-}23)$$

Equation (14-11) represents an ellipse, as shown in Fig. 14-11(b).

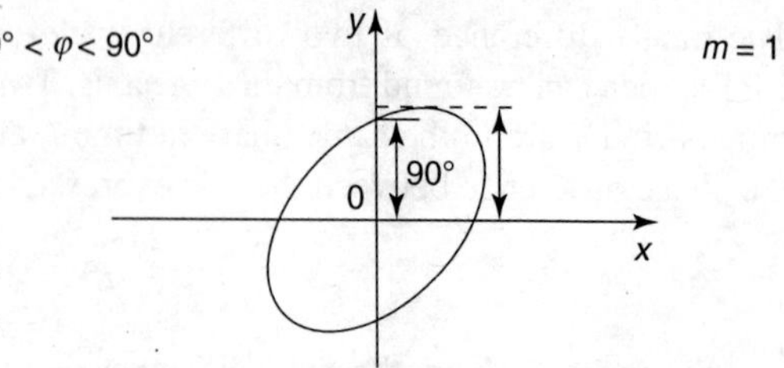


**Figure 14-11(b)** Lissajous figure at $0° < \varphi < 90$ takes the shape of an ellipse

**Case III: When $\varphi = 90°$, $\cos\varphi = 0$, $\sin\varphi = 1$**

Then Eq. (14-21) reduces to:

$$v_x^2 + v_y^2 = A^2 \qquad (14\text{-}24)$$

Equation (14-24) represents a circle shown in Fig. 14-12.

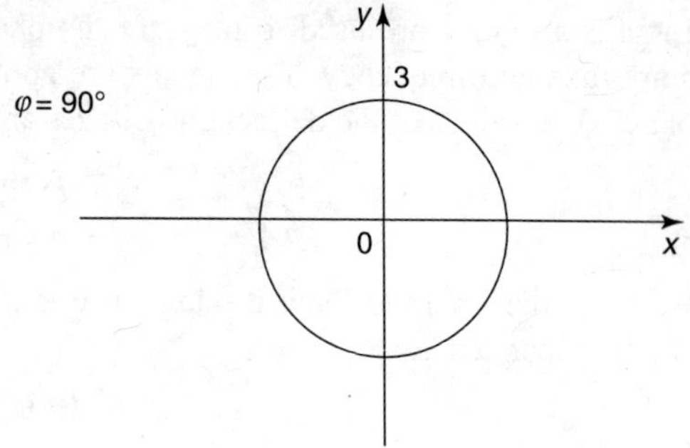


**Figure 14-12** Lissajous figure at $\varphi = 90°$: it forms a circle

**Case IV: When $90 < \varphi < 180$; say $\varphi = 135°$,**

$$\cos\varphi = -\frac{1}{\sqrt{2}}, \quad \sin\varphi = \frac{1}{\sqrt{2}}$$

Then Eq. (14-21) reduces to:

$$v_x^2 + v_y^2 + \sqrt{2}v_x y_y = \frac{A^2}{2} \qquad (14\text{-}25)$$

Equation (14-25) represents an ellipse shown in Fig. 14-13.

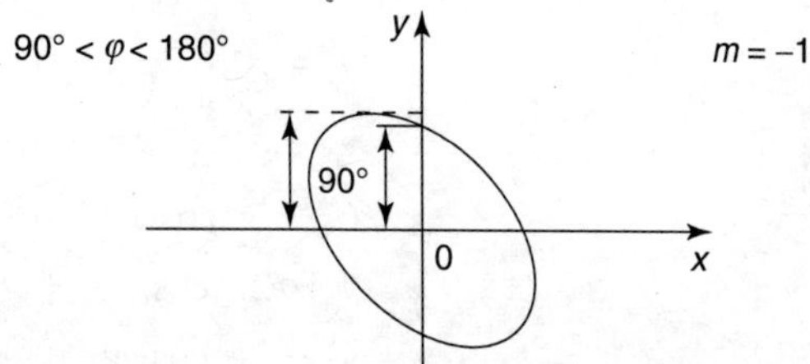


**Figure 14-13** Lissajous figure when $90 < \varphi < 180$

**Case V: $\varphi = 180°$, $\cos\varphi = -1$, $\sin\varphi = 0$**

Then Eq. (14-21) reduces to:

$$v_x^2 + v_y^2 + 2v_x v_y = 0$$

$$(v_x + v_y)^2 = 0$$

$$v_x = -v_y \qquad (14\text{-}26)$$

Equation (14-26) represents a straight line with slope $m = -1$; a slope of 45° in the negative direction of the $x$-axis, as shown in Fig. (14-14).

The maximum $y$-displacement, $A$, and the vertical displacement, $V_y$, at time $t = 0$ can be measured from the vertical scale of the CRO. Putting $t = 0$ in Eq. (14-17), we get:

$$v_{y0} = A\sin\varphi \qquad (14\text{-}27)$$

$$\sin\varphi = \frac{v_{y0}}{A} \qquad (14\text{-}28)$$

Thus, the phase angle can be found from Eq. (14-28) using any form of the Lissajous figure.

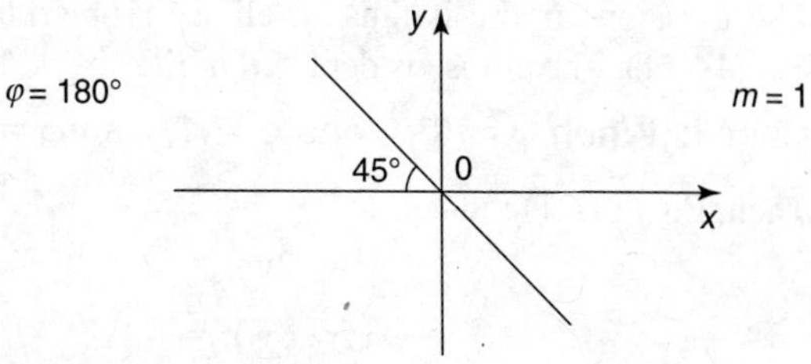


**Figure 14-14** Lissajous figure at $\varphi = 180°$ with negative slope $m = -1$

## Solved Examples

**Example 14.8** A certain Lissajous pattern is produced by applying sinusoidal voltages to the vertical and horizontal input terminals of a CRO. The pattern makes five tangencies with the vertical three with the horizontal. Calculate the frequency of the signal applied to the vertical amplifier if the frequency of the input voltage is 3 kHz.

**Solution:**
A feature that is common to all Lissajous figures is that the horizontal and the vertical lines are tangent to the pattern at a number of points. The number depends on the frequency applied to the vertical and to the horizontal deflecting plates and is given by:

$$\frac{\text{Number of points at which figure is tangent to vertical line}}{\text{Number of points at which figure is tangent to horizontal line}} = \frac{\omega_x}{\omega_y}$$

According to the data given in the problem, we have:

$$\frac{5}{3} = \frac{3}{\omega_y}\ \text{kHz}$$

Hence the frequency of the signal applied to the vertical amplifier is:

$$\omega_y = \frac{9}{5} = 1.8\ \text{kHz}$$

**Example 14-9** A CRT is designed to have a deflection sensitivity of 0.5 mm/V. The deflecting plates are 3 cm long and 6 mm apart. *T* distance of the screen from the centre of the plates is 20 cm. Calculate the required voltage to be applied to the final anode?

**Solution:**

Deflection sensitivity: $$S = \frac{lL}{2dV_a}$$

Therefore, the necessary anode voltage: $$V_a = \frac{lL}{2dS}$$

where, $l = 3$ cm, $L = 20$ cm, d = 0.6 cm, $S = 0.05$ cm/V.

Therefore, the anode voltage $V_a = 1000$ V

**Example 14-10** Two ac signals of same frequency but having a phase difference are displayed simultaneously on a dual-trace CRO. If the horizontal separation between two neighbouring peaks of the displayed waveform corresponds to 2 divisions and the distance between two consecutive peaks of a signal wave corresponds to 14 divisions of the horizontal scale, calculate the phase difference between the two signals.

**Solution:**
The phase difference $$\phi = \frac{2\pi}{D} \times d$$

Here, $d = 2$ divs and, $D = 14$ divs

Therefore, $$\phi = \frac{2\pi}{D} \times 2 = \frac{\pi}{3} = 60°$$

**Example 14-11** A certain Lissajous pattern is produced by applying sinusoidal voltages to the vertical and horizontal input terminals of a CRO. The pattern makes five tangencies with the vertical and three with the horizontal. Calculate the frequency of the signal applied to the vertical amplifier if the frequency of the input voltage is 14 kHz.

**Solution:**
A feature that is common to all Lissajous figures is that the horizontal line and the vertical line are tangent to the pattern at a number of points. The number depends on the frequency applied to the vertical and to the horizontal deflecting plates and is given by:

$$\frac{\text{Number of points at which figure is tangent to vertical line}}{\text{Number of points at which figure is tangent to horizontal line}} = \frac{\omega_x}{\omega_y}$$

According to the data given in the problem, we have:

$$\frac{5}{3} = \frac{12}{\omega_y}\text{ kHz}$$

Hence the frequency of the signal applied to the vertical amplifier is:

$$\omega_y = \frac{36}{5} = 7.2\text{ kHz}$$

■

## 14-6 TYPES OF CATHODE-RAY OSCILLOSCOPE

The categorization of CROs is done on the basis of whether they are digital or analog. Digital CROs can be further classified as storage oscilloscopes. Different types of CRO models are explained in brief in the following sections.

### 14-6-1 Analog CRO

In an analog CRO, the amplitude, phase and frequency are measured from the displayed waveform, through direct manual reading.

### 14-6-2 Digital CRO

A digital CRO offers digital read-out of signal information, i.e., the time, voltage or frequency along with signal display. It consists of an electronic counter along with the main body of the CRO.

### 14-6-3 Storage CRO

A storage CRO retains the display up to a substantial amount of time after the first trace has appeared on the screen. The storage CRO is also useful for the display of waveforms of low-frequency signals. The displayed waveform can also be stored or saved in the memory section of the CRO.

### 14-6-4 Dual-Beam CRO

In the dual-beam CRO two electron beams fall on a single CRT. The dual-gun CRT generates two different beams. These two beams produce two spots of light on the CRT screen which make the simultaneous observation of two different signal waveforms possible. The comparison of input and its corresponding output becomes easier using the dual-beam CRO.

## 14-7 SWEEP FREQUENCY GENERATOR

A sweep frequency generator is a signal generator which can automatically vary its frequency smoothly and continuously over an entire frequency range. Figure 14-15 shows the basic block diagram of a sweep frequency generator. The sweep frequency generator has the ramp generator and the voltage-tuned oscillator as its basic components.

The output of the ramp generator is a linear ramp voltage which serves as the input to the voltage-tuned oscillator. The basic circuit of a voltage-tuned oscillator is similar to that of a frequency modulator circuit.

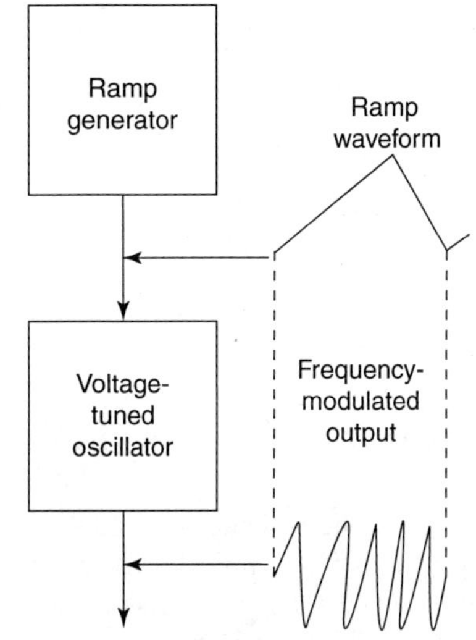


**Figure 14-15** Block diagram of a sweep frequency generator

The resonant frequency of the tank circuit is given by:

$$f = \frac{1}{2\pi\sqrt{LC}} \tag{14-29}$$

With the increase of the voltage level of the ramp output, the reverse-bias on the diode of the oscillator circuit increases. This, is in turn, reduces the capacitance $C_d$ and the resonance frequency of the tank circuit increases. As the ramp voltage returns to its zero level, the diode capacitance and the output frequency of the oscillator decrease to return to their starting levels. The frequency range over which the oscillator frequency is swept is predetermined by choosing appropriate values of $L$ and $C$. Figure 14-16 shows an oscillator tank circuit.

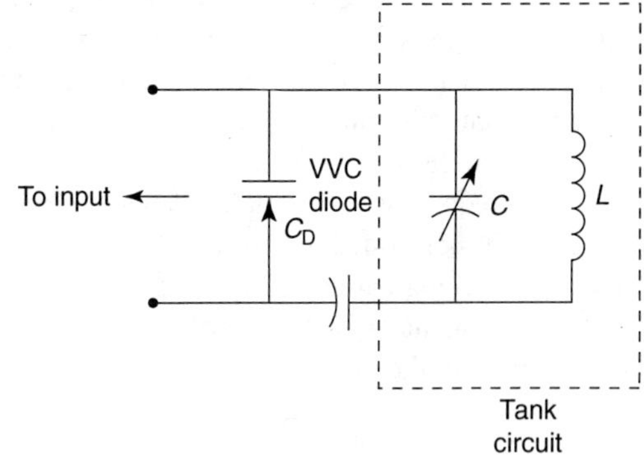


**Figure 14-16** Oscillator tank circuit

### 14-7-1 Applications of the Sweep Frequency Generator

1. Sweep frequency generators are used to display the response curve of the various stages of frequency of television or radio receivers.
2. Sweep frequency generators can be used to determine the characteristics of a device over a wide continuous range of frequencies.

## 14-8 FUNCTION GENERATOR

A function generator provides a variety of output waveforms. It can produce sine, square, ramp, pulse and triangular waveforms. The output amplitudes and frequencies are variable and a dc-offset adjustment is possible. The output frequency of a frequency generator may vary from a small fraction of a hertz to several

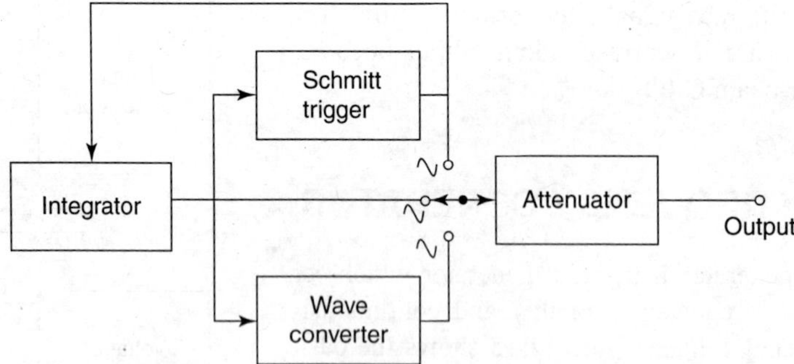


**Figure 14-17** Block diagram of a function generator

hundred kilohertz. Output amplitude is usually 0–20 V *p*-to-*p* and 0–2 V *p*-to-*p*, while the output impedance is typically a few ohms. The accuracy of frequency selection of any function generator is around ±2% of full scale of a given range, and distortion is less than 1%. Figure 14-17 gives the basic block diagram of a function generator. The basic components of a function generator are:

(i) Integrator
(ii) Schmitt trigger circuit
(iii) Sine wave converter
(iv) Attenuator

The output from the Schmitt trigger circuit is a square wave. The integrator produces a triangular waveform. The sine wave converter is used to convert a square or a triangular waveform into a sine wave. This instrument generates square or triangular waves as a primary waveform, which can then be applied to appropriate circuitry to produce the remaining waveform. Figure 14-18 gives the circuit diagram of a function generator.

The frequency of the function generator is controlled by the capacitor in the *LC* or *RC* circuit. The frequency is controlled by varying the magnitude of current which drives the integrator. The instrument produces sine, triangular and square waves with a frequency range of 0.01 Hz to 100 kHz.

The frequency controlled voltage regulates two current sources. The upper current source supplies constant current to the integrator whose output voltage increases linearly with time. The output of the integrator

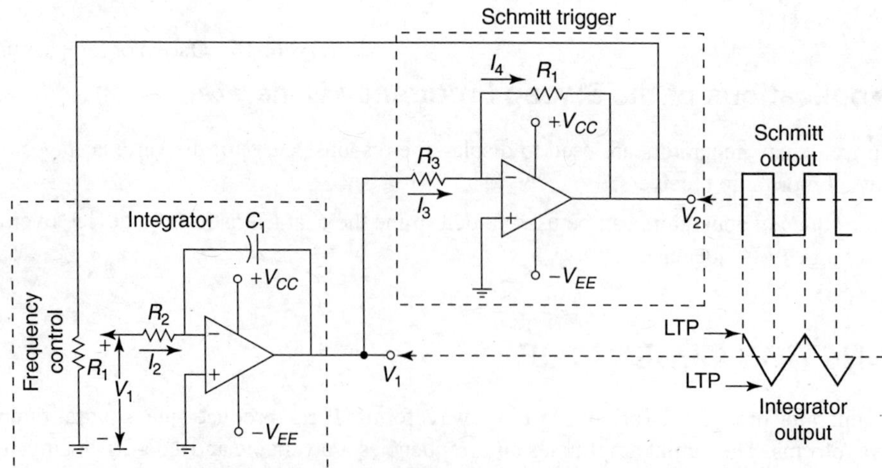


**Figure 14-18** Circuit diagram of a function generator

is a triangular waveform whose frequency is determined by the magnitude of the current supplied by the constant current source.

Attenuators are designed to change the magnitude of the input signal seen at the input stage, while presenting constant impedance on all ranges at the attenuator input. The attenuator is required to attenuate all frequencies equally.

## 14-9 SINE WAVE GENERATOR

Generally a sine wave is not produced from the function generator as a primary waveform. This is because at low frequencies amplitude and frequency distortions are introduced. A sine wave is produced by converting a triangular wave, applying proper circuits. The triangular wave is produced by employing an integrator and a Schmitt trigger circuit. This triangular wave is then converted to a sine wave using the diode loading circuit, as shown in Fig. 14-19.

Resistors $R_1$ and $R_2$ behave as the voltage divider. When $VR_2$ exceeds $+V_1$, the diode $D_1$ becomes forward-biased. There is more attenuation of the output voltage levels above $+V_1$ than levels below $+V_1$. With the presence of the diode $D_1$ and resistor $R_3$ in the circuit, the output voltage rises less steeply. The output voltage falls below $+V_1$ and the diode stops conducting, as it is in reverse-bias. The circuit behaves as a simple voltage-divider circuit. This is also true for the negative half-cycle of the input $V_i$. If $R_3$ is carefully chosen to be the same as $R_4$, the negative and the positive cycles of the output voltage will be the same. The output is an approximate sine wave.

The approximation may be further improved by employing a six-level diode loading circuit, as shown in Fig. 14-20(a). All the diodes are connected to different bias voltage levels by appropriate values of resistors. As there are six diodes, there will three positive- and three negative-bias voltage levels. Therefore, at each half-cycle of the output voltage, the slope changes six times and the output wave shape is a better approximation of the sine wave.

The triangular to sine wave converter is an amplifier whose gain varies inversely with the amplitude of the output voltage.

$R_1$ and $R_3$ set the slope of $V_0$ at low amplitudes near the zero crossing. As $V_0$ increases, the voltage across $R_3$ increases to begin forward biasing $D_1$ and $D_3$ for positive outputs or $D_2$ and $D_4$ for negative output. When these diodes conduct, they shunt feedback resistance $R_3$ lowering the gain. This tends to shape the triangular output into the sine wave. In order to get the rounded tops for sine waves, output $R_2$ and diodes $D_1$ and $D_2$ are adjusted to make the amplifier gain approach zero at the peak. The circuit is shown in Fig. 14-20(b).

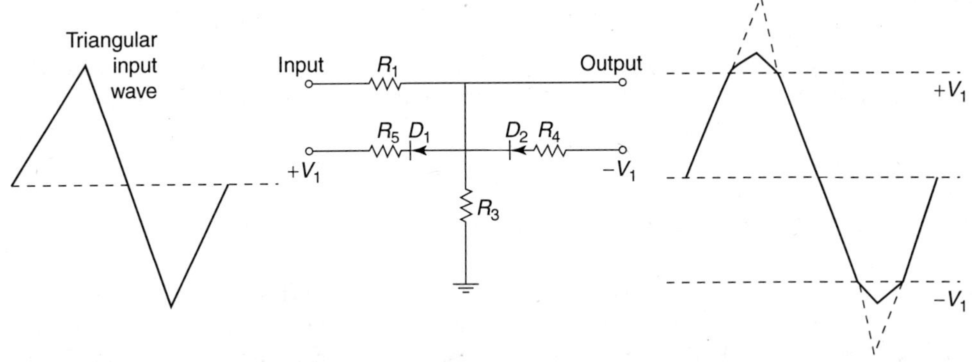


**Figure 14-19** Two-level diode loading circuit

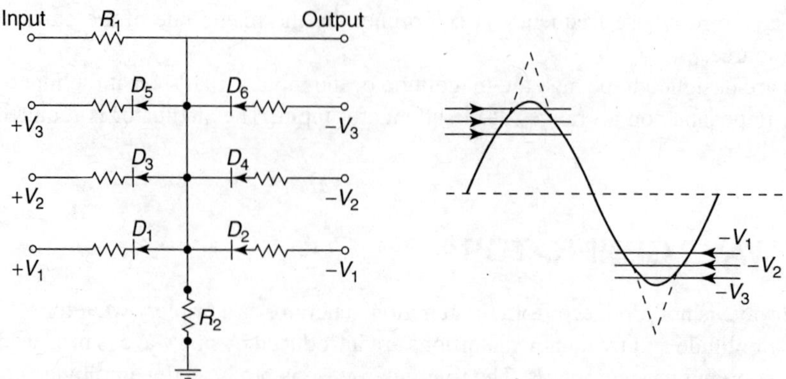


**Figure 14-20(a)** Diagram for the six-level diode loading circuit

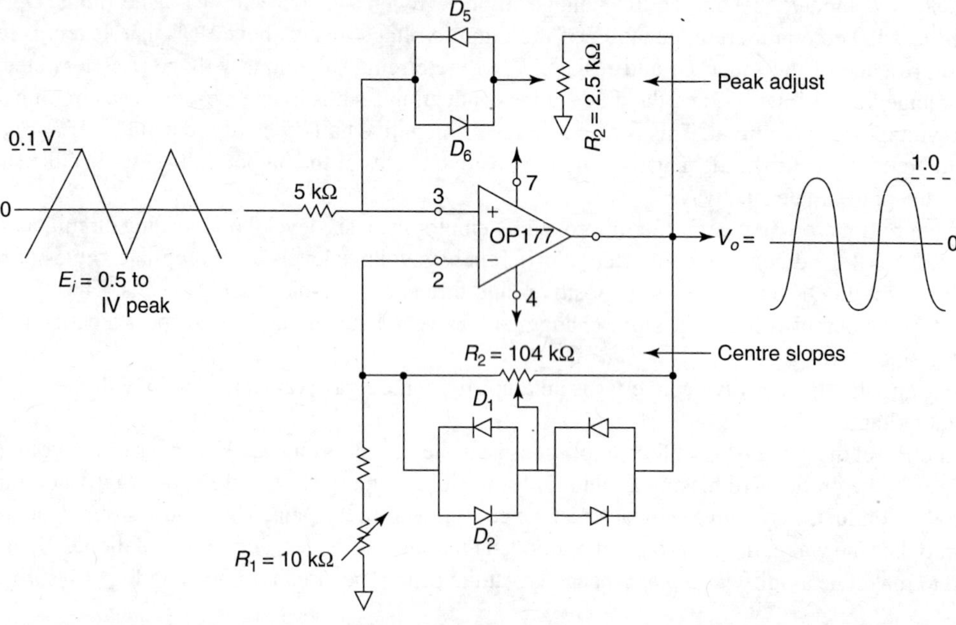


**Figure 14-20(b)** Triangular to sine wave generator using op-amp

The circuit is adjusted by comparing a 1 kHz sine wave and the output of the triangular/sine wave converter on a dual-track CRO. $R_1$, $R_2$, $R_3$ and the peak amplitude of $E_i$ are adjusted in sequence for the best sinusoidal shape.

## 14-10 SQUARE WAVE GENERATOR

A square wave can be most easily obtained from an operational amplifier astable multi-vibrator. An astable multi-vibrator has no stable state—the output oscillates continuously between high and low states.

In Fig. 14-21, the block comprising the op-amp, resistors $R_2$ and $R_3$ constitutes a Schmitt trigger circuit. The capacitor $C_1$ gets charged through the resistor $R_1$. When the voltage of the capacitor reaches the upper trigger

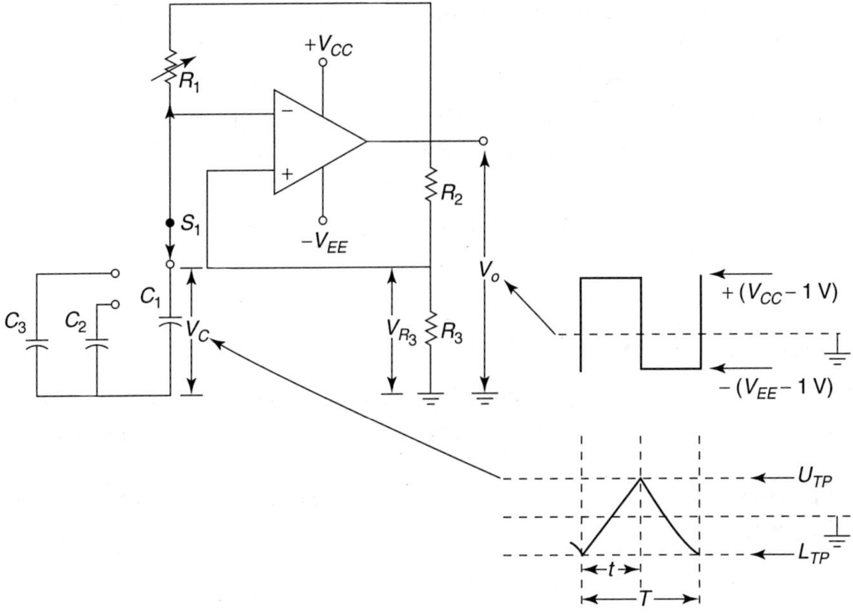


**Figure 14-21** Sine wave generator

point of the Schmitt trigger circuit, the output of the op-amp switches to output low. This is because the Schmitt trigger is a non-inverting type. Now, when the op-amp output is low, the capacitor $C_1$ starts getting discharged. As the capacitor discharges and the capacitor voltage reaches the lower trigger point of the Schmitt trigger, the output of the op-amp switches back to the output high state. The capacitor charges through the resistor again and the next cycle begins. The process is repetitive and produces a square wave at the output. The frequency of the output square wave depends on the time taken by the capacitor to get charged and discharged when the capacitor voltage varies from UTP (upper trigger point) and LTP (lower trigger point).

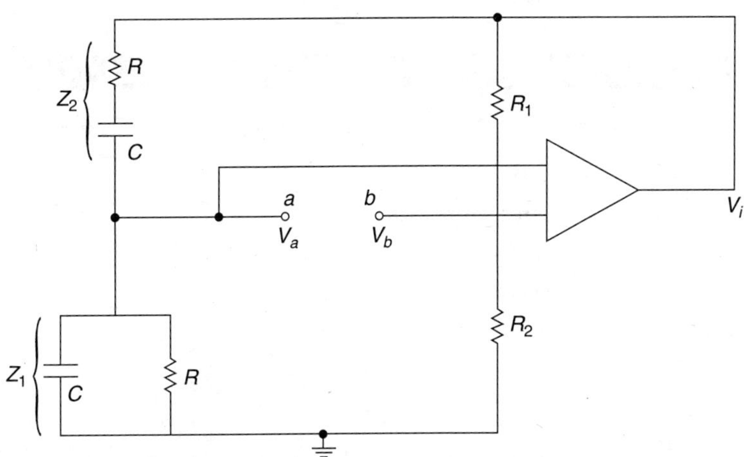


**Figure 14-22** Wien-bridge feedback network with an amplifier

The frequency of the output square wave is given by:

$$f = \frac{1}{T} = \frac{1}{2t} \tag{14-30}$$

where, $t$ is the time taken by the capacitor to get charged or discharged. The UTP and LTP values for the Schmitt trigger can be fixed by choosing appropriate values of $R_2$ and $R_3$.

$$|\text{UTP}| = |\text{LTP}| = V_0 \frac{R_3}{R_2 + R_3} \tag{14-31}$$

## 14-11 AF SIGNAL GENERATOR

An AF signal generator generally uses an oscillator which is regulated by a controlled phase shift through a resistor and capacitor network. The Wien-bridge oscillator produces sine waves using an RC network as a feedback.

The amplifier is connected as an oscillator in order to determine at what frequency the Wien-bridge provides the required criterion for oscillation. With respect to ground, the voltage at $A$ is given by:

$$V_a = \frac{Z_1}{Z_1 + Z_2} V_i \tag{14-32}$$

and

$$V_b = \frac{R_1}{R_1 + R_2} V_i \tag{14-33}$$

Since $V_a$ and $V_b$ are the same, from Eqs. (14-32) and (14-33) we can write:

$$\frac{R_1}{R_1 + R_2} = \frac{Z_1}{Z_1 + Z_2} \tag{14-34}$$

At a frequency $f_0 = 1/2\pi RC$, the phase angle between $V_a$ and the output is zero. The Wien-bridge oscillator is tuned with a variable capacitance and the oscillator is band-switched using the resistance. The Wien-bridge oscillator is usually the heart of a general purpose AF signal generator. Harmonic distortion is then less than a few tenths of a percent.

## POINTS TO REMEMBER

1. CRO is used to study waveforms.
2. CRT is the main component of a CRO.
3. Prosperous P31 is used for the fluorescent screen of a CRO.
4. A CRO has the following components:
   (a) Electron gun
   (b) Deflecting system
   (c) Florescent screen
5. Lissajous figures are used to measure frequency and phase of the waves under study.
6. A time-base generator produces saw-tooth voltage.
7. An oscilloscope amplifier is used to provide a faithful representation of input signal applied to its input terminals.
8. Function generators can produce sine, square, ramp, pulse and triangular waveforms.
9. A sine wave is produced by converting a triangular wave, applying proper circuits.
10. Wien-bridge oscillator is usually the heart of a general purpose AF signal generator.

## IMPORTANT FORMULAE

1. The deflection sensitivity of the CRT is:

$$S = \frac{l_{total}}{V_d} = \frac{lL}{2sV_a} \text{ m/V}$$

2. The deflection factor of the CRT is:

$$G = \frac{1}{S} = \frac{2sV_a}{lL} \text{ V/m}$$

3. Phase angle is given by:

$$\varphi = \frac{2\pi}{T}(t_1 - t_2)$$

4. Lissajous equation is given by:

$$v_x^2 + v_y^2 - 2v_x v_y \cos\varphi = A^2 \sin^2\varphi$$

## OBJECTIVE QUESTIONS

1. Input impedance of CRO is:
   (a) 1 MΩ (b) 1 kΩ
   (c) 100 Ω (d) 1 Ω
2. CRO displays:
   (a) AC signals
   (b) DC signals
   (c) Both ac and dc signals
   (d) None of the above
3. CRO uses:
   (a) Electrostatic deflection
   (b) Magnetic deflection
   (c) Electro-magnetic deflection
   (d) None of the above
4. CRO fluroscent screen uses the phosphorous isotope:
   (a) P31 (b) P32
   (c) P30 (d) None of the above
5. Lissajous figure is used in CRO for:
   (a) Phase measurement
   (b) Frequency measurement
   (c) Amplitude measurement
   (d) None of the above
6. Lissajous figure at a 45 degree angle means:
   (a) Straight line (b) Circle
   (c) Oval (d) None of the above
7. The difference between the spectrum analyser (SA) and CRO is:
   (a) CRO and SA both measure time domain signal
   (b) CRO and SA both measure frequency domain signal
   (c) CRO measures time domain signal and SA measures frequency domain
   (d) CRO measures frequency domain signal and SA measures time domain
8. Digital storage oscilloscope is more preferable because of:
   (a) Digital display
   (b) Storage
   (c) Both (a) and (b)
   (d) None of the above
9. Saw-tooth voltage of a CRO means:
   (a) Sweep time + fly back time
   (b) Fly back time + sweep time
   (c) Only fly back time
   (d) Only sweep time
10. Defection sensitivity of CRO depends on:
    (a) Deflection voltage, separation between the plates and plate length
    (b) Only deflection voltage
    (c) Only separation between plates
    (d) Electron density

## REVIEW QUESTIONS

1. Draw the block diagram of a CRO and explain the function of each block.
2. What is CRT? How it is used?
3. Calculate the deflection sensitivity of a CRO? Why is electrostatic deflection used in a CRO?
4. How are different signal parameters measured by a CRO?
5. Explain the significance of Lissajous figures in CRO measurement.
6. How can you measure the frequency of a signal voltage using a CRO?
7. How can the phase difference between two ac voltages be measured by a CRO?
8. How can a waveform be displayed in a CRO?
9. State the applications of CRO.
10. Explain the following terms:
    (a) Sweep voltage
    (b) Synchronization
    (c) Time-base
11. Distinguish between a CRT and a CRO.
14. Why is Aquadag coating neccessary for a CRO?
13. Derive the expression for deflection sensitivity.
14. Draw a sketch to illustrate the electrostatic deflection.
15. Define deflection factor and deflection sensitivity.
16. Discuss the factors that affect brightness of the display.
17. Briefly discuss the screen of a CRO.
18. Explain the sweep circuit of a CRO.
19. Describe the procedure of making amplitude and time measurement on an oscilloscope.
20. How are Lissajous figures obtained?
21. Derive the expression for obtaining Lissajous figures.
22. Describe the generation of a saw-tooth waveform.
23. What must be done to obtain a steady oscillogram?
24. How can a CRO measure the phase difference between two ac voltages?
25. What is post-acceleration and why is it used in a CRO?

## PRACTICE PROBLEMS

1. In a CRT, an electron beam is magnetically deflected after being accelerated through a potential difference of 1000 V. The deflecting magnetic field acts over an axial length of 1.8 cm. If a deflection sensitivity of 2.2 mm/gauss is to be attained, calculate the distance of the CRT screen from the centre of the deflection system?
2. A Lissajous pattern is obtained on a CRO screen when the sinusoidal voltages are applied to the two sets of deflecting plates. The figure makes three tangencies with the horizontal and five tangencies with the vertical. If the frequency of the horizontal signal is 2.7 kHz, find the frequency of the vertical signal.
3. The accelerating voltage of a CRT is 1 kV. A sinusoidal voltage is applied to a set of deflecting plates of axial length 2.1 cm. Calculate the frequency of the deflecting voltage if the electrons remain between the plates for one half-cycles.
4. A deflection amplifier has the following components:

$$R_1 = R_7 = 10\ \text{k}\Omega$$

$$R_2 = R_8 = 5.7\ \text{k}\Omega$$

$$R_3 = R_6 = 15\ \text{k}\Omega$$

$$R_5 = 2.3\ \text{k}\Omega$$

$$R_9 = R_{11} = 15.2\ \text{k}\Omega$$

$$R_{10} = 2.2\ \text{k}\Omega$$

If the supply voltage is ±15 V, determine the dc voltage levels throughout the circuit when the input level is zero and the moving contact of $R_{10}$ is at its centre position.

5. A 1 kHz triangular wave with peak amplitude of 11 V is applied to the vertical deflecting plates of a CRT. A 1 kHz saw-tooth wave with peak amplitude of 22 V is applied to the horizontal deflecting plates. The CRT has a vertical deflection sensitivity of 0.4 cm/V and a horizontal deflection sensitivity of 0.25 cm/V. Assuming that the two inputs are synchronized, determine the waveform displayed on the screen.

6. Repeat Problem 5 with the triangular-wave frequency changed to 2.2 kHz

7. If the vertical amplifier of the oscilloscope has the bandwidth of 17 MHz, what is the fastest rise time that an input may have to be displayed without distortion?

8. A certain Lissajous pattern is produced by applying sinusoidal voltages to the vertical and horizontal inputs of a CRO. The pattern makes five tangencies with the vertical and three with the horizontal. If the frequency of the horizontal input is 3 kHz, determine the frequency of the signal applied to the vertical input.

10. In a CRT the length of the deflecting plates is 1.6 cm, the spacing of the plates is 5 mm and the distance of the screen from the centre of the plates is 20 cm. If the final anode voltage is 1400 V, what would be the deflection sensitivity?

11. The deflection system in a CRT employs a magnetic field of $10^{-4}$ T acting over an axial length of 4 cm and is placed 24 cm from the screen. If the accelerating voltage is 600 V, find the deflection of the spot on the fluorescent screen.

12. If the time/div control of a waveform is set to 10 ms and the volts/div control is at 5.5 V, determine the peak amplitude and frequency of each waveform.

13. Two waveforms (A and B), each occupying five horizontal divisions for one cycle, are displayed on an oscilloscope. Wave B commences 1.6 divisions after commencement of wave A. Calculate the phase difference between the two.

14. If the volts/div of waveforms is set to 0.1 V and the time/div control is at 22 $\mu$s, determine the pulse amplitude, the pulse frequency, the delay time, the rise time and the fall time.

15. A signal with an amplitude of $V_s$ = 450 mV and a source resistance of 1 kilo ohms is connected to an oscilloscope with $R_i$ = 1 MΩ in parallel with $C_i$ = 45 pF. The coaxial cable of the 1:1 probe used has a capacitance of $C_{cc}$ = −88 pF. Calculate the signal voltage level $V$ at the oscilloscope terminals when the signal frequency is 150 Hz. Also calculate the signal frequency at which $V_t$ is 3 dB below $V_v$.

16. In a CRT the distance between the deflecting plates is 1.0 cm, the length of the deflecting plate is 4.0 cm and the distance of the screen from the centre of the deflecting plate is 30 cm. If the accelerating voltage supply is 350 V, calculate the deflecting sensitivity of the tube.

17. An electrostatically deflected CRT has deflecting plate which is 3 cm long and 0.5 cm apart and the distance from the centre to the screen is 22 cm. The electron beam is accelerated by 2000 V and is projected centrally between the plates. Calculate the deflecting voltage required to cause the beam to strike the screen. Also, find the corresponding deflection.

18. In a neon tube time-base generator $R$ = 100 k and $C$ = 0.01 $\mu$F. The extinguishing and striking potentials are 90 V and 140 V respectively. If the supply voltage is 180 V, determine the frequency of the time base.

19. Consider a circuit with $R_1$ = 100 kΩ, $R_2$ = 50 kΩ and $V_s$ = 150 V. The voltage across $R_2$ is measured by the 0–50 V scale of a voltmeter of sensitivity 10 kΩ/V. Find the percentage error in measurement due to loading effect of the voltmeter.

20. Consider two 1 MΩ resistors connected in series and supplied by a source of 150 V. A multi-meter having a sensitivity 20 kΩ/V is used to measure voltages across one of the resistors. The scale range used is 50 V. What will be the reading on the screen?

21. For the waveforms illustrated in the following diagram, the time/div control is at 50 ms, and

the volts/div control is set to 22 mV. Determine the peak amplitude and frequency of each waveform, and calculate the phase difference.

22. Use the diagram provided for Problem 21, to calculate the number of horizontal divisions between the beginning of each waveform cycle for a phase difference of 35°.

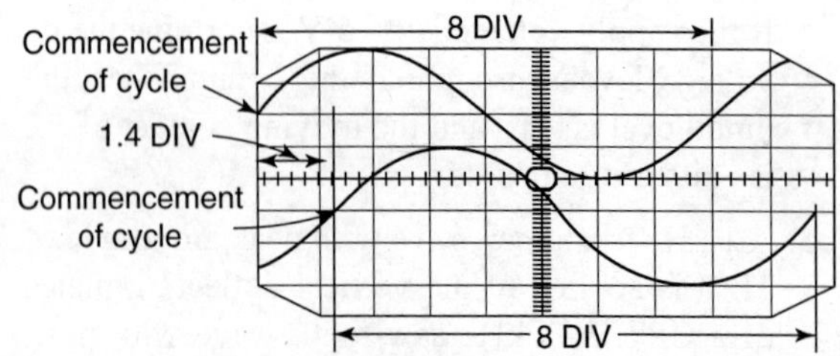


## SUGGESTED READINGS

Kalsi, K. L. 2000. *Electronic Measurement*. New Delhi: Tata McGraw Hill.

Harris, F.K. 2000. *Electrical Measurements*. New York: John Wiley and Sons.

Golding E.W. and F.C.Wides. 1999. *Electrical Measuring Instruments*. A. H. Wheeler and Co.

Bell, David A. 1994. *Electronic Instrumentation and Measurements*. New Jersey: Prentice Hall.

Helfrick, Albert D. and William D. Cooper. 2000. *Modern Electronic Instrumentation and Measurement Techniques*. New Delhi: PHI.

Chattopadhyay, D. and P.C.Rakshit. 2006. *Electronics Fundamentals and Applications*. New Delhi: New Age International Publishers.

# INDEX

## D

## E



## F

## G



## H



## I



## J

## L



## M



## N



## O

## P



## Q

## R



## S

## T



## U



## V



## W



## X



## Z